Library of America, a nonprofit organization,
champions our nation's cultural heritage
by publishing America's greatest writing in
authoritative new editions and providing resources
for readers to explore this rich, living legacy.

EDWARD O. WILSON

EDWARD O. WILSON

Biophilia

The Diversity of Life

Naturalist

David Quammen, *editor*

THE LIBRARY OF AMERICA

E. O. Wilson: Biophilia, The Diversity of Life, Naturalist
is published and kept permanently in print
with support from

THE GIORGI FAMILY FOUNDATION

Contents

Introduction

BY DAVID QUAMMEN

THE ONE-EYED BOY

Edward O. Wilson was born on June 10, 1929, in Birmingham, Alabama; spent much of his boyhood in the piney forests and bayous and swamps of the Southeast; found his way to Harvard to do a doctorate on ants; and rose to become one of the most famous, celebrated, and controversial biologists of the late twentieth and early twenty-first centuries. It has been a long and serpentine path. He recounts portions of this improbable journey in some of his writings, but nowhere more pithily than in a single sentence from his 1994 autobiography, *Naturalist*: "I am blind in one eye and cannot hear high-frequency sounds; therefore I am an entomologist."

The blinding, which he has described as "a fortuitous constriction of physiological ability," occurred when he was seven. He was spending a summer on the Florida panhandle coast, boarded out among strangers as his parents drifted toward divorce, and doing the sorts of things that a solitary child who finds solace and fascination in nature would do. Wandering the strand. Wading in the shallows. Gazing at stingrays. That particular day he was fishing from a dock. A small spiny animal called a pinfish touched his bait, and he struck back, too forcefully. "It flew out of the water and into my face. One of its spines pierced the pupil of my right eye." Characteristically for Wilson, when he tells this story he identifies the pinfish to species: *Lagodon rhomboides*. His eye never healed well, the pupil clouded over, and months later he squirmed through an old-fashioned ether-drip anaesthetization so that a surgeon could remove his lens. For years afterward, the smell of ether made him nauseous.

But he still had the left eye and, fortunately for him, its short-range acuity was exceptionally high. He could count tiny hairs on the thorax of a very small insect—which is how, sometimes, you tell one species from another—though he couldn't spot the brown bird amid the brown limbs in a distant tree. Sometime during adolescence, for reasons that may have involved genetic predisposition (anyway, not angling), Wilson also lost most of his hearing in the high registers. So he couldn't detect the brown bird by its song either. This is how lives are shaped. He focused his natural history enthusiasms at relatively close range, developing a passionate interest in butterflies, which could be caught in a cheesecloth net, killed and mounted, their markings studied in fine detail. He also, less conventionally for a nature boy, became fascinated with ants. They were available but mysterious. You could find them by ripping bark off a dead tree, and then they would scurry away—to go where, do what? You could read about them in *National Geographic*. Ants could be captured live, their tunneling behavior observed in a jar of sand. Let me repeat that, because it's important: *their behavior observed*. Not many years after the eye injury, precociously early in fact, his love of nature began florescing into another dimension: an aspiration toward science.

This is a step not every ten-year-old lepidopterist takes. Little Ed got hold of a classic text, R. E. Snodgrass's *Principles of Insect Morphology*, and pored over it with half-comprehending reverence. He wanted to learn exactly what the title advertised: *principles*, not just the names of pretty creatures. He wanted to pose questions, seek reasons, and discern patterns. Eventually he arrived at what, with adolescent seriousness and no small bit of ambition, he saw as a decision point. He should chart his career in entomology. "The time had come to select a group of insects on which I could become a world authority." He was sixteen years old, still one year from entering the University of Alabama as a freshman.

As Wilson tells this story on himself, in *Naturalist*, there were several well-calculated considerations. Butterflies were out: too well-known and much studied by other scientists. Flies were more enticing—true flies, the order Diptera—especially the so-called long-legged flies, known as dolichopodids to the cognoscenti (of which teenage Wilson was one), a diverse family of delicate, stilt-walking forms, many of them decorated with a metallic blue-green sheen, like a prized 1958 Chevy. But to collect, mount, and study flies, Wilson would need certain tools, including a very special sort of long, black insect pin, for mounting and drying and labeling specimens. Those pins were made mainly in Czechoslovakia. But it was 1945 now, just after the Second World War, and Soviet-occupied Czechoslovakia wasn't exporting insect pins.

Ants, on the other hand, being wingless (except the sexual individuals, during their brief mating flights) and therefore slightly less fragile, could be pickled and stored in little glass bottles of alcohol. Rubbing alcohol and small vials for holding medicinal potions—you could get these at a drugstore in Decatur, Alabama, where Wilson was finishing high school. "I quickly hit upon ants."

Although he makes it sound almost larkish or random, that choice was vastly consequential. I invite you to ponder how the intellectual landscape of modern biology might be different today—notably, the ferment of ideas about behavior among social animals, from ants and termites to naked mole-rats, baboons, and humans, and the evolutionary origins of such behavior, the field sometimes known as sociobiology—if young Ed Wilson had committed his life to the study of butterflies.

∾

There's another factor that made this one-eyed boy exceptional as he grew into a mature scientist, a man with broad intellectual scope as well as a keen focus on ants, a fearless theorist as well as an eminent taxonomist, restless to detect patterns in the scientific data and push the boundaries of conventional thinking: he could write.

Wilson was, and remained, very good with the English language. I don't favor overzealous exercises in hindsight to see the shape of the oak revealed in the curves of the acorn, but in this case there is one acorn's curve worth noting. It's another little datum from his childhood, reported in *Naturalist*. Rummaging backward through family papers as he wrote that book, Wilson found an old letter from his fifth-grade teacher to his parents, sent in 1940, which said: "Ed has genuine writing ability, and when he combines this with his great knowledge of insects, he produces fine results." That's probably far more prophetic than anything your fifth-grade teacher or mine ever said about us.

Wilson at his best is a writer of great skill and aplomb. His books for the general public, including the three collected in this volume (and more that will appear in a future Library of America Wilson volume), flow with high narrative skill and expository eloquence, offering many passages of keen character portrait, self-deprecating humor, emotional impact, and elegant description. He has a voice and he knows how to use it. He has the chops. For that reason, reading Wilson is more than just informative, stimulating, often provocative, sometimes aggravating, and always challenging to comfortable assumptions. It's also a literary pleasure.

This skill was first broadly recognized in 1979, when his incendiary book *On Human Nature* received the Pulitzer Prize for General Non-Fiction. (By nice coincidence, the prize for poetry that year went to another writer who had been a one-eyed boy closely observing nature in the South, Robert Penn Warren. It was Warren's third of three Pulitzers and Wilson's first of two, but alas they never had occasion to compare notes.) *On Human Nature* was the final book in a trilogy, on the ideas and applications of sociobiology, that Wilson himself hadn't consciously seen as trilobate until he had almost finished writing. It was also the crescendo of what science historians have called "the sociobiology controversy," of which Wilson was chief provocateur and central figure. His work in that vein will constitute much of Library of America's second Wilson volume, but it bears mentioning here as an important phase in his evolution as a scientist, a public figure, and a writer. Among the most vivid accounts of that controversy—persuasive and lucid, though it can't be called objective—is Wilson's own, in the later chapters of *Naturalist*.

Sociobiology is the study of the ecological, evolutionary, and genetic dimensions of social behavior among animals, including humans. Wilson came to it naturally from his study of social insects, publishing an ambitious overview of ants, termites, wasps, and bees as *The Insect Societies* in 1971. In the last chapter of that book, he wrote: "When the same parameters and quantitative theory are used to analyze both termite colonies and troops of rhesus macaques, we will have a unified science of sociobiology." He set to work toward advancing that unification himself. Four years later, in the first chapter of his tome *Sociobiology*, which is subtitled *The New Synthesis*, he picked up where *The Insect Societies* left off, noting that the task of sociobiology is to understand forms of social behavior with the help of Darwinian logic, and adding: "For the present it focuses on animal societies, their population structure, castes, and communication, together with all of the physiology underlying the social adaptations." Then he dropped a pebble in that clear pool: "But the discipline is also concerned with the social behavior of early man and the adaptive features of organization in the more primitive contemporary human societies."

There followed twenty-five chapters, more than five hundred double-column pages, describing a vast body of work by various researchers on the principles of this science and their application to many kinds of animals, many differing forms and degrees of sociality. The *tour d'horizon* proceeded from ant and termite societies to schooling among fish, territorial competition among frogs, "mob" assemblages and dominance hierarchies among wallabies, pride formation among lions, troop structure among ringtail lemurs, and the intricate forms of transactional cooperation that Jane Goodall and others had observed among chimpanzees. In the final chapter, as he had done with *The Insect Societies*, Wilson bridged to the next big idea: "Let us now consider man in the free spirit of natural history, as though we were zoologists from another planet completing a catalog of social species on Earth." Altruism, bonding, division of labor, vocal communication, tribalism, ritual, even religion: Was it possible that these unusual forms of behavior originated from factors that *included* (note: he didn't say *were limited to*) natural selection acting on genetic variation within human populations? Flash: the fat hit the fire.

More on all that in volume two. Suffice it here to say that a loose collection of excitable scientists and other academics in Boston, including several of Wilson's Harvard colleagues, constituted themselves into what they called the Sociobiology Study Group, devoted to denouncing Wilson's ideas and writings as well as studying them, and that if the 1979 Pulitzer judges had viewed *On Human Nature* through the same lens that the Sociobiology Study Group did, that book would never have received the prize. But it did, and even as it did, Wilson was pivoting toward the next phase of his extraordinary career.

∽

Again a hint of the new interests appeared in late pages of one book, heralding the next. In the last chapter of *On Human Nature* he postulated the existence and evolutionary origin of certain "cardinal values" in the collective human population. One of those values is the very survival of the human species, the human genetic line. We care, more than members of *Tyrannosaurus rex* could have consciously cared, about our own collective continuation. Another value is the diversity of the human gene pool. We sense (though certain politicians tend to forget) that we are well served by our own variousness. A third cardinal value, Wilson wrote, is universal human rights. We may disagree about just what they are, but the concept is deeply ingrained. Beyond these three, he wrote, are some secondary values that may also have arisen and persisted in part because they have helped our survival and reproductive success: the thrill of discovery, triumph in battle, satisfaction from acts of generosity, ethnic pride, and finally but not least "the secure biophilic pleasure from the nearness of animals and growing plants." As far as I can tell, that was Wilson's first allusion to what he would soon develop as the concept of *biophilia*.

The word itself wasn't his invention. It could be traced back to writings of the social psychologist Erich Fromm, from the 1960s and early 1970s, who defined biophilia as "the passionate love of life and of all that is alive." Fromm's meaning was a bit different from Wilson's idea, more a matter of clinging to life and furthering growth than an innate inclination to affiliate oneself with other living forms. "The tendency to preserve life and fight against death," Fromm wrote, in 1964, "is the most elementary form of the biophilous orientation, and is common to all living substance." Wilson may have been unaware of Fromm's prior use and put together this portmanteau word independently, as his recollection in *Naturalist* suggests. Anyway, in early 1979 he introduced it with a new meaning within a new context—more evolutionary than Fromm's—by way of a fragment of essay published in *The New York Times Book Review*. That fragment, somewhat oddly, served as text of an advertisement for Harvard University Press. "Capital Ideas from People Who Publish with Harvard" was the ad's headline, with Wilson's capital idea as exemplar: "Our deepest needs stem from ancient and still poorly understood biological adaptations. Among them is biophilia, the rich, natural pleasure that comes from being surrounded by living organisms, not just other human beings but a diversity of plants and animals that live in gardens and woodlots, in zoos, around the home, and in the wilderness." True to promise, Harvard University Press published this capital idea five years later in his book *Biophilia*.

Something important had happened, in the meantime, to Wilson's scientific focus and personal sense of imperatives: his long-standing devotion to biological diversity, and his concern over the rate of species extinctions, population declines, and habitat losses on Earth, due to human actions, had converged to a critical point.

∾

Human destruction of biological diversity was nothing new, in the early 1980s, either to Wilson or to other biologists and attentive nonscientists. The British ecologist Charles Elton had published a lucidly detailed warning, back in 1958, in his book *The Ecology of Invasions by Animals and Plants*. Elton focused on the impacts of invasive species, and of chemically enforced homogeneity, not on the other foremost diversity-reducing factors, such as habitat loss, habitat fragmentation, and unsustainable "harvest" by humans; but he called out the net effects. "On the exploited lands of the world," Elton wrote,

> we see a decrease in richness and variety of species: monocultures with rigid spraying programmes, pastures of pure grass populations, pure stands of trees, the replacement of stratified and mature deciduous woods by quick-growing conifers with their relatively barren structure and poor inhabitants, the cleaning up of waste patches, the hormone spraying of roadsides, and the planting of exotic

species many of which may literally be quite sterile of animal life—at first. We might sum up this stream of events in the words of Isaiah: "Woe unto them that join house to house, that lay field to field, till there be no place, that they may be placed alone in the midst of the earth!"

Others, biologists and writers, followed with similar warnings during the next couple of decades: Peter Matthiessen (*Wildlife in America*, 1959), Rachel Carson (*Silent Spring*, 1962), James Greenway (*Extinct and Vanishing Birds of the World*, especially the second edition, 1967), Norman Myers (*The Sinking Ark*, 1979), plus various works of Daniel Janzen, Alwyn Gentry, Peter Raven, Paul and Anne Ehrlich, Robert Michael Pyle, and many more.

Wilson, by his own account, came abashedly late to this cause, in about 1980, moving only then from a state of quiet discomfort at what he had seen since his Alabama boyhood—rampant destruction of habitats, loss of diversity—to a state of abhorrent alarm. That was just after he read Norman Myers's systematic estimates of the rate at which tropical rainforests were being cut and burned, and at a time when Wilson's friend Peter Raven, an eminent botanist, was chairing a National Research Council study of the problem and its possible remedies. Wilson tells this story in *Naturalist* too. Raven's was a leading voice, urging field biologists to fight against the obliteration of biological diversity, the very matter they studied. Wilson called Raven and said: Count me in.

The shift from sociobiology and its principles to biological diversity and its preservation was a panning movement along a spectrum of issues for Wilson, not a vault over some intellectual wall. The concept of biophilia, for instance, like the concept of altruism as a heritable tendency, involves speculation about how human behavioral motifs have been shaped, partly, by natural selection over evolutionary stretches of time, acting upon genetic variation.

Still, there was something distinctly different. *Biophilia* the book, even more than biophilia the idea, marked a new stage in Wilson's career and his reputation. He revealed himself as a graceful writer and a compelling memoirist who could mix storytelling with big ideas, delivering a potpourri of scientific, philosophic, and narrative material in a very human, companionable voice. He drew on old journals from his field travels, especially to Suriname in 1961—a respite from Harvard, after his Melanesian journey and before his partnership with Robert MacArthur, a mathematical ecologist with whom he coauthored his first important, very technical book. (That was *The Theory of Island Biogeography*, 1967.) The anecdotes and observations from Suriname, the central Amazon, New Guinea, and elsewhere served him, like shovels, as he dug toward deeper insights.

And he didn't confine his attention to ants. Here in the Suriname forest was a three-toed sloth, upon which lived tiny moths of the species *Cryptoses choloepi*, innocuous passengers amid the mammal's fur. When the sloth climbed down from the treetops to the forest floor to defecate, a weekly routine, female moths would emerge from its fur, flying out (or jumping off?) to lay their eggs on its fresh, warm dung, which would nourish the hatchling caterpillars from those eggs. When the caterpillars grew big and metamorphosed into adults, they would fly back into the canopy, find another sloth to ride, and the cycle would repeat. By living directly and patiently on the sloths, rather than flapping around in the understory looking for sloth dung, the females of *C. choloepi* saved themselves energy while ensuring that their young would get immediate access to choice food amid a very competitive forest.

"The study of every kind of organism matters," Wilson wrote, because "nothing in the whole system makes sense until the natural history of the constituent species becomes known." (This was an echo, probably conscious, of the Russian-American geneticist Theodosius Dobzhansky's famous dictum: "Nothing in biology makes sense except in the light of evolution.") That is, the science of ecology must be built by beginning from individual life histories. Wilson's statement was hyperbolic, maybe, but hyperbole containing a truth to which he has remained committed: details count. Work from the ground up. Even the big principles of theoretical ecology—and Wilson has ever been hungry to find big principles—cannot be extracted except from the assemblage of multifarious data about how and where particular creatures live. Robert MacArthur, the brilliant mathematical modeler, did his dissertation on warblers.

Charles Darwin devoted eight years to barnacle taxonomy. Wilson knew ants, first of all for their own sake. That's why he later titled his autobiography *Naturalist*, not *Theorist*.

The other change that manifested in *Biophilia* was Wilson's awakened concern over the global trend of biodiversity loss. The whole book, in a sense, was an argument that we humans *need* and we *want*—more deeply than we even know, as deep as our genomes, shaped by millions of years of evolution—to live surrounded by other living creatures of all sorts, wriggling, furry, large, tiny, brown, yellow, green. In its penultimate chapter, titled "The Conservation Ethic" (his variant on Aldo Leopold's "The Land Ethic"), Wilson places human-caused destruction of biological diversity second only below nuclear war on the list of regrettable blunders that humans might inflict upon the planet and their own descendants. He judges it worse than "energy depletion, economic collapse, conventional war, or even the expansion of totalitarian governments," as an irreparable tragedy. (Interestingly, he doesn't mention climate change, or even "global warming," on this list, which suggests how little that problem occupied attentions at the time. Nowadays he would cite climate change as a rising contributor to the losses of biological diversity—among corals and other marine species, for instance.) What's worse than everything except nuclear Armageddon? "The one process now going on that will take millions of years to correct is the loss of genetic and species diversity by the destruction of natural habitats. This is the folly our descendants are least likely to forgive us."

Those words, published in the slim, anomalous *Biophilia* in 1984, defined much of his working agenda for the next thirty-five years. Most immediately, they led to his encyclopedic but dazzling 1992 book, *The Diversity of Life*, also included here. Beyond that, they guided him toward more wild places, more efforts of persuasion to save those wild places, more scientific studies, more tireless acts of hope and determination, more explosions of boyish wonderment, and (best for us) more literary projects, as he moved across planet Earth, with his forceps, and his collecting bottles, and his hand lens, and his eye wide open.

BIOPHILIA

Soft, to your places, animals,
Your legendary duty calls.
THOMAS KINSELLA

Contents

Prologue

O N MARCH 12, 1961, I stood in the Arawak village of Bernhardsdorp and looked south across the white-sand coastal
forest of Surinam. For reasons that were to take me twenty
years to understand, that moment was fixed with uncommon
urgency in my memory. The emotions I felt were to grow more
poignant at each remembrance, and in the end they changed
into rational conjectures about matters that had only a distant
bearing on the original event.

The object of the reflection can be summarized by a single
word, biophilia, which I will be so bold as to define as the
innate tendency to focus on life and lifelike processes. Let me
explain it very briefly here and then develop the larger theme
as I go along.

From infancy we concentrate happily on ourselves and other
organisms. We learn to distinguish life from the inanimate
and move toward it like moths to a porch light. Novelty and
diversity are particularly esteemed; the mere mention of the
word *extraterrestrial* evokes reveries about still unexplored
life, displacing the old and once potent *exotic* that drew earlier generations to remote islands and jungled interiors. That
much is immediately clear, but a great deal more needs to be
added. I will make the case that to explore and affiliate with
life is a deep and complicated process in mental development.
To an extent still undervalued in philosophy and religion, our
existence depends on this propensity, our spirit is woven from
it, hope rises on its currents.

There is more. Modern biology has produced a genuinely
new way of looking at the world that is incidentally congenial
to the inner direction of biophilia. In other words, instinct
is in this rare instance aligned with reason. The conclusion I
draw is optimistic: to the degree that we come to understand
other organisms, we will place a greater value on them, and on
ourselves.

Bernhardsdorp

A T BERNHARDSDORP on an otherwise ordinary tropical morning, the sunlight bore down harshly, the air was still and humid, and life appeared withdrawn and waiting. A single thunderhead lay on the horizon, its immense anvil shape diminished by distance, an intimation of the rainy season still two or three weeks away. A footpath tunneled through the trees and lianas, pointing toward the Saramacca River and far beyond, to the Orinoco and Amazon basins. The woodland around the village struggled up from the crystalline sands of the Zanderij formation. It was a miniature archipelago of glades and creekside forest enclosed by savanna—grassland with scattered trees and high bushes. To the south it expanded to become a continuous lacework fragmenting the savanna and transforming it in turn into an archipelago. Then, as if conjured upward by some unseen force, the woodland rose by stages into the triple-canopied rain forest, the principal habitat of South America's awesome ecological heartland.

In the village a woman walked slowly around an iron cooking pot, stirring the fire beneath with a soot-blackened machete. Plump and barefoot, about thirty years old, she wore two long pigtails and a new cotton dress in a rose floral print. From politeness, or perhaps just shyness, she gave no outward sign of recognition. I was an apparition, out of place and irrelevant, about to pass on down the footpath and out of her circle of required attention. At her feet a small child traced meanders in the dirt with a stick. The village around them was a cluster of no more than ten one-room dwellings. The walls were made of palm leaves woven into a herringbone pattern in which dark bolts zigzagged upward and to the onlooker's right across flesh-colored squares. The design was the sole indigenous artifact on display. Bernhardsdorp was too close to Paramaribo, Surinam's capital, with its flood of cheap manufactured products to keep the look of a real Arawak village. In culture as in name, it had yielded to the colonial Dutch.

A tame peccary watched me with beady concentration from beneath the shadowed eaves of a house. With my own, taxonomist's eye I registered the defining traits of the collared species, *Dicotyles tajacu:* head too large for the piglike body, fur coarse and brindled, neck circled by a pale thin stripe, snout tapered, ears erect, tail reduced to a nub. Poised on stiff little dancer's legs, the young male seemed perpetually fierce and ready to charge yet frozen in place, like the metal boar on an ancient Gallic standard.

A note: Pigs, and presumably their close relatives the peccaries, are among the most intelligent of animals. Some biologists believe them to be brighter than dogs, roughly the rivals of elephants and porpoises. They form herds of ten to twenty members, restlessly patrolling territories of about a square mile. In certain ways they behave more like wolves and dogs than social ungulates. They recognize one another as individuals, sleep with their fur touching, and bark back and forth when on the move. The adults are organized into dominance orders in which the females are ascendant over males, the reverse of the usual mammalian arrangement. They attack in groups if cornered, their scapular fur bristling outward like porcupine quills, and can slash to the bone with sharp canine teeth. Yet individuals are easily tamed if captured as infants and their repertory stunted by the impoverishing constraints of human care.

So I felt uneasy—perhaps the word is embarrassed—in the presence of a captive individual. This young adult was a perfect anatomical specimen with only the rudiments of social behavior. But he was much more: a powerful presence, programed at birth to respond through learning steps in exactly the collared-peccary way and no other to the immemorial environment from which he had been stolen, now a mute speaker trapped inside the unnatural clearing, like a messenger to me from an unexplored world.

I stayed in the village only a few minutes. I had come to study ants and other social insects living in Surinam. No trivial task: over a hundred species of ants and termites are found within a square mile of average South American tropical forest. When all the animals in a randomly selected patch of woodland are collected together and weighed, from tapirs and parrots down to the smallest insects and roundworms, one third of the weight is found to consist of ants and termites. If you close your eyes and lay your hand on a tree trunk almost anywhere in the tropics until you feel something touch it, more times than not the crawler will be an ant. Kick open a rotting log and termites pour out. Drop a crumb of bread on the ground and within minutes ants of one kind or another drag it down a nest hole. Foraging ants are the chief predators of insects and other small animals in the tropical forest, and termites are the key animal decomposers of wood. Between them they form the conduit for a large part of the energy flowing through the forest. Sunlight to leaf to caterpillar to ant to anteater to jaguar to maggot to humus to termite to dissipated heat: such are the links that compose the great energy network around Surinam's villages.

I carried the standard equipment of a field biologist: camera; canvas satchel containing forceps, trowel, ax, mosquito repellent, jars, vials of alcohol, and notebook; a twenty-power hand lens swinging with a reassuring tug around the neck; partly fogged eyeglasses sliding down the nose and khaki shirt plastered to the back with sweat. My attention was on the forest; it has been there all my life. I can work up some appreciation for the travel stories of Paul Theroux and other urbanophile authors who treat human settlements as virtually the whole world and the intervening natural habitats as troublesome barriers. But everywhere I have gone—South America, Australia, New Guinea, Asia—I have thought exactly the opposite. Jungles and grasslands are the logical destinations, and towns and farmland the labyrinths that people have imposed between them sometime in the past. I cherish the green enclaves accidentally left behind.

Once on a tour of Old Jerusalem, standing near the elevated site of Solomon's Throne, I looked down across the Jericho Road to the dark olive trees of Gethsemane and wondered which native Palestinian plants and animals might still be found in the shade underneath. Thinking of "Go to the ant, thou sluggard; consider her ways," I knelt on the cobblestones to watch harvester ants carry seeds down holes to their subterranean granaries, the same food-gathering activity that had impressed the Old Testament writer, and possibly the same species at the very same place. As I walked with my host back past the Temple Mount toward the Muslim Quarter, I made inner calculations of the number of ant species found within the city walls. There was a perfect logic to such eccentricity: the million-year history of Jerusalem is at least as compelling as its past three thousand years.

AT BERNHARDSDORP I imagined richness and order as an intensity of light. The woman, child, and peccary turned into incandescent points. Around them the village became a black disk, relatively devoid of life, its artifacts adding next to nothing. The woodland beyond was a luminous bank, sparked here and there by the moving lights of birds, mammals, and larger insects.

I walked into the forest, struck as always by the coolness of the shade beneath tropical vegetation, and continued until I came to a small glade that opened onto the sandy path. I narrowed the world down to the span of a few meters. Again I tried to compose the mental set—call it the naturalist's trance, the hunter's trance—by which biologists locate more elusive organisms. I imagined that this place and all its treasures were mine alone and might be so forever in memory—if the bulldozer came.

In a twist my mind came free and I was aware of the hard workings of the natural world beyond the periphery of ordinary attention, where passions lose their meaning and history is in another dimension, without people, and great events pass without record or judgment. I was a transient of no consequence in this familiar yet deeply alien world that I had come to love. The uncounted products of evolution were gathered there for purposes having nothing to do with me; their long Cenozoic history was enciphered into a genetic code I could not understand. The effect was strangely calming. Breathing and heartbeat diminished, concentration intensified. It seemed to me that something extraordinary in the forest was very close to where I stood, moving to the surface and discovery.

I focused on a few centimeters of ground and vegetation. I willed animals to materialize, and they came erratically into view. Metallic-blue mosquitoes floated down from the canopy in search of a bare patch of skin, cockroaches with variegated wings perched butterfly-like on sunlit leaves, black carpenter ants sheathed in recumbent golden hairs filed in haste through moss on a rotting log. I turned my head slightly and all of them vanished. Together they composed only an infinitesimal fraction of the life actually present. The woods were a biological maelstrom of which only the surface could be scanned by the naked eye. Within my circle of vision, millions of unseen organisms died each second. Their destruction was swift and silent; no bodies thrashed about, no blood leaked into the ground. The microscopic bodies were broken apart in clean biochemical chops by predators and scavengers, then assimilated to create millions of new organisms, each second.

Ecologists speak of "chaotic regimes" that rise from orderly processes and give rise to others in turn during the passage of life from lower to higher levels of organization. The forest was a tangled bank tumbling down to the grassland's border. Inside it was a living sea through which I moved like a diver groping across a littered floor. But I knew that all around me bits and pieces, the individual organisms and their populations, were working with extreme precision. A few of the species were locked together in forms of symbiosis so intricate that to pull out one would bring others spiraling to extinction. Such is the consequence of adaptation by coevolution, the reciprocal genetic change of species that interact with each other through many life cycles. Eliminate just one kind of tree out of hundreds in such a forest, and some of its pollinators, leafeaters, and woodborers will disappear with it, then various of their parasites and key predators, and perhaps a species of bat or bird that depends on its fruit—and when will the reverberations end? Perhaps not until a large part of the diversity of the forest collapses like an arch crumbling as the keystone is pulled away. More likely the effects will remain local, ending with a minor shift in the overall pattern of abundance among the numerous surviving species. In either case the effects are beyond the power of present-day ecologists to predict. It is enough to work on the assumption that all of the details matter in the end, in some unknown but vital way.

After the sun's energy is captured by the green plants, it flows through chains of organisms dendritically, like blood spreading from the arteries into networks of microscopic capillaries. It is in such capillaries, in the life cycles of thousands of individual species, that life's important work is done. Thus nothing in the whole system makes sense until the natural history of the constituent species becomes known. The study of every kind of organism matters, everywhere in the world. That conviction leads the field biologist to places like Surinam and the outer limits of evolution, of which this case is exemplary:

> The three-toed sloth feeds on leaves high in the canopy of the lowland forests through large portions of South and Central America. Within its fur live tiny moths, the species *Cryptoses choloepi*, found nowhere else on Earth. When a sloth descends to the forest floor to defecate (once a week), female moths leave the fur briefly to deposit their eggs on the fresh dung. The emerging caterpillars build nests of silk and start to feed. Three weeks later they complete their development by turning into adult moths, and then fly up into the canopy in search of sloths. By living directly on the bodies of the sloths, the adult *Cryptoses* assure their offspring first crack at the nutrient-rich excrement and a competitive advantage over the myriad of other coprophages.

At Bernhardsdorp the sun passed behind a small cloud and the woodland darkened. For a moment all that marvelous environment was leveled and subdued. The sun came out again and shattered the vegetative surfaces into light-based niches. They included intensely lighted leaf tops and the tops of miniature canyons cutting vertically through tree bark to create shadowed depths two or three centimeters below. The light filtered down from above as it does in the sea, giving out permanently in the lowermost recesses of buttressed tree trunks and penetralia of the soil and rotting leaves. As the light's intensity rose and fell with the transit of the sun, silverfish, beetles, spiders, bark lice, and other creatures were summoned from their sanctuaries and retreated back in alternation. They responded according to receptor thresholds built into their eyes and brains, filtering devices that differ from one kind of animal to another. By such inborn controls the species imposed a kind of prudent self-discipline. They unconsciously halted their population growth before squeezing out competitors, and others did the same. No altruism was needed to achieve this balance, only specialization. Coexistence was an incidental by-product of the Darwinian advantage that accrued from the avoidance of competition. During the long span of evolution the species divided the environment among themselves, so that now each tenuously preempted certain of the capillaries of energy flow. Through repeated genetic changes they sidestepped competitors and built elaborate defenses against the host of predator species that relentlessly tracked them through matching genetic countermoves. The result was a splendid array of specialists, including moths that live in the fur of three-toed sloths.

NOW TO THE very heart of wonder. Because species diversity was created prior to humanity, and because we evolved within it, we have never fathomed its limits. As a consequence, the living world is the natural domain of the most restless and paradoxical part of the human spirit. Our sense of wonder grows exponentially: the greater the knowledge, the deeper the mystery and the more we seek knowledge to create new mystery. This catalytic reaction, seemingly an inborn human trait, draws us perpetually forward in a search for new places and new life. Nature is to be mastered, but (we hope) never completely. A quiet passion burns, not for total control but for the sensation of constant advance.

At Bernhardsdorp I tried to convert this notion into a form that would satisfy a private need. My mind maneuvered through an unending world suited to the naturalist. I looked in reverie down the path through the savanna woodland and imagined walking to the Saramacca River and beyond, over the horizon, into a timeless reconnaissance through virgin forests to the land of magical names, Yékwana, Jívaro, Sirionó, Tapirapé, Siona-Secoya, Yumana, back and forth, never to run out of fresh jungle paths and glades.

The same archetypal image has been shared in variations by others, and most vividly during the colonization of the New World. It comes through clearly as the receding valleys and frontier trails of nineteenth-century landscape art in the paintings of Albert Bierstadt, Frederic Edwin Church, Thomas Cole, and their contemporaries during the crossing of the American West and the innermost reaches of South America.

In Bierstadt's *Sunset in Yosemite Valley* (1868), you look down a slope that eases onto the level valley floor, where a river flows quietly away through waist-high grass, thickets, and scattered trees. The sun is near the horizon. Its dying light, washing the surface in reddish gold, has begun to yield to blackish green shadows along the near side of the valley. A cloud bank has lowered to just beneath the tops of the sheer rock walls. More protective than threatening, it has transformed the valley into a tunnel opening out through the far end into a sweep of land. The world beyond is obscured by the blaze of the setting sun into which we are forced to gaze in order to see that far. The valley, empty of people, is safe: no fences, no paths, no owners. In a few minutes we could walk to the river, make camp, and afterward explore away from the banks at leisure. The ground in sight is human-sized, measured literally by foot strides and strange new plants and animals large enough to be studied at twenty paces. The dreamlike quality of the painting rolls time forward: what might the morning bring? History is still young, and human imagination has not yet been chained by precise geographic knowledge. Whenever we wish, we can strike out through the valley to the unknown terrain beyond, to a bor-derland of still conceivable prodigies—bottomless vales and boundless floods, in Edgar Allan Poe's excited imagery, "and chasms, and caves and Titan woods with forms that no man can discover." The American frontier called up the old emo-tions that had pulled human populations like a living sheet over the world during the ice ages. The still unfallen western world, as Melville wrote of the symbolizing White Steed in *Moby-Dick*, "revived the glories of those primeval times when Adam walked majestic as a god."

Then a tragedy: this image is almost gone. Although perhaps as old as man, it has faded during our own lifetime. The wilder-nesses of the world have shriveled into timber leases and threat-ened nature reserves. Their parlous state presents us with a dilemma, which the historian Leo Marx has called the machine in the garden. The natural world is the refuge of the spirit, remote, static, richer even than human imagination. But we cannot exist in this paradise without the machine that tears it apart. We are killing the thing we love, our Eden, progenitrix, and sibyl. Human beings are not captive peccaries, natural crea-tures torn from a sylvan niche and imprisoned within a world of artifacts. The noble savage, a biological impossibility, never existed. The human relation to nature is vastly more subtle and ambivalent, probably for this reason. Over thousands of generations the mind evolved within a ripening culture, creating itself out of symbols and tools, and genetic advantage accrued from planned modifications of the environment. The unique operations of the brain are the result of natural selection operating through the filter of culture. They have suspended us between the two antipodal ideals of nature and machine, forest and city, the natural and the artifactual, relentlessly seeking, in the words of the geographer Yi-Fu Tuan, an equilibrium not of this world.

So at Bernhardsdorp my own thoughts were inconstant. They skipped south to the Saramacca and on deep into the Amazon basin, the least spoiled garden on Earth, and then swiftly back north to Paramaribo and New York, greatest of machines. The machine had taken me there, and if I ever seriously thought of confronting nature without the conveniences of civilization, reality soon regained my whole attention. The living sea is full of miniature horrors designed to reduce visiting biologists to their constituent amino acids in quick time. Arboviruses visit the careless intruder with a dismaying variety of chills and diarrhea. Breakbone fever swells the joints to agonizing tightness. Skin ulcers spread remorselessly outward from thorn scratches on the ankle. Triatoma assassin bugs suck blood from the sleeper's face during the night and leave behind the fatal microorganisms of Chagas' disease—surely history's most unfair exchange. Leishmaniasis, schistosomiasis, malignant tertian malaria, filariasis, echinococcosis, onchocerciasis, yellow fever, amoebic dysentery, bleeding bot-fly cysts . . . evolution has devised a hundred ways to macerate livers and turn blood into a parasite's broth. So the romantic voyager swallows chloraquin, gratefully accepts gamma globulin shots, sleeps under mosquito netting, and remembers to pull on rubber boots before wading in freshwater streams. He hopes that enough fuel was put into the Land Rover that morning, and he hurries back to camp in time for a hot meal at dusk.

The impossible dilemma caused no problem for ancestral men. For millions of years human beings simply went at nature with everything they had, scrounging food and fighting off predators across a known world of a few square miles. Life

was short, fate terrifying, and reproduction an urgent priority: children, if freely conceived, just about replaced the family
members who seemed to be dying all the time. The population flickered around equilibrium, and sometimes whole bands
became extinct. Nature was something out there—nameless
and limitless, a force to beat against, cajole, and exploit.

If the machine gave no quarter, it was also too weak to break
the wilderness. But no matter: the ambiguity of the opposing
ideals was a superb strategy for survival, just so long as the
people who used it stayed sufficiently ignorant. It enhanced the
genetic evolution of the brain and generated more and better
culture. The world began to yield, first to the agriculturists
and then to technicians, merchants, and circumnavigators.
Humanity accelerated toward the machine antipode, heedless
of the natural desire of the mind to keep the opposite as well.
Now we are near the end. The inner voice murmurs *You went
too far*, and disturbed the world, and gave away too much for
your control of Nature. Perhaps Hobbes's definition is correct, and this will be the hell we earned for realizing truth
too late. But I demur in all this. I suggest otherwise: the same
knowledge that brought the dilemma to its climax contains the
solution. Think of scooping up a handful of soil and leaf litter
and spreading it out on a white ground cloth, in the manner
of the field biologist, for close examination. This unprepossessing lump contains more order and richness of structure,
and particularity of history, than the entire surfaces of all the
other (lifeless) planets. It is a miniature wilderness that can take
almost forever to explore.

Tease apart the adhesive grains with the aid of forceps, and
you will expose the tangled rootlets of a flowering plant, curl-
ing around the rotting veins of humus, and perhaps some larger
object such as the boat-shaped husk of a seed. Almost certainly
among them will be a scattering of creatures that measure the
world in millimeters and treat this soil sample as traversable:
ants, spiders, springtails, armored oribatid mites, enchytraeid
worms, millipedes. With the aid of a dissecting microscope,
proceed on down the size scale to the roundworms, a world
of scavengers and fanged predators feeding on them. In the
hand-held microcosm all these creatures are still giants in a
relative sense. The organisms of greatest diversity and numbers

are invisible or nearly so. When the soil-and-litter clump is progressively magnified, first with a compound light microscope and then with scanning electron micrographs, specks of dead leaf expand into mountain ranges and canyons, soil particles become heaps of boulders. A droplet of moisture trapped between root hairs grows into an underground lake, surrounded by a three-dimensional swamp of moistened humus. The niches are defined by both topography and nuances in chemistry, light, and temperature shifting across fractions of a millimeter. Organisms now come into view for which the soil sample is a complete world. In certain places are found the fungi: cellular slime molds, the one-celled chitin-producing chytrids, minute gonapodyaceous and oomycete soil specialists, Kickxellales, Eccrinales, Endomycetales, and Zoopagales. Contrary to their popular reputation, the fungi are not formless blobs, but exquisitely structured organisms with elaborate life cycles. The following is a recently discovered extreme specialization, the example of the sloth moth repeated on a microscopic scale:

In water films and droplets, attack cells of an oomycete, *Haptoglossa mirabilis*, await the approach of small, fat wormlike animals the biologists call rotifers. Each cell is shaped like a gun; its anterior end is elongated to form a barrel, which is hollowed out to form a bore. At the base of the bore is a complicated explosive device. When a rotifer swims close, the attack cell detects its characteristic odor and fires a projectile of infective tissue through the barrel and into its body. The fungal cells proliferate through the victim's tissues and then metamorphose into a cylindrical fruiting body, from which exit tubes sprout. Next tiny spores separate themselves inside the fruiting body, swim out the exit tubes with the aid of whip-shaped hairs, and settle down to form new attack cells. They await more rotifers, prepared to trigger the soundless explosion that will commence a new life cycle.

Still smaller than the parasitic fungi are the bacteria, including colony-forming polyangiaceous species, specialized predators that consume other bacteria. All around them live rich mixtures of rods, cocci, coryneforms, and slime azotobacteria. Together these microorganisms metabolize the entire spectrum of live and dead tissue. At the moment of discovery some are actively growing and fissioning, while others lie dormant in wait for the right combination of nutrient chemicals. Each species is kept at equilibrium by the harshness of the environment. Any one, if allowed to expand without restriction for a few weeks, would multiply exponentially, faster and faster, until it weighed more than the entire Earth. But in reality the individual organism simply dissolves and assimilates whatever appropriate fragments of plants and animals come to rest near it. If the newfound meal is large enough, it may succeed in growing and reproducing briefly before receding back into the more normal state of physiological quiescence.

Biologists, to put the matter as directly as possible, have begun a second reconnaissance into the land of magical names. In exploring life they have commenced a pioneering adventure with no imaginable end. The abundance of organisms increases downward by level, like layers in a pyramid. The handful of soil and litter is home for hundreds of insects, nematode worms, and other larger creatures, about a million fungi, and ten billion bacteria. Each of the species of these organisms has a distinct life cycle fitted, as in the case of the predatory fungus, to the portion of the microenvironment in which it thrives and reproduces. The particularity is due to the fact that it is programed by an exact sequence of nucleotides, the ultimate molecular units of the genes.

The amount of information in the sequence can be measured in bits. One bit is the information required to determine which of two equally likely alternatives is chosen, such as heads or tails in a coin toss. English words average two bits per letter. A single bacterium possesses about ten million bits of genetic information, a fungus one billion, and an insect from one to ten billion bits according to species. If the information in just one insect—say an ant or beetle—were to be translated into a code of English words and printed in letters of standard size, the string would stretch over a thousand miles. Our lump of earth contains information that would just about fill all fifteen editions of the *Encyclopaedia Britannica*.

To see what such molecular information can do, consider a column of ants running across the floor of a South American forest. Riding on the backs of some of the foragers are minute workers of the kind usually confined to duties within the underground nursery chambers. The full significance of hitchhiking is problematic, but at the very least the act helps to protect the colony against parasites. Tiny flies, members of the family Phoridae, hover above the running foragers. From time to time a fly dives down to thrust an egg into the neck of one of them. Later the egg hatches into a maggot that burrows deeper into the ant's body. The maggot grows rapidly, transforms into a pupa, and eventually erupts through the cuticle as an adult fly to restart the life cycle. The divebombers find the runners easy targets when they are burdened with a fragment of food. But when one also carries a hitchhiker, the smaller ant is able to chase the intruder away with its jaws and legs. It serves as a living fly whisk.

The brain of the fly or of the fly-whisk ant, when dissected out and placed in a drop of saline solution on a glass slide, resembles a grain of sugar. Although barely visible to the naked eye, it is a complete command center that choreographs the insect's movements through its entire adult cycle. It signals the precise hour for the adult to emerge from the pupal case; it processes the flood of signals transduced to it by the outer sensors; and it directs the performance of about twenty behavioral acts through nerves in the legs, antennae, and mandibles. The fly and the ant are hardwired in a manner unique to their respective species and hence radically different from each other, so that predator is implacably directed against prey, flier against runner, solitaire against colony member.

With advanced techniques it has been possible to begin mapping insect nervous systems in sufficient detail to draw the equivalent of wiring diagrams. Each brain consists of somewhere between a hundred thousand and a million nerve cells, most of which send branches to a thousand or more of their neighbors. Depending on their location, individual cells appear to be programed to assume a particular shape and to transmit messages only when stimulated by coded discharges from neighbor units that feed into them. In the course of evolution, the entire system has been miniaturized to an extreme. The fatty sheaths surrounding the axon shafts of the kind found in larger animals have been largely stripped away, while the cell bodies are squeezed off to one side of the multitudinous nerve connections. Biologists understand in very general terms how the insect brain might work as a complete on-board computer, but they are a long way from explaining or duplicating such a device in any detail.

The great German zoologist Karl von Frisch once said of his favorite organism that the honeybee is like a magic well: the more you draw from it, the more there is to draw. But science is in no other way mystical. Its social structure is such that anyone can follow most enterprises composing it, as observer if not as participant, and soon you find yourself on the boundaries of knowledge.

You start with the known: in the case of the honeybee, where it nests, its foraging expeditions, and its life cycle. Most remarkable at this level is the waggle dance discovered by von Frisch, the tail-wagging movement performed inside the hive to inform nestmates of the location of newly discovered flower patches and nest sites. The dance is the closest approach known in the animal kingdom to a true symbolic language. Over and over again the bee traces a short line on the vertical surface of the comb, while sister workers crowd in close behind. To return to the start of the line, the bee loops back first to the left and then to the right and so produces a figure-eight. The center line contains the message. Its length symbolically represents the distance from the hive to the goal, and its angle away from a line drawn straight up on the comb, in other words away from twelve o'clock, represents the angle to follow right or left of the sun when leaving the hive. If the bee dances straight up the surface of the comb, she is telling the others to fly toward the sun. If she dances ten degrees to the right, she causes them to go ten degrees right of the sun. Using such directions alone, the members of the hive are able to harvest nectar and pollen from flowers three miles or more from the hive.

The revelation of the waggle-dance code has pointed the way to deeper levels of biological investigation, and a hundred new questions. How does the bee judge gravity while on the darkened comb? What does it use for a guide when the sun goes behind a cloud? Is the waggle dance inherited or must it be learned? The answers create new concepts that generate still more mysteries. To pursue them (and we are now certainly at the frontier) investigators must literally enter the bee itself, exploring its nervous system, the interplay of its hormones and behavior, the processing of chemical cues by its nervous system. At the level of cell and tissue, the interior of the body will prove more technically challenging than the external workings of the colony first glimpsed. We are in the presence of a biological machine so complicated that to understand just one part of it—wings, heart, ovary, brain—can consume many lifetimes of original investigation.

And if that venture were somehow to be finished, it will merely lead on down into the essence of the machine, to the interior of cells and the giant molecules that compose their distinctive parts. Questions about process and meaning then take center stage. What commits an embryonic cell to become part of the brain instead of a respiratory unit? Why does the mother's blood invest yolk in the growing egg? Where are the genes that control behavior? Even in the unlikely event that all this microscopic domain is successfully mapped, the quest still lies mostly ahead. The honeybee, *Apis mellifera*, is the product of a particular history. Through fossil remains in rock and amber, we know that its lineage goes back at least 50 million years. Its contemporary genes were assembled by an astronomical number of events that sorted and recombined the constituent nucleotides. The species evolved as the outcome of hourly contacts with thousands of other kinds of plants and animals along the way. Its range expanded and contracted across Africa and Eurasia in a manner reminiscent of the fortunes of a human tribe. Virtually all this history remains unknown. It can be pursued to any length by those who take a special interest in *Apis mellifera* and seek what Charles Butler called its "most sweet and sov'raigne fruits" when he launched the modern scientific study of the honeybee in 1609.

Every species is a magic well. Biologists have until recently been satisfied with the estimate that there are between three and ten million of them on Earth. Now many believe that ten million is too low. The upward revision has been encouraged by the increasingly successful penetration of the last great

unexplored environment of the planet, the canopy of the tropical rain forest, and the discovery of an unexpected number of
new species living there. This layer is a sea of branches, leaves,
and flowers crisscrossed by lianas and suspended about one
hundred feet above the ground. It is one of the easiest habitats
to locate—from a distance at least—but next to the deep sea
the most difficult to reach. The tree trunks are thick, arrow-
straight, and either slippery smooth or covered with sharp
tubercles. Anyone negotiating them safely to the top must then
contend with swarms of stinging ants and wasps. A few athletic
and adventurous younger biologists have begun to overcome
the difficulties by constructing special pulleys, rope catwalks,
and observation platforms from which they can watch high
arboreal animals in an undisturbed state. Others have found a
way to sample the insects, spiders, and other arthropods with
insecticides and quick-acting knockdown agents. They first
shoot lines up into the canopy, then hoist the chemicals up
in canisters and spray them out into the surrounding vegetation by remote control devices. The falling insects and other
organisms are caught in sheets spread over the ground. The
creatures discovered by these two methods have proved to be
highly specialized in their food habits, the part of the tree in
which they live, and the time of the year when they are active.
So an unexpectedly large number of different kinds are able to
coexist. Hundreds can fit comfortably together in a single tree
top. On the basis of a preliminary statistical projection from
these data, Terry L. Erwin, an entomologist at the National
Museum of Natural History, has estimated that there may be
thirty million species of insects in the world, most limited to
the upper vegetation of the tropical forests.

Although such rough approximations of the diversity of life
are not too difficult to make, the exact number of species is
beyond reach because—incredibly—the majority have yet to
be discovered and specimens placed in museums. Furthermore,
among those already classified no more than a dozen have been
studied as well as the honeybee. Even *Homo sapiens*, the focus
of billions of dollars of research annually, remains a seemingly
intractable mystery. All of man's troubles may well arise, as
Vercors suggested in *You Shall Know Them*, from the fact that
we do not know what we are and do not agree on what we want

to become. This crucial inadequacy is not likely to be remedied until we have a better grasp of the diversity of the life that created and sustains us. So why hold back? It is a frontier literally at our fingertips, and the one for which our spirit appears to have been explicitly designed.

I WALKED on through the woodland at Bernhardsdorp to see what the day had to offer. In a decaying log I found a species of ant previously known only from the midnight zone of a cave in Trinidad. With the aid of my hand lens I identified it from its unique combination of teeth, spines, and body sculpture. A month before I had hiked across five miles of foothills in central Trinidad to find it in the original underground habitat. Now suddenly here it was again, nesting and foraging in the open. Scratch from the list what had been considered the only "true" cave ant in the world—possessed of workers pale yellow, nearly eyeless, and sluggish in movement. Scratch the scientific name *Spelaeomyrmex*, meaning literally cave ant, as a separate taxonomic entity. I knew that it would have to be classified elsewhere, into a larger and more conventional genus called *Erebomyrma*, ant of Hades. A small quick victory, to be reported later in a technical journal that specializes on such topics and is read by perhaps a dozen fellow myrmecologists. I turned to watch some huge-eyed ants with the formidable name *Gigantiops destructor*. When I gave one of the foraging workers a freshly killed termite, it ran off in a straight line across the forest floor. Thirty feet away it vanished into a small hollow tree branch that was partly covered by decaying leaves. Inside the central cavity I found a dozen workers and their mother queen—one of the first colonies of this unusual insect ever recorded. All in all, the excursion had been more productive than average. Like a prospector obsessed with ore samples, hoping for gold, I gathered a few more promising specimens in vials of ethyl alcohol and headed home, through the village and out onto the paved road leading north to Paramaribo.

Later I set the day in my memory with its parts preserved for retrieval and closer inspection. Mundane events acquired the raiment of symbolism, and this is what I concluded from them: That the naturalist's journey has only begun and for all intents and purposes will go on forever. That it is possible to spend a lifetime in a magellanic voyage around the trunk of a single tree. That as the exploration is pressed, it will engage more of the things close to the human heart and spirit. And if this much is true, it seems possible that the naturalist's vision is only a specialized product of a biophilic instinct shared by all, that it can be elaborated to benefit more and more people. Humanity is exalted not because we are so far above other living creatures, but because knowing them well elevates the very concept of life.

The Superorganism

IN MARCH 1983 I returned to South America to begin a new program of study on tropical ants. I was interested in the way the communication systems and division of labor of these insects adapt them to their environment. My first stop was the field site of the Minimum Critical Size Project of the World Wildlife Fund, located in Amazonian forest sixty miles north of Manaus, Brazil. I was accompanied by Thomas Lovejoy, the young and vigorous vice-president for science of WWF-US, who had conceived the project in the late 1970s. We joined an assortment of researchers, students, and assistants who were working back and forth between Manaus and the site on weekly tours. Camaraderie came easily and was genuine; our shared values were implicit, forming a bond too strong to allow much discussion of why we had come together in this unlikely place. We talked only about organisms, in endless technical detail.

My hosts were not ordinary field biologists. They did not affect the verbal delicacy and critical reserve of typical academics encountered on leave from comfortable bases in Berkeley, Ann Arbor, and Cambridge. Their manner was self-confident and achievement-oriented, tough in a pleasant way, and they reminded me a bit of settlers I had met in Australia and New Guinea (and the Israeli biologist who pointed out the house where he had commanded a company during the 1967 war, as we returned from a field trip to the Dead Sea). Even though the World Wildlife Fund is operating on a slender budget, its Amazon Project is truly pioneering on a large scale. Planned to run into the next century, it is designed to answer one of the key questions of ecology and conservation practice: how extensive does a wildlife preserve have to be to sustain permanently most or all of the kinds of plants and animals protected within its boundaries?

We know that when a species loses part of its range, it is in greater danger of extinction. Expressed in loosely mathematical terms, the chance that a population of organisms will go extinct in a given year increases as its living space is cut back and its numbers are held at correspondingly lower numbers.

All populations fluctuate in size to some extent, but those kept at a low maximum are more likely to zigzag all the way down to zero than those permitted to fluctuate at higher levels. For example, a population of ten grizzly bears living on one hundred square miles of land will probably disappear much earlier than a population of a thousand grizzly bears living on ten thousand square miles of similar land; the thousand could persist for centuries or, so far as ordinary human awareness is concerned, forever.

This simple fact of nature bears heavily on the design of nature reserves. When a piece of primeval forest is set aside and the surrounding forest cleared, it becomes an island in an agricultural sea. Like wave-lapped Puerto Rico or Bali, it has lost most of its connections with other natural land habitats from which new immigrations can occur. Over a period of years the number of plant and animal species will fall to a new and predictable level. Some impoverishment of diversity is inevitable even if men never put an ax to a single tree in the reserve. This natural decline presents biologists with a problem that is technically difficult and soluble only through the chancing of risky compromises. The reserve they recommend must be small enough to be economically reasonable. They cannot ask that an entire country be set aside. But they have the obligation to insist that the patch be made large enough to sustain the fauna and flora. It is their job to prove that a certain minimum area is required, to list as completely as possible which species will be saved in the reserve, and for approximately how long.

The tropical rain forest north of Manaus, like that in many other parts of the Amazon basin, is being clear-cut from the edge inward. It is being lifted up from the ground entire like a carpet rolled off a bare floor, leaving behind vast stretches of cattle range and cropland that need artificial fertilization to sustain even marginal productivity for more than two or three years. A rain forest in Brazil differs fundamentally from a deciduous woodland in Pennsylvania or Germany in the way its key resources are distributed. A much greater fraction of organic matter is bound up in the tissues of the standing trees, so that the leaf litter and humus are only a few inches deep. When the forest is felled and burned, the hard equatorial downpours quickly wash away the thin blanket of top soil.

Although I had this general information in advance, I was still shaken by the sight of newly cleared land around Manaus. The pans and hillocks of lateritic clay, littered with blackened tree stumps, bore the look of a freshly deserted battlefield. Spherical termite mounds sprouted from the fallen wood in an ill-fated population explosion, while vultures and swifts wheeled overhead in representation of the mostly vanished bird fauna. Bony white cattle, forlorn replacements of a magnificent heritage, clustered in small groups around the scattered watersheds. Near midday the heat of the sun bounced up from the bare patches of soil to hit with an almost tactile force. It was another world altogether from the shadowed tunnels of the nearby forest, and a constant reminder of what had happened: tens of thousands of species had been scraped away as by a giant hand and will not be seen in that place for generations, if ever. The action can be defended (with difficulty) on economic grounds, but it is like burning a Renaissance painting to cook dinner.

The Brazilian authorities have sanctioned the opening of the wilderness on the basis of a logical formula: the impoverished Northeast has people and no land, the Amazon has land and no people; join them together and build a nation. But they are also well aware of the problems of environmental degradation. More recently, influenced by biologists such as Warwick Kerr, Paulo Nogueira Neto, and Paulo Vanzolini, they have begun to forge a policy of preservation. Now by law, honored at least in principle, half of the forest must be left standing. Of equal importance, more than twenty Amazonian reserves and parks have been set aside in key areas where the greatest number of species of plants and animals are thought to exist. Most are over a thousand square miles in extent, the minimum area that, according to Princeton University's John Terborgh and other experts on the subject, is needed to hold the number of disappearing species to less than 1 percent of the initial complement over the next century. In other words, the formula is meant to ensure that 99 out of every 100 kinds of organisms will still be present in the year 2100. With reserves of this size there is hope even for the harpy eagle and the jaguar, of which single individuals need three square miles or more to survive, as well as the flourish of rare orchids, monkeys, river fish, and

brilliant toucans and macaws that symbolize the admirable élan
of Brazil itself.

But this leaves the smaller reserves elsewhere in Brazil and
in the more densely populated countries of South and Central
America. The American and Brazilian scientists at the Manaus
project are addressing the problem in the following way. At the
edge of the cutting, where the rain forest begins and continues
virtually unbroken north to Venezuela, they have marked off a
series of twenty plots ranging in size from one to a thousand
hectares (a hectare is 100 meters on the side and equals 2.47
acres). In each plot they survey the most easily classified and
monitored of the big organisms, including the trees, butterflies, birds, monkeys, and other large mammals. Then, with the
cooperation of the landowners, they supervise the clearing of
the surrounding land, leaving the plots behind as forest islands
in a newly created agricultural sea. The surveys were begun
in 1980 and will be continued for many years. Eventually the
data should reveal how much faster the smaller island-reserves
lose their species than larger ones, which kinds of animals and
plants decline the most rapidly, why they become extinct, and,
most crucially, the minimum area needed to hold on to the
greater part of the diversity of life. No process being addressed
by modern science is more complicated or, in my opinion,
more important.

WE RODE in a World Wildlife Fund truck to a camp just inside
the forest border at Fazenda Esteio, where the biologists were
conducting one of the initial surveys. True to the philosophy
of the sponsoring organization, the camp was a tiny clearing
with a temporary shelter just large enough to hold a few ham-
mocks, plus a cook's shed and fireplace, and nothing more. To
my delight I found that I could roll out of my hammock in the
morning, take twenty steps, and be in virgin rain forest. For
five days I stayed in the woods except for meals and as little
sleep as I needed to keep going.

I savored the cathedral feeling expressed by Darwin in 1832
when he first encountered tropical forest near Rio de Janeiro
("wonder, astonishment & sublime devotion, fill & elevate the
mind"). And once again I could hold still for long intervals
to study a few centimeters of tree trunk or ground, finding

some new organism at each shift of focus. The intervals of total silence, often prolonged, became evidence of the intensity of the enveloping life. Several times a day I heard what may be the most distinctive sound of the primary tropical forest: a sharp *crack* like a rifle shot, followed by a whoosh and a solid thump. Somewhere a large tree, weakened by age and rot and top heavy from layers of vines, has chosen that moment to fall and end decades or centuries of life. The process is random and continuous, a sprinkling of events through the undisturbed portion of the forest. The broad trunk snaps or keels over to lever up the massive root system, the branches plow down through the canopies of neighboring trees at terrifying speed, and the whole thunders to the ground in a cloud of leaves, trailing lianas, and fluttering insects. There may be a hundred thousand trees within earshot of any place the hiker stands in the forest, so that the odds of hearing one coming down on a given day are high. But the chance of being close enough to be struck by any part of the tree is remote, comparable to that of stepping on a poisonous snake or coming round the bend of a trail one day to meet a mother jaguar with cubs. Still, the lifetime risk builds up cumulatively, like that in daily flights aboard single-engined airplanes, so that those who spend years in the forest count falling trees as an important source of danger.

Most of the time I worked with a restless energy to get ahead on several research projects I had in mind. I opened logs and twigs like presents on Christmas morning, entranced by the endless variety of insects and other small creatures that scuttled away to safety. None of these organisms was repulsive to me; each was beautiful, with a name and special meaning. It is the naturalist's privilege to choose almost any kind of plant or animal for examination and be able to commence productive work within a relatively short time. In the tropical forest, with thousands of mostly unknown species all around, the number of discoveries per investigator per day is probably greater than anywhere else in the world.

As if to dramatize the point, an insect I most wanted to find made its appearance soon after my arrival at Fazenda Esteio, with no effort of my own and literally at my feet. It was the leafcutter ant (*Atta cephalotes*), one of the most abundant and visually striking animals of the New World tropics. The *saúva*, as it is called locally, is a prime consumer of fresh vegetation, rivaled only by man, and a leading agricultural pest in Brazil. I had devoted years of research to the species in the laboratory but never studied it in the field. At dusk on the first day in camp, as the light failed to the point where we found it difficult to make out small objects on the ground, the first worker ants came scurrying purposefully out of the surrounding forest. They were brick red in color, about a quarter inch in length, and bristling with short, sharp spines. Within minutes, several hundred had arrived and formed two irregular files that passed on either side of the hammock shelter. They ran in a nearly straight line across the clearing, their paired antennae scanning right and left, as though drawn by some directional beam from the other side. Within an hour, the trickle expanded to twin rivers of tens of thousands of ants running ten or more abreast. The columns could be traced easily with the aid of a flashlight. They came up from a huge earthen nest a hundred yards from the camp on a descending slope, crossed the clearing, and disappeared again into the forest. By climbing through tangled undergrowth we were able to locate one of their main targets, a tall tree bearing white flowers high in its crown. The ants streamed up the trunk, scissored out pieces of leaves and petals with their sharp-toothed mandibles, and headed home carrying the fragments over their heads like little parasols. Some floated the pieces to the ground, where most were picked up and carried away by newly arriving nestmates. At maximum activity, shortly before midnight, the trails were a tumult of ants bobbing and weaving past each other like miniature mechanical toys.

For many visitors to the forest, even experienced naturalists, the foraging expeditions are the whole of the matter, and individual leafcutter ants seem to be inconsequential ruddy specks on a pointless mission. But a closer look transforms them into beings of another order. If we magnify the scene to human scale, so that an ant's quarter-inch length grows into six feet, the forager runs along the trail for a distance of about ten miles at a velocity of 16 miles an hour. Each successive mile is covered in three minutes and forty-five seconds, about the current (human) world record. The forager picks up a burden of 750 pounds and speeds back toward the nest at 15 miles an hour—hence, four-minute miles. This marathon is repeated many times during the night and in many localities on through the day as well.

From research conducted jointly by biologists and chemists, it is known that the ants are guided by a secretion paid onto the soil through the sting, in the manner of ink being drawn out of a pen. The crucial molecule is methyl-4-methylpyrrole-2-carboxylate, which is composed of a tight ring of carbon and nitrogen atoms with short side chains made of carbon and oxygen. The pure substance has an innocuous odor, judged by various people to be faintly grassy, sulphurous, or fruitlike with a hint of naphtha (I'm not sure I can smell it at all). But whatever the impact on human beings, it is an ichor of extraordinary power for the ants. One milligram, a quantity that would just about cover the printed letters in this sentence, if dispensed with theoretical maximum efficiency, is enough to excite billions of workers into activity or to lead a short column of them three times around the world. The vast difference between us and them has nothing to do with the trail substance itself, which is a biochemical material of unexceptional structure. It lies entirely in the unique sensitivity of the sensory organs and brains of the insects.

One millimeter above the ground, where ants exist, things are radically different from what they seem to the gigantic creatures who peer down from a thousand times that distance. The ants do not follow the trail substance as a liquid trace on the soil, as we are prone to think. It comes up to them as a cloud of molecules diffusing through still air at the ground surface. The foragers move inside a long ellipsoidal space in which the gaseous material is dense enough to be detected. They sweep their paired antennae back and forth in advance of the head to catch the odorant molecules. The antennae are the primary sensory centers of the ant. Their surfaces are furred with thousands of nearly invisible hairs and pegs, among which are scattered diminutive plates and bottle-necked pits. Each of these sense organs is serviced by cells that carry electrical impulses into the central nerve of the antenna. Then relay cells take over and transmit the messages to the integrating regions of the brain. Some of the antennal organs react to touch, while others are sensitive to slight movements of air, so that the ant responds instantaneously whenever the nest is breached by intruders. But most of the sensors monitor the chemicals that swirl around the ant in combinations that change through each second of its life. Human beings live in a world of sight and sound, but social insects exist primarily by smell and taste. In a word, we are audiovisual where they are chemical.

The oddness of the insect sensory world is illustrated by the swift sequence of events that occurs along the odor trail. When a forager takes a wrong turn to the left and starts to run away from the track, its left antenna breaks out of the odor space first and is no longer stimulated by the guiding substance. In a few thousandths of a second, the ant perceives the change and pulls back to the right. Twisting right and left in response to the vanishing molecules, it follows a tightly undulating course between the nest and tree. During the navigation it must also dodge moment by moment through a tumult of other runners. If you watch a foraging worker from a few inches away with the unaided eye, it seems to touch each passerby with its antennae, a kind of tactile probe. Slow-motion photography reveals that it is actually sweeping the tips of the antennae over parts of the other ant's body to smell it. If the surface does not present exactly the right combination of chemicals—the colony's unique odor signature—the ant attacks at once. It may simultaneously spray an alarm chemical from special glands located in the head capsule, causing others in the vicinity to rush to the site with their mandibles gaping.

An ant colony is organized by no more than ten or twenty such signals, most of which are chemical secretions leaked or sprayed from glands. The workers move with swiftness and precision through a life that human beings have come to understand only with the aid of mathematical diagrams and molecular formulas. We can also simulate the behavior. Computer technology has made it theoretically possible to create a mechanical ant that duplicates the observed activity. But the machine, if for some reason we chose to build one, would be the size of a small automobile, and even then I doubt if it would tell us anything new about the ant's inner nature.

At the end of the trail the burdened foragers rush down the nest hole, into throngs of nestmates and along tortuous channels that end near the water table fifteen feet or more below.

The ants drop the leaf sections onto the floor of a chamber, to be picked up by workers of a slightly smaller size who clip them into fragments about a millimeter across. Within minutes still smaller ants take over, crush and mold the fragments into moist pellets, and carefully insert them into a mass of similar material. This mass ranges in size between a clenched fist and a human head, is riddled with channels, and resembles a gray cleaning sponge. It is the garden of the ants: on its surface a symbiotic fungus grows which, along with the leaf sap, forms the ants' sole nourishment. The fungus spreads like a white frost, sinking its hyphae into the leaf paste to digest the abundant cellulose and proteins held there in partial solution.

The gardening cycle proceeds. Worker ants even smaller than those just described pluck loose strands of the fungus from places of dense growth and plant them onto the newly constructed surfaces. Finally, the very smallest—and most abundant—workers patrol the beds of fungal strands, delicately probing them with their antennae, licking their surfaces clean, and plucking out the spores and hyphae of alien species of mold. These colony dwarfs are able to travel through the narrowest channels deep within the garden masses. From time to time they pull tufts of fungus loose and carry them out to feed their larger nestmates.

The leafcutter economy is organized around this division of labor based on size. The foraging workers, about as big as houseflies, can slice leaves but are too bulky to cultivate the almost microscopic fungal strands. The tiny gardener workers, somewhat smaller than this printed letter I, can grow the fungus but are too weak to cut the leaves. So the ants form an assembly line, each successive step being performed by correspondingly smaller workers, from the collection of pieces of leaves out of doors to the manufacture of leaf paste to the cultivation of dietary fungi deep within the nest.

The defense of the colony is also organized according to size. Among the scurrying workers can be seen a few soldier ants, three hundred times heavier than the gardener workers. Their sharp mandibles are powered by massive adductor muscles that fill the swollen, quarter-inch-wide head capsules. Working like miniature wire clippers, they chop enemy insects into pieces and easily slice through human skin. These behemoths are especially adept at repelling large invaders. When entomologists digging into a nest grow careless, their hands become nicked all over as if pulled through a thorn bush. I have occasionally had to pause to stanch the flow of blood from a single bite, impressed by the fact that a creature one-millionth my size could stop me with nothing but its jaws.

No other animals have evolved the ability to turn fresh vegetation into mushrooms. The evolutionary event occurred only once, millions of years ago, somewhere in South America. It gave the ants an enormous advantage: they could now send out specialized workers to collect the vegetation while keeping the bulk of their populations safe in subterranean retreats. As a result, all of the different kinds of leafcutters together, comprising fourteen species in the genus *Atta* and twenty-three in *Acromyrmex*, dominate a large part of the American tropics. They consume more vegetation than any other group of animals, including the more abundant forms of caterpillars, grasshoppers, birds, and mammals. A single colony can strip an orange tree or bean patch overnight, and the combined populations inflict over a billion dollars' worth of damage yearly. It was with good reason that the early Portuguese settlers called Brazil the Kingdom of the Ants.

At full size, a colony contains three to four million workers and occupies three thousand or more underground chambers. The earth it excavates forms a pile twenty feet across and three to four feet high. Deep inside the nest sits the mother queen, a giant insect the size of a newborn mouse. She can live at least ten years and perhaps as long as twenty. No one has had the persistence to determine the true longevity. In my laboratory I have an individual collected in Guyana fourteen years ago. When she reaches eighteen, and breaks the proved longevity record of the seventeen-year locusts, my students and I will open a bottle of champagne to celebrate. In her lifetime an individual can produce over twenty million offspring, which translates into the following: a mere three hundred ants, a small fraction of the number emerging from a single colony in a year, can give birth to more ants than there are human beings on Earth.

The queen is born as a tiny egg, among thousands laid daily by the old mother queen. The egg hatches as a grublike larva, which is fed and laved incessantly throughout its month-long existence by the adult worker nurses. Through some unknown treatment, perhaps a special diet controlled by the workers, the larva grows to a relatively huge size. She then transforms into a pupa, whose waxy casement is shaped like an adult queen in fetal position, with legs, wings, and antennae folded tightly against the body. After several weeks the full complement of adult organs develops within this cuticle, and the new queen emerges. From the beginning she is fully adult and grows no more in size. She also possesses the same genes as her sisters, the colony workers. Their smaller size and pedestrian behavior is not due to heredity but rather to the different treatment they received as larvae.

In bright sunshine following a heavy rain, the virgin queen comes to the surface of the nest and flies up into the air to join other queens and the darkly pigmented, big-eyed males. Four or five males seize and inseminate her in quick succession, while she is still flying through the air. Their sole function now completed, they die within hours without returning to the home nest. The queen stores their sperm in her spermatheca, a tough muscular bag located just above and behind her ovaries. These reproductive cells live like independent microorganisms for years, passively waiting until they are released into the oviduct to meet an egg and create a new female ant. If the egg passes through the oviduct and to the outside without receiving a sperm, it produces a male. The queen can control the sex of her offspring, as well as the number of new workers and queens she produces, by opening or shutting the passage leading from her sperm-storage organ to the oviduct.

The newly inseminated queen descends to the ground. Raking her legs forward, she breaks off her wings, painlessly because they are composed of dead, membranous tissue. She wanders in a random pattern until she finds a patch of soft, bare soil, then commences to excavate a narrow tunnel straight down. Several hours later, when the shaft has been sunk to a depth of about ten inches, the queen widens its bottom into a small room. She is now set to start a garden and a colony of her own. But there is a problem in this life-cycle strategy. The queen has completely separated herself from the mother colony. Where can she obtain a culture of the vital symbiotic fungus to start the garden? Answer: she has been carrying it all along in her mouth. Just before leaving home, the young queen gathered a wad of fungal strands and inserted it into a pocket in the floor of her oral cavity, just back of the tongue. Now she passes the pellet out onto the floor of the nest and fertilizes it with droplets of feces.

As the fungus proliferates in the form of a whitish mat, the queen lays eggs on and around its surface. When the young larvae hatch, they are fed with other eggs given to them by the queen. At the end of their development, six weeks later, they transform into small workers. These new adults quickly take over the ordinary tasks of the colony. When still only a few days old, they proceed to enlarge the nest, work the garden, and feed the queen and larvae with tufts of the increasingly abundant fungus. In a year the little band has expanded into a force of a thousand workers, and the queen has ceased almost all activity to become a passive eating and egg-laying machine. She retains that exclusive role for the rest of her life. The measure of her Darwinian success is whether some of her daughters born five or ten years down the line grow into queens, leave on nuptial flights, and—rarest of all achievements—found new colonies of their own. In the world of the social insects, by the canons of biological organization, colonies beget colonies; individuals do not directly beget individuals.

People often ask me whether I see any human qualities in an ant colony, any form of behavior that even remotely mimics human thought and feeling. Insects and human beings are separated by more than 600 million years of evolution, but a common ancestor did exist in the form of one of the earliest multicellular organisms. Does some remnant of psychological continuity exist across that immense phylogenetic gulf? The answer is that I open an ant colony as I would the back of a Swiss watch. I am enchanted by the intricacy of its parts and the clean, thrumming precision. But I never see the colony as anything more than an organic machine.

Let me qualify that metaphor. The leafcutter colony is a superorganism. The queen sits deep in the central chambers, the vibrant growing tip from which all the workers and new queens originate. But she is not in any sense the leader or the repository of an organizational blueprint. No command center directs the colony. The social master plan is partitioned into the brains of the all-female workers, whose separate programs fit together to form a balanced whole. Each ant automatically performs certain tasks and avoids others according to its size and age. The superorganism's brain is the entire society; the workers are the crude analogue of its nerve cells. Seen from above and at a distance, the leafcutter colony resembles a gigantic amoeba. Its foraging columns snake out like pseudopods to engulf and shred plants, while their stems pull the green pieces down holes into the fungus gardens. Through a unique step in evolution taken millions of years ago, the ants captured a fungus, incorporated it into the superorganism, and so gained the power to digest leaves. Or perhaps the relation is the other way around: perhaps the fungus captured the ants and employed them as a mobile extension to take leaves into the moist underground chambers.

In either case, the two now own each other and will never pull apart. The ant-fungus combination is one of evolution's master clockworks, tireless, repetitive, and precise, more complicated than any human invention and unimaginably old. To find a colony in the South American forest is like coming upon some device left in place ages ago by an extraterrestrial visitor for a still undisclosed purpose. Biologists have only begun to puzzle out its many parts.

Because of modern science the frontier is no longer located along the retreating wall of the great rain forest. It is in the bodies and lives of the leafcutters and thousands of other species found for the most part on the other side of that tragic line.

The Time Machine

YOU CAN ENVISION the full sweep of biology best by imagining that you own a motion-picture projector of magical versatility. The image it projects can be slowed to explode seconds into hours and days or speeded up to condense years and centuries into a few minutes. The image can be magnified to reveal microscopic detail or compressed to take in broad vistas from a distance. The projector serves as the scientist's time machine. It performs what Einstein called thought experiments.

Begin with a moment in history, any moment. Appropriate to our theme is the late evening of May 12, 1859, when Louis Agassiz and Benjamin Peirce are strolling in the spring air along a street in Cambridge, Massachusetts, conversing on the war between France and Austria and the threat to Switzerland's neutrality. It is a notable pair. Agassiz is the most celebrated American scientist of his generation, pioneer in the study of glaciers, leading authority on fishes and the general classification of animals, much-sought-after lecturer, professor at Harvard, founder of the Museum of Comparative Zoology, close friend of Emerson, Longfellow, and other literati, and about to become the most bitter and effective American opponent of Charles Darwin's theory of evolution. Peirce is a prominent mathematician and professor of astronomy at Harvard, already shaping up into one of Agassiz's strongest allies in the country's young intellectual community. The two are returning from a dinner given at the home of Asa Gray, professor of botany and Darwin's principal American supporter. The occasion was one of the fortnightly meetings of the Cambridge Scientific Club, consisting of about a dozen members of the Harvard faculty and others in the town with more than a passing interest in science. The event was one of the few that can be correctly called historic in the world of ideas: Gray has just presented the essentials of the Darwinian theory for the first time in the western hemisphere. Earlier in the year, he and Agassiz had circled each other warily at a meeting of the American

Academy of Arts and Sciences, also held in Cambridge, sparring lightly on the vicarious distribution of plant species and other evidences of evolution but without addressing the central issue of the evolutionary process itself. Gray was too cautious to argue Darwin's theory of natural selection in a forthright manner before a large group of scholars with the formidable and popular Agassiz sitting there. Now, in the more relaxed company of the Cambridge Scientific Club, he has done so.

Few at the meeting sensed the importance of the Darwinian theory. The issue was cut between the two men alone. Gray enjoyed himself, letting the ideas and evidence tumble out in good spirits. Agassiz was disturbed. He said: "Gray, we must stop this." Indeed, much of the remainder of his career was spent trying to do just that. Now, at the moment on which we have chosen to focus the time machine, the conversation turns to current events, to war in Europe—a polite diversion from harder subjects about which close friends cannot afford to disagree. The historian A. Hunter Dupree has said of the two strollers, "Did they know that they stood on the knife-edge between two epochs in the intellectual history of the Western world? Did they know that the hesitant, eager tones of plain, familiar Asa Gray carried a message of more importance than Napoleon III, Franz Joseph, and all their legions rolled together?"

Picture the motion of the two men and the quiet flow of their words. Your effort consumes several seconds of time. Agassiz and Peirce and we, and all larger organisms, live in such *organismic time*, where most critical actions cover seconds or minutes. The reason for this deceptively simple fact is that human beings are constructed of billions of cells that communicate across their membranes by means of chemical surges and electrical impulses. A sentence is spoken: "Agassiz, I am much concerned." In a millisecond—one thousandth of a second— the compressed air strikes Agassiz's eardrum and transfers its energy inward to a row of three bony levers, which relay it instantly to the inner ear, an organ shaped like a snail shell; rows of sensitive cells deployed across the spiral resonate to the varying pitch of the vibration and trigger the discharge of equivalent nerve cells leading into the auditory nerve; as more milliseconds tick away, the coded electrical signals race into the hindbrain, cascade into predetermined pathways of the midbrain, the auditory cortex of the forebrain, and finally the seats of consciousness of the cerebral cortex—and Agassiz hears the sentence spoken by Peirce. Coordinated pulses of the neurons change their pattern through the cerebral cortex and special memory and emotive centers of the limbic system, generating new and quickly changing linkages of concepts and words; Agassiz is thinking. The brain combines new information from the banks of long-term memory into the temporary circuits of short-term memory. In a process consuming additional tenths of a second, the relevant images are pieced together and valuated by the emotive circuits they activate. Without pause the integrating centers of speech along the parietal cortex—Broca's and Wernicke's areas—are fully engaged, commands are issued through the cells of the relay stations of the motor cortex out to the tongue, lips, and larynx, and Agassiz responds: "Peirce, we must await developments." Four seconds have elapsed.

Now slow the reel in the time machine a thousand times. Agassiz and Peirce appear to freeze in their tracks. Their movements actually continue but are too slow to be perceived with unaided vision. Next magnify Agassiz until his individual nerve fibers come into view, then his cells, and finally molecules and atoms. Once again activity is normally paced and easily followed. The cell constituents swarm in passage through their appointed rounds, like the inhabitants of a city—like strollers in Cambridge. Enzyme molecules lock on to proteins and cleave them neatly into parts. A nerve cell discharges: along the length of its membrane, the voltage drops as sodium ions flow inward. At each point on the shaft of the nerve cell, these events consume several thousandths of a second, while the electrical signal they create—the voltage drop—speeds along the shaft at thirty feet per second. If we were to magnify the cell without slowing its action, the events would occur too swiftly to be seen. An electric discharge on the cell membrane would cross the field of vision faster than a rifle bullet. In order to understand such events at the molecular level, we must think in thousandths of seconds or less, the units of chemical reactions. That is why we slowed the action. We are in *biochemical time*. The magic projector allows us to picture its passage clearly by translating milliseconds into seconds, long enough for the myriad of our brain cells to interact and to recreate the image of the microscopic events that underlie the formation of the image now unfolding on the screen.

Spin the reel faster and return to organismic time. The biochemical reactions occur too swiftly to be comprehended, even if the magnification remains exactly right. So we reduce the magnification while moving back from Agassiz's body. As a consequence the projected atoms and molecules multiply and coalesce into vast aggregates, first as cells, then as tissues and organs. Once again, at these higher levels of biological organization, the action is slow enough to consume seconds—and hence to be understood by the brain. The diaphragm rises and falls, the heart pulsates, the leg muscles contract. Agassiz resumes his walk and conversation with Peirce.

Keep going. Speed the action still more—minutes and then hours pass within seconds—and pull away from Agassiz and Peirce. Like comic figures in an early silent movie, they speed jerkily out of the picture. As the reel turns ever faster, we rise above Cambridge to view the countryside of Massachusetts, then the full northeastern seaboard. Day and night pass in quickening succession. When the alternation between them reaches the flicker-fusion frequency, ten or more in a second of viewing time, they merge in our brains, so that the landscape is suffused by a continuous but dimmer light. Individual people and other organisms are no longer distinguishable except for a few long-lived trees that spring into existence and enlarge briefly before evaporating. But something new has appeared. We are aware of the presence of whole populations of species, say all of the sugar maples and red-eyed vireos, as they pass through cycles of expansion and retreat across the New England landscape. Ecosystems, formed of combinations of these species, have become the creatures of our vision. A pond is fringed with larch, fills up with waterweed, and then congeals into a bog. A sand dune sprouts beach grass, then wild rose and other low shrubs, which yield to jack pine and finally hardwood forest. We have entered *ecological time*. Biochemical events have been compressed beyond reckoning. Organisms are no more than ensembles defined by the mathematical laws of birth and death, competition, and replacement.

Where have Agassiz and Peirce and the other organisms of 1859 gone, during the acceleration of time? Dissolved into the gene pool of their species. Broken into tiny fragments by the shearing action of meiosis and fertilization. Erased as individuals, but preserved in perpetuity as DNA. They contributed half the genes of each of their children, one fourth of each grandchild, one eighth of each great-grandchild. Balancing this attrition was the multiplication of descendants through each successive generation. In a steady-state population, the average person has twice as many grandchildren as children, four times as many great-grandchildren, and so on up in a geometric progression. So the genes of one individual diffuse steadily outward through the population. Across a thousand years, the approximate threshold interval of *evolutionary time*, individuals lose most of their relevance as biological units. Families divide into multiplying lines of descent until they become coextensive across a large part of the population. Racial distinctions are blurred and eventually rendered meaningless. In the course of a thousand years, populations are even capable of splitting into entirely new species—although they have not done so in the case of the human line since the dawn of *Homo sapiens* half a million years ago.

The modern biological vision sweeps from microseconds to millions of years and from micrometers to the biosphere. But it is merely ordinary vision expanded by the electron microscope, earth-scan satellite, and other prosthetic devices of science and technology. The precise discipline is defined by the point of entry. Organismic biology explores the way we walk and speak; cell biology, the assembly and structure of our tissues; molecular biology, the ultimate chemical machinery; and evolutionary biology, the genetic history of our whole species. The modes of study depend upon the levels of organization chosen, which ascend in a hierarchical fashion: molecules compose cells, cells tissues, tissues organisms, organisms populations, and populations ecosystems. To understand any given species and its evolution requires a knowledge of each of the levels of organization sufficient to account for the one directly above it. Molecular biology is at the bottom (as its practitioners are always keen to point out) because everything depends upon the ultramicroscopic building blocks. Yet molecular biology on its own is a helpless giant. It cannot specify the parameters of space, time, and history that are crucial to and define the higher levels of organization. Consider the elementary fact that an embryo's development depends not just on its genes but on the way its cells are deployed in the surrounding environment. Or that an organism's behavior is shaped in part by learning, in other words by the alteration of its nerve cells by external stimuli. In a still deeper dimension, the very genes that comprise the central interest of molecular biology were assembled through a long history of mutation and selection within changing environments. When this last relationship became too obvious to ignore any longer, in the 1970s, molecular and evolutionary biology began to fuse, and the other branches of biology were realigned accordingly. The Darwinian conception approached its high watermark, more than a century after the publication of the *Origin of Species*.

AGASSIZ'S APPREHENSION over Darwin's theory deepened as he began to read *On the Origin of Species* late in 1859. "Agassiz, when I saw him last, had read but part of it," Asa Gray wrote J. D. Hooker in England in January 1860. "He says it is *poor—very poor*!! (*entre nous*). The fact is that he is very much annoyed by it . . . Tell Darwin all this."

The impact of the *Origin* began to exceed that of Agassiz's own masterwork, the "Essay on Classification" in volumes of *Contributions to the Natural History of the United States.* The American zoologist had promoted his own theory of the origin of species: they are creations in the mind of God, brought to life when the creator thinks of them and extinguished when he ceases to think of them. It seemed a perfect conception, uniting science and religion in a form consistent with the transcendentalist beliefs then ruling America's intellectual scene. Its rejection by the Darwinians was something that Agassiz could not understand, no matter how hard he tried. Near the end of his life he complained:

What of it, if it were true? Have those who object to repeated acts of creation ever considered that no progress can be made in knowledge without repeated acts of thinking? And what are thoughts but specific acts of mind?

Why should it then be unscientific to infer that the facts of nature are the result of a similar process, since there is no evidence of any other cause?

Darwin wrote Asa Gray: "Agassiz's name, no doubt, is a heavy weight against us." But not his logic and evidence, which in letters to friends Darwin dismissed as wild, paradoxical, and religiously inspired. When Agassiz composed an article on the geology of the Amazon with arguments against evolution, Darwin told Charles Lyell he was glad to read it but "chiefly as a psychological curiosity."

Agassiz and Darwin were type specimens in a fundamental classification that far exceeded their conflict and history. There have always been two kinds of scientists, two kinds of natural philosophers. The first look upon the Creator, or at least the ineffability of the human spirit, as the ultimate explanation of first choice. The second follow the venerable dictum attributed to Polybius that, whenever it is possible to find out the cause of what is happening, one should not have recourse to the gods. The historian Loren Graham has given names to the two camps: restrictionists and expansionists. The first believe that science can go only so far, after which new forms of explanation and understanding have to be devised. The second acknowledge no intrinsic limits. They favor Bertrand Russell's definition of science as the things we know, distinct from philosophy as the things we do not know.

Darwin was the great expansionist. He shocked the world by arguing convincingly that life is the creation of an autonomous process so simple that it can be understood with just a moment of reflection. No equations, photons, or computer read-outs required. It can all be summarized in a couple of lines: new variations in the hereditary material arise continuously, some survive and reproduce better than others, and as a result organic evolution occurs. And even more briefly as follows: natural selection acting on mutations produces evolution. Given enough time (and the Earth is over four billion years old) even radically new kinds of organisms can be assembled this way, insects from myriapods, amphibians from lungfish, birds from small dinosaurs, and even life itself from inanimate matter.

Such a proposition was shocking in 1859 because before then almost everyone had worked under the opposite assumption, that great effects imply great causes. The eye of the eagle, the human hand, the whale's giant heart—such feats of engineering are so extraordinary as to suggest a designing power, if not God then at least an Idea of divine profundity. It had been difficult to think of the world in any other way. But Darwin showed that even the most complicated organism can be self-assembled through a series of small steps. God—and, yes, philosophy—was excused from the living world so that biology might seek its independent destiny.

And if that much were to be granted, what about the mind itself? The brain can also evolve by natural selection. If the mind is the creation of the brain, then it must be subject to material explanation. In 1838, shortly after he had conceived the principle of natural selection, Darwin wrote in his *N* Notebook that "to study Metaphysics, as they have always been studied, appears to me to be like puzzling at astronomy without mechanics.—Experience shows that the problem of the mind cannot be solved by attacking the citadel itself."

It certainly seemed to follow. The mind cannot understand its own workings and ultimate meaning merely by thinking about itself. If truly material in origin, the citadel is not to be entered directly but roundabout, through an exploration of the brain. The brain must lose some of its magic when it is regarded as a product of evolution through natural selection, like other organs of the body. So Darwin wrote in his 1838 *M* Notebook: "Origin of man now proved.—Metaphysics must flourish.—He who understand[s] baboon would do more toward metaphysics than Locke."

MODERN BIOLOGY has been built upon two great ideas. The first, a product of the nineteenth century, is that all life descended from elementary, single-celled organisms by means of natural selection. The second, perfected in the twentieth century, is that organisms are entirely obedient to the laws of physics and chemistry. No extraneous "vital force" runs the living cell. Each of the two ideas supports the other in compelling fashion. On the one side, the argument that organisms are physicochemical entities makes the universal operation of natural selection all the more plausible. On the other hand, the proof of natural selection in even a limited number of cases helps to explain why organisms are physicochemical mechanisms rather than the vessels of a mystic life force.

That is why expansionism has prevailed so far, passing beyond the boundaries of physics and chemistry into the domain of life and the mind, enabling its proponents to crank out knowledge at an accelerating rate. The biologist's time machine has grown into an awesome device that searches across centuries and down inside molecules. The vistas it has opened are the enchanted terrain of the new age.

But wait—a *machine*? Opening virgin territories? That has a familiar ring, and indeed we have arrived at the core of the fear of science, the cause of its historic alienation from the humanities. Science rampant is resisted as "scientism," an unpleasant doctrine. The dilemma of the machine in the garden exists in the realm of the spirit as surely as it does in the shrinking wilderness.

"Science grows and Beauty dwindles," wrote Tennyson. In the 1800s the romantic movement in poetry came to flourish as a fierce assault by free minds against the philosophy of the Enlightenment. They rejected the idea that all nature and human affairs are open to rational investigation, or that Newtonian law can be spread beyond physics. Keats warned in *Lamia*: "Philosophy will clip an Angel's wings. / Conquer all mysteries by rule and line, / Empty the haunted air, and gnomed mine— / Unweave a rainbow."

The romantic world view has been kept alive in sophisticated arguments by such modern theologians and philosophers as John Bowker, Theodore Roszak, and William Irwin Thompson. Their bill of indictment can be summarized: "Science reduces and oversimplifies / Condenses and abstracts, drives toward generality / Presumes to break the insoluble / Forgets the spirit / Imprisons the spark of artistic genius."

The distinction between the two cultures of science and the humanities made famous by C. P. Snow thus persists. Until that fundamental divide is closed or at least reconciled in some congenial manner, the relation between man and the living world will remain problematic.

The Bird of Paradise

COME WITH ME NOW to another part of the living world. The role of science, like that of art, is to blend exact imagery with more distant meaning, the parts we already understand with those given as new into larger patterns that are coherent enough to be acceptable as truth. The biologist knows this relation by intuition during the course of field work, as he struggles to make order out of the infinitely varying patterns of nature.

Picture the Huon Peninsula of New Guinea, about the size and shape of Rhode Island, a weathered horn projecting from the northeastern coast of the main island. When I was twenty-five, with a fresh Ph.D. from Harvard and dreams of physical adventure in far-off places with unpronounceable names, I gathered all the courage I had and made a difficult and uncertain trek directly across the peninsular base. My aim was to collect a sample of ants and a few other kinds of small animals up from the lowlands to the highest part of the mountains. To the best of my knowledge I was the first biologist to take this particular route. I knew that almost everything I found would be worth recording, and all the specimens collected would be welcomed into museums.

Three days' walk from a mission station near the southern Lae coast brought me to the spine of the Sarawaget range, 12,000 feet above sea level. I was above treeline, in a grassland sprinkled with cycads, squat gymnospermous plants that resemble stunted palm trees and date from the Mesozoic Era, so that closely similar ancestral forms might have been browsed by dinosaurs 80 million years ago. On a chill morning when the clouds lifted and the sun shone brightly, my Papuan guides stopped hunting alpine wallabies with dogs and arrows, I stopped putting beetles and frogs in bottles of alcohol, and together we scanned the rare panoramic view. To the north we could make out the Bismarck Sea, to the south the Markham Valley and the more distant Herzog Mountains. The primary forest covering most of this mountainous country was broken into bands of different vegetation according to elevation. The

zone just below us was the cloud forest, a labyrinth of inter-
locking trunks and branches blanketed by a thick layer of moss,
orchids, and other epiphytes that ran unbroken off the tree
trunks and across the ground. To follow game trails across this
high country was like crawling through a dimly illuminated
cave lined with a spongy green carpet.

A thousand feet below, the vegetation opened up a bit and
assumed the appearance of typical lowland rain forest, except
that the trees were denser and smaller and only a few flared
out into a circle of blade-thin buttresses at the base. This is the
zone botanists call the mid-mountain forest. It is an enchanted
world of thousands of species of birds, frogs, insects, flowering plants, and other organisms, many found nowhere else.
Together they form one of the richest and most nearly pure
segments of the Papuan flora and fauna. To visit the mid-
mountain forest is to see life as it existed before the coming of
man thousands of years ago.

The jewel of the setting is the male Emperor of Germany bird
of paradise (*Paradisaea guilielmi*), arguably the most beautiful
bird in the world, certainly one of the twenty or so most striking in appearance. By moving quietly along secondary trails
you might glimpse one on a lichen-encrusted branch near the
tree tops. Its head is shaped like that of a crow—no surprise
because the birds of paradise and crows have a close common
lineage—but there the outward resemblance to any ordinary
bird ends. The crown and upper breast of the bird are metallic
oil-green and shine in the sunlight. The back is glossy yellow,
the wings and tail deep reddish maroon. Tufts of ivory-white
plumes sprout from the flanks and sides of the breast, turning
lacy in texture toward the tips. The plume rectrices continue
on as wirelike appendages past the breast and tail for a distance
equal to the full length of the bird. The bill is blue-gray, the
eyes clear amber, the claws brown and black.

In the mating season the male joins others in leks, common
courtship arenas in the upper tree branches, where they display
their dazzling ornaments to the more somberly caparisoned
females. The male spreads his wings and vibrates them while
lifting the gossamer flank plumes. He calls loudly with bub-
bling and flutelike notes and turns upside down on the perch,
spreading the wings and tail and pointing his rectrices skyward.

The dance then reaches a climax as he fluffs up the green breast feathers and opens out the flank plumes until they form a brilliant white circle around his body, with only the head, tail, and wings projecting beyond. The male sways gently from side to side, causing the plumes to wave gracefully as if caught in an errant breeze. Seen from a distance his body now resembles a spinning and slightly out-of-focus white disk.

This improbable spectacle in the Huon forest has been fashioned by millions of generations of natural selection in which males competed and females made choices, and the accouterments of display were driven to a visual extreme. But this is only one trait, seen in physiological time and thought about at a single level of causation. Beneath its plumed surface, the Emperor of Germany bird of paradise possesses an architecture culminating an ancient history, with details exceeding those that can be imagined from the naturalist's simple daylight record of color and dance.

Consider one such bird for a moment in the analytic manner, as an object of biological research. Encoded within its chromosomes is the developmental program that led with finality to a male *Paradisaea guilielmi*. The completed nervous system is a structure of fiber tracts more complicated than any existing computer, and as challenging as all the rain forests of New Guinea surveyed on foot. A microscopic study will someday permit us to trace the events that culminate in the electric commands carried by the efferent neurons to the skeletal-muscular system and reproduce, in part, the dance of the courting male. This machinery can be dissected and understood by proceeding to the level of the cell, to enzymatic catalysis, microfilament configuration, and active sodium transport during electric discharge. Because biology sweeps the full range of space and time, there will be more discoveries renewing the sense of wonder at each step of research. By altering the scale of perception to the micrometer and millisecond, the laboratory scientist parallels the trek of the naturalist across the land. He looks out from his own version of the mountain crest. His spirit of adventure, as well as personal history of hardship, misdirection, and triumph, are fundamentally the same.

Described this way, the bird of paradise may seem to have been turned into a metaphor of what humanists dislike most about science: that it reduces nature and is insensitive to art, that scientists are conquistadors who melt down the Inca gold. But bear with me a minute. Science is not just analytic; it is also synthetic. It uses artlike intuition and imagery. In the early stages, individual behavior can be analyzed to the level of genes and neurosensory cells, whereupon the phenomena have indeed been mechanically reduced. In the synthetic phase, though, even the most elementary activity of these biological units creates rich and subtle patterns at the levels of organism and society. The outer qualities of *Paradisaea guilielmi*, its plumes, dance, and daily life, are functional traits open to a deeper understanding through the exact description of their constituent parts. They can be redefined as holistic properties that alter our perception and emotion in surprising and pleasant ways.

There will come a time when the bird of paradise is reconstituted by the synthesis of all the hard-won analytic information. The mind, bearing a newfound power, will journey back to the familiar world of seconds and centimeters. Once again the glittering plumage takes form and is viewed at a distance through a network of leaves and mist. Then we see the bright eye open, the head swivel, the wings extend. But the familiar motions are viewed across a far greater range of cause and effect. The species is understood more completely; misleading illusions have given way to light and wisdom of a greater degree. One turn of the cycle of intellect is then complete. The excitement of the scientist's search for the true material nature of the species recedes, to be replaced in part by the more enduring responses of the hunter and poet.

What are these ancient responses? The full answer can only be given through a combined idiom of science and the humanities, whereby the investigation turns back into itself. The human being, like the bird of paradise, awaits our examination in the analytic-synthetic manner. As always by honored tradition, feeling and myth can be viewed at a distance through physiological time, idiosyncratically, in the manner of traditional art. But they can also be penetrated more deeply than ever was possible in the prescientific age, to their physical basis in the processes of mental development, the brain structure, and indeed the genes themselves. It may even be possible to trace them back through time past cultural history to the evolutionary origins of human nature. With each new phase of synthesis to emerge from biological inquiry, the humanities will expand their reach and capability. In symmetric fashion, with each redirection of the humanities, science will add dimensions to human biology.

The Poetic Species

ONE OF THE most dramatic events of this century was the setting of the first Viking probe on Mars. The landing was scheduled for July 4, 1976, to coincide with the bicentennial of the United States, but it was actually accomplished on July 20. Scientists have anticipated few events with such electric excitement. There was an outside chance that Martian organisms might be detected quickly; a new biology could be created in one stroke. I know that feeling was shared by many following the event in the news, but I had a somewhat more personal interest. In 1964 I had attended a conference about Mars chaired by the ever ebullient Carl Sagan. We examined the best telescopic data available, speculating in every direction on the possible existence of life on the red planet and ways that it might be analyzed. I served as the instant "exo-ecologist," trying without much seriousness or effect to put a biological construction on the dark zones that spread and retreat through the middle latitudes (they were later proved to be sand storms). The conferees went home with no firm conclusions but high hopes for the NASA program then being charted.

Now at last, after twelve years of waiting, everyone was to have a close look at the surface of Mars, as though standing there in person, in a place where life might be found. The cameras scanned the Chryse Plain from the foot of the lander to the horizon and transmitted color pictures with a maximum resolution of less than a millimeter. The result was disappointing: no bushes dotted the landscape, no animals walked past the lens. A mechanical arm scooped up soil and analyzed it chemically to reveal reactions that were lifelike but not proof of the presence of microorganisms. The overall scene was nonetheless compelling: it was a landfall on another world remarkably like Earth in many respects. An outwardly familiar desert reached to the horizon where at sunset the thin atmosphere briefly glowed pink and turquoise. Every half-buried pebble, every wind pocket in the soil seen from a few feet away held the attention, hammered the imagination.

Then it was over. A giddying potential had been reduced to the merely known. The cold dust of the desert plain was committed to photographs in magazines, then to technical monographs, textbooks, and encyclopedias. The adventure became a set of facts, somehow mundane, to be looked up by students and recalled during leisure reading. A fundamental trait of science was exemplified: the magic had been consumed quickly and the action moved elsewhere. While a great deal remained to be learned, the high tide of research had swept on past an entire planet in less than a year.

Such brilliance fades quickly because newly discovered truths, and not truth in some abstract sense alone, are the ultimate goal and yardstick of the scientific culture. Scientists do not discover in order to know, they know in order to discover. That inversion of purpose is more than just a trait, it is the essence of the matter. Humanists are the shamans of the intellectual tribe, wise men who interpret knowledge and transmit the folklore, rituals, and sacred texts. Scientists are the scouts and hunters. No one rewards a scientist for what he knows. Nobel Prizes and other trophies are bestowed for the new facts and theories he brings home to the tribe. One great discovery and the scientist himself is great forever, no matter how foolish the rest of his deeds and pronouncements. No discovery, and he will probably be forgotten, even if he is learned and wise in matters scientific. The humanist grows in stature as he grows in wisdom. He can gain immortality as a critic, and justly so. But this vocational opportunity is not open—not yet—to the scientist. The most memorable critics among scientists are those who served as foils for the discoverers, helping them to clear error from the path. Thus the great Agassiz, cherished by Emerson and Longfellow, the idol of lecture halls along the Atlantic seaboard, is remembered most often today for having been wrong about Darwin.

Scientists therefore spend their productive lives struggling to reach the edge of knowledge in order to make discoveries. David Hilbert, the most successful mathematician of the early 1900s, stated the rules very well: "So long as a branch of science offers an abundance of problems, it will stay alive; a lack of problems foreshadows extinction or the cessation of independent development."

The scientist is not a very romantic figure. Each day he goes into the laboratory or field energized by the hope of a great score. He is brother to the prospector and treasure hunter. Every little discovery is like a gold coin on the ocean floor. The professional's real business, the bone and muscle of the scientific endeavor, amounts to a sort of puttering: trying to find a good problem, thinking up experiments, mulling over data, arguing in the corridor with colleagues, and making guesses with the aid of coffee and chewed pencils until finally something—usually small—is uncovered. Then comes a flurry of letters and telephone calls, followed by the writing of a short paper in an acceptable jargon. The great majority of scientists are hard-working, pleasant journeymen, not excessively bright, making their way through a congenial occupation.

Einstein spoke on the occasion of Max Planck's sixtieth birthday. He said that three types of people occupy the temple of science. There are those who enter for purely utilitarian reasons, to have a calling and invent things useful to mankind. Others are attracted by the sport in science, satisfying their ambition through the exercise of superior intellectual power. If an angel of the Lord were to come, Einstein said, and drive all belonging to these first two categories from the temple, a few people would be left, including Planck—"and that is why we love him."

The scientists most esteemed by their colleagues are those who are both very original and committed to the abstract ideal of truth in the midst of clamoring demands of ego and ideology. They pass the acid test of promoting new knowledge even at the expense of losing credit for it. They can face a fact, in accordance with the prayer of Thomas Henry Huxley, even though it slays them. Their principal aim is to discover natural law marked by *elegance*, the right mix of simplicity and latent power. The theory they accept is the one that defeats rival schemes by uniquely explaining the experiments of numerous independent investigators. It is a sleek instrument forged by repeated exposure to stubborn and sometimes inconvenient data. Conversely, the ideal experiment is the one that settles the rival claims of competing theories. Both the dominant theory and its patron data endure only if they fit the explanations of other disciplines through a network of logically tight and quantitative arguments.

This tidy conception is made the more interesting by the deep epistemological problem it creates and the biological process it implies. Elegance is more a product of the human mind than of external reality. It is best understood as a product of organic evolution. The brain depends upon elegance to compensate for its own small size and short lifetime. As the cerebral cortex grew from apish dimensions through hundreds of thousands of years of evolution, it was forced to rely on tricks to enlarge memory and speed computation. The mind therefore specializes on analogy and metaphor, on a sweeping together of chaotic sensory experience into workable categories labeled by words and stacked into hierarchies for quick recovery. To a considerable degree science consists in originating the maximum amount of information with the minimum expenditure of energy. Beauty is the cleanness of line in such formulations, along with symmetry, surprise, and congruence with other prevailing beliefs. This widely accepted definition is why P. A. M. Dirac, after working out the behavior of electrons, could say that physical theories with some physical beauty are also the ones most likely to be correct, and why Hermann Weyl, the perfecter of quantum and relativity theory, made an even franker confession: "My work always tried to unite the true with the beautiful; but when I had to choose one or the other, I usually chose the beautiful."

Einstein offered the following solution to the dilemma of truth versus beauty: "God does not care about our mathematical difficulties. He integrates empirically." In other words, a mind with infinite memory store and calculating ability could compute any system as the sum of all its parts, however minute and numerous. Mathematics and beauty are devices by which human beings get through life with the limited intellectual capacity inherited by the species. Like a discerning palate and sexual appetite, these esthetic contrivances give pleasure. Put in more mechanistic terms, they play upon the circuitry of the brain's limbic system in a way that ultimately promotes survival and reproduction. They lead the scientist adventitiously into the unexplored fractions of space and time, from which he returns to report his findings and fulfill his social role. Riemannian geometry is declared beautiful no less than the bird of paradise, because the mind is innately prepared to receive its symmetry and power. Pleasure is shared, triumph ceremonies held, and the communal hunt resumed. In a memorial tribute to Hermann Minkowski, David Hilbert described this perpetual cycle with gentle botanic images:

> Our Science, which we loved above everything, had brought us together. It appeared to us as a flowering garden. In this garden there were wellworn paths where one might look around at leisure and enjoy oneself without effort, especially at the side of a congenial companion. But we also liked to seek out hidden trails and discovered many an unexpected view which was pleasing to our eyes; and when the one pointed it out to the other, and we admired it together, our joy was complete.

Scientific innovation sometimes sounds like poetry, and I would claim that it is, at least in the earliest stages. The ideal scientist can be said to think like a poet, work like a clerk, and write like a journalist. The ideal poet thinks, works, and writes like a poet. The two vocations draw from the same subconscious wellsprings and depend upon similar primal stories and images. But where scientists aim for a generalizing formula to which special cases are obedient, seeking unifying natural laws, artists invent special cases immediately. They transmit forms of knowledge in which the knower himself is revealed. Their works are lit by a personal flame and above all else they identify, in Roger Shattuck's expression, "the individual as the accountable agent of his action and as the potential seat of human greatness."

The aim of art is not to show how or why an effect is produced (that would be science) but literally to produce it. And not by just any cry from the heart—it requires mental discipline no less than in science. In poetry, T. S. Eliot explained, the often-quoted criterion of sublimity misses the mark. What counts is not the greatness of the emotion but the intensity of the artistic process, the pressure under which the fusion takes place. The great artist touches others in surgical manner with the generating impulse, transferring feeling precisely. His work is personal in style but general in effect.

Ideally art is powerful enough to cross cultures; it reads the code of human nature. Octavio Paz's poem "The Broken Waterjar" (*El cántaro roto*) accomplishes this result with splendid effectiveness. Paz is torn by the contradiction in the Mexican experience. He says that the minds of his people are capable of long flights of imagination and visions of piercing beauty. The people look at the sky and add torches, wings, and "bracelets of flaming islands." But they also look down to a desiccated landscape, symbolizing physical and spiritual poverty. A potentially great nation has been divided by the Conquest and stifled by oppression:

> Bare hills, a cold volcano, stone and a sound of panting
> > under such splendor, and drouth, the taste of dust,
> the rustle of bare feet in the dust, and one tall tree in the
> > middle of the field like a petrified fountain!

The resolution is not offered in the form of practical advice, which might easily prove wrong, but in the poet's vision of unity in a search back through time, "más allá de las aguas del bautismo," to a more secure metaphysical truth, and Paz says:

> vida y muerte no son mundos contrarios, somos un solo
> > tallo con dos flores gemelas

Mexico is a single stem with twin flowers, united by the continuity of time.

The essence of art, no less than of science, is synecdoche. A carefully chosen part serves for the whole. Some feature of the subject directly perceived or implied by analogy transmits precisely the quality intended. The listener is moved by a single, surprising image. In "The Broken Waterjar" the rustle of bare feet in the dust conveys the pauperization of Mexico. The artist knows which sensibilities shared by his audience will permit the desired impact.

Picasso defined art as the lie that helps us to see the truth. The aphorism fits both art and science, since each in its own way seeks power through elegance. But this inspired distortion is only a technique of thinking and communication. There is a still more basic similarity: both are enterprises of discovery. And the binding force lies in our biology and in our relationship to other organisms. In art, the workings of the mind are explored, whereas in science the domain is the world at large and now, increasingly, the workings of the mind as well. Of equal importance, both rely on similar forms of metaphor and analogy, because they share the brain's strict and peculiar limitations in the processing of information.

Most scientists have been self-conscious at one time or another about their own procedures of discovery. The stakes are high: a major advance can be made by a single insight consuming just a few seconds. Scientific theory is the last and greatest of the cottage industries and the principal source of vitality in every discipline. Is there a secret, a hidden meta-formula by which the mind creates these visible formulas? This question has been addressed in studies of creativity by cognitive psychologists. Their work has advanced rapidly during the past ten years as the iron grip of behaviorist philosophy relaxed and studies of the mind became more respectable. Of equal weight is the testimony of scientists about their own steps to discovery. Essays by Freeman Dyson, J. B. S. Haldane, Werner Heisenberg, Willard Libby, Henri Poincaré, John Wheeler, Chen Ning Yang, and others form a veritable psychologist's case book in this most elusive of mental operations.

In my own search for a meta-formula I have had the good fortune of working with gifted mathematicians on subjects for which there was little or no previous theory—no framework of well-defined ideas on which information might be deployed and linked in explanatory chains. Early in my life I discovered that I have very little talent for mathematics. It is simply one of those things you either have or not, and no amount of training or effort will bring it to you, just as most people have little capacity for playing the violin or running a fast mile. Through hard work at college and while a young professor I became mathematically semiliterate. I can puzzle through articles on pure theory in the journals and advanced textbooks, but I cannot write the rows of original equations that transport the mind from one or two easy propositions by some miracle to a new, counterintuitive truth. My ability lies in seeing the problem in the first place, envisioning what a subject might look like if a proper theoretical scaffolding and beautiful facts were put in place. In other words scout, not architect. Nothing is more attractive to me than a muddled domain awaiting its first theory. I feel most at home with a jumble of glittering data and the feeling that they might be fitted together for the first time into some new pattern. This inclination made me especially compatible with mathematicians. I became fascinated with the way they think, why they should be so much better at quantitative reasoning than I, what difference it made in the end, why I should be the one so often to suggest moving in a particular direction, but then even more frequently not be able to do so, and finally how different everything looked after a little progress had been made.

From this personal experience and the impressions recorded by others, let me offer the following rough map of innovation in science. You start by loving a subject. Birds, probability theory, explosives, stars, differential equations, storm fronts, sign language, swallowtail butterflies—the odds are that the obsession will have begun in childhood. The subject will be your lodestar and give sanctuary in the shifting mental universe.

A pioneer in molecular biology (still young, because most pioneering work was done after 1950) once told me that his fascination with the replication of DNA molecules began when he was given an erector set as a child. Playing with the toy, he saw the possibilities of creation by the multiplication and rearrangement of identical units. The great metallurgist Cyril Smith owed his devotion to alloys to the fact that he was color blind. The impairment caused him to turn his attention at an early age to the intricate black-and-white patterns to be seen everywhere in nature, to swirls, filigree, and banding, and eventually to the fine structure of metal. Albert Camus spoke for all such innovators when he said that "a man's work is nothing but this slow trek to rediscover, through the detours of art, those two or three great and simple images in whose presence his heart first opened."

The subject we love is probably also well known to others. So we have to travel away from it into regions deliberately chosen for their lack of previous attention. Science has flourished in western cultures because this difficult step was recognized by society as valuable, and rewarded. Nothing comes harder than original thought. Even the most gifted scientist spends only a tiny fraction of his waking hours doing it, probably less than one tenth of one percent. The rest of the time his mind hugs the coast of the known, reworking old information, adding lesser data, giving reluctant attention to the ideas of others (what use can *I* make of them?), warming lazily to the memory of successful experiments, and looking for a problem—always looking for a problem, something that can be accomplished, that will lead somewhere, anywhere.

There is in addition an optimal degree of novelty in problem-seeking, difficult to measure and follow. Stick to the coast too tightly and only minor new data will follow. Venture out of sight and you risk getting lost at sea. Years of effort might then be wasted, competitors will hint that the enterprise is pseudoscience, grants and other patronage will be cut off, and tenure and election to the academies denied. The fate of the overly daring is to sail off the rim of the world.

On one point both psychologists and successful voyagers agree. The key instrument of the creative imagination is analogy. Hideki Yukawa, who reflected on this matter for forty years while working on the nuclear binding force, explained it as follows:

> Suppose that there is something which a person cannot understand. He happens to notice the similarity of this something to some other thing which he understands quite well. By comparing them he may come to understand the thing which he could not understand up to that moment. If his understanding turns out to be appropriate and nobody else has ever come to such an understanding, he can claim that his thinking was really creative.

We have returned to the common, human origin of science and art. The innovator searches for comparisons that no one else has made. He scrambles to tighten his extension by argument, example, and experiment. Important science is not just any similarity glimpsed for the first time. It offers analogies that map the gateways to unexplored terrain. The comparisons meet the criterion of principal metaphor used by art critics: one commanding image synthesized from several units, such that a single complex idea is attained not by analysis but by the sudden perception of an objective relation.

Theoretical scientists, inching away from the safe and known, skirting the point of no return, confront nature with a free invention of the intellect. They strip the discovery down and wire it into place in the form of mathematical models or other abstractions that define the perceived relation exactly. The now-naked idea is scrutinized with as much coldness and outward lack of pity as the naturally warm human heart can muster. They try to put it to use, devising experiments or field observations to test its claims. By the rules of scientific procedure it is then either discarded or temporarily sustained. Either way, the central theory encompassing it grows. If the abstractions survive they generate new knowledge from which further exploratory trips of the mind can be planned. Through the repeated alternation between flights of the imagination and the accretion of hard data, a mutual agreement of the workings of the world is written, in the form of natural law.

The scientist entrepreneur can pick a subject virtually at random and soon be on the edge of discovery, if he is at all lucky. In 1962 Robert H. MacArthur and I, both in our early thirties, decided to try something new in biogeography. The discipline, which studies the distribution of plants and animals around the world, was ideal for theoretical research. Biogeography was intellectually important, replete with poorly organized information, underpopulated, and almost devoid of quantitative models. Its borders with ecology and genetics, specialties in which we also felt well prepared, were blank swaths across the map.

MacArthur was then an associate professor of biology at the University of Pennsylvania, the same rank I held at Harvard. He later moved to Princeton, where he spent the remainder of his short life. He was medium tall and thin, with a handsomely angular face. He met you with a level gaze supported by an ironic smile and widening of eyes. He spoke with a thin baritone voice in complete sentences and paragraphs, signaling his more important utterances by tilting his face slightly upward and swallowing. He had a calm understated manner, which in intellectuals suggests tightly reined power. Because very few professional academics can keep their mouths shut long enough to be sure about anything, MacArthur's restraint gave his conversation an edge of finality he did not intend. In fact he was basically shy and reticent. He was not a mathematician of the first class—very few scientists are, otherwise they would become pure mathematicians—but he joined superior talent in that field with an extraordinary creative drive, decent ambition, and a love of the natural world, birds, and science, in that order.

By general agreement MacArthur was the most important ecologist of his generation. His use of evolutionary theory in the explanation of population growth and competition was so original and productive that biologists today refer informally to the MacArthur school of ecology, or more justly to the Hutchinson-MacArthur school, in order to include his influential teacher at Yale, G. Evelyn Hutchinson. MacArthur died of renal cancer in 1972. Hours before he died in his sleep, I talked with him at length over the telephone, from Cambridge to Princeton. It was the same as ten years before. We touched on familiar subjects: the future of ecology, the key unsolved problems of evolution, and the merits of various colleagues. MacArthur's easy concentration on these matters, as if he had a hundred years to live, was but one more measure of his intellectual integrity.

In 1960, when we first met, I was finishing a ten-year stint of field work and knew the distribution of animals quite well. I had worked out the classification of hundreds of species of ants throughout the Pacific region and elsewhere. I had the sense that there was some general order within the exciting chaos, some powerful process to be uncovered, but only a vague idea of its outline. In our first laconic discussion (MacArthur had the effect on me of shortening my sentences), we quickly realized that something of value lay close to the surface. In the following exchange I have telescoped our conversations and letters on the subject in order to convey the crucial steps in the origin of species-equilibrium theory—almost, as it were, out of the air.

Wilson: I think biogeography can be made into a science. There are striking regularities no one has explained. For example, the larger the island, the more the species of birds or ants that live on it. Look at what happens when you go from little islands, such as Bali and Lombok, to big ones like Borneo and Sumatra. With every tenfold increase in area, there is roughly a doubling of the number of species found on the island. That appears to be true for most other kinds of animals and plants for which we have good data. Here's another piece in the puzzle. I've found that as new ant species spread out from Asia and Australia onto the islands between them, such as New Guinea and Fiji, they eliminate other ones that settled there earlier. At the level of the species this fits in pretty well with the views of Philip Darlington and George Simpson. They proved that in the past major groups of mammals, such as all the deer or all the pigs taken together, have tended to replace other major groups in South America and Asia, filling the same niches. So there seems to be a balance of nature down to the level of the species, with waves of replacement spreading around the world.

MacArthur: Yes, a species equilibrium. It looks as though each island can hold just so many species, so if one species colonizes the island, an older resident has to go extinct. Let's treat the whole thing as if it were a kind of physical process. Think of the island filling up with species from an empty state up to the limit. That's just a metaphor, but it might get us somewhere. As more species settle, the rate at which they are going extinct will rise. Let me put it another way: the probability that any given species will go extinct increases as more species crowd onto the island. Now look at the species arriving. A few colonists of each are making it each year on the wind or floating logs or, like birds, flying in on their own power. The more species that settle on the island, the fewer *new* ones that will be arriving each year, simply because there are fewer that aren't already there. Here's how a physicist or economist would represent the situation. As the island fills up, the rate of extinction goes up and the rate of immigration goes down, until the two processes reach the same level. So by definition you have a dynamic equilibrium. When extinction equals immigration, the *number* of species stays the same, even though there may be a steady change in the particular species making up the fauna.

Look what happens when you play around a little with the rising and falling curves. Let the islands get smaller. The extinction rates have to go up, since the populations are smaller and more liable to extinction. If there are only ten birds of a kind sitting in the trees, they are more likely to go to zero in a given year than if there are a hundred. But the rate at which new species are arriving won't be affected very much, because islands well away from the mainland can vary a lot in size without changing much in the amount of horizon they present to organisms traveling toward them. As a result, smaller islands will reach equilibrium sooner and end up with a smaller number of species at equilibrium. Now look at pure distance as a factor. The farther the island is from the source areas, say the way Hawaii is farther out in the Pacific than New Guinea, the fewer new species that will be arriving each year. But the rate of extinction stays the same because, once a species of plant or animal is settled on an island, it doesn't matter whether the island is close or far. So you expect the number of species found on distant islands to be fewer. The whole thing is just a matter of geometry.

Weeks pass. We are sitting next to the fireplace in MacArthur's living room, with notes and graphs spread out on a coffee table.

Wilson: So far so good. The numbers of bird and ant species *do* go down as islands get smaller and farther from the mainland. We'll label the two trends the *area effect* and the *distance effect.* Let's take them both as given for the moment. How do we know that they prove the equilibrium model? I mean, other people are almost certainly going to come up with a rival theory to explain the area and distance effects. If we claim that the results prove the model that predicted them, we will commit what logicians call the Fallacy of Affirming the Consequent. The only way we can avoid that impasse is to get results that are uniquely predicted by our model and no one else's.

MacArthur: All right, we've gone this far with pure abstraction—let's go on. Try the following: line up the extinction and emigration curves so that where they cross and create the equilibrium, they are straight lines and tilted at approximately the same angle. As an exercise in elementary differential calculus, you can show that the number of years an island takes to fill up to 90 percent of its potential should just about equal the number of species at equilibrium divided by the number going extinct every year.

Wilson: Let's look at Krakatoa.

KRAKATOA is the small island in the Sunda Strait between Java and Sumatra that exploded with an equivalent force of approximately 100 megatons of TNT on August 27, 1883. As a wave raced out across the Indian Ocean (eventually to cause ships to tug at their anchor chains in the English Channel) a blanket of glowing hot pumice covered most of what was left of the island and killed the last remnants of life. Scientists realized that they had a once-in-a-century opportunity to witness the recolonization of a dead island. Between 1884 and 1936 several principal expeditions were mounted under the auspices of the Dutch colonial government to follow the return of plants and animals to Krakatoa. The data were published in a scattering of articles and books, but very little use had been made of them in the years that followed, largely because no quantitative theory of island biogeography existed.

The Dutch reported that the vegetation returned to Krakatoa quickly. The first plants sprouted in the rain-moistened ash within a year, and a luxuriant forest covered most of its surface by 1920. Large numbers of animal species simultaneously colonized the island. The journal data were especially good for birds. We got Krakatoa's area and fitted it on our own curve relating area to species. Krakatoa should have 30 bird species at equilibrium. The Dutch surveys indicated that it reached about 90 percent of that number in 30 years. The elementary equilibrium equation predicted that by the late 1920s about one species per year (30 species divided by 30 years) should be going extinct, to be replaced by one new species coming in. We scanned the pages of the reports eagerly. Would the Dutch scientists mention extinction? They did. They were impressed, they said, by a remarkably high turnover in bird species. We calculated that they saw an average of one extinction every five years. That rate was five times lower than our prediction, but nevertheless much higher than most naturalists expect to find on such islands. And when extinctions and immigrations occurring unseen in the intervals between the surveys were provisionally figured in, the fit to the theory was closer.

Other biologists were encouraged by this mathematically simple attempt to define a dynamic equilibrium. They were intrigued by the very idea that the diversity of life rises to a certain level and stays there, with species coming and going at a predictable rate. The theory could be applied not just to islands in the ocean but also to "habitat islands," such as woodlots in a sea of grass, ponds and streams in a sea of land, and in fact to any habitat enclosed by a different environment hostile to its organisms. It might even be used to predict the fate of parks and nature reserves over a period of years or centuries.

Species-equilibrium theory, in other words, was heuristic. It promoted further study, a quality highly valued in science. In giving answers to a few old questions, it raised new ones and suggested techniques for their solution. New, more elaborate studies soon proliferated. Mountain tops, lakes, coral reefs, and bottles of water were added to the list of habitat islands under study, while guidelines were suggested for the design of parks. The World Wildlife Fund used some of the models in planning its rain-forest reserve project near Manaus. It was all very exciting, but the first models that MacArthur and I had fashioned were too crude to fit these additional cases. A whole new canon of theory was invented, and appropriate experiments followed. The study of species equilibria grew into a rich and sophisticated branch of ecology. Twenty years later our particular contributions were no longer clearly distinguishable from those of equal or superior merit contributed by other biologists. The surviving fragments had been absorbed into the mainstream, which continues to broaden and shift each year.

That is the way of science. The scientist may think like a poet, but the products of his imagination are seldom preserved in their original state. It is often said that a discipline is successful according to how quickly its founders are forgotten—or, more precisely, how soon they are replaced in the textbooks and vade mecums of the trade. No original MacArthurs hang in galleries; no biologists return to his original texts in the *Proceedings of the National Academy of Sciences* to sift for nuance and symbolism. MacArthur lives on as he would have liked, in the irreversible change he caused in an important branch of science.

AT THE MOMENT the spark ignites, when intuition and metaphor are all-important, the artist most closely resembles the scientist. But he does not then press on toward natural law and self-dissolution within the big picture. All his skills are aimed at the instant transference of images and control of emotions in others. For purposes of craft, he carefully avoids exact definitions or the display of inner logic.

In 1753 Bishop Lowth made the correct diagnosis in *Lectures on the Sacred Poetry of the Hebrews:* the poetic mind is not satisfied with a plain and exact description but seeks to heighten sensation. "For the passions are naturally inclined to amplifications; they wonderfully magnify and exaggerate whatever dwells upon the mind, and labour to express it in animated, bold, and magnificent terms."

The essential quality can be rephrased in more modern terms. The mind is biologically prone to discursive communication that expands thought. Mankind, in Richard Rorty's expression, is the poetic species. The symbols of art, music, and language freight power well beyond their outward and literal meanings. So each one also condenses large quantities of information. Just as mathematical equations allow us to move swiftly across large amounts of knowledge and spring into the unknown, the symbols of art gather human experience into novel forms in order to evoke a more intense perception in others. Human beings live—literally live, if life is equated with the mind—by symbols, particularly words, because the brain is constructed to process information almost exclusively in their terms.

I have spoken of art as a device for exploration and discovery. Its practitioners and expert observers, whose authority is beyond question, have stressed other functions as well. In Samuel Johnson's definition, to instruct by pleasing. According to Keats, to uplift by the refinement of shared feelings. No—moral, the role of art is moral, according to D. H. Lawrence. A spell against death, to create and preserve the self, in the formulation of Richard Eberhart. For the more prosaic cultural anthropologists, art above all else expresses the purposes of a society. Indeed, affirmation may have been the original evolutionary driving force behind Paleolithic cave art. It was certainly served by the early oral poets of Europe, including the illiterate Homeric bards who recited the *Iliad* and *Odyssey* at festivals and thus transmitted the central myths and legends of ancient Greece. When this cohesive function fails, and tradition and taste fragment as part of culture's advance, criticism becomes a necessary and honored profession. Then we also witness revolutionary art, which goes beyond innovation to promote a different society and culture.

All these functions are variously filled according to circumstances. Nevertheless, art generally considered to be important appears to be marked by one consistent quality: it explores the unknown reaches of the mind. The departure is both calculated and tentative, as in science. The poet focuses on the inward search itself and attracts us to his distant constructions. Something moves on the edge of the field of vision, a new connection is glimpsed, holds for a moment. Words pour in and around, and the image takes substantial form, at first believed familiar, then seen as strikingly new. It is something, as in Thomas Kinsella's "Midsummer,"

> that for this long year
> Had hid and halted like a deer
> Turned marvellous,
> Parted the tragic grasses, tame,
> Lifted its perfect head and came
> To welcome us.

But the poet refuses to take us any farther. If he goes on the precise image will melt into abstract descriptions; light and beauty will congeal into rows of formulas. In this essential way art differs from science. The world of interest is the mind, not the physical universe on which mental process feeds. Richard Eberhart, a keen observer of nature, listened to the same New England birds that led Robert MacArthur to mathematical theories of ecology, and I daresay that the first swirl of imagery, the first tensile pleasures were the same, but the two then diverged, the poet inward and the scientist outward into separate existences:

No, may the thrush among our high pine trees
Be ambiguous still, elusive in true song,
Never or seldom seen, and if never seen
May it to my imperious memory belong.

He holds back, on himself and on us, in order to cast his spell. Again we see that the dilemma of the machine-in-the-garden exists in the rain forests of the mind as surely as it does in the American continent. Our intrinsic emotions drive us to search for fresh habitats, to cross unexplored terrain, but we still crave the sense of a mysterious world stretching infinitely beyond. The free-living birds (thrush, nightingale, bird of paradise), being rulers of the blank spaces on the map and negligent of human existence, are worthy symbols of both art and science.

I have emphasized the expansive role of poetry to argue that, whereas art and science are basically different in execution, they are convergent in what they might eventually disclose about human nature. Until recently science was minimally concerned with the mind. Even those who granted mental process a material origin classified it as an ephemeron, the proper subject of some other occupation, a different way of thinking, a separate literature—in short, the humanities. Now all that has changed. Cognitive psychology has emerged as a strong discipline. Parallel studies on the nervous system and artificial intelligence are contributing further insights. Scientists view the human brain as one of the last remaining frontiers of empirical research. They are beginning to pour in from genetics, molecular biology, and other neighboring disciplines to join its settlement.

The subject of greatest immediate interest is long-term memory. We are essentially what we remember or can remember at some time in the future. People build memory by linking new images and concepts to old ones. In its size, in the space it fills, the mind expands like a coral reef, adding new branches and cross-ties out from the edge of those parts already established and anchored, while its central body settles and coalesces. A widely used theory of learning uses a more abstract but equivalent metaphor, the node-link model. The nodes are concepts such as dog, red, bark, aboveground, running, and teeth. Each node is linked to certain others, so that the activation of one in a person's memory tends to pull in a whole group. The image of a dog (or the mere word "dog") may evoke red, running, aboveground, fur, teeth, and a great deal more: concatenations of memories, frames of node-link structures riffled back and forth through time, and emotional nodes that can only be labeled by broad generic words such as fear and affection. The mind probes and adapts by a process that some psychologists have called spreading activation. Given new images or altered circumstances, it encompasses widening circles of nodes and links in a search for similarities, finally settling on the best categories and analogies previously stored in long-term memory.

Suppose that a strange new animal walks out of the tangled undergrowth and into our view. It might be compared to a dog, or a monkey, or something else. Perhaps the mind, over-loaded with its novelty, will simply abandon the search, giving the creature a new name and a fuller description than usual in order to establish a new node-link cluster. *This dark furry beast—let us call it X—is smaller than a dog. It has batlike ears, round luminous eyes, and ratlike teeth. It creeps about slug-gishly while picking at objects with long spidery fingers. At night it prowls through the treetops and inspires superstitious terror in the few natives who glimpse it by torchlight along the forest trails.* (This particular animal happens to be the aye-aye of Madagascar.)

SO NOW there can be a more explicit description of what theo-retical scientists and artists, dreamers of a kind, accomplish during the first stages of original thought. It is controlled growth, a disciplined spread of the mind into hidden recesses where concepts and linkages are still embryonic or nonexistent.

Genius is this kind of expertise born aloft on the wings of energy, daring, and luck. The combination came together in a famous letter sent in 1913 from Srinivasa Ramanujan, a Hindu clerk, to Godfrey H. Hardy, the English mathematician. By the age of twenty-five and with the aid of only one obscure text on higher mathematics, Ramanujan had independently solved some of the problems that previously occupied the best mathematicians of Europe for over a century. A few of the equations were known already: number (1.8) in Ramanujan's series was a formula of Laplace first proved by Jacobi, while (1.9) had already been published by Rogers in 1907. Equations (1.5) and (1.6) looked somewhat familiar and could be confirmed, according to Hardy, but only with a surprising amount of effort. Then came something really new:

> The formulae (1.10)–(1.13) are on a different level and obviously both difficult and deep . . . (1.10)–(1.12) defeated me completely; I had never seen anything in the least like them before. A single look at them is enough to show that they could only be written down by a mathematician of the highest class. They must be true because, if they were not true, no one would have had the imagination to invent them.

They must be true: the same can be said of outstanding achievements in literature and the arts, which pull the rest of us along until the construction becomes self-evident. Eliot wrote that "unless we have those few men who combine an exceptional sensibility with an exceptional power over words, our own ability, not merely to express, but even to feel any but the crudest emotions, will degenerate." The difference in power is one of degree rather than kind, but it crosses a threshold to create a qualitative new result in the same way that a critical speed lifts a glider off the ground into flight.

Research on cognitive development has shown that in the course of its growth the mind probes certain channels much more readily than others. Some of the responses are automatic and can be measured by physiological changes of which the individual is mostly or entirely unaware. For example, using electroencephalograms in the study of response to graphic designs, the Belgian psychologist Gerda Smets found that maximal arousal (measured by the blockage of the alpha wave) occurs when the figure contains about 20 percent redundancy. That is the amount present in a spiral with two or three turns, or a relatively simple maze, or a neat cluster of ten or so triangles. Less arousal occurs when the figure consists of only one triangle or square, or when the design is more complicated than the optimum—as in a difficult maze or an irregular scattering of twenty rectangles. The data are not the result of a chance biochemical quirk. When selecting symbols and abstract art, people actually gravitate to about the levels of complexity observed in Smets's experiments. Furthermore, the preference has its roots in early life. Newborn infants gaze longest at visual designs containing between five and twenty angles. During the next three months their preference shifts toward the adult pattern measured with electroencephalograms. Nor is there anything foreordained or otherwise trivial in the aesthetic optimum of human beings. It is easy to imagine the evolution of some other intelligent species in another time or on some other planet, possessing different eyes, optic nerves, and brain—and thus distinct optimal complexity and artistic standards.

We can reasonably suppose that the compositions of artists play upon the rules of mental development that are now beginning to receive the objective attention of experimental psychology. The distinction between science and art can be understood more clearly from this different perspective. The abstracted qualities of the developmental rules of the mind are the principal concern of science. In contrast, the node-link structures themselves, their emotional color, tone, cadence, fidelity to personal experience, and the images they fleetingly reveal, are more the domain of art. Of equal importance to both enterprises are the symbols and myths that evoke the mental structures in compelling fashion. Certain great myths—the origin of the world, cataclysm and rebirth, the struggle between the powers of light and darkness, Earth Mother, and a few others—recur dependably in cultures around the world. Lesser, more personal myths appear in crisis poems and romantic tales, where they blend imperceptibly into legend and history. Through the deep pleasures they naturally excite, and the ease with which they are passed from one person to another, these stories invade the developing mind more readily than others, and they tend to converge to form the commonalities of human nature. Yeats in his 1900 essay on Shelley distinguished between the theoretician who seeks abstract truth and the naturalist-poet who celebrates detail. In the universe of the mind, Yeats said, no symbol tells all its meanings to any generation. Only by discovering the ancient symbols can the artist express meanings that cross generations and open the full abundance of nature.

We need not worry about the extravagances of visionary artists, so long as they reveal the deeper channels of their minds in a manner that gives meaning to our own. Each human mindscape is idiosyncratic and yet ultimately obedient to biological law. Like the forest of some newly discovered island, it possesses unique contours and previously undescribed forms of life, treasures to be valued for their own sake, but the genetic process that spawned them is the same as elsewhere. Continuity is essential for comprehension; the imagery chosen by the artist must draw on common experience and values, however tortuous the manner of presentation. Thus in 1919 the American modernist Joseph Stella created "Tree of My Life" to translate his own vaulting optimism into a physical paradise within the mind. Bright tropical plants and animals served as the symbols. He described his feelings that led to the painting:

> AND one clear morning in April I found myself in the midst of joyous singing and delicious scent—the singing and the scent of the birds and flowers ready to celebrate the baptism of my new art, the birds and the flowers already enjewelling the tender foliage of the new-born tree of my hopes.

We are in the fullest sense a biological species and will find little ultimate meaning apart from the remainder of life. The fiery circle of disciplines will be closed if science looks at the inward journey of the artist's mind, making art and culture objects of study in the biological mode, and if the artist and critic are informed of the workings of the mind and the natural world as illuminated by the scientific method. In principle at least, nothing can be denied to the humanities, nothing to science.

The Serpent

SCIENCE and the humanities, biology and culture, are bridged in a dramatic manner by the phenomenon of the serpent. The snake's image enters the conscious mind with ease during dreams and reverie, fabricated from symbols and bearing portents of magic. It appears without warning and departs abruptly, leaving behind not the perception of any real snake but the vague memory of a more powerful creature, the serpent, surrounded by a mist of fear and wonderment.

These qualities are dominant in one particular dream I have experienced often through my life, for reasons I will try to clarify later. I find myself in a locality that is wooded and aquatic, silent and drawn wholly in shades of gray. As I walk into this somber environment I am gripped by an alien feeling:

> The terrain before me is mysterious, on the rim of the unknown, at once calm and forbidding. I am required to be there but in the dream state cannot grasp the reasons. Suddenly the Serpent appears. It is not a snake of the ordinary kind, a literal reptile, but much more, a threatening presence with unusual powers. Its size is indeterminately large. While I watch its muscular coils slide into the water, beneath prop roots, and back onto the bank, protean in size and shape, armored, irresistible. The poisonous head radiates a cold, inhuman intelligence. The Serpent is somehow the spirit of that shadowed place and guardian of the passage into deeper reaches. I sense that if I could capture or control or even just evade it, a great change in the ambience would follow. The change cannot be defined immediately, but its anticipation stirs old and still unnamed emotions. The risk is also vaguely felt, like that emanating from a knife blade or high cliff. The Serpent is life-promising and life-threatening, seductive and treacherous. It now slips close to me, turning importunate, ready to strike. The dream ends uneasily, without clear resolution.

The snake and the serpent, flesh-and-blood reptile and demonic dream-image, reveal the complexity of our relation to nature and the fascination and beauty inherent in all forms of organisms. Even the deadliest and most repugnant creatures bring an endowment of magic to the human mind. Human beings have an innate fear of snakes or, more precisely, they have an innate propensity to learn such fear quickly and easily past the age of five. The images they build out of this peculiar mental set are both powerful and ambivalent, ranging from terror-stricken flight to the experience of power and male sexuality. As a consequence the serpent has become an important part of cultures around the world.

There is a principle of many ramifications to consider here, which extends well beyond the ordinary concerns of psychoanalytic reasoning about sexual symbols. Life of any kind is infinitely more interesting than almost any conceivable variety of inanimate matter. The latter is valued chiefly to the extent that it can be metabolized into live tissue, accidentally resembles it, or can be fashioned into a useful and properly animated artifact. No one in his right mind looks at a pile of dead leaves in preference to the tree from which they fell.

What is it exactly that binds us so closely to living things? The biologist will tell you that life is the self-replication of giant molecules from lesser chemical fragments, resulting in the assembly of complex organic structures, the transfer of large amounts of molecular information, ingestion, growth, movement of an outwardly purposeful nature, and the proliferation of closely similar organisms. The poet-in-biologist will add that life is an exceedingly improbable state, metastable, open to other systems, thus ephemeral—and worth any price to keep.

Certain organisms have still more to offer because of their special impact on mental development. I have suggested that the urge to affiliate with other forms of life is to some degree innate, hence deserves to be called biophilia. The evidence for the proposition is not strong in a formal scientific sense: the subject has not been studied enough in the scientific manner of hypothesis, deduction, and experimentation to let us be certain about it one way or the other. The biophilic tendency

is nevertheless so clearly evinced in daily life and widely distributed as to deserve serious attention. It unfolds in the predictable fantasies and responses of individuals from early childhood
onward. It cascades into repetitive patterns of culture across
most or all societies, a consistency often noted in the literature of anthropology. These processes appear to be part of
the programs of the brain. They are marked by the quickness
and decisiveness with which we learn particular things about
certain kinds of plants and animals. They are too consistent to
be dismissed as the result of purely historical events working
on a mental blank slate.

Perhaps the most bizarre of the biophilic traits is awe and
veneration of the serpent. The dreams from which the dominant images arise are known to exist in all those societies where
systematic studies have been conducted on mental life. At least 5
percent of the people at any given time remember experiencing
them, while many more would probably do so if they recorded
their waking impressions over a period of several months. The
images described by urban New Yorkers are as detailed and
emotional as those of Australian aboriginals and Zulus. In all
cultures the serpents are prone to be mystically transfigured.
The Hopi know Palulukon, the water serpent, a benevolent but
frightening godlike being. The Kwakiutl fear the *sisiutl*, a kind
of three-headed serpent with both human and reptile faces,
whose appearance in dreams presages insanity or death. The
Sharanahua of Peru summon reptile spirits by taking halluci-
nogenic drugs and stroking the severed tongues of snakes over
their faces. They are rewarded with dreams of brightly colored
boas, venomous snakes, and lakes teeming with caimans and
anacondas. Around the world serpents and snakelike creatures
are the dominant elements of dreams in which animals of any
kind appear. Inspiring fear and veneration, they are recruited
as the animate symbols of power and sex, totems, protagonists
of myths, and gods.

These cultural manifestations may seem at first detached and
mysterious, but there is a simple reality behind the ophidian
archetype that lies within the experience of ordinary people.
The mind is primed to react emotionally to the sight of
snakes, not just to fear them but to be aroused and absorbed
in their details, to weave stories about them. This distinctive

predisposition played an important role in an unusual experience of my own. Let me tell a childhood story about an encounter with a large and memorable snake, a creature that actually existed.

I GREW UP in the Panhandle of northern Florida and the adjacent counties of Alabama, in circumstances that eventually turned me into a field biologist. Like most boys in that part of the country set loose to roam the woods, I enjoyed hunting and fishing and made no clear distinction between these activities and life at large. But I also cherished natural history for its own sake and decided very early to become a biologist. I had a secret ambition to find a Real Serpent, a snake so fabulously large or otherwise different that it would exceed the bounds of imagination.

Certain peculiarities in the environment encouraged this adolescent fantasy. It helped at the outset that I was an only child with indulgent parents, encouraged to develop my own interests and hobbies, however farfetched. In other words, I was spoiled. And except for issues pertaining to the literal claims of the King James Bible, our neighbors were equally tolerant of eccentric kids. They had to be: we all knew, even though we did not discuss openly, that certain families in the neighborhood kept very unusual children in their homes instead of placing them in institutions. It was a time in the South, about to come to a close, when family obligations and loyalties were unquestioned and spoken about mostly in oblique, ritual terms.

The physical surroundings inclined youngsters toward an awe of nature. That part of the country had been covered, four generations back, by a wilderness as formidable in some respects as the Amazon. Dense thickets of cabbage palmetto descended into meandering spring-fed streams and cypress sloughs. Carolina parakeets and ivory-billed woodpeckers flashed overhead in the sunlight, where wild turkeys and passenger pigeons could still be counted on as game. On soft spring nights after heavy rains a dozen varieties of frogs croaked, rasped, bonged, and trilled their love songs in mixed choruses. Much of the Gulf Coast fauna had been derived from species that spread north from the tropics over millions of years and adapted to the local, warm temperate conditions. Columns of miniature army ants, close replicas of the large marauders of South America, marched mostly unseen at night over the forest floor. *Nephila* spiders the size of saucers spun webs as wide as garage doors across the woodland clearings.

From the stagnant pools and knothole sinks, clouds of mosquitoes rose to afflict the early immigrants. They carried the Confederate plague, malaria and yellow fever, which periodically flared into epidemics and reduced the populations along the coastal lowlands. This natural check is one of the reasons the strip between Tampa and Pensacola remained sparsely settled for so long and why even today, long after the diseases have been eradicated, it is still the relatively natural "other Florida."

Snakes abounded. The Gulf Coast has a greater variety and denser populations than almost any other place in the world, and they are frequently seen. Striped ribbon snakes hang in Gorgonlike clusters on branches at the edge of ponds and streams. Poisonous coral snakes root through the leaf litter, their bodies decorated with warning bands of red, yellow, and black. They are easily confused with their mimics, the scarlet kingsnakes, banded in a different sequence of red, black, and yellow. The simple rule recited by woodsmen is: "Red next to yellow will kill a fellow, Red next to black is a friend of Jack." Hognoses, harmless thick-bodied sluggards with upturned snouts, are characterized by an unsettling resemblance to venomous African gaboon vipers and a habit of swallowing toads live. Pygmy rattlesnakes two feet long contrast with diamondbacks seven feet long or more. Watersnakes are a herpetologist's medley told apart by size, color, and the arrangement of body scales. They comprise ten species of *Natrix, Seminatrix, Agkistrodon, Liodytes,* and *Farancia.*

Of course limits to the abundance and diversity exist. Because snakes feed on frogs, mice, fish, and other animals of similar size, they are necessarily scarcer than their prey. You can't just go out on a stroll and point to one individual after another. An hour's careful search will often turn up none at all. But I can testify from personal experience that on any given day you are ten times more likely to meet a snake in Florida than in Brazil or New Guinea.

There is something oddly appropriate about the abundance of snakes. Although the Gulf wilderness has been largely

converted into macadam and farmland, and the sound of television and company jets is heard in the land, a remnant of the
old rural culture remains, as if the population were still pitted
against the savage and the unknown. "Push the forest back
and fill the land" remains a common sentiment, the colonizer's
ethic and tested biblical wisdom (the very same that turned the
cedar groves of Lebanon into the fought-over desert they are
today). The prominence of snakes lends symbolic support to
this venerable belief.

In the back country during a century and a half of settlement, the common experience of snakes was embroidered into
the lore of serpents. Cut off a rattlesnake's head, one still hears,
and it will live on until sundown. If a snake bites you, open the
puncture wounds with a knife and wash them with kerosene
to neutralize the poison (I never met anyone who claimed to
have done that and survived). If you believe with all your heart
in Jesus, you can hang rattlers and copperheads around your
neck without fear. If one strikes you just the same, accept it as
a sign from the Lord and find peace in whatever follows. The
hognose snake, on the other hand, is always death in the shape
of a slithery S. Those who get too close to one will have venom
sprayed in their eyes and be blinded; the very breath from the
snake's skin is lethal. This species is the beneficiary of its dreadful legend: I never heard of one being killed.

Deep in the woods live creatures of startling power. (*That* is
what I most wanted to hear.) Among them is the hoop snake.
Skeptics, who used to be found hunkered down in a row along
the county courthouse guardrail on a Saturday morning, say
it is only mythical; on the other hand it might be the familiar
coachwhip racer turned vicious by special circumstances. Thus
transformed, it puts its tail in its mouth and rolls down hills
at great speed to attack its terrified victims. Then there were
reports of the occasional true monsters: a giant snake believed
to live in a certain swamp (used to be there anyway, even if
no one's seen it in recent years); a twelve-foot diamondback
rattler a farmer killed on the edge of town a few years back;
some unclassifiable prodigy recently glimpsed as it sunned itself
along the river's edge.

It is a wonderful thing to grow up in southern towns where
animal fables are taken half seriously, breathing into the

adolescent mind a sense of the unknown and the possibility that something extraordinary might be found within a day's walk of where you live. No such magic exists in the environs of Schenectady, Liverpool, and Darmstadt, and for all children dwelling in such places where the options have finally been closed, I feel a twinge of sadness. I found my way out of Mobile, Pensacola, and Brewton to explore the surrounding woods and swamps in a languorous mood. I formed the habit of quietude and concentration into which I still pass my mind during field excursions, having learned to summon the old emotions as part of the naturalist's technique.

Some of these feelings must have been shared by my friends. In the mid-1940s during the hot season between spring football practice and the regular schedule of games in the fall, working on highway cleanup gangs and poking around outdoors were about all we had to do. But there was some difference: I was hunting snakes with a passionate intensity. On the Brewton High School football team of 1944–45 most of the players had nicknames, leaning toward the infantilisms and initials favored by southerners: Bubba Joe, Flip, A. J., Sonny, Shoe, Jimbo, Junior, Snooker, Skeeter. As the underweight third-string left end, allowed to play only in the fourth quarter when the foe had been crushed beyond any hope of recovery, mine was Snake. And while of this measure of masculine acceptance I was inordinately proud, my main hopes and energies had been invested elsewhere. There are an incredible forty species of snakes native to that region, and I managed to capture almost all of them.

One kind became a special target just because it was so elusive: the glossy watersnake *Natrix rigida*. The adults lay on the bottom of shallow ponds well away from the shore and pointed their heads out of the alga-green water in order to breathe and scan the surface in all directions. I waded out toward them very carefully, avoiding the side-to-side movements to which snakes are most alert. I needed to get within three or four feet in order to manage a diving tackle, but before I could cover the distance they always pulled their heads under and slipped silently away into the opaque depths. I finally solved the problem with the aid of the town's leading slingshot artist, a taciturn loner my age, proud and quick to anger, the sort of boy who in an earlier time might have distinguished himself at Antietam or Shiloh. Aiming pebbles at the heads of the snakes, he was able to stun several long enough for me to grab them underwater. After recovering, the captives were kept for a while in homemade cages in our backyard, where they thrived on live minnows placed in dishes of water.

Once, deep in a swamp miles from home, half lost and not caring, I glimpsed an unfamiliar brightly colored snake disappearing down a crayfish burrow. I sprinted to the spot, thrust my hand after it and felt around blindly. Too late: the snake had squirmed out of reach into the lower chambers. Only later did I think about the possibilities. Suppose I had succeeded and the snake was poisonous? My reckless enthusiasm did catch up with me on another occasion when I miscalculated the reach of a pygmy rattlesnake, which struck out faster than I thought possible and hit me with startling authority on the left index finger. Because of the small size of the reptile, the only result was a temporarily swollen arm and a fingertip that still grows a bit numb at the onset of cold weather.

But I digress. I found my Serpent on a still July morning in the swamp fed by the artesian wells of Brewton, while working toward higher ground along the course of a weed-choked stream. Without warning a very large snake crashed away from under my feet and plunged into the water. Its movement startled me even more than it would have in other circumstances, because I had grown accustomed through the day to modestly proportioned frogs and turtles silently tensed on mudbanks and logs. This snake was more nearly my size as well as violent and noisy—a colleague, so to speak. It sped with wide body undulations to the center of the shallow watercourse and came to rest on a sandy riffle. It was not quite the monster I had envisioned but nevertheless unusual, a water moccasin (*Agkistrodon piscivorus*), one of the poisonous pit vipers, more than five feet long with a body as thick as my arm and a head the size of a fist. It was the largest snake I had ever seen in the wild. I later calculated it to be just under the published size record for the species. The snake now lay quietly in the shallow clear water completely open to view, its body stretched along the fringing weeds, its head pointed back at an oblique angle to watch my approach. Moccasins are like that. They don't always keep going until they are out of sight, in the manner of ordinary watersnakes. Although no emotion can be read in the frozen half-smile and staring yellow cat's eyes, their reactions and posture make them seem insolent, as if they see their power reflected in the caution of human beings and other sizable enemies.

I moved through the snake handler's routine: pressed the snake stick across the body in back of the head, rolled it forward to pin the head securely, brought one hand around to grasp the neck just behind the swelling masseteric muscles, dropped the stick to seize the body midway back with the other hand, and lifted the entire animal clear of the water. The technique almost always works. The moccasin, however, reacted in a way that took me by surprise and put my life in immediate danger. Throwing its heavy body into convulsions, it twisted its head and neck slightly forward through my gripped fingers, stretched its mouth wide open to unfold the inch-long fangs and expose the dead-white inner lining in the intimidating "cottonmouth" display. A fetid musk from its anal glands filled the air. At that moment the morning heat became more noticeable, the episode turned manifestly frivolous, and at last I wondered why I should be in that place alone. Who would find me? The snake began to turn its head far enough to clamp its jaws on my hand. I was not very strong for my age and I was losing control. Without thinking I heaved the giant out into the brush and it thrashed frantically away, this time until it was out of sight and we were rid of each other.

I sat down and let the adrenaline race my heart and bring tremors to my hand. How could I have been so stupid? What is there in snakes anyway that makes them so repellent and fascinating? The answer in retrospect is deceptively simple: their ability to remain hidden, the power in their sinuous limbless bodies, and the threat from venom injected hypodermically through sharp hollow teeth. It pays in elementary survival to be interested in snakes and to respond emotionally to their generalized image, to go beyond ordinary caution and fear. The rule built into the brain in the form of a learning bias is: become alert quickly to any object with the serpentine gestalt. *Overlearn* this particular response in order to keep safe.

Other primates have evolved similar rules. When guenons and vervets, the common monkeys of the African forest, see a python, cobra, or puff adder, they emit a distinctive chuttering call that rouses other members in the group. (Different calls are used to designate eagles and leopards.) Some of the adults then follow the intruding snake at a safe distance until it leaves the area. The monkeys in effect broadcast a dangerous-snake alert, which serves to protect the entire group and not solely the individual who encountered the danger. The most remarkable fact is that the alarm is evoked most strongly by the kinds of snakes that can harm them. Somehow, apparently through the routes of instinct, the guenons and vervets have become competent herpetologists.

The idea that snake aversion on the part of man's relatives can be an inborn trait is supported by other studies on rhesus macaques, the large brown monkeys of India and surrounding Asian countries. When adults see a snake of any kind, they react with the generalized fear response of their species. They variously back off and stare (or turn away), crouch, shield their faces, bark, screech, and twist their faces into the fear grimace, in which the lips are retracted, the teeth are bared, and the ears are flattened against the head. Monkeys raised in the laboratory without previous exposure to snakes show the same response to them as those brought in from the wild, although in weaker form. During control experiments designed to test the specificity of the response, the rhesus failed to react to other, nonsinuous objects placed in their cages. It is the form of the snake and perhaps also its distinctive movements that contain the key stimuli to which the monkeys are innately tuned.

Grant for the moment that snake aversion does have a hereditary basis in at least some kinds of nonhuman primates. The possibility that immediately follows is that the trait evolved by natural selection. In other words, individuals who respond leave more offspring than those who do not, and as a result the propensity to learn fear quickly spreads through the population—or, if it was already present, is maintained there at a high level.

How can biologists test such a proposition about the origin of behavior? They turn natural history upside down. They search for species historically free of forces in the environment believed to favor the evolutionary change, to see if in fact the organisms do *not* possess the trait. The lemurs, primitive relatives of the monkeys, offer such an inverted opportunity. They are indigenous inhabitants of Madagascar, where no large or poisonous snakes exist to threaten them. Sure enough, lemurs presented with snakes in captivity fail to display anything resembling the automatic fear responses of the African and Asian monkeys. Is this adequate proof? In the chaste idiom of scientific discourse, we are permitted to conclude only that the evidence is consistent with the proposal. Neither this nor any comparable hypothesis can be settled by a single case. Only further examples can raise confidence in it to a level beyond the reach of determined skeptics.

Another line of evidence comes from studies of the chimpanzee, a species thought to have shared a common ancestor with prehumans as recently as five million years ago. Chimps raised in the laboratory become apprehensive in the presence of snakes, even if they have had no previous experience. They back off to a safe distance and follow the intruder with a fixed stare while alerting companions with the *Wah!* warning call. More important, the response becomes gradually more marked during adolescence.

This last quality is especially interesting because human beings pass through approximately the same developmental sequence. Children under five years of age feel no special anxiety over snakes, but later they grow increasingly wary. Just one or two mildly bad experiences, such as a garter snake seen writhing away in the grass, a playmate thrusting a rubber model at them, or a counselor telling scary stories at the campfire, can make children deeply and permanently fearful. The pattern is unusual if not unique in the ontogeny of human behavior. Other common fears, notably of the dark, strangers, and loud noises, start to wane after seven years of age. In contrast, the tendency to avoid snakes grows stronger with time. It is possible to turn the mind in the opposite direction, to learn to handle snakes without apprehension or even to like them in some special way, as I did—but the adaptation takes a special effort and is usually a little forced and self-conscious. The special sensitivity will just as likely lead to full-blown ophidiophobia, the pathological extreme in which the mere appearance of a snake brings on a feeling of panic, cold sweat, and waves of nausea. I have witnessed these events:

At a campsite in Alabama, on a Sunday afternoon, a four-foot-long black racer glided out from the woods across the clearing and headed for the high grass along a nearby stream. Children shouted and pointed. A middle-aged woman screamed and collapsed to the ground sobbing. Her husband dashed to his pickup truck to get a shotgun. But black racers are among the fastest snakes in the world, and this one made it safely to cover. The onlookers probably did not know that the species is nonvenomous and harmless to any creature larger than a cotton rat.

Halfway around the world, at the village of Ebabaang in New Guinea, I heard shouting and saw people running down a path. When I caught up with them they had formed a circle around a small brown snake that was essing leisurely across the front yard of a house. I pinned the snake and carried it off to be preserved in alcohol for the museum collections at Harvard. This seeming act of daring earned either the admiration or the suspicion of my hosts—I couldn't be sure which. The next day children followed me around as I gathered insects in the nearby forest. One brought me an immense orb-weaving spider gripped in his fingers, its hairy legs waving and the evil-looking black fangs working up and down. I felt panicky and sick. It so happens that I suffer from mild arachnophobia. To each his own.

Why should serpents have such a strong influence during mental development? The direct and simple answer is that throughout the history of mankind a few kinds have been a major cause of sickness and death. Every continent except Antarctica has poisonous snakes. Over large stretches of Asia and Africa the known death rate from snake bite is 5 persons per 100,000 each year, or higher. The local record is held by a province in Burma, with 36.8 deaths per 100,000 a year. Australia has an exceptional abundance of deadly snakes, a majority of which are relatives of the cobras. Among them the tiger snake is especially feared for its large size and tendency to strike without warning. In South and Central America live the bushmaster, fer-de-lance, and jaracara, among the largest and most aggressive of the pit vipers. With backs colored like rotting leaves and fangs long enough to pass through a human hand, they lie in ambush on the floor of the tropical forest for the small warm-blooded animals that form their major prey. Few people realize that a complex of dangerous snakes, the "true" vipers, are still relatively abundant throughout Europe. The common adder *Viperus berus* ranges to the Arctic Circle. The number of people bitten in such improbable places as Switzerland and Finland is still high enough, running into the hundreds annually, to keep outdoorsmen on a sort of yellow alert. Even Ireland, one of the few countries in the world lacking snakes altogether (thanks to the last Pleistocene glaciation and not Saint Patrick), has imported the key ophidian symbols and traditions from other European cultures and preserved the fear of serpents in art and literature.

HERE, THEN, is the sequence by which the agents of nature appear to have been translated into the symbols of culture. For hundreds of thousands of years, time enough for the appropriate genetic changes to occur in the brain, poisonous snakes have been a significant source of injury and death to human beings. The response to the threat is not simply to avoid it, in the way that certain berries are recognized as poisonous through a process of trial and error. People also display the mixture of apprehension and morbid fascination characterizing the non-human primates. They inherit a strong tendency to acquire the aversion during early childhood and to add to it progressively, like our closest phylogenetic relatives, the chimpanzees. The mind then adds a great deal more that is distinctively human. It feeds upon the emotions to enrich culture. The tendency of the serpent to appear suddenly in dreams, its sinuous form, and its power and mystery are the natural ingredients of myth and religion.

Consider how sensation and emotional states are elaborated into stories during dreams. The dreamer hears a distant thunderclap and changes an ongoing episode to end with the slamming of a door. He feels a general anxiety and is transported to a schoolhouse corridor, where he searches for a classroom he does not know in order to take an examination for which he is unprepared. As the sleeping brain enters its regular dream periods, marked by rapid eye movement beneath closed eyelids, giant fibers in the lower brainstem fire upward into the cortex. The awakened mind responds by retrieving memories and fabricating stories around the sources of physical and emotional discomfort. It hastens to recreate the elements of past real experience, often in a jumbled and antic form. And from time to time the serpent appears as the embodiment of one or more of these feelings. The direct and literal fear of snakes is foremost among them, but the dream-image can also be summoned by sexual desire, a craving for dominance and power, and the apprehension of violent death.

We need not turn to Freudian theory in order to explain our special relationship to snakes. The serpent did not originate as the vehicle of dreams and symbols. The relation appears to be precisely the other way around and correspondingly easier to study and understand. Humanity's concrete experience with poisonous snakes gave rise to the Freudian phenomena after it was assimilated by genetic evolution into the brain's structure. The mind has to create symbols and fantasies from something. It leans toward the most powerful preexistent images or at least follows the learning rules that create the images, including that of the serpent. For most of this century, perhaps overly enchanted by psychoanalysis, we have confused the dream with the reality and its psychic effect with the ultimate cause rooted in nature.

Among prescientific people, whose dreams are conduits to the spirit world and snakes a part of ordinary experience, the serpent has played a central role in the building of culture. There are magic incantations for simple protection as in the hymns of the Atharva Veda:

> With my eye do I slay thy eye, with poison do I slay thy poison. O Serpent, die, do not live; back upon thee shall thy poison turn.

"Indra slew thy first ancestors, O Serpent," the chant continues, "and since they are crushed, what strength forsooth can be theirs?" And so the power can be controlled and even diverted to human use through iatromancy and the casting of magic spells. Two serpents entwine the caduceus, which was first the winged staff of Mercury as messenger of the gods, then the safe-conduct pass of ambassadors and heralds, and finally the universal emblem of the medical profession.

Balaji Mundkur has shown how the inborn awe of snakes matured into rich productions of art and religion around the world. Serpentine forms wind across stone carvings from paleo-lithic Europe and are scratched into mammoth teeth found in Siberia. They are the emblems of power and ceremony for the shamans of the Kwakiutl, the Siberian Yakut and Yenisei Ostyak, and many of the tribes of Australian aboriginals. Styl-ized snakes have often served as the talismans of the gods and spirits who bestow fertility: Ashtoreth of the Canaanites, the demons Fu-Hsi and Nu-kua of the Han Chinese, and the pow-erful goddesses Mudammā and Manasā of Hindu India. The ancient Egyptians venerated at least thirteen ophidian deities ministering to various combinations of health, fecundity, and vegetation. Prominent among them was the triple-headed giant Nehebkau who traveled widely to inspect every part of the river kingdom. Amulets in gold inscribed with the sign of a cobra god were placed in the wrappings of Tutankhamen's mummy. Even the scorpion goddess Selket bore the title "mother of serpents." Like her offspring she prevailed simultaneously as a source of evil, power, and goodness.

The Aztec pantheon was a phantasmagoria of monstrous forms among whom serpents were given pride of place. The calendrical symbols included the ophidian *olin nahui* and *cipactli*, the earth crocodile that possessed a forked tongue and rattlesnake's tail. The rain god Tlaloc consisted in part of two coiled rattlesnakes whose heads met to form the god's upper lip. *Coatl*, serpent, is the dominant phrase in the names of the divinities. Coatlicue was a threatening chimera of snake and human parts, Cihuacoatl the goddess of childbirth and mother of the human race, and Xiuhcoatl the fire serpent over whose body fire was rekindled every fifty-two years to mark a major division in the religious calendar. Quetzalcoatl, the plumed serpent with a human head, reigned as god of the morning and evening star and thus of death and resurrection. As inventor of the calendar, deity of books and learning, and patron of the priesthood, he was revered in the schools where nobles and priests were taught. His reported departure over the eastern horizon upon a raft of snakes must have been the occasion of consternation for the intellectuals of the day, something like the folding of the Guggenheim Foundation.

A contradiction of ophidian images was a feature of Greek religion as well. Among the early forms of Zeus was the serpent Meilikhios, god of love, gentle and responsive to supplication, and god of vengeance, whose sacrifice was a holocaust offered at night. Another great serpent protected the lustral waters at the spring of Ares. He coexisted with the Erinyes, demons of the underworld so horrible they could not be pictured in early mythology. They were given the form of serpents when brought to stage by Euripides in the *Iphigeneia in Tauris:* "Dost see her, her the Hades-snake who gapes / To slay me, with dread vipers, open-mouthed?"

Slyness, deception, malevolence, betrayal, the implicit threat of a forked tongue flicking in and out of the masklike head, all qualities tinged with miraculous powers to heal and guide, forecast and empower, became the serpent's prevailing image in western cultures. The serpent in the Garden of Eden, appearing as in a dream to serve as Judaism's evil Prometheus, gave humankind knowledge of good and evil and with it the burden of original sin, for which God repaid in kind:

> I will put enmity between you and the woman,
> between your brood and hers.
> They shall strike at your head,
> and you shall strike at their heel.

TO SUMMARIZE the relation between man and snake: life gathers human meaning to become part of us. Culture transforms the snake into the serpent, a far more potent creation than the literal reptile. Culture in turn is a product of the mind, which can be interpreted as an image-making machine that recreates the outside world through symbols arranged into maps and stories. But the mind does not have an instant capacity to grasp reality in its full chaotic richness; nor does the body last long enough for the brain to process information piece by piece like an all-purpose computer. Rather, consciousness races ahead to master certain kinds of information with enough efficiency to survive. It submits to a few biases easily while automatically avoiding others. A great deal of evidence has accumulated in genetics and physiology to show that the controlling devices are biological in nature, built into the sensory apparatus and brain by particularities in cellular architecture.

The combined biases are what we call human nature. The central tendencies, exemplified so strikingly in fear and veneration of the serpent, are the wellsprings of culture. Hence simple perceptions yield an unending abundance of images with special meaning while remaining true to the forces of natural selection that created them.

How could it be otherwise? The brain evolved into its present form over a period of about two million years, from the time of *Homo habilis* to the late stone age of *Homo sapiens*, during which people existed in hunter-gatherer bands in intimate contact with the natural environment. Snakes mattered. The smell of water, the hum of a bee, the directional bend of a plant stalk mattered. The naturalist's trance was adaptive: the glimpse of one small animal hidden in the grass could make the difference between eating and going hungry in the evening. And a sweet sense of horror, the shivery fascination with monsters and creeping forms that so delights us today even in the sterile hearts of the cities, could see you through to the next morning. Organisms are the natural stuff of metaphor and ritual. Although the evidence is far from all in, the brain appears to have kept its old capacities, its channeled quickness. We stay alert and alive in the vanished forests of the world.

The Right Place

THE NATURALIST is a civilized hunter. He goes alone into a field or woodland and closes his mind to everything but that time and place, so that life around him presses in on all the senses and small details grow in significance. He begins the scanning search for which cognition was engineered. His mind becomes unfocused, it focuses on everything, no longer directed toward any ordinary task or social pleasantry. He measures the antic darting of midges in a conical mating swarm, the slant of sunlight by which they are best seen, the precise molding of mosses and lichens on the tree trunk on which they spasmodically alight. His eye travels up the trunk to the first branch and out to a spray of twigs and leaves and back, searching for some irregularity of shape or movement of a few millimeters that might betray an animal in hiding. He listens for any sound that breaks the lengthy spells of silence. From time to time he translates his running impressions of the smell of soil and vegetation into rational thought: the ancient olfactory brain speaks to the modern cortex. The hunter-in-naturalist knows that he does not know what is going to happen. He is required, as Ortega y Gasset expressed it, to prepare an attention of a different and superior kind, "an attention that does not consist in riveting itself to the presumed but consists precisely in not presuming anything and avoiding inattentiveness."

Every practicing naturalist has favorite stories to tell about the rewards of chance in the field. I once went out with Jesse Nichols, a professional animal collector, to a woodlot in central Alabama late at night in a cold rain to look for frogs and salamanders. I had been to the site several times before on sunny days and seen nothing. That night, as soon as we walked into the woods, we found a teeming population of a pygmy salamander in the genus *Desmognathus*, recently described by zoologists as a new species. The delicately built amphibians, which resembled shiny popeyed lizards, were climbing up grass and low bushes. They jumped agilely from branch to branch in search of prey. It had been our good fortune to encounter them under the most favorable environmental conditions at the height of their activity, and it occurred to us that as a result we had made a worthwhile discovery about desmognaths in general. This class of salamanders usually lives at the edge of water or concealed in litter and soil. We now knew that one of the species is also partly arboreal and behaves a bit like tree frogs. It follows that desmognath salamanders as a whole are more ecologically diverse than originally believed. They have undergone a moderate evolutionary expansion near the center of their range, a broad area in the southeastern United States that includes Alabama. We discussed these weighty matters as we shivered in the rain, plucking enough pygmy salamanders off bushes to give to museums around the country.

Field research consists of hard physical work broken by moments of happy surprise. In his autobiography William Mann, until 1958 Director of the National Zoological Park in Washington but by training an entomologist, tells of a trip he made as a young man into the Sierra de Trinidad of central Cuba. When he lifted a rock to see what animals were hiding underneath (there are always animals of some kind, usually very small, under every rock), it split down the middle to expose a half-teaspoonful of metallic-green ants living in a small cavity deep inside. Mann went on to name this remarkable creature *Macromischa wheeleri*, in honor of William Morton Wheeler, his major professor at Harvard and the reigning world authority on ants. Thirty-six years later, with his discovery a romantic image in my head, I was climbing a steep slope in the same mountains, another young man at the start of a career in entomology. I had begun an ant-hill odyssey around the world remarkably similar to Mann's. A rock I grabbed for support split in my hand, exposing a half-teaspoonful of the same glittering green species. I accepted the event as one of the rites of passage.

Within hours on the same hillside I had another piece of luck. I obtained a live adult of the rare giant anole lizard, *Chamaeleolis chamaeleontides*, found only in Cuba and previously something of a mystery to biologists. It belongs to a group of species sometimes called the false chameleons because of an ability, shared with the true chameleons of Africa, to change skin color according to background and mood. The foot-long lizard also had a naturally wrinkled skin and a tired-looking expression, and I named my specimen Methuselah. During the remainder of my travels in Cuba and Mexico that summer of 1953 and after we came back to Cambridge in the fall, Methuselah spent much of his time riding on my shoulders. By watching him almost daily over a six-month's period as I fed him live mealworms and other insects, I came to realize that *Chamaeleolis* closely resembles the African chameleons in behavior as well as appearance. Both hunt with a slowness and deliberateness unusual for lizards, swivel their partly fused eyelids around to change the field of vision, and capture prey by flicking out long sticky tongues at nearly invisible speed. The similarity provided one more textbook example of evolutionary convergence between separate lines of animals that originated in the Old and New Worlds, in this case Africa as opposed to Cuba. The facts (which I published in a short technical article) were less than earth-shaking, but solid and satisfying—at least they will outlive Methuselah and me.

INSOFAR as organisms have been scrutinized, the naturalist can place them: their linkage in the ecosystem, life cycle, behavior, genetics, evolutionary history, physiology, and from all this information something of their general significance by whatever philosophy guided the naturalist to his life's pursuit in the first place. He is conducting a hunt in another mode, not for the animal's body but for discoveries, new information that will become part of the permanent record about the species viewed as an enduring entity. The pursuit is peculiarly satisfying because it enters that part of the real world, largely unrecognized, where humanity evolved during most of its two-million-year history. The vivifying eye of the naturalist is the orderly response to the original human environment.

What was that environment? To answer the question, we must turn natural history partly into an exercise in aesthetic judgment. The more habitats I have explored, the more I have felt that certain common features subliminally attract and hold my attention. Is it unreasonable to suppose that the human mind is primed to respond most strongly to some narrowly defined qualities that had the greatest impact on survival in the past? I am not suggesting the existence of an instinct. There is no evidence of a hereditary program hardwired into the brain.

We learn most of what we know, but some things are learned much more quickly and easily than others. The hypothesis of biased learning is at least worth examining, and the logical point of departure is a pair of derived questions. What was the prevailing original habitat in which the brain evolved? Where would people go if given a completely free choice?

The whole matter may seem imponderable at first, but a workable approach can be found in this generalization from ecology: the crucial first step to survival in all organisms is habitat selection. If you get to the right place, everything else is likely to be easier. Prey become familiar and vulnerable, shelters can be put together quickly, and predators are tricked and beaten consistently. A great many of the complex structures in the sense organs and brain that characterize each species serve the primary function of habitat selection. They determine the sounds, sights, and smells individuals receive and the sequence of responses these stimuli evoke.

Following inborn rules of behavior, animals turn to the special routes and crannies for which the remainder of their anatomy and physiology is particularly well suited. A few make crucial choices in the first few minutes of their lives. In what may be the ultimate case among mammals, the newborn kangaroo travels over its mother's belly from her genital opening to the nipples located deep in the pouch. Because it is completely blind, the peanut-sized creature must rely on a precise instinctual reading of the odor and feel of every centimeter of fur. In order to duplicate an equal feat of orientation, a human infant would have to emerge unaided from the womb, crawl down onto the carpet, make its way directly through the house to the nursery and into the crib, seize a bottle and start feeding.

So precise is habitat selection by many animals that closely related species can often be told apart more quickly by where they are found than by any obvious physical trait. The North American flycatchers, for example, are relatively small, inconspicuously colored birds that flit in and out of trees to snatch insects from the air. Only an expert can separate the species readily by outward appearance alone, but even a beginner can make a reliable identification if the habitat is added in. The alder flycatcher lives primarily in swamps and wet thickets, while each of the other species chooses a special combination of sites from among coniferous forests, cold bogs, farmland, and open mixed woodland.

Even more instructive is the case of the prairie deer mouse of the central United States. Wild populations remain strictly in open terrain, avoiding all kinds of forests, even those with grassy floors. When biologists raised individuals in outdoor enclosures simulating the principal natural environments, they found that the orientation is inborn, although it can be additionally reinforced by early exposure to open places. They were also able to breed the trait out of captive deer mouse populations in less than twenty generations, so that afterward individual mice were just as likely to enter woods as fields.

Salamanders, frogs, and insects make finer discriminations appropriate to their smaller size. They settle on precisely defined sites beneath stones or on vegetation that offer the optimal combination of moisture, light, and temperature for their species. Even colon bacteria swim skillfully to the position in a drop of water where nutrients are most concentrated—but in a decidedly peculiar way. They move by spinning the whip-like flagellum at the end of the body like a ship's propeller. If the effort takes a particular bacterium from a higher to a lower concentration, in other words away from the nutrients, the organism changes course by reversing the spin, forcing the filaments of its flagellum to fly apart. This action makes it tumble through the water. When the tumbling stops, the filaments come together again, allowing the bacterium to swim in a new direction. Eventually, by trial and error, it reaches a zone of concentration high enough to let it feed. Microbiologists have succeeded in locating the genes and sensitive proteins that guide this simplest of all known orientation devices. They have identified mutations that change the structure of the control-ling molecules and hence the direction in which bacteria swim. An important test of evolutionary theory has been passed: it is possible to alter an organism so that it automatically chooses the wrong habitat and condemns itself to death.

The question of interest is the preferred habitat of human beings. It is often said that *Homo sapiens* is the one species that can live anywhere—on top of ice floes, inside caves, under the sea, in space, anywhere—but this is just a half truth. People must jigger their environment constantly in order to keep it within a narrow range of atmospheric conditions. And once they have managed to rise above the level of bare subsistence, they invest large amounts of time to improve the appearance of their immediate surroundings. Their aim is to make the habitat more "livable" according to what are usually called aesthetic criteria.

With aesthetics we return to the central issue of biophilia. It is interesting to inquire about the prevalent direction of this vector in cultural evolution, in other words the ideal toward which human beings unconsciously strive no less relentlessly than flycatchers and deer mice. For if animals choose habitats by orientation devices and prepared learning built in during generations of natural selection, it is possible that people do the same. If certain human feelings are innate, they might not be easily expressed in rational language. A more promising approach is to explore the nature of the environment in which the brain evolved. The logical hypothesis I raised earlier can then be more precisely expressed. It is that certain key features of the ancient physical habitat match the choices made by modern human beings when they have a say in the matter.

The archeological evidence seems clear on the question of the original environment. For most of two million years human beings lived on the savannas of Africa, and subsequently those of Europe and Asia, vast, parklike grasslands dotted by groves and scattered trees. They appear to have avoided the equatorial rain forests on one side and the deserts on the other. There was nothing foreordained about this choice. The two extreme habitats have no special qualities that deny them to primates. Most monkeys and apes flourish in the rain forest, and two species, the hamadryas baboon and gelada, are specialized for life in the relatively barren grasslands and semideserts of Africa. The prehistoric species of *Homo* can be viewed both as the progenitors of modern human beings and as one more product among many within the great primate radiation of the Old World. In the latter role they belong to the minority of species that hit upon an intermediate topography, the tropical savanna. Most students of early human evolution agree that the bipedal locomotion and free-swinging arms fitted these ancestral forms very well to the open land, where they were able to exploit an abundance of fruits, tubers, and game.

THE BODY—YES. But is the *mind* predisposed to life on the savanna, such that beauty in some fashion can be said to lie in the genes of the beholder? Three scientists, Gordon Orians, Yi-Fu Tuan, and the late René Dubos, have independently suggested that this is indeed the case. They point out that people work hard to create a savanna-like environment in such improbable sites as formal gardens, cemeteries, and suburban shopping malls, hungering for open spaces but not a barren landscape, some amount of order in the surrounding vegetation but less than geometric perfection. Orians in particular has elaborated the idea according to modern evolutionary theory and added a small but suggestive body of supporting evidence. According to his formulation, the ancestral environment contained three key features.

First, the savanna by itself, with nothing more added, offered an abundance of animal and plant food to which the omnivorous hominids were well adapted, as well as the clear view needed to detect animals and rival bands at long distances. Second, some topographic relief was desirable. Cliffs, hillocks, and ridges were the vantage points from which to make a still more distant surveillance, while their overhangs and caves served as natural shelters at night. During longer marches, the scattered clumps of trees provided auxiliary retreats sheltering bodies of drinking water. Finally, lakes and rivers offered fish, mollusks, and new kinds of edible plants. Because few natural enemies of man can cross deep water, the shorelines became natural perimeters of defense.

Put these three elements together: it seems that whenever people are given a free choice, they move to open tree-studded land on prominences overlooking water. This worldwide tendency is no longer dictated by the hard necessities of hunter-gatherer life. It has become largely aesthetic, a spur to art and landscaping. Those who exercise the greatest degree of free choice, the rich and powerful, congregate on high land above lakes and rivers and along ocean bluffs. On such sites they build palaces, villas, temples, and corporate retreats. Psychologists have noticed that people entering unfamiliar places tend to move toward towers and other large objects breaking the skyline. Given leisure time, they stroll along shores and river banks. They look along the water and up, to the hills beyond or to high buildings, expecting to see the sacred and beautiful places, the sites of historic events, now the seats of government, museums, or the homes of important personages. And they often do, in such landmarks as the Zähringen-Kyburg fortress of Thun, the Belvedere palace of Vienna, the cathedral of Saint Etienne, the chateau of Angers, and the Potala, and among the more imposing sites from past eras, Thingvellir, location of the ancient parliament of Iceland, the Parthenon, and the great plaza at Tenochtitlán.

The most revealing manifestation of the triple criterion occurs in the principles of landscape design. When people are confined to crowded cities or featureless land, they go to considerable lengths to recreate an intermediate terrain, something that can tentatively be called the savanna gestalt. At Pompeii the Romans built gardens next to almost every inn, restaurant, and private residence, most possessing the same basic elements: artfully spaced trees and shrubs, beds of herbs and flowers, pools and fountains, and domestic statuary. When the courtyards were too small to hold much of a garden, their owners painted attractive pictures of plants and animals on the enclosure walls—in open geometric assemblages. Japanese gardens, dating from the Heian period of the ninth to twelfth centuries (and hence ultimately Chinese in origin), similarly emphasize the orderly arrangement of trees and shrubs, open space, and streams and ponds. The trees have been continuously bred and pruned to resemble those of the tropical savanna in height and crown shape. The dimensions are so close as to make it seem that some unconscious force has been at work to turn Asiatic pines and other northern species into African acacias.

I will grant at once the strangeness of the comparison and the possibility that the convergence is merely a large coincidence. It is also true that individuals often yearn to retain the dominant and sometimes peculiar qualities of the environment in which they were raised. But entertain for a while longer the idea that the landscape architects and gardeners, and we who enjoy their creations without special instruction or persuasion, are responding to a deep genetic memory of mankind's optimal environment. That given a completely free choice, people gravitate statistically toward a savanna-like environment. The theory accommodates a great many seemingly disconnected facts from other parts of the world.

Far away, on the western frontier of the United States, explorers were given a brief opportunity to select the landscape to which their hearts led them. In their journals and memoirs they made clear the habitat they most valued. Not the dark forest, waiting to be cut back and replaced with a pastoral landscape of crops and hedges. Not the empty desert flats, good only if irrigated and planted in grass and trees. But the intermediate habitat already in place, a terrain that we ourselves can instantly appreciate: a savanna, rolling gold and green, dissected by a sharp tracery of streams and lake, with clean dry air and clouds dappling a blue sky. When Captain R. B. Marcy, on a United States government expedition through the southern plains in 1849, encountered the land around the headwaters of the Clear Fork of the Brazos River, he declared it to be "as beautiful a country for eight miles as I ever beheld."

> It was a perfectly level grassy glade, and covered with a growth of large mesquite trees at uniform distances, standing with great regularity, and presenting more the appearance of an immense peach orchard than a wilderness. The grass is of the short buffalo variety and as uniform and even as new mown meadow, and the soil is as rich, and very similar to that of the Red River Bottoms.

W. P. Parker, Marcy's companion, agreed: "The view was the most extensive and glowing in the sunset, the most striking that we had enjoyed during the whole trip, combining the grandeur of immense space—the plain extending to the horizon on every side from our point of view—with the beauty of the contrast between the golden carpet of buffalo grass and the pale green of the mesquite trees dotting its surface."

A note on botany: the trees are mesquite, a mimosaceous tree-shrub. The Brazos country is a passable replica of the tropical savanna with a dominant life form closely related to the African acacias, which are also members of the Mimosaceae. I have felt a similar attraction while traveling through the sawgrass and buttonbush flats of the Florida Everglades, the eucalyptus woodland of Queensland, and most compellingly the immense virgin savannas of South America.

Not long ago I joined a group of Brazilian scientists on a tour of the upland savanna, the *cerrado*, around the capital city of Brasilia. We went straight to one of the highest elevations as if following an unspoken command. We looked out across the rippled terrain of high grass, parkland, and forest enclaves and watched birds circling in the sky. We scanned the cumulus clouds that tower like high mountains above the plains during the wet season and found a gray curtain of rain descending into a valley behind some distant hills. We traced gallery forests, groves of trees that wind along the banks of the widely spaced streambeds. We studied Brasilia itself, now almost at the horizon, to admire the shining buildings and monuments that rise like well-spaced terraced cliffs and giant trees, and discussed the green belt and artificial lake that were designed and executed to make existence more livable—more human. It was, all agreed, very beautiful. Of such feelings Melville wrote, "Were Niagara but a cataract of sand, would you travel your thousand miles to see it?"

The practical-minded will argue that certain environments are just "nice" and there's an end to it. So why dilate on the obvious? The answer is that the obvious is usually profoundly significant. Some environments are indeed pleasant, for the same general reason that sugar is sweet, incest and cannibalism repulsive, and team sports exhilarating. Each response has its peculiar meaning rooted in the distant genetic past. To understand why we have one particular set of ingrained preferences, and not another, out of the vast number possible remains a central question in the study of man.

It might still be argued that people are just tracking ideal features of an environment sought out by other creatures as well. If that were true, the whole issue would be trivialized. If the most general properties of human nature are shared with lower organisms in a manner similar to eating and elimination, they could be studied more efficiently in simple animals such as squirrels and bobolinks. But such is not the case. Although the rules of sexual choice, diet selection, and social behavior are to some extent shared with a few other species, the overall pattern is particular to *Homo sapiens*. Not only symbolization and language, but also most of the basic cognitive specializations are unique. Among them appears to be biophilia, which is richly structured and quite irrational, in conformity with a primate genetic history played out in the warm climates of the Old World. Arcturian zoologists visiting this planet could make no sense of our morality and art until they reconstructed our genetic history—nor can we.

THERE IS ANOTHER way to measure the strength of human biophilia. Visualize a beautiful and peaceful world, where the horizon is rimmed by snowy peaks reaching into a perfect sky. In the central valley, waterfalls tumble down the faces of steep cliffs into a crystalline lake. On the crest of the terminal bluff sits a house containing food and every technological convenience. Artisans have worked across the terrain below to create a replica of one of Earth's landscape treasures, perhaps a formal garden from late eighteenth-century England, or the Garden of the Golden Pavilion at Kyoto, marked by an exquisite balance of water, copse, and trail. The setting is the most visually pleasing that human imagination can devise. Except for one thing—it contains no life whatever. This world has always been dead. The vegetation of the garden is artificial, shaped from plastic and colored by master craftsmen down to the last blade and stem. Not a single microbe floats in the lake or lies dormant in the ground. The only sounds are the broken rhythms of the falling water and an occasional whisper of wind through the plastic trees.

Where are we? If the ultimate act of cruelty is to promise everything and withhold just the essentials, the locality is a department of hell. It is a tomb built on a lunar landscape with air and elaborate contrivances added. This is a world (and more than a theoretical possibility in the age of space travel) where people would find their sanity at risk. Without beauty and mystery beyond itself, the mind by definition is deprived of its bearings and will drift to simpler and cruder configurations. Artifacts are incomparably poorer than the life they are designed to mimic. They are only a mirror to our thoughts. To dwell on them exclusively is to fold inwardly over and over, losing detail at each translation, shrinking with each cycle, finally merging into the lifeless facade of which they are composed.

When exceptions occur, they are incomplete and temporary. A few people can escape for a time into a world consisting exclusively of themselves and their machines and exist there without noticeable loss, providing they have strong character and a clearly defined goal. When Cyril Smith began his career in metallurgy as an employee of the American Brass Company, he treated the fire and clangor of the foundries as an aesthetic experience:

> I still have vivid sensual memories of that time: The smell of burning lard oil. Streams of molten brass in the casting shop. Some of the last coke-fired pit furnaces in operation, and men drawing crucibles, skimming and pouring the metal. The magnificent row of rolling mills, all driven continuously by a Corliss engine with a huge flywheel and a shaft running the full length of the large shop. The dance and clangor of drop and screw presses . . . To this day a frequent dream is of wandering through complex assemblies of industrial buildings full of such machines, in search of something I never find.

But Smith was no satanic apprentice adapted to an artifactual world. He was attracted to the most swiftly changing and visually dramatic events, in other words to quasi-life, one might say ultimately back toward life itself. Even in his anxiety dreams he searched for new and undefined experiences of similar kind. When he expanded these themes in his autobiographical *A Search for Structure*, he compared the most attractive patterns of the physical world and technology to artistic representations of plants and animals. People react more quickly and fully to organisms than to machines. They will walk into nature, to explore, hunt, and garden, if given the chance. They prefer entities that are complicated, growing, and sufficiently unpredictable to be interesting. They are inclined to treat their most formidable contraptions as living things or at least to adorn them with eagles, floral friezes, and other emblems representative of the peculiar human perception of true life. The ultimate machine of the futurist's imagination is a self-replicating robot that is benignly independent of its creators, hence in key respects quasi-alive. Mechanophilia, the love of machines, is but a special case of biophilia.

These qualities should impart a certain reserve about man's destiny out there among the stars. Let me qualify that remark at once. As a scientist and hence professional optimist, I am inspired perhaps more than most by the exploration of space. Our knowledge and self-understanding have been greatly expanded by orbiting scanners, probes, soft landings; and the technical spinoff seems to have no limit. If we can stripmine the moon, sweep rare elements from a comet's tail, and change the atmosphere of Venus to resemble Earth's ("terraforming it" is the favored expression), we should not hesitate—so long as the practical and scientific benefits are commensurate to the costs.

But the actual colonization of space by human beings is another matter altogether. No one doubts that the venture has compelling virtues. It would vault the historic expansion of the species around the world out to the unlimited frontiers beyond the planet, feeding the best in the human spirit. It would blast surplus populations from the source of their (more important, *our*) problems. The pioneer of this dream, Gerard O'Neill, and other experts including NASA engineers have explored the technical aspects of the project and are sure it can be done. The gigantic cylinders and toroids they envision are admirable in scope and ingenuity. The interiors will be lined with agricultural fields, parks, and lakes, already depicted persuasively in preliminary layout paintings. These visualizations clearly reflect the designers' unconscious concession to the pull of the primitive human environment. And therein lies the problem as I see it.

For tugging at the bottom of the minds of the planners is an awareness that the mental health of the colonists is as important as their physical well-being. The whole enterprise is afflicted by an unsolved problem of unknown magnitude: can the psychic thread of life on Earth be snapped without eventually fatal consequences? A stable ecosystem can probably be created from an eternal cycling of microorganisms and plants. But it would still be an island of minute dimensions desperately isolated from the home planet, and simpler and less diverse by orders of magnitude than the environment in which human beings evolved. The tedium in such a reduced world would be oppressive for highly trained people aware of the grandeur of the original biosphere.

Even more painful would be the responsibility for keeping the station alive. There is a fundamental difference between the projected mental life of space colonies and ordinary mental life on Earth. It is far more frightening to know that only expert human intervention prevents the whole world from collapsing than merely to know that human beings can destroy it if they try. The comparison is similar to maintaining a patient in intensive care as opposed to watching him walk down the street in good health. People cannot be expected to carry such a burden; they were not built to be godlike in this particular sense. So when we dream of human populations expanding through the solar system and beyond, I believe we dream too far.

The chief significance of the life-in-space debate is symbolic rather than practical. Space colonies are very far down on the list of public priorities and not likely to be undertaken for generations—being part of the agenda, as it were, of the twenty-first century. They are useful right now for what they reveal about the poverty of our self-knowledge. The audaciously destructive tendencies of our species run deep and are poorly understood. They are so difficult to probe and manage as to suggest an archaic biological origin. We run a risk if we continue to diagnose them as by-products of history and suppose that they can be erased with simple economic and political remedies. At the very least, the Sophoclean flaws of human nature cannot be avoided by an escape to the stars. If people perform so badly on Earth, how can they be expected to survive in the biologically reduced and more demanding conditions of space?

Surely we would be better advised to invest the money on the workings of the mind. We should pay more attention to the quality of our dependence on other organisms. The brain is prone to weave the mind from the evidences of life, not merely the minimal contact required to exist, but a luxuriance and excess spilling into virtually everything we do. People can grow up with the outward appearance of normality in an environment largely stripped of plants and animals, in the same way that passable looking monkeys can be raised in laboratory cages and cattle fattened in feeding bins. Asked if they were happy, these people would probably say yes. Yet something vitally important would be missing, not merely the knowledge and pleasure that can be imagined and might have been, but a wide array of experiences that the human brain is peculiarly equipped to receive. Of that much I feel certain, and I will offer it in the form of a practical recommendation: on Earth no less than in space, lawn grass, potted plants, caged parakeets, puppies, and rubber snakes are not enough.

The Conservation Ethic

WHEN VERY LITTLE is known about an important subject, the questions people raise are almost invariably ethical. Then as knowledge grows, they become more concerned with information and amoral, in other words more narrowly intellectual. Finally, as understanding becomes sufficiently complete, the questions turn ethical again. Environmentalism is now passing from the first to the second phase, and there is reason to hope that it will proceed directly on to the third.

The future of the conservation movement depends on such an advance in moral reasoning. Its maturation is linked to that of biology and a new hybrid field, bioethics, that deals with the many technological advances recently made possible by biology. Philosophers and scientists are applying a more formal analysis to such complex problems as the allocations of scarce organ transplants, heroic but extremely expensive efforts to prolong life, and the possible use of genetic engineering to alter human heredity. They have only begun to consider the relationships between human beings and organisms with the same rigor. It is clear that the key to precision lies in the understanding of motivation, the ultimate reasons why people care about one thing but not another—why, say, they prefer a city with a park to a city alone. The goal is to join emotion with the rational analysis of emotion in order to create a deeper and more enduring conservation ethic.

Aldo Leopold, the pioneer ecologist and author of *A Sand County Almanac*, defined an ethic as a set of rules invented to meet circumstances so new or intricate, or else encompassing responses so far in the future, that the average person cannot foresee the final outcome. What is good for you and me at this moment might easily sour within ten years, and what seems ideal for the next few decades could ruin future generations. That is why any ethic worthy of the name has to encompass the distant future. The relationships of ecology and the human mind are too intricate to be understood entirely by unaided intuition, by common sense—that overrated capacity composed of the set of prejudices we acquire by the age of eighteen.

Values are time-dependent, making them all the more difficult to carve in stone. We want health, security, freedom, and pleasure for ourselves and our families. For distant generations we wish the same but not at any great personal cost. The difficulty created for the conservation ethic is that natural selection has programed people to think mostly in physiological time. Their minds travel back and forth across hours, days, or at most a hundred years. The forests may all be cut, radiation slowly rise, and the winters grow steadily colder, but if the effects are unlikely to become decisive for a few generations, very few people will be stirred to revolt. Ecological and evolutionary time, spanning centuries and millennia, can be conceived in an intellectual mode but has no immediate emotional impact. Only through an unusual amount of education and reflective thought do people come to respond emotionally to far-off events and hence place a high premium on posterity.

The deepening of the conservation ethic requires a greater measure of evolutionary realism, including a valuation of ourselves as opposed to other people. What do we really owe our remote descendants? At the risk of offending some readers I will suggest: Nothing. Obligations simply lose their meaning across centuries. But what do we owe ourselves in planning for them? Everything. If human existence has any verifiable meaning, it is that our passions and toil are enabling mechanisms to continue that existence unbroken, unsullied, and progressively secure. It is for ourselves, and not for them or any abstract morality, that we think into the distant future. The precise manner in which we take this measure, how we put it into words, is crucially important. For if the whole process of our life is directed toward preserving our species and personal genes, preparing for future generations is an expression of the highest morality of which human beings are capable. It follows that the destruction of the natural world in which the brain was assembled over millions of years is a risky step. And the worst gamble of all is to let species slip into extinction wholesale, for even if the natural environment is conceded more ground later, it can never be reconstituted in its original diversity. The first rule of the tinkerer, Aldo Leopold reminds us, is to keep all the pieces.

This proposition can be expressed another way. What event likely to happen during the next few years will our descendants

most regret? Everyone agrees, defense ministers and environmentalists alike, that the worst thing possible is global nuclear
war. If it occurs the entire human species is endangered; life
as normal human beings wish to live it would come to an end.
With that terrible truism acknowledged, it must be added that
if no country pulls the trigger the worst thing that will *probably*
happen—in fact is already well underway—is not energy depletion, economic collapse, conventional war, or even the expansion of totalitarian governments. As tragic as these catastrophes
would be for us, they can be repaired within a few generations.
The one process now going on that will take millions of years
to correct is the loss of genetic and species diversity by the
destruction of natural habitats. This is the folly our descendants are least likely to forgive us.

Extinction is accelerating and could reach ruinous proportions during the next twenty years. Not only are birds and
mammals vanishing but such smaller forms as mosses, insects,
and minnows. A conservative estimate of the current extinction
rate is one thousand species a year, mostly from the destruction
of forests and other key habitats in the tropics. By the 1990s
the figure is expected to rise past ten thousand species a year
(one species per hour). During the next thirty years fully one
million species could be erased.

Whatever the exact figure—and the primitive state of evolutionary biology permits us only to set broad limits—the current rate is still the greatest in recent geological history. It is
also much higher than the rate of production of new species by
ongoing evolution, so that the net result is a steep decline in
the world's standing diversity. Whole categories of organisms
that emerged over the past ten million years, among them the
familiar condors, rhinoceros, manatees, and gorillas, are close
to the end. For most of their species, the last individuals to
exist in the wild state could well be those living there today.
It is a grave error to dismiss the hemorrhaging as a "Darwinian" process, in which species autonomously come and go and
man is just the latest burden on the environment. Human
destructiveness is something new under the sun. Perhaps it
is matched by the giant meteorites thought to smash into the
Earth and darken the atmosphere every hundred million years
or so (the last one apparently arrived 65 million years ago and

contributed to the extinction of the dinosaurs). But even that interval is ten thousand times longer than the entire history of civilization. In our own brief lifetime humanity will suffer an incomparable loss in aesthetic value, practical benefits from biological research, and worldwide biological stability. Deep mines of biological diversity will have been dug out and care-lessly discarded in the course of environmental exploitation, without our even knowing fully what they contained.

The time is late for simple answers and divine guidance, and ideological confrontation has just about run its course. Little can be gained by throwing sand in the gears of industrialized society, even less by perpetuating the belief that we can solve any problem created by earlier spasms of human ingenuity. The need now is for a great deal more knowledge of the true biological dimensions of our problem, civility in the face of common need, and the style of leadership once characterized by Walter Bagehot as agitated moderation.

Ethical philosophy is a much more important subject than ordinarily conceded in societies dominated by religious and ideological orthodoxy. It faces an especially severe test in the complexities of the conservation problem. When the time scale is expanded to encompass ecological events, it becomes far more difficult to be certain about the wisdom of any particular decision. Everything is riddled with ambiguity; the middle way turns hard and general formulas fail with dispiriting consis-tency. Consider that a man who is a villain to his contempo-raries can become a hero to his descendants. If a tyrant were to carefully preserve his nation's land and natural resources for his personal needs while keeping his people in poverty, he might unintentionally bequeath a rich, healthful environment to a reduced population for enjoyment in later, democratic generations. This caudillo will have improved the long-term welfare of his people by giving them greater resources and more freedom of action. The exact reverse can occur as well: today's hero can be tomorrow's destroyer. A popular political leader who unleashes the energies of his people and raises their standard of living might simultaneously promote a population explosion, overuse of resources, flight to the cities, and poverty for later generations. Of course these two extreme examples are caricatures and unlikely to occur just so, but they suffice to illustrate that, in ecological and evolutionary time, good does not automatically flow from good or evil from evil. To choose what is best for the near future is easy. To choose what is best for the distant future is also easy. But to choose what is best for both the near and distant futures is a hard task, often internally contradictory, and requiring ethical codes yet to be formulated.

AN ENDURING CODE of ethics is not created whole from absolute premises but inductively, in the manner of common law, with the aid of case histories, by feeling and consensus, through an expansion of knowledge and experience, influenced by the epigenetic rules of mental development, during which well-meaning and responsible people sift the opportunities and come to agree upon norms and directions.

The conservation ethic is evolving according to this pattern. It started centuries ago as a scattering of incidental thoughts and actions. The first biological preserves around the world were the by-products of selfish interests created, like most early art and learning, for the pleasure of the ruling classes. Among them were the gardens of the Kandy kings in Sri Lanka, the royal hunting reserves of Europe, and a few islands, such as Niihau in the Hawaiian group and Lignumvitae Key in Florida Bay, cordoned off for the use of private families.

I have visited all of these places, except Niihau, and many others as well, drawn by the opportunity offered for original biological research. In Cuba, on June 25, 1953, a month before Fidel Castro's assault on the Moncada barracks in Santiago de Cuba, I arrived on a far more modest mission in a jeep at a place called Blanco's Woods, near Cienfuegos. The tract was owned by a wealthy family who lived in Spain and declined to develop the land. All the surrounding forest had been cut down and converted into pasture and agricultural fields, leaving Blanco's Woods a rare refuge of native plants and animals of the coastal lowlands. To walk into that otherwise unprepossessing wood-lot was to travel back into Cuba's geologic past, into the Pleistocene age before the coming of man—all thanks to what some would rightfully call the selfish actions of one family. Over 50 million years the Greater Antillean Islands, Cuba among them, had broken apart and drifted away from Central America eastward across the Caribbean Sea. In countless episodes the forests of Cuba were seeded with plants and animals from the mainland and surrounding islands. Many of the populations became extinct; others hung on to evolve during thousands of generations into distinct genera and species, found nowhere else, woven together into intricate systems of competitors, predators, and prey. Biologists have given many of the organisms formal scientific names reflecting their origin and exclusive stronghold, such as *cubaensis, antillana, caribbaea,* and *insularis.* Now it has come down to this: in a negligible interval of evolutionary time, within the lifespan of Fidel Castro and one unheroic entomologist of approximately the same age visiting a nonstrategic part of the island, much of the woodland and hence a large part of Cuba's history have vanished. In 1953, on trial in Batista's court, Castro declared that history would absolve him. I wonder if it will, whether Blanco's Woods have since been cleared for the "good of the people"— meaning one or two generations—and to what degree the Cuban people will someday treasure such places as part of their national heritage, when heroes and political revolutions are dim in their memory.

Advances in conservation elsewhere in the world have been equally subordinate to whim and short-term social needs. The ginkgo tree, a relict of the ancient Asiatic forests and sole surviving species of an entire order of gymnospermous plants, was saved only because it was planted as an ornamental in Chinese and Japanese temple gardens over a period of centuries, long after it became extinct in the wild. Père David's deer held on for generations as an inhabitant of the imperial compound at Peking, after being hunted out over the rest of its once extensive range in China. In 1898, just before this final herd was destroyed, a new population was established by the Duke of Bedford on the grounds of Woburn Abbey. The stock has since been used to populate other reserves and parks. The great value of such by-the-fingernails species preservation is that it keeps alive the possibility of reconstituting original faunas and floras. Individuals can be transferred back to the original habitats and allowed to breed up to stable levels. Père David's deer itself may someday roam fresh in the relict woodlands of China.

Some kinds of organisms survive as the accidental beneficiaries of religion and magic. In Israel rare plants, largely exterminated in the surrounding agricultural land, grow in and around the Tel Dan, tombs of the holy men located near the sources of the Jordan River. When the biologist Michael J. D. White set out to analyze the genetic constitution of a group of interesting Australian grasshoppers called the Morabinae, he found them in sufficient numbers only in cemeteries and along railroad tracks. In the Western Ghats of India, sacred groves dating back to hunter-gatherer times today contain the best-preserved remnants of the original flora and fauna. Madhav Gadgil, one of India's foremost biologists and recipient of a gold medal in science from Prime Minister Indira Gandhi, has recommended that the groves serve as the nuclei of a system of national biotic reserves.

The modern practice of conservation has moved steadily forward from such primitive beginnings, but its philosophical foundations remain shaky. It still depends almost entirely on what may be termed surface ethics. That is, our relationship to the rest of life is judged on the basis of criteria that apply to other, more easily defined categories of moral behavior. This mode of reasoning is approximately the same as promoting literature because good writing helps to sell books, or art because it is useful for portraiture and scientific illustration. Of course the criteria are not in error—just spectacularly incomplete.

Thus we favor certain animals because they fill the superficial role of surrogate kin. It is the most disarming reason for nurturing other forms of life, and only a churl could find fault. Dogs are especially popular because they live by humanlike rituals of greeting and subservience. The family to whom they belong is part of their pack. They treat us like giant dogs, automatically alpha in rank, and clamor to be near us. We in turn respond warmly to their joyous greetings, tail wagging, slavering grins, drooped ears, groveling, bristling fur, and noisy indignation at territorial trespass. (Just as I write this line I have to pause to calm down my own cocker spaniel, who is barking at a passing jogger. I say without thinking, "Quiet! Good *boy*!") The key to the compatibility of the two species is that dogs are descended with little behavioral modification from wolves. Like human beings, they and their wild cousins are happy carnivores, specialized to hunt large, swift, or otherwise unusually difficult prey in tightly coordinated groups. The wolf pack can catch mice and other small animals easily enough, but its real distinction is that it is also a superb instrument for bringing down a moose. The adaptation entails an extreme sensitivity to the moods of others. Dogs (domesticated wolves) are always ready for the communal hunt. They are primed to charge out the door with members of the human family in attendance, perhaps to chase down and slaughter a squirrel or rabbit, which, after an appropriate amount of fussing about and posturing in reconfirmation of status, will be shared with others. When not on the run or its equivalent (being carried along ecstatically in the family automobile), they follow the wolf's primal custom of spraying urine onto tree trunks and bushes (fireplugs and telephone poles will suffice) in order to mark out territory. At home, they metamorphose into children. The King Charles spaniel was bred to be an extreme specialist in this role. The adult possesses the small size, round head, and pug face of a puppy—also, let us acknowledge it frankly, of a baby—and is meant to be held in the lap.

Kinship affects emotion in other, unexpected ways. One of the most oddly disquieting events of my life was an encounter with Kanzi, a young pygmy chimpanzee. I was a guest of Sue Savage-Rumbaugh at the Language Research Center outside Atlanta, waiting in her office, when Kanzi was led in by a young woman who is helping to raise him. It was the first time I had seen this rare primate in life. I had a more than ordinary interest in it as an evolutionary biologist. The pygmy chimpanzee is arguably a distinct species from the ordinary chimpanzee. It appears to be somewhat less modified for arboreal existence than its sister species, and of the two it is the closer to man in certain key features of anatomy and behavior. The arms are longer and the legs shorter relative to the body. The head is more rounded, the forehead higher, and the jaw and brow less protruding. Overall the pygmy chimpanzee is remarkably similar in skeletal structure to "Lucy," the type specimen of *Australopithecus afarensis*, one of the probable direct precursors of man. It is the most humanlike of all animals. Its existence lends weight to the belief of many biologists that the evolutionary lines leading to human beings and chimpanzees split from a common stock in Africa as recently as five million years ago. There are also a few equally impressive similarities in behavior. The pygmy chimpanzee walks erect much of the time, and it learns many tasks more quickly and vocalizes more freely than the common chimpanzee. In sexual behavior it is closer than any other nonhuman primate to human beings. Females remain sexually receptive through most of their cycle, and they take a face-to-face position with the male in about a third of the couplings.

The pygmy chimpanzee is also endangered as a species. Wild populations are found only in one remote area in the Lomoko forest of Zaire, where a German lumber company has begun to conduct logging operations (in 1983, the time of writing). Only several dozen of the animals exist in captivity. Realizing the unique importance and threatened status of the species, scientists such as Savage-Rumbaugh, Adrienne Zihlman, and Jeremy Dahl are engaged in intensive studies of its biology and social behavior. Among the perhaps thirty million species of organisms on Earth, this is one that in my opinion deserves the highest priority in research and preservation.

Kanzi walked into the office and spotted me sitting in a chair on the far side of the room. He went into a frenzy of excitement, yelping and gesticulating to the two women with him in a way that seemed to exclaim, "That's a stranger! Why is he here? What are we going to do about him?" After a few minutes he calmed down and walked cautiously over to me, flicking glances from side to side as though plotting an emergency escape route. When he came near I brought my left hand up slowly and held it out, palm down and fingers slightly crumpled. I thought this was the very essence of humility and friendly intention, but he slapped my hand hard and backed off with a loud cry. The trainer murmured, "Oh, you're such a brave little boy!" (He *was* a brave little boy.) I didn't mind that my hand stung a bit. At that moment Kanzi's comfort and well-being seemed much more important than my own.

The trainer gave him a cup of grape juice, and he climbed into her lap to drink it and be cuddled. After a short wait he slid down to the floor and drifted back over to me. This time, having been coached by Sue Savage-Rumbaugh, I imitated the flutelike conciliatory call of the species, *wu-wu-wu-wu-wu* . . . with my lips pursed and what this time I believed to be a sincere, alert expression on my face. Now Kanzi reached out and touched my hand, nervously but gently, and stepped back a short distance to study me once again. The trainer gave me a cup of grape juice of my own. I flourished the cup as if offering a toast and took a sip, whereupon Kanzi climbed into my lap, took the cup, and drank most of the juice. Then we cuddled. Afterward everyone in the room had a good time playing ball and a game of chase with Kanzi.

The episode was unnerving. It wasn't the same as making friends with the neighbor's dog. I had to ask myself: was this really an animal? As Kanzi was led away (no farewells), I realized that I had responded to him almost exactly as I would to a two-year-old child—same initial anxieties, same urge to communicate and please, same gestures and food-sharing ritual. Even the conciliatory call was not very far off from the sounds adults make to comfort an infant. I was pleased that I had been accepted, that I had proved adequately human (was that the word?) and *sensitive* enough to get along with Kanzi.

WE ARE LITERALLY KIN to other organisms. The common and pygmy chimpanzees constitute the extreme case, the two species closest to human beings out of the contemporary millions. About 99 percent of our genes are identical to the corresponding set in chimpanzees, so that the remaining 1 percent accounts for all the differences between us. The chromosomes, the rodlike structures that carry the genes, are so close that only high-resolution photography and expert knowledge can tell many of them apart. Bishop Wilberforce's darkest thoughts might well be true; the creationists are justified in spending restless nights. The genetic evidence suggests that we resemble the chimpanzees in anatomy and a few key features of social behavior by virtue of a common ancestry. We descended from something that was more like a modern ape than a modern human being, at least in brain and behavior, and not very long ago by the yardstick of evolutionary time. Furthermore, the greater distances by which we stand apart from the gorilla, the orangutan, and the remaining species of living apes and monkeys (and beyond them other kinds of animals) are only a matter of degree, measured in small steps as a gradually enlarging magnitude of base-pair differences in DNA.

The phylogenetic continuity of life with humanity seems an adequate reason by itself to tolerate the continued existence of

apes and other organisms. This does not diminish humanity—
it raises the status of nonhuman creatures. We should at least
hesitate before treating them as disposable matter. Peter Singer,
a philosopher and animal liberationist, has gone so far as to
propose that the circle of altruism be expanded beyond our
own species to all animals with the capacity to feel and suffer,
just as we have extended the label of brotherhood steadily until
most people now feel comfortable with an all-inclusive phrase,
the family of man. Christopher D. Stone, in *Should Trees
Have Standing?*, has examined the legal implications of this
enlarged generosity. He points out that until recently women,
children, aliens, and members of minority groups had few or
no legal rights in many societies. Although the policy was
once accepted casually and thought congenial to the prevailing ethic, it now seems hopelessly barbaric. Stone asks why we
should not extend similar protection to other species and to the
environment as a whole. People still come first—humanism has
not been abandoned—but the rights of owners should not be
the exclusive yardstick of justice. If procedures and precedents
existed to permit legal action to be taken on behalf of certain
agreed-upon parts of the environment, the argument continues, humanity as a whole would benefit. I'm not sure I agree
with this concept, but at the very least it deserves more serious
debate than it has received. Human beings are a contractual
species. Even religious dogma is hammered out as a system
of mutual agreements. The working principles of ownership
and privilege are arrived at by slow mutual consent, and legal
theorists are a long way from having explored their limits.

If nobility is defined as reasoned generosity beyond expedi-
ence, animal liberation would be the ultimate ennobling act.
Yet to force the argument entirely inside the flat framework of
kinship and legal rights is to trivialize the case favoring conservation, to make it part of the surface ethic by justifying one
criterion on the basis of another. It is also very risky. Human
beings, for all their professed righteousness and brotherhood,
easily discriminate against strangers and are content to kill
them during wars declared for relatively frivolous causes. So
it is much easier to find an excuse to exterminate another species. A stiffer dose of biological realism appears to be in order.
We need to apply the first law of human altruism, ably put by

Garrett Hardin: never ask people to do anything they consider contrary to their own best interests. The only way to make a conservation ethic work is to ground it in ultimately selfish reasoning—but the premises must be of a new and more potent kind.

An essential component of this formula is the principle that people will conserve land and species fiercely if they foresee a material gain for themselves, their kin, and their tribe. By this economic measure alone, the diversity of species is one of Earth's most important resources. It is also the least utilized. We have come to depend completely on less than 1 percent of living species for our existence, with the remainder waiting untested and fallow. In the course of history, according to estimates recently made by Norman Myers, people have utilized about 7,000 kinds of plants for food, with emphasis on wheat, rye, maize, and about a dozen other highly domesticated species. Yet at least 75,000 exist that are edible, and many of these are superior to the crop plants in use. The strongest of all arguments from surface ethics is a logical conclusion about this unrealized potential: the more the living world is explored and utilized, the greater will be the efficiency and reliability of the particular species chosen for economic use. Among the potential star species are these:

• The winged bean (*Psophocarpus tetragonolobus*) of New Guinea has been called a one-species supermarket. It contains more protein than cassava and potato and possesses an overall nutritional value equal to that of soybean. It is among the most rapidly growing of all plants, reaching a height of fifteen feet within a few weeks. The entire plant can be eaten, tubers, seeds, leaves, flowers, stems, and all, both raw and ground into flour. A coffeelike beverage can be made from the liquefied extract. The species has already been used to improve the diet in fifty tropical countries, and a special institute has been set up in Sri Lanka to study and promote it more thoroughly.

• The wax gourd (*Benincasa hispida*) of tropical Asia grows an inch every three hours over the course of four days, permitting multiple crops to be raised each year. The fruit attains a size of up to 1 by 6 feet and a weight of 80 pounds. Its crisp white flesh can be eaten at any stage, as a cooked vegetable, a base for soup, or a dessert when mixed with syrup.

• The Babussa palm (*Orbigyna martiana*) is a wild tree of the Amazon rain forest known locally as the "vegetable cow." The individual fruits, which resemble small coconuts, occur in bunches of up to 600 with a collective weight of 200 pounds. Some 70 percent of the kernel mass is composed of a colorless oil, used for margarine, shortening, fatty acids, toilet soap, and detergents. A stand of 500 trees on one hectare (2.5 acres) can produce 125 barrels of oil per year. After the oil has been extracted the remaining seedcake, which is about one-fourth protein, serves as excellent animal fodder.

Even with limited programs of research, biologists have compiled an impressive list of such candidate organisms in the technical literature. The vast majority of wild plants and animals are not known well enough (certainly many have not yet been discovered) even to guess at those with the greatest economic potential. Nor is it possible to imagine all the uses to which each species can be put. Consider the case of the natural food sweeteners. Several species of plants have been identified whose chemical products can replace conventional sugar with negligible calories and no known side effects. The katemfe (*Thaumatococcus danielli*) of the West African forests contains two proteins that are 1,600 times sweeter than sucrose and are now widely marketed in Great Britain and Japan. It is outstripped by the well-named serendipity berry (*Dioscoreophyllum cumminsii*), another West African native whose fruit produces a substance 3,000 times sweeter than sucrose.

Natural products have been called the sleeping giants of the pharmaceutical industry. One in every ten plant species contains compounds with some anticancer activity. Among the leading successes from the screening conducted so far is the rosy periwinkle, a native of the West Indies. It is the very paradigm of a previously minor species, with pretty five-petaled blossoms but otherwise rather ordinary in appearance, a roadside casual, the kind of inconspicuous flowering plant that might otherwise have been unknowingly consigned to extinction by the growth of sugarcane plantations and parking lots. But it also happens to produce two alkaloids, vincristine and vinblastine, that achieve 80 percent remission from Hodgkin's disease, a cancer of the lymphatic system, as well as 99 percent remission from acute lymphocytic leukemia. Annual sales of the two drugs reached $100 million in 1980.

A second wild species responsible for a medical breakthrough is the Indian serpentine root (*Rauwolfia serpentina*). It produces reserpine, a principal source of tranquilizers used to relieve schizophrenia as well as hypertension, the generalized condition predisposing patients to stroke, heart malfunction, and kidney failure.

The natural products of plants and animals are a select group in a literal sense. They represent the defense mechanisms and growth regulators produced by evolution during uncounted generations, in which only organisms with the most potent chemicals survived to the present time. Placebos and cheap substitutes were eliminated at an early stage. Nature has done much of our work for us, making it far more efficient for the medical researcher to experiment with extracts of living tissue than to pull chemicals at random off the laboratory shelf. Very few pharmaceuticals have been invented from a knowledge of the first principles of chemistry and medicine. Most have their origin in the study of wild species and were discovered by the rapid screening of large numbers of natural products.

For the same reason, technical advances utilizing natural products have been achieved in many categories of industry and agriculture. Among the most important have been the development of phytoleum, new plant fuels to replace petroleum; waxes and oils produced from indefinitely renewing sources at more economical rates than previously thought possible; novel kinds of fibers for paper manufacture; fast-growing siliceous plants, such as bamboo and elephant grass, for economical dwellings; superior methods of nitrogen fixation and soil reclamation; and magic-bullet techniques of pest control, by which microorganisms and parasites are set loose to find and attack target species without danger to the remainder of the ecosystem. Even the most conservative extrapolation indicates that many more such discoveries will result from only a modest continuing research effort.

Furthermore, the direct harvesting of free-living species is only a beginning. The favored organisms can be bred over ten to a hundred generations to increase the quality and yield of their desired product. It is possible to create strains that do well in new climates and the special environments required for mass production. The genetic material comprising them is an additional future resource; it can be taken apart gene by gene and distributed to other species. Thomas Eisner, one of the pioneers of chemical ecology, has used a striking analogy to explain these two levels of utilization of wild organisms. Each of the millions of species can be visualized as a book in a library. No matter where it originates, it can be transferred and put to use elsewhere. No matter how rare in its original state, it can be copied many times over and disseminated to become indefinitely abundant. An orchid down to the last hundred individuals in a remote valley of the Peruvian Andes, which also happens to be the source of a medicinal alkaloid, can be saved, cultured, and converted into an important crop in gardens and greenhouses around the world. But there is much more to the species than the alkaloid or other useful material that it happens to package. It is not really a conventional book but more like a looseleaf notebook, in which the genes are the equivalent of detachable pages. With new techniques of genetic engineering, biologists will soon be able to lift out desirable genes from one species or strain and transfer them to another. A valuable food plant, for example, can be given DNA from wild species conferring biochemical resistance to its most destructive disease. It can be altered by parallel procedures to grow in desert soil or through longer seasons.

A notable case in point is the primitive form of maize, *Zea diploperennis*, recently discovered in a mountain forest of southwestern Mexico. It is still known from three patches covering a mere ten acres (at any time a bulldozer might easily have extinguished the entire species, within hours). *Zea diploperennis* possesses genes for perennial growth, making it unique among all other known varieties of corn. It is thus the potential source of a hereditary trait that could reduce growing time and labor costs, making cultivation more feasible in ecologically marginal areas.

There are few countries in the world that do not harbor unique species and genetic strains still unknown to the people who live there. There is no country that would fail to benefit from the importation of such undiscovered organisms. With these facts in mind I find it astonishing that so little attention is being given to the exploration of the living world. The set of disciplines collectively called evolutionary biology, including initial field surveys, taxonomy, ecology, biogeography, and comparative biochemistry, remains among the most poorly funded in science. The amount spent globally in 1980 on such research in the tropics, where the great majority of organisms live, was $30 million—somewhat less than the cost of two F-15 Eagle fighter-bombers, approximately 1 percent of the grants for health-related research in the United States, or a few weeks' liquor bill for the populace of New York City.

Let us postpone for the moment moral arguments of the conventional kind. It would be to the direct economic advantage of most governments to invest more in the study of their own living resources. Because evolutionary biology exists so close to the poverty line, it offers society what economists call increasing returns to scale: a modest absolute expenditure in dollars will yield large relative benefits. The reason is that the existing low level of activity causes most opportunities to remain unmet, with the result that the marketplace stays largely empty. Museums, meant at their founding to be national research centers, are everywhere understaffed. Taxonomy, the principal occupation of museum scientists, is a declining profession through lack of support. The neglect is all the more puzzling because the value of the research is widely appreciated within the scientific community. Any biologist who tries to get an identification of an organism in order to facilitate its further study knows that he may be in for a long wait. Even when the research has considerable economic potential, it is often at risk because of delays and inadequate data.

The diversity of species is so immense that the Linnaean enterprise of describing the living world remains by force a part of modern science. In addition to more and better staffed museums, we (scientists, individual countries, the world) would benefit from institutes for the extended study of the organisms once they have been classified. There the previously unknown species can be screened for economic and medical potential, their ecology and physiological traits probed. The accumulating data will also reveal the complex processes by which species originate and go extinct, information needed to guide the practice of conservation.

A few such institutions of high quality exist today, among them Brazil's National Institute for Research on Amazonia, the Marine Biological Laboratory at Woods Hole, Massachusetts, and the Smithsonian Tropical Research Institute in Panama. But even if these pioneering organizations were operated at full current capacity, they could handle only a minute fraction of the different kinds of organisms around the world. The most urgent need is for an increased research capacity in the tropics, where perhaps 90 percent of species exist.

I will now add a note of optimism that I know is shared by many biologists. The exploration of natural resources is the kind of research most readily justified in the underdeveloped countries, especially those in the tropics. It is also the kind they can most easily afford. These nations occasionally need accelerators, satellites, mass spectrometers, and the other accouterments of big science, but such equipment can be borrowed during cooperative ventures with the richer countries. The economically less developed countries can do better with skilled and semiskilled workers who make expeditions into the wild, collect and prepare specimens, culture promising varieties, and spend the long hours of close observation needed to understand growth and behavior. This kind of science is labor-intensive, best performed by people who love the land and organisms for their own sake. Its results will gain worldwide recognition and serve as a source of national pride.

Can there be an Ecuadoran biology, a Kenyan biology? Yes, if they focus on the uniqueness of indigenous life. Will such efforts be important to international science? Yes, because evolutionary biology is a discipline of special cases woven into global patterns. Nothing makes sense except in the light of the histories of local faunas and floras. It is further true that all of biology, from biochemistry to ecology, is moving toward a greater emphasis on evolution and its resultant particularity.

Finally, the efforts of generations to come will be frustrated unless they are safeguarded with national reserve systems of the kind recently pioneered by Brazil, Costa Rica, and Sri Lanka, where the parcels of land set aside are chosen to achieve a maximum protection of organic diversity. Otherwise hundreds of species will continue to vanish each year without so much as the standard double Linnaean names to record their existence. Each takes with it millions of bits of genetic information, a history ages long, and potential benefits to humanity left forever unmeasured.

TO SUMMARIZE: a healthful environment, the warmth of kinship, right-sounding moral strictures, sure-bet economic gain, and a stirring of nostalgia and sentiment are the chief components of the surface ethic. Together they are enough to make a compelling case to most people most of the time for the preservation of organic diversity. But this is not nearly enough: every pause, every species allowed to go extinct, is a slide down the ratchet, an irreversible loss for all. It is time to invent moral reasoning of a new and more powerful kind, to look to the very roots of motivation and understand why, in what circumstances and on which occasions, we cherish and protect life. The elements from which a deep conservation ethic might be constructed include the impulses and biased forms of learning loosely classified as biophilia. Ranging from awe of the serpent to the idealization of the savanna and the hunter's mystique, and undoubtedly including others yet to be explored, they are the poles toward which the developing mind most comfortably moves. And as the mind moves, picking its way through the vast number of choices made during a lifetime, it grows into a form true to its long, unique evolutionary history.

I have argued in this book that we are human in good part because of the particular way we affiliate with other organisms. They are the matrix in which the human mind originated and is permanently rooted, and they offer the challenge and freedom innately sought. To the extent that each person can feel like a naturalist, the old excitement of the untrammeled world will be regained. I offer this as a formula of reenchantment to invigorate poetry and myth: mysterious and little known organisms live within walking distance of where you sit. Splendor awaits in minute proportions.

Why then is there resistance to the conservation ethic? The familiar argument is that people come first. After their problems have been solved, we can enjoy the natural environment as a luxury. If that is indeed the answer, the wrong question was asked. The question of importance concerns purpose. Solving practical problems is the means, not the purpose. Let us assume that human genius has the power to thread the needles of technology and politics. Let us imagine that we can avert nuclear war, feed a stabilized population, and generate a permanent supply of energy—what then? The answer is the same all around the world: individuals will strive toward personal fulfillment and at last realize their potential. But what is fulfillment, and for what purpose did human potential evolve?

The truth is that we never conquered the world, never understood it; we only think we have control. We do not even know why we respond a certain way to other organisms, and need them in diverse ways, so deeply. The prevailing myths concerning our predatory actions toward each other and the environment are obsolete, unreliable, and destructive. The more the mind is fathomed in its own right, as an organ of survival, the greater will be the reverence for life for purely rational reasons.

Natural philosophy has brought into clear relief the following paradox of human existence. The drive toward perpetual expansion—or personal freedom—is basic to the human spirit. But to sustain it we need the most delicate, knowing stewardship of the living world that can be devised. Expansion and stewardship may appear at first to be conflicting goals, but they are not. The depth of the conservation ethic will be measured by the extent to which each of the two approaches to nature is used to reshape and reinforce the other. The paradox can be resolved by changing its premises into forms more suited to ultimate survival, by which I mean protection of the human spirit.

Surinam

E TERNAL SURINAM: the image of the land I kept for many years symbolized the tangle of dreams and boyhood adventures from which I had originally departed, the home country of all naturalists, and the quiet refuge from which personal beliefs might someday be redeemed in a permanent and more nearly perfect form. It is appropriate, then, to describe the reality of that particular place before returning a final time to its image.

Surinam is a sovereign country with a fertile coastal plain, interior wilderness, and one of the richest forest reserves in the world. It is often called the ornithologist's paradise for the variety of neotropical bird species seen more easily there than in most of the rest of South America. Parrots flock among the palms within the city limits of Paramaribo. Over a hundred kinds of hummingbirds and cotingas flash through the flowering canopies of the nearby forests. A short drive and boat trip to the south will bring you to guans, tinamous, manakins, bellbirds, ant-thrushes, and toucans, and perhaps provide a glimpse of the harpy eagle, the giant predator of monkeys and sloths and apex of the arboreal energy pyramid. It is a general rule that, when the bird fauna stays intact, so does the rest of the fauna and flora. The interior of Surinam is a fragment of tropical America as it was ten thousand years ago, or at least approximately so, when the first Indian colonists walked in from the Panamanian land bridge.

Location: north coast of South America, bracketed by French Guiana to the east and Guyana to the west, with Brazil sharing the southern border. Population: 350,000, mostly concentrated on the coast, especially in and around Paramaribo. Agriculture is mixed and moderately successful, with emphasis on rice as the principal export crop. One of the largest hydroelectric plants in South America is located at the Brokopondo Dam; it delivers the bulk of the power used by a highly productive bauxite operation, still mostly foreign-owned. The Surinamese people are courteous and friendly, adding considerably to the potential of tourism as an economic resource. They react with special warmth to visitors who struggle with Takki-Takki, the national Creole dialect, although Dutch or English will serve you well almost anywhere in the country.

Climate: sweltering. Education: valued and improving. Roads: few. The Netherlands gave Surinam independence in 1975 and a promised allowance of $100 million annually for fifteen years. By 1982 the per capita income was $2,500, one of the highest among developing countries. One in three persons owned an automobile, while refrigerators and television sets were routinely stocked in private homes. The long-term future seems bright for this little country, which has a bountiful environment, a small population, and hence a period of grace that was not granted to most of the Third World when colonialism came to an end.

Bernhardsdorp has changed strikingly since my visit in 1961. Touched at last by the population sprawl originating in Paramaribo and Lelydorp, it grew from a tiny Arawak village into a town of about five hundred people of Javanese, Chinese, Amerindian, and Creole ancestry, an ethnic microcosm of larger Surinam. Today the scene is classic tropical-rural. The thatched huts are outnumbered by conventional one- or two-room dwellings built on basement pilings from plank sidings and sheetmetal roofs. The lush pastureland and gardens, crisscrossed by drainage ditches, yield an abundance of vegetables, dairy products, and poultry for local consumption and nearby markets. In the center of town, by the main dirt road, is a small store run by a Chinese family. Someone has erected Coca-Cola signs and a billboard with the national coat of arms, featuring two armed Arawak warriors, a circular shield emblazoned with sailing ship, star, and palm tree, and beneath these figures, on the flying scroll, the motto "Justitia, Pietas, Fides." The bulldozer came: the forest has been mostly cleared, leaving behind a scattering of palms and second-growth edge thickets. There is also a tall tree with tear-shaped oropendola nests hanging in military rows beneath its horizontal branches. The town is not yet on any map I have been able to find. A carefully lettered sign at the turnoff from the paved Lelydorp-Zanderij road proudly proclaims its existence: BERNHARDSDORP.

In 1980 all this bright picture was darkened by the advent of barbarism. The democratically elected government of Henk

Arron was overthrown by Revolutionary Leader Dési Bouterse, a military physical-education instructor with scant education. At first suspicious of socialism, but then schooled in Marxism-Leninism by his teacher and mistress, Bouterse drifted leftward and began to court Fidel Castro and the Soviet Union. In December 1982 Bouterse, without warning, ordered the arrest and execution of fifteen of the country's leading citizens, including lawyers, journalists, and union leaders. The next morning all but one were dead. With a substantial fraction of the old leadership erased and hundreds of citizens soon joining the tens of thousands already in exile, Bouterse announced "the building of a new Surinam."

As I write, it is a Surinam of silence and fear. The modest tourist trade has ended, aid from the Netherlands and United States has been cut off, unemployment is rising, and the once substantial foreign reserves are quickly drying up. The state university has been closed, and the key radio stations and union headquarters have been either burned or blown up. Plain-clothes police arrest citizens randomly for questioning. People rarely speak about the government for fear of the informers they now believe to be everywhere. In the words of one exile, Surinam has become a "country of mutes." Its anxiety is the projected mood of a frightened, paranoid ruler. Rumors are circulating of a planned military coup to overthrow Bouterse, perhaps backed clandestinely by the United States. This has of course been denied. Cuba has been rebuffed, its ambassador expelled, through fear of a coup from the left. In contrast, Brazil has managed to expand contact with Bouterse's regime with the expressed hope of meliorating it. All are wrestling with a problem as old as recorded history: how to deal with the kingdoms of Caliban.

There is a way for the mind to find some ease in such matters. Whatever its denouement the Bouterse episode, the national tragedy, can be seen as no more than a tick in what will be the ultimate history of Surinam. Its people will survive to see ecological and then evolutionary change, within which biography and political events become cyclical and shrink steadily in proportion.

The swiftness of human change and the transience of power are well remembered in the words of a man who saw it all on a

grander stage and far enough back in time to gain convincing
authority: the wise Stoic emperor Marcus Aurelius. Take the
distant view, he said, and observe that those who praise and
those praised both endure for a short season and are gone, and
"all this too in a tiny corner of this continent, and not even
there are all in accord, no nor a man with himself."

> The persons men wish to please, the objects they wish to
> gain, the means they employ—think of the character of
> all these! How soon will Time hide all things! How many
> a thing has it already hidden!

I wish I could ask him: Marcus Aurelius, do you agree that
tragedy, like value, is dependent upon the scale of time? If you
could be a philosopher king in this century, and sail to some
new Ionia in search of wisdom, would you turn to conservation? Is it possible that humanity will love life enough to save it?

I will remember Bernhardsdorp as a special place, a portal to
far-reaching dreams. To the south stretches Surinam eternal,
Surinam serene, a living treasure awaiting assay. I hope that
it will be kept intact, that at least enough of its million-year
history will be saved for the reading. By today's ethic its value
may seem limited, well beneath the pressing concerns of daily
life. But I suggest that as biological knowledge grows the ethic
will shift fundamentally so that everywhere, for reasons that
have to do with the very fiber of the brain, the fauna and flora
of a country will be thought part of the national heritage as
important as its art, its language, and that astonishing blend
of achievement and farce that has always defined our species.

Acknowledgments

F OR TECHNICAL HELP and sound advice not always taken I thank the following colleagues and friends: Freeman J. Dyson, Gerald Holton, Kathleen M. Horton, Jehoshua Kugler, William N. Lipscomb, Thomas Lovejoy, Charles J. Lumsden, Peter Marler, Marvin L. Minsky, Eviatar Nevo, Gordon H. Orians, Raymond A. Paynter, Jr., Richard Prum, Glenn Rowe, Joshua Rubenstein, Michael Ruse, Sue Savage-Rumbaugh, Lyle K. Sowls, J. Gray Sweeney, James H. Tumlinson, Barry D. Valentine, Ernest E. Williams, and Renee Wilson. This being a more personal book than my other ones, I feel it also appropriate to acknowledge with warm gratitude the staff of Harvard University Press for the effort and trust they have placed in my work during fifteen years of collaboration. Our relationship made the pleasures of authorship lasting and its pains already mostly forgotten.

THE DIVERSITY OF LIFE

To my mother
Inez Linnette Huddleston
in love and gratitude

Contents

VIOLENT NATURE,
RESILIENT LIFE

CHAPTER ONE

Storm over the Amazon

IN THE AMAZON BASIN the greatest violence sometimes begins as a flicker of light beyond the horizon. There in the perfect bowl of the night sky, untouched by light from any human source, a thunderstorm sends its premonitory signal and begins a slow journey to the observer, who thinks: the world is about to change. And so it was one night at the edge of rain forest north of Manaus, where I sat in the dark, working my mind through the labyrinths of field biology and ambition, tired, bored, and ready for any chance distraction.

Each evening after dinner I carried a chair to a nearby clearing to escape the noise and stink of the camp I shared with Brazilian forest workers, a place called Fazenda Dimona. To the south most of the forest had been cut and burned to create pastures. In the daytime cattle browsed in remorseless heat bouncing off the yellow clay and at night animals and spirits edged out onto the ruined land. To the north the virgin rain forest began, one of the great surviving wildernesses of the world, stretching 500 kilometers before it broke apart and dwindled into gallery woodland among the savannas of Roraima.

Enclosed in darkness so complete I could not see beyond my outstretched hand, I was forced to think of the rain forest as though I were seated in my library at home, with the lights turned low. The forest at night is an experience in sensory deprivation most of the time, black and silent as the midnight zone of a cave. Life is out there in expected abundance. The jungle teems, but in a manner mostly beyond the reach of the human senses. Ninety-nine percent of the animals find their way by chemical trails laid over the surface, puffs of odor released into the air or water, and scents diffused out of little hidden glands and into the air downwind. Animals are masters of this chemical channel, where we are idiots. But we are geniuses of the audiovisual channel, equaled in this modality only by a few odd groups (whales, monkeys, birds). So we wait for the dawn, while they wait for the fall of darkness; and because sight and sound are the evolutionary prerequisites of intelligence, we alone have come to reflect on such matters as Amazon nights and sensory modalities.

I swept the ground with the beam from my headlamp for signs of life, and found—diamonds! At regular intervals of several meters, intense pinpoints of white light winked on and off with each turning of the lamp. They were reflections from the eyes of wolf spiders, members of the family Lycosidae, on the prowl for insect prey. When spotlighted the spiders froze, allowing me to approach on hands and knees and study them almost at their own level. I could distinguish a wide variety of species by size, color, and hairiness. It struck me how little is known about these creatures of the rain forest, and how deeply satisfying it would be to spend months, years, the rest of my life in this place until I knew all the species by name and every detail of their lives. From specimens beautifully frozen in amber we know that the Lycosidae have survived at least since the beginning of the Oligocene epoch, forty million years ago, and probably much longer. Today a riot of diverse forms occupy the whole world, of which this was only the minutest sample, yet even these species turning about now to watch me from the bare yellow clay could give meaning to the lifetimes of many naturalists.

The moon was down, and only starlight etched the tops of the trees. It was August in the dry season. The air had cooled enough to make the humidity pleasant, in the tropical manner, as much a state of mind as a physical sensation. The storm I guessed was about an hour away. I thought of walking back into the forest with my headlamp to hunt for new treasures, but was too tired from the day's work. Anchored again to my chair, forced into myself, I welcomed a meteor's streak and the occasional courtship flash of luminescent click beetles among the nearby but unseen shrubs. Even the passage of a jetliner 10,000 meters up, a regular event each night around ten o'clock, I awaited with pleasure. A week in the rain forest had transformed its distant rumble from an urban irritant into a comforting sign of the continuance of my own species.

But I was glad to be alone. The discipline of the dark envelope summoned fresh images from the forest of how real organisms look and act. I needed to concentrate for only a second and they came alive as eidetic images, behind closed eyelids, moving across fallen leaves and decaying humus. I sorted the memories this way and that in hope of stumbling on some pattern not obedient to abstract theory of textbooks. I would have been happy with *any* pattern. The best of science doesn't consist of mathematical models and experiments, as textbooks make it seem. Those come later. It springs fresh from a more primitive mode of thought, wherein the hunter's mind weaves ideas from old facts and fresh metaphors and the scrambled crazy images of things recently seen. To move forward is to concoct new patterns of thought, which in turn dictate the design of the models and experiments. Easy to say, difficult to achieve.

The subject fitfully engaged that night, the reason for this research trip to the Brazilian Amazon, had in fact become an obsession and, like all obsessions, very likely a dead end. It was the kind of favorite puzzle that keeps forcing its way back because its very intractability makes it perversely pleasant, like an overly familiar melody intruding into the relaxed mind because it loves you and will not leave you. I hoped that some new image might propel me past the jaded puzzle to the other side, to ideas strange and compelling.

Bear with me for a moment while I explain this bit of personal esoterica; I am approaching the subject of central interest. Some kinds of plants and animals are dominant, proliferating new species and spreading over large parts of the world. Others are driven back until they become rare and threatened by extinction. Is there a single formula for this biogeographic difference, for all kinds of organisms? The process, if articulated, would be a law or at least a principle of dynastic succession in evolution. I was intrigued by the circumstance that social insects, the group on which I have spent most of my life, are among the most abundant of all organisms. And among the social insects, the dominant subgroup is the ants. They range 20,000 or more species strong from the Arctic Circle to the tip of South America. In the Amazon rain forest they compose more than 10 percent of the biomass of all animals. This means that if you were to collect, dry out, and weigh every animal in a piece of forest, from monkeys and birds down to mites and roundworms, at least 10 percent would consist of these insects alone. Ants make up almost half of the insect biomass overall and 70 percent of the individual insects found in the treetops. They are only slightly less abundant in grasslands, deserts, and temperate forests throughout the rest of the world.

It seemed to me that night, as it has to others in varying degrees of persuasion many times before, that the prevalence of ants must have something to do with their advanced colonial organization. A colony is a superorganism, an assembly of workers so tightly knit around the mother queen as to act like a single, well-coordinated entity. A wasp or other solitary insect encountering a worker ant on its nest faces more than just another insect. It faces the worker and all her sisters, united by instinct to protect the queen, seize control of territory, and further the growth of the colony. Workers are little kamikazes, prepared—eager—to die in order to defend the nest or gain control of a food source. Their deaths matter no more to the colony than the loss of hair or a claw tip might to a solitary animal.

There is another way to look at an ant colony. Workers foraging around their nest are not merely insects searching for food. They are a living web cast out by the superorganism, ready to congeal over rich food finds or shrink back from the most formidable enemies. Superorganisms can control and dominate the ground and treetops in competition with ordinary, solitary organisms, and that is surely why ants live everywhere in such great numbers.

I heard around me the Greek chorus of training and caution: *How can you prove that is the reason for their dominance? Isn't the connection just another shaky conclusion that because two events occur together, one causes the other? Something else entirely different might have caused both. Think about it—greater individual fighting ability? Sharper senses? What?*

Such is the dilemma of evolutionary biology. We have problems to solve, we have clear answers—too many clear answers. The difficult part is picking out the right answer. The isolated mind moves in slow circles and breakouts are rare. Solitude is better for weeding out ideas than for creating them. Genius is the summed production of the many with the names of the few attached for easy recall, unfairly so to other scientists. My mind drifted into the hourless night, no port of call yet chosen.

The storm grew until sheet lightning spread across the western sky. The thunderhead reared up like a top-heavy monster in slow motion, tilted forward, blotting out the stars. The forest erupted in a simulation of violent life. Lightning bolts broke to the front and then closer, to the right and left, 10,000 volts dropping along an ionizing path at 800 kilometers an hour, kicking a countersurge skyward ten times faster, back and forth in a split second, the whole perceived as a single flash and crack of sound. The wind freshened, and rain came stalking through the forest.

In the midst of chaos something to the side caught my attention. The lightning bolts were acting like strobe flashes to illuminate the wall of the rain forest. At intervals I glimpsed the storied structure: top canopy 30 meters off the ground, middle trees spread raggedly below that, and a lowermost scattering of shrubs and small trees. The forest was framed for a few moments in this theatrical setting. Its image turned surreal, projected into the unbounded wildness of the human imagination, thrown back in time 10,000 years. Somewhere close I knew spear-nosed bats flew through the tree crowns in search of fruit, palm vipers coiled in ambush in the roots of orchids, jaguars walked the river's edge; around them eight hundred species of trees stood, more than are native to all of North America; and a thousand species of butterflies, 6 percent of the entire world fauna, waited for the dawn.

About the orchids of that place we knew very little. About flies and beetles almost nothing, fungi nothing, most kinds of organisms nothing. Five thousand kinds of bacteria might be found in a pinch of soil, and about them we knew absolutely nothing. This was wilderness in the sixteenth-century sense, as it must have formed in the minds of the Portuguese explorers, its interior still largely unexplored and filled with strange, myth-engendering plants and animals. From such a place the pious naturalist would send long respectful letters to royal patrons about the wonders of the new world as testament to the glory of God. And I thought: there is still time to see this land in such a manner.

The unsolved mysteries of the rain forest are formless and seductive. They are like unnamed islands hidden in the blank spaces of old maps, like dark shapes glimpsed descending the far wall of a reef into the abyss. They draw us forward and stir strange apprehensions. The unknown and prodigious are drugs to the scientific imagination, stirring insatiable hunger with a single taste. In our hearts we hope we will never discover everything. We pray there will always be a world like this one at whose edge I sat in darkness. The rain forest in its richness is one of the last repositories on earth of that timeless dream.

That is why I keep going back to the forests forty years after I began, when I flew down to Cuba, a graduate student caught up in the idea of the "big" tropics, free at last to look for something hidden, as Kipling had urged, something lost behind the Ranges. The chances are high, in fact certain, of finding a new species or phenomenon within days or, if you work hard, hours after arrival. The hunt is also on for rare species already discovered but still effectively unknown—represented by one or two specimens placed in a museum drawer fifty or a hundred years ago, left with nothing but a locality and a habitat note handwritten on a tiny label ("Santarém, Brazil, nest on side of tree in swamp forest"). Unfold the stiff yellowing piece of paper and a long-dead biologist speaks: I was there, I found this, now you know, now move on.

There is still more to the study of biological richness. It is a microcosm of scientific exploration as a whole, refracting hands-on experience onto a higher plane of abstraction. We search in and around a subject for a concept, a pattern, that imposes order. We look for a way of speaking about the rough unmapped terrain, even just a name or a phrase that calls attention to the object of our attention. We hope to be the first to make a connection. Our goal is to capture and label a process, perhaps a chemical reaction or behavior pattern driving an ecological change, a new way of classifying energy flow, or a relation between predator and prey that preserves them both, almost anything at all. We will settle for just one good question that starts people thinking and talking: Why are there so many species? Why have mammals evolved more quickly than reptiles? Why do birds sing at dawn?

These whispering denizens of the mind are sensed but rarely seen. They rustle the foliage, leave behind a pug mark filling with water and a scent, excite us for an instant and vanish. Most ideas are waking dreams that fade to an emotional residue.

A first-rate scientist can hope to capture and express only several in a lifetime. No one has learned how to invent with any consistent success the equations and phrases of science, no one has captured the metaformula of scientific research. The conversion is an art aided by a stroke of luck in minds set to receive them. We hunt outward and we hunt inward, and the value of the quarry on one side of that mental barrier is commensurate with the value of the quarry on the other side. Of this dual quality the great chemist Berzelius wrote in 1818 and for all time:

> All our theory is but a means of consistently conceptualizing the inward processes of phenomena, and it is presumable and adequate when all scientifically known facts can be deduced from it. This mode of conceptualization can equally well be false and, unfortunately, presumably is so frequently. Even though, at a certain period in the development of science, it may match the purpose just as well as a true theory. Experience is augmented, facts appear which do not agree with it, and one is forced to go in search of a new mode of conceptualization within which these facts can also be accommodated; and in this manner, no doubt, modes of conceptualization will be altered from age to age, as experience is broadened, and the complete truth may perhaps never be attained.

The storm arrived, racing from the forest's edge, turning from scattered splashing drops into sheets of water driven by gusts of wind. It forced me back to the shelter of the corrugated iron roof of the open-air living quarters, where I sat and waited with the *mateiros*. The men stripped off their clothing and walked out into the open, soaping and rinsing themselves in the torrential rain, laughing and singing. In bizarre counterpoint, leptodactylid frogs struck up a loud and monotonous honking on the forest floor close by. They were all around us. I wondered where they had been during the day. I had never encountered a single one while sifting through the vegetation and rotting debris on sunny days, in habitats they are supposed to prefer.

Farther out, a kilometer or two away, a troop of red howler monkeys chimed in, their chorus one of the strangest sounds to be heard in all of nature, as enthralling in its way as the songs of humpback whales. A male opened with an accelerating series of deep grunts expanding into prolonged roars and was then joined by the higher-pitched calls of the females. This far away, filtered through dense foliage, the full chorus was machine-like: deep, droning, metallic.

Such raintime calls are usually territorial advertisements, the means by which the animals space themselves out and control enough land to forage and breed. For me they were a celebration of the forest's vitality: *Rejoice! The powers of nature are within our compass, the storm is part of our biology!*

For that is the way of the nonhuman world. The greatest powers of the physical environment slam into the resilient forces of life, and nothing much happens. For a very long time, 150 million years, the species within the rain forest evolved to absorb precisely this form and magnitude of violence. They encoded the predictable occurrence of nature's storms in the letters of their genes. Animals and plants have come to use heavy rains and floods routinely to time episodes in their life cycle. They threaten rivals, mate, hunt prey, lay eggs in new water pools, and dig shelters in the rain-softened earth.

On a larger scale, the storms drive change in the whole structure of the forest. The natural dynamism raises the diversity of life by means of local destruction and regeneration.

Somewhere a large horizontal tree limb is weak and vulnerable, covered by a dense garden of orchids, bromeliads, and other kinds of plants that grow on trees. The rain fills up the cavities enclosed by the axil sheaths of the epiphytes and soaks the humus and clotted dust around their roots. After years of growth the weight has become nearly unsupportable. A gust of wind whips through or lightning strikes the tree trunk, and the limb breaks and plummets down, clearing a path to the ground. Elsewhere the crown of a giant tree emergent above the rest catches the wind and the tree sways above the rain-soaked soil. The shallow roots cannot hold, and the entire tree keels over. Its trunk and canopy arc downward like a blunt ax, shearing through smaller trees and burying understory bushes and herbs. Thick lianas coiled through the limbs are pulled along. Those that stretch to other trees act as hawsers to drag down still more vegetation. The massive root system heaves up to create an instant mound of bare soil. At yet another site, close to the river's edge, the rising water cuts under an overhanging bank to the critical level of gravity, and a 20-meter front collapses. Behind it a small section of forest floor slides down, toppling trees and burying low vegetation.

Such events of minor violence open gaps in the forest. The sky clears again and sunlight floods the ground. The surface temperature rises and the humidity falls. The soil and ground litter dries out and warms up still more, creating a new environment for animals, fungi, and microorganisms of a different kind from those in the dark forest interior. In the following months pioneer plant species take seed. They are very different from the young shade-loving saplings and understory shrubs of the prevailing old-stand forest. Fast-growing, small in stature, and short-lived, they form a single canopy that matures far below the upper crowns of the older trees all around. Their tissue is soft and vulnerable to herbivores. The palmate-leaved trees of the genus *Cecropia*, one of the gap-filling specialists of Central and South America, harbor vicious ants in hollow internodes of the trunk. These insects, bearing the appropriate scientific name *Azteca*, live in symbiosis with their hosts, protecting them from all predators except sloths and a few other herbivores specialized to feed on *Cecropia*. The symbionts live among new assemblages of species not found in the mature forest.

All around the second-growth vegetation, the fallen trees and branches rot and crumble, offering hiding places and food to a vast array of basidiomycete fungi, slime molds, ponerine ants, scolytid beetles, bark lice, earwigs, embiopteran webspinners, zorapterans, entomobryomorph springtails, japygid diplurans, schizomid arachnids, pseudoscorpions, real scorpions, and other forms that live mostly or exclusively in this habitat. They add thousands of species to the diversity of the primary forest.

Climb into the tangle of fallen vegetation, tear away pieces of rotting bark, roll over logs, and you will see these creatures teeming everywhere. As the pioneer vegetation grows denser, the deepening shade and higher humidity again favor old-forest species, and their saplings sprout and grow. Within a hundred years the gap specialists will be phased out by competition for light, and the tall storied forest will close completely over.

In the succession, pioneer species are the sprinters, old-forest species the long-distance runners. The violent changes and a clearing of space bring all the species briefly to the same starting line. The sprinters dash ahead, but the prolonged race goes to the marathoners. Together the two classes of specialists create a complex mosaic of vegetation types across the forest which, by regular tree falls and landslides, is forever changing. If square kilometers of space are mapped over decades of time, the mosaic turns into a riotous kaleidoscope whose patterns come and go and come again. A new marathon is always beginning somewhere in the forest. The percentages of successional vegetation types are consequently more or less in a steady state, from earliest pioneer species through various mixes of pioneer and deep-forest trees to stands of the most mature physiognomy. Walk randomly on any given day for one or two kilometers through the forest, and you will cut through many of these successional stages and sense the diversity sustained by the passage of storms and the fall of forest giants.

It is diversity by which life builds and saturates the rain forest. And diversity has carried life beyond, to the harshest environments on earth. Rich assemblages of animals swarm in the shallow bays of Antarctica, the coldest marine habitats on earth. Perch-like notothenioid fishes swim there in temperatures just above the freezing point of salt water but cold enough to turn ordinary blood to ice, because they are able to generate glycopeptides in their tissues as antifreeze and thrive where other fish cannot go. Around them flock dense populations of active brittlestars, krill, and other invertebrate animals, each with protective devices of its own.

In a radically different setting, the deep unlighted zone of caves around the world, blind white springtails, mites, and beetles feed on fungi and bacteria growing on rotting vegetable matter washed down through ground water. They are eaten in turn by blind white beetles and spiders also specialized for life in perpetual darkness.

Some of the harshest deserts of the world are home to unique ensembles of insects, lizards, and flowering plants. In the Namib of southwestern Africa, beetles use leg tips expanded into oarlike sandshoes to swim down through the shifting dunes in search of dried vegetable matter. Others, the swiftest runners of the insect world, race over the baking hot surface on bizarre stilt legs.

Archaebacteria, one-celled microorganisms so different from ordinary bacteria as to be candidates for a separate kingdom of life, occupy the boiling water of mineral hot springs and volcanic vents in the deep sea. The species composing the newly discovered genus *Methanopyrus* grow in boiling vents at the bottom of the Mediterranean Sea in temperatures up to 110°C.

Life is too well adapted in such places, out to the edge of the physical envelope where biochemistry falters, and too diverse to be broken by storms and other ordinary vagaries of nature. But diversity, the property that makes resilience possible, is vulnerable to blows that are greater than natural perturbations. It can be eroded away fragment by fragment, and irreversibly so if the abnormal stress is unrelieved. This vulnerability stems from life's composition as swarms of species of limited geographical distribution. Every habitat, from Brazilian rain forest to Antarctic bay to thermal vent, harbors a unique combination of plants and animals. Each kind of plant and animal living there is linked in the food web to only a small part of the other species. Eliminate one species, and another increases in number to take its place. Eliminate a great many species, and the local ecosystem starts to decay visibly. Productivity drops as the channels of the nutrient cycles are clogged. More of the biomass is sequestered in the form of dead vegetation and slowly metabolizing, oxygen-starved mud, or is simply washed away. Less competent pollinators take over as the best-adapted bees, moths, birds, bats, and other specialists drop out. Fewer seeds fall, fewer seedlings sprout. Herbivores decline, and their predators die away in close concert.

In an eroding ecosystem life goes on, and it may look superficially the same. There are always species able to recolonize the impoverished area and exploit the stagnant resources, however clumsily accomplished. Given enough time, a new combination of species—a reconstituted fauna and flora—will reinvest the habitat in a way that transports energy and materials somewhat more efficiently. The atmosphere they generate and the composition of the soil they enrich will resemble those found in comparable habitats in other parts of the world, since the species are adapted to penetrate and reinvigorate just such degenerate systems. They do so because they gain more energy and materials and leave more offspring. But the restorative power of the fauna and flora of the world as a whole depends on the existence of enough species to play that special role. They too can slide into the red zone of endangered species.

Biological diversity—"biodiversity" in the new parlance—is the key to the maintenance of the world as we know it. Life in a local site struck down by a passing storm springs back quickly because enough diversity still exists. Opportunistic species evolved for just such an occasion rush in to fill the spaces. They entrain the succession that circles back to something resembling the original state of the environment.

This is the assembly of life that took a billion years to evolve. It has eaten the storms—folded them into its genes—and created the world that created us. It holds the world steady. When I rose at dawn the next morning, Fazenda Dimona had not changed in any obvious way from the day before. The same high trees stood like a fortress along the forest's edge; the same profusion of birds and insects foraged through the canopy and understory in precise individual timetables. All this seemed timeless, immutable, and its very strength posed the question: how much force does it take to break the crucible of evolution?

Krakatau

K RAKATAU, earlier misnamed Krakatoa, an island the size of Manhattan located midway in the Sunda Strait between Sumatra and Java, came to an end on Monday morning, August 27, 1883. It was dismembered by a series of powerful volcanic eruptions. The most violent occurred at 10:02 A.M., blowing upward like the shaped explosion of a large nuclear bomb, with an estimated force equivalent to 100–150 megatons of TNT. The airwave it created traveled at the speed of sound around the world, reaching the opposite end of the earth near Bogotá, Colombia, nineteen hours later, whereupon it bounced back to Krakatau and then back and forth for seven recorded passages over the earth's surface. The audible sounds, resembling the distant cannonade of a ship in distress, carried southward across Australia to Perth, northward to Singapore, and westward 4,600 kilometers to Rodriguez Island in the Indian Ocean, the longest distance traveled by any airborne sound in recorded history.

As the island collapsed into the subterranean chamber emptied by the eruption, the sea rushed in to fill the newly formed caldera. A column of magma, rock, and ash rose 5 kilometers into the air, then fell earthward, thrusting the sea outward in a tsunami 40 meters in height. The great tidal waves, resembling black hills when first sighted on the horizon, fell upon the shores of Java and Sumatra, washing away entire towns and killing 40,000 people. The segments traversing the channels and reaching the open sea continued on as spreading waves around the world. The waves were still a meter high when they came ashore in Ceylon, now Sri Lanka, where they drowned one person, their last casualty. Thirty-two hours after the explosion, they rolled in to Le Havre, France, reduced at last to centimeter-high swells.

The eruptions lifted more than 18 cubic kilometers of rock and other material into the air. Most of this tephra, as it is called by geologists, quickly rained back down onto the surface, but a residue of sulfuric-acid aerosol and dust boiled upward as high

as 50 kilometers and diffused through the stratosphere around
the world, where for several years it created brilliant red sunsets
and "Bishop's rings," opalescent coronas surrounding the sun.

Back on Krakatau the scene was apocalyptic. Throughout
the daylight hours the whole world seemed about to end for
those close enough to witness the explosions. At the climac-
tic moment of 10:02 the American barque *W. H. Besse* was
proceeding toward the straits 84 kilometers east northeast of
Krakatau. The first officer jotted in his logbook that "terrific
reports" were heard, followed by

> a heavy black cloud rising up from the direction of Krakatoa
> Island, the barometer fell an inch at one jump, suddenly rising
> and falling an inch at a time, called all hands, furled all sails
> securely, which was scarcely done before the squall struck the
> ship with terrific force; let go port anchor and all the chain
> in the locker, wind increasing to a hurricane; let go starboard
> anchor, it had gradually been growing dark since 9 A.M. and by
> the time the squall struck us, it was darker than any night I ever
> saw; this was midnight at noon, a heavy shower of ashes came
> with the squall, the air being so thick it was difficult to breathe,
> also noticed a strong smell of sulfur, all hands expecting to be
> suffocated; the terrible noises from the volcano, the sky filled
> with forked lightning, running in all directions and making
> the darkness more intense than ever; the howling of the wind
> through the rigging formed one of the wildest and most awful
> scenes imaginable, one that will never be forgotten by any one
> on board, all expecting that the last days of the earth had come;
> the water was running by us in the direction of the volcano at
> the rate of 12 miles per hour, at 4 P.M. wind moderating, the
> explosions had nearly ceased, the shower of ashes was not so
> heavy; so was enabled to see our way around the decks; the ship
> was covered with tons of fine ashes resembling pumice stone, it
> stuck to the sails, rigging and masts like glue.

In the following weeks, the Sunda Strait returned to outward normality, but with an altered geography. The center of
Krakatau had been replaced by an undersea crater 7 kilometers
long and 270 meters deep. Only a remnant at the southern end
still rose from the sea. It was covered by a layer of obsidian-laced
pumice 40 meters or more thick and heated to somewhere

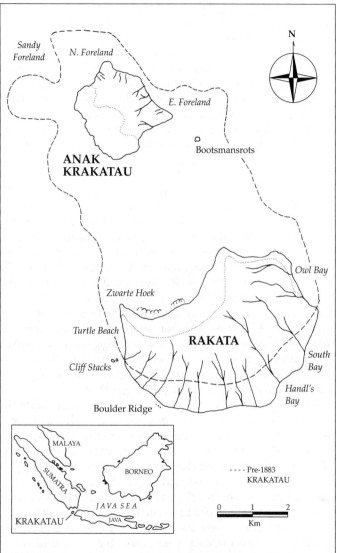

Old Krakatau was destroyed by a volcanic eruption in 1883, leaving only Rakata, a lifeless remnant, at the southern end. Anak Krakatau emerged from the sea as a volcanic cone in 1930.

between 300° and 850°C, enough at the upper range to melt lead. All traces of life had, of course, been extinguished.

Rakata, the ash-covered mountain of old Krakatau, survived as a sterile island. But life quickly enveloped it again. In a sense, the spinning reel of biological history halted, then reversed, like a motion picture run backward, as living organisms began to return to Rakata. Biologists quickly grasped the unique opportunity that Rakata afforded: to watch the assembly of a tropical ecosystem from the very beginning. Would the organisms be different from those that had existed before? Would a rain forest eventually cover the island again?

The first search for life on Rakata was conducted by a French expedition in May 1884, nine months after the explosions. The main cliff was eroding rapidly, and rocks still rolled down the sides incessantly, stirring clouds of dust and emitting a continuous noise "like the rattling of distant musketry." Some of the stones whirled through the air, ricocheting down the sides of the ravines and splashing into the sea. What appeared to be mist in the distance turned close up into clouds of dust stirred by the falling debris. The crew and expedition members eventually found a safe landing site and fanned out to learn what they could. After searching for organisms in particular, the ship's naturalist wrote that "notwithstanding all my researches, I was not able to observe any symptom of animal life. I only discovered one microscopic spider—only one; this strange pioneer of the renovation was busy spinning its web."

A baby spider? How could a tiny wingless creature reach the empty island so quickly? Arachnologists know that a majority of species "balloon" at some point in their life cycle. The spider stands on the edge of a leaf or some other exposed spot and lets out a thread of silk from the spinnerets at the posterior tip of its abdomen. As the strand grows it catches an air current and stretches downwind, like the string of a kite. The spider spins more and more of the silk until the thread exerts a strong pull on its body. Then it releases its grip on the surface and soars upward. Not just pinhead-sized babies but large spiders can occasionally reach thousands of meters of altitude and travel hundreds of kilometers before settling to the ground to start a new life. Either that or land on the water and die. The voyagers have no control over their own descent.

Ballooning spiders are members of what ecologists, with the accidental felicity that sometimes pops out of Greek and Latin sources, have delightfully called the aeolian plankton. In ordinary parlance, plankton is the vast swarm of algae and small animals carried passively by water currents; aeolian refers to the wind. The creatures composing the aeolian plankton are devoted almost entirely to long-distance dispersal. You can see some of it forming over lawns and bushes on a quiet summer afternoon, as aphids use their feeble wings to rise just high enough to catch the wind and be carried away. A rain of planktonic bacteria, fungus spores, small seeds, insects, spiders, and other small creatures falls continuously on most parts of the earth's land surface. It is sparse and hard to detect moment by moment, but it mounts up to large numbers over a period of weeks and months. This is how most of the species colonized the seared and smothered remnant of Krakatau.

The potential of the planktonic invasion has been documented by Ian Thornton and a team of Australian and Indonesian biologists who visited the Krakatau area in the 1980s. While studying Rakata they also visited Anak Krakatau ("Child of Krakatau"), a small island that emerged in 1930 from volcanic activity along the submerged northern rim of the old Krakatau caldera. On its ash-covered lava flows they placed traps made from white plastic containers filled with seawater. This part of the surface of Anak Krakatau dated from localized volcanic activity from 1960 to 1981 and was nearly sterile, resembling the condition on Rakata soon after the larger island's violent formation. During ten days the traps caught a surprising variety of windborne arthropods. The specimens collected, sorted, and identified included a total of 72 species of spiders, springtails, crickets, earwigs, barklice, hemipterous bugs, moths, flies, beetles, and wasps.

There are other ways to cross the water gaps separating Rakata from nearby islands and the Javan and Sumatran coasts. The large semiaquatic monitor lizard *Varanus salvator* probably swam over. It was present no later than 1899, feasting on the crabs that crawl along the shore. Another long-distance swimmer was the reticulated python, a giant snake reaching up to 8 meters in length. Probably all of the birds crossed over by powered flight. But only a small percentage of the species of Java and Sumatra were represented because it is a fact, curiously, that many forest species refuse to cross water gaps even when the nearest island is in full view. Bats, straying off course, made the Rakata landfall. Winged insects of larger size, especially butterflies and dragonflies, probably also traveled under their own power. Under similar conditions in the Florida Keys, I have watched such insects fly easily from one small island to another, as though they were moving about over meadows instead of salt water.

Rafting is a much less common but still important means of transport. Logs, branches, sometimes entire trees fall into rivers and bays and are carried out to sea, complete with microorganisms, insects, snakes, frogs, and occasional rodents and other small mammals living on them at the moment of departure. Blocks of pumice from old volcanic islands, riddled with enough closed air spaces to keep them afloat, also serve as rafts.

Once in a great while a violent storm turns larger animals such as lizards or frogs into aeolian debris, tearing them loose from their perches and propelling them to distant shores. Waterspouts pick up fish and transport them live to nearby lakes and streams.

Swelling the migration further, organisms carry other organisms with them. Most animals are miniature arks laden with parasites. They also transport accidental hitchhikers in soil clinging to the skin, including bacteria and protozoans of immense variety, fungal spores, nematode worms, tardigrades, mites, and feather lice. Seeds of some species of herbs and trees pass live through the guts of birds, to be deposited later in feces, which serves as instant fertilizer. A few arthropods practice what biologists call phoresy, deliberate hitchhiking on larger animals. Pseudoscorpions, tiny replicas of true scorpions but lacking stings, use their lobster-like claws to seize the hairs of dragonflies and other large winged insects, then ride these magic carpets for long distances.

The colonists poured relentlessly into Rakata from all directions. A 100-meter-high electrified fence encircling the island could not have stopped them. Airborne organisms would still have tumbled in from above to spawn a rich ecosystem. But the largely happenstance nature of colonization means that flora and fauna did not return to Rakata in a smooth textbook manner, with plants growing to sylvan thickness, then herbivores proliferating, and finally carnivores prowling. The surveys made on Rakata and later on Anak Krakatau disclosed a far more haphazard buildup, with some species inexplicably going extinct and others flourishing when seemingly they should have quickly disappeared. Spiders and flightless carnivorous crickets persisted almost miraculously on bare pumice fields; they fed on a thin diet of insects landing in the aeolian debris. Large lizards and some of the birds lived on beach crabs, which subsisted in turn on dead marine plants and animals washed ashore by waves. (The original name of Krakatau was Karkata, or Sanskrit for "crab"; Rakata also means crab in the old Javanese language.) Thus animal diversity was not wholly dependent on vegetation. And for its part vegetation grew up in patches, alternately spreading and retreating across the island to create an irregular mosaic.

If the fauna and flora came back chaotically, they also came back fast. In the fall of 1884, a little more than a year after the eruption, biologists encountered a few shoots of grass, probably *Imperata* and *Saccharum*. In 1886 there were fifteen species of grasses and shrubs, in 1897 forty-nine, and in 1928 nearly three hundred. Vegetation dominated by *Ipomoea* spread along the shores. At the same time grassland dotted with *Casuarina* pines gave way here and there to richer pioneer stands of trees and shrubs. In 1919 W. M. Docters van Leeuwen, from the Botanical Gardens at Buitenzorg, found forest patches surrounded by nearly continuous grassland. Ten years later he found the reverse: forest now clothed the entire island and was choking out the last of the grassland patches. Today Rakata is covered completely by tropical Asian rain forest typical in outward appearance. Yet the process of colonization is far from complete. Not a single tree species characterizing the deep, primary forests on Java and Sumatra has made it back. Another hundred years or more may be needed for investment by a forest fully comparable to that of old, undisturbed Indonesian islands of the same size.

Some insects, spiders, and vertebrates aside, the earliest colonists of most kinds of animals died on Rakata soon after arrival. But as the vegetation expanded and the forest matured, increasing numbers of species took hold. At the time of the Thornton expeditions of 1984–85, the inhabitants included thirty species of land birds, nine bats, the Indonesian field rat, the ubiquitous black rat, and nine reptiles, including two geckos and *Varanus salvator*, the monitor lizard. The reticulated python, recorded as recently as 1933, was not present in 1984–85. A large host of invertebrate species, more than six hundred in all, lived on the island. They included a terrestrial flatworm, nematode worms, snails, scorpions, spiders, pseudoscorpions, centipedes, cockroaches, termites, barklice, cicadas, ants, beetles, moths, and butterflies. Also present were microscopic rotifers and tardigrades and a rich medley of bacteria.

A first look at the reconstituted flora and fauna of Rakata, in other words Krakatau a century after the apocalypse, gives the impression of life on a typical small Indonesian island. But the community of species remains in a highly fluid state. The number of resident bird species may now be approaching an equilibrium, the rise having slowed markedly since 1919 to settle close to thirty. Thirty is also about the number on other islands of Indonesia of similar size. At the same time, the *composition* of the bird species is less stable. New species have been arriving, and earlier ones have been declining to extinction. Owls and flycatchers arrived after 1919, for example while several old residents such as the bulbul (*Pycnonotus aurigaster*) and gray-backed shrike (*Lanius schach*) disappeared. Reptiles appear to be at or close to a similar dynamic equilibrium. So are cockroaches, nymphalid butterflies, and dragonflies. Flightless mammals, represented solely by the two kinds of rats, are clearly not. Nor are plants, ants, or snails. Most of the other invertebrates are still too poorly explored on Rakata over sufficiently long periods of time to judge their status, but in general the overall number of species appears to be still rising.

Rakata, along with Panjang and Sertung, and other islands of the Krakatau archipelago blasted and pumice-coated by the 1883 explosion, have within the span of a century rewoven a semblance of the communities that existed before, and the diversity of life has largely returned. The question remains as to whether endemic species, those found only on the archipelago prior to 1883, were destroyed by the explosion. We can never be sure because the islands were too poorly explored by naturalists before Krakatau came so dramatically to the world's attention in 1883. It seems unlikely that endemic species ever existed. The islands are so small that the natural turnover of species may have been too fast to allow evolution to attain the creation of new species, even without volcanic episodes.

In fact the archipelago has suffered turbulence that destroyed or at least badly damaged its fauna and flora every few centuries. According to Javanese legend, the volcano Kapi erupted violently in the Sunda Strait in 416 A.D.: "At last the mountain Kapi with a tremendous roar burst into pieces and sunk into the deepest of the earth. The water of the sea rose and inundated the land." A series of smaller eruptions, burning at least part of the forest, occurred during 1680 and 1681.

Today you can sail close by the islands without guessing their violent history, unless Anak Krakatau happens to be smoldering that day. The thick green forest offers testimony to the ingenuity and resilience of life. Ordinary volcanic eruptions are not enough, then, to break the crucible of life.

CHAPTER THREE

The Great Extinctions

W^HAT WAS the greatest blow ever suffered by life through all time? Not the 1883 explosions at Krakatau, which were not even the worst in recorded history. An 1815 eruption at Tambora, 1,400 kilometers to the east of Krakatau on the Indonesian island of Sumbawa, lifted five times as much rock and ash as Krakatau. It inflicted more environmental destruction and killed tens of thousands of people. About 75,000 years ago a still greater eruption occurred in the center of northern Sumatra. It blew out a phenomenal 1,000 cubic kilometers of solid material, creating an oval depression 65 kilometers long that filled with fresh water and persists to this day as Lake Toba. Paleolithic people lived on the island then. We can only imagine what they felt in the presence of an eruption one hundred times the magnitude of Krakatau, and what stories of gods and apocalypse proliferated in the culture afterwards.

Great eruptions are likely to have occurred repeatedly across long stretches of geological time. A simple form of statistical reasoning leads to this conclusion. The frequency curve of the intensity of volcanic eruptions around the world, like so many chance phenomena, peaks near the low end and tapers off for a long distance toward the high end. This means that most eruptions are relatively minor perturbations, consisting of a plume of vapor from a fumarole here, a minor lava flow there. Lava fountains and big flows, the next step up, are less common but still occur on a yearly basis somewhere in the world. An event the size of the Krakatau explosion happens once or twice a century. An eruption as big as the one at Toba is far rarer but, over millions of years, probably inevitable.

The same statistical reasoning applies to the fall of meteorites. A large number ranging in size from dust particles to pebbles reach the earth's surface each year, streaking in at 15 to 75 kilometers a second. A much smaller number range in size from baseballs to soccer balls. They account for the majority of the thirty or so meteorites worldwide that can be seen traveling all the way down and are then located by searchers on foot. A very few are much more massive. The largest ever observed in the United States was a 5,000-kilogram meteorite that fell in Norton County, Kansas, on February 18, 1948. Over millions of years only a few truly gigantic meteorites reach the earth's surface. One with a diameter of 1,250 meters gouged out Canyon Diablo in Arizona. Another monster, 3,200 meters in diameter, created the Chubb Depression at Ungava, Quebec.

By extrapolation upward along the scale of violence, it is conceivable and even likely that a volcanic eruption or a meteorite strike occurs once every 10 million or 100 million years so great as to literally shake the earth, drastically change its atmosphere, and as a result extinguish a substantial portion of the species then living. Something of that kind might have happened at the end of the Mesozoic era 66 million years ago, when dinosaurs and a few other prevailing groups of animals were set back or extinguished altogether. So concluded Luis Alvarez and three other physical scientists at Berkeley in 1979. They found abnormally high concentrations of iridium, an element of the platinum group, in a thin geological deposit separating the older Mesozoic era from the younger Cenozoic era. More precisely, the layer separates rocks of the Cretaceous period, youngest of the Mesozoic periods, and the Tertiary period, oldest of the Cenozoic. Proceeding upward across this thin line, called the K-T boundary (the two single-letter symbols given the two periods respectively), the fossils shift from a prevalence of dinosaurs and a few small mammals to no dinosaurs and a prevalence of mammals. Iridium has a strong affinity for iron; as a result, during the formation of the planet most of it had been drawn deep into the iron-rich core of the earth. Its presence in the K-T boundary, so close to the surface, was a mystery.

The Berkeley team noticed that iridium is also abundant in some meteorites. This anomaly and some mathematical modeling led them to the following scenario: 66 million years ago a meteorite 10 kilometers in diameter crashed into the earth at 72,000 kilometers an hour. The impact conveyed a force greater than the detonation of all the nuclear weapons in the world. It rang the earth like a bell, ignited wildfires, washed the shores with giant tsunamis, and kicked up an immense dust cloud that enshrouded the planet and then either cooled the atmosphere by blocking out the sun or else warmed it by trapping heat as in a greenhouse. As the dust settled it formed a silt layer half a centimeter thick, laced with iridium. Afterward acid rain washed the surface residue for months or years. All these effects combined, according to the Alvarez scenario, to kill the dinosaurs and a medley of other plants and animals.

If a strike anywhere near this magnitude did occur, it should have left telltale signs in addition to iridium enrichment of the earth's surface. In the intense discussion and research that followed Alvarez's proposal, one key piece of new evidence came to light. Geochemists know that when quartz is subjected to extreme pressures, such as those at an impact site, it is "shocked": the crystal lattice is disrupted so that irregular planes appear in thin sections of the mineral examined microscopically between crossed polarizing filters. Such planes were indeed found in quartz grains in some parts of the K-T boundary. At this point the case for the meteorite hypothesis looked very good.

First rule of the history of science: when a big, new, persuasive idea is proposed, an army of critics soon gathers and tries to tear it down. Such a reaction is unavoidable because, aggressive yet abiding by the rules of civil discourse, this is simply how scientists work. It is further true that, faced with adversity, proponents will harden their resolve and struggle to make the case more convincing. Being human, most scientists conform to the psychological Principle of Certainty, which says that when there is evidence both for and against a belief, the result is not a lessening but a heightening of conviction on both sides. During the 1980s, hundreds of experts wrote over two thousand articles for and against the meteorite hypothesis. Tensions rose at scientific conferences, arguments and counterarguments flowed through the pages of *Science*, a small industry grew up in the laboratories and seminar rooms of research universities.

Rule number two: the new idea will, like mother earth, take some serious hits. If good it will survive, probably in modified form. If bad it will die, usually at the time of death or retirement of the last original proponent. As Paul Samuelson once said of the science of economics: funeral by funeral, theory

advances. In this case, the antimeteorite critics had a powerful competing hypothesis. They said that every few tens of
millions of years huge volcanic eruptions, either one-time gar-
gantuan Krakataus or concerted volleys of ordinary Krakataus,
could produce the effects observed at the K-T boundary. Some
present-day volcanoes do bring up elevated levels of iridium in
their ash. They might also generate enough pressure to shock
quartz, although field tests in progress (as I write) have not
resolved the matter either way.

The volcanists and other critics raised another even more
troubling piece of evidence to undermine the meteorite
hypothesis: many extinctions did occur at the end of the Cretaceous period, no question about that, but not all at once. The
times of extinction of various groups were smeared out over
millions of years on either side of the K-T boundary. Dinosaurs,
for example, declined noticeably during the last ten million
years before the end of the Cretaceous period. In Montana and
southern Alberta, about thirty species were present 10 million
years before the end. The number decreased gradually to thirteen just before the end, with the horned dinosaur *Triceratops*
remaining most abundant in the final group. A similar pattern
was followed by the ammonoids, mollusks with a chambered
shell like that of the modern pearly nautilus. It was also followed
by inoceramid pelecypods, bivalve mollusks that included giant
species with shells a meter wide, and by rudists, other bivalves
that built reefs from the mass of their shells. Many groups
of foraminiferans, amoeba-like marine creatures that secrete
elaborate and exquisitely designed siliceous skeletons, phased
out in steps across a million years. Some disappeared before the
end of the Cretaceous, others later at differing times, all to be
replaced by new kinds of foraminiferans emerging during several hundred thousand years. Insects passed through the K-T
boundary relatively unscathed. All of the orders, the highest-
ranking taxonomic groups, survived, including the Coleoptera
(beetles), Diptera (flies), Hymenoptera (bees, wasps, and ants),
and Lepidoptera (moths and butterflies). Most if not all of the
families, the next highest-ranking groups, also came through,
including the Formicidae (ants), Curculionidae (weevils), and
Stratiomyidae (soldier flies). The fossil record is still too poor
on the Cretaceous side to estimate extinctions at the species

level, of particular kinds such as the modern-day house fly (*Musca domestica*) or the cabbage white butterfly (*Pieris rapae*).

In order to accommodate the staggered schedule of extinctions across the K-T boundary brought into focus by the controversy, some paleontologists conceived of a series of violent eruptions over millions of years toward the end of the Cretaceous, creating periodic global dust veils, wildfires, acid rain, and climatic cooling. These malign events conspired to reduce population levels of all kinds of organisms and to shrink their geographical ranges to limited regions of the world. A few kinds of animals, such as the dinosaurs, ammonoids, and foraminiferans, were hit hard. Insects and plants persisted more or less intact, perhaps because of their ability to function at low physiological levels for months or even years at a time.

Some of the scientists favoring the meteorite hypothesis, but impressed by the new evidence on extinction, adjusted their model by abandoning the hypothesis of a single cataclysmic event. They postulated a series of smaller meteorite strikes across the million-year transition. Many such events, they said, could have smeared out the timing of the extinctions widely on either side of the K-T boundary.

Not all paleontologists were so ready to abandon both the mega-Krakatau and the single-strike hypotheses. They redoubled their efforts to locate fossils close to the K-T boundary in order to fix more precisely the time of the mass extinctions. Now the balance appears to be shifting somewhat to the single-event hypothesis. With more fossils in hand, it does seem more plausible that dinosaurs and ammonoids were cut down suddenly at the time of the supposed meteorite strike or the mega-Krakatau. The data on foraminiferans remain ambiguous and disputed. Plants offer clearer evidence of a single catastrophe. Their fossils are more abundant and easily interpreted, especially the pollen grains incorporated into lake-bottom silt year by year. Western North America saw a sudden and severe reduction in the pollen grains of flowering plants at the K-T boundary, followed by an equally abrupt jump in fern spores— the "fern spike" of the fossil record—followed soon afterward by a return of the pollen of flowering plants, this time representing a different assemblage of species. The temporary decline of flowering plants and the rise of ferns is consistent with a boundary-event winter, a darkening and cooling of the climate caused by clouds of dust and smoke and lasting for a year or two. Some plant species became extinct, especially broad-leaved evergreens in the general category represented today by magnolias and rhododendrons. Others came back after a time, the descendants of scattered survivors, but as part of a different mix of the post-Mesozoic era. In the southern hemisphere the effect on the vegetation was less severe.

Most paleontologists are now leaning cautiously toward the hypothesis of a sudden, catastrophic closure of the Mesozoic era. Meanwhile, the search goes on for the kind of evidence most coveted during all scientific odysseys: one easily under-stood discovery that definitely implicates a single major cause while dismissing the alternatives. The most obvious candidate is the smoking gun from a great meteorite strike, a giant crater somewhere on earth that could be dated precisely to the K-T boundary. Since two-thirds of the earth is covered by water, the remnant of a large blowout might lie undiscovered on the ocean floor. In 1990 two candidate craters were proposed on the basis of the distribution of shocked quartz and distinctive geological formations in accessible strata: one in the Caribbean southwest of Haiti, the other just south of western Cuba 1,350 kilometers from the first site. The evidence is not yet strong enough to be accepted. The geological conformations are under study, and the search continues in other ocean basins.

A compromise may be in order. Both the meteorite and the volcano explanations could be correct. The two events might have occurred at the same time. A meteorite 10 kilometers in diameter striking the surface at thousands of kilometers an hour would not only shake the earth's surface and darken the atmosphere but also trigger volcanic eruptions over all the planet. Alternatively, an unprovoked volcanic activity might be the key, with a meteorite strike delivering the coup de grace to the dinosaurs and the most sensitive marine animals at the time we have come to call the K-T boundary.

This brings us to the important fact that the Cretaceous extinction was only one of five such catastrophes that occurred over the last half-billion years, and it was not the most severe. Furthermore, the earlier spasms appear not to have been asso-ciated with meteorite strikes or unusually heavy volcanism.

The five mass extinctions occurred in this order, according to geological period and time before the present: Ordovician, 440 million years; Devonian, 365 million years; Permian, 245 million years; Triassic, 210 million years; and Cretaceous, 66 million years. There have been a great many second- and third-order dips and rises, but these five are at the far end of the curve of violence, and they stand out. They are to other episodes as a catastrophe is to a misfortune, a hurricane to a summer squall.

The organisms that most clearly display extinction rates were animals that lived in the sea, from mollusks and arthropods to fishes, for the simple artifactual reason that their remains settled quickly to the bottom to be silted over and turned into fossils before they fully decomposed. It is further true that the units of taxonomic measure are families of related species because if any species in the group were alive at the time of the deposit, there is a good chance that at least one will now come to light in fossil form. To rely on individual species, many of which were likely to be rare or spottily distributed at any given time, is to introduce a large statistical error.

Consider the large amount of data on marine animals collected and analyzed by John Sepkoski and David Raup of the University of Chicago and others. The loss of families on which reliable data are available was just about the same, roughly 12 percent, in each of the spasms except the Permian, which saw a staggering loss of 54 percent. Statistical methods exist by which it is possible to count the number of extinguished families and make a sound educated guess as to the loss of the species that composed the families. The great Permian crash is estimated to have resulted in the loss of between 77 and 96 percent of all marine animal species. Raup has remarked that "if these estimates are reasonably accurate, global biology (for higher organisms at least) had an extremely close brush with total destruction." Trilobites and placoderm fishes, two highly distinctive and dominant groups in earlier periods, did in fact come to an end. On land, the mammal-like reptiles, distant ancestors of humanity, were devastated, with only a few survivors squeezing through. Insects and plants were less affected; they somehow acquired the invisible shield that was to surround them through all the later episodes.

No iridium has been found in deposits dating to the time of the first four spasms. Thus there was evidently no meteorite strike of sufficient magnitude to cause first-order extinction spasms. There were massive volcanic eruptions in north-central Siberia about the time of the Permian extinctions, perhaps sufficient to alter the global climate, but the connection to the decline of life is far from proved. So what did happen? In the view of Steven Stanley and some other paleobiologists, the primary agent of destruction was long-term climatic change. The evidence is circumstantial but persuasive. It includes a general retreat of tropical organisms toward the equator, reaching a peak at the time of the crises. Reef-building organisms, including algae and calcareous sponges, were especially vulnerable. They vanished over large portions of the earth. The nonliving, skeletal parts of the reefs were then either eroded by wave action or silted over. (One fossil reef, formed in Western Australia 350 million years ago, somehow resisted erosion and is still a prominent feature of the landscape.) The ranges of surviving tropical organisms were compressed toward the equator during the crises. Glaciation was more extensive.

The earth thus appears to have cooled dramatically during the first four crises, eliminating many species and forcing others into smaller ranges, rendering them more vulnerable to extinction from other causes.

I have begged the question of ultimate causation. If global cooling was the killing event, what caused the cooling? The most likely answer deduced by geologists is the movement of land masses and the fringing seas during continental drift. At the time of the first major extinction spasms—Ordovician, Devonian, and Permian—the land was united as a single supercontinent called Pangaea. As its southern block, Gondwanaland, edged over the South Pole in Late Ordovician and Devonian times, extensive glaciation resulted, and the biological crises occurred more or less concurrently. During Permian times, Pangaea drifted farther northward, and glaciers spread over both the northern and southern ends. As ice formed, the sea level fell, drastically reducing the extent of the warm inland seas in which most of marine life then lived.

Continental drift does not appear to have been a cause of global cooling at the end of the Mesozoic period, so our attention is justifiably fixed on meteorites and volcanoes. Today the land mass of the world is arrayed in a configuration that favors high levels of diversity: widely separated continents with long shorelines and stretches of shallow tropical water dotted with lots of islands. No evidence exists of meteorite showers or volcanic blasts of world-altering proportions for the past 66 million years, at least none strong enough to have collapsed the house of cards we know as biodiversity.

To summarize: life was impoverished in five major events, and to lesser degree here and there around the world in countless other episodes. After each downturn it recovered to at least the original level of diversity. How long did it take for evolution to restore the losses after the first-order spasms? The number of families of animals living in the sea is as reliable a measure as we have been able to obtain from the existing fossil evidence. In general, five million years were enough only for a strong start. A complete recovery from each of the five major extinctions required tens of millions of years. In particular the Ordovician dip needed 25 million years, the Devonian 30 million years, the Permian and Triassic (combined because they were so close together in time) 100 million years, and the Cretaceous 20 million years. These figures should give pause to anyone who believes that what *Homo sapiens* destroys, Nature will redeem. Maybe so, but not within any length of time that has meaning for contemporary humanity.

In the chapters to follow I will describe the formation of life's diversity as it is understood—with traversals—by most biologists. I will give evidence that humanity has initiated the sixth great extinction spasm, rushing to eternity a large fraction of our fellow species in a single generation. And finally I will argue that every scrap of biological diversity is priceless, to be learned and cherished, and never to be surrendered without a struggle.

BIODIVERSITY RISING

CHAPTER FOUR

The Fundamental Unit

THE MOST WONDERFUL mystery of life may well be the means by which it created so much diversity from so little physical matter. The biosphere, all organisms combined, makes up only about one part in ten billion of the earth's mass. It is sparsely distributed through a kilometer-thick layer of soil, water, and air stretched over a half billion square kilometers of surface. If the world were the size of an ordinary desktop globe and its surface were viewed edgewise an arm's length away, no trace of the biosphere could be seen with the naked eye. Yet life has divided into millions of species, the fundamental units, each playing a unique role in relation to the whole.

For another way to visualize the tenuousness of life, imagine yourself on a journey upward from the center of the earth, taken at the pace of a leisurely walk. For the first twelve weeks you travel through furnace-hot rock and magma devoid of life. Three minutes to the surface, five hundred meters to go, you encounter the first organisms, bacteria feeding on nutrients that have filtered into the deep water-bearing strata. You breach the surface and for ten seconds glimpse a dazzling burst of life, tens of thousands of species of microorganisms, plants, and animals within horizontal line of sight. Half a minute later almost all are gone. Two hours later only the faintest traces remain, consisting largely of people in airliners who are filled in turn with colon bacteria.

The hallmark of life is this: a struggle among an immense variety of organisms weighing next to nothing for a vanishingly small amount of energy. Life operates on only 10 percent of the sun's energy reaching earth's surface, that portion fixed by the photosynthesis of green plants. The free energy is then sharply discounted as it passes through the food webs from one organism to the next: very roughly 10 percent passes to the caterpillars and other herbivores that eat the plants and bacteria, 10 percent of that (or 1 percent of the original) to the spiders and other low-level carnivores that eat the herbivores, 10 percent of the residue to the warblers and other middle-level carnivores that eat the low-level carnivores, and so on upward to the top carnivores, which are consumed by no one except parasites and scavengers. Top carnivores, including eagles, tigers, and great white sharks, are predestined by their perch at the apex of the food web to be big in size and sparse in numbers. They live on such a small portion of life's available energy as always to skirt the edge of extinction, and they are the first to suffer when the ecosystem around them starts to erode.

A great deal can be learned quickly about biological diversity by noticing that species in the food web are arranged into two hierarchies. The first is the energy pyramid, a straightforward consequence of the law of diminishing energy flow as noted: a relatively large amount from the sun's energy incident on earth goes into the plants at the bottom, tapering to a minute quantity to the big carnivores on top. The second pyramid is composed of biomass, the weight of organisms. By far the largest part of the physical bulk of the living world is contained in plants. The second largest amount belongs to the scavengers and other decomposers, from bacteria to fungi and termites, which together extract the last bit of fixed energy from dead tissue and waste at every level in the food web, and in exchange return degraded nutrient chemicals to the plants. Each level above the plants diminishes thereafter in biomass until you come to the top carnivores, which are so scarce that the very sight of one in the wild is memorable. Let me stress that point. No one looks twice at a sparrow or squirrel, or even once at a dandelion, but a peregrine falcon or mountain lion is a lifetime experience. And not just because of their size (think of a cow) or ferocity (think of a house cat), but because they are rare.

The biomass pyramid of the sea is at first glance puzzling: it is turned upside down. The photosynthetic organisms still capture almost all the energy, which is discounted in steps by the 10 percent rule, but they have less total bulk than the animals that eat them. How is this inversion possible? The answer is that the photosynthetic organisms are not plants in the traditional landbound sense. They are phytoplankton, microscopic single-celled algae carried passively by water currents. Cell for cell, planktonic algae fix more solar energy and manufacture more protoplasm than plants on the land, and they grow, divide, and die at an immensely faster pace. Small animals, particularly copepods and other small crustaceans carried in the sea currents, hence called zooplankton, consume the algae. They harvest huge quantities without exhausting the standing photosynthetic crop in the water. Zooplankton in turn are eaten by larger invertebrate animals and fish, which are then eaten by still larger fish and marine mammals such as seals and porpoises, which are hunted by killer whales and great white sharks, the top carnivores. The inversion of the biomass pyramid is why the waters of the open ocean are so clear, why you can look into them and spot an occasional fish but not the green plants—algae—on which all the animals ultimately depend.

We have arrived at the question of central interest. The larger organisms of earth, composing the visible superstructures of the energy and biomass pyramids, owe their existence to biological diversity. Of what then is biodiversity composed? Since antiquity biologists have felt a compelling need to posit an atomic unit by which diversity can be broken apart, then described, measured, and reassembled. Let me put the matter as strongly as this important issue merits. Western science is built on the obsessive and hitherto successful search for atomic units, with which abstract laws and principles can be derived. Scientific knowledge is written in the vocabulary of atoms, subatomic particles, molecules, organisms, ecosystems, and many other units, including species. The metaconcept holding all of the units together is hierarchy, which presupposes levels of organization. Atoms bond into molecules, which are assembled into nuclei, mitochondria, and other organelles, which aggregate into cells, which associate as tissues. The levels then progress on upward as organs, organisms, societies, species, and ecosystems. The reverse procedure is decomposition, the breaking of ecosystems into species, species into societies and organisms, and so on downward. Both theory and experimental analysis in science are predicated on the assumption—the trust, the faith—that complex systems can be cleaved into simpler systems. And so the search proceeds relentlessly for natural units until, like the true grail, they are found and all rejoice.

Scientific fame awaits those who discover the lines of fracture
and the processes by which lesser natural units are joined to
create larger natural units.

So the species concept is crucial to the study of biodiversity.
It is the grail of systematic biology. Not to have a natural unit
such as the species would be to abandon a large part of biology into free fall, all the way from the ecosystem down to the
organism. It would be to concede the idea of amorphous variation and arbitrary limits for such intuitively obvious entities
as American elms (species: *Ulmus americana*), cabbage white
butterflies (*Pieris rapae*), and human beings (*Homo sapiens*).
Without natural species, ecosystems could be analyzed only
in the broadest terms, using crude and shifting descriptions
of the organisms that compose them. Biologists would find it
difficult to compare results from one study to the next. How
might we assess, for example, the thousands of research papers
on the fruit fly, which form much of the foundation of modern
genetics, if no one could tell one kind of fruit fly from another?

I will try to cut to the heart of the matter with the "biological-species concept": *a species is a population whose members are
able to interbreed freely under natural conditions.* This definition is an idea easily stated but filled with exceptions and difficulties, all interesting, all reflective of the range of complexity
in evolutionary biology itself. My opinion is that the grail,
though nicked and tarnished, is in our possession. The chalice
sits on the shelf. I must add at once that not all biologists accept
the biological-species concept as sound or as the pivotal unit
on which the description of biological diversity can be based.
They look to the gene or the ecosystem to play these roles, or
they are happy to live with conceptual anarchy. I think they are
wrong, but in any case will return shortly to the difficulties of
the biological species to give voice to their misgivings.

For the moment let me go on to expand the definition, which
is accepted at least provisionally by a majority of evolutionary
biologists. Notice the qualification it contains, "under natural
conditions." This says that hybrids bred from two kinds of animals in captivity, or two kinds of plants cultivated in a garden,
are not enough to classify them as a member of a single species.
To take the most celebrated example, zookeepers have for years
crossed tigers with lions. The offspring are called tiglons when

the father is a tiger and ligers when the father is a lion. But the existence of these creatures proves nothing, except perhaps that lions and tigers are genetically closer to each other than they are to other kinds of big cats. The still unanswered question is, do lions and tigers hybridize freely where they meet under natural conditions?

Today the two species do not meet in the wild, having been driven back by the expansion of human populations into different corners of the Old World. Lions occur in Africa south of the Sahara and in one small population in the Gir Forest of northwestern India. Tigers live in small, mostly endangered populations from Sumatra north through India to southeastern Siberia. In India, no tigers are found near the Gir Forest. It would seem at first that the test of the biological-species concept, free interbreeding in nature, cannot be applied. But this is not so: during historical times the two big cats overlapped across a large part of the Middle East and India. To learn what happened in these earlier days is to find the answer.

At the height of the Roman Empire, when North Africa was covered by fertile savannas—and it was possible to travel from Carthage to Alexandria in the shade of trees—expeditions of soldiers armed with net and spear captured lions for display in zoos and in colosseum spectacles. A few centuries earlier, lions were still abundant in southeastern Europe and the Middle East. They preyed on humans in the forests of Attica while being hunted themselves for sport by Assyrian kings. From these outliers they ranged eastward to India, where they still thrived during British rule in the nineteenth century. Tigers ranged in turn from northern Iran eastward across India, thence north to Korea and Siberia and south to Bali. To the best of our knowledge, no tiglons or ligers were recorded from the zone of overlap. This absence is especially notable in the case of India, where under the British Raj trophies were hunted and records of game animals kept for more than a century.

We have a good idea why the two species of big cats, despite their historical proximity, failed to hybridize in nature. First, they liked different habitats. Lions stayed mostly in open savanna and grassland and tigers in forests, although the segregation was far from perfect. Second, their behavior was and is radically different in ways that count for the choice of mates.

Lions are the only social cats. They live in prides, whose enduring centers are closely bonded females and their young. Upon
maturing, males leave their birth pride and join other groups,
often as pairs of brothers. The adult males and females hunt
together, with the females taking the lead role. Tigers, like all
other cat species except lions, are solitary. The males produce
a different urinary scent from that of lions to mark their territories and approach one another and the females only briefly
during the breeding season. In short, there appears to have
been little opportunity for adults of the two species to meet
and bond long enough to produce offspring.

Each biological species is a closed gene pool, an assemblage
of organisms that does not exchange genes with other species. Thus insulated, it evolves diagnostic hereditary traits
and comes to occupy a unique geographic range. Within the
species, particular individuals and their descendants cannot
diverge very far from others because they must reproduce sexually, mingling their genes with those of other families. Over
many generations all families belonging to the same biological
species are by definition tied together. Linked as one by the
chains of ancestry and future descent, they all evolve in the
same general direction.

The biological-species concept works best if used in a single
locality, such as a state or county or small island, over a short
period of time. Consider any group of organisms in such a
place and time. Select one haphazardly: say the hawks of Harris
County, Texas. Walk through the remaining natural habitats
around the city of Houston looking for accipiters, buzzard
hawks, harriers, ospreys, and falcons, and you will eventually
find sixteen species. Some, such as the red-shouldered hawk
(*Buteo lineatus*) and kestrel (*Falco sparverius*), are relatively
common. Others, Harlan's hawk (*Buteo harlani*) and the prairie falcon (*Falco mexicanus*), are rare. In the end, after enough
visits to fields, scrub-pine woods, and timbered swamps, you
will have compiled the same list as other veteran birdwatchers,
and your characterizations will coincide with those in Roger
Tory Peterson's *Field Guide to the Birds of Texas and Adjacent
States*. Each hawk species possesses a diagnostic combination
of anatomical traits, call, favored prey, flight pattern, and
geographical range. Some of these qualities, such as mating

behavior, can be seen to contribute to the reproductive isolation of the sixteen species. Hybrids in nature are next to none.

It might immediately have come to your mind that the general agreement on hawk species is only a cultural artifact, a convention about anatomy and scientific names that arose in the same fashion as the autonomous evolution of common law, from intuition and historical accident—dependent on who first used plumage color to classify types, who first applied a Latin name to some recognizable form or other, and so on until a classification emerged with which a sufficient number of people were able to feel comfortable and, finally, to which Roger Tory Peterson gave his imprimatur. You would be wrong. There is a test that can distinguish cultural artifacts from natural units: the comparison of classifications made by human societies that have never been in contact. In 1928 the great ornithologist Ernst Mayr traveled as a young man to the remote Arfak Mountains of New Guinea to make the first thorough collection of birds, including hawks. Before departing, he visited key bird collections already deposited in European museums. By studying specimens gathered from western New Guinea, he estimated that a little more than one hundred bird species would be likely to occur in the Arfak Mountains. His species concept was that of a European scientist looking at dead birds, who then sorts the specimens in piles according to their anatomy, as a bankteller stacks nickels, dimes, and quarters. Once settled in camp, after a long and hazardous trek, Mayr hired native hunters to help him collect all the birds of the region. As the hunters brought in each specimen, he recorded the name they used in their own classification. In the end he found that the Arfak people recognized 136 bird species, no more, no less, and that their species matched almost perfectly those distinguished by the European museum biologists. The only exception was a single pair of closely similar species that Mayr, the trained scientist, was able to separate but that Arfak mountain people, although practiced hunters, lumped together.

Many years later, when I was twenty-five years old, about Mayr's age at the time of his Arfak adventure, I made a long trek through the Saruwaget Mountains of northeastern New Guinea to collect ants. I repeated the cross-cultural test and found that the Saruwaget people could not tell one ant from another. An ant was an ant was an ant. This should have come as no surprise. It was not that Saruwaget ants and natives failed the test, only that Papuans have no practical need to classify ants. The Arfak people are hunters who use their knowledge of bird diversity to make a living, just as European ornithologists do. In Mayr's time at least, wild birds were their principal source of meat.

To the same end, the Amerindian tribes of the Amazon and Orinoco basins have an intimate knowledge of the plants of the rain forest. A few shamans and tribal elders are able to put names on a thousand or more species of plants. Not only do the botanists of Europe and North America generally agree with these species distinctions, but they have learned a great deal from their Amerindian colleagues about the habitat preferences, flowering seasons, and practical uses of the different plants. It is a remarkable fact that the only crop used widely by developed countries not already known to native peoples is the macadamia nut, which originated in Australia. Unfortunately, much of the indigenous knowledge is being lost as European culture continues to intrude and the last preliterate native cultures in tropical countries weaken and disappear. We are losing irretrievably what is in a real sense scientific knowledge.

In all cultures, taxonomic classification means survival. The beginning of wisdom, as the Chinese say, is calling things by their right names. Following the discovery in 1895 that malaria is carried by *Anopheles* mosquitoes, governments around the world set out to eradicate those insect vectors by draining wetlands and spraying infested areas with insecticides. In Europe the relation between the malarial agent, protozoan blood parasites of the genus *Plasmodium*, and the vector mosquito, *Anopheles maculipennis*, seemed at first inconsistent, and control efforts lacked pinpoint accuracy. In some localities the mosquito was abundant but malaria rare or absent, while in others the reverse was true. In 1934 the problem was solved. Entomologists discovered that *A. maculipennis* is not a single species but a group of at least seven. In outward appearance the adult mosquitoes seem almost identical, but in fact they are marked by a host of distinctive biological traits, some of which prevent them from hybridizing. The first such "characters"

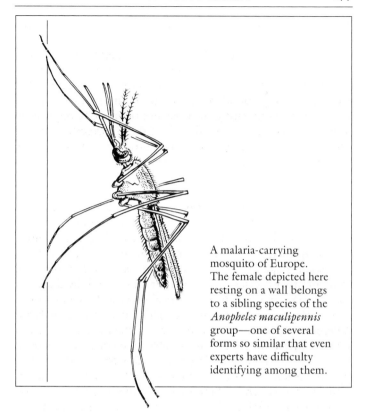

A malaria-carrying mosquito of Europe. The female depicted here resting on a wall belongs to a sibling species of the *Anopheles maculipennis* group—one of several forms so similar that even experts have difficulty identifying among them.

detected were the size and shape of the rafts of eggs laid by female mosquitoes on the water surface. Those of two species were found to produce no clusters at all, instead laying each egg separately. Entomologists were alerted, and other parts of the puzzle soon fell into place. More characters were quickly added: color of the eggs, gross structure of the chromosomes, hibernation versus continuous breeding in winter, and geographical distribution. Most important of all, some of the species distinguished by these traits were found to feed on human blood and thus to carry the malarial parasites. Once identified, the dangerous members of the *A. maculipennis* complex could be targeted and eradicated. Malaria virtually disappeared from Europe.

Systematists are often able to solve biological problems in this manner by breaking species into characters of sibling species. In the reverse direction, they also frequently lump other forms previously considered good species into larger, variable species by demonstrating the existence of only one population of freely breeding individuals. When done correctly, splitting and lumping open the door to a secure interpretation of the organisms on which the analysis was performed.

Still, the biological-species concept has chronic deep problems. From the beginning of its first clear formulation at the turn of the century, it has been corroded by exceptions and ambiguities. The fundamental reason is that each species defined as a reproductively isolated population or set of populations is in some stage or other of evolution that makes it different from all other species. It is moreover a unique individual, not merely one unit in a class of identical units such as a hydrogen atom or a molecule of benzene. This qualification makes it different from a concept in physics or chemistry, which is a summary term for a set of measurable quantities. An electron, for example, is a postulated unit with 4.8×10^{-10} units of negative charge and 9.1×10^{-28} grams of mass. Of course no one has actually seen electrons, but physicists believe in them because the properties attributed to them permit a precise explanation of cathode rays, electromagnets, the photoelectric effect, electricity, and chemical bonds. A large part of physics and chemistry depends on a precise visualization of the way electrons are stripped from atoms and molecules to create positive ions and free electrons. In the language of physics, they are virtual; they have undoubted corporeal existence. At Cambridge University in the 1930s Lord Rutherford and his research group sang of these invisible bodies at the annual Cavendish dinner, to the tune of the popular song "My Darling Clementine":

> There the atoms in their glory,
> Ionize and recombine.
> Oh my darlings, oh my darlings,
> Oh my darlings, ions mine.

But all the members of a given class are identical, and the class is forever absolute and unalterable.

If an electron is truly an electron and an ion is an ion, and all in a class are interchangeable, a species is a thing-unto-itself that merely shares certain properties with most other species, most of the time. For species are always evolving, which means that each one perpetually changes in relation to other species. In some cases sibling species are so similar that only biochemical tests or mating experiments can tell them apart, bringing despair to practicing biologists who have to sort organisms quickly. In the eastern United States slipper animalcules, the familiar little protozoans of the genus *Paramecium* widely used in high-school biology classes, used to be separated into three common "species," *P. aurelia*, *P. bursaria*, and *P. caudatum*, on the basis of anatomical differences that can be readily seen under the light microscope. Close study, however, has disclosed that no fewer than 20 species exist, separable at the least by their mating behavior and hence constituting independently evolving populations. There is a strong temptation just to ignore the biological complexity and stay with the three old, easy species, but the malaria example counsels otherwise. Biologists know in their hearts that there can be no compromise on matters of such importance, that they must soldier on until all the true closed-gene pools have been defined. Every one of the atomic units must be given a name.

Sibling species present no more than technical problems. They threaten no crisis for biological theory. More serious conceptual problems are created by "semispecies," populations that partially interbreed—not enough to constitute one big freely interbreeding gene pool, but enough to produce a good many hybrids under natural conditions. The problem is acute in many plants, especially in those that are pollinated by the wind so that pollen is broadcast scattershot and often settles on flowers of the wrong species. Along the Pacific coast of North America, about a third of the oak and pine species are actually semispecies. Yet somehow the semispecies stay apart as breeding systems. They are recognizable as discrete entities in the field even while exchanging genes by occasional hybridization. They can be distinguished by the anatomy of their foliage and flowers and the habitats in which they preferentially grow. Alan

Whittemore and Barbara Schaal, after completing a study of
DNA differences among the native white oaks of the eastern
United States, concluded:

> The genus *Quercus* (the oaks) is outstanding for the very poor
> development of sterility barriers between its species. Oak spe-
> cies are interfertile in many combinations, and natural hybrids
> may be formed between pairs of species that are very different
> from one another both morphologically and physiologically.
> Although some pairs of interfertile species show strong eco-
> logical separation, many interfertile species pairs show extensive
> ecological overlap.

Yet somehow the white oaks remain distinct. Hybridization
between the species is much less frequent than breeding within
them, and as a result the gene pools stay partly closed.

It is further true that an abundance of hybrids, and with it
the maintenance of semispecies in a state of ambiguous tension, may not be a global phenomenon in plants. Tropical
species appear to exchange genes less extensively than those
in the temperate zones. In other words, they "behave" more
like animals in the maintenance of a stricter pattern of species
diversity. Since the great majority of plant species occur in the
tropics, this evolutionary conservatism may prove to be a more
general botanical trait than hybridization at the intensity displayed by oaks. A conspicuous exception is the large tree genus
Erythrina, whose species commonly hybridize. But the genetic
study of hybridization and species formation in tropical plants
has hardly begun, and caution is advised.

It is the nature of species formation, to be explained in the
next chapter, that for a while after the splitting of a single
species into two species—call them A and B—some members
of species A can be more closely related to some members of
species B than they are to other As, and vice versa. These relatives in A and B have a close common ancestry, but they have
acquired one or a few crucial differences that prevent them
from exchanging genes. They are like sisters who live in different countries and are unable to cross the national boundary. Some biologists have argued that such individuals should
be put in the same species, a single "phylogenetic species,"

regardless of their inability to interbreed. Other biologists, and I am one of them, firmly disagree. The idea of the phylogenetic species is an interesting and useful but not fatal challenge to the biological species. To seek genealogical information within and between populations does not require the overthrow of reproductive isolation as the salient process of diversification at the population level. Sisterhoods are important, but nations more so.

We must now confront an even more serious conceptual difficulty for the biological-species concept. The idea of the closed-gene pool has no meaning at all for the minority of organisms that are either obligatorily hermaphroditic—those that have both ovaries and testes and fertilize themselves—or parthenogenetic, producing offspring from unfertilized eggs. By one or the other of these devices, various microorganisms, fungi, plants, mites, tardigrades, crustaceans, insects, and even lizards simply forgo the inconvenience and perils and gambler's excitement of sexual reproduction.

How to solve the dilemma? Asexual and self-fertilizing forms tend to maintain a remarkable integrity. The vast majority, even though freed from the evolutionary constraints of sexual compatibility, do not vary in all directions, do not fan out to create wide continuous variation and taxonomic confusion. The gene combinations of the organisms are prone to exist in clusters, enabling systematists to place most specimens with ease. It is widely believed that the clustering is due to the lower survival and reproduction rate of vagrant intermediate forms. Only those organisms with anatomy and behavior close to the norm are able to do well. Also, many of the nonsexual species have recently evolved from sexually reproducing ancestors and therefore have not had enough time to diverge or expand. In the end, however, the lines drawn by the biologist around such species must be arbitrary.

The closed-pool concept also loses its meaning in the case of chronospecies, which are stages in the evolution of the same species through time. Consider our own species, *Homo sapiens*, which evolved in a straight line from *Homo erectus*, found through Africa and Eurasia roughly a million years ago. Obviously we cannot know whether *H. sapiens* and *H. erectus* would interbreed freely if brought together in nature. The question is hollow when presented away from context; it is a scientist's koan, the equivalent of the sound of one hand clapping, the length of a string. Yet paleontologists, driven by practical necessity, go on distinguishing and naming chronospecies. They are right to do so. It would be irresponsible to call *H. sapiens* and *H. erectus* the same species, and even more so to add their immediate antecedent *H. habilis* and—still further back—the primitive australopithecine man-apes.

Searching for an anchor, willing to compromise in order to find some process shared by a large fraction of organisms, biologists keep returning to the biological-species concept. In spite of its difficulties, regardless of the fact that it can never be employed as an abstract entity like the electron to make exact quantitative calculations, the concept is likely to continue to hold center stage for the simple reason that it works well enough in enough studies on most kinds of organisms, most of the time.

The vast majority of species are in fact sexual; they do exist as closed-gene pools. The biological-species concept works very well in the study of local faunas and floras, such as the hawks of Texas and the mosquitoes of Europe and the primates of the Old World including *Homo sapiens*, and maximally so in well-demarcated communities on islands and isolated habitat patches, circumstances, in short, found in a large part of the real world.

For years I have suffered through seminars and hallway debates on the biological species. I have read through a library of opinions and witnessed the concept wax and wane in the minds of evolutionary biologists. The core problem seems to be the democratic process of science, in the sense that the concept has a weak constituency: people really don't need it most of the time. Systematists perform their sorting largely on the basis of differences they find between museum specimens. Asked if the differences are maintained by reproductive isolation, they respond: probably. But they don't care enough to go to the trenches. They are satisfied that the anatomical gaps exist and leave it to population biologists to find out why this is so. Population biologists for their part are enchanted by the dynamism of the speciation process and the many problems thrown up to embarrass the biological-species concept in the early stages of species separation. What is wrong, many ask, with disorder or even chaos, for a while? Why not play around with many species concepts, each designed to fit ad hoc circumstances? Content with the thrill of the race they are running, and with scattered applause along the way, few see profit in sprinting to the finish line.

Unlike the systematists, population biologists do not have to classify a million species. And they forget that reproductive isolation between breeding populations is the point of no return in the creation of biological diversity. During the earliest stages of divergence, there may be less difference between the two species than exists as variation within them. A surge of hybrids may yet occur to erase the barrier and confuse the picture even more. But in most cases the two species are embarked on an endless journey that will carry them further and further apart. The differences between them will in time far exceed anything possible among the members of their own breeding populations. In the real world, the great range of biological diversity has been generated by the divergence of species that were created in turn by the defining step expressed in the biological-species concept.

Someday biologists may come up with a single concept that unifies sexual species, nonsexual species, and chronospecies into a single, theoretically powerful natural unit. But I doubt it. The dynamism of the evolutionary process and the individuality of species make it unlikely that a completely universal species definition will ever be fashioned. Instead, two to several concepts will continue to be recognized, like the waves and particles of physics, as optimal in different circumstances. Of these, the biological species is likely to remain central to the explanation of global diversity. But whatever the outcome, the imperfections of the concept, and thereby of our classificatory system, reflect the idiosyncratic essence of biological diversity. They give even more reason to cherish each species in turn as a world unto itself, worthy of lifetimes of study.

CHAPTER FIVE

New Species

W HAT IS THE ORIGIN of biological diversity? This profoundly important problem can be most quickly solved
by recognizing that evolution creates two patterns across time
and space. Think of a butterfly species with blue wings as it
evolves into another species with purple wings. Evolution has
occurred but leaves only one kind of butterfly. Now think of
another butterfly species, also with blue wings. In the course of
its evolution it splits into three species, bearing purple, red, and
yellow wings respectively. The two patterns of evolution are
vertical change in the original population and speciation, which
is vertical change plus the splitting of the original population
into multiple races or species. The first blue butterfly experienced pure vertical change without speciation. The second blue
butterfly experienced pure vertical change plus speciation. Speciation requires vertical evolution, but vertical evolution does
not require speciation. The origin of most biological diversity,
in a phrase, is a side product of evolution.

Vertical change is mostly what Darwin had in mind when he
published his 1859 masterwork. The full title tells the story: *On
the Origin of Species by Means of Natural Selection, or the Preservation of Favoured Races in the Struggle for Life*. In essence,
Darwin said that certain hereditary types within a species
(the "favoured races") survive at the expense of others and
in so doing transform the makeup of the entire species across
generations. A species can be altered so extensively by natural
selection as to be changed into a different species, said Darwin.
Yet no matter how much time elapses, no matter how much
change occurs, only one species remains. In order to create
diversity beyond mere variation among the competing organisms, the species must split into two or more species during the
course of vertical evolution.

Darwin understood in a general way the difference between
vertical evolution and the splitting of species, but he lacked a
biological-species concept based on reproductive isolation. As

a result he did not discover the process by which multiplication occurs. His thinking on diversity remained fuzzy. In that sense, the short title *On the Origin of Species* is misleading.

The distinction between the two patterns of evolution is illustrated in concrete detail by human evolution. The earliest known hominid in the fossil record is the man-ape *Australopithecus afarensis*. A hominid, to be technically explicit, is a member of the family Hominidae, which includes modern *Homo sapiens* and earlier human and humanlike species. When *A. afarensis* lived in the savannas and woodlands of Africa five to three million years ago, it was, so far as the evidence shows, the only species of its kind. The scant fossils reveal that adults walked in a bipedal manner broadly similar to that of *Homo sapiens*. This posture was and is unique among all the mammals. It freed the man-apes to carry burdens in their arms and hands. They or their descendants were able to transport infants, tools, and food for long distances. Perhaps they established campsites (although evidence so far back is lacking), and from that habit divided labor among those who stayed at home to tend the camp and those who foraged abroad. In brains *A. afarensis* was not conspicuously endowed. Its cranial capacity was no greater than that of modern chimpanzees, about 400 cubic centimeters. But the stage was set for the evolutionary advance toward humankind.

As time passed, to no later than two million years ago, the early man-ape populations both evolved and split into at least three distinct species. Two of the species, the advanced man-apes *Australopithecus boisei* and *Australopithecus robustus*, were 1.5 meters (five feet) tall, with a gorilla-like bony crest down the midline of the skull that anchored enormous jaw muscles. They were probably vegetarians, using molar teeth up to 2 centimeters long to crush seeds and shred tough vegetation, much as gorillas do today. The third species derived from the primitive man-apes, *Homo habilis*, was more nearly a true human in the contemporary sense, enough so for anthropologists to remove it from the genus *Australopithecus* and place it in *Homo*. It stood just under 1.5 meters and weighed about 45 kilograms (100 pounds). Its form was essentially that of modern *H. sapiens*, with one outstanding exception: the brain volume was between 600 and 800 cubic centimeters, still only half as much as in modern *H. sapiens* but considerably larger than in chimpanzees.

During the next million years the man-apes disappeared, and with them most hominid diversity. *H. habilis* survived and continued to evolve in total size and cranial capacity, slowly at first and then at an accelerating rate. It metamorphosed into the intermediate species *Homo erectus*, reaching that level, or "evolutionary grade" as biologists like to call it, about 1.5 million years ago. Sometime during its early history *H. erectus* expanded its range from Africa to Europe and Asia.

Fossils found at Zhoukoudian, near Beijing, China, bear witness to the ensuing evolution over a quarter million years. The brain size rose steadily from 915 to 1,140 cubic centimeters, while the stone tools designed by these "Peking men" grew increasingly sophisticated. Across the far-flung *H. erectus* populations of the Old World, the brain size and dentition progressed to essentially modern levels, by half a million years before the present. The chronospecies *H. sapiens*, or the modern species derived from the archaic species by vertical evolution, had arrived. Its taxonomic diagnosis is extraordinary: brain 3.2 times larger than in an ape of human size, housed in a wobbly spherical skull; jaw and teeth feeble; body borne erect on elongated hindlegs; skin mostly hairless except for patches that warm the head and display the genitalia; internal organs supported by a basin-shaped pelvis; thumb abnormally long for a primate, turning the hand into a specialized device for handling tools; mind fashioned from symbolic language and semantic memory with the aid of elaborate speech-control centers located in the parietal cortex.

Now to recapitulate human evolution as a display of the two patterns of evolution: the hominids experienced a modest amount of speciation during the early man-ape period, followed by extinction of all the lines except the rapidly evolving single species of *Homo*. The ancestral *Australopithecus afarensis* probably had a generalized diet. The hominids that succeeded it combined vertical evolution and full species formation enforced by reproductive isolation. The species deployed into different niches in the manner typical of successful, expanding animal groups. The man-apes *Australopithecus boisei* and *A.*

robustus became increasingly vegetarian. *Homo habilis*, which had already diverged enough to be placed in a genus of its own, added meat to its diet by means of both hunting and scavenging. At the same time, it evidently consumed enough vegetable material to retain what today would be called a balanced diet. Then the man-apes disappeared, possibly driven to extinction by *H. habilis* or its descendant, *H. erectus*. What followed was mostly the evolution of a single species, traversing the series *H. habilis* to *H. erectus* to *H. sapiens*.

Most animal and plant species, including *H. sapiens*, retain their identity simply by not breeding with other species. How does this segregation come about in the first place? The process as we understand it is surprisingly simple. Any evolutionary change whatsoever that reduces the chances of producing a fertile hybrid can yield a new species. The reason is that the launching of fertile hybrids is a complicated and delicate procedure. It is somewhat like putting a space vehicle into orbit. A vast number of parts must work correctly, and the timing of their operation must be nearly perfect. Otherwise the mission fails. Consider a male of species A and a female of species B trying to create a fertile hybrid offspring. Because they are genetically different from one another, things can go wrong. The two individuals might want to mate in different places. They might try to breed at different seasons or times of the day. Their courtship signals could be mutually incomprehensible. And even if the representatives of the two species actually mate, their offspring might fail to reach maturity or, attaining maturity, turn out to be sterile. The wonder is not that hybridization fails but that it ever works. *The origin of species is therefore simply the evolution of some difference—any difference at all—that prevents the production of fertile hybrids between populations under natural conditions.*

Biologists speak weightily of all the things that fail as "intrinsic isolating mechanisms." By intrinsic they mean hereditary, in other words differences prescribed by the genes of the opposing populations. They do not mean something extrinsic, such as a river or mountain range, that keeps populations A and B apart. Step by step through the reproductive process, no intrinsic isolating mechanism must intrude if two populations are to remain in the same species. Step by step, the appearance of even one such mechanism will cleave them into two distinct species. Distinct, that is, if you accept the biological-species concept, which you must do if we are to avoid chaos in general discussions of evolution.

By all means let us avoid chaos. Take any set of sexually reproducing species that live together in the same geographical region. They are reproductively isolated from one another by their own distinctive isolating mechanisms. This is just a formal way of saying that some hereditary difference between the species, examined pair by pair, prevents them from producing large numbers of fertile hybrids.

Consider flycatchers of the genus *Empidonax*, little birds that perch on tree limbs or powerlines and dart out from time to time to snatch flying insects. Five species occur together in the northern United States. They remain genetically distinct in part because they prefer different habitats. For example:

Least flycatcher (*E. minimus*), open woods and farmland

Alder flycatcher (*E. alnorum*), alder swamps and wet thickets

Yellow-bellied flycatcher (*E. flaviventris*), coniferous woods and cold bogs

In addition, each species uses its own identifying call during the breeding season, so distinctive in combination with habitat choice as to leave little room for mistakes and the creation of hybrids.

The possibility for error has no limit, and so intrinsic isolating mechanisms are endless in variety. Examples worked out by field researchers are not merely the substance of academic biology. They also explain a great deal of wondrous and otherwise indecipherable natural history. Some examples:

• The giant silkworm moths of North America (family Saturniidae) fly and mate at various times during late afternoon and through the night. The females call in the males over distances of up to several kilometers by releasing a powerful chemical scent. They pop out an eversible sac folded into the rear tips of their bodies, exposing the attractant to the air and allowing it to evaporate and disperse into a downwind plume. The males are extremely responsive to the sex attractant. When they detect only a few molecules, they fly upwind, bringing them close to the females. Each species of giant silkworm moth is sexually active only during a limited span of time in each daily cycle, as follows:

Females of promethea moths (*Callosamia promethea*) call from about 1600 to 1800 hours

Females of polyphemus moths (*Antheraea polyphemus*) call from about 2200 to 0400 hours

Females of cecropia moths (*Hyalophora cecropia*) call from about 0300 to 0400 hours

And so on through the 69 species of North American Saturniidae, each of which so far as we know has its own *heure d'amour*. Agitated in flight, the males select females of their own species not only by the time of the calls but by the chemical composition of the sex attractants. Few if any mistakes are made thanks to the combination of timing and scent, and thus few if any hybrids come into being.

• Male jumping spiders (family Salticidae) recognize females of their own species by sight. The courting male faces the female. His visage, which is strikingly colored according to species, is instantly recognizable to the sharp-eyed female spider and to human observers alike. In the woods and fields of New England lives a species with a red brow, white face, and black fangs. Another can be found bearing gray brow, red face, and white fangs; still another with black brow and face and snowy white hair that envelop the fangs like ermine muffs; and yet another with tufts of hair elevated behind the head like black-speckled fairy wings. The males posture and dance before the females. Those of one species supplement the yellow and black vertical striping of their fangs by raising black-tipped yellow forelegs above their heads in a gesture resembling surrender. Those of others variously bob up and down, weave side to side, bend their abdomens over their heads or twist them to the side, or raise their forelegs and wave them side to side like semaphore flags. When biologists give females a choice between males, the spiders use the colors and movements to select mates of their own species.

• The heliconias (family Heliconiaceae) are mostly tropical American plants pollinated by hummingbirds, which they attract with huge flower-like structures called bracts and copious offerings of nectar. Hummingbirds respond to this extravagant generosity by carrying the pollen of the heliconias with speed and efficiency. They serve the plants well, but there is a problem: the birds like to visit more than one heliconia species, creating a risk of hybridization. The heliconias have solved the problem by the evolution of flower parts of differing length. Each species plants its pollen on a particular zone of the hummingbird's body. The hummingbird in turn deposits the pollen only on the stigmas of flowers of the same length, in other words only on flowers belonging to the same species.

It is possible to proceed down the catalogue of evolved procedures known to biologists by which species avoid hybridization and seldom see one repeated in exact detail. Many of the most elaborate and beautiful displays of nature function as intrinsic isolating mechanisms, from bright colors to beguiling odors and melodious songs.

But wait: I have been speaking of the origin of species in paradoxical language. In the traditional language of biology, the "mechanisms" have "functions." Yet they represent whatever can go wrong, not what can go right. In other words, beauty arises from error. How can both of these apparently contradictory perceptions be true? The answer, based on studies of many populations in the wild, is this: *the differences between species ordinarily originate as traits that adapt them to the environment, not as devices for reproductive isolation*. The adaptations may also serve as intrinsic isolating mechanisms, but the result is accidental. Speciation is a by-product of vertical evolution.

To see why this strange relationship holds, consider the special but widespread mode of diversification called *geographical speciation*. Start with an imaginary population of birds—say, flycatchers—that was split by the last glacial advance in North America. Over several thousand years, the population living in what would today be the southwestern United States adapted to life in open woodland, while the other population, in the southeastern United States, adapted to life in swamp forests. These differences were independently acquired and functional. They allowed the birds to survive and reproduce better in the habitats most readily available to them south of the glacial front. With the retreat of the ice, the two populations expanded their ranges until they met and intermingled across the northern states. One now breeds in open woodland, the other in swamps. The differences in their preferred habitats, based on hereditary differences acquired during the period of enforced geographical separation, make it less likely that the two newly evolved populations will closely associate during the breeding season and hybridize. The adaptive difference in habitat thus accidentally came to serve as an isolating mechanism.

Other traits might well have diverged between two such bird populations during their period of geographical separation, including songs used by the males to attract females and places within the forest favored for the construction of nests. Any of these hereditary differences might lessen the chances that adults of the woodland and swampland species mate. If mating does occur, the hybrids will be intermediate in the newly evolved traits. They will not be well suited to either open woodland or swampland, and hence have less chance to survive. Any kind of barrier, from habitat differences to maladapted hybrids, will work as an intrinsic barrier if it is strong enough. The two populations have turned into distinct species because they are reproductively isolated where they meet under natural conditions. The single ancestral, pre-glaciation species has been split into two species, an entirely incidental result of the vertical evolution of its populations while they were separated by a geographical barrier.

The fission of the flycatcher species is a simplified composite of real speciation events that occurred in North America during the last glacial advance. Something like it has been repeated in many parts of the world, among many kinds of plants and animals, over hundreds to thousands of years. The process is initiated by geographical barriers that rise and fall, promoting the origin of a random array of hereditary isolating mechanisms sufficient to keep newly formed species from interbreeding when they do come in contact. Evolutionary

A species multiplies into two or more daughter species when its populations are isolated from one another long enough by geographical barriers. In this example, based on a composite of real cases, the parental species is at first widely distributed across terrain composed mostly of grassland (*top*). The local climate grows wetter, and a river splits the species in two, so that now one population lives in grassland and the other in woodland (*bottom*).

In time the two populations evolve apart, until they attain the level of new species (*top*). When the river barrier disappears, the two forms are able to live together without interbreeding (*bottom*).

biologists have discovered an extreme diversity of such barriers, as seen in these examples:

• In the Amazon Basin, droughts break up continuous forest into scattered patches. Some of the plant and animal populations newly isolated within them start to diverge. Other populations are fragmented by shifts in river courses that repeatedly open and close rain-forest corridors linking the patches.

• Along the coast of New Guinea, stretches of the continental shelf are partially submerged as the sea level rises. Those parts with the highest elevation remain above water as islands, and populations restricted to them start to diverge.

• The Hawaiian islands are colonized by birds, crickets, wasps, damselflies, beetles, snails, flowering plants, and other kinds of organisms arriving as occasional windblown waifs. As the first colonists multiply and spread, they evolve in response to the distinctive environment of the islands and hence diverge from the ancestral populations left behind on the mainlands of North America and Asia. They also spread from island to island within the archipelago and from valley to valley and along the ridge tops on each island's mountainous interior, generating new isolated and diverging populations as they go. A single species that colonized Hawaii 100,000 years ago could easily have given rise to hundreds of species, each endemic at the present time to a particular island, valley, or mountain ridge.

I have stressed the imperfections of the species as a natural unit. The flaws that plague it are the inevitable consequence of the particularities of history. Every population of animals exists in a highly dynamic state, growing in size if it can, thrusting into new terrain when permitted, evolving in new directions as opportunity arises. Raw chance weighs heavily in guiding its evolutionary trajectory.

Consider a biologically diverse environment such as a forested valley on Kauai, a shallow-water shelf along the shores of Lake Victoria, or a cypress swamp in northern Florida. Some of the resident species are individually specialized for narrow niches and limited to small geographical ranges. Their dispersal ability is poor, their close relatives few or nonexistent. Their vertical evolution creeps along, and speciation is stalled. They have no geographic races and little prospect for multiplication. At the opposite extreme, other species possess flexible food habits and are excellent dispersers. They form new populations readily and evolve quickly into new niches, from diet to habitat to season of activity. Their potential for diversification is high; and they pile up species in the same localities by repeated cycles of dispersal and reinvasion.

In focusing on this last group of actively evolving populations, we are most likely to encounter all the stages of geographical speciation as interpreted by prevailing theory. At the earliest stage, the population is spread continuously across its range, and all the organisms in it freely interbreed. Few if any differences occur from one end of the range to the other. At the next stage, the population is still continuously distributed but divided into subspecies. Where the subspecies meet, they freely crossbreed. Imagine such a race of butterflies in Texas with large spots on their wings, and another in Mississippi lacking spots. Where the two subspecies meet in Louisiana and interbreed, the butterflies have spots of intermediate size. Those near Texas have larger spots, tending toward full Texas size; those near Mississippi have very small spots, tending toward the unadorned Mississippi condition.

Time passes, and at a more advanced stage the subspecies are still able to interbreed freely if they meet, but by now they have diverged in many genetic qualities. Butterflies from the same populations might differ not only in wing pattern but also in size, preferred food plant, growth rate of the caterpillars, and so on through any combination of hundreds of traits subject to genetic variation. The divergence of the subspecies will be hastened if some physical barrier, such as a broad river or dry grassland corridor, separates the two populations and restricts the flow of genes between them.

Finally, the two populations have diverged so far that they do not interbreed when they meet. They have been transformed into full-fledged biological species. Our two butterfly species now coexist in Louisiana, kept apart by differences in breeding season, courtship behavior, or some other intrinsic isolating mechanism singly or in combination—an inherited failure, in other words, to mesh their reproductive activity. Few if any small-spotted hybrids occur in the zone of overlap.

This classical model of geographical speciation makes a tidy picture. It has a core of truth, but real evolution is much

messier. In fact, evolution is so messy that a faithful description
of real cases converts the science into natural history, in which
unique details are as important as the principles by which they
are explained.

Consider the subspecies. The category seems an inevitable
intermediate step in an Aristotelian progression running
from no subspecies to subspecies to species. What exactly is
a subspecies? The textbooks define it as a geographical race, a
population with distinctive traits occupying part of the range
of the species.

What then is a population? We are in immediate trouble. It is
easy to say that a clearly defined population, one recognizable
by everyone at a glance, occupies an exclusive part of the range
of a species. And geneticists like to add, for purposes of mathematical clarity but not as an absolute requirement, that the
population is a "deme": its members interbreed at random, and
any member is equally likely to mate with any other member in
the population, regardless of its location.

Few such objectively definable populations exist in nature.
Most that do look like textbook examples are endangered species, with so few organisms left that there is no doubt as to the
boundaries of the population they compose. The last surviving ivory-billed woodpeckers, found in one mountain forest
of eastern Cuba, belong to this parlous category. So do the
Devil's Hole pupfish, barely hanging on in a tiny desert spring
at Ash Meadows, Nevada. You can stand at the entrance to
Devil's Hole, look down 15 meters to where the water laps over
a sunlit ledge, and see the entire species swimming around like
goldfish in a bowl.

Most species are not confined this stringently, which is fortunate for conservation and unfortunate for textbook theory. Take
the redbacked salamander (*Plethodon cinereus*), one of the most
widespread and abundant salamanders of North America. The
species ranges from Nova Scotia and Ontario south to Georgia
and Louisiana. Redbacks occur almost continuously through
the northern three quarters of this range. It is tempting to
classify the whole northern ensemble as one huge population.
Salamander taxonomists do just that and call it a subspecies,
Plethodon cinereus cinereus (a formal rule of nomenclature: to

designate a subspecies, add a third name). But redbacks are far from continuously distributed. They are largely confined to moist lowland forest, which is not a continuous habitat but an irregularly broken filigree laid on the land. Even within habitable forest, the population is divided into local aggregates that slowly expand, contract, and reform into new configurations across the generations. The rate of interbreeding among local demes in forested valleys and woodlots is unresearched and unknown. In short, with more data biologists might be able to distinguish thousands of populations across the vast range of *P. cinereus cinereus.* A diligent taxonomist might legitimately break the one formally recognized subspecies into large numbers of subspecies with smaller ranges.

To the south, in the mountains of northern Georgia and Alabama, there is another generally recognized subspecies,

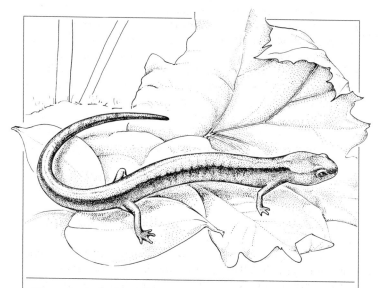

The redbacked salamander (*Plethodon cinereus*) of eastern North America, a widespread species with a pattern of racial variation that is typically ambiguous, such that subspecies can be defined only by subjective criteria.

Plethodon cinereus polycentratus, separated from *P. cinereus cinereus* by 80 kilometers of redback-free terrain. A third subspecies, *P. cinereus serratus,* occurs in several widely separated localities in the hill country of Arkansas, Oklahoma, and Louisiana. These two additional races offer the same difficulties as the main northern subspecies. Their triple names are a convenient shorthand, the statement of a rough truth. The classification works so long as we recognize that dicing up the whole species geographically is imprecise and to a large degree arbitrary. Depending on the criteria used, there could be one subspecies of *P. cinereus,* or there could be hundreds.

An even more fundamental difficulty of subspecies is the discordance of the traits by which the subspecies are defined. Suppose that we agreed to ignore the population problem for the moment. Imagine that easily definable populations exist in an idealized species (which I will continue to call the redback salamander for clarity). One species comprises thousands of small populations across North America. Individuals from the *southern* half, Georgia to Virginia, have stripes over most of the body; those from the *northern* half, Maryland to Canada, lack stripes. On the basis of this one character, there are two subspecies, two geographic races: striped southern redbacks and plain northern redbacks. We notice, however, that *western* individuals of the species are larger. These two characters, stripedness and size, are obviously discordant—they break along different geographic lines. They can be used to define four subspecies: big striped in the *southwest,* little striped in the *southeast,* little unstriped in the *northeast,* and big unstriped in the *northwest.* Next we find that the eyes of juveniles are amber southwest of a line running from the Great Lakes to Georgia and yellow to the northeast of the line. Two more subspecies are added to produce a total of six overall. Looking still more closely we find . . .

Here is the point of this exercise in geometry: most traits varying geographically in a given species are discordant. They change at different places and in different directions. It follows that subspecies are recognized according to whatever traits taxonomists choose to study. It also follows that the greater the number of traits, the larger the number of the subspecies that must be recognized.

The uncertainty of the limits of populations combined with the discordance of traits means that the subspecies is an arbitrary unit of classification. That uncertainty is reflected in the confusion over human races. In past years anthropologists struggled hopelessly with attempts to define human races. Estimates of the number of races made by researchers during the 1950s ranged from six to more than sixty. The variation in numbers is due precisely to the fact that *Homo sapiens* is a typical evolving species.

Anthropologists, like biologists, have now largely abandoned the formal subspecies concept. They prefer a convenient shorthand to designate a certain part of a population with reference to one or two traits. They say, for example, "northern Asians tend to have more prominent canthal folds," knowing full well that canthal folds differ in geography from blood types, which deviate in turn from average height, lactose intolerance, Tay-Sachs syndrome, eye color, hair structure, infant passivity, and so on through hundreds of other more or less discordant traits prescribed or at least influenced by 200,000 or so human genes scattered through 46 chromosomes. The emphasis in research in anthropology and biology has passed from the description of subspecies to the analysis of the geography of separate traits and their respective contributions, singly or in combination, to survival and reproduction.

The demotion of the subspecies should carry with it a word of caution, in the service of moderation. Real populations do exist, however difficult to define. Genetic traits still vary. It may be artificial to divide up and label redback salamanders from the southern United States as subspecies, but they nevertheless differ in many genetic traits and compose a reservoir of unique genes. It is further true that some populations of widespread animals and plants are sufficiently isolated and genetically distinct to compose objective subspecies even in the abstract textbook sense. It is useful to label such populations formally as subspecies. Stephen O'Brien and Ernst Mayr, for example, have proposed guidelines to that end for use by conservation biologists and policy makers. They suggest that subspecies be defined as individuals occupying a particular part of the range of the species, with genes and natural history distinct from those of other subspecies. Members of different subspecies can freely interbreed. They can arise either as populations adapting to local conditions or as hybrids between subspecies.

The delimitation of subspecies, an occasional bureaucratic necessity when the U.S. Endangered Species Act or its equivalent is invoked, will usually be difficult and even controversial. Evolution, to repeat, is messy. The Florida panther offers a case in point. Once the panther, also known as the mountain lion or cougar, occurred throughout the southern United States. Now it is down to about fifty individuals in southern Florida, the subspecies *Felis concolor coryi*. Biochemical tests have revealed that this tiny population was derived from two stocks: the final survivors of the original North American panthers that once roamed Florida, and seven animals of mixed North American and South American origin released from captivity into the Everglades between 1957 and 1967. The present population is thus of hybrid origin, but it contains a unique ensemble of genes of partial North American origin and deserves protection as a native mammal.

The ambiguity of the subspecies as a taxonomic unit creates an interesting dilemma in evolutionary reasoning. We have before us an idealized sequence that starts with a geographically isolated population still identical to other populations of the same species. The population then evolves into a subspecies, still capable of interbreeding with the other populations, if they could somehow breach the geographical barrier and meet along the edges of their ranges. The subspecies finally evolves into a full species, meaning that if it meets the other populations it would no longer freely interbreed with them. The dilemma is this: if subspecies are usually amorphous and cannot be defined by a single objective criterion, how can such an arbitrary unit give rise to the species, which is sharply defined and objective?

The answer to the puzzle tells us a great deal about the origin of diversity. In order to spring forth as a species, a group of breeding individuals need only acquire one difference in one trait in their biology. This difference, the innate isolating mechanism, prevents them from freely interbreeding with other groups. It does not matter if the limits of the populations as a whole are poorly marked. Nor does it matter if all the other traits vary in a crazy-quilt manner across the populations that are splitting off as species. What counts is that somehow a group of individuals occupying some part of the total range evolves a different sex attractant, nuptial dance, mating season, or any other hereditary trait that prevents them from freely interbreeding with other populations. When that happens, a new species is born. The truly objective unit, the closed-gene pool of future generations, is the group of individual organisms that acquires the isolating trait. This new species can be defined by a single isolating trait. Other characters that vary geographically—hairiness, color, cold hardiness, whatever— can show any pattern of geographical variation whatever, either concordant with the isolating trait or totally different, without changing the outcome. Once segregated in this manner, the species will inevitably evolve away from other species, becoming ever more different in a steadily enlarging suite of traits as time passes.

The decisive isolating change can be further based on only a slight alteration in the genes or chromosomes. Some species of leafroller moths in the family Tortricidae, for instance, are segregated by relatively minor deviations in the chemistry of their female sex attractants. The variation they manifest from one species to the next is the kind commonly based on mutations in only one gene. The segregation can in fact be even more elementary. Some leafroller species are separated by close to the absolute conceivable minimum: a difference not in the organic structure of the substances composing the sex attractants but in the percentages of the substances, in this case a variety of acetates, that go into the blend. The females of each moth species have their own delicate bouquet that the males sample, and from this signal alone they choose either to press forward or to depart. In theory at least, it is possible for large numbers of tortricid moth species to arise quickly through minor changes in their genes that alter either the chemistry or the proportions of the sex attractants.

Geographical speciation is supplemented in nature by a rich medley of other modes of speciation. By far the best documented is polyploidy, an increase in the number of chromosomes. More precisely, a polyploid individual has twice as many chromosomes in each cell as an ordinary individual—or three times or four times or any other exact multiple. Polyploidy is virtually instantaneous in its effect, potentially isolating a group of individuals from its ancestors in one generation. This immediate isolation is caused by the inability of the hybrids of polyploid individuals and nonpolyploid individuals to develop in a normal manner or, if full development is achieved, to reproduce. Polyploidy is responsible for the origin of almost half of the living species of flowering plants and of a smaller number of animal species.

The stage for polyploid speciation is set by the passage of all sexually reproducing species through a two-part life cycle. In the haploid phase of the cycle, there is a single set of chromosomes in each cell. That is followed by the diploid phase, in which there are two sets of chromosomes in each cell. The diploid phase is ended by a reduction of the chromosomes to a single set, returning the organisms to the haploid phase, and so on. The haploid phase includes the sperm and eggs, each of which is a little organism with a single set of chromosomes. In higher plants and animals, the haploid phase consists exclusively of this ephemeral period. When a sperm and an egg unite, the chromosome number is doubled and the diploid phase begins. Two little organisms become one little organism, which is then capable of growing into a very big organism consisting of billions of diploid cells. The haploid number of chromosomes in human beings (the number in each sex cell) is 23; the diploid number, found in the remaining tissues after fertilization, is 46. If the base number is somehow tripled, thus creating a triploid organism (a triploid human being, for example, would have 69 chromosomes), the organism is marked for trouble. Difficulties are likely to be encountered during development of the embryo and beyond, into adult life. Down's syndrome is one of many defects caused by triple numbers, in this case three copies of chromosome number 21 (so called because biologists give each of the 23 chromosomes its own number for quick labeling).

When a triploid starts to manufacture sex cells, it will encounter difficulty. The two chromosomes of a normal diploid plant or animal can go through meiosis—the reduction to one chromosome per cell—with relative ease. In the first cell division of meiosis, when reduction takes place, the two related chromosomes pair up (in human beings, for example, all the chromosomes together create 23 pairs), then separate to different cells each with the haploid chromosome number. The three chromosomes of a triploid get tangled up during the pairing and separation (in this, the ultimate sense, three *is* a crowd), and either abort the process or produce large numbers of abnormal sex cells.

Triploids play the pivotal role in species isolation by polyploidy, simply because they are the unworkable products created when polyploid organisms try to breed with their diploid relatives. In full, the process of speciation by polyploidy proceeds as follows:

• Start with a polyploid plant that has newly originated within a diploid population. It is usually a tetraploid, in which the ordinary diploid number was accidentally doubled in early embryonic development. Each ordinary cell therefore has four chromosomes of each kind in each ordinary cell, rather than the usual two. As a result the tetraploid plant—the new species—places two chromosomes of a kind in each sex cell instead of the usual one.

• Suppose that plants of the ancestral species have 10 chromosomes in each ordinary cell and 5 in each sex cell. The polyploid plants have 20 and 10, respectively, in the two kinds of cell. Diploid plants of the ancestral species can interbreed with one another, and polyploid plants can interbreed with other polyploid plants.

• Some of the diploid and polyploid plants cross-breed to produce hybrids. When an ordinary sex cell (5 chromosomes) is fused with a polyploid sex cell (10 chromosomes), the hybrid is a triploid (15 chromosomes). As a result the hybrid plant may experience difficulties during growth. Even if it is able to attain sexual maturity, however, it cannot produce normal sex cells and is sterile. The diploid ancestor and polyploid derivative are reproductively isolated, and hence the polyploid derivative is a new species, created in a single generation.

There is another, even more innovative way that polyploidy can create species: by multiplying the number of chromosomes in hybrids of two species already in existence. The ordinary hybrids of many species of plants are sterile, even when they have the same number of chromosomes and growth is untroubled up to the time of flowering. The reason for the sterility is the incompatibility of the parental chromosomes during the formation of the sex cells. Let us call the two species that hybridize A and B. When a chromosome from species A attempts to line up with its counterpart from species B, a normal procedure for the exchange of blocks of genes during the production of sex cells, the A and B chromosomes differ too much from each other to complete the maneuver.

The way out of the impasse is to double the number of chromosomes in the hybrid. Then during the production of its sex cells, each A chromosome can be matched with an identical A chromosome, and each B chromosome can be matched with an identical B chromosome. Now the hybrid polyploid is fertile. It can breed with other hybrid polyploids of identical type, but not with either of the diploid parents from which it arose. Such hybridization followed by doubling does occur spontaneously in nature from time to time.

New species can be created in the laboratory and garden by joining old species in this Frankensteinian manner. The most celebrated example is the polyploid hybrid of the radish (*Raphanus sativus*) and the cabbage (*Brassica oleracea*), which are genetically similar but reproductively isolated members of the mustard family Brassicaceae. Radishes and cabbages both have 9 chromosomes in their sex cells and 18 chromosomes in their diploid tissues. Hybrid radish-cabbage plants can be readily produced by cross-fertilization. They too have 18 chromosomes in their diploid tissues, 9 from each parent. But the two sets of 9 cannot pair with each other and complete the formation of sex cells during the reduction division, a defect that renders the hybrids sterile. When the chromosome numbers of the hybrids are doubled, bringing the diploid number to 36, the plants are fertile. Now each radish chromosome and each cabbage chromosome has an exact counterpart with which it can pair, and the production of sex cells can proceed normally. The rabbage or cabbish—choose the name you like—is self-sustaining as a species. It cannot be bred back to either of the parent species.

In spite of its clear importance in the origin of plant diversity, polyploidy may not be the most prevalent rapid process for organisms as a whole. Another, possibly even more widespread mode is nonpolyploid sympatric speciation. The term *sympatric speciation* refers to the origin of a new species in the same place as the parent species (literally, "of the same country"). It is contrasted with geographic speciation or, more formally, allopatric ("of different countries") speciation, in which the new species originates in a different place while isolated by a physical barrier.

Speciation by polyploidy is sympatric because the new polyploid form arises as a few plants directly from diploid plants, in one generation. Nonpolyploid sympatric speciation is merely sympatric origin by some other means. The most persuasively modeled and documented process in this category is through the intermediate host races of insects that feed on plants. Here are the key steps suggested by Guy Bush and others who have developed and tested the theory in recent years:

• Members of the parental insect species live and mate on one kind of plant. This degree of specificity is widespread among insects. It might characterize millions of species of plant-eating and parasitic insects and other small creatures that spend most or all of their lives on single plants.

• Some individuals of a given species then move to a second kind of plant, where they begin to feed and mate. The new host plant grows in the immediate vicinity of the old, so closely that individual plants of the two might even be intermingled. The insect shift is accompanied by enough genetic change to alter preference to the new plant species and improve the chance of survival on it. Later the evolved insects seek out the new host species whenever they disperse from plant to plant.

• When the evolution of the new insect strain proceeds far enough to settle it firmly on the adopted host, but not enough to isolate it reproductively in full from the old strain, it becomes by definition a host race. When the host race diverges still further, picking up enough differences to forestall even the potential of interbreeding, it becomes by definition a full species.

Host races can originate and then evolve to species rank within a few generations. Some fruit flies of the genus *Rhago-letis* appear to make the transition that quickly. In North America, species that infest hawthorns have occasionally spread to domestic fruit trees. The colonizing flies are prone to turn into host races because they breed only while the host fruit is available, and different tree species come into fruit at different times of the growing season. In 1864 *Rhagoletis pomonella,* which lives on native hawthorns, invaded apples in the Hudson River Valley and subsequently spread over most of the apple-growing regions of North America. A second host race of *R. pomonella* colonized cherries in Door County, Wisconsin, in the mid-1960s. The three races are partially separated by the season in which their host fruit matures, in the following order through the spring and summer: cherry, apple, hawthorn.

Speciation in the sympatric mode, transiting no more than an eyeblink of geological time, might easily have created vast numbers of insects and other invertebrates known to be specialized on particular plant species. It could also have underwritten the proliferation of parasite species that spend most or all of their lives on one kind of animal host. The theory looks good, but we can only guess at the purity of its truth. The early stages are difficult to detect, and few studies have been initiated in the invertebrates most likely to display them. *Rhagoletis* fruit flies are exceptional in the attention they receive, because of the economic damage they cause.

In conclusion, species can be created quickly, and diversity can therefore expand explosively. Our knowledge of evolution, though imperfect, tells us at the very least why life has that potential. Given the right circumstances, a new species can arise in one to several generations.

This vision of the origin of diversity raises a troubling question with ethical overtones: if evolution can occur rapidly, with the number of species quickly restored, why should we worry about species extinction? The answer is that new species are usually cheap species. They may be very different in outward traits, but they are still genetically similar to the ancestral forms and to the sister species that surround them. If they fill a new niche, they probably do so with relative inefficiency. They have not yet been fine-tuned by the vast number of mutations and episodes of natural selection needed to insert them solidly into the community of organisms into which they were born. Pairs of newly created sister species are often so close in their diet, nest preference, susceptibility to particular diseases, and other biological traits that they cannot coexist. Each tends to push out the other by competition. They then occupy different ranges, so that local communities are not enriched by the presence of both.

Great biological diversity takes long stretches of geological time and the accumulation of large reservoirs of unique genes. The richest ecosystems build slowly, over millions of years. It is further true that by chance alone only a few new species are poised to move into novel adaptive zones, to create something spectacular and stretch the limits of diversity. A panda or a sequoia represents a magnitude of evolution that comes along only rarely. It takes a stroke of luck and a long period of probing, experimentation, and failure. Such a creation is part of deep history, and the planet does not have the means nor we the time to see it repeated.

CHAPTER SIX

The Forces of Evolution

WHAT DRIVES EVOLUTION? This is the question that Darwin answered in essence and twentieth-century biologists have refined to produce the synthesis, called neo-Darwinism, with which we now live in uneasy consensus. To answer it in modern idiom is to descend below species and subspecies to the genes and chromosomes, and thence to the ultimate sources of biological diversity.

The fundamental evolutionary event is a change in the frequency of genes and chromosome configurations in a population. If a population of butterflies shifts through time from 40 percent blue individuals to 60 percent blue individuals, and if the color blue is hereditary, evolution of a simple kind has occurred. Larger transformations are accomplished by a great many such statistical changes in combination. Shifts can occur purely in the genes, with no effect on wing color or any other outward trait. But whatever their nature or magnitude, the changes in progress are always expressed in percentages of individuals within or among populations. Evolution is absolutely a phenomenon of populations. Individuals and their immediate descendants do not evolve. Populations evolve, in the sense that the proportions of carriers of different genes change through time. This conception of evolution at the population level follows ineluctably from the idea of natural selection, which is the core of Darwinism. There are other causes of evolution, but natural selection is overwhelmingly dominant.

Evolution by natural selection as we understand it today is a continuous cycle that can be stopped only by the death of the entire population. The starting point is the origin of variation by mutations, which are random changes in the chemical makeup of the genes, in the positions of the genes on the chromosomes, and in the numbers of chromosomes themselves. Genes are the portions of the DNA that ultimately prescribe outward traits, as simple as the color of wings and as complex as the power of flight. Each gene is composed of up to several thousand nucleotide pairs, or genetic "letters." Three nucleotide pairs in a row specify an amino acid. The amino acids in turn are assembled into proteins; proteins are the building blocks of the cells, and cells are the building blocks of organisms.

The number of genes in a typical larger organism, such as a human being, is on the order of 100,000. At least five genes on different chromosome positions affect variation in quantitative traits such as the date of flowering in plants, fruit size, the eye diameter of fish, and skin color in human beings. As many as one hundred genes work together to prescribe traits as complex as ear structure or skin texture.

A large number of molecular steps translate the nucleotide code into the assembly of the distinctive qualities of a species. The precise order of march leads from the triplet letters of DNA to messenger RNA and from there in sequence to transfer RNA and amino acids; the amino acids bond together into proteins; some of the proteins assemble into cell structures, and others into enzymes that catalyze the cell construction itself; additional enzymes accelerate metabolism; and finally the whole self-organizing ensemble projects to the world those properties of anatomy, physiology, and behavior by which the organism lives or dies, reproduces or not.

The commonest and most elementary kind of mutation is an alteration in the chemistry of a gene, specifically a chance substitution of one nucleotide pair for another. Sickle-cell anemia in human beings is one of the most thoroughly researched of all examples of evolution at this ultimate molecular level. The sickle-cell condition originates in perhaps one out of 100,000 persons each generation, as the alteration of a single gene. When present in double dose (unopposed by a normal counterpart in each cell), the mutant gene causes full-blown anemia. Recall that each cell contains two chromosomes of each type and hence two positions in which either a normal or a sickle-cell gene is located. The sickle-cell gene prescribes a change in the chemistry of the hemoglobin molecules that crystallizes them into an elongate form when the oxygen level falls in the surrounding blood. Hemoglobin is carried in the red blood cells, which are normally disk-shaped with a thin center. As the modified hemoglobin molecules of the sickle-cell carriers lengthen, they stretch the red blood cells into the shape of a sickle. The change in form causes the cells to block the smallest blood vessels in tissues throughout the body, slowing circulation downstream and thereby inducing ischemia, or localized anemia. In spite of its debilitating effect, the mutant sickle-cell gene has spread widely in some human populations. Biologists have pieced together the full sequence of this small bit of human evolution from gene chemistry to ecology to tell the following story.

The sickle-cell mutation was the accidental, random substitution of one nucleotide pair for another at one of the billion nucleotide positions strung along the 46 human chromosomes.

This change in a genetic letter translates into the replacement of one amino acid (glutamic acid) by another (valine) in two of the positions on the hemoglobin molecule. There are 574 such positions, or 574 amino acids that make up each hemoglobin molecule.

The substitution of valine for glutamic acid causes the hemoglobin molecules to line up in long spindles when the hemoglobin gives up its oxygen to surrounding tissues. The realignment distorts the red blood cells into a sickle shape.

A double dose of the gene causes sickling in more than a third of the cells, and severe anemia. A single dose results in less than 1 percent sickling and at most mild anemia. But, and this is the most important part, the carrier of one or two sickle-cell genes is also protected from malignant malaria. This second deadly disease is caused by the amoeba-like parasite *Plasmodium falciparum*, which invades and consumes red blood cells. The sickle form of hemoglobin is less vulnerable to the *Plasmodium* invaders.

As a result of the resistance it confers, a half dose of sickle-cell genes, one per cell, is advantageous in those parts of the world where malignant malaria is common. Until recent historical times, this danger zone included tropical Africa, the eastern Mediterranean, the Arabian peninsula, and India. Over most of this area, natural selection favored the sickle-cell gene. Its

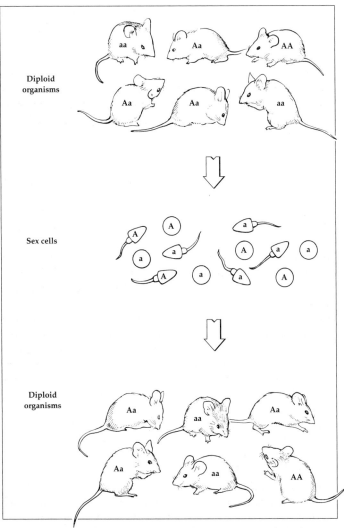

The life cycle and the gene pool. The diploid organisms, each with two genes of a kind per cell, produce sex cells that individually carry only one gene of a kind. The sex cells, composed of sperm and eggs, are the haploid generation in the cycle. Sperm combine with eggs to create the next crop of diploid organisms. Thus the genes in a population—collectively the gene pool—repeatedly separate and recombine to create new variations to be acted on by natural selection. (The animal modeled here is the salt-marsh harvest mouse, a threatened species of California's wetlands.)

frequency commonly hovered above 5 percent, and it rose to as high as 20 percent in a few regions of Mozambique, Tanzania, and Uganda. The natural selection is balanced. When the gene becomes common, more people acquire a double dose and die off from hereditary anemia. When it becomes rare, more people die of malaria, a parasite-induced anemia. Over centuries the percentages of the gene observed in Africa and elsewhere have moved or down according to the frequency with which they encounter malignant malaria.

Of the multitude of gene mutations and chromosomal re-arrangements that arise throughout a population in each generation, many are so minor as to be neutral in effect, neither favoring nor disfavoring survival and reproduction. Either that, or they affect such quantitative traits as height and longevity, adding or subtracting from the traits in ways that are difficult to detect. The vast majority of genetic changes whose effects are large enough to be easily detected are also harmful. By definition they are opposed by natural selection and therefore scarce. In human beings these genetic defects are called genetic diseases. They include Down's syndrome, Tay-Sachs disease, cystic fibrosis, hemophilia, and sickle-cell anemia, as well as thousands of other abnormalities. When on the other hand a new mutant or novel combination of rare preexisting alleles (different forms of the same gene) happens to be superior to the ordinary, "normal" allele, it tends to spread through the population over many generations. In time it becomes the new genetic norm. If human beings were to move into a new environment that somehow gave sickle-cell hemoglobin a total (as opposed to just partial) Darwinian advantage over ordinary hemoglobin, then in time the sickle-cell trait would predominate and be regarded as the norm.

The sickle-cell trait puts a twist on moral reasoning that is worth a moment's reflection. Natural selection, it reminds us, is ethically neutral. Malarial anemia is balanced by hereditary anemia through the mindless agency of differential survival. Those who die of malaria are victims of a harsh environment. Those who die of a double dose of sickle genes are Darwinian wreckage, cast off as the accidental side product of a chance mutation. The tragedy of the hereditary loss is repeated relentlessly in high numbers because in this case natural selection happens to have been balanced rather than directional. No gods decreed it, no moral precept emerges from it. The sickle-cell gene happens to be common in a few parts of the world because the hemoglobin molecule defeats a parasite through the agency of one of its conveniently available mutant forms, and it does so in an inept manner.

The process of evolution by natural selection can be summarized as follows. Random nucleotide substitutions in the gene yield corresponding changes in anatomy, physiology, or behavior. The process sprinkles multiple forms of the gene created this way through the population. Genetic change is also initiated when genes shift position on the chromosomes, or when the number of chromosomes (and hence the number of genes) is raised or lowered. In biological phrasing, the genotype has been altered by one of these forms of mutation or another, and as a result there is now a different phenotype. New phenotypes, or the altered traits in anatomy, physiology, or behavior, usually have some effect on survival and reproduction. If the effect is favorable, if they confer higher rates of survival and reproduction, the mutant genes prescribing them proceed to spread through the population. If the effect is unfavorable, the prescribing genes decline and may disappear altogether.

It can be easily seen why Darwinism is both the greatest idea in nineteenth-century science and the simplest. Its power arises from the fact that natural selection is protean in form. In some cases selection is lethal, mediated by predation, disease, and starvation. In others it is benign, arising from differences in family size, without increasing mortality in the least. Its products range in magnitude from fixing the number of hairs on a fly's wing to the creation of the human brain. Like the old god Proteus it is endless in the form it takes and is therefore filled with the information of realized Nature. Natural selection has these near magical properties because, in one sense, it is a creation of our language. It is nothing more than the active-voice metaphor of all the differences in survival and reproduction among genotypes arising from the effects of the genotypes on organisms. But what it represents is real, and very powerful.

The environment is the theater, as the ecologist G. Evelyn Hutchinson once remarked, and evolution is the play. And more: genetic prescription of the developmental process is the language, and mutation invents the words—but like an idiot spouting gibberish. Finally, natural selection is the editor and principal driving, creative force. Guided by no vision, bound to no distant purpose, evolution composes itself word by word to address the requirements of only one or two generations at a time.

Evolution is blinkered still more by the fact that the frequency of genes and chromosomes can be shifted by pure chance. The process, an alternative to natural selection called genetic drift, occurs most rapidly in very small populations. It proceeds faster when the genes are neutral, having little or no effect on survival and reproduction. Genetic drift is a game of chance. Suppose that a population of organisms contained 50 percent A genes and 50 percent B genes at a particular chromosome site, and that in each generation it reproduced itself by passing on A and B genes at random. Imagine that the population comprises only five individuals and hence 10 genes on the chromosome site. Draw out 10 genes to make the next generation. They can all come from one pair of adults or as many as five pairs of adults. The new population could end up with exactly 5 A and 5 B genes, duplicating the parental population, but there is a high probability that in such a tiny sample the result instead will be 6 A and 4 B, or 3 A and 7 B, or something else again. Thus in very small populations the percentages of alleles can change significantly in one generation by the workings of chance alone. That in a nutshell is genetic drift, about which mathematicians have published volumes of sophisticated and usually incomprehensible calculations.

But let us go on. Population size is critical in genetic drift. If the population were 500,000 individuals with 500,000 A genes and 500,000 B genes respectively, the picture could be entirely different. At this large number, and given that even a small percentage of the adults reproduced—say 1 percent reproduced—the sample of genes drawn would remain very close to 50 percent A and 50 percent B in each generation. In such large populations genetic drift is therefore a relatively minor factor in evolution, meaning that it is weak if opposed by natural selection. The stronger the selection, the more quickly the perturbation caused by drift will be corrected. If drift leads to a high percentage of B genes but A genes are superior in nature to B genes, the selection will tend to return the B genes to a lower frequency.

An important version of genetic drift is the founder effect, believed by some evolutionary biologists to accelerate the formation of new species. Suppose that we started with the same large population containing a mix of A and B alleles. Again, for simplicity make it 50 percent of each. A small group of individuals strays to an offshore island or some other remote locality previously unoccupied by the species. Say a mated pair of birds flies to the new location. The genes they carry are four in number at each chromosome position, including the one that carries gene forms A and B. By sheer accident the founder population may turn out to contain 2 A and 2 B, preserving the ancestral ratios. But there is also an excellent chance that they will contain 3 A and 1 B or 1 A and 3 B, or all A or all B. In other words, because founder populations are so likely to be small in size, they are also likely to differ genetically from the parent population by chance alone. That initial difference, combined with geographical isolation and the exigencies of a new and different environment, can propel populations into new ways of life, new adaptive zones. It can also lead them more quickly to the formation of reproductive barriers and full species status.

Three features of evolution conspire to give it great creative potential. The first is the vast array of mutations, including nucleotide-pair substitutions, shifts in positions of genes on chromosomes, changes in the numbers of chromosomes, and transfers of pieces of chromosomes. All populations are subject to a continuous rain of such new genetic types that test the old.

A second source of evolutionary creativity is the speed at which natural selection can act. Selection does not need geological time, spanning thousands or millions of years, to transform a species. The point is best argued with explicit examples from the theory of population genetics. Take a dominant gene, one whose expression overrides recessive genes at the same chromosome position. For example, the dominant gene for normal blood clotting overrides the one for hemophilia, and the dominant gene for the ability to roll the tongue into a tube overrides the one for lack of that ability. When a dominant

gene occurs in the same cells as the recessive gene, a combination called the *heterozygous condition*, it is the one expressed
in the phenotype. Only when the recessive gene is alone, in
other words the double-dose or homozygous condition, does
its phenotype find expression. A dominant gene whose phenotype enjoys a 40 percent advantage in survival or reproduction
over the recessive phenotype can largely replace it within the
population in twenty generations, passing from a frequency of
5 percent to a frequency of 80 percent in that interval. Twenty
generations amount to as little as four or five hundred years
in human beings, forty years or less in dogs, and one year in
fruit flies. A recessive gene with the same degree of advantage
requires sixty generations to traverse the same frequency range,
still a very short period by geological standards. If dominance
is incomplete—both genes are expressed when together—and
if the advantage of the winning gene is total, the changeover
can, in theory and laboratory populations at least, be accomplished in a single generation.

The final creative feature of natural selection is the ability
to assemble complicated new structures and physiological processes, including new patterns of behavior, with no blueprint
and no force behind them other than natural selection itself
acting on chance mutations. This is a key point missed by creationists and other critics of evolutionary theory, who like to
argue that the probability of assembling an eye, a hand, or life
itself by genetic mutations is infinitesimally small—in effect,
impossible. But the following thought experiment shows that
the opposite is true. Suppose that a new trait emerges if two
new mutations, call them C and D, occur simultaneously on
different chromosome sites. The chance of C occurring is one
in a million per individual organism, a typical mutation rate in
the real world, and the chance of D occurring is also one in a
million. Then the chances of both C and D occurring in the
same individual is a million times a million, or a trillion, a near
impossibility—as the critics have insisted. But natural selection subverts the process. If C confers even a slight advantage
all by itself, it will become the prevailing gene through the
population at its chromosome position. Now the chance of CD
appearing is one in a million. In moderate to large populations

of plants and animals, which often contain more than a million individuals, the changeover to CD is a virtual certainty.

The emerging picture of evolution at the level of the gene has altered our conception of both the nature of life and the human place in nature. Before Darwin it was customary to use the vast complexity of living organisms as proof of the existence of God. The most famous exposition of this "argument from design" came from the Reverend William Paley, who in his 1802 *Natural Theology* introduced the watchmaker analogy: the existence of a watch implies the existence of a watchmaker. In other words, great effects imply great causes. Common sense would seem to dictate the truth of this deduction, but common sense is merely unaided intuition, and unaided intuition is reasoning performed in the absence of instruments and the tested knowledge of science. Common sense tells us that massive satellites cannot hang suspended 36,000 kilometers above one point on the earth's surface, but they do, in synchronous equatorial orbits.

Phenotypic evolution, based on gene action but expressed in the outward traits of organisms, can be correspondingly swift. If a single gene is easily substituted in fewer than a hundred generations by moderate selection pressure, a single gene thus inserted might also exercise profound effects on the biology of a species. One gene can change the shape of a skull. It can lengthen lifespans, restructure the color pattern on a wing, or create a race of giants.

This point is made most compellingly by allometry, when parts of the body grow at different rates. A familiar example is the slower growth of the human skull relative to the body in children, causing adults to end up with heads not much larger than those of infants but atop large muscular bodies. If the allometry of a species is strong, small adults can be strikingly different from large adults in many biological traits, even if they are all identical genetically in the trait under consideration. Among animals the process can be taken to bizarre extremes. In some stag-beetle species, such as the European *Lucanus cervus*, little males have relatively short, simple mandibles, while large males have more massive mandibles half as long as the rest of the body, an armament that gains them superiority in combat. What is inherited in the males is not one of a series of body types, and not necessarily even a particular body size, but rather the allometric growth pattern common to all the males. Males that obtain less food or terminate growth early end up small and feminized; those that reach large size become hulking, top-heavy supermales. The allometry itself is relatively simple, dependent only on differences in rates of growth among certain patches of tissue. It is easy to imagine a rapid switch of a magnitude often associated with the origin of species that is nevertheless based on the simplest hereditary change. Minor mutations in one or several genes might easily alter the allometric pattern, so that all of the males come to more closely resemble females. Alternatively, the change could push the pattern the other way, so that all stag-beetle males sprout huge mandibles.

The social systems of ants illustrate the power of allometry even more dramatically. The caste system of each ant colony, from queens to big-headed soldiers to small-headed workers, is based on a single allometric pattern common to all female members of the colony. Depending on the food and chemical stimuli she receives as a larva, a female ant becomes a queen or a soldier or a minor worker. All fall within the same allometric framework. Genes have nothing to do with the caste determination of the female, but they do determine the allometry of the colony and thus the characteristics of the caste system as a whole. If the allometry is changed even a little by gene muta-tions, a different caste system emerges.

Natural selection is, then, the wellspring of biological diver-sity. The allelic differences that occur among individuals of the same species, across all the chromosomes and the genes they bear, together with differences in the number and structure of the chromosomes themselves, constitute genetic variation. Furthermore, genetic variation is the material from which new species originate, by giving rise to the hereditary reproductive barriers that split old species. Thus there are two basic levels in the diversity of life: genetic variation within species and differ-ences among species.

The two levels of biological diversity are paralleled roughly by microevolution, the small changes that can be tracked down

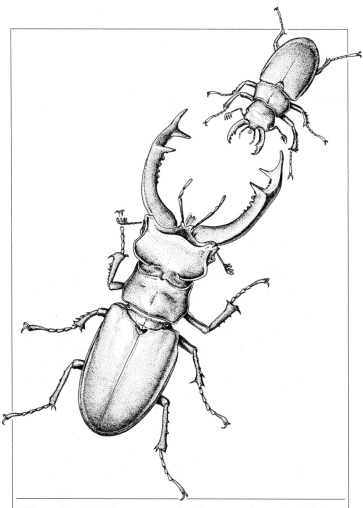

Two males of the European stag beetle prepare for combat. Their different body forms are due to allometry, the more rapid growth of certain parts of the body relative to others, in this species the head and mandibles grow fastest, so that the larger male is much more heavily armored than its smaller opponent.

to the level of the gene and chromosome, and macroevolution, the more complex and profound changes less susceptible to immediate genetic analysis. The origin of blue eye color is a case of microevolution; the origin of color vision is a case of macroevolution. The rise and spread of sickle-cell anemia is microevolution; the formation of the circulatory system in which it is expressed is macroevolution. The splitting of a bird species into two similar daughter species is microevolution; the splitting of a single bird species into a wide array of species from warbler-like to finch-like is macroevolution.

Some paleontologists, impressed by the striking evolutionary changes displayed in fossils, have from time to time suggested that macroevolution is too complex or occurs too rapidly, or on occasion too slowly, to be explained by conventional evolutionary theory. The most recent version of this expostulation, punctuated equilibrium, was made by Niles Eldredge and Stephen Jay Gould in 1972 and developed by them and others in later publications. It claims that not only does evolution periodically bound forward but it tends to slow to a virtual halt at other times. Species emerge quickly and fully formed after a rapid burst of evolution, then persist almost unchanged for millions of years. And, conversely, rapid evolution is driven mostly or entirely during species formation. The alternation between leaps and pauses creates a jerky pattern, a punctuated equilibrium, so extreme as to point to novel processes of evolution beyond the natural selection of genes and chromosomes. Macroevolution, the reasoning in its most radical form concludes, is in some fashion unique, not the same as microevolution.

The punctuated-equilibrium thesis received a great deal of attention because it was at first promoted as a challenge to the neo-Darwinian theory of evolution; in effect, a new theory of evolution. That claim has been abandoned by most of its proponents. The fossil evidence for the widespread occurrence of jerky patterns proved weak, and most examples put forth at the outset were discredited.

More to the point, the possibility of swift evolution was already a cornerstone of traditional evolutionary theory and therefore in no sense a challenge to it. The models of population genetics, the foundation of quantitative theory, predict that evolution by natural selection can be so rapid as to seem nearly instantaneous in geological time. The models also allow for stasis, or long periods with little or no evolution of a kind detectable in fossils. These predictions of population genetics have been upheld by decades of careful study in the laboratory and field, across a wide range of animals, plants, and micro-organisms. They have illustrated gradual transitions between closely related species from a single tick in microevolution to large advances in macroevolution, from the earliest beginnings of geographical variation to the origin of species and panoramic expansions of species into multiple adaptive zones.

In general, the continuity between microevolution and macroevolution has been upheld. Neo-Darwinian theory was not challenged in substance, only semantically—a renaming, so to speak, as opposed to a reinventing of the wheel. Punctu-ated equilibrium is now used mostly as a descriptive term for a pattern of alternating rapid and slow evolution, especially when the rapid phase is accompanied by species formation. Its fate illustrates the principle that in science failed ideas live on as ghosts in the glossaries of the survivors. The value of the punctuated-equilibrium challenge lay not in its claims but in the research it stimulated on evolutionary rates and in the favorable public attention it brought to evolution studies as a whole.

To say that microevolution shades into macroevolution by degree instead of kind does not mean, however, that all we know about evolution is written in the narrow script of modern genetics. It only affirms that nothing learned so far is inconsis-tent with that canon, in the same sense that nothing learned so far about the molecular processes of the cell is inconsistent with contemporary physics and chemistry. There is still a great deal more to evolution than its genetic mechanisms.

A case in point is species selection, a process that has begun to be profitably explored by paleontologists, who study fossils, as well as by neontologists, who work on modern organisms. The subject has been muddled somewhat by the fact that the two groups favor different vocabularies, but the process itself is easily explained. A newly evolved species, like a newborn organism, comes into the world bearing its own distinctive set of traits. Depending on which traits it possesses, the species can last a long time or a short time before it becomes extinct. It is also prone either to split into multiple species or to remain intact as a single species throughout its lifespan. These influential hereditary traits are emergent properties of species but are not invested by some mysterious macroevolutionary process. They are the properties evolved by the organisms that compose the species. They originate by microevolution, in other words by changes in the frequencies of genes and chromosome arrays, and are translated upward into the species-level patterns we call macroevolution.

This translation has two key properties. It is blind, and it echoes back down to speed or to slow the evolution of organisms. Organisms, in the course of their struggle for survival and reproduction, are unconcerned in a Darwinian sense with the persistence of the species as a whole. They are also unaffected by the degree to which the species multiplies. Thus their genes are inserted into the next generation or perish by their own idiosyncratic actions, no matter whether the species is expanding and multiplying or dwindling toward extinction. The traits they possess nevertheless cause the species to last a long time or a short time and to remain a single species or to multiply. That influence is what has been identified as the upward translation from microevolution to macroevolution. Conversely, and this is the essence of species selection, the longevity of a species, together with its tendency to form new species, affects how rapidly the crucial traits are spread through the fauna or flora as a whole. This is the downward echo that makes the theory of species selection more than just a boring statement of the obvious.

Consider a set of species that might be subject to selection at the higher level. By a *set* I mean multiple species of common ancestry, such as the cichlid fishes of Lake Victoria or the lycaenid butterflies of tropical America. Natural selection among the species can reinforce organism-level selection that occurs within each species in turn. Hence the evolution of the trait proceeds more rapidly within the set of species as a whole. The character of the fauna and flora shifts accordingly. Alternatively, species selection can oppose organism-level selection, slowing the rate of spread.

How important is species selection? If the group is defined broadly enough, such as all vascular plants or all land vertebrates, it is of overwhelming importance. In the late Mesozoic era, cycads and conifers gave way to the poleward-spreading flowering plants. After the catastrophic end of the Mesozoic, mammals took over from dinosaurs and crocodilians. But we knew all that; the fact sheds little light on the biological process of species selection. To link species selection to natural selection at the level of individuals and populations, we need smaller groups of species and sharper insights into the fine details of ecology and adaptation. It is easy to imagine the existence of such groups but difficult to find them in nature. We have been limited so far to a scattering of cases, of which the following appear to be the most promising.

• Among insects, a shift from predatory behavior or scavenging to plant feeding increases the rate of species formation. The reason is that more species can specialize on particular kinds of plants or even different parts of the same plant. They can radiate into these niches even more rapidly by forming host races, which are believed to be precursors to full-fledged species. In such insects, to put the matter as simply as possible, individual selection and species selection conspire to increase the rate of evolution.

• During the latter half of the Mesozoic era 100 to 66 million years ago, oysters, clams, and other mollusks varied from one set of species to another in dispersal ability and hence in the breadth of their geographical ranges. Those with large ranges also survived for longer periods of geological time. The study of present-day mollusk species shows that dispersal ability is probably the result of natural selection at the organism level. If true, it follows that individual and species selection worked together to increase the average geographical range and longevity of mollusks.

• Ants, beetles, lizards, and birds display what has been called the *taxon cycle*. Some of the species adapt—more precisely, their member organisms adapt—to habitats from which it is easy to disperse. Such places include the seashore, edges of rivers, and windswept grasslands. Quite coincidentally, these habitats are also the best staging areas for long-distance dispersal. The species concentrated there attain the widest geographical spread and potential for speciation. When some of the far-flung populations penetrate more sheltered habitats nearby, they "settle down"—lose dispersal power and thereby become more prone to species formation. Eventually, they decline to extinction. The question arises: do they decline to extinction more rapidly, as was the case in the Mesozoic mollusks? If so, organisms that adapt for life in restricted habitats can be said to improve their own Darwinian fitness at the expense of the longevity of their species. In short, the two levels of selection, individual and species, are countervailing.

• A process similar to the taxon cycle has been occurring for millions of years in the rich fauna of antelopes, buffalo, and other bovid mammals of Africa. Species that are generalized, able to occupy more than one habitat—to shift from forest to grassland, say, and back again—survive for longer periods of time. Those specialized to live in particular habitats are more likely to be trapped there and to decline to extinction as the climate changes and the forest alternately advances and retreats. With their populations prone to fragmentation, the bovid specialists are also more likely to generate new species, and thus they gain and lose species more rapidly than bovid generalists. Overall, natural selection of individual animals leads to the natural selection of species, by either increasing their longevity or decreasing it according to circumstances.

• Desert plants such as *Dedeckera eurekensis* of the Mojave Desert may experience another category of individual-level natural selection that conflicts with species selection. During droughts, few seeds have a chance to germinate. Natural selection might easily lead to a strategy in which individual plants shut down seed production and concentrate their resources on survival. (The alternative strategy, not used by *Dedeckera*, is to produce many seeds, which then wait for rainfall.) If hard times drag on, a premium is put on longevity rather than on reproductive ability. The species whose members are pushed by natural selection to adopt the longevity strategy will end up limited to a small number of long-lived but nearly sterile individuals. The surviving individual plants are winners in the game of organism-level selection, but their success brings the species to the edge of extinction.

The picture emerging of natural selection at the level of the individual, whether or not it is enhanced by species selection, is one of exuberance, power, and a potential for quickness. If enough raw hereditary material exists in the first place, and if the selection pressure (differences in survival and reproduction) are strong, one gene or chromosome type can be substituted for another in fewer than a hundred generations. The possibility is there for rapid microevolution and even the early stages of macroevolution.

This capacity is well understood in theory and has been realized in laboratory experience. It is also displayed in wild populations when species are subjected to new selection pressures, such as the threat from a new parasite or access to a new food source. There has been time enough and more for natural selection alone to create radical new types of organisms. Consider: the Age of Reptiles lasted for 100 million reptilian generations, and the Age of Mammals succeeding it passed more than 10 million mammalian generations before the human species emerged. Earth took hundreds of millions of years to produce the first unicellular organisms, which were assembled out of an astronomical number of potent molecules.

What we understand best about evolution is mostly genetic, and what we understand least is mostly ecological. I will go further and suggest that the major remaining questions of evolutionary biology are ecological rather than genetic in content. They have to do with selection pressures from the environment as revealed by the histories of particular lineages, not with genetic mechanisms of the most general nature. I could be very wrong. Molecular biology is so vigorous and rapidly growing that new mechanisms driving evolution one way or the other may be discovered. We have a vast amount to learn about the way functional genes, the exons of DNA, originated and how they were reshuffled and elaborated to set the basis for the full flowering of biological diversity. It is further possible that extragenetic constraints on embryonic development, such as fundamental physical limits on cell size and tissue organization, play a guiding role. Competition and interference among cells and tissues might entail still undiscovered principles of a novel kind. Many surprises await us in the study of development. Sometime soon discoveries in the two key domains of the genetic code and embryonic development could shake neo-Darwinism to its foundations. But I doubt it. I think the greatest advances in evolutionary biology will be made in ecology, explaining more fully in time why the diversity of life is of such and such a nature and not some other.

CHAPTER SEVEN

Adaptive Radiation

E VOLUTION on a large scale unfolds, like much of human history, as a succession of dynasties. Organisms possessing common ancestry rise to dominance, expand their geographic ranges, and split into multiple species. Some of the species acquire novel life cycles and ways of life. The groups they replace retreat to relict status, being diminished in scattershot fashion by competition, disease, shifts in climate, or any other environmental change that serves to clear the way for the newcomers. In time the ascendant group itself stalls and begins to fall back. Its species vanish one by one until all are gone. Once in a while, in a minority of groups, a lucky species hits upon a new biological trait that allows it to expand and radiate again, reanimating the cycle of dominance on behalf of its phylogenetic kin.

When viewed in one slice of geological history, all contemporary dynastic successions taken together present a complex and strikingly beautiful pattern across the surface of the earth. Now the comparison is to a palimpsest, an ancient parchment on which the current dominant groups are boldly spread and past rulers survive as faded traces in spaces between the lines, in shrunken niches. Mammals, the dominant large vertebrates on the land today, are accompanied by turtles and crocodilians, among the last survivors of yesteryear's ruling reptiles. Forests of flowering plants shelter scattered ferns and cycads, remnants of the prevailing vegetation of the Age of Reptiles. And on a smaller scale the air is filled with flies, wasps, moths, and butterflies, relative newcomers of insect evolution. They are preyed upon by dragonflies, Paleozoic relics that still possess stiff outstretched wings and other archaisms that date back to the dawn of flight. Dragonflies are the Fokkers and Sopwith Camels of the insect world that somehow stayed aloft all those years.

Adaptive radiation is the term applied to the spread of species of common ancestry into different niches. *Evolutionary convergence* is the occupation of the same niche by products of different adaptive radiations, especially in different parts of the world. The Tasmanian wolf of Australia, a marsupial, outwardly resembles the "true" wolf of Eurasia and North America, a placental mammal. The first is a product of adaptive radiation in Australia, the second the product of a parallel adaptive radiation in the northern hemisphere. The two species converged to occupy similar niches within independent adaptive radiations on the different continents.

Adaptive radiation and evolutionary convergence are displayed in textbook fashion in distant archipelagoes around the world, including the Galápagos, Hawaii, and the Mascarenes. They are also sharply defined in ancient lakes such as Baikal and the Great Lakes of East Africa's Rift Valley. These places are so isolated that only a few kinds of plants and animals have been able to reach them. The fortunate colonists originated in large, crowded faunas and floras, pressed by competition, predators, and disease, restricted in habitat and diet. They arrived in a new and mostly empty world where, initially at least, opportunity was spread before them in abundance.

The archipelagoes and lakes are not only isolated but also small and young enough, in comparison with the continents and oceans, to keep the patterns of adaptive radiation and convergence simple and hence decipherable. That is why biologists see Hawaii as one of the prime laboratories of evolution. It is an archipelago rather than a single island, setting the stage for the splitting of populations into full-blown species. It is geographically the most remote of all archipelagoes, so that relatively few colonists have reached its shores. It is large enough to provide niches for the radiation of large numbers of new species, yet small enough to constrict and clearly display the patterns of speciation and adaptive radiation. Finally, although youthful relative to continents, it is still old enough, about 5 million years in the case of Kauai, for adaptive radiation to have attained an impressive degree of maturity.

The 10,000 known endemic species of insects on Hawaii are believed to have evolved from only about 400 immigrant species. Some have made unique shifts in habitat and lifestyle. Around the world, for example, almost all larvae of damselflies (small, delicate relatives of dragonflies) are aquatic, feeding on insects around them in ponds and other bodies of fresh water. But on Hawaii the nymphs of one species, *Megalagrion oahuense*, have left the water entirely and now hunt insect prey on the floor of wet mountain forests. Displaying an even more radical shift, caterpillars of the moth genus *Eupithecia* have abandoned their habit of feeding on plants to become ambush predators. These bizarre wormlike larvae lie in wait on vegetation for passing insects and seize their victims with sudden strikes of their forelegs. A cricket in the genus *Caconemobius* has gone from life on the land to a partial marine existence, living among boulders in the wave-splash zone and feeding on flotsam that washes ashore. Another *Caconemobius* species lives on bare lava flows, where it browses on windblown vegetation debris. Still other crickets of the same genus, all blind, live in caves. The killer caterpillars and entrepreneurial crickets have all been discovered within the past twenty years. Hawaii, familiar as it may seem to the casual visitor, is still a paradise full of surprises for the explorer naturalist.

Radiation and convergence on distant archipelagoes are marked by disharmony, defined in evolutionary biology as the wildly disproportionate representation of some major groups and the absence of others. When a few species split swiftly and massively in response to exceptional opportunity, they and their descendants seize a large part of the environment and hold it thereafter with a lion's share of the total diversity. The fauna and flora as a whole are thus unbalanced in comparison with those of continents, whose great biological diversity originated from many stocks over longer spans of time.

Hawaii harbors the most disharmonic bird fauna in the world. Until recent historical times, more than one hundred of the known species were endemic, by which is meant they are native and found nowhere else in the world. Their numbers include sixty species extinguished successively by the Polynesian and then the European colonists, and forty still surviving. More than half are or were honeycreepers, composing a unique tribe of their own, the Drepanidini, which in formal classifications is a branch of the finch subfamily Carduelinae, which in turn is a branch of the larger finch and sparrow family Fringillidae. All honeycreepers are descended from a single pair or small flock of colonists, most likely blown to the islands

by a storm many thousands of years ago. This ancestral species was a relatively primitive cardueline bird, probably small,
slender, and with a bill resembling that of a goldfinch. Its diet
probably consisted of seeds and insects. Non-Hawaiian car-
duelines include goldfinches, canaries, and crossbills, which
occur around the northern hemisphere but are concentrated
in temperate Europe and Asia. It seems likely, then, that the
first Hawaiian colonists flew or were blown in by a storm from
either North America or eastern Asia. The honeycreepers, their
populations expanding, then achieved an explosive adaptive
radiation. They penetrated many new niches and diversified
their anatomy and behavior in concert. As ecological conquerors of the first rank, they offer a textbook display of radiation
and convergence on a scale small enough to be dissected and
explained with reasonable certitude.

I should say that their *memory* offers such a display. Before
the coming of the Polynesians 2,000 years ago, and before the
arrival of European merchantmen and settlers eighteen centuries later, the forests of Hawaii swarmed with a riot of the spar-
row-sized honeycreepers. The many species were distinguished
variously by red and yellow and olive green plumages, slashed
with wing bands in black, gray, and varying nuances of white.
Even today the scarlet arapane (*Himatione sanguinea*) occurs
in populations of one thousand to a square kilometer in some
localities. Walk among them in a grove of ohia lehua trees,
watch for their scintillas of bright color, listen for their thin
whistling songs, and you will be granted a last snapshot of old
Hawaii, as it was before the first Tahitian canoes touched shore.

Most of the honeycreepers are gone now. They retreated and
vanished under pressure from overhunting, deforestation, rats,
carnivorous ants, and malaria and dropsy carried in by exotic
birds introduced to "enrich" the Hawaiian landscape. They
disappeared as vanished species usually do, not in a dramatic
cataclysm but unnoticed, at the end of a decline when those
who knew them could acknowledge that none had been seen
around for a while, that perhaps there were still a few to be
found in such-and-such a valley, where in fact a predator had
already snatched the last living individual, a lonely male, let
us suppose, from its nocturnal perch. In old Polynesian times
several generations might pass, the last tattered feather panache

would be replaced in a ceremonial headdress laid aside for good, and the species would be as consigned and as forgotten, in the words of the Catholic liturgy, as the unremembered dead.

Still the sweep of the radiation even among the surviving honeycreepers is the greatest of any closely related group of birds in the world. The Maui parrotbill (*Pseudonestor xanthophrys*) has an anatomy somewhat resembling that of a true parrot but feeds on insects rather than fruit and seeds. It wields its stout bill to chew and rip open twigs and branches in order to reach beetle grubs and other insects burrowing through the wood. The ou (*Psittirostra psittacea*), the finch analogue, possesses a thick bill with which it feeds on seeds primarily and insects secondarily, in the generalized finch manner. The akepa (*Loxops coccinea*) represents a partial approach to crossbills of the northern hemisphere. The tips of its bill reach past each other in a lateral direction, allowing it to twist open leaf buds and legume pods in search of insect prey. Other species of *Loxops* and *Himatione* resemble warblers, with small delicate bodies and short thin bills. Like typical warblers, which are dominant birds on most continents, they hunt insects that fly in the open or rest exposed on vegetation. The iiwi (*Vestiaria coccinea*) and several species of *Hemignathus* are closely convergent to the sunbirds of tropical Africa and Asia. They use long, slender, downward curving bills like siphons to extract nectar from flowers.

The species of *Hemignathus* have achieved a miniature adaptive radiation within the larger radiation, a secondary deployment into major niches. In addition to the primary nectar feeders with full, curved bills, they include the nukupuu, *Hemignathus lucidus*, whose lower bill is shortened to only a little more than half the length of the upper bill. This curious form has evolved partway to the status of a woodpecker. In addition to nectar feeding, accomplished with the upper bill, it uses the lower bill to tap tree trunks and branches, pry up pieces of bark, probe into crevices, and chase out insect prey caught unaware by the maneuver.

A second, even more remarkable species, the akiapolaau (*Hemignathus wilsoni*), has traveled almost the full distance into the woodpecker niche. It uses its lower bill, which is short and completely straight, to hammer and chisel open bark and

wood. This woodpeckerish behavior is a straightforward extension of the gentler tapping behavior of the nukupuu. Walter
Bock has described it: "The decurved upper jaw is raised out of
the way of the mandible when the bird is pecking. After a hole
has been cut and the insect exposed, the longer decurved upper
jaw is then used to probe for the insects. The combination of
straight chisel-like mandible and a long decurved upper jaw
for probing is an unusual, and perhaps unique, example of the
two jaws in a single avian species being adapted for two quite
different actions, both of which are essential for the feeding
method of this species."

The step from sunbird-like *Hemignathus* to the nukupuu
and then akiapolaau levels of woodpecker design is an instructive example of an important evolutionary change arising as
part of species formation. Their coexistence in contemporary
Hawaii is a freeze-frame of microevolution attaining the scale
of macroevolution. Macroevolution, seen in the two steps away
from the sunbird level, is microevolution writ large, with species multiplication added.

The ersatz woodpeckers of Hawaii deserve further attention
as exemplars of imperfect convergent evolution that results
from bold adaptive radiation given too little time to mature.
They contrast with members of the bird family Picidae, which
we are entitled for several reasons to call the true woodpeckers.
The Picidae, a well-knit group whose common ancestry is far
removed from that of the Hawaiian Drepanidini, comprises
about 200 species worldwide. The 19 species of the United
States include the familiar northern flicker (*Colaptes auratus*),
downy woodpecker (*Picoides pubescens*), and sapsuckers of
the genus *Sphyrapicus.* They also include two newly extinct
victims of deforestation in North America, the ivory-billed
woodpecker (*Campephilus principalis*), largest of all the picids
of the Nearctic realm, and the closely related imperial woodpecker of Mexico (*Campephilus imperialis*), largest woodpecker
in the world.

The picids are called true woodpeckers simply because they
are widespread and common enough to be the birds to which
the vernacular name was originally applied. But they also have
the right stuff to be the nominal standard bearers. They are
the premier specialists in their ecological class. Many other

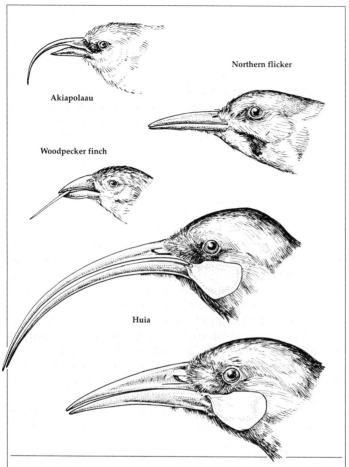

Woodpeckers and woodpecker-like forms illustrate the dual patterns of adaptive radiation and evolutionary convergence. During radiations of birds in different parts of the world, separate lines evolved to fill the woodpecker niche: the akiapolaau, a honeycreeper of Hawaii; the common flicker of North America, one of many "true" woodpeckers; the woodpecker finch of the Galápagos; and the huia (female above, male below) of New Zealand.

kinds of birds peck and pry wood to expose insects, but none do it with the élan and precision of a picid. The details of the hunting strike of one typical species, the acorn woodpecker (*Melanerpes formicivorus*) of California, can be seen with the aid of slow-motion photography. The awl-shaped bill hits the wood at between 20 and 25 kilometers an hour, whereupon it instantaneously decelerates at 1000G, where 1G is the acceleration needed to counteract earth's gravity and 4G is what an astronaut experiences on liftoff. An ordinary brain jarred hundreds of times daily by a blow to the head of this magnitude would be reduced to pulp. The woodpecker survives because it possesses two unusual features. The brain case is made of unusually dense spongy bone joined with sets of opposing muscles that appear to act as shock absorbers. And the woodpecker brings its head up and down like a metronome in a single plane, avoiding the rotational forces that would skew the brain from side to side and tear it loose from its moorings.

Jackhammer feeding is only one adaptation of the picid woodpeckers. Many of the species have a stiff wedge-shaped tail that braces them against tree trunks and bristlelike feathers over the nostrils that shield the air passages from wood dust. They employ cylindrical, sticky tongues that can be extruded beyond the tip of the bill by as much as twenty centimeters and snaked through the insect burrows to seize their prey, then retracted and coiled in a cavity encircling the inner surface of the skull.

On the other hand, picid woodpeckers are not very good at traveling over open water, and therein lies our tale. They never colonized Hawaii during the millions of years its bird fauna was evolving. The honeycreepers were free to fill the woodpecker niche, and they did so with the ingenious innovation of the akiapolaau. The species has a jerry-built look to it when put alongside one of the sophisticated, wood-pulverizing picids. The akiapolaau could never have survived in competition, or originated in the first place, had native Hawaiian picids been hammering in the forests when the first honeycreepers flew ashore. To exist as a woodpecker with full dependence on dead and dying trees as a hunting ground requires space: an ivorybill breeding pair used about 8 square kilometers of old-growth swamp forest. When its habitat was drastically reduced by lumbering in the southern United States, the species was doomed to extinction. The ivorybill population was never very large. It declined precipitously, and the last authenticated sightings were made in the 1970s. Today a tiny remnant population lives on in the mountain forests of eastern Cuba. Woodpeckers therefore compete intensely for the relatively scarce resource on which they depend, and they would almost certainly displace any akiapolaaus they encountered.

Picid woodpeckers are also absent from the Galápagos. This volcanic archipelago, located in deep water 800 kilometers west of Ecuador, is the site of extensive adaptive radiations in many kinds of plants and animals. The productions are not as rich as those on Hawaii, but they are conspicuous enough to have inspired Darwin with the idea of evolution. Among those he found most interesting were Darwin's finches, or the subfamily Geospizinae in technical classification. A single colonizing ancestor expanded into thirteen contemporary species overall, which fill some of the same feeding niches as the Hawaiian honeycreepers. They shout the truth of evolution, and a naturalist of Darwin's caliber could not have missed it. He wrote in his 1842 *Journal of Researches* the words that foretold his theory: "The most curious fact is the perfect gradation in the size of the beaks of the different species of Geospiza—Seeing this gradation and diversity of structure in one small, intimately related group of birds, one might really fancy that, from an original paucity of birds in this archipelago, one species had been taken and modified for different ends."

Some of the Darwin's finch species are warbler-like, using their slender bills to capture insects and drink nectar. Others act like "true" finches, wielding relatively thick bills to tear apart fruit and crack open seeds. The larger the bird and thicker the beak, the wider the range of food objects consumed. In hard times, those with the thickest bills are able to specialize on the largest and toughest fruits and seeds.

Adaptive radiation is never complete, neither on archipelagoes nor on continents. Perhaps because there are fewer flowers in the dry Galápagos forests, Darwin's finches have not entered the sunbird niche, so expertly filled by several species of Hawaiian honeycreepers. None possesses a long, curved bill or long tongue of the kind needed to collect nectar from the deep recesses of flowers. On the other hand, the Galápagos radiation has produced an adaptive type unique among all the birds of the world: vampires. On the small and remote islands of Darwin and Wolf, ground finches alight on the backs of boobies, large seabirds of the genus *Sula*, and peck at the feather roots on the wings and tail, drinking the blood that flows out. Not content with this villainy, the vampire finches also crack open the eggs of seabirds by pushing them against rocks, then drink the contents.

Two geospizines, the woodpecker finch (*Cactospiza pallida*) and the mangrove finch (*Cactospiza heliobates*), have entered the woodpecker niche, again in a manner wholly new for birds. Their bills are shaped like those of conventional insect feeders. The birds peck at the surface of trunks and branches and pull up loose pieces of bark, but they do not strike the wood with vertical hammer blows. In this respect they are ordinary Darwin's finches, close to a few other species of similar mien. They neither pluck insects out with a long curved bill, like Hawaii's akiapolaau, nor fish for them with an extrusible tongue in the picid woodpecker manner. Their innovation is purely behavioral. Woodpecker finches pick up a cactus spine, twig, or leaf petiole, adjust it so that it sticks out in front of their head like a stiff extruded tongue, then insert this makeshift probe into crevices here and there to stir up insects and winkle them out for capture. This trick is one of the few uses of tools known among animals. Seeing them in action, it is hard not to think of the woodpecker finches as intelligent. They have in fact been observed to correct mistakes during their hunting procedure. One individual was seen trying to break in two a stick that had proved too long to use as a tool. Another picked up a forked stick, tried unsuccessfully to probe with the forked tip, then turned the stick around to work with the unforked tip, this time successfully.

How did the birds conceive such an innovation? Peter Grant, who has watched Darwin's finches in the wild longer than any other person, believes that the birds achieved tool use by accident rather than by thought and then relied on operant conditioning. "I can imagine," he wrote, "a frustrated Woodpecker Finch failing to drop a piece of bark it had just removed from around the entrance to a crevice in a branch, accidentally pushing it into the crevice and touching the prey, and being rewarded by the prey moving toward the entrance and within reach of the bird's beak." Evolution by genetic assimilation might follow. Birds with a greater capacity for such trial-and-error learning would imitate those who invented the technique, and thus survive at a higher rate. In time the population would then contain not only brighter birds but those with an instinct hardened to pick up and manipulate sticks in the first place. Evolutionary biologists believe that genetic assimilation of this kind can on occasion greatly accelerate evolution, with behavioral flexibility leading the way.

If necessity is the mother of invention, then opportunity is its mother's milk. Tool use in Darwin's finches, no less than the fantastical double-functioned bill of Hawaii's akiapolaau, arose in a remote place in the absence of competition from the dominant picid woodpeckers. There is an even more bizarre twist to illustrate this principle, provided by the huia (*Heteralocha acutirostris*) on New Zealand. In the absence of competition from native picids, this odd crowlike species evolved a division of labor between the male and female that enabled them to work together as a kind of compound woodpecker. The species is extinct now, having been last seen on North Island in 1907, but enough observations were made in its last days to give a picture of the foraging techniques, again unique among birds. The male was armed with a straight, stout bill, similar in shape to a picid's. He chiseled open both dead wood and green saplings, then snatched up the first beetle grubs and other insects exposed. His mate, in contrast, had a long, slender, curved bill like that of many Hawaiian honeycreepers. She worked closely with the male, probing into deeper crevices and plucking out insects beyond his reach.

The archives of natural history are filled with other cases of species formation exploding as a response to ecological opportunity. On the Galápagos, Rarotonga, Juan Fernandez, and other remote oceanic islands, members of the plant family Compositae have radiated repeatedly to fill a large fraction of the niches available to vegetation. Composites as a whole rank among the most diverse and widespread flowering plants in the world. They include such familiar plants as the asters, sunflowers, thistles, marigolds, and lettuce. Their flowers are actually flower heads, tight clusters of many small flowers surrounded by leaflike structures (bracts). In addition to gracing gardens and banks of wildflowers, they are encountered everywhere as pretty weeds, such as dandelions and goldenrods, indomitable in the summer, gone with the winter frost.

On the most distant forested islands, many composite species also dominate the native shrubs and trees, having evolved from small herbaceous plants into what may be loosely called tree-asters and tree-lettuces. St. Helena is one of the most isolated islands in the world, located in the South Atlantic midway between Africa and South America. Before being completely settled first by Dutch and then British colonists, a process completed by the late 1800s, St. Helena's volcanic slopes were covered by forests of woody composites. Among them grew additional composite species and other plants of herbaceous form, the whole flora comprising thirty-six endemic species of flowering plants. Living in the forests were 157 or more St. Helenan beetle species, evolved from as few as twenty colonizing stocks, feeding on vegetation, dead wood, fungi, and each other. Seventy percent of these insects were weevils, a proportion totally out of line with coleopterous faunas in the rest of the world. Yet the strange ensemble worked. St. Helena was a nearly closed ecosystem, a biosphere functioning in great isolation, one step removed from a satellite colony in space.

The floras of each such composite-invested islands around the world contain among them every principal step in the transition from herbs to shrubs to trees. Each island is a contemporary laboratory of macroevolution, its flora an independent evolutionary experiment in progress, waiting for evolutionary biologists to pick up the clues and tell the story. The experiments are made even more persuasive by the fact that they have been replicated in still other groups of herbaceous plants, including especially members of the family Lobeliaceae. "The metamorphosis of these lettuces into shrubs or trees," Sherman Carlquist wrote in an account of island biology, "invites comparison with what has happened on other islands, to other plants. The Hawaiian lobeliads provide an almost exact parallel . . . Each growth form and leaf type can be roughly matched, showing that islands with a particular climate and a particular degree of isolation tend to promote these forms, these sizes."

What selection force drives the herbs to larger dimensions and assembles the island forests? Evidence from many sources suggests that it is ecological opportunity afforded by the absence of conventional trees. The vast majority of both temperate and tropical tree species have limited dispersal powers. Beech mast, dipterocarp seeds, and citrus fruits cannot travel far from the mother trees or survive immersion in salt water. But composites, among the dominant weedy plants of the world, are superb dispersers. When islands such as St. Helena and Oahu emerged as volcanic cones from the sea, these plants along with grasses were evidently among the first to arrive. They were also among the pioneers of Krakatau after the 1883 explosion. Around the world the long-distant emigrants entered an environment mostly or entirely devoid of shrubs and trees. They had the chance to evolve into shrubs and trees and preempt the land before traditional woody plants arrived, assuming this later event were even possible. Darwin correctly deduced the process in *On the Origin of Species*, using the new idiom of natural selection:

> Trees would be little likely to reach distant oceanic islands; and an herbaceous plant, though it would have no chance of successfully competing in stature with a fully developed tree, when established on an island and having to compete with herbaceous plants alone, might readily gain an advantage by growing taller and taller and overtopping the other plants. If so, natural selection would often tend to add to the stature of herbaceous plants when growing on an island, to whatever order they belonged, and thus convert them first into bushes and ultimately into trees.

The arborescence of island weeds raises the larger question of why certain groups of organisms undergo radiation, and others do not. The example of the Compositae shows that superior dispersal ability empowers at least some organisms some of the time. A species able to invade a new island, lake, or other empty environment, to fill it, and to divide into multiple specialized species is likely to control the land by preempting invasion and diversification by other species. On the Galápagos

Islands a small assemblage of flycatchers, mockingbirds, and warblers coexists with the thirteen species of Darwin's finches, but none has achieved a comparable adaptive radiation. Is it possible that the finches, or more precisely the ancestral finch, simply got to the Galápagos first and closed off opportunities for later arrivals? Dominance could have been the reward of nothing more than superior dispersal power. Since we do not know the date of arrival of these birds, we cannot say for sure.

Alternatively, perhaps the ancestral Darwin's finch had qualities that allowed it to evolve and radiate more decisively than its rivals, no matter when it arrived. It might have possessed a generalized anatomy and behavior that adapted it quickly to a partially empty environment. If that is true, would it be possible to deduce the nature of the original species in this respect? Not with certainty, but we can make a good guess because, amazingly, something like the ancestral species still exists. One more Darwin's finch, number fourteen, lives on Cocos Island, a 47-square-kilometer speck of land 580 kilometers to the northeast of the Galápagos. The island, owned by Costa Rica, is hilly, uninhabited by human beings, and densely clothed in tropical rain forest. The Cocos Island finch, *Pinaroloxias inornata*, coexists with only three other breeding species of landbirds—a cuckoo, a flycatcher, and a yellow warbler. This scarcity of competitors has allowed it to engage in what biologists call *ecological release*, the expansion of a single species into multiple habitats.

Ecological release is a common phenomenon on remote islands with small faunas and floras—I have observed it many times in ants, for example—but it occurred to spectacular degree in the Cocos Islands finch. The birds, all members of a single freely breeding species, occupy niches ordinarily divided among species, genera, and even entire families of birds. They range from shore to hilltop, forage within the forest from ground to canopy, and feed variously on insects, spiders, and other arthropods, mollusks, small lizards, seeds, fruit, and nectar. In this respect the Cocos finch far exceeds any one species of Darwin's finch in the Galápagos. Most striking of all, each individual bird specializes on a particular kind of food and maintains the habit for at least several weeks, perhaps for its entire life. This microcosmic adaptive radiation appears to be based on observational learning. During a ten months' visit to Cocos, Tracey Werner and Thomas Sherry, the biologists who discovered the release phenomenon, watched juvenile finches approach and imitate the distinctive feeding behavior of yellow warblers and sandpipers. No doubt the young birds also copy older birds of their own species. Like medieval apprentices selecting masters within specialized guilds, the young birds appear to prosper with personalized instruction.

Cocos finches have bills roughly intermediate in size and shape to those of warblers and finches. With the tendency of the birds to diversify in feeding habits individually, the stage is set for rapid species formation and adaptive radiation in the Galápagos mode, if circumstances permitted it. But circumstances do not permit it. The place is too small and too remote from other islands to allow the formation of new species. The Cocos finch radiation therefore remains stalled in an embryonic state, one species to the island.

Certain kinds of plants and animals, by virtue of distinctive biological traits they already possess, seem poised to expand and preempt many niches in sparsely inhabited environments. If the new home is complex enough to allow species formation and ecological specialization, the radiation proceeds to term. A second example of radiation proneness, as graphic as that of Darwin's finches, is found in the freshwater Cichlidae. This prolific fish family occurs from Texas to South America in the New World, from Egypt to the Cape Province in the Old World. A suite of primitive species also lives on Madagascar, and three more species are endemic to southern India and Sri Lanka.

The Great Lakes of East Africa, the necklace of fresh water strung along the Rift Valley from Uganda to Mozambique, swarm with cichlids. These fishes dominate the aquatic fauna, having radiated to fill almost all major niches available to freshwater fishes as a whole. Cichlids are the lacustrine equivalents of the Hawaiian honeycreepers. The three hundred or more species of Lake Victoria alone, for example, include the following major adaptive types:

Astatotilapia elegans, perch-like in shape, a generalized bottom feeder

Paralabidochromis chilotes, large mouth with thickened lips; preys on insects

Macropleurodus bicolor, small mouth; uses pebble-shaped pharyngeal teeth to crush snails and other mollusks

Lipochromis obesus, possesses a heavier body and somewhat enlarged mouth; preys on the young of other fish

Prognathochromis macrognathus, resembles a pike, with slender body, disproportionately large head and jaws, sharp teeth; preys on other fish

Pyxichromis parorthostoma, constricted head, upturned mouth with thickened lips; probably a feeding specialist but habits still unknown

Haplochromis obliquidens, teeth expanded and flat at the tips; grazes on algae

The Lake Victoria array is the greatest found anywhere in the world in a single group of fishes limited to a single body of water. Equally remarkable are the graduated steps that link the species in each of the adaptive classes, from the earliest stages of anatomical modification to the most extreme specialization in body form. Among the mollusk feeders, to illustrate the point, are found some species with only a few slightly enlarged pharyngeal teeth used by the fishes to crush the shells of their prey. Others, somewhat more advanced, possess a greater number of such teeth, many pebble-shaped, which are ground together on the mollusk shells with somewhat enlarged throat muscles. Still others, the extreme mollusk specialists, use pharyngeal bones packed with pebble teeth and powered by heavier throat muscles. Comparable morphoclines—series of species arrayed from the most generalized to the most specialized—occur among the cichlid algal feeders and predators on other fish.

All of the Lake Victoria cichlids appear to have descended from a single ancestral species that colonized Lake Victoria from nearby older lakes. The evidence, presented by Axel Meyer and his coworkers in 1990, is based on the degree of similarity in the genetic codes of the fishes. More specifically, fourteen of the species, representing nine genera, display very little variation in the nucleotide sequences of mitochondrial DNA, less diversity than occurs within the entire human species.

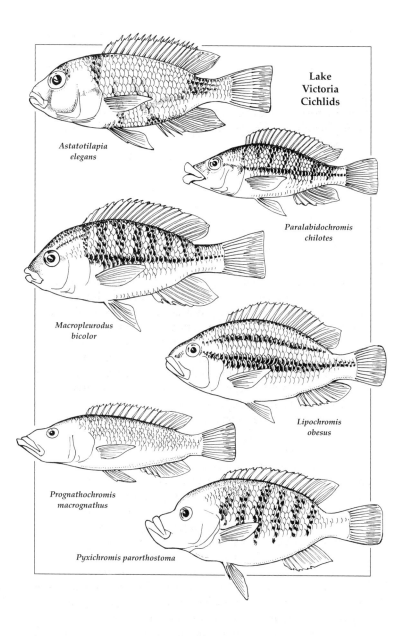

Lake Victoria Cichlids

Astatotilapia elegans

Paralabidochromis chilotes

Macropleurodus bicolor

Lipochromis obesus

Prognathochromis macrognathus

Pyxichromis parorthostoma

Most of the Lake Victoria cichlids belong to a larger group called *haplochromines*, an informal designation used in past years to suggest a common recent ancestry, a hypothesis now supported by the molecular data. Other haplochromines occur in Lake Malawi and Lake Tanganyika. They resemble the Lake Victoria species in mitochondrial DNA sequences, but are not nearly so close to them as the Victoria species are to one another.

Another remarkable feature of the Lake Victoria cichlids is the youth of their radiation. The lake is estimated to be between 250,000 and 750,000 years old. Using DNA sequences from the cytochrome *b* gene, where steady evolution serves as a "molecular clock" for animals generally, Meyer and his research team calculated that the full cichlid evolution was achieved in no more than 200,000 years.

The Victoria cichlids fall in the special category of adaptive radiation called *species flocks*: they comprise relatively numerous species of immediate common ancestry and are limited to a single well-isolated area such as a lake, river basin, island, or mountain range. The chief theoretical puzzle created by species flocks is the process by which they grow. How can populations split repeatedly into species within a closed habitat that has no geographical barriers? If haplochromines are typical of fishes and other vertebrates, they would seem to need intervening barriers, such as isthmuses of dry land that alternately rise and fall, to break the populations up and give the fragments time to diverge to species level. Lake Victoria appears on first scrutiny to have experienced too few such cycles in its history to have generated three hundred species from a single ancestor. We are pressed by the evidence to conclude that the cichlids speciate by sympatric means, in other words by splitting in two without first having been divided by a physical barrier. On the other hand, perhaps not. Recall that only a single trait, such as a change in courtship behavior or a shift in the mating season, is enough to create a new species. Consider also that Lake Victoria is a large body of water, almost 70,000 square kilometers in area, larger than the combined countries of Rwanda and Burundi nearby, and home for millions of small fishes. It is rimmed by a twisting shoreline over 24,000 kilometers long and occupied by numerous local habitats of widely varying character, from wave-lapped inlets to deep offshore basins whose bottoms never see sunlight. On many occasions during the hundred or more millennia of their history, cichlid populations must have contracted their ranges along the shoreline, breaking into local, temporarily isolated populations. In theory at least, differences in courtship or habitat preferences could be fixed within tens or hundreds of generations, a process fast enough many times over to have generated three hundred cichlid species during the lifespan of Lake Victoria.

The evolutionary explosion might have occurred all the more easily if cichlid fishes are prone to rapid evolution in the demonstrated manner of Darwin's finches. The clue to look for is a cichlid species somewhere that is the equivalent of the Cocos Island finch, that is, opposed by few if any competitors, highly variable, and multipurpose in its life habits. And, as if created for the delectation of life scientists and textbook writers, such an example exists. It does not occur in the crowded waters of the African Great Lakes, where competition and specialization in the species flock has reached a level of near saturation. To find the right conditions we must journey to the waters of Cuatro Ciénegas, in the state of Coahuila, northern Mexico. There the prescribed species, *Cichlasoma minckleyi*, lives in streams, ponds, and canals. A small, spotted perch-like fish, it coexists with several other fishes of similar size, including one additional cichlid. Its populations contain two radically different feeding types: the papilliform morph, with more slender jaws and slender teeth, and the molariform morph, with thicker jaws and pebble-shaped teeth. They look like entirely different species, but they are not. Both types freely breed together, thus constituting a single species. Both feed readily in the same places on the same wide range of small prey, including insects, crustaceans, and worms. When food grows scarce, however, the molariform cichlasomas—though not their papilliform associates—switch increasingly to snails, which they alone are able to crush with their stronger jaws and thick, flat teeth. By expanding their diet, the molariforms reduce competition with the papilliforms and persist through the hard times. It is easy to imagine such a species as *Cichlasoma minckleyi* invading a new body of water of the Lake Victoria class and then radiating into many niches in short time. The first step would almost certainly be a division into two full, reproductively isolated species, a papilliform *Cichlasoma* concentrating on insects and other soft-bodied prey, and a molariform *Cichlasoma* preying on snails and other mollusks.

Biologists have begun to search more systematically for such species on the threshold of adaptive radiation and hence macroevolution. What may prove to be the most dramatic case discovered so far, exceeding even the Cocos finch and the Mexican cichlasoma, is the arctic char, *Salvelinus alpinus*, a salmon-like fish found in lakes and rivers around the northern polar region. Very few other fish species occur with it, and the char has a relatively wide range of uncontested feeding niches to exploit. Many of the local populations contain several anatomically distinct forms with different food habits and growth rates. In Iceland's Thingvallavatn (*vatn*, lake), there are four such morphs: a large bottom feeder, a small bottom feeder, a predator on other fish, and a vegetarian that grazes algae. Skúli Skúlason and his fellow researchers at the University of Iceland found that these specialists differ one from another genetically but still interbreed freely, forming a single, highly plastic species. Like the finch and cichlasoma, the arctic char seems to be an adaptive radiation waiting to happen, or perhaps certain to happen given more time. The Arctic lakes in which the char lives were created by the retreat of the continental glaciers only a few thousand years ago.

Natural history becomes all the more pleasing and interesting when we look at it through the lens of evolutionary theory and search for the starbursts of adaptive radiation—and all the more foreboding when we learn how quickly these creations can be extinguished. The vast majority of radiated groups stay near the peak of diversity from thousands to millions of years. The cichlid fishes of Lake Victoria, in contrast, are disappearing almost instantaneously by this standard. They are being extinguished en masse by the giant Nile perch, a voracious predator introduced as a game fish by Ugandan officials in the 1920s. This "elephant of the water," reaching 2 meters in length and 180 kilograms, is literally eating its way through the cichlids southward from its northern point of introduction.

Where it has become dominant, more than half of the cichlid species have disappeared.

The confinement of groups like the African cichlid fishes and the Hawaiian honeycreepers to a single lake or archipelago renders them extremely vulnerable to environmental change, and they can be obliterated by a swipe of the human hand. In their company are groups of higher taxonomic order and broader geographic distribution that hang on as diminished remnants of a flamboyant past: cycads, crocodilians, lungfishes, rhinoceros, and other so-called living fossils. These too are being pushed by human activity to the brink of extinction, after tenures lasting millions of years. At the opposite extreme are a few select groups that have remained fully radiated for an equivalent period of time. These species display an astonishing array of radically different body forms and life cycles, and are widely distributed and abundant throughout the world. Among these old-money dynasties are ciliated protozoans, spiders, isopod crustaceans, and beetles, as well as one group that in my opinion deserves special attention in any serious talk about diversity and natural history: the sharks.

Sharks, fishes that compose the three superorders Squatinomorphii, Squalomorphii, and Galeomorphii of the class Chondrichthyes, shadows in the sea of our nightmares, lone-wolf predators of frightening quickness, questioners of the Darwinian importance of intelligence, have been on earth for 350 million years. They began as small, stiff-bodied cladodonts in the late Devonian period, then radiated and maintained high diversity in seas throughout the world until the beginning of the Permian period. At that time, 290 million years ago, they declined to a low level of diversity lasting for 100 million years. The survivors recovered, expanded a second time, and somehow passed full-blown through the great extinction spasm at the end of the Age of Dinosaurs. Today they are at least as diverse as they have ever been.

Sharks seen at a distance (fin and back roiling the surface for a heart-stopping moment, then the vaguely torpedo shape slipping into deeper water) may not seem to differ much from one species to the next except in size. In fact the 350 species found in the world vary immensely, so much so as to stretch the very definition of the word *shark*; we must infer their common ancestry from traits of internal anatomy in order to place them all within the same group. The archaic radiation of sharks is marked by differences among species far greater than those in the still youthful radiations of Darwin's finches and Lake Victoria cichlids. It is tempting to think that age has fine-tuned their specializations, pitted them against more competitors, extinguished a higher number over longer periods of time, and produced a generally tougher, more durable group of contemporary species.

If ever there was a prototypical shark in the popular imagination, it is most likely the tiger shark (*Galeocerdo cuvier*), the great fish sometimes called the garbage can of the sea. Reaching 6 meters in length, weighing up to a ton, tiger sharks are often attracted to harbors, where they consume almost anything sizable even hinting of animal protein. From the stomachs of such specimens have been retrieved fish, boots, beer bottles, bags of potatoes, coal, dogs, and parts of human bodies. One dissected giant contained three overcoats, a raincoat, a driver's license, one cow's hoof, the antlers of a deer, twelve undigested lobsters, and a chicken coop with feathers and bones still inside. Tiger sharks are man-eaters in a casual way, by which I mean taking swimmers not by design but by happenstance as part of their catholic diet.

Not so the great white shark, *Carcharodon carcharias*, famed killing machine and, with the saltwater crocodile and Sundarbans tiger, the last expert predator of man still living free. Great whites are by all odds the most frightening animals on earth—swift, relentless, mysterious (no one knows where they come from and where they go), and unpredictable. They are, in my admittedly emotional judgment, the full possessors of shark *arete*, the essence of sharkness. They are more completely predators, less scavengers, than tiger sharks, consuming a wide range of bony fish, other sharks, sea turtles, and—this is the salient trait as far as human beings are concerned—marine mammals such as porpoises, seals, and sea lions. The best place to encounter *Carcharodon carcharias* is in colder water around seal and sea-lion rookeries, such as those at California's Farallon Islands and at Dangerous Reef off South Australia. Great whites are dangerous simply because they fail to make a clear distinction between sea mammals and human swimmers. Divers in rubberized suits and swimmers on surf boards, lying prone with their arms out, are more than passable imitations of seals and sea lions. The shark sees what it thinks is the silhouette of its familiar prey, noses about for a while, makes up its mind, and sprints toward the swimmer at speeds over 40 kilometers an hour. At the last moment it rolls its eyes backward to protect them from impact. It opens its huge mouth wide, raising the head to project the tooth-ringed maw forward, and bites down hard for a second. Then it waits for the victim to bleed to death. In this interval, as it circles nearby, rescuers are often able to carry victims to safety without great danger to themselves.

I have been enchanted by both the reality and the image of the great white shark for years, back to a time when its natural history was little known and before it became a mythic horror in popular culture. There is much to admire about the species. It is the decathlon champion of the sea, wonderfully designed for the speed and strength to hunt big prey and endurance needed for long journeys through open water. Adults grow to immense size, reaching a known maximum of seven meters (23 feet) in length and 3,300 kilograms (3.6 tons) in weight. Its eyes are disproportionately large, an accommodation to the dark waters in which it hunts much of the time. Great whites have something like the classic tuna shape associated with fast pelagic fishes: body spindle-shaped and rigidly muscular, nose pointed to cut the water like the prow of a submarine. Ridges to the rear along each side of the trunk guide the flow of water evenly past the body. The powerful tail sweeps smoothly from side to side. The mouth, lined with parallel rows of serrated triangular teeth, hangs partly open in a fixed clown's grin, affirming the impression of human divers that the fish is glad to see them. Water flows continuously through the mouth and back over the gills, part of a ramjet system that feeds oxygen efficiently to the large, active body. The great white is warm-blooded, allowing it to cruise the colder waters of most of the world's oceans and to forage from the surface down to at least 1,300 meters.

In 1976 the naturalist Hugh Edwards, who watched for great whites from a shark cage in waters off the Albany whaling

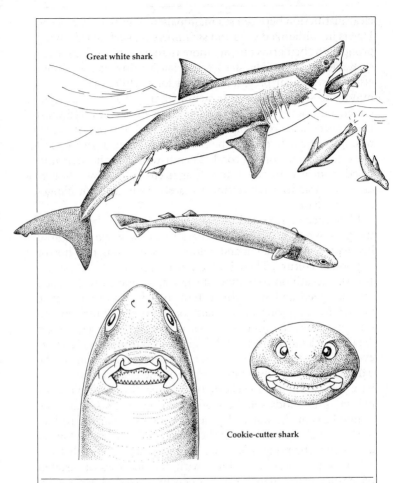

Great white shark

Cookie-cutter shark

The adaptive radiation of sharks has gone to extreme lengths, exemplified by the great white shark (*Carcharodon carcharias*), which preys extensively on seals and other marine mammals. Another, more bizarre form is the cookie-cutter shark (*Isistius brasiliensis*), a parasite that carves plugs of flesh from the bodies of marine mammals and large fish without killing them.

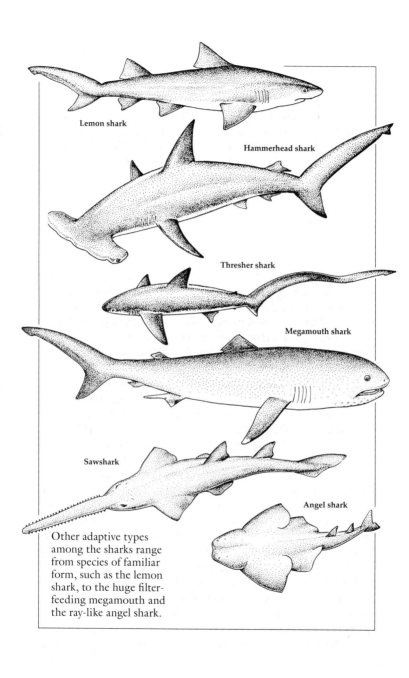

Lemon shark

Hammerhead shark

Thresher shark

Megamouth shark

Sawshark

Angel shark

Other adaptive types among the sharks range from species of familiar form, such as the lemon shark, to the huge filter-feeding megamouth and the ray-like angel shark.

station of Western Australia, turned to see a large male suspended two meters away. He later wrote:

> In all our lives there are milestones, important moments we
> remember long after. This was one of them. For the brief time
> of his appearance I drank in every detail of the shark—his eyes,
> black as night; the magnificent body; the long gill slits slightly
> flaring; the wicked white teeth; the pectoral fins like the wings
> of a large aeroplane; and above all the poise and balance in the
> water and the feeling conveyed of strength, power, and intel-
> ligence. To see the shark alive was a revelation. He was strong,
> he was beautiful. No dead shark or second-hand account could
> convey the vitality and presence of the live creature. A few sec-
> onds face to face were worth more than all the years of hearsay,
> pictures, and slack-jawed corpses.

Working off the image of these classic sharks, I will now argue that, given enough time, evolution can fine-tune and harden adaptive types to create the most extreme radiations. Of greatly different anatomy and biology from the tiger and great white sharks, for example, is the cookie-cutter shark (*Isistius brasiliensis*). It is not a predator at all but a parasite, of porpoises, whales, bluefin tuna, and even other sharks. Only half a meter in length, shaped like a cigar, the cookie-cutter has a curving row of huge teeth on its lower jaw. It thrusts its maw into the bodies of its victims and twists to slice out 5-centimeter-wide conical plugs of skin and flesh. For many years the circular scars on porpoises and whales were a mystery, attributed variously to bacterial infection or an unknown invertebrate parasite, until the true habits of the small sharks were discovered in 1971. Cookie-cutters also attack nuclear submarines, taking unnu-tritious bites from the neoprene coating of sonar domes and hydrophone arrays. The cookie-cutter shark passes what I like to call the test of a complete adaptive radiation: the existence of a species specialized to feed on other members of its own group, other products of the same adaptive radiation.

Equally specialized in a wholly different direction are the filtering sharks, gigantic fishes that cruise placidly near the surface of the open ocean, seining out and swallowing huge quantities of copepod crustaceans and other small planktonic animals in the manner of the baleen whales. The whale shark

(*Rhincodon typus*), reaching 13 meters in length and many tons, may be the largest fish that ever lived. At the opposite extreme is the green lanternshark (*Etmopterus virens*), which at 23 centimeters—the size of a large goldfish—is the smallest of all sharks.

Other major adaptive types expand the cavalcade of the living sharks:

Horn sharks (example, *Heterodontus japonicus*), inshore bottom dwellers that use their hard, molar-like teeth to feed on mollusks

Frilled sharks (example, *Chlamydoselachus anguineus*), deep-sea dwellers with elongated, eel-shaped bodies and fins, teeth shaped like fishhooks

Angel sharks (example, *Squatina dumerili*), squat bottom dwellers outwardly resembling rays more than sharks, but anatomically sharks

Thresher sharks (example, *Alopias vulpinus*), large pelagic forms that sometimes run in pairs and stun smaller fish by lashing them with long, whiplike tails

Unknown species of sharks almost certainly swim the seas. Some are likely to be very large. I base this conjecture on the megamouth shark (*Megachasma pelagios*), discovered in 1976. The first specimen was hauled up from deep water off Hawaii by the United States Navy after it became entangled in a cargo parachute used as a sea anchor. It was almost 5 meters long and weighed 750 kilograms. To the surprise of the navy and consulting ichthyologists alike, it was unlike any other shark seen to that time. Four more individuals of the same species have been encountered since. Two were caught in gill nets off California, and one each washed ashore at Japan and Western Australia.

Megamouth is so different in anatomy from all previously known sharks that it has been placed in a taxonomic family of its own, the Megachasmidae. Its most striking feature is an enormous maw, used to draw in water and filter copepods, euphausiid shrimps, and other small planktonic animals on which it feeds. Megamouth is thus in the same ecological guild as the whale shark, as well as the huge basking shark of northern seas. Its body is cylindrical and flabby, its eyes small, its movements stiff and slow. It flees into deep water at small disturbances. Its upper jaw and palate are covered with a silvery, iridescent lining, possibly a deposit of guanine or other reflective waste material. When the Los Angeles specimen became entangled in a gill net, scuba-suited researchers were able to implant transmitters into its body and track it at sea for two days. In that time the shark cruised about 10 to 15 meters below the surface at night and descended to 200 meters during the day. This vertical migration is typical of fishes of the deep scattering layer, the thick concentrations of organisms, detectable by sonar, that travel up and then back down again through each twenty-four hour cycle. Megamouth's deep submergence during the day and shy, aversive behavior overall perhaps explain why the species remained undiscovered for so long.

The metaphor of dynastic succession I chose at the beginning to describe the turnover of adaptively radiated groups implies a balance of nature. One dynasty, this conception holds, cannot tolerate another dynasty of closely similar kind. A limit to organic diversity exists so that when one group radiates into a part of the world, another group must retreat. Because evolution is so dismayingly idiosyncratic, the balance of nature cannot be ranked as a law of biology. But at least it is a rule, a statistical trend: dominant, expanding groups tend to replace those groups found in the same places that are ecologically most similar to them.

The displacement of one group by another is seldom if ever a blitzkrieg. Almost always it is a sitzkrieg, with the newer group pushing gradually into the terrain of the older group, enveloping its rival slowly, and replacing it species by species. Just as often the replacement is favored by decimation of the older dynasty through climatic change or loss of food supply. The rise of the mammals after the fall of the dinosaurs is the textbook case, but examples exist among corals, mollusks, archosaur reptiles, ferns, conifers, and other organisms following the demise of their competitors in one of the major extinction spasms. These temporary winners seized opportunities provided by vacant niches just as the Hawaiian honeycreepers and Lake Victoria haplochromine fishes invaded newly created environments. Their success was global in scope, however, not limited to an archipelago or lake.

We now come to the interesting question implied by the balance of nature: what happens when two full-blown, closely similar dynasties meet head on? If it were possible to play God with geological spans of time to wait and watch, the ideal experiment would be this: allow two isolated parts of the world to fill up with independent adaptive radiations of plants and animals, so that the majority of species in each theater have close ecological equivalents in the other theater; then connect the two regions with a bridge and see what happens. When the organisms intermingle, would those from one theater replace the other, so that a single biota comes to occupy the entire range?

The experiment has in fact been performed once in relatively recent geological time, and we can deduce a great deal of what happened by comparing fossil and living species. Two and a half million years ago the Panama isthmus rose above the sea, allowing the mammals of South America to mix with the mammals of North and Central America. First I should explain that the contemporary mammals of the world are primarily the products of three great adaptive radiations, and three only. The reason is that it takes an entire continent to spawn a mammalian radiation. For insects, a single island is adequate. Beetle species have proliferated luxuriously on St. Helena in the South Atlantic, Rapa in the South Pacific, and Mauritius in the Indian Ocean. Had flightless mammals reached these same specks of land—which they did not and probably could not prior to the coming of man—it is doubtful that the species would have multiplied. Mammals, even rats and mice, are simply too big, active, and wide-ranging. To produce an adaptive radiation of the honeycreeper or cichlid magnitude, their species need a continent.

The first of the three continents on which mammalian radiation did attain full expression is Australia. In biogeographical terms Australia is only an extremely large island, having been isolated from the rest of the world since the breakup of the Gondwanaland super continent more than 200 million years

ago. The second body of land spacious enough for mammalian radiation is the "World Continent," comprising Africa,
Europe, Asia, and North America as far south as the southern
rim of the Mexican plateau. The World Continent has been
more or less cohesive throughout the Age of Mammals, during
the past 66 million years, because the closeness of its parts has
allowed many kinds of plants and animals to emigrate from one
to the next. North America, the most isolated of the elements,
was joined to Europe across present-day Greenland and Scan-
dinavia during the early Age of Mammals. Alaska and northeastern Siberia have been connected off and on by land bridges,
the most recent about 10,000 years ago. The third continental
center of mammalian evolution is South America, which was
isolated during the breakup of Gondwanaland, drifted north,
and was finally linked solidly to North and Central America
2.5 million years ago.

For most people, the mammals of the World Continent are
"typical" and "true" mammals, simply because they are the
most familiar. These are the animals we were born and raised
with. The mammals of Australia and South America are nevertheless highly evolved in their own right.

Today three major groups contribute to the composition of
the native, prehuman fauna of Australia. The first are mono-
tremes or egg-laying mammals, remnants of an ancient and
largely superseded radiation—faded lines on the palimpsest.
They include the duck-billed platypus, an aquatic form that
looks as if it were fabricated from the head of a duck and the
body of a web-footed muskrat; and the short-beaked echidna,
a land dweller best described as a porcupine with the tapered
cylindrical snout of an anteater. The second group are placental
mammals, which carry their young attached to a placenta in the
uterus. Relative newcomers to Australia, yet already composing a third of the species, they comprise a wide array of bats
and rodents. Their immediate ancestors island-hopped across
Indonesia to reach northern Australia and then spread through
parts of the continent.

The third native group of Australia are marsupials, mammals
that give birth to their young as tiny fetuses and carry them to
an advanced stage in a belly pouch (marsupium). It is this third,

relatively ancient and still dominant, group that has converged with the greatest fidelity to the World Continent placental fauna. Here are the chief mammalian analogues across the two regions and the adaptive roles they fill:

Australian marsupial mammals	World Continent placental mammals	Adaptive type
Dibbler (*Parantechinus apicalis*)	Mice	Small, secretive omnivores
Jerboa marsupial mouse (*Antechinomys spenceri*)	Jerboas, kangaroo rats	"Jumping mice" of desert areas; insect feeders in Australia
Bandicoots (*Macrotis lagotis, etc.*)	Rabbits, hares	Long, saltatory hind legs; diet of grass and other plant materials; some omnivorous
Quolls or dasyures (*Dasyurus geoffroii* and *D. viverrinus*)	Small cats	Predators of small mammals, reptiles, and birds
Gliders (*Petaurus sciureus, etc.*)	Flying squirrels	Arboreal gliders, using membranes on sides of body; mostly herbivores
Anteater or numbat (*Myrmecobius fasciatus*)	Anteaters	Feed on termites with long, flexible, sticky tongues
Tree wallabies (*Dendrolagus lumholtzi, etc.*)	Catarrhine monkeys	Arboreal, mostly herbivores
Marsupial mole (*Notoryctes typhlops*)	Moles	Subterranean, feed on insects and worms
Wombats (*Lasiorhinus krefftii, etc.*)	Woodchuck	Secretive, burrowing herbivores

Australian marsupial mammals	World Continent placental mammals	Adaptive type
Large kangaroos (*Macropus robustus,* etc.)	Horses, antelopes, other ungulates	Grazers, using chisel-like front teeth and broad grinding molars
Tasmanian devil (*Sarcophilus harrisi*)	Wolverine	Predators of small animals
Tasmanian wolf or thylacine (*Thylacinus cynocephalus*)	Wolves, big cats	Predators of kangaroos, other mammals, and birds

The stage set, we have arrived at the moment of the experiment. The mammalian radiation in South America was as expansive as that in Australia, and its convergence to the World Continent fauna was even closer. Yet the look-alike species are much less familiar—toxodonts, marsupial cats, macrauchenians, glyptodonts—because so few lived to be seen by human beings. They disappeared at about the time the Panama land bridge rose and elements of the World Continent fauna poured into South America. Others that survived failed to diversify at the same rate as the northern invaders. In the exchange North and Central America contributed more to South America than the reverse.

Prior to the overland migrations back and forth, known as the Great American Interchange, the old endemic mammals of South America had been assembled during two waves of radiation and partial extinction. The first began near the end of the Mesozoic era, about 70 million years ago, and climaxed during the succeeding 40 million years. The early stocks of these archaic mammals had risen even earlier in Mesozoic times, in the remnants of Gondwanaland, when South America was still close to Africa and Antarctica and dinosaurs prevailed. Now relieved of the constricting influence of dinosaurs, the mammals expanded to fill the abandoned niches. In the grasslands lived litopterns superficially similar to the "true" horses of the World Continent, members of the family Equidae with which human beings were to evolve in close intimacy. Litopterns possessed fully developed hooves and skulls equipped for grazing long before these specializations evolved in the equids. Other litopterns were more like camels. Toxodonts variously resembled rhinos and hippopotamuses, while some astrapotheres and pyrotheres came passably close to tapirs and elephants. Argyrolagids, good imitations of kangaroo rats but with enormous eyes set far back on the skull, bounced about on springy hind legs. Borhyaenids, whose species resembled shrews, weasels, cats, and dogs, were among the main predators of other mammals. A sabertooth marsupial cat, *Thylacosmilus*, bore an amazing resemblance to the sabertooth tigers of the World Continent fauna.

The herbivores of old South America were mostly placentals, and the carnivores were marsupials. Paleontologists are not sure why this was so, or why in contrast the mammals were primarily marsupials in Australia and placentals in the World Continent. It might have been no more than the luck of the draw: which group penetrated the major adaptive zones first, radiated, and preempted the other. We may never know, for the number of continents with which to test the hypothesis ran out at three. (Being limited to one planet and a small number of continents and archipelagoes is the curse of evolutionary biology.)

About 30 million years ago a long, slow second wave worked its way into South America, this time from the north across island stepping stones. North and South America were still separated by a broad seaway running through the Bolivar Trough. Present-day Central America consisted then of islands scattered across the Trough, with the newly formed West Indies close by and drifting eastward. A few mammal species were able to expand their ranges southward from one such island to the next and eventually onto the continent of South America itself. These island hoppers included an early monkey-like species of primate, which proceeded to proliferate into howler monkeys, spider monkeys, marmosets, titis, tamarins, capuchins, sakis, and other dwellers of the forest canopy. Many possessed prehensile tails, the hallmark of the New World species (if a monkey can hang from its tail, it is from the American tropics). Even more successful members of the second wave were ancient ancestors of the guinea pigs, the rabbitty viscachas,

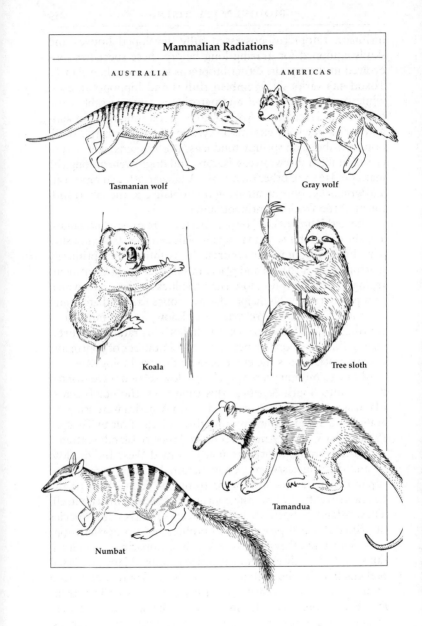

Mammalian Radiations

AUSTRALIA

AMERICAS

Tasmanian wolf

Gray wolf

Koala

Tree sloth

Tamandua

Numbat

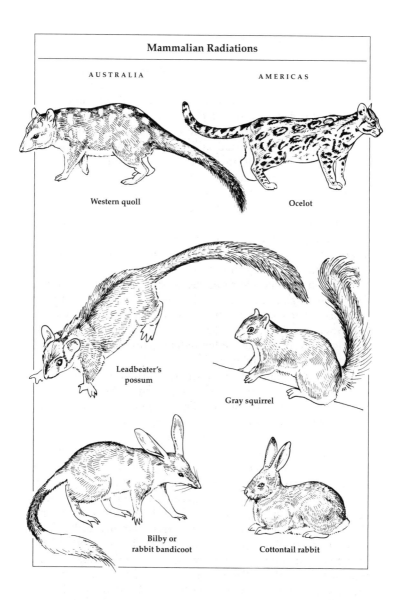

Mammalian Radiations

AUSTRALIA   AMERICAS

Western quoll   Ocelot

Leadbeater's
possum

Gray squirrel

Bilby or
rabbit bandicoot   Cottontail rabbit

the porcupines, and the horsefaced aquatic capybaras, largest rodents on earth.

A thousand ages in thy sight are like an evening gone. If we could travel back in time to the mid-Cenozoic savanna of South America, when the continent was still surrounded by straits and oceans, we might think ourselves on safari somewhere in a national park in modern Africa. Everything would be off somewhat, distorted and out of focus, like a picture studied through an astigmatic lens, yet it would seem *almost* normal. Say we are there on the edge of a lake, early one sunny morning, turning our gaze slowly through a full circle. The vegetation looks much like modern savanna. Out in the water a crash of rhinoceros-like animals browse belly deep through a bed of aquatic plants. On the shore something resembling a large weasel drags an odd-looking mouse into a clump of shrubs and disappears into a hole. A creature vaguely like a tapir watches immobile from the shadows of a nearby copse. Out of the high grass a big, cat-like animal suddenly charges a herd of—what?—animals that are not quite horses. Its mouth is thrown open nearly 180 degrees, knife-shaped canines projecting forward. The horse look-alikes panic and scatter in all directions. One stumbles, and . . .

The simulacrum that was ancient South America is all the more remarkable because its mammals had nothing to do with those in the rest of the world. They were evolved in this replicate mega-experiment along different lines out of different stocks, yet roughly to the same effect.

If you limit your scrutiny to faunas isolated for a long time, and if you are willing to settle for loose standards in scoring similarity of anatomy and niche, evolution is predictable. But wild cards always break the pattern. Back to Cenozoic South America, we turn at the sound of tree branches snapping and crashing to the ground, as some large mammal pulls them down to feed. We expect elephants but find ground sloths, immense, ungainly animals covered with thick reddish fur, who gather in the foliage with clawed hands and chew the leaves and tender branches with vaguely horse-like heads. They fill the elephant niche but use different tools. And now a stunning surprise. A *Titanis* appears, a flightless carnivorous bird standing 3 meters (10 feet) tall, its eagle head tipped by a massive hooked beak 38 centimeters (15 inches) long. Loping along on stilt legs like a malevolent ostrich, it flicks its head to left and right in search of prey, which can be as large as a deer. *Titanis* was only the largest of a variety of phororhacoid ground birds; some were as small as geese. Never before or since have mammals faced anything like the phororhacoids, except during their earliest evolution in the Age of Dinosaurs. In South America *Titanis* and its relatives must have been serious rivals to the borhyaenids and other carnivorous marsupials. Since anatomists consider birds as a whole to be direct descendants of dinosaurs, close enough to be called dinosaurs (although that is stretching it), the phororhacoids might be called the final echo of the ruling reptiles.

Phororhacoids, saber-toothed marsupial cats, toxodont rhinoceroids—all that splendid assemblage is gone. We will never ride a litoptern or feed peanuts to a long-trunked pyrothere in the zoo. Although biological history is a flow of events with causes and effects that can in principle be rationally joined, one extraneous accident can change everything. When the Bolivar Trough disappeared and the Panamanian land bridge rose across its center less than 3 million years ago, the final wave of mammals rolled swiftly into South America. Many of the World Continent mammals, which had been blocked for millions of years by the Bolivar straits, now simply walked onto the continent. Most traveled along corridors of grassland, which at the time extended southward along the eastern slopes of the Andes all the way to Argentina.

So successful was this incursion that about half of the most familiar South American mammals today are of geologically recent World Continent origin: jaguars, ocelots, margays, peccaries, tapirs, coatis, kinkajous, bush dogs, giant otters, alpacas, vicunas, llamas, and the recently extinct mastodons. South America's autochthons flowed in the opposite direction. For a while at least, North America was home to giant sloths, armadillos, possums, glyptodonts, porcupines, anteaters, and toxodonts. *Titanis* spread all the way to Florida.

The Great American Interchange resulted in a sharp increase for a time in the mammalian diversity on both continents. Consider first the taxonomic level of the family. Examples of mammalian families are the Felidae, or cats; Canidae, dogs and their relatives; Muridae, the common mice and rats; and of course Hominidae, human beings. The number of mammalian families in South America before the interchange was thirty-two. It rose to thirty-nine soon after the isthmus connection and then subsided gradually to the present-day level of thirty-five. The history of the North American fauna was closely comparable: about thirty before the interchange, rising to thirty-five, and subsiding to thirty-three. The number of families crossing over was about the same from both sides.

When biologists see a number go up following a disturbance and then fall back to the original level, whether body temperature, density of bacteria in a flask, or biological diversity on a continent, they suspect an equilibrium. The restoration of the numbers of mammalian families in both North and South America points to such a balance of nature. In other words, there appears to be a limit to diversity, in the sense that two very similar major groups cannot coexist in their fully radiated condition. A closer examination of the ecological equivalents on both continents, dwellers in the same broad niche, reinforces this conclusion. In South America marsupial big cats and smaller marsupial predators were replaced by their placental equivalents. Toxodonts gave way to tapirs and deer. Still some unusual specialists—the wild cards—were able to persist. Anteaters, tree sloths, and monkeys continue to flourish in South America, while armadillos are not only abundant throughout tropical America but are represented by one species that has expanded its range throughout the southern United States.

In general, where close ecological equivalents met during the interchange, the North American elements prevailed. They also attained a higher degree of diversification, as measured by the number of genera. A genus is a group of related species less well demarcated than those composing a family. The genus *Canis*, for example, comprises domestic dogs, wolves, and coyotes; other genera in the dog family Canidae include *Vulpes* (foxes), *Lycaon* (African wild dogs), and *Speothos* (South American bush dogs). During the interchange, the number of genera rose sharply in both North and South America and remained high thereafter. In South America it began at about 70 and has reached 170 at the present time. The swelling of numbers has come principally from speciation and radiation of the World Continent mammals after they arrived in South America. The old, pre-invasion South American elements were not able to diversify significantly in either North or South America. So the mammals of the western hemisphere as a whole now have a strong World Continent cast. Nearly half of the families and genera of South America belong to stocks that have immigrated from North America during the past 2.5 million years.

Why did the World Continent mammals prevail? No one knows for sure. The answer has been largely concealed by complex events imperfectly preserved in the fossil record—the paleontologist's equivalent of the fog of war. The question remains before us, part of the larger unsolved problem toward which our understanding of dynastic succession is directed. Evolutionary biologists keep coming back to it compulsively, as I did while waiting for the night storm at Fazenda Dimona, in the Brazilian Amazon, surrounded by mammals of World Continent origin. What comprises success and dominance? Before returning a last time to the Great American Interchange, let me try to rephrase these important terms into more useful conceptions.

Success in biology is an evolutionary idea. It is best defined as the longevity of a species with all its descendants. The longevity of the Hawaiian honeycreepers will eventually be measured from the time the ancestral finch-like species split off from other species, through its dispersal to Hawaii, and finally to that time when the last honeycreeper species ceases to exist.

Dominance, in contrast, is both an ecological and an evolutionary concept. It is best measured by the relative abundance of the species group in comparison with other, related groups and by the relative impact it has on the life around it. In general, dominant groups are likely to enjoy greater longevity. Their populations, simply by being larger, are less prone to sink all the way to extinction in any given locality. With greater numbers, they are also better able to colonize more localities, increasing the number of populations and making it less likely that every population will suffer extinction at the same time. Dominant groups often are able to preempt the colonization of potential competitors, reducing still further the risk of extinction.

Because dominant groups spread farther across land and sea, their populations tend to divide into multiple species that adopt different ways of life: dominant groups are prone to experience adaptive radiations. Conversely, dominant groups that have diversified to this degree, such as the Hawaiian honeycreepers and placental mammals, are on average better off than those composed of only a single species; as a purely incidental effect, highly diversified groups have better balanced their investments and will probably persist longer into the future. If one species comes to an end, another occupying a different niche is likely to carry on.

The mammals of North American origin proved dominant as a whole over the South American mammals, and in the end they remained the more diverse. Over two million years into the interchange, their dynasty prevails. To explain this imbalance, paleontologists have forged a widely held theory, an evolutionary-biologist kind of theory, in other words a rough consensus that violates the minimal number of facts. The fauna of North America, they note, was not insular and discrete like that of South America. It was and remains part of the World Continent fauna, which extends beyond the New World to Asia, Europe, and even Africa. The World Continent is by far the larger of the two land masses. It has tested more evolutionary lines, built tougher competitors, and perfected more defenses against predators and disease. This advantage has allowed its species to win by confrontation. They have also won by insinuation: many were able to penetrate sparsely occupied niches more decisively, radiating and filling them quickly. With both confrontation and insinuation, the World Continent mammals gained the edge.

The testing of this theory has just begun. Right or wrong, whether decisive in empirical support or not, its pursuit alone promises to link paleontology in interesting new ways to ecology and genetics. That synthesis will continue as the study of biological diversity expands in widening circles of inquiry to other disciplines, to other levels of biological organization and farther reaches of time.

CHAPTER EIGHT

The Unexplored Biosphere

IN 1983 a previously unknown creature, *Nanaloricus mysticus*, which vaguely resembles an ambulatory pineapple, was described as a new species, new genus, new family, new order, and new phylum of animals. Barrel-shaped, a quarter of a millimeter long (one-hundredth of an inch), sheathed in neat rows of scales and spines, it possesses a snout up front and, when young, a pair of flippers shaped like penguin wings at the rear. *Nanaloricus mysticus* lives in the gravel and coarse sand 10 to 500 meters deep on ocean bottoms around the world. Almost nothing is known about its ecology and behavior, but we can guess from the body shape and armament that it burrows like a mole in search of microscopic prey.

To place a species in its own phylum, the decision taken in this case by the Danish zoologist Reinhardt Kristensen, is a bold step. He said, and other zoologists agreed, that *Nanaloricus mysticus* is anatomically distinct enough to deserve placement alongside major groups such as the phylum Mollusca, comprising all the snails and other mollusks, and phylum Chordata, consisting of all the vertebrates and their close relatives, it is like ranking Liechtenstein with Germany, Bhutan with China. Kristensen named the new phylum Loricifera, from the Latin *lorica* (corset) and *ferre* (to bear). The corset in this case is the cuticular sheath that encases most of the body.

The loriciferans—now a larger group, since about thirty other species have been discovered in the past decade—live among a host of other tiny bizarre animals found in spaces between grains of sand and gravel on the ocean bottom. They include gnathostomulids (raised to phylum status in 1969), rotifers, kinorhynchs, and cephalocarid crustaceans. This Lilliputian fauna is so poorly known that most of the species lack a scientific name. They are nevertheless cosmopolitan and extremely abundant. And they are also almost certainly vital to the healthy functioning of the ocean's environment.

The existence of loriciferans and their submicroscopic associates is emblematic of how little we know of the living world, even that part necessary for our own existence. We dwell on a largely unexplored planet. Consider that our earth is a planet of a certain size, its continents and seas are arrayed in such and such a way, and all its life is based on a single nucleic-acid code, in the same sense that all of written English is based on twenty-six letters. In the universe there must exist a vast array of life-bearing planets of other sizes and geographies, and perhaps different codes as well, each combination fixing a particular level of natural biodiversity. Several lines of evidence, including the history of adaptive radiations, suggest that earth is at or close to its own particular capacity. But what exactly is that capacity? No one has the faintest idea; it is one of the great unsolved problems of science.

In the realm of physical measurement, evolutionary biology is far behind the rest of the natural sciences. Certain numbers are crucial to our ordinary understanding of the universe. What is the mean diameter of the earth? It is 12,742 kilometers (7,913 miles). How many stars are there in the Milky Way, an ordinary spiral galaxy? Approximately 10^{11}, 100 billion. How many genes are there in a small virus? There are 10 (in ϕX174 phage). What is the mass of an electron? It is 9.1×10^{-28} grams. And how many species of organisms are there on earth? We don't know, not even to the nearest order of magnitude. The number could be close to 10 million or as high as 100 million. Large numbers of new species continue to turn up every year. And of those already discovered, over 99 percent are known only by a scientific name, a handful of specimens in a museum, and a few scraps of anatomical description in scientific journals. It is a myth that scientists break out champagne when a new species is discovered. Our museums are glutted with new species. We don't have time to describe more than a small fraction of those pouring in each year.

With the help of other systematists, I recently estimated the number of known species of organisms, including all plants, animals, and microorganisms, to be 1.4 million. This figure could easily be off by a hundred thousand, so poorly defined are species in some groups of organisms and so chaotically organized is the literature on diversity in general. More to the point, evolutionary biologists are generally agreed that this estimate is less than a tenth of the number that actually live on earth.

To see why the biodiversity audit is so far short of reality, consider the phylum Arthropoda, which includes all the insects, spiders, crustaceans, centipedes, and related organisms with jointed, chitinous exoskeletons. About 875,000 arthropod species have been described, or more than half the total for all organisms. Insects in particular, with 750,000 species known, compose the unchallenged dynasty of animals in the small to small-medium range on the land, and they have been thus installed since late Carboniferous times, more than 300 million years ago. Their terrestrial corulers in the plant kingdom for the past 150 million years have been the angiosperms, or flowering plants, constituting about a quarter million species, 18 percent of the total known for all organisms.

The immense diversity of the insects and flowering plants combined is no accident. The two empires are united by intricate symbioses. The insects consume every anatomical part of the plants, while dwelling on them in every nook and cranny. A large fraction of the plant species depend on insects for pollination and reproduction. Ultimately they owe them their very lives, because insects turn the soil around their roots and decompose dead tissue into the nutrients required for continued growth.

So important are insects and other land-dwelling arthropods that if all were to disappear, humanity probably could not last more than a few months. Most of the amphibians, reptiles, birds, and mammals would crash to extinction about the same time. Next would go the bulk of the flowering plants and with them the physical structure of most forests and other terrestrial habitats of the world. The land surface would literally rot. As dead vegetation piled up and dried out, closing the channels of the nutrient cycles, other complex forms of vegetation would die off, and with them all but a few remnants of the land vertebrates. The free-living fungi, after enjoying a population explosion of stupendous proportions, would decline precipitously, and most species would perish. The land would return to approximately its condition in early Paleozoic times, covered by mats of recumbent wind-pollinated vegetation, sprinkled

Number of Living Species of All Kinds of Organisms Currently Known

(According to Major Group)

ALL ORGANISMS: TOTAL SPECIES, 1,413,000

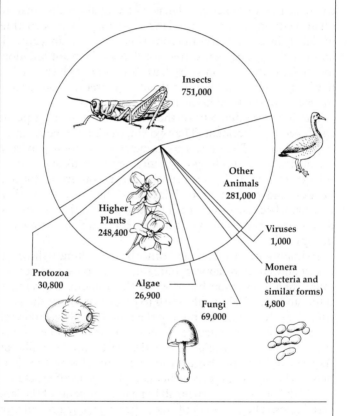

Insects
751,000

Other
Animals
281,000

Higher
Plants
248,400

Viruses
1,000

Protozoa
30,800

Algae
26,900

Fungi
69,000

Monera
(bacteria and
similar forms)
4,800

Insects and higher plants dominate the diversity of living organisms known to date, but vast arrays of species remain to be discovered in the bacteria, fungi, and other poorly studied groups. The grand total for all life falls somewhere between 10 and 100 million species.

Number of Living Species of Higher Plants Currently Known

(According to Major Group)

HIGHER PLANTS: TOTAL SPECIES, 248,000

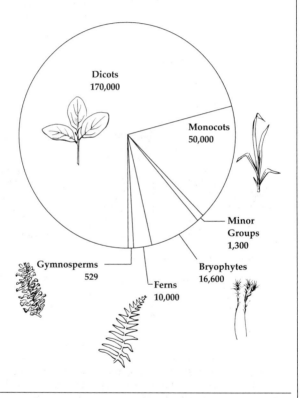

Dicots
170,000

Monocots
50,000

Minor
Groups
1,300

Gymnosperms
529

Ferns
10,000

Bryophytes
16,600

The plant diversity of the world consists primarily of angiosperms (flowering plants), which in turn make up grasses and other monocots and a huge variety of dicots, from magnolias to asters and roses. Most flowering plants live on the land; algae (26,900 known species) prevail in the sea.

Number of Living Animal Species Currently Known

(According to Major Group)

ANIMALS: TOTAL SPECIES, 1,032,000

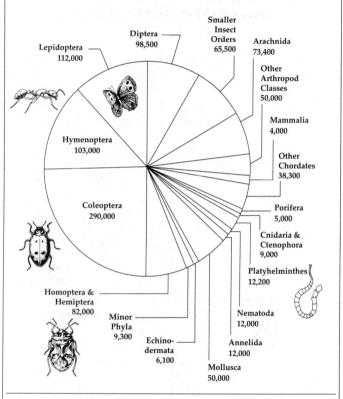

Diptera
98,500

Smaller
Insect
Orders
65,500

Arachnida
73,400

Lepidoptera
112,000

Other
Arthropod
Classes
50,000

Mammalia
4,000

Hymenoptera
103,000

Other
Chordates
38,300

Coleoptera
290,000

Porifera
5,000

Cnidaria &
Ctenophora
9,000

Platyhelminthes
12,200

Homoptera &
Hemiptera
82,000

Nematoda
12,000

Minor
Phyla
9,300

Echino-
dermata
6,100

Annelida
12,000

Mollusca
50,000

Among animals known to science, the insects are overwhelming
in number. Because of this imbalance, most animal species live on
the land; but most phyla (Echinodermata, etc.), the highest units
of classification, are found in the sea.

with clumps of small trees and bushes here and there, largely devoid of animal life.

Arthropods are thus all around us, life-giving, and we have never taken their measure. There are far more species than the 875,000 given a scientific name to date. In 1952 Curtis Sabrosky, working at the U.S. Department of Agriculture, conjectured on the basis of the flood of new species pouring continuously into museums that there are about 10 million kinds of insects among an unknown diversity of other arthropods. In 1982 Terry Erwin of the National Museum of Natural History raised the ante threefold, estimating that there are 30 million species of arthropods in the tropical forest alone, of which the great majority are insects. Most of the variety, he said, is concentrated in the crowns of rain-forest trees. This layer of leaves and branches, which conducts most of the photosynthesis for the forest, was already known to be rich in animal diversity. Yet it has been inaccessible because of the height of the trees, 30–40 meters, the slick surface of the trunks, and the swarms of stinging ants and wasps awaiting human climbers at all levels.

To overcome these difficulties, entomologists developed the "bug bomb," a method for blowing fogs of a rapidly acting insecticide from the ground up into the treetops, enveloping the arthropods and chasing them out of their hiding places even as it kills them. The specimens are then collected as they fall dying to the ground. The particular fogging procedure used by Erwin and his research team in South and Central America is conducted mostly at night. Walking into the rain forest in the evening, they select a tree for sampling and lay out a grid of 1-meter-wide funnels beneath it. The funnels feed into bottles partly filled with 70 percent alcohol, the specimen preservative of choice. In the predawn hours next morning, when the wind in the treetops dies down to a minimum, the crew forces the insecticide upward into the canopy from a motor-driven "cannon." They continue the treatment for several minutes. Then they stand by for five hours while the dead and dying arthropods rain down in the thousands, many falling into the funnels. Finally, the collected specimens are then sorted, roughly classified to major taxonomic group (such

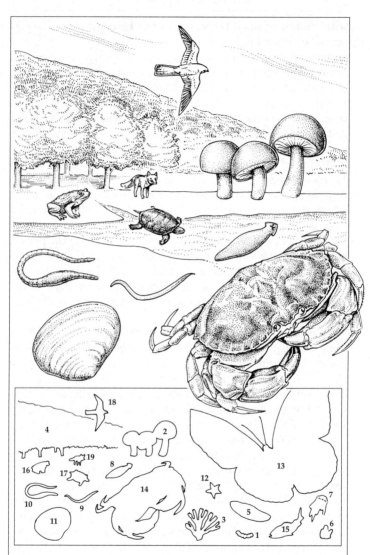

The species-scape. The size of the representative organism in each group has been made to be roughly proportional to the number of species currently known to science. The code and number of species are given below. Viruses and some minor invertebrate groups have been omitted.

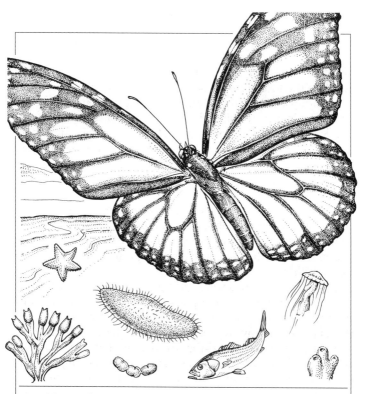

1. Monera (bacteria, cyanobacteria), 4,800
2. Fungi, 69,000
3. Algae, 26,900
4. Higher plants, 248,400
5. Protozoa, 30,800
6. Porifera (sponges), 5,000
7. Cnidaria and Ctenophora (corals, jellyfish, comb jellies, and relatives), 9,000
8. Platyhelminthes (flatworms), 12,200
9. Nematoda (roundworms), 12,000
10. Annelida (earthworms and relatives), 12,000
11. Mollusca (mollusks), 50,000
12. Echinodermata (starfish and relatives), 6,100
13. Insecta, 751,000
14. Noninsectan arthropods (crustaceans, spiders, etc.), 123,400
15. Fishes and lower chordates, 18,800
16. Amphibians, 4,200
17. Reptiles, 6,300
18. Birds, 9,000
19. Mammals, 4,000

as ants, leaf beetles, or jumping spiders), and sent to specialists for further study.

Erwin himself studied the beetles of the canopy. He made some counts in a small sample in a Panamanian rain forest, then proceeded by an arithmetical progression to arrive at a guess of the total number of arthropod species in tropical forests worldwide. Erwin first estimated that there are 163 species of beetles living exclusively in the canopies of one tree species, the leguminous *Luehea seemannii*. There are about 50,000 tropical tree species in all, so that if *Luehea seemannii* is typical, the total number of canopy-dwelling tropical beetle species is 8,150,000. Beetles represent about 40 percent of all species of insects, spiders, and other arthropods. If that proportion also exists in the tropical canopy, the number of arthropod species in the habitat comes to about 20 million. There are about twice as many arthropod species in the rain-forest canopy as on the ground, so that the total number of tropical species might well be 30 million.

Erwin's calculations were an important step forward in the study of biodiversity. The explicit figure he arrived at initially, however, is somewhat like an upside-down pyramid balanced on its point. At any step on the road to the final total of 30 million tropical-forest arthropods, the number of species can be shifted drastically up or down by changing assumptions. If the true total is within 10 million of that number either way, it will be sheer luck.

So are there really such a great number of beetle species on each tree species worldwide? Data are very sparse, but legumes such as *Luehea seemannii* seem to support a greater variety of insects than most other kinds of trees. This could move the total species number down by millions of species. Are the arthropods found on a tree species the same everywhere that particular tree species occurs? A great deal of evidence suggests that there is often a change in the kinds of beetles found on the same kind of tree from one locality to the next. This could move the total number back up. Are 10 percent of beetle species found on a particular tree species restricted to that tree? A change in this parameter, on which little exact information is available in the tropics, could shift the total strongly up or down.

Nigel Stork, by reevaluating Erwin's estimates and leavening them with other data from Borneo, England, and South Africa, concluded that the total number of tropical arthropod species is indeed very large, but likely to be lower than projected by Erwin, perhaps anywhere from 5 to 10 million. Kevin Gaston interviewed specialists on different kinds of insects and found them to be conservative, also pointing to a total of 5 to 10 million species. Illumination by these and other studies has been only partial. In a sense we are back to square one: the number of species of organisms on earth is immense but still cannot be placed to the nearest order of magnitude.

The great naturalist-explorer William Beebe said of the rain-forest canopy in 1917: "Yet another continent of life remains to be discovered, not upon the earth, but one to two hundred feet above it." The subsequent decades have revealed a second unexplored continent 1,000 meters and more beneath the surface of the oceans, on the floor of the deep sea. This vast domain, 300 million square kilometers in extent, is with the possible exception of Antarctica's valleys the least hospitable habitat on earth, bitterly cold, crushed by pressure from the water columns above, and pitch-dark except for rare pinpoints of light from passing luminescent organisms. Biologists of the early nineteenth century thought the deep sea to be lifeless. They were proved wrong by dredging operations conducted during the Challenger expedition of 1872–1876, whose mud samples disclosed a wide array of previously unknown organisms. Thus was the abyssal benthos discovered, the community of organisms living on or close to the ocean floor. In the 1960s a major advance was made by the introduction of the epibenthic sled, which rakes the top layer of the floor with fine mesh nets and traps the residue with a closing door to prevent the winnowing and loss of smaller organisms. The new sample yielded animal life diverse to a degree beyond even the boldest imaginings of the biologists. From these collections, and from photographs and more recent selective sampling made from deep-sea vehicles, we know that the abyssal benthos contains swarms of polychaete worms, peracarid crustaceans, mollusks, and other animals found nowhere else on earth. Many of the invertebrate animals are minuscule and subsist at low metabolic rates for life spans that may last decades. Bacteria are present that can grow and divide only in cold water under very high pressure. The abyssal benthos is a subdued miniaturized world. There is no way of guessing the full number of species present, but it certainly ranges into the hundreds of thousands and probably beyond. J. Frederick Grassle, after reviewing data on all the samples taken up to 1991, conjectured that the number of animal species could range in the tens of millions. The diversity of bacteria and other microorganisms cannot even be guessed to order of magnitude.

In an ecological sense, the animals of rain forests and the abyssal benthos occupy opposite ends of the earth; one could say that they dwell on two planets. Their environments are as physically different as possible, and their biotas share not a single species of plant or animal. Yet all the diversity they contain may be dwarfed by that of the bacteria, organisms that saturate the two extreme environments and every other place on earth. It is a common misperception among both biologists and nonbiologists that bacteria are relatively well known because they are so important in medicine, ecology, and molecular genetics. The truth is that the vast majority of bacterial types remain completely unknown, with no name and no hint of the means needed to detect them. Take a gram of ordinary soil, a pinch held between two fingers, and place it in the palm of your hand. You are holding a clump of quartz grains laced with decaying organic matter and free nutrients, and about 10 billion bacteria. How many bacterial species are present? Take one-millionth of the pinch of soil and spread it evenly over nutrients poured into standard culture dishes. If each and every one of the bacteria in this near-invisible soil sample could multiply, we would expect to see over 10,000 little colonies growing on the nutrient surface, one from each bacterium. But they cannot, and we do not. We will get only between 10 and 100 colonies.

Some of the bacterial cells that failed to respond were dead at the moment of implantation, but most simply did not find the conditions of the culture medium congenial for division and colony formation. Such species are incommunicado—they refuse to respond to the microbiologists who use standard techniques. They await the right temperature, acidity, air pressure, and combination of sugars, fats, proteins, and minerals appropriate to their genetically dictated metabolic needs. Moreover, each of these silent species may be represented in the pinch of soil by as few as one or two individuals per million. To find them, microbiologists must offer one culture medium and ambient environment after another until they hit on exactly the right combination. Then a colony proliferates and enough of the bacteria become available to be sorted and analyzed by standard microscopy and biochemical techniques.

Microbiologists rarely try to find silent bacteria. They are interested only in the select group of species already proved to be of some scientific or practical interest. One of the most famous species in the world, the colon bacterium *Escherichia coli*, is the key experimental organism of molecular biology. All beginning biology textbooks celebrate the knowledge won from its short life cycle and ease of culture. But from an evolutionary biologist's point of view, *E. coli* is only a somewhat peculiar symbiont of the large intestine of mammals, one that helps to convert exhausted food into feces. The bacterial proletariat, that vast majority of other species representing three billion years of adaptive radiation, remains unstudied and unheralded.

How many species of bacteria are there in the world? *Bergey's Manual of Systematic Bacteriology*, the official guide updated to 1989, lists about 4,000. There has always been a feeling among microbiologists that the true number, including the undiagnosed species, is much greater, but no one could even guess by how much. Ten times more? A hundred? Recent research suggests that the answer might be at least a thousand times greater, with the total number ranging into the millions.

Jostein Goksøyr and Vigdis Torsvik went in search of silent bacterial species in a natural environment. They chose to cut the Gordian knot of selective culturing by separating and comparing the DNA of the bacteria directly. They took small quantities of soil from a Norwegian beech forest near their laboratory. Using a succession of steps in extraction and centrifugation, they separated the bacteria from the soil and removed and purified the DNA of these organisms in a single common batch. They employed extreme high pressure to shear the double-stranded DNA molecules into fragments of uniform size. When heated, DNA molecules separate into their constituent single strands.

To divide DNA into single strands means that the letters of the DNA code, the base pairs, have been split apart. In most DNA, the base pairs are adenine-thymine and cytosine-guanine, AT and CG for short. As you proceed down the DNA helix, any of the four bases can be on the right or left in each nucleotide, so that in reading off the genetic code four permutations are possible: AT, TA, CG, GC. A typical sequence might then be TA-CG-CG-AT-GC, and so on for thousands or millions of such letters per cell. When the DNA helix is split, the two complementary single strands will read, for the segment just cited, T-C-C-A-G and A-G-G-T-C respectively.

When cooled to about 25°C below the melting temperature, the DNA strands are easily brought together again to form a double helix; they are "annealed," as the molecular biologists say. The higher the concentration of complementary strands in a solution, the faster the annealing will occur. If there are mixes of different species and strains of species, like those in the Norwegian soil bacteria, the concentration of complementary single strands will be much lower than is the case when DNA from only one species is present; the annealing process will be slowed to a corresponding degree. The rate at which annealing occurs can be measured exactly and calibrated against standards consisting of DNA single strands from a bacterium (the familiar *E. coli*) with known quantities of DNA. By this means, it is possible indirectly to estimate the overall percentage of matches among the diverse single strands of DNA in an entire bacterial community—that is, in all the bacteria living in a pinch of soil.

The percentage of DNA matching can be used as an indirect means to calculate the number of bacterial species. In so doing, microbiologists cannot use the biological-species concept directly. It is beyond their power to observe which bacterial cells exchange DNA, as if these organisms were so many birds and oak trees in the Norwegian forest. They are forced to rely on the similarity in DNA of one cell to the next. The arbitrary standard proposed by classifiers of bacteria is the following: a bacterial species consists of all those cells whose nucleotides are at least 70 percent identical, and hence are at least 30 percent different from the nucleotides of other species. This cutoff point is actually conservative; many higher plant and animal species are separated by far less than a 30 percent difference.

I have provided so much technical detail in order to reveal the difficulties faced by microbiologists and emphasize why it has taken so long to make inroads into bacterial diversity. Here is the result of the Norwegian research group: between 4,000 and 5,000 bacterial species were found in a single gram of beech-forest soil. A similar number of species, with little or no overlap, was found in a gram of sediment from shallow water off the Norway coast.

"It is obvious," Jostein Goksøyr has written, "that microbiologists will not run out of work for a couple of centuries." If over 10,000 microbial types exist in two pinches of substrate from two localities in Norway, how many more await discovery in other, radically different habitats? Again, no one has the faintest idea. It seems inevitable that entire new complexes of bacteria await discovery on the floor of the deep sea, in the axils of rain-forest orchids, amid the algal scum of mountain lakes, and so on around the world in thousands of sites that ordinarily escape our notice. Recent drilling of deep aquifers in South Carolina revealed large numbers of peculiar bacteria down to at least 500 meters below ground. The species changed from one stratum to the next. More than 3,000 forms, all new to science, were found in the early probes.

Still another uncharted world of bacteria and other microorganisms exists in and on the bodies of larger organisms. Some of the species are neutral guests, neither harming nor helping their hosts. Others have been characterized, however, that assist their hosts in digestion, excretion, and even the production of light by luminescent chemical reactions within their tiny bodies. So useful—indeed vital—are many that their hosts maintain specialized cells and tissues to carry them, while resorting to elaborate steps in physiology and behavior to pass the symbionts between the sexes and from parent to offspring. The phenomenon is well illustrated in the transmission of bacteria and yeasts by the scale insect *Rastrococcus iceryoides*. The microorganisms move to the offspring by an elegantly choreographed *pas de deux* within the developing egg. It has been described by the great authority on symbiosis, Paul Buchner:

> Both types of symbionts infect at the same place, thus forming in the mature egg a roundish ball at the upper pole. When the germ band approaches them, the two partners, which were united until now, separate and it is interesting to see how differently the host treats each of them. First, it shows interest only in the yeasts. While the yolk nuclei migrate toward them and soon penetrate them on all sides, the bacteria, which in the meantime have markedly increased, glide toward the periphery of the embryo in irregular groups, yet without allying themselves with nuclei. Yeasts and bacteria are soon separated. By the time the extremities sprout, cell limits have been formed around the yeasts, whereas the bacteria groups, unchanged, are situated here and there in the plasma.

Although hundreds of such peculiar partnerships have been discovered, they are described in the literature in only fragmentary manner. Very few of the bacterial species have been given a scientific name or described beyond adjectives such as "rod-shaped" or "vesicle-shaped."

To plumb the depth of our ignorance, consider that there are millions of insect species still unstudied, most or all of which harbor specialized bacteria. There are millions of other invertebrate species, from corals to crustaceans to starfish, in similar state. Consider that each bacterial type, each species if we employ the DNA-matching rule, can utilize at most a hundred carbon sources, such as different sugars or fatty acids. Most can in fact metabolize only one to several such compounds. Consider further that bacteria can evolve rapidly to exploit these sources. Different strains and even species readily exchange genes, especially during periods of food shortage and other forms of environmental stress. Their generations are extremely short, allowing natural selection to act on new assortments of genes within days or even hours, shifting the heredity this way or that, perhaps creating new species.

Consider, finally, a centimeter-wide patch of soil, the area of a fingernail, at a randomly chosen spot on the forest floor. A decaying splinter of wood lying on the surface contains one set of bacterial forms, leached sand grains a millimeter away another flora, and specks of humus a centimeter down yet another. All told there are thousands of species. Now assemble all such microfloras across an entire forest, then across all forests and habitats for the entire world, and we might expect to find many millions of hitherto unstudied species. The bacteria await biologists as the black hole of taxonomy. Few scientists have even tried to dream of how all that diversity can be assayed and used.

As exploration of the natural world continues, new species of even the largest, most conspicuous organisms continue to turn up. In Colombia's Chocó Region, embracing the mountainous rain forests west of Medellín, half the plant species remain unregistered, and of these a large portion still lacks a scientific name. An average of two new bird species a year are discovered somewhere in the world, usually in remote mountain valleys and recesses of the last surviving tropical forests. Even new kinds of mammals are discovered from time to time. In the banner year 1988 the following novelties were published: Tattersall's sifaka (*Propithecus tattersalli*), a new lemur from Madagascar; the sun-tailed guenon (*Cercopithecus solatus*), a monkey from Gabon, central Africa; and a new muntjak deer from the mountains of western China. In 1990 a previously unknown primate, the black-faced lion tamarin, was discovered on the small coastal island of Superaqui, only 65 kilometers from the city of São Paulo. It was, in the words of Russell Mittermeier, "one of the most amazing primatological discoveries of this century." And caught in the nick of time, I might add, because the species is represented by only several dozen individuals. One hunter could have extinguished the species in a matter of days.

Not even the order Cetacea, containing the largest animals on earth, the whales and porpoises, is fully known. It is true that the species of the very largest, the baleen whales, including the blue, right, and humpback whales, had all been described by 1878. In contrast, the toothed whales, which include the giant sperm whale, killer whales, and their smaller relatives, the beaked whales and porpoises, have continued to yield new species at an average rate of one a decade during the twentieth

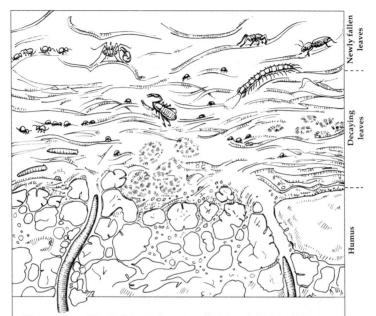

Newly fallen leaves

Decaying leaves

Humus

The teeming life of a North American deciduous forest appears to exist on a two-dimensional plane when viewed in a typically human perspective from above (*left*). In this assemblage, the lithobiid centipede in the center is surrounded, from the top clockwise, by a greenbottle fly, social wasp, long-snouted acorn weevil, bess beetle, termite, wood cockroach, carpenter ant, sowbug, ground beetle, tick, ichneumon wasp, aphid, earwig, and harvestman (daddy-longlegs). When the litter and soil are cut vertically and viewed from the side (*right*), a different, three-dimensional world is revealed. The dead leaves piled loosely at the top provide a dry, airy living space, inhabited in this example by small globular springtails, tiny turtle-shaped oribatid mites, harvestman (feeding on a snail), jumping spider, centipede, and ground beetle. A few centimeters deeper, in denser and moister litter amid piles of arthropod and earthworm fecal matter, are scattered more springtails and mites, pseudoscorpion (claws but no stinger), and two slug-shaped crane fly larvae. Deeper still in the now tightly packed humus and soil, two earthworms rest in their burrows.

century. Here are the eleven discovered since 1908, composing 13 percent or more of all the living cetaceans known:

Andrews' beaked whale, *Mesoplodon bowdonini* Andrews, 1908

Spectacled porpoise, *Australophocaena dioptrica* (Lahille), 1912

True's beaked whale, *Mesoplodon mirus* True, 1913

Chinese river porpoise (baiji), *Lipotes vexillifer* Miller, 1918

Longman's beaked whale, *Mesoplodon pacificus* Longman, 1926

Tasman beaked whale, *Tasmacetus shepherdi* Oliver, 1937

Fraser's porpoise, *Lagenodelphis hosei* Fraser, 1956

Vaquita (Gulf of California harbor porpoise), *Phocoena sinus* Norris and McFarland, 1958

Ginkgo-toothed beaked whale, *Mesoplodon ginkgodens* Nishiwaki and Kamiya, 1958

Hubbs' beaked whale, *Mesoplodon carlhubbsi* Moore, 1963

Pygmy beaked whale, *Mesoplodon peruvianus* Reyes, Mead, and Van Waerebeek, 1991

Many of the smaller whales and porpoises have been recorded solely from scattered carcasses or body parts washed ashore in remote parts of the world, and their natural history remains a mystery. Of the Tasman beaked whale, the cetacean expert Willem Mörzer Bruyns wrote in 1971: "A total 6 specimens washed up on beaches of Stewart Island, Bank's Peninsula and Cook's strait, east coast New Zealand." Of Hector's beaked whale, discovered in 1871: "Originally described from three skulls of very immature, perhaps neonatal calves, found in New Zealand waters . . . In 1967 the skull of an adult female was found in Tasmania." And of Longman's beaked whale: "Described from a skull found near Mackay, Queensland, Australia. In March 1968 Dr. Maria Louise Azzaroli described a second skull found in 1955 near Mogadiscio (Somali) which establishes the identity of a separate species." The rarity and elusiveness of these species suggest that other ocean giants await discovery. Individuals of at least one distinctive new species of beaked whale have in fact been sighted several times in the waters of the eastern tropical Pacific, but none has been captured.

A large part of species diversity stares us in the face but goes unrecognized. Previously I defined sibling species as two or more populations reproductively isolated from one another and yet so similar in outward appearance as to be lumped together even by expert taxonomists. Only a careful study of fine details in anatomy, cell structure, biochemistry, and behavior brings the differences to light and allows systematists to define the species with certainty. Early in my career as a classifier of ants, I combined all the slave-making ants of eastern North America into two species, thinking that there were only two reproductively isolated populations. I was wrong. A second entomologist, William Buren, took a closer look and broke the slave makers into five species, on the basis of small differences in their hair patterns, the shapes and color of their bodies, and the other species of ants they kidnap as slaves. There is little doubt that all are in fact reproductively isolated populations, each with a unique genetic constitution.

Groups exist, such as the protozoans and fungi, that are rife with sibling species for a purely technical reason: they have few outward traits by which species can be separated even with refined microscopical techniques. Owing to the frailties of the human sensory apparatus, their species are hidden. The recorded diversity of such groups can be expected to rise sharply as the DNA sequences and physiological requirements of more and more species are teased apart. It is further true that closer examination will turn many more subspecies into species. When the exact geographic limits of populations are mapped, many of those previously thought to be widely distributed species are revealed to comprise multiple species with exclusive ranges.

Most biological diversity, however, in the old-fashioned way, awaits discovery by foot, net, and scuba gear. To confront diversity, biologists continue to go out of the laboratory and into the world. They count species in three ways, according to the breadth of the geographic area surveyed. Alpha diversity is the number of species at one habitat in one locality. Two of my coworkers, Stefan Cover and John Tobin, and I recently set out to break the world record for alpha diversity in ants. We got it: 275 species collected in 8 hectares of rain forest near Puerto Maldonado, Peru. Beta diversity, the second measure, is the rate at which the species number increases as nearby habitats are added. If the Puerto Maldonado study were extended to swamp forest, river banks, and grassland patches, our catalog would almost certainly increase to over 350. Finally, gamma diversity is the totality of species in all habitats across a broad area. To survey thoroughly all the ants of Peru, valley by valley and across all the tributaries of the Amazon, might easily yield 2,000 species. It is gamma diversity, of course, that biologists have assayed with least precision. Knowing this, they press on into unexplored mountain ridges, river headwaters, and coral reefs. For most countries in the world, especially the tropics, the plumb line is still being let out; we have no idea where it will all end. The rewards of physical adventure, the excitement of grimy, sweat-soaked exploration into remote corners of the earth, still beckon in science.

But imagine for a moment that all the diversity of the world were finally revealed and then described, say one page to a species. The description would contain the scientific name, a photograph or drawing, a brief diagnosis, and information on where the species is found. If published in conventional book form, with pages bound into ordinary thousand-page volumes 17 centimeters wide inside cloth covers, this Great Encyclopedia of Life would occupy 60 meters of library shelf per million species. If there are 100 million species of organisms on earth, they would extend through 6 kilometers of shelving, the size of a medium-sized public library. Of course biodiversity studies will never come to that. Long before all the species are discovered, long before we put away our butterfly nets and plant presses, descriptions will be electronically recorded, so that the Great Encyclopedia can reside on disks placed in a box on one side of an office desk. Far more information will be added for each species as it becomes available, from its genetic code to its role in ecosystems, and these data will be readily available through networks arrayed around international and regional biodiversity centers.

The Great Encyclopedia of Life will record additional measures of diversity in current use by biologists. One is equitability, or the evenness of the abundance of species. Up to this
point I have spoken of the measure of diversity only as the
number of species: so many bacteria in a pinch of soil, so many
ants in a stretch of rain forest. What also matters very much is
the relative abundance of species. Suppose that we encounter a
fauna of butterflies consisting of 1 million individuals, divided
into 100 species. Say one of the species is extremely abundant,
represented by 990,000 individuals, and each of the other species therefore comprises an average of about 100 individuals.
One hundred species are present but, as we walk along the
forest paths and across the fields, we encounter the abundant
butterfly almost all the time and each of the other species only
rarely. This is a fauna of low equitability. Then in a nearby
locality we encounter a second butterfly fauna, comprising
the same 100 species, but this time all are equally abundant,
represented by 10,000 individuals each. This is a fauna of high
equitability, in fact the highest possible. Intuitively we feel that
the high-equitability fauna is the more diverse of the two, since
each butterfly encountered in turn is less predictable and therefore gives us more information on average, just as each word in
a rich vocabulary fully employed gives more information. To
study a highly diverse fauna is to be continually informed—
and therefore pleased in an ultimate aesthetic sense. Diversity
in this dimension also has practical importance in ecology.
A fauna of high equitability is likely to have a very different
impact on an ecosystem than one of low equitability, sustaining
a larger variety of plants and other animals dependent on it.

Biologists measure the diversity of life not only by species, but
by genera, families, and other higher categories of classification
up to and including phyla and kingdoms. Each such higher unit
is a cluster of species that resemble one another and are thought
to share a common ancestry. In particular, a genus is a group
of species placed together in the classification because they are
very similar and of more or less immediate common ancestry.
A family is a group of similar, related genera (its species overall
are more distantly related than those within a genus); an order
is a group of similar, related families; and so on through the

hierarchy of classification all the way up to kingdoms, which embrace plants as a whole, animals as a whole, and so on. Here in briefest form is the complete taxonomic placement of the domestic cat, *Felis domestica*:

Species: *domestica*
  Genus: *Felis*
    Family: Felidae
      Order: Carnivora
        Class: Mammalia
          Phylum: Chordata
            Kingdom: Animalia

The basic principles of classification follow a transparent logic that can be stated in a few words. First principle: the species is the pivotal unit. Second principle: two definitions are used in setting up the hierarchical classification, as follows. A category is an abstract level of classification used universally in classification. The categories are the species, the genus, the family, and so on. A taxon, in contrast, is a concrete group of organisms, a particular set of populations given the rank of one or the other of the categories. Examples of taxa include the species *Felis domestica* and the family Felidae. Categories are the abstraction, taxa the reality. Third principle: a higher taxon such as the genus *Felis* is a group of species that are all descended from a single ancestral species. The species of a different taxon of equal rank, such as the great cats of the genus *Panthera*, are all descended from *another* ancestral species. When two genera are placed together to form a family, however, in this case the Felidae, they are considered to be descended from an even older ancestral species; this earlier progenitor gave rise to the two more recent ancestral species, which in turn gave rise to the species composing the two respective genera. Fourth principle: as these last examples make clear, the higher categories are a mental construct invented for convenience. They are based on a conception of species splitting into new species through time, and reflect the branching pattern that the successive splittings produce. The construction of branching patterns to map evolutionary change is called *cladistics*, and the devising of higher classifications (genus and up) to conform with the results of cladistics is called *phylogenetic systematics.* Classifications are expected to be consistent with phylogeny: in other words, the family trees of species.

Fifth and final principle: the exact limits of the higher taxa are arbitrary. The species themselves, the atomic units, are natural—more or less. So are their phylogenetic (ancestral) trees, if we have deduced them correctly. But the *limits* of genera, families, and still higher taxa, those lines drawn around clusters of species, are arbitrary. This statement may seem paradoxical because I just finished saying that the whole point of cladistics is to create a natural classification at the level of the genus and above. That much is true. Cladistics does allow us to judge which species are most likely to share a common ancestor, validating their placement in the same genus or family or higher taxon. What is arbitrary are the limits of each higher taxon. Should *Felis* and *Panthera* be kept as separate genera, or should they be lumped together in the single genus *Felis?* Either classification is correct by the standards of cladistics. Again: should the Felidae be allowed to stand as the only cat family, or should it be divided into two families, say the Felidae, the "true" cats, and Acinonycidae, the cheetahs? Cladistics is silent on this matter.

Systematists look at the reconstructed evolutionary trees. They see which species are descended from which common ancestors and can be bound into higher taxa, that is, into clusters of related species. They use criteria—common sense for the most part—to decide how to divide the clusters into smaller clusters. If all the species are very similar, it is sensible to place them in a single genus. If one species is very different from the others, even if it shares the same ancestor, the best procedure is to erect a new genus, thus drawing attention to its unusual properties. To recognize one genus or two, in the large number of borderline cases, is a judgment call. Systematics is mostly science but also a bit of art.

This fuzzy resolution is the right compromise to reach. The subjective nature of higher taxonomic categories reflects the chaotic nature of organic evolution. Like stars in an expanding universe, species are always evolving away from all other species, until they become extinct—or, in a few cases, break down the reproductive barriers between them and hybridize. This principle of evolution stems in turn from the immense variety made possible by rearrangements in the sequence of nucleotide letters of the genetic code. The code contains about a million nucleotide pairs in bacteria and between 1 to 10 billion nucleotide pairs in higher plants and animals. Evolution proceeds mostly by the accidental substitution of one or more of the letters, followed by the winnowing of these mutations and their combinations through natural selection. Because mutations occur at random, and because natural selection is affected by idiosyncratic changes in the environment that differ from one place and time to the next, no two species ever follow exactly the same path for more than a couple of steps. The real world, then, consists of species that differ from one another in infinitely varying directions and distances. So far as we know, no way exists to lump or to split them into groups except by what the human mind finds practical and aesthetically pleasing.

There is another consequence of evolution-as-expanding-universe that affects the taxonomic rank of species and their perceived value to humanity. Every species born, given enough time to evolve and proliferate into multiple species, is a potential genus or taxon of still higher rank. The longer this assemblage survives and evolves, the more it comes to differ genetically from the remainder of life. Because extinction is all but inevitable, the assemblage ordinarily dwindles until only a few relict species remain. These survivors are old, unique, and precious. Think of a species that has lived a very long time. Either its sister species have been stripped away by extinction, or else it is the sole occupant of an ancient line that never proliferated into multiple species in the first place. It stands alone now, accorded the rank of genus, family, or still higher category. It deserves special consideration by the human race for the story it tells. Such is the status of the giant panda, sole member of the genus *Ailuropoda*; the coelacanth fish *Latimeria chalumnae*, most celebrated of living fossils; and the tuatara, *Sphenodon punctatus*, a lizard-like reptile limited to small islands off New Zealand and only one of two members of the order Rhyncho-cephalia to survive from the Mesozoic era.

The known diversity of life has expanded within each level of the taxonomic hierarchy; so many species per genus, so many genera per family, and on up. At the very top, a total of 89 living phyla are distributed among the kingdoms of life. According to one widely used but highly subjective classification, the kingdoms are five in number:

Plantae: multicellular plants, from algae to flowering plants

Fungi: mushrooms, molds, and other fungi

Animalia: multicellular animals from sponges and jellyfish to vertebrates

Protista: single-celled eukaryotic organisms (protozoans and other single-celled organisms)

Monera: single-celled prokaryotic organisms (such as bacteria and cyanobacteria)

To describe diversity by organizing species into clusters according to how closely they resemble one another was a fundamental advance of eighteenth-century biology. A later and equally important way of describing diversity is by level of biological organization. The organizational levels of importance to biological diversity are arrayed in this hierarchy:

Ecosystem
   Community
      Guild
         Species
            Organism
               Gene

The idea is best conveyed by a concrete example that can be followed down through the full stack. The example I choose: *A goshawk* (Accipiter gentilis) *hunts for songbirds through the Black Forest of Germany, flying fast and low among the fir trees, shifting direction abruptly. It sees a wood warbler* (Phylloscopus sibilatrix) *resting on a pine branch. With a few short wingbeats and a long silent glide, it closes on its prey.*

The goshawk lives in a particular ecosystem, the upland fir forest of the Black Forest. The land is covered by granitic soil weathered from round low hills, laced with streams that form the headwaters of the Danube and Neckar rivers. The ecosystem consists of this physical base plus all the organisms living in the habitats of the forest, in the woods, the glades, and small bodies of fresh water. The combined physical and biotic elements, from rock and streams to trees, hawks, and warblers, are tightly bound to one another. Energy is carried as in a leaky bucket from one species to another through the food webs of organisms. Nutrients flow through the organisms, soil, water, and air, and back in unending biogeochemical cycles. The character of the soil cover and water drainage depends intimately on the organisms living in the forest. The Black Forest ecosystem is unique in its particular combination of physical environment and tenant organisms. We look up and across southern Germany, at Europe as a whole, and finally around the entire world to measure the diversity of ecosystems in existence. We find that the possibilities are astronomical in number—the combinations of millions of species that can live in all of the distinguishable physical settings is a number beyond practical calculation. This incapacity is interesting but not important. The real number of ecosystems is what counts. Each ecosystem has intrinsic value. Just as a country treasures its finite historical episodes, classic books, works of art, and other measures of national greatness, it should learn to treasure its unique and finite ecosystems, resonant to a sense of time and place.

Within the Black Forest ecosystem, the goshawk belongs to a particular community of organisms, defined as all the species connected in the food web and by any other activity that influences the life cycles of the species. The fir tree is in the goshawk's food web because it nourishes the moth larvae that feed the songbirds that feed the hawk. The common buzzard, a European buteo hawk, is a member of the same community by virtue of competition and accidental symbiosis. The occasional small bird it kills reduces the goshawk's larder. The nest it abandons and makes available to the goshawk improves the breeding chances of that less discriminating bird. The diversity of communities is measured within a particular ecosystem. More accurately, it is subjectively assayed, since the limits of a community can seldom be drawn with exactitude.

Inside the community, the goshawk is a member of a guild, a set of species that live in the same place and harvest the same food by similar means. Strictly speaking, the goshawk shares its guild with only one species in the Black Forest community, the sparrow hawk (*Accipiter nisus*). They are both accipiters, possessed of short rounded wings and long tails. They hunt small birds by rapid, twisting flight through the forest, occasionally ascending to brief soars above the trees. Other guilds of the Black Forest include the aster-eating flower insects, the warblers, and the forest-dwelling shrews and small mice. Because guilds tell us something about ecology, they are as valuable a measure of diversity within an ecosystem as the number of species.

We are approaching the lowest levels of biological diversity. The goshawk is a species, a nexus of ill-defined local populations that altogether range from continental Europe across Asia to Canada and the northern and western United States. The individual birds composing it are the repositories of genetic diversity, the differences that exist among the chromosomes and genes and the level below species diversity. This is a level most quickly grasped by using familiar examples in human heredity. A single gene difference determines whether the ear has a free-hanging lobe or is attached all the way along its base. Possession of an earlobe is a dominant trait. If either one of the two genes in each cell is an earlobe gene, the person acquires a fully developed earlobe. Only if both genes are for the recessive, lobeless condition does the person develop that alternative trait. The earlobe genes, the particular pair of genes determining the presence or absence of earlobes, occur at just one of the 200,000 or more sites arrayed in a row along the 46 chromosomes. Other examples of human variation based on single genes are the blood types, the ability to roll the tongue into a tube, the presence of a widow's peak in the hairline, whether the last segment of the thumb angles back sharply when the thumb is stretched out (the condition called "hitchhiker's thumb"), and a myriad of hereditary diseases from sickle-cell anemia to albinism, hemophilia, and Huntington's chorea. Many other traits, such as height, skin color, and predisposition to diabetes, are affected by combinations of genes at many chromosome sites that work together, the so-called polygenes.

By counting such variations in outward traits known to arise from mutations in single genes and polygenes, it is possible to

arrive at a figure of total genetic diversity. The estimate, however, would fall short by orders of magnitude. The reason is that
differences among alternative genes at the same chromosome
position cause differences that are often invisible; they prescribe
variations in proteins detectable only by chemical analysis. In
the 1960s an advance was made in resolving power with the
introduction of gel electrophoresis, a technique that allows
the rapid purification and identification of enzymes. When
molecules are placed in a charged field on material through
which they can move, such as a porous gel, they migrate at
a rate proportionate to their own electric charge. As a result
they space out like runners with varying abilities. Enzymes
are protein molecules whose design, including their electric
charge, is prescribed by genes.

Even small differences in the genes caused by mutations
translate to variations in the enzymes, often (but not always)
translating into differences in electric charge, which cause the
enzymes to travel at different velocities and separate on the
electrified gel plates. Geneticists follow a straightforward procedure to take advantage of this chain of events. They crush
tissues from the organisms to be studied, extract materials
containing enzymes, and place the extracts at one end of a
gel plate. They let the enzymes run in the electrical field for a
while, then stain them with dyes to reveal their positions. They
count the stained enzymes that have separated on the plate.
They are then able to determine the number and identity of
the enzymes and hence infer the number and identity of the
prescribing genes. By sampling many individuals in a species
and by proceeding from one kind of enzyme to another—
thus from one set of genes to another—they can estimate the
amount of overall genetic diversity in the species.

Surveys using electrophoresis have ranged widely over many
kinds of organisms, from flowering plants and insects to fishes,
birds, and mammals. Of all the discoveries made, one stands
out: the amount of genetic diversity revealed is very large,
much greater than had been expected in pre-electrophoresis
days when researchers relied principally on such visible traits
as earlobes and skin color. In order to express the diversity
as a number, geneticists use the concept of polymorphism.

A gene is said to be polymorphic when it occurs in multiple forms, or multiple alleles as they are more technically called. The rarer alleles are not counted unless they exceed some arbitrarily selected frequency, usually 1 percent of the total for that particular gene. In other words, only if earlobe alleles were more than 1 percent in the human population would they be included in the count (in fact, they are 45 percent), and only then would the controlling gene be called polymorphic (it is in this case). The electrophoretic studies have shown that in the great majority of species, somewhere between 10 and 50 percent of genes are polymorphic. A typical figure is roughly 25 percent.

High levels of polymorphism per gene through the population also produce high levels of polymorphism within the bodies of individual organisms. In each individual on the average, again according to species, between 3 and 20 percent of the genes are polymorphic. This means that each organism is heterozygous for that number of genes; in human beings it means possessing one gene for earlobes and one for the lobeless state in each cell, or one for A blood type and one for B, and so on through the 200,000 or more genes that make up the total hereditary composition of a human being.

Yet even the unexpectedly high electrophoretic numbers are minimal estimates and almost certainly too low. Some enzyme variants have no special electrical charge or molecular arrangement by which they can be separated, so they stay silent on the electrophoretic field. In order to take an exact and final measure of genetic diversity, it is necessary to go past the proteins and straight to the genes themselves and learn the nucleotide sequence, the letters of the genetic code. The true, ultimate measure of genetic diversity is nucleotide diversity. It must be determined base pair by base pair through a large part of the chromosomes, and in many individuals belonging to the same species.

During the 1980s rapid advances were made in DNA sequencing. The human genome project was born, aiming at nothing less than a complete nucleotide map of one human being. A similar project is planned for a species of fruit fly. When sequencing becomes cheap enough, and reading genetic codes is as routine as counting feathers and molar teeth, we will be technically prepared in full to address the question of how much biodiversity exists on earth.

Meanwhile I will venture a guess at the final outcome: to the nearest order of magnitude, or powers of ten, 10^8 (100 million) species, multiplied by 10^9 (1 billion) nucleotide pairs on average per species; hence a total of 10^{17} (100 quadrillion) nucleotide pairs specifying the full genetic diversity among species. Nucleotide diversity, it should be noted in passing, is limited to a maximum of four kinds of nucleotide per site and hence does not add as much as an order of magnitude.

That figure, 10^{17}, is in one sense the entire diversity of life. Yet it still does not take into account the differences among individuals belonging to the same species. When that dimension is added, the potential grows still more. Consider that, within a typical sexually reproducing species, two nucleotides occurring at the same site on different chromosomes can generate three combinations; the letters AT and CG, for example, can generate the combinations (AT) (AT), (AT) (CG), and (CG) (CG). If only one out of a thousand sites had two such variants somewhere in the species, then with 10^6 positions (in other words, one-thousandth of 10^9 positions in the whole genetic makeup of the species) there would be 10^{18} possible combinations for each species. This immense figure is still an underestimate. Whatever the true figure, it represents the potential of biodiversity at the level of the organism, the great field of possible genetic combinations through which each species travels with the raw materials it possesses, guided by natural selection and now, increasingly, the ignorant hands of humanity.

CHAPTER NINE

The Creation of Ecosystems

THE BALD EAGLE, one species, flies above the Chippewa National Forest of Minnesota. A thousand species of plants compose the vegetation below. Why does this particular combination obtain rather than a thousand eagles and one plant? Or a thousand eagles and a thousand plants? It is natural to ask whether the numbers that do exist are governed by mathematical laws. If there are such laws, it follows that we can someday predict diversity in other places, in other groups of organisms. To master complexity by such an economical means would be the crowning achievement of ecology.

There are no laws unfortunately, at least none that biologists have hit upon yet, not in the sense ordained by physicists and chemists. But, as in any study of evolution, there are principles that can be written in the form of rules or statistical trends. The discipline formulating these weaker statements, community ecology, is still youthful and rapidly growing, which is a polite way of saying that it is a long way behind the physical sciences—but there is progress, and ambition.

Before us now is the overwhelmingly important problem of how biodiversity is assembled by the creation of ecosystems. We can address it by recognizing two extreme possibilities. One is that a community of organisms, like that occupying the Chippewa National Forest, is in total disorder. The species come and go as free spirits. Their colonization and extinction are not determined by the presence or absence of other species. Consequently, according to this extreme model, the amount of biodiversity is a random process, and the habitats in which the various species live fail to coincide except by accident. The second extreme possibility is perfect order. The species are so closely interdependent, the food webs so rigid, the symbioses so tightly bound that the community is virtually one great organism, a superorganism. This means that if only one of the species were named, say the Acadian flycatcher, marbled salamander, or goblin fern, the thousands of other species could be ticked off without further information about that particular community.

Ecologists dismiss the possibility of either extreme. They envision an intermediate form of community organization, something like this: whether a particular species occurs in a given suitable habitat is largely due to chance, but for most organisms the chance is strongly affected—the dice are loaded—by the identity of the species already present.

In such loosely organized communities there are little players and big players, and the biggest players of all are the keystone species. As the name implies, the removal of a keystone species causes a substantial part of the community to change drastically. Many other species decline to near or total extinction or else rise to unprecedented abundance. Sometimes other species previously excluded from the community by competition and lack of opportunity now invade it, altering its structure still more. Put the keystone species back in and the community typically, but not invariably, returns to something resembling its original state.

The most potent keystone species known in the world may be the sea otter (*Enhydra lutris*). This wonderful animal, large and supple in body, cousin to the weasels, whiskered like a cat, staring with a languorously deadpan expression, once thrived among the kelp beds close to shore from Alaska to southern California. It was hunted by European explorers and settlers for its fur, so that by the end of the nineteenth century it was close to extinction. In places where sea otters disappeared completely, an unexpected sequence of events unfolded. Sea urchins, normally among the major prey of the otters, exploded in numbers and proceeded to consume large portions of the kelp and other inshore seaweeds. In otter times, the heavy kelp growth, anchored on the sea bottom and reaching to the surface, was a veritable forest. Now it was mostly gone, literally eaten away. Large stretches of the shallow ocean floor were reduced to a desert-like terrain, called sea-urchin barrens.

With strong public support, conservationists were able to restore the sea otter and with it the original habitat and biodiversity. A small number of the animals had managed to survive at far opposite ends of the range, in the outer Aleutian Islands to the north and a few localities along the southern

California coast. Some of these were now transported to scattered intermediate sites in the United States and Canada, and strict measures were taken to protect the species throughout its range. The otters waxed and the sea urchins waned. The kelp forests grew back to their original luxuriance. A host of lesser algal species moved in, along with crustaceans, squid, fishes, and other organisms. Gray whales migrated closer to shore to park their young in breaks along the kelp edge while feeding on the dense concentrations of animal plankton.

Ecologists, like the organisms they study, cannot make nature conform to their perfect liking. They search for openings and seize opportunity, exploiting the occasional discovery of keystone species like sea otters to gain insight into the organization of communities in different environments. Other examples have been found. In the undisturbed forests of Central and South America—more precisely, in the dismayingly few such forests remaining—jaguars and pumas prey on a wide variety of small animals encountered on the ground. They are "searchers," taking whatever animals they meet, as opposed to "pursuers" like cheetahs and wild dogs, which select only a few kinds of animals and then chase them down. The big cats are especially fond of coatis, members of the raccoon family with elongated bodies and tapered noses, and agoutis and pacas, outsized rodents variously resembling jackrabbits and small deer. When jaguars and pumas disappeared from Barro Colorado Island in Panama, because the forest was no longer extensive enough to support them, the prey species soon increased tenfold. Effects from this shift in balance now appear to be rippling downward through the food chain. Coatis, agoutis, and pacas feed on large seeds that fall from the rain-forest canopy. When they become superabundant, as on Barro Colorado Island, they reduce the reproductive ability of the particular tree species that produce these seeds. Other tree species whose seeds are too small to be of interest to the animals benefit by the lessened competition. Their seeds set and their seedlings flourish, and a larger number of the young trees reach full height and reproductive age. Over a period of years, the composition of the forest shifts in their favor. It seems inevitable that the animal species specialized to feed on them also prosper, the predators that attack these animals increase, the fungi and bacteria that parasitize the small-seed trees and associated animals spread, the microscopic animals feeding on the fungi and bacteria grow denser, the predators of these creatures increase, and so on outward across the food web and back again as the ecosystem reverberates from the removal of the keystone species.

In a very different way, elephants, rhinoceros, and other big herbivores rank as keystone species in the savannas and dry woodlands of Africa. When allowed to reach natural high densities, they control the entire physical structure of these habitats. "Modern African elephants," Norman Owen-Smith has written, "push over, break, or uproot trees,"

> altering vegetation physiognomy and hence habitat conditions for other animal species. Trees killed by elephants are replaced by regenerating shrubs or grasses that offer more accessible foliage for consumption by smaller herbivores. The leaves of rapidly growing woody plants are less strongly defended chemically than those of the slower-growing trees they replace. Rates of nutrient cycling are also accelerated. Grazing pressure from white rhinoceroses and hippopotamuses transforms medium-tall grasslands into a mosaic of short and tall grass patches. Short, creeping grasses are generally less fibrous and more nutrient rich than taller grasses. As a result of such vegetation changes, food quality is improved for smaller, more selective grazers. Animal species dependent upon a dense cover of woody vegetation or tall grasses for predator evasion may persist in areas of low impact.

For millions of years the great herbivores of sub-Saharan Africa ranged freely across the vast parklands, creating a mosaic of habitats, a swath of short grassland here, an acacia copse or remnant of riverine forest there, reed-lined pools grown from mud wallows scattered widely about. The total effect was a huge enrichment of biological diversity.

Focusing now from the kilometer reach of elephants down to grassroot level, we find a wholly different class of keystone species. Where big mammals control the vegetation structure, a colony of driver ants at their feet captures millions of victims each day and alters the nature of the community of small animals. Viewed a few meters away, a driver-ant raiding column seems a living thing, a giant pseudopodium reaching out to engulf its prey. The victims are snared with hook-shaped jaws, stung to death, and carried to the bivouac, a labyrinth of underground tunnels and chambers housing the queen and immature forms. Each expeditionary force comprises several million workers who flow out of this retreat. The hungry legions emerging from the bivouac are like an expanding sheet that lengthens into a treelike formation. The trunk grows from the nest, the crown expands as an advancing front, and numerous branches pour back and forth between the two. The swarm is shaped but leaderless. Excited workers rush back and forth throughout its length at an average speed of a centimeter per second. Those in the van press forward for a short distance and then fall back to yield their front position to other runners. The feeder columns resemble thick black ropes laid along the ground, slowly writhing from side to side. The front, advancing at 20 meters an hour, blankets all the ground and low vegetation in its path. The columns expand into it like a river entering a delta, where the workers race back and forth in a feeding frenzy, consuming most of the insects, spiders, and other invertebrates in their path, attacking snakes and other large animals unable to move away.

Day after day the driver ants scythe through the animal life around their bivouac. They reduce its biomass and change the proportions of species. The most active flying insects escape. So do invertebrate animals too small to be noticed by the ants, particularly roundworms, mites, and springtails. Other insects and invertebrates are hit hard. One driver-ant colony, comprising as many as 20 million workers—all daughters of a single mother queen—is a heavy burden for the ecosystem to bear. Even the insectivorous birds must fly to a different spot to find enough food.

It has become clear that an elite group of species exercises an influence on biological diversity out of all proportion to its numbers. Scientists are drawn to such strong cases, not just in ecology but in other fields as disparate as astrophysics and neurobiology, because they yield quick information and an entry into systems that are otherwise intractable. But they can be misleading if overgeneralized. There comes a time, in all science, when it is profitable to move away from the bold

A keystone species at the grassroots level: a swarm of driver ants marches across a savanna in Kenya. The ant armies drastically alter the abundance of insects and other small animals in the habitats through which they pass.

and obvious and circle around a bit, inventing more subtle approaches to search for concealed phenomena. In the study of communities, this strategy requires greater attention to context, history, and chance.

One successful recent approach has been to deduce the assembly rules of faunas and floras. Although the attempt to identify keystone species takes a community pretty much as it is and figures out what happens when the candidate species is removed, the assembly rules reconstruct the sequence in which species were added when the community came into being. It does more: it postulates the sequences that are possible and those that are not. Let me express the idea with an imaginary example chosen for clarity. A certain plant species establishes itself, say on a mountainous island. Its presence allows the colonization of the habitat by a beetle species that feeds only on it. A wasp species that parasitizes the beetle is added next. In another dimension, entailing competition, a second assembly rule is manifested. A woodpecker species arrives; call it A. It multiplies to such abundance and dominates so much of the food supply that when two more woodpecker species arrive, B and C, either one of them but not both can squeeze into the community. Now we have a woodpecker fauna consisting of either AB or AC, depending on which of the latecomers

Assembly rules determine which species can coexist in a community of organisms (such as the bird species occupying a forest patch). The rules also determine the sequence in which species are able to colonize the habitat. A set of imaginary rules is represented here as pieces of a jigsaw puzzle that can be fitted together in one of two combinations, ABD or AC.

arrived first. Finally, woodpecker species D appears. Occupying a distinctive niche of its own—say foraging on large conifers—it can squeeze in with the other species if the preexisting combination is AB but not if the combination is AC. So the first stable woodpecker fauna in the community is either ABD or AC.

Ecologists deduce assembly rules by observing which species actually live together in nature. One approach, used by Jared Diamond in pioneering work on the birds of New Guinea, compares communities in many different localities to see which combinations of species occur and which ones rarely or never occur. The preliminary conclusions reached in this way can then be tested further by detailed studies of habitat preferences of the individual species. Suppose, in the woodpecker case just cited, B and C almost never coexist in the same localities because they compete until one or the other is extinguished. Suppose further that additional studies refine the pattern: B and C occur on some mountains together but almost always at different elevations, so in fact they are seldom members of the same community. On mountains where both occur, B ranges from 200 to 1,000 meters and C from 1,000 to 2,000 meters. Where only one of the species occurs on a mountain, it spreads all the way from 200 to 2,000 meters. This expansion in the absence of a competitor is the same phenomenon we have already met, ecological release. The constriction in the presence of the competitor is called *ecological displacement*. The existence of ecological release and displacement is considered strong presumptive evidence that even when B and C occupy the same geographical range they cannot live closely together in the same habitat and community. They withdraw to elevations where each in turn is the superior competitor, in this case B in the lowlands and C higher up.

It will be interesting now to return to Krakatau and to recall the example it offers of the assembly of species. A community does not arrive on the shores of such an island as a finished product. Instead, it is stacked like a house of cards, one species on another, loosely obedient to assembly rules. Most propagules, whether plant seeds or wandering bird flocks, are doomed to failure. For them the soil is wrong, the forest glades are still too small, the prey species have not yet arrived, or formidable competitors wait at the shore. Even many of the species established earlier cannot hold on as conditions inevitably change: grassy swales are closed out by forest growth, disease strikes, a stronger competitor invades, chance fluctuations in members bounce the population to zero. The community shifts continuously, and by an unconscious trial and error, through innumerable fits and starts, its biodiversity slowly rises. Species excluded earlier at last find room, symbiotic pairs and trios are fitted together, the forest grows deeper and richer, new niches are prepared. The community thus approaches a mature state, actually a dynamic equilibrium with species forever arriving and disappearing and the total species numbers bobbing up and down inside narrow limits.

Throughout the process of colonization, accommodations are made. Species in collision sometimes compromise through ecological displacement. They yield part of the environment to their competitors and survive. Fire ants, for example, are among the most aggressive territorial animals known, and it is unusual to find more than two or three species coexisting in the same community. Their colonies, made up of a mother queen and thousands of stinging, biting workers, engage each other in organized combat. They seek out and destroy smaller colonies and settle territorial boundaries with larger ones by continuous *combat d'usure*, pushing at one another until a balance of power is attained. Sometime during the 1930s a South American species, the imported fire ant *Solenopsis invicta*, was accidentally introduced into the port of Mobile, Alabama. It was successful from the start, needing just forty years to spread across the southern United States from the Carolinas to Texas. Throughout most of that range it confronted a native fire ant, *Solenopsis geminata*, which up to that time had been a dominant ant in both woodland and open habitats. The native fire ant is still abundant, but it has been largely forced back into scattered woodland sites. The habitats most favored by fire ants generally, pastureland, lawns, and roadsides, are now owned by the newcomers. If the imported fire ant could somehow be removed (an event fervently but hopelessly desired by southerners), the native fire ant would almost certainly reoccupy its old haunts.

The case of the fire ants illustrates the well-documented principle that closely similar species can fit together when their requirements are elastic. Elasticity is the hallmark of Darwin's finches of the Galápagos, for the simple reason that their long-term survival depends on it. They live on volcanic desert islands with harsh, variable environments, changing in the quality of life they offer from month to month and year to year. During the wet season, when most plant growth occurs and food is relatively abundant, the birds enjoy a broad diet. Species that live on the same island and are anatomically similar to one another feed to substantial degree on the same items. In the dry season, food grows scarce and the species come to differ in the items they select. Some become specialists, while others broaden their diets.

The tiny island of Daphne Major is home to two resident species, the medium ground finch *Geospiza fortis* and the cactus finch *Geospiza scandens*. Both live in dense stands of opuntia cactus. In the wet season, when the cactus is in full bloom, the two species consume much the same food. They take the nectar and pollen of the flowers, and they also feast on various kinds of seeds and insects. In the dry season, as the food supply drops, *scandens* narrows its diet to concentrate on edible parts of the opuntia plant. *Fortis* broadens its diet to include an even broader range of items than before, wherever it can find them.

Imagine a case in which two such species have been squeezed together in the same communities long enough for evolution to occur. When they first came into contact, they were elastic and could diverge in their habits enough to lessen competition. The differences were phenotypic, the result of environment and not genes. The compression occurred in traits that were relatively easy to change, most likely by a retreat from parts of the habitat and diet by one or both of the species. As the generations passed, genetic differences arose and hardened the distinction between the two species. Individual birds found it advantageous to excel in those portions of the niche to which they had been driven. The success of those genetically predisposed to do so caused the population as a whole to specialize— to consume a certain food or to build nests in one or another habitat. The differences between the two species next extended to anatomy and physiology. Then the two species competed less with one another, most likely at the price of some of their original elasticity. They experienced the evolutionary change called *character displacement*.

The classic example of character displacement is the change in bill size and food habits of Darwin's finches. Adaptive radiation among the thirteen Galápagos species was based to a large degree on variations in thickness of the bill, and this trait has been engineered in part by character displacement. The selection pressure behind the evolution is improved efficiency during specialization. The deeper the bill at the plane of its attachment to the head, the more power it can exert along the cutting edges and at the tip. Finches with thick bills are well equipped to rip open tougher fruits and to crush bigger and more brittle seeds. Finches with thin bills are limited to softer fare, but they are compensated by an ability to probe narrow crevices and manipulate small objects. A rough analogy from human technology is the adaptive radiation of pliers. To turn a bolt or twist a thick wire with dispatch calls for either linesman's pliers or parrot-head gripping pliers. To manipulate fine bolts and wires you need nose pliers, which are thinner and proportionately longer.

The shape of the bill is not the whole story of displacement and radiation in Darwin's finches. The size of the jaw muscles, the stereotyped movement of the birds during feeding, and perhaps even the chemistry of the digestive traits have been altered as part of dietary specialization among the species. But bill depth remains the most obvious and easily measured among all the traits. It is a convenient proxy by which the larger syndrome of specialized changes can be studied.

The surest test of character displacement as an engine of adaptive radiation is the demonstration of a certain two-part geographic pattern: species have evolved away from each other in places where they are in contact, but they have failed to do so, or have even converged, where they live alone. In the special case of Darwin's finches, we look for enhanced differences among the species on islands where they live together, and particularly in those traits such as bill shape that allow them to specialize and reduce competition. And we need a control: on other islands harboring only a single species, the competitors should more closely resemble each other, again in those traits believed to be most subject to competition. If this dual pattern is strong and convincing, we may reasonably conclude that where the species has been forced to compete, it evolved away from its opponent to fill a special niche, and where it lacked competition it stayed put—or else evolved in the direction of the opponent to fill both niches.

In testing for character displacement in Darwin's finches, Peter Grant put to use the fact that some of the species occur on many islands in the Galápagos. He looked at thirteen cases where pairs of closely related species occur together on various islands. In eleven such instances he found them to differ more in beak depth than when they occur alone, on islands of their own. The evidence was nevertheless short of decisive. Grant recognized that there is another way in which such a pattern can arise in the absence of competition. Character displacement could also occur by reproductive reinforcement of the differences that isolate species as distinct gene pools. If two species hybridize to some extent when they meet, and the hybrids are inferior or sterile, it is to the advantage of both species to avoid interbreeding altogether. One device might be to evolve traits such as distinctive bill shapes that allow individuals to select members of their own species with greater accuracy. Using stuffed female birds, which in spite of their immobility are courted by unsuspecting males, Grant discovered that the males prefer females with the right bill shape on islands where similar species live together. They are much less selective, however, where the same species lives alone. In other words, bill shape *is* used as a cue by male finches to choose females of their own species, and reproductive reinforcement does occur as an evolutionary process. By closely weighing the factors, however, Grant showed that character displacement occurs primarily through competition, and reproductive reinforcement is hooked onto it as a secondary effect. This means that once bills evolve apart as a consequence of competition, related species of Darwin's finches also use the differences to avoid hybridization.

Character displacement has been persuasively documented in a few other groups of organisms, including frogs, fruit flies, ants, and snails, but it is far from a universal biological process.

It allows a bit of compression here and there, and enables a few more species to squeeze into local communities. It represents one process by which communities can be organized to some degree, mediating a rise in general biological diversity.

To the forces that increase biodiversity, add predators. In a celebrated experiment on the seacoast of Washington state, Robert Paine discovered that carnivores, far from destroying their prey species, can protect them from extinction and thereby salvage diversity. The starfish *Pisaster ochraceus* is a keystone predator of mollusks living in rock-bound tidal waters, including mussels, limpets, and chitons. It also attacks barnacles, which look like mollusks but are actually shell-encased crustaceans that remain rooted to one spot. Where the *Pisaster* starfish occurred in Paine's study area, fifteen species of the mollusk and barnacle species coexisted. When Paine removed the starfish by hand, the number of species declined to eight. What occurred was unexpected but in hindsight logical. Free of the depredations of *Pisaster*, mussels and barnacles increased to abnormally high densities and crowded out seven of the other species. In other words, the predator in this case was less dangerous than the competitors. The assembly rule is this: insert a certain predator, and more species of sedentary animals can invade the community later.

Still another dimension of complexity is added by symbiosis, defined broadly as the intimate association of two or more species. Biologists recognize three classes of symbiosis. In parasitism, the first, the symbiont is dependent on the host and harms but does not kill it. Put another way, parasitism is predation in which the predator eats the prey in units of less than one. Being eaten one small piece at a time and surviving, often well, a host organism is able to support an entire population of another species. It can also sustain many species simultaneously. A single unfortunate and unmedicated human being might, theoretically at least, support head lice (*Pediculus humanus capitis*), body lice (*Pediculus humanus humanus*), crab lice (*Pthirus pubis*), human fleas (*Pulex irritans*), human bot flies (*Dermatobia hominis*), and a multitude of roundworms, tapeworms, flukes, protozoans, fungi, and bacteria, all metabolically adapted for life on the human body. Each species of organism, especially each kind of larger plant or animal, is host to such a customized fauna and flora of parasites. The gorilla, for example, has its own crab louse, *Pthirus gorillae*, which closely resembles the one on *Homo sapiens*. A mite has been found that lives entirely on the blood it sucks from the hind feet of the soldier caste of one kind of South American army ant. Tiny wasps are known whose larvae parasitize the larvae of still other kinds of wasps that live inside the bodies of the caterpillars of certain species of moths that feed on certain kinds of plants that live on other plants.

Raising diversity still more are the commensals, symbiotic organisms that live on the bodies of other species or in their nests but neither harm nor help them. Without any awareness of the fact, most human beings carry around on their foreheads two kinds of mites, slender creatures with wormlike bodies and spidery heads so small as to be almost invisible to the naked eye. One (*Demodex folliculorum*) dwells in the hair follicles, the other (*Demodex brevis*) in the sebaceous glands. You can get to know your own forehead mites the following way: stretch the skin tight with one hand, carefully scrape a spatula or butter knife over the skin in the opposite direction, squeezing out traces of oily material from the sebum glands. (Avoid using too sharp an object, such as a glass edge or sharpened knife.) Next scrape the extracted material off the spatula with a cover slip and lower the slip face down onto a drop of immersion oil previously placed on a glass microscope slide. Then examine the material with an ordinary compound microscope. You will see the creatures that literally make your skin crawl.

People would never notice their forehead mites in any other way. These acarines and other commensals slip the thin wedge in, sip small amounts of nutrients and energy virtually useless to their hosts, and live secure lives of flawless modesty. Their biomass is small to microscopic, their diversity immense. They are everywhere, but it takes a special eye to find them. On the leaves of trees in the tropical rain forests grow flat, centimeter-wide gardens of lichens, mosses, and liverworts. Among the epiphylls—plants that live on the leaves—thrive a host of tiny mites, springtails, and barklice. Some of the animals browse on the epiphylls, others prey on the epiphyll browsers. Thus a single leaf of a tree, often composing less than one part in

10,000 of that single large organism, is home to an entire miniaturized fauna and flora.

The tightest bond of all among species, the one that gives
the word *community* more than metaphorical meaning, is
mutualism. This third kind of relationship, often considered
the true symbiosis and employed that way in less formal prose,
is an intimate coexistence of two species benefiting both. A
large part of dead wood is decomposed by termites—not by
the termites really, but by protozoans and bacteria that live in
the hind guts of the termites. And not entirely by these microorganisms either, since they need the termites to provide them
a home and a steady stream of wood chewed into digestible
pulp. So the right way to put the original phrase is: a large
part of wood is decomposed by the termite-microorganism
symbiosis. The termites harvest the wood but cannot digest it;
the microorganisms digest the wood but cannot harvest it. It
might be said that over millions of years the termites domesticated the microorganisms to serve their special needs. That,
however, would be big-organism chauvinism. It is equally correct to say that termites have been harnessed to the needs of the
microorganisms. Such is the nature of mutualistic symbiosis:
to attain the highest level of intimacy, the partners are melded
into a single organism.

Mutualistic symbioses are more than simply curiosities for
the delectation of biologists. Most life on land depends ultimately on one such relationship: the mycorrhiza (literally from
the Greek for fungus-root), the intimate and mutually dependent coexistence of fungi and the root systems of plants. Most
kinds of plants, from ferns to conifers and flowering plants,
harbor fungi that are specialized to absorb phosphorus and
other chemically simple nutrients from the soil. The mycorrhizal fungi give up part of these vital materials to their plant
hosts, and the plants repay them with shelter and a supply of
carbohydrates. Plants deprived of their fungi grow slowly;
many die.

According to species, the fungi either enter the outer root
cells of their plant hosts or envelop the entire roots to form
dense webs. A plant pulled up almost anywhere in the world
reveals a tangle of delicate fibers clutching masses of soil particles. Some of the extensions are likely to be rootlets of the

plant, but others are the moldlike hyphae of the symbiotic fungi. In many kinds of plants, fungal hyphae have completely replaced the rootlets during evolution.

Without the plant-fungus partnership, the very colonization of the land by higher plants and animals, 450 to 400 million years ago, probably could not have been accomplished. The barren, rain-lashed soil of that time was not hospitable to organisms more complex than bacteria, simple algae, and mosses. The earliest vascular plants were leafless, seedless forms that superficially resembled modern-day horsetails and quillworts. By allying themselves with fungi, they took hold of the land. Some of the pioneers evolved into the lycophyte trees and seed ferns of the great Paleozoic coal forests. They also gave rise to the ancestors of modern conifers and flowering plants, whose vegetation came in the fullness of time to harbor the largest array of animal life that has ever existed. Today the tropical rain forests, which may contain more than half the species of plants and animals on earth, grow on a mat of mycorrhizal fungi.

Coral reefs, the marine equivalents of rain forests, are also built on a platform of mutualistic symbiosis. The living coral organisms, which cover the carbonaceous bulk of the reef, are the polyps, close relatives of the jellyfish. Like the jellyfish and other coelenterates, they use feathery tentacles to capture crustaceans and other small animals. They also depend on the energy provided by single-celled algae, which they shelter within their tissues and to which they donate some of the nutrients extracted from their prey. In most coral species, each individual polyp lays down a skeletal container of calcium carbonate that surrounds and protects its soft body. Coral colonies grow by the budding of individual polyps, with the skeletal cups being added one on another in a set geometric pattern particular to each species. The result is a lovely, bewildering array of skeletal forms that mass together to make the whole reef—a tangled field of horn corals, brain corals, staghorn corals, organ pipes, sea fans, and sea whips. As the colony grows, the older polyps die, leaving their calcareous shells intact beneath; and in time the living members form a layer on top of a growing reef of skeletal remains. These massive remains, many of which are thousands of years old, play a major role in the formation of tropical islands, in particular the fringing reefs of volcanic islands and the atolls left behind when the volcanoes erode away. They create the physical basis and photosynthetic energy for tightly packed communities of thousands of species, from sea hornets and mantis shrimps to carpet sharks.

What, in summarizing to this point, do we understand of the assembly of communities? Obviously we know that there is a large amount of organization in the connections among species. But how much? The answer is unknown for any kind of community—all the organisms in a patch of hardwood forest, for example, or in a coral reef or desert spring. We know some keystone species, some assembly rules, some processes of competition and symbiosis that serve as a weak gravitational force.

We know how some species fit together in twos and threes, but not how the whole community fits together. There are a few hints of what is to come as research grows more sophisticated. Think of the community as a food web, a connection of species that prey on other species. Consider what might happen when a species is extinguished, simply plucked out of the food web as were the sea otters. What is the effect? With field studies and mathematical models, ecologists have pieced together a few of the most general properties of food webs that bear on the result of such an experiment. They have learned that the food chains making up the web are very short. If you track who eats whom in different parts of the web, you will usually find the number of links in the chain to be five or fewer. For example: in a marshy glade of the north central states, reedgrass is eaten by short-horned grasshoppers, the grasshoppers are eaten by orb-weaver spiders, the spiders are eaten by palm warblers, and the warblers are eaten by marsh hawks. Because the grass eats no one and the hawks are eaten *by* no one (except by bacteria and other decomposers when they die), these two species form the ends of the chain. A second rule is that the number of links in the food web does not increase as the size of the community increases. No matter how many species manage to persist in the community, the average number of links from a given plant species to a given top predator does not increase.

I cite these two generalizations to illustrate the more solid principles of community ecology. But I cite them also to show

how incomplete and tenuous those principles are. Imagine
that you excise the palm warblers from the marsh food web.
That food chain is broken, but the ecosystem remains intact,
more or less. The reason is that each species in the chain is
linked to additional chains. Other species of birds still present in the marsh will eat more spiders, and the marsh hawks
will turn, almost imperceptibly, to a larger number of birds,
rodents, snakes, and other creatures. Feather mites, bird lice,
and other symbionts found only on palm warblers, part of yet
other chains, disappear with their host, but their loss has a
negligible effect on the community at large.

Expand the thought experiment to extirpate two warbler
species, then all warbler species, and finally all the songbirds in
the community. As the knife cuts deeper, its effects will spread
with increasing severity through a large but indeterminate part
of the community. Take out the ants, the principal predators
and scavengers of insects and other small animals, and the
effects will intensify—yet the details are even less predictable.
Most species of birds, ants, and other plants and animals are
linked to multiple chains in the food web. It is very difficult to
assess which survivors will fill in for the extinguished species
and how competently they will perform in that role. Physicists
can chart the behavior of a single particle; they can predict with
confidence the interaction of two particles; they begin to lose
it at three and above. Keep in mind that ecology is a far more
complex subject than physics.

The reverse of the extinction process is species packing.
Ecologists are unable for the most part to predict which species
can still invade the community and add to its diversity. Select
a habitat at random. How tightly packed are the species? What
is the upper limit of stable diversity, the highest number of
species that can be maintained without human intervention? It
is easy to enhance local diversity by the artificial introduction
of more and more species—orchids affixed to tree trunks, zoo-
bred tigers released into the jungle—but most would eventually perish. Without constant and intrusive manipulation, most
overloaded communities will revert to a lower state of diversity,
perhaps resembling the original, perhaps not.

The indeterminacy of community structure is increased by
the existence of connections between species lying beyond the

conventional food webs, and for which few reliable laws or rules exist. Competition—especially that resulting in the exclusion of one species by another—is especially difficult to call. So are the effects of removing scavengers and symbionts. Most difficult of all to assess is the impact of species that alter the physical environment over many years. Dominant tree species overgrow and change the temperature and humidity regimes in which other plants and animals must live. Mound-building termites turn and enrich the soil; they alter the composition of chemical elements and determine the species of plants that can grow near their underground tunnels. Populations of mites and springtails bloom, and fungus spores and humus correspondingly decline—all to indeterminate degree.

The unpredictability of ecosystems is a consequence of the particularity of the species that compose them. Each species is an entity with a unique evolutionary history and set of genes, and so each species responds to the rest of the community in a special way. I will finish with my own favorite example of law-destroying idiosyncrasy. Tree holes often fill with rainwater, creating small aquatic habitats for animals and microorganisms. On the west coast of the United States live larvae of a tree-hole mosquito species, *Aedes sierrensis.* They feed on microscopic ciliated protozoans, *Lambornella clarki,* which resemble the familiar paramecia used in biology courses. The protozoans in turn feed on bacteria and other microorganisms breeding in the tree-hole water. After the protozoans have been exposed to the odor of the mosquito larvae for one to three days, they turn the tables on their tormentors. Some of them metamorphose into parasitic forms that invade the bodies of the larvae and start to feed on their tissues and blood. Thus a segment of the food chain is flipped upside down, creating a food cycle where each species is simultaneously predator and prey of the other.

The mosquito-protozoan cycle of predation and counterpredation is emblematic of the direction that community ecology must take: analyze ecosystems in detail, from the bottom up. Biologists are returning to natural history with a new sense of mission. They cannot expect to learn much more from the top down, from the properties of whole ecosystems (energy flow, nutrient cycles, biomass) interpolated to the properties of communities and species. Only with a detailed knowledge of the life cycles and biology of large numbers of constituent species will it be possible to create principles and methods that can precisely chart the future of ecosystems in the face of the human onslaught.

Then there might be an answer to the question I am asked most frequently about the diversity of life: if enough species are extinguished, will the ecosystems collapse, and will the extinction of most other species follow soon afterward? The only answer anyone can give is: possibly. By the time we find out, however, it might be too late. One planet, one experiment.

CHAPTER TEN

Biodiversity Reaches the Peak

THREE BILLION years ago the land was virtually devoid of life. More than that, it was uninhabitable. No ozone layer existed in the stratosphere, and the progenitor molecules of oxygen in the air below were too thin to create it. Short-length ultraviolet radiation traveled unimpeded to the earth and beat down on the dry basaltic rocks. It assaulted organisms venturing there out of the sea, shutting down their enzymatic synthesis, opening their membranes to ambient poisons, and rupturing their cells. But in the water, safe from the lethal rays, microscopic organisms swarmed. They were close to modern cyanobacteria (sometimes called blue-green algae) and a mélange of bacteria and bacteria-like species. Most were single-celled and prokaryotic, and a few were composed of cells strung together in filaments. These simple organisms were devoid of nuclear membranes, mitochondria, chloroplasts, and the other organelles that give structural complexity to the cells of higher plants and animals.

A large portion of the early life forms were concentrated in thin scummy sheets called microbial mats. Under the mats they built up distinctive rock formations called stromatolites, resembling stacks of mattresses (*stroma*, mattress) strewn about the shallow sea bottoms like packages on a warehouse floor. Modern versions of these organism-topped rocks still grow in subtidal marine waters in a few scattered localities such as Baja California and northwestern Australia. Some are soft enough to be cut by a hunting knife. Others have been infiltrated by enough calcium carbonate to make them as hard as the fossil stromatolites. The formations grow by accretion. The living organisms on them are periodically covered by silt and debris carried by the tide and storms. They multiply upward, pressing through the fouling cover to touch once again clear water and sunlight, and by this means they add height to the stromatolite foundations year by year.

Not all modern microbial mats have thick columns under them. Many form thin, unsupported sheets in marginal habitats where physical conditions are severe and predators and competing organisms scarce, such as hot springs, salty lagoons, Antarctic lakes, deep-sea sediments, and damp rock surfaces on the land. They are scarce and scattered in comparison with most ecosystems. But three billion years ago all available space in the shallow seas was probably covered by a variety of such microbial formations, each kind specialized for a particular niche of light, temperature, and acidity.

Since the beginning of life, the denizens of microbial mats have gathered into communities of considerable complexity. The plain appearance of the outer coat viewed with the naked eye is misleading. When a mat is sliced vertically and examined under the microscope, it is seen to be packed with photosynthetic organisms from the surface to a depth of a millimeter. Across that short distance, half the height of a capital letter on this page, sunlight attenuates to 1 percent of the intensity it has in the water above. That is about the same amount of energy lost by sunlight in traveling from tree crown to floor in a dense forest. And the analogy runs deeper: the mat community is even organized somewhat like a forest. Cyanobacteria, which capture solar energy, are distributed in succession from top to bottom like different kinds of trees, with least shade-tolerant species near the surface and most shade-tolerant species toward the bottom. They use the energy to combine water and carbon dioxide into organic molecules, giving off oxygen in the process. Farther down, in the miniature equivalent of the dark forest interior (or deep sea, below the upper lighted waters), live sulfur-oxidizing bacteria. These archaic organisms, of a kind that may have preceded cyanobacteria in evolution, are not photosynthesizers. They do not decompose water into hydrogen and oxygen with the aid of solar energy, but instead split the weaker sulfide bonds unaided by sunlight.

Swimming and drifting in open water around the ancient microbial mats were almost certainly populations of cyanobacteria and other prokaryotic forms different from the mat organisms. Some lived by photosynthesis, others by preying on prokaryotes or scavenging their dead cells. Life must have been already diverse at the microscopic level, appropriating relatively large quantities of energy and nutrients. Yet the early organisms were not all that diverse, not when compared with

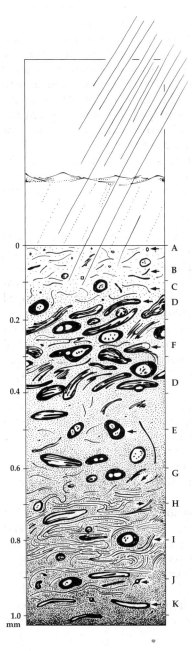

Among the most primitive ecosystems on earth are the microbial mats, thin assemblages of microscopic organisms that date back in geological time almost to the beginning of life. In this living mat from shallow marine water, organisms are arrayed by species according to their depth across 1 millimeter of the mat and hence by the amount of light and kind of nutrients reaching them.

A diatoms (microscopic algae)
B *Spirulina* (cyanobacteria or blue-green algae)
C *Oscillatoria* (cyanobacteria)
D *Microcoleus* (cyanobacteria)
E nonphotosynthesizing bacteria
F mixed single-celled cyanobacteria
G bacterial mucilage
H *Chloroflexus* (green bacteria)
I *Beggiatoa* (sulfide-oxidizing bacteria)
J unidentified grazing organisms
K discarded sheaths of cyanobacteria

present-day biotas. No forests and grasslands succoring millions of animal species blanketed the land; no kelp beds choked the ocean margins; no flocks of terns hunted fish across blue waters. If you and I could travel back in time to wade along the shore of the ancient sea, searching for plants and animals with unaided vision, we would find nothing certifiably alive, only unprepossessing pond-scum smudges of brown and green and slimy rock surfaces of uncertain provenance. Visible organisms and high diversity were to come much later.

Biological diversity has increased a thousandfold since the early days of the microbial mats, pulled along by evolutionary progress, measured in turn by four great steps that mark the passage of eons:

• The origin of life itself, spontaneously from prebiotic organic molecules, about 3.9 to 3.8 billion years ago. The first organisms were single-celled and hence microscopic. Stromatolite ecosystems appeared no later than 3.5 billion years ago.

• The origin of eukaryotic organisms—"higher organisms"—about 1.8 billion years ago. Their DNA was enveloped in membranes, and the remainder of the cell contained mitochondria and other well-formed organelles. At first eukaryotes were single-celled, in the manner of modern protozoans and the simpler forms of algae, but soon they gave rise to more complex organisms composed of many eukaryotic cells organized into tissues and organs.

• The Cambrian explosion, 540 to 500 million years ago. Newly abundant macroscopic animals, large enough to be seen with the naked eye, evolved in a radiative pattern to create the major adaptive types that exist today.

• The origin of the human mind, in later stages of evolution in the genus *Homo*, probably from a million to 100,000 years ago.

Some biologists and philosophers have trouble with that term, "evolutionary progress." The expression is inexact and loaded with humanistic nuance, granted, but I use it just the same to identify a paradox pivotal to the understanding of biological diversity. In the strict sense, the concept of progress implies a goal, and evolution has no goal. Goals are not inherent in DNA. They are not implied by the impersonal forces of natural selection. Rather, goals are a specialized form of behavior, part of the outer phenotype that also includes bones, digestive enzymes, and the onset of puberty. Once assembled by natural selection, human beings and other sentient organisms formulate goals as part of their survival strategies. Because goals are the ex-post-facto responses of organisms to the necessities imposed by the environment, life is ruled by the immediate past and the present, not by the future. In short, evolution by natural selection has nothing to do with goals, and so it would seem to have nothing to do with progress.

And yet there is another meaning of "progress" that does have considerable relevance to evolution. Biological diversity embraces a vast number of conditions that range from the simple to the complex, with the simple appearing first in evolution and the more complex later. Many reversals have occurred along the way, but the overall average across the history of life has moved from the simple and few to the more complex and numerous. During the past billion years, animals as a whole evolved upward in body size, feeding and defensive techniques, brain and behavioral complexity, social organization, and precision of environmental control—in each case farther from the nonliving state than their simpler antecedents did. More precisely, the overall averages of these traits and their upper extremes went up. Progress, then, is a property of the evolution of life as a whole by almost any conceivable intuitive standard, including the acquisition of goals and intentions in the behavior of animals. It makes little sense to judge it irrelevant. Attentive to the adjuration of C. S. Peirce, let us not pretend to deny in our philosophy what we know in our hearts to be true.

An undeniable trend of progressive evolution has been the growth of biodiversity by increasing command of earth's environment. New methods to detect microscopic fossils in billion-year-old sedimentary rocks, chemical analyses of ancient environments, and statistical estimates of the relative abundances of extinct species have allowed geochemists and paleontologists during the past decade to bring this history into sharper focus.

By two billion years before the present, a large fraction of earth's organisms were generating oxygen through photosynthesis. But this element, so vital to life as we know it today, did not accumulate in the water and atmosphere. It was captured by ferrous iron, which dissolves in water and was abundant enough to saturate the early seas. The two elements combined to form ferric oxides, insoluble in water, which settled to the ocean floor. As J. William Schopf neatly summarized the situation, the world rusted.

Denied oxygen by the ferrous sink, the organisms were forced to remain anaerobic. The aerobic pathways of metabolism, which are highly efficient means to obtain and deploy free energy, could at most have evolved as an auxiliary adaptation. By 2.8 billion years ago, the sink had partially filled, and a few local habitats sustained low levels of molecular oxygen. Aerobic organisms, still single-celled prokaryotes, appeared about this time. During the next billion years, oxygen levels rose worldwide to constitute about 1 percent of the atmosphere. By 1.8 billion years ago, the first eukaryotic organisms appeared: alga-like forms, forerunners of the dominant photosynthesizers of the modern seas. By no later than 600 million years ago, near the end of the Proterozoic era, the first animals evolved. Members of this Ediacaran fauna, named after the Ediacara Hills of South Australia where many of the first specimens were found, were soft-bodied and typically flat. They vaguely resembled jellyfish, annelid worms, and arthropods, and some may have been members of those surviving groups.

Approximately 540 million years ago, near the beginning of the Cambrian period, earliest of the time segments of the Phanerozoic eon in which we now live, a seminal event occurred in the history of life. Animals increased in size and diversified explosively. The supply of free oxygen in the atmosphere was by this time near the 21 percent level of today. The two trends are probably linked, for the simple reason that large, active animals need aerobic respiration and a rich supply of oxygen. Within a few million years, the fossil record held almost every modern phylum of invertebrate animals a millimeter or more in length and possessed of skeletal structures, hence easily preserved and detectable later. A large portion of present-day classes and orders had also come on stage. Thus occurred the Cambrian explosion, the big bang of animal evolution. Bacteria and single-celled organisms had long since attained comparable levels of biochemical sophistication. Now, in a dramatic new

Full History of Life

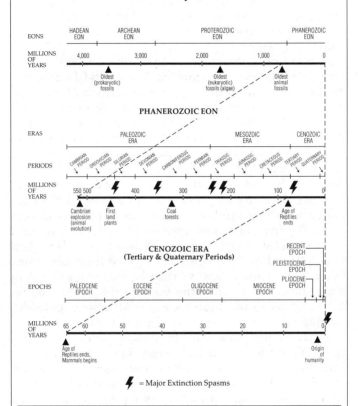

= Major Extinction Spasms

The full geological history of life goes back more than 3.5 billion years, when the first single-cell organisms appeared. Key episodes in evolution are placed within the divisions of geological time: eons divided into eras, eras into periods, and periods into epochs. Biodiversity was sharply reduced by the great extinction spasms, indicated here by lightning flashes.

radiation, they augmented their niches to include life on the bodies and waste materials of the newly evolved animals. They created a new, microscopic suzerainty of pathogens, symbionts, and decomposers. In broad outline at least, life in the sea attained an essentially modern aspect no later than 500 million years ago.

By this time a strong ozone layer existed as well, screening out lethal short-wave radiation. The intertidal reaches and dry land were safe for life. By the late Ordovician period, 450 million years ago, the first plants, probably derived from multicellular algae, invaded the land. The terrain was generally flat, lacking mountains, and mild in climate. Animals soon followed: invertebrates of still unknown nature burrowed and tunneled through the primitive soil. Paleontologists have found their trails but still no bodies. Within 50 or 60 million years, early into the Devonian period, the pioneer plants had formed thick mats and low shrubbery widely distributed over the continents. The first spiders, mites, centipedes, and insects swarmed there, small animals truly engineered for life on the land. They were followed by the amphibians, evolved from lobe-finned fishes, and a burst of land vertebrates, relative giants among land animals, to inaugurate the Age of Reptiles. Next came the Age of Mammals and finally the Age of Man, amid continuing tumultuous change at the level of class and order.

By 340 million years before the present, the pioneer vegetation had given way to the coal forests, dominated by towering lycophyte trees, seed ferns, tree horsetails, and a great variety of ferns. Life was close to the attainment of its maximal biomass. More organic matter was invested in organisms than ever before. The forests swarmed with insects, including dragonflies, beetles, and cockroaches. By late Paleozoic and early Mesozoic times, close to 240 million years ago, most of the coal vegetation had died out, with the exception of the ferns. Dinosaurs arose among a newly constituted, mostly tropical vegetation of ferns, conifers, cycads, and cycadeoids. From 100 million years on, the flowering plants swept to domination of the land vegetation, reconstituting the forests and grasslands of the world. The dinosaurs died out during the hegemony of this essentially modern vegetation, at a time when tropical rain forests were assembling the greatest concentration of biodiversity of all time.

For the past 600 million years, the thrust of biodiversity, mass extinction episodes notwithstanding, has been generally upward. In the sea the orders of marine animals climbed slightly above a hundred during the Cambrian and Ordovician periods and stayed in place for the remaining 450 million years. Families, genera, and species closely traced the same pattern to the end of the Paleozoic era, 245 million years ago. They were knocked down sharply by the extinction catastrophe that closed the Paleozoic era, followed in only 50 million years by a smaller spasm in the Triassic period. Thereafter they climbed steeply, with a dip at the end of the Mesozoic, reaching unprecedented levels during the past several million years. Plant and animal diversity on the land, after a delay of 100 million years during which colonization took place, followed the same trajectory to our time.

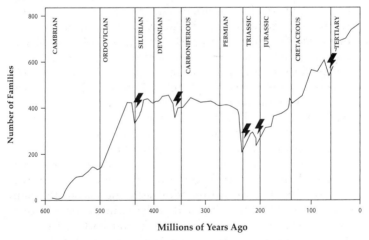

Biological diversity has increased slowly over geological time, with occasional setbacks through mass global extinctions. There have been five such extinctions so far, indicated here by lightning flashes. The data given are for families (groups of related species) of marine organisms. A sixth major decline is now underway as a result of human activity.

Each extinction spasm reduced the numbers of species most and the numbers of classes and phyla least. The lower the taxonomic category, the more it was diminished. At the end of the Paleozoic era, as many as 96 percent of the species of marine animals and foraminiferans vanished, compared with 78 to 84 percent of the genera and 54 percent of the families. Apparently no phyla came to an end.

This descending vulnerability by taxonomic rank is an artifact, a straightforward consequence of the hierarchical manner in which biologists classify organisms. But it is an interesting and technically useful artifact, for reasons most clearly expressed by what has been called the Field of Bullets Scenario. Imagine advance infantrymen walking forward in eighteenth-century manner, posting arms in one serried rank into a field of fire. Each man represents a species, belonging to a platoon (genus), which is a unit of a company (family), in turn a unit of a battalion (order), and so on up to the corps (phylum). Each man has the same chance each moment of being hit by a bullet. When he falls the species he represents is extinguished, but other members of the platoon-genus march on, so that even though diminished in size the genus survives. In time all the members of the genus may perish, yet remnants of other genera are still standing down the line, and so the company-family presses forward. At the end of the long lethal march, the vast majority of species, genera, families, orders, and even classes may be gone, but so long as one species out of the multitude remains alive the phylum-corps survives.

Across 600 million years of Phanerozoic evolution, the turnover of species was nearly total. More than 99 percent of all species that ever lived in each period perished, to be replaced by even larger numbers drawn from the descendants of the survivors. Such is the nature of dynastic successions through the history of life, often initiated by the extinction spasms that bring down entire company-families and battalion-orders. Ninety-nine percent is not a surprising figure. Imagine a group such as the archaic amphibians of the Paleozoic. A thousand species die, and one survives to produce the primitive reptiles. A thousand of those also die, but one survivor carries on to become the ancestor of the Mesozoic dinosaurs. The survival rate of species in this sequence is 1 in 2,000. In other words, only a single line persists out of the 2,000 created, and yet life flourishes as diversely as before.

This brings us to the notable possibility that no phylum has ever been extinguished. Let me phrase this conjecture in a more operational manner: no major group that has gone extinct can be definitely assigned a taxonomic rank as high as the phylum. A great many battalions and regiments have disappeared in full, but we cannot be sure that any one of them constitutes an entire corps. If any true phylum has vanished, it is most likely to have been one that originated during the Cambrian explosion of animal diversity. Something happened to the environment then, most likely the availability of atmospheric oxygen, that opened the sea to large animals. The waters of the world became a new continent on which animals above a centimeter long could evolve and adaptively radiate; and so they did, creating most or all of the remaining phyla we know today.

It has been widely believed that the Cambrian explosion was a period of wild experimentation during which basic body plans never seen before or afterward were invented and discarded. If this view is correct, some of the most extreme of the short-lived species must qualify as phyla that went extinct. It would also follow that at the phylum level biodiversity reached its peak during the Cambrian explosion and subsided soon afterward to modern levels. This interpretation is supported by the existence of well-preserved fossils in the Burgess Shale of British Columbia, early to middle Cambrian in age, which appear to fit no established phylum. Other Burgess-type fossils, equally bizarre, have been found in Europe, China, and Australia.

These fossils taken together leave little doubt that many unusual animal types arose during Cambrian times and disappeared after brief tenures. In taxonomic language, orders and classes originated that lasted only a few million years. But the fossils do not yet tell us definitively that radical new body types—innovations that confer phylum rank—were created and discarded. In 1989 Simon Conway Morris, a leading authority on Burgess Shale faunas, recognized eleven modern phyla present in those ancient assemblages, together with "19 distinct body plans that for the most part are as different from each other as any of the remaining phyla in the fauna." The swift radiation, Conway Morris continued, is reflected in the still-living phylum Arthropoda, in which the Burgess Shale "cavalcade of morphologies seems to be almost inexhaustible. The overall impression is of an enormous mosaic, individual species being assembled according to differences in number and types of jointed appendages, number of segments and extent of tergite fusion, and overall proportions of the body."

Still, the totality of known arthropod diversity in the Cambrian fossils does not exceed that in living arthropods, and it probably falls far short. The array of classes and orders in the sea remains vast. The insects had not yet arisen to shower the land and fresh water with burrowers, swimmers, and flying machines of fantastical design. I cannot help thinking that if we were to take just four living species of this single class (the Insecta)—say the maggot of a blackfly, the giant bottlenosed lantern fly, a female coccid scale, and the water penny—and preserve them with the same fidelity as the Burgess rock fossils, the remains would be classified erroneously as four distinct phyla. They display what appear superficially to be wholly different body designs.

The paleontologists working on the Burgess faunas have approached their subject with admirable caution. They refer to the fossils with body types not assignable to modern phyla as "problematica." When better-preserved fossils turned up, as in the Chengjiang beds of southern China, and improved methods were developed to study the old specimens, the problematica were whittled down. The armored creatures of the genera *Hallucigenia*, *Microdictyon*, and *Xenusion* have recently been placed among the Onychophora, the living phylum of caterpillar-like animals thought to be intermediate between arthropods and annelid worms. The essence of weirdness, *Wiwaxia corrugata*, thought to resemble a scaly slug with long spikes sprouting from its back, has proved to be fragments of a polychaete worm, member of an extant phylum, the Annelida.

To summarize the grand parade to date: the number of living animal phyla, all of which have representatives in the sea, is about thirty-three. Of these, approximately twenty comprise animals large and abundant enough to leave fossils of the kind preserved in beds of the Burgess Shale type. The number of Cambrian animal phyla identified with confidence remains at eleven. To the best of our knowledge, no phylum has yet gone extinct. The level of diversity in the sea inched upward after the Cambrian explosion, and it also increased on the land after the assembly of the coal forests and their insect and amphibian inhabitants. The greatest rise in general has occurred during the past 100 million years.

We may now ask why, in spite of major and minor temporary declines along the way, in spite of the nearly complete turnover of species, genera, and families on repeated occasions, the trend in biodiversity has been consistently upward. Part of the answer is that the continental land masses have changed in a way that enhances species formation. During late Paleozoic times, the earth's land surface was composed of the single supercontinent Pangaea. By the early Mesozoic, Pangaea had split into two great fragments, Laurasia to the north and Gondwana to the south; India had broken off as a smaller fragment and was crawling northward toward its rendezvous with the Himalayan arc. Around 100 million years ago, the modern continents were

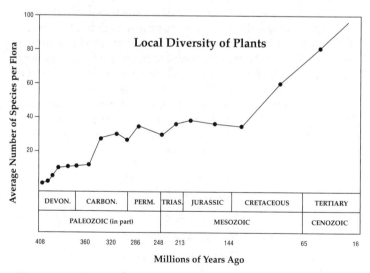

The average number of plant species found in local floras has risen steadily since the invasion of the land by plants 400 million years ago. The increase reflects a growing complexity in terrestrial ecosystems around the world.

in place, with the waters between them progressively widening. Major faunas and floras evolved in a state of deepening isolation. The length of coastline increased everywhere, along with the area available for inshore bottom-dwelling organisms. Shallow seas spread and retreated back and forth over the land, creating and abolishing new habitats and assemblages of organisms adapted to them. These inhabited worlds, the "faunistic and floristic provinces" as they are called, have waxed and waned, but what we live with today is a peak of enrichment.

Global biodiversity was brought to its Cenozoic pinnacle first by the creation of the aerobic environment and second by the fragmentation of the land masses. But that is far from the whole story. The number of species living together within particular habitats, shallow bays and tropical forests for example, has also increased by fits and starts. The number at least doubled in marine organisms and more than tripled in land plants during the past 100 million years. The trends imply that more species are now packed into local communities—that species either originate more quickly or die more slowly.

To see why local communities have grown richer in diversity through geological time, we must return to living faunas and floras, where the biology of individual species can be examined in finer detail. First we can compare living communities poor in species with those that are rich in species. The major clue before us is the latitudinal diversity gradient, the increase in species (or any other taxonomic category) encountered while traveling from the poles to the equator. Here is a slice of the northern hemispheric gradient in breeding bird species, using land areas of roughly the same size:

Greenland	56
Labrador	81
Newfoundland	118
New York State	195
Guatemala	469
Colombia	1,525

Some 30 percent of the world's 9,040 bird species occur in the Amazon Basin, another 16 percent in Indonesia. Most of these faunas are limited to rain forests and closely associated habitats, such as riverine and swamp woodland.

In fact, a large part of the gradient is due to the extraordinary richness of tropical rain forests. This major habitat type, or *biome* as ecologists call it, is defined as forest growing in regions with at least 200 centimeters of annual rainfall spread evenly enough through the year to allow a heavy growth of broad-leaved evergreen trees. The forest is arrayed in multiple layers, from upper canopy 30 meters or more in height, broken by scattered emergent trees soaring over 40 meters, down through ragged middle levels to chest-high understory shrubs. Lianas, strangler vines, and creepers coil around the tree trunks and dangle from the high limbs straight to the ground. Gardens of orchids and other epiphytes festoon the thicker branches. Palms are common in the lower and middle layers of many rain forests, adding lush beauty and a deceptive feel of benignity for the visitor on foot. So effective are the staggered canopy layers at intercepting sunlight that the energy-starved lower vegetation is as sparse as in a juniper woodland. You can walk through it easily, pushing the crisscrossing fronds and branches aside, stepping around the trunks of the great trees, bending to pass under lianas and low tree limbs. There is almost never a need to slash a path with a machete through tangled vegetation, in accord with the popular image of jungles. Machetes are for second growth and forest borders, the true jungles. Rain forests are green cathedrals. They are like the gentle temperate forests familiar to most, except that they tower high and have somehow stayed mysterious and wild.

In the deep rain forest, sunlight dapples the ground, which is covered in spots by a thin layer of leaves and humus and is completely bare in others. Away from the sunlit patches, the ground is so dark that a flashlight is needed to study it closely, to see the insects, spiders, sowbugs, millipedes, daddy-longlegs, and other small creatures swarming there, together forming the cemetery squads of scavengers and the predators that hunt them in turn.

Tropical rain forests, though occupying only 6 percent of the earth's land surface, are believed to contain more than half the species of organisms on earth. I say "believed" because no exact estimates of diversity have been made either for the world as a whole or for rain forests in particular. The more-than-half figure has simply emerged as a consensus from technical reports and conversations among experts, bolstered by educated guesses and logical extensions by the theoreticians of biological diversity. It is largely based, I admit, on anecdotes and piecemeal analysis. Yet in the aggregate this circumstantial evidence grows more persuasive with time.

Here are the elements of reasoning behind the more-than-half argument. The latitudinal diversity gradient, which I illustrated with birds, is a true general principle of biology: the largest numbers of species occur in the equatorial regions of South America, Africa, and Asia. Another explicit example is offered by the vascular plants, which include the flowering plants, ferns, and a mix of lesser groups such as the club mosses, horsetails, and quillworts. These groups together compose more than 99 percent of land vegetation. Of the approximately 250,000 species known, 170,000 (68 percent) occur in the tropics and subtropics, especially in the rain forests. The peak of global plant diversity is the combined flora of the three Andean countries of Colombia, Ecuador, and Peru. There over 40,000 species occur on just 2 percent of the world's land surface. The world record for tree diversity at one site was set by Alwyn Gentry in the rain forest near Iquitos, Peru. He found about 300 species in each of two 1-hectare (2.5-acre) plots. Peter Ashton discovered over 1,000 species in a combined census of ten selected hectare plots in Borneo. These numbers are to be compared with 700 native species found in all of the United States and Canada, in every major habitat from the mangrove swamps of Florida to the coniferous forests of Labrador.

Butterflies are even more disproportionately rich in the rain forests. The largest documented faunas in the world occur in the Río Madre de Dios drainage in southeastern Peru. To date Gerardo Lamas and his coworkers have recorded 1,209 butterfly species within the 55 square kilometers of the Tambopata Reserve. In a close race, Thomas Emmel and George Austin have identified 800 species in a forest patch several square kilometers in extent at Fazenda Rancho Grande, near the center of the state of Rondonia, western Brazil. By adding probable species numbers from still poorly researched groups, they estimate the total inventory to fall between 1,500 and 1,600 species.

Nearby, at Jaru, on October 5, 1975, one entomologist sighted an astonishing 429 butterfly species within twelve hours (this site has since been cleared for agriculture and almost all its butterflies are gone). In contrast, there are only about 440 species in all of eastern North America and 380 in Europe and the Mediterranean coast of North Africa combined.

The ants rival the butterflies in the steepness of their latitudinal gradient. At the Tambopata Reserve, Terry Erwin used a bug bomb to collect all the insects from a single leguminous tree in the rain forest. I identified the ants in his sample and found 43 species in 26 genera, approximately equal to the entire ant fauna of the British Isles. The ants are in turn dwarfed by the beetles. Erwin estimated that over 18,000 species occur in 1 hectare of Panamanian rain forest, with most previously unknown to science—in other words, still lacking a scientific name. To date, only 24,000 beetle species are known from all of the United States and Canada, 290,000 from the entire world.

Thus it goes, biodiversity pyramiding southward, group by group on the land. A few kinds of plants and animals, including conifers, aphids, and salamanders, are more diverse in the temperate zones. These are the exceptions and in any case are not unusually diverse. There are, for example, fewer than 400 species of salamanders known in the entire world. Other groups of plants and animals are mostly tropical but specialized to live in deserts, grassland, and dry forest. They too are generally less diverse than the inhabitants of nearby rain forests.

The organisms of the shallow-water marine environments follow the same latitudinal trend: plankton and bottom dwellers increase in diversity toward the tropics, and the densest concentrations of all are found in the coral reefs, the marine equivalent of rain forests. They abound in absolute diversity, most of it unexplored. Hundreds of species of crustaceans, annelid worms, and other invertebrates can be found in a single coral head, the equivalent of a rain-forest tree.

To summarize the present global pattern, latitudinal diversity gradients rising toward the tropics are an indisputable general feature of life. And on the land biodiversity is heavily concentrated in the tropical rain forests. So immense are the insect faunas alone in these forests, comprising possibly tens of millions of species, overwhelming even the opulence of the coral reefs, that on this basis alone it is reasonable to suppose that over half of all species are found there.

The cause of tropical preeminence poses one of the great theoretical problems of evolutionary biology. Biologists have focused variously on climate, solar energy, amount of habitable terrain, variety of habitats available, amount and frequency of environmental disturbance, degree of isolation of the faunas and floras, and the mostly intangible idiosyncrasies of history. Many have called the problem intractable, supposing its solution to be lost somewhere in an incomprehensible web of causes or else dependent on past geological events that have faded beyond recall. Yet a light glimmers. Enough solid analyses and theory have locked together to suggest a relatively simple solution, or at least one that can be easily understood: the Energy-Stability-Area Theory of Biodiversity, or ESA theory for short. In a nutshell, the more solar energy, the greater the diversity; the more stable the climate, both from season to season and from year to year, the greater the diversity; finally, the larger the area, the greater the diversity.

The evidence for this theory has come from several directions and tells us a great deal not only about biological diversity but also about the importance of physical environment in the organization of ecosystems. David Currie, for example, studied the effects of a wide range of environmental variables on the number of tree and vertebrate species in different parts of North America. This continent offers an excellent laboratory for such a multifactorial analysis. It lies entirely within the temperate zone and hence possesses the same well-marked seasons everywhere, yet varies greatly in precipitation and topography from east to west. Under these conditions—no tropics to worry about for the moment—the overwhelming factor is the amount of solar energy and humidity available to organisms throughout the year. The measurement that captures both variables is evapotranspiration, the quantity of water evaporating from a saturated surface. This amount depends in turn on the energy available to evaporate water, and that energy comes from the heat of the sunlight combined with the temperature of the ambient air and the movement of drying air currents.

It depends to a lesser degree on humidity. In North America at least, warm and humid environments support more species of trees. The diversity of land vertebrates, including mammals, birds, reptiles, and amphibians, rises with solar energy but is less dependent on humidity. Put as briefly as possible, dry spots are bad news for trees, much less so for vertebrates. For both kinds of organisms, however, more solar energy means more diversity.

The parts of the world with the highest year-round temperatures are the equatorial tropics, and the habitats with the greatest combined heat and humidity are the tropical rain forests. Given an equal amount of nutrients, the hottest, most humid places are also the most productive in terms of the quantity of plant and animal tissue grown each year. It would seem to follow that the higher the production of living matter, the greater the number of species that can coexist in the same community. Put another way, the larger the pie, the greater the number of possible slices big enough to sustain the lives of individual species.

But energy and biomass production alone cannot explain the tropical dominance of biological diversity. What is to prevent one superbly adapted species in each broad category— one flowering plant, one frog, one wood-boring beetle—from taking over the entire habitat? Something like that has in fact occurred in red mangrove swamps and cordgrass marshes, two of the most productive wetlands in the world. In each of these habitats one species of plant composes more than 90 percent of the vegetation. But simple ecosystems are the rare exceptions worldwide, and diverse ecosystems are the rule. A fuller explanation of latitudinal gradients requires shifting the analysis to look at the role of the seasons. In the temperate and polar regions, organisms experience wild swings in temperature through the year. They must adapt to a wide range of physical and biological environments as part of their life cycles. In winter they variously hibernate, die off after setting seed, shed leaves, move down the mountainside, descend trees to the ground, burrow deeper in the soil, change their diet to cold-hardy prey, change activity from nocturnal to diurnal peaks, or, in the case of migratory birds and monarch butterflies, flee the country altogether. In the spring animals enjoy a flush of fresh vegetation, which tapers off with the dryness of late summer and forces a shift to new foods and habitats.

Because animal and plant species of cold climates are therefore adapted to a greater breadth of local environments, they also occupy larger geographical ranges. In particular, they are distributed across a wider range of latitudes. If a butterfly can thrive in the cool, wet spring of New England, it can endure the winter of Florida. This trend has been called Rapoport's rule, after the Argentine ecologist Eduardo Rapoport, who first suggested it in 1975. It means that as you travel southward down North America or northward up temperate South America, the ranges of individual species shrink steadily the closer you come to the equator. Of equal importance, the altitudinal range of species along the sides of mountains also contract. Thus more species are packed into the same amount of space in the tropics than in the colder temperate zones.

Higher energy, greater biomass production, the narrowing of geographical ranges within a less varying environment—all these properties elevate the level of biodiversity in the tropics over long stretches of evolutionary time. But there is still more to the engine of tropical exuberance. Stable climates with muted seasons allow more kinds of organisms to specialize on narrower pieces of the environment, to outcompete the generalists around them, and so to persist for longer periods of time. Species are packed more tightly. No niche, it seems, goes unfilled. Specialization is likely to be pushed to bizarre, beautiful extremes. In the sunlit clearings of Central American rain forests there are giant helicopter damselflies, drifting back and forth in the still air, with bands on their beating transparent wings that seem to spin around their bodies. Their nymphs are not to be found in ponds and streams, the conventional haunts of damselflies elsewhere, but in the water-filled axils of epiphytes high in the canopy. The adults feed by plucking spiders from their webs. On the hindfeet of army-ant soldiers nearby are fastened parasitic mites found nowhere else in the world. While sucking the blood of the ants, they allow themselves to be used as artificial feet; the ants walk on the bodies of the parasites with no sign of discomfort on either side. The mite covers the claws of the ant by which the ant hangs while nesting and renders it useless, but no matter: the mite has curved hindlegs the size of the claws, and the ant uses them instead. On the vegetation of mountain rain forests of Papua New Guinea live weevils half the size of a human thumb, sluggish and long-lived, their backs covered with algae, lichens, and mosses. In this miniature traveling garden dwell distinctive species of tiny mites and nematodes. I could go on with this bestiary, moving from one country to the next—the literature of tropical biology never runs out of surprises. Where conventional niches have already been filled, it seems that enterprising species invent new ones.

Walk the floor of a tropical rain forest, searching for specimens of almost any group, whether orchids, frogs, or butterflies. You will find that the species change subtly every hundred or thousand meters. A certain kind is common at one spot, then fades and disappears, to be replaced by closely similar species not encountered farther back. Then a stroke of luck: there appears a single individual of a species never before encountered in the entire area. Collect or at least photograph it carefully because you will never see it again. In Central American rain forests, the nymphaline butterfly *Dynamine hoppi* is a pretty species distinguished by large white spots on the forewing and metallic blue fringes on the hindwing bands. It has been encountered three times in history. One female was collected by the lepidopterist Philip DeVries in a clearing at Finca La Selva, Costa Rica, in July. It was the only individual he found during a six-month study of the butterflies of this particular forest. A second female was taken the following year in the same spot, also in July. Then no more. If you return to the forest day after day and from one year to the next, quartering it on foot with net and binoculars, your list of orchids or frogs or butterflies will grow—and grow.

At first the picture of diversity in the forest seems hopelessly confusing, but in time a general pattern emerges: there are a few common species, most distributed in scattered patches, and a great many rare species, including some like *Dynamine hoppi* that are extremely rare. How has this skewed statistical curve come into existence?

Some of the rare species are on the edge of extinction, especially where the forest has been disturbed or cut back, but there is another, more likely explanation. Most species are specialized for a particular set of conditions within the forest. Trees of a certain kind grow best when they are touched by direct sunlight for more hours of the day (or fewer hours), when the slope in which they are rooted is well drained (or poorly drained), and when symbiotic root fungi of the needed species are available (or not). If any of these three conditions changes, the trees are likely to yield to other species. Particular ensembles of insect species flourish when logs at a certain stage of decomposition are available (for example, wood firm but soft enough to break apart by hand, with the bark still in place), and vanish when the logs rot still further (wood crumbling, the bark sloughing off from its own weight). The condition of particular rotting shifts from spot to spot.

The rain forest may look uniform through the windows of an airplane passing over, but on foot it is seen to be endlessly heterogeneous, a daunting labyrinth of transient local physical environments and overlapping species distributions. Individual species are most abundant in places that are just right for them. Their populations breed luxuriantly there, grow in size, and send out colonists in all directions. Such locations are called *source areas* for the favored species. The colonists often land in less-suitable sites, where they may survive and even reproduce for a while, but not well enough to be self-sustaining. These places are called *sink areas*. In the source-sink model of ecology, the successful populations subsidize the failing populations. If you rope off a sample area at random—one hectare, a hundred hectares, whatever—you will have within it sources for some species, which are relatively common, and sinks for others, which are relatively rare. Sources and sinks occur in all kinds of habitats, but they contribute far more to biodiversity in the tropical forests, where the environmental requirements of species have been rigidly set in the course of evolution.

A balance between source and sink was one of the key features discovered by Stephen Hubbell and Robin Foster during their superb study of tree diversity in a 50-hectare plot on Barro Colorado Island, Panama. The investigators and their diligent assistants tracked more than 238,000 trees and shrubs belonging to 303 species for a period of years. From their data Hubbell and Foster concluded:

Many (at least one third) of the rare species (fewer than 50 total individuals) do not appear to have self-maintaining populations in the plot. Their presence appears to be the result of immigration from population centers outside the plot, and their numbers are probably kept low by a combination of unfavorable regeneration conditions, lack of appropriate habitat, or both, in the plot.

There is yet another way that abundant solar energy and the evenness of climate have contributed to the rise of biodiversity: the piggybacking of species. Benign, less variable environments permit the existence of larger life forms that cannot survive in harsher climates. Other, smaller species live on these large organisms, often in great variety. In the tropical forests, but not in the temperate deciduous and coniferous forests, woody lianas abound. They sprout on the forest floor as herbs, then send long shoots up the trunks of nearby trees and any other available vegetation. At maturity all traces of their conventional origins are gone. They are like heavy anchored ropes, reaching all the way from their roots in the ground to their branches and leaves intertwined high above with those of the supporting trees. They create a supplemental form of vegetation, a source of food and hiding places for animals that could not otherwise survive. Alongside them grows another class of vines, the stem climbers, which fasten themselves onto tree trunks by sucker-like roots. One prominent group is the arums, including the philodendrons and monsteras. They have large heart-shaped leaves and a tolerance for deep shade, qualities that make them favorite house plants. In the rain forests stem climbers are plastered in dense clusters over the surfaces of the tree trunks. Their stems and roots accumulate layers of soil and decaying organic matter, home to yet another unique assemblage of tiny plants, insects, scorpions, sowbugs, and less common invertebrates. These serial forms are adapted to a way of life virtually absent in the temperate zones.

The greatest multipliers of tropical diversity, however, are the epiphytes, plants growing on trees that do not extract water or nutrients from them. Orchids make up a majority of the species of epiphytes, but they are accompanied by vast arrays of ferns, cacti, gesneriads, arums, members of the pepper family, and others, together comprising 28,000 known species in 84 families, or a little less than 10 percent of all the higher plants. These arboricolous plants turn tree limbs into Babylonian hanging gardens. Each one is a small habitat, complete with soil captured from the dust of the air and home to animals that can range from mites and roundworms to snakes and small mammals. The tank bromeliads of the American tropics hold as much as a liter of water in their stiff, upturned leaves. In these catchments live aquatic animals found nowhere else, including tadpoles of tree-dwelling frogs and the specialized larvae of mosquitoes and damselflies.

At the Monteverde Cloud Forest Reserve of Costa Rica, Nalini Nadkarni and other botanists have encountered what may be the ultimate piggyback phenomenon and the physically most complex arboreal ecosystem in the world. The epiphyte gardens on some of the heavier horizontal limbs are so profuse and tangled as to resemble a thicket of miniature woodland. Even small trees, of a kind usually found only on the ground, sprout from the dense clusters. The ecological house of cards has become a tower, an emblem of the prodigiousness of life on earth: big trees supporting orchids and other epiphytes, epiphytes supporting smaller trees with their root masses, lichens and other tiny plants growing on the leaves of the smaller trees, mites and small insects browsing among these leaf-top plants, and protozoans and bacteria living in the tissues of the insects.

Area counts in the buildup of diversity: the larger the forest or desert or ocean or any other definable habitat, the greater the number of species. As a rule of thumb, a tenfold increase in area results in a doubling of the number of species. If a forest-covered island 1,000 square kilometers in area has 50 species of butterflies, a nearby forest-covered island of 10,000 square kilometers can be expected to have about double that number, or 100 butterfly species. The reasons for this logarithmic increase are complicated, but two factors stand out. The larger island can support a greater overall population, say of butterflies, and so more rare species can be squeezed into the same forest. And the larger island is more likely to have additional habitats in which species can find refuge. It may have a central mountain, offering the beginnings of a closed zone—higher rainfall and lower temperatures for butterflies specialized for life in that climatic regime. The tropics contain vast areas of both land and shallow water that serve as theaters for the evolution of extreme diversity.

And time counts, by which I mean evolutionary time, enough time for the large piggyback organisms to evolve, symbiotic bargains to be struck, competition to be moderated, extinction rates to drop, and species thereby to assemble in respectable numbers. We have come back to climatic stability as a factor, but on a larger scale. The tropical rain forests, unlike large portions of the temperate forests and grasslands, were not obliterated by the continental glaciers of the Ice Age. They were never overridden by ice sheets or forced into new lands hundreds of kilometers from their original ranges. In the prolonged droughts that accompanied glacial cycles at higher latitudes, the lowland rain forests did retreat and were replaced by grasslands and in a few places by semideserts. The change was especially drastic in equatorial Africa. There were nevertheless abundant refuges into which species assemblages could persist more or less intact, along river courses, in regional pockets of persisting moderate rainfall, and in remnant tracts partway up the sides of cloud-enveloped mountains. Each time year-round rains came back to the equatorial river basins, the tropical rain forests expanded to carpet the earth again. The historical circumstance of interest is that the forests have persisted over broad parts of the continents since their origins as strongholds of the flowering plants 150 million years ago. Just prior to the coming of man they occupied more than 10 percent of the land surface, about 20 million square kilometers, and in earlier times much more than that. During the Eocene epoch, 60–50 million years ago, the marginal land mass that is now the British Isles was covered by forests similar in broad features to those of present-day Vietnam.

Let us put this notion of the importance of climatic stability to a test. If stability over a large area through evolutionary time is a prerequisite for high biological diversity, we should expect to find the most diversity anywhere in the world where stability prevails, not only in tropical forests. The ideal testing ground would be a very stable environment with small amounts of energy. Energy can then be discounted and the role

of stability more confidently identified. The floor of the deep
sea has the right geographical and historical qualifications. It
covers an area of over 200 million square kilometers, has lain
relatively undisturbed in most places for millions of years (no
winters, no dry seasons), and is devoid of all energy except for
widely scattered volcanic vents and a thin rain of organic debris
from the lighted zones above. The animals are for the most
part small annelid worms, starfish and other echinoderms,
and clams and other bivalve mollusks. Compared with similar
forms inhabiting shallow, lighted sea bottoms, they are sparse
in numbers, sluggish, and long-lived. But in accord with the
stable-environment hypothesis, they are extremely diverse.
Their species number into the hundreds of thousands, perhaps
millions. The stability part of the general biodiversity theory
is thus upheld, to surprising degree.

   Where are the niches on the deep-sea floor into which species can be packed? It has no forests or rivers. Large stretches of
the terrain appear flat and outwardly barren, like a desert. Yet
the floor is far from uniform in a biological sense. If examined
millimeter by millimeter, the scale at which small animals and
microorganisms live, it is seen to offer finely divided niches on
which deep-sea life can specialize. Sediments pile up in small
mounds, while the burrows of digging worms and bivalves
create ridges and depressions. The concentration of food varies
immensely from spot to spot. Almost all the energy comes
drifting down in the form of dead animals and plant matter.
Each piece—fish head, waterlogged fragment of driftwood,
strand of seaweed—is a bonanza on which animals gather to
feed and bacteria and other microscopic organisms proliferate.
Their predators also assemble, and in time a tiny local community is created, often different in composition from other
communities existing a few meters away. The bonanzas vary
not just on a local scale but also regionally across thousands of
square kilometers of ocean floor. Sections near the mouths of
great rivers receive logs and tree branches carried downstream
and out to sea, as well as more nutrient-rich sediments washed
from the land by rain. In the horse latitudes of the North
Atlantic, the benthic deeps are the cemetery of the sargassum
beds, receiving dead vegetation and animals directly from that
unique clearwater ecosystem high above.

In every habitat on land or in the sea, whether richly diverse or impoverished, the size of an organism exercises an important influence on the number of species in its group. Very small plants and animals are more diverse by far than very large organisms. Herbs and epiphytes exceed trees, and insects exceed vertebrates. The rule also holds within finer taxonomic divisions: among the 4,000 species of mammals found throughout the world, a thousandfold decrease in weight means (very roughly) a tenfold increase in the number of species. This translates to about ten times as many species the size of mice as species the size of deer.

The reason for the pyramid of diversity stacked by size is that small organisms can divide the environment into smaller niches than large organisms can. In 1959 the ecologists G. Evelyn Hutchinson and Robert MacArthur suggested that the number of species increases directly with a decrease in the area of the body of animals, or the square of the decrease in their weight. The reason for this rule, they proposed, is that animals living on surfaces require spaces that are the square of their body lengths. In other words, the animals move around not along a straight line, and not up and down in the three-dimensional space of midair, but over a surface, so that with each millimeter increase in their length they need a square millimeter more to find new roles, open new niches, and split into new species. Hence the more millimeters in an animal's length, the fewer species by the square of that length.

Though this mathematical exercise is beguiling, it is not very accurate. Nature is always too devious to obey simple formulations in any but a slovenly manner. To see why this is so, and to come closer to the truth, picture in your mind a large beetle 50 millimeters long, living on the side of a tree. As it walks around the trunk, browsing on lichens and fungi, it measures a trunk circumference of 5 meters. But it cannot take account of the much smaller world at its feet. The beetle is scarcely aware of the many dips and hollows in the bark only a millimeter across. In that irregularity live other species of beetles small enough to make it home. They exist in an entirely different scale of space. To them, irregularities are not trivial. As they crawl down the sides of the crevices and up again, the circumference of the tree trunk is about ten times what it is for the giant beetle, which knows nothing of tiny crevices. The surface of the trunk is a hundred times greater for the small beetles, the square of the difference in the circumference perceived by them and that perceived by the big beetle. The disparity translates into more niches. Different crevices contain their own regimes of humidity and temperature and a variety of combinations of algae and fungi on which insects can feed. Hence small beetles have many more dwelling places and foods on which they can specialize, and as a result a correspondingly larger number of species can evolve.

Let us descend deeper into the microscopic. At the feet of the small beetles are still smaller crevices and patches of algae and fungi too narrow for them to enter. Living there, however, are the smallest of all insects together with armored oribatid mites, measuring under a millimeter in length. A close scan of the surface geometry reveals that the species of this diminutive fauna live as if the surface of the tree trunk were a hundred times or more greater than the surface embraced by beetles the next size up, and thousands of times greater than the titan beetle looming over the whole ensemble. Finally, the tiny insects and mites stand on grains of sand lodged in algal films and the rhizoids of mosses, and on a single grain of sand may grow colonies of ten or more species of bacteria.

I have dwelled on this tree-trunk microcosm to stress that in the real world, where species multiply until halted, space is not measured in ordinary Euclidean dimensions but in fractal dimensions. Size depends on the span of the measuring stick or, more precisely, on the size and foraging ambit of the organisms dwelling on the tree.

In the evolution of biodiversity, smaller size means more species. Within particular groups of animals, such as the insects, the smallest organisms are able to exploit more niches and thus pack more species into local communities. In the mountain rain forests of Papua New Guinea, the large weevil (*Gymnopholus lichenifer*) carries a garden of lichens on its back, a microhabitat that supports several species of mites and springtails. At its feet, in a world of their own, are miniature anobiid beetles of an unknown species.

In the fractal world, an entire ecosystem can exist in the plumage of a bird. Among the prominent organisms living in that peculiar environment are feather mites, spidery organisms apparently subsisting on oily secretions and cellular detritus. Individuals are so small and territorial that they can spend most of their lives on one part of one feather. Each species is specialized on a feather type and feather position, such as the outer quill of a primary wing feather, or the vane of a body contour feather, or the interior of a downy feather, and so on through what to feather mites is the equivalent of a forest of trees and shrubs. A single parrot species, the green conure of Mexico, is host to as many as thirty species, each with four life stages, making a total of over a hundred life forms. Each of these forms in turn has its own preferred site and pattern of behavior. A single conure harbors fifteen or more species of feather mites, with seven occupying different sites on the same individual feather. Tila Pérez of the National University of Mexico recently collected six species from the plumage of museum specimens of the extinct Carolina parakeet. If this near-microscopic fauna was indeed unique to the parakeet, which seems likely, then the mite species also vanished when the last bird died in the Santee Swamp of South Carolina in the late 1930s.

Statistical studies have shown that the most diverse animals are not only small in size but also highly mobile, giving them access to the most bountiful variety of foods and other resources. The ultimate exemplars of this principle are the insects, so diverse and abundant that they project a popular image of near invincibility. *(In the nuclear aftermath a cockroach surveys the scorched landscape atop a blasted beer can.)* Entomologists are often asked whether insects will take over if the human race extinguishes itself. This is an example of a wrong question inviting an irrelevant answer: insects have already taken over. They originated on the land nearly 400 million years ago. By Carboniferous times, 100 million years later, they had radiated into forms nearly as diverse as those existing today. They have dominated terrestrial and freshwater habitats around the world ever since. They easily survived the great extinction spasm at the end of the Paleozoic era, when life survived more than the equivalent of a total nuclear war.

Today about a billion billion insects are alive at any given time around the world. At nearest order of magnitude, this amounts to a trillion kilograms of living matter, somewhat more than the weight of humanity. Their species, most of which lack a scientific name, number into the millions. The human race is a newcomer dwelling among the six-legged masses, less than two million years old, with a tenuous grip on the planet. Insects can thrive without us, but we and most other land organisms would perish without them.

Richard Southwood has explained the preeminence and hyperdiversity of insects with three words: size, metamorphosis, and wings. Size for the small niches to be defined and the many species therein generated. Metamorphosis for the transition from one life stage to another—from larva or nymph to adult—that allows the penetration of more than one habitat and the fabrication of still more niches. And wings, for dispersal to far corners of the land environment, across lakes and desert corridors, to outermost leaf tips and distant sanctuaries, putting insects within easy reach of additional food sources and places to mate and to escape from enemies. To this may be added preemption: because insects were the first to expand into all of the terrestrial niches, including the air, they were no doubt too well entrenched to be evicted by newcomers.

The human species came into the world as a late product of the radiations that, 550 million years into the Phanerozoic, lifted global biodiversity to its all-time high. In a more than biblical sense, humanity was born in the Garden of Eden and Africa was the cradle. During most of its recent geological history, from the Mesozoic era to approximately 15 million years ago, that continent was cut off from Europe to the north and Asia to the east by the Tethys Sea, a body of shallow tropical water connecting the Atlantic and Indian Oceans. As the Tethys dwindled to its remnant as the Mediterranean Sea, Africa was joined to Europe and Asia and became part of the World Continent, the loosely united biogeographical realm through which major groups of plants and animals were able to spread. Before that time Africa was an island continent similar in bulk and isolation to Australia and South America. Like those severed land masses, it developed a distinctive mammalian fauna: elephants, hyracoids, giraffes, barytheres, elephant shrews, and, not least, man-apes and the earliest true humans. Some of the groups were indigenous to Africa. Others, including the big cats and the primates, flourished all across Europe and Asia and periodically invaded Africa, where occasional lines then branched into multiple species during secondary bursts of evolution. The man-apes and early men were one of the final products of the post-Tethyean secondary radiation of the primates. They walked upright onto the stage, bearing Promethean fire—self-awareness and knowledge taken from the gods—and everything changed.

THE HUMAN IMPACT

CHAPTER ELEVEN

The Life and Death of Species

EVERY SPECIES lives a life unique to itself, and every species dies a different way. The New Zealand mistletoe *Trilepidea adamsii* was a pretty plant with pale green glabrous leaves, red tubular flowers tinged with yellowish green, and bright-red ellipsoidal fruits. It disappeared from its last stronghold on North Island in 1954. The species grew as a parasite on shrubs and small trees in the understory of native forest. Never common, it was limited at the time of the first European botanical explorations to a few localities in the northern peninsula, around Auckland.

Trilepidea adamsii came to an end by a combination of circumstances that no one could have foreseen a hundred years ago. Its habitat was reduced by deforestation, first by the original Maoris during a thousand years of occupation and then at an accelerating rate by British settlers in the late nineteenth century. Placed at risk, the population was reduced still further by collectors eager to secure specimens of what was recognized as a rare and desirable plant. The dispersal of the mistletoe dropped still more as bird populations declined in the area, depressed by clearing of their forest habitats and predation from introduced mammals. Birds are necessary for the transport of the seed from one host tree or shrub to another. By the early 1950s *Trilepidea adamsii* was close to extinction. The nature of its final days are unknown. The last few plants may have been eaten by brush-tailed possums, a species of arboreal browsing mammal deliberately introduced from Australia during the 1860s to establish a fur trade. The possums were never abundant enough to destroy the mistletoe while it flourished, but they could have tipped it into extinction at the very end.

Consider this familiar paradox of biological diversity: almost all the species that ever lived are extinct, and yet more are alive today than at any time in the past. The solution of the paradox is simple. The life and death of species have been spread across more than three billion years. If most species last an average of, say, a million years, then it follows that most have expired across that vast stretch of geological time, in the same sense that all the people who ever lived during the past 10,000 years are dead though the human population is larger than it has ever been. The turnover would have been even greater if the grand pattern were dynastic, with one species giving rise to many species, most or all of which yielded to later ascendant groups.

Evolution is indeed dynastic, and million-year longevities are close to the mark for many kinds of organisms. The precise measure of interest is not the longevity of species but of the clade, composed of the species and all its descendants, taken from the time the ancestral species first splits off from other species to the moment the last organism belonging to that species and all of its descendants disappears. Chronospecies extinction, or pseudo-extinction as it is also called, does not count. If a population of organisms evolves so much that biologists declare it to be a new species, or a chronospecies, the species did not go extinct; it just changed a great deal. The life of the clade goes on, and that particular lineage of genes endures.

Each major group of organisms appears to have a characteristic clade longevity. Because of the relative richness of fossils in shallow marine deposits, the duration of fish and invertebrate clades living there can often be determined with a modest degree of confidence. During Paleozoic and Mesozoic times, the average persistence of most lay between 1 and 10 million years—for example, 6 million years for starfish and other echinoderms, 1.9 million years for graptolites (colonial animals distantly related to the vertebrates), and 1.2 to 2 million years for ammonites (shelled mollusks resembling modern nautiluses). On the land, the longevities of clades of flowering plants during Cenozoic times also appear to fall within the range of 1 to 10 million years. Those of mammals vary from .5 million to 5 million years, depending on the geological epoch.

The probability of extinction of species within clades is more or less constant through time. As a result the frequency of species in a clade surviving to a greater and greater age falls off as an exponential decay function. If, to use an oversimplified example, one half the species are alive at the end of a million years, about one half of those (or one quarter of the original) persist 2 million years, one half of those again (or one eighth of

The extinct New Zealand mistletoe (*Trilepidea adamsii*).

the original) last 3 million years, and so on. The progression is often accelerated by shifts in climate that cause waves of extinction and subsequent rebirth—not only the great catastrophes that ended the Paleozoic and Mesozoic eras but smaller, more frequent, and more local events. Clades of buffalos and antelopes in Africa south of the Sahara have persisted from 100,000 to several million years. But about 2.5 million years ago many came to an end, and others first appeared almost simultaneously. The controlling event was apparently a period of cooling and diminished rainfall that caused grasslands to spread over a large part of the African continent.

Local climatic instability is only one of the reasons not to generalize too quickly about the lifespan of species from the fossil record. Sibling species, so similar in anatomical detail as not to be traceable in fossils, could come and go in rapid succession without being detected. Small local species might also turn over at a high rate in places where fossilization seldom occurs, such as desert valleys and the interiors of small islands, leaving no evidence whatever of their existence.

We know that contemporary species formation in the northern Andean cloud forests is both profuse and resistant to the formation of fossils. In the mountain habitats of Colombia, Ecuador, and Peru, populations of plants and animals are prone to fast evolution and early extinction by geographical location alone. The ridges on which they live are isolated and differ from one another in temperature, rainfall, and the species composing local communities. The populations are small. Alwyn Gentry and Calaway Dodson estimate that in these places some orchid species could multiply in only fifteen years. By implication, the longevity of species might be short as well, measured in decades or centuries. Orchids are by far the most diverse of living plants, comprising at least 17,000 species or 8 percent of all flowering plants. Many are rare and local like the Andean endemics, and they could originate and die at a high rate without leaving a trace. The general biology of orchids is also such as to erase their history. They live mostly in the tropics, where the fossil record is poor. Most grow as epiphytes in the crowns of forest trees, a habitat not conducive to the fossilization of plant parts. And unlike the vast majority of other flowering plants, they do not scatter their pollen as simple grains, letting much of it fall into lakes and streams where it can form easily studied microfossils. Instead, orchid plants bind pollen together in solid bodies, the pollinia, which are carried from flower to flower by insects. These two traits taken together, rapid speciation and the difficulty of fossilization, mean that orchid floras leave almost no record by which we can hope to measure the longevity of species.

The orchids are not alone. They merely instruct us that, in addition to the species whose fossils suggest a longevity of 1 to 10 million years, there is a large hidden group of species that appear and disappear at a far higher rate. New species occupy small ranges on the average and are often started by small numbers of pioneers that land on island shores or distant mountain ridges. If extinction of such young, vulnerable populations were high, the equivalent of infant mortality among organisms struggling in a harsh environment, a large percentage of species would die young with no record of their existence. The birth and death of most species may therefore lie behind a veil of artifact. Only the more widespread populations in or near bodies of water are fossilized with consistency for direct measurement. Behind the veil lie vast numbers of species that once lived in restricted habitats and are forever beyond direct access.

To pull the veil back, to visualize how rare species live and die, we must take a less direct approach by returning to the principles of ecology and to natural history, which is ecology expressed in the details of the biology of individual species that still live or have recently perished. Consider first the laws of ecology. They are written in the equations of demography. The number of plants or animals belonging to a particular species is exactly determined by the rate at which new individuals are born, the age at which they reproduce, and the age at which they die. The distribution of the population by age (how many newborn, how many juvenile, how many young and old adults) is set by these schedules of birth and death. The schedules themselves are influenced by the size of the population or, more precisely, by its density. The number of birds crowded into a woodlot or the number of algal cells living on a wet stone affects food supply, how heavily predators and disease pathogens strike, to what degree reproduction is delayed, how long individuals live, which competitors can force themselves

into the same community. All this has an important consequence: if ecology is ultimately a matter of demography, then
demography eventually must turn into natural history, with
parameters expressed as a function of particular time and place.
The equations of demography are specified by context.

So it is when we move on up to the life and death of particular
species. The laws of biological diversity are written in the equations of speciation and extinction. Ecologists and paleontologists have begun to pursue these laws, aware of the importance
of data on the birth rates of species and the longevity of the
clades they spawn. The equations are beginning to resemble
those of ecology, and they too are being fleshed out by the
details of natural history.

Consider a newly formed island in the sea, devoid of life, say
Krakatau in 1883, Surtsey off Iceland in 1963, or Kauai 5 million
years ago. Plants and animals soon arrive, showering down as
aerial plankton or blown ashore by storms. Next focus on a
particular group, say landbirds, reptiles, or grasses. At first the
rate of arrival of new species in the group is relatively high, but
it inevitably drops because the strong dispersers become established early. On islands and continents nearby, there are other
species that can cross the sea, but they compose a less able pool
of potential colonists. As the island fills up with more and more
species of the group of interest—birds, reptiles, grasses—the
rate of arrival of species not already established keeps dropping.
It might start with an average of one new species a year and
decline during the next century to one every ten years. At the
same time, the rate of extinction rises as more and more species
contend for the available space and resources.

In time the rate of extinction of species already on the island,
say in species per year, will just about equal the rate of immigration of new species onto the island, again in species per year.
The number of species is in a dynamic equilibrium. New species are arriving, old species are disappearing, the composition
of the fauna and flora is constantly changing, but the number
of species present at any moment in time stays the same.

This very simple model of a balance between immigration
and extinction is the basis of the theory of island biogeography that Robert MacArthur and I developed in 1963. We had

noticed that faunas and floras of islands around the world show a consistent relation between the area of the islands and the number of species living on them. The larger the area, the more the species. Cuba has many more kinds of birds, reptiles, plants, and other organisms than does Jamaica, which has a larger fauna and flora in turn than Antigua. The relation showed up almost everywhere, from the British Isles to the West Indies, Galápagos, Hawaii, and archipelagoes of Indonesia and the western Pacific, and it followed a consistent arithmetical rule: the number of species (birds, reptiles, grasses) approximately doubles with every tenfold increase in area. Take an actual case, the landbirds of the world. There is an average of about 50 species on islands of 1,000 square kilometers, and about twice that many, 100 species, on islands of 10,000 square kilometers. In more exact language, the number of species increases by the area-species equation $S = CA^z$, where A is the area and S is the number of species. C is a constant and z is a second, biologically interesting constant that depends on the group of organisms (birds, reptiles, grasses). The value of z also depends on whether the archipelago is close to source areas, as in the case of the Indonesian islands, or very remote, as with Hawaii and other archipelagoes of the eastern Pacific. In short, z is a parameter. It holds constant for a given group of organisms and set of islands, such as the birds of the West Indies, but can change when we proceed to other organisms on other islands, such as the grasses of Indonesia. It ranges among faunas and floras around the world from about 0.15 to 0.35. To give the rule of thumb that a tenfold increase in area doubles the fauna and flora is the same as saying that $z = 0.30$, or $\log_{10} 2$. Notice that we can if we wish, and this is very important for conservation, state the rule in reverse: a tenfold *decrease* in area cuts the number of species in half.

The rise of biodiversity with the size of islands is called the *area effect*, and it follows from the equilibrium model in a straightforward way. Think of a row of newly emerged islands along the edge of a continent, all located an equal distance from the shore of that larger body of land but varying in size. As they fill up with species, the fringing islands will all have about the same immigration rate—the same number of new species arriving each year—since they are all equally far from the continent. On the other hand, the extinction rates will rise more slowly on the larger islands. The reason is that more area means more space, more space means larger populations for each species, and finally larger populations mean an expectation of longer life for the species. You are less likely to go completely broke if you are rich at the start, and more people can be crowded onto large tracts of land before they become poor. So the overall extinction rate comes to equal the immigration rate on a large island only after many species have colonized the island, and larger islands have more species at equilibrium than smaller islands do.

The *distance effect* is this: the farther the island from continents and other islands, the fewer the species living on it. Like the area effect, this trend in biogeography can be explained by the basic equilibrium model in a straightforward way. Merely reverse the arrangement of islands, so that now all are the same size but located at varying distances from the continent. As they fill up with species of birds, reptiles, or grasses, the extinction rate on all the islands increases at about the same rate (because they are the same size). But the distant islands fill up more slowly; organisms have farther to travel, and their immigration rate (new species arriving each year) is lower. The extinction rate comes to equal the immigration rate when fewer species are present. Distant islands thus end up at equilibrium with fewer species than nearby islands.

Still theory, even if airtight and plausible, is not enough to put the seal on such complex ecological processes as the growth of species numbers. There must be experiments to confirm the predictions of theory and, in the case of ecology particularly, to expose them to the exhilarating intrusions of natural history. But how can experiments be performed on archipelagoes and entire faunas and floras?

The answer is to miniaturize. In the early 1960s I spent a great deal of time poring over maps of the United States, daydreaming, searching for little islands that might be visited often and somehow manipulated to test the models of island biogeography. I thought a lot about insects, creatures small enough to maintain large populations in tight places. A full-blown bird or mammal fauna might require an island the size of Guernsey or Martha's Vineyard, but aphids and bark beetles

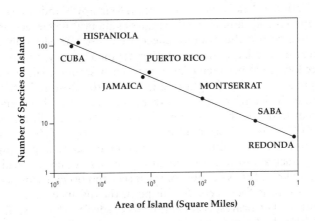

The number of species living on islands increases or decreases with the area of the island. The diversity of reptiles and amphibians in the West Indies, depicted here, is typical: a reduction of 90 percent in area from one island to the next results in a 50 percent loss of species.

can flourish in large numbers on a single tree. I finally settled on the Florida Keys, especially the tiny red mangrove islets that dot the shallow waters of Florida Bay immediately westward. Clumps of the salt-tolerant mangroves vary from a single tree to forests hundreds of hectares in extent. They are present in huge numbers, from the Ten Thousand Islands group of the upper bay to abundant miniature archipelagoes strung along the northern rim of the Lower Keys.

In 1966 Daniel Simberloff, then a graduate student at Harvard University and now a distinguished professor of ecology at Florida State University, joined me in an attempt to turn the mangrove islets into an outdoor laboratory. We needed a series of little Krakataus, islands that could be completely cleared of insects, spiders, and other arthropods and then monitored month by month. To accomplish that much would be to witness recolonization from scratch and to learn unambiguously whether biodiversity is at equilibrium. With the permission of the National Park Service, we selected four tiny islands for our experiment, all clumps of red mangroves about 15 meters across. In order to test the distance effect, we chose one islet only 2 meters from a large island, another 533 meters away, and two more located in between. The 2 meters may seem trivial, being no greater than the height of a guard on the Boston Celtics, but it is the length of 1,000 worker ants which, translated into human terms, is about a mile. We set to work. Crawling over each of the islets from mud bottom to the tops of the trees, examining every millimeter of leaf and bark surface, probing every crack and seam, photographing and collecting, we made as complete a list as possible of the insect and other arthropod species on the four islets. At this point the distance effect was manifest. The nearest islet held the most species, the most distant islet the fewest, and the two intermediate islets numbers in between.

Next we hired a pest-control company in Miami to destroy all the arthropods on the islands, using a method ordinarily employed to fumigate entire buildings. The workmen first covered the islands with a rubberized nylon tent. Then they fumigated the interior with methyl bromide gas at a concentration and at a preselected period of time that was fatal to arthropods but not to mangrove trees. When the tents were removed, we had four empty islands, four little Krakataus.

The recolonization began within days. In less than a year, the faunas had reattained their original levels on the islands. They lined up once again in accordance with the distance effect: from 43 species back to 44 species on the near islet, from 25 back to 22 species on the distant islet, and an equally close return on the middle islets to the numbers prevailing earlier. The numbers held remarkably constant to the end of the second year, when the experiment was discontinued. The equilibrium was also dynamic, with many of the arthropod species colonizing a given islet, vanishing after a month or two, then making a second appearance or else giving way to one or two similar species. The fauna tracked through time was kaleidoscopic, with the total numbers more or less balanced but the composition changing constantly, like travelers in an air terminal. Depending on the island, only between 7 and 28 percent of the species of the new colonists were the same halfway through the experiment as those present before the fumigation.

The Florida Keys experiment yielded new information on the abilities of different groups of small organisms to immigrate and endure. Spiders rained down on the islets, some of them large, no doubt ballooning over the water on silken threads. But many of the species went quickly extinct. Their distant relatives, the mites, were slower to arrive, blown in air currents like random particles of dust—literally part of the dust—but their individual species persisted longer. Cockroaches, crickets, moths, and ants arrived early and colonized solidly. Centipedes and millipedes, although well established before the fumigation, never made it back during the two years we visited the islets.

The mangrove experiment was inspired by Krakatau and the scientific interest of a land swept clean of all animals. A second method of assessing the balance of diversity would be to reduce the size of islands and observe the decline in numbers of species from a higher equilibrium to a lower one. In the late 1970s Thomas Lovejoy adopted this approach in what was to become the largest biological experiment in history. He took advantage of a Brazilian law that required owners of rain-forest tracts in the Amazon region to leave at least 50 percent of their land covered in forest; the rest they were free to convert into pastureland and farms. With the support of the World Wildlife Fund and the Brazilian government, Lovejoy set out to observe the fate of diversity in patches of forest left behind as the clearing proceeded. He persuaded owners of land along the Boa Vista road north of Manaus to leave square patches of forests varying from 1 hectare to 1,000 hectares in size. A fellow ornithologist, Richard Bierregaard, joined him as field director, and other experts were invited in as guest investigators to move the huge project along. The biologists set out to survey diversity at the experimental sites while the plots were in pristine condition and, later, after they had been turned into islands by the forest clearing. (One of these tracts was Fazenda Dimona, where I sat to watch a storm.)

The enterprise was first called the Minimum Critical Size of Ecosystems Project, MCS for short, because its ultimate goal was to determine the smallest size a rain-forest reserve must be in order to sustain the plant and animal species native to the immediate vicinity. How much land is needed, say, to sustain 99 percent of all the original species for a hundred years? Later the study became part of the Biological Dynamics of Forest Fragments Project, meant eventually to cover habitats all over Brazil. Staff members refer to it as the Forest Fragments Project for short, and many Brazilians call it Projeto Lovejoy. The monitoring near Manaus was begun just before the land was clear-cut in the late 1970s, and it is expected to carry over into the next century.

A mountain of data must be sifted from the Manaus experiment, but even in the first decade, 1979–1989, useful new facts have emerged. The diversity of the smaller "islands" is decreasing the most rapidly, as expected. The extinction of species has been accelerated by the unexpectedly deep penetration of daytime winds, which dry out the forest from the edge in and kill deep-forest trees and shrubs for distances inward of 100 meters or more. Many plant and animal species have disappeared from the smaller plots, but a few others are increasing their numbers. The reasons for the changes are sometimes obvious, but just as often baffling. Army-ant colonies, which require more than 10 hectares to maintain their worker force, quickly disappeared from the 1- and 10-hectare plots. With them went five species of ant birds that make their living by following ant swarms and feeding on the insects driven forward by the 10-meter-wide raiding front. Shade-loving butterflies of the deep forest declined quickly from the wind-drying effect, but other species specialized to live around forest edges and second growth flourished. Large, metallic green and blue euglossine bees, which are premier pollinators of orchids and other plants, were hit hard in the plots of up to 100 hectares. Saki monkeys, which eat fruit, dropped out of the 10-hectare plots. But red howler monkeys, which are leaf eaters and hence able to harvest more food, stayed on. Larger ground-dwelling mammals, including margay cats, jaguars, pumas, pacas, and peccaries, simply walked away from the smaller plots and out of the fauna altogether.

By the late 1980s second-order effects could be seen spreading through the food web. With the peccaries gone, there were no wallows in which temporary forest pools could form. Without the pools, three species of *Phyllomedusa* frogs failed to breed and disappeared. As the mammal and bird populations declined, dung and carrion became scarcer. Scarab beetles that feed on these materials dropped in numbers of species and individuals. The average size of the surviving beetles grew smaller. Bert Klein, who documented these changes, prophesied still more reverberations through a broad segment of the animal community, including carnivorous mites that ride on the beetles and attack fly maggots, with still deeper effects on the disease organisms of mammals and birds:

> A first-order disruption in abundance of some scarabs will undoubtedly result in second-order changes in mite dispersal, which may trigger third-order changes in populations of dung- and carrion-breeding flies. What fourth-order changes could occur because of changes in fly abundance needs further study. By eating and burying dung and carrion, the Scarabaeinae kill nematode larvae and other gastrointestinal parasites of vertebrates. Thus, a change in dung beetle communities may alter the incidence of parasites and disease in some isolated forest fragments or biological reserves.

Through perturbations spreading to third rings of interaction among the species and possibly beyond, the diversity of the smaller forest plots spirals downward to new and still unpredictable levels. We know at least this much: an Amazon forest chopped into many small fragments will become no more than a skeleton of its former self.

Theory confirms common sense with the following theorem: the smaller the average population size of a given species through time, and the more the size fluctuates from generation to generation, the sooner will the population drift all the way down to zero and go extinct. Think of an island with a thousand sparrows on average, varying by chance alone by a hundred individuals either way once or twice a century. Another island holds a hundred sparrows of the same species, and this population also varies by a hundred individuals once or twice in a century. The second population, which is both smaller and experiences a higher degree of fluctuation, faces a shorter life. More precisely, many such populations go extinct sooner than many otherwise comparable larger populations.

The prediction has been proved correct in a meticulous study of one hundred species of landbirds on small islands off the English and Irish coasts conducted by Stuart Pimm, Lee Jones, and Jared Diamond. They found that the lifespan of local island populations of birds does indeed grow shorter as the population size decreases. It also drops when the population fluctuates more widely over time.

In order to see the importance of population size in a broader perspective, imagine that we could protect a local population from catastrophic destruction. The habitat is kept intact, a steady food source is ensured, and no devastating diseases or predators are allowed to sweep the area. The fluctuations in the number of individuals in the population are then based on pure chance in the events of birth and death—how many females are mated that year, how many young survive infancy, and so on. Chance itself is the summed outcome of many other, largely unpredictable events in rainfall, temperature, food supply, and enemy assault. Mathematical models of the history of such steady-state populations reveal that population size and fluctuations can have enormous effects on longevity. A tenfold rise in average population size, say from ten to a hundred individuals, might increase average longevity thousands of times. Expressed in more practical terms, there is threshold below which the population is in eminent danger of extinction from one year to the next. The bright side is that endangered species can often be rescued from this red zone by a relatively modest increase in habitat area and hence average population size.

Because extinction is forever, rare species are the focus of conservation biology. Specialists in this young scientific discipline conduct their studies with the same sense of immediacy as doctors in an emergency ward. They look for quick diagnoses and procedures that can prolong the life of species until more leisurely remedial work is possible. They understand that the populations of a species can be habitually small and frequently vanishing, but if others are born through colonization of new sites at the same rate, and if there are many such populations, then the species as a whole is in no particular danger. Rareness, therefore, requires a many-layered definition in order to be addressed realistically. The essential idea can be expressed by diagnosing three of the most endangered species of birds in North America.

Bachman's warbler. A species is endangered if it occurs over a wide area but is scarce throughout its range. Such is the case of Bachman's warbler (*Vermivora bachmanii*), which is the rarest bird in North America in numbers of individuals per square kilometer of its geographical range. Small, yellow-breasted, olive-green on the back, and black-throated in the male, the warbler once bred in thicket-grown river swamps from Arkansas to South Carolina. Its present breeding range and population size are unknown, and it appears to be close to extinction if not already lost.

Kirtland's warbler. A species is rare if it is densely concentrated but limited to a few small populations restricted to tiny ranges. Kirtland's warbler (*Dendroica kirtlandii*), with lemon-yellow breast, bluish-gray back streaked with black, and dark mask in the male, is such a case. It is loosely colonial, with a breeding range restricted to jack-pine country in the north-central part of the lower peninsula of Michigan. Between 1961 and 1971 the known population plunged from 1,000 to 400 birds. The decline was apparently due to increased nest parasitism by brown-headed cowbirds (*Molothrus ater*), which place their eggs in the warbler's nest. Kirtland's warblers are as dense as ever in the localities where they occur, but the progressive restriction of their range has brought them close to extinction.

Red-cockaded woodpecker. A species can be rare even if it has a broad range and is locally numerous, but is specialized to occupy a scarce niche. The red-cockaded woodpecker (*Picoides borealis*), with zebra back, white breast speckled with black, and each white cheek touched by a carmine speck, is the outstanding example. It ranges across most of the southeastern United States but requires pine forests at least eighty years old. The birds live in small societies composed of a breeding pair and up to several offspring, with the latter helping their parents to protect and rear the younger siblings. Each group requires an average of 86 hectares of woodland to produce an adequate harvest of insect prey. To nest, red-cockaded woodpeckers hollow out cavities in living, mature longleaf pines eighty to one hundred and twenty years old, in which the heartwood has already been destroyed by fungus. These exacting conditions are no longer easy to find in the piney woods of the south. The total size of the woodpecker breeding population was estimated in 1986 to be only 6,000. It was falling steadily, by as much as 10 percent a year in Texas and probably just as fast elsewhere. The species appears doomed unless the cutting of the oldest pine forests is stopped immediately.

Species trapped by specialization and pressed by shrinking habitat form the largest endangered class. The scarcity of Bachman's warbler across the southern United States is no mystery, despite the abundance of riverine swampland in which it can breed. It winters (or wintered) exclusively in the forests of western Cuba and the nearby Isle of Pines, where virtually all the forests have been cleared to grow sugar cane. The bottleneck is the loss of wintering ground and starvation for even the remnant of warblers produced in the lusher summer environment of the United States.

John Terborgh has given a poignant account of his own experience with one of the last Bachman's warblers. In May

1954, as an eighteen-year-old birder (now a foremost ornithologist), he learned of the sighting of a male Bachman's on
Pohick Creek in Virginia not far from his home. The song of
the Bachman's had been described to him as resembling that
of a black-throated green warbler with a downward sweep at
the end: *zee-zee-zee-zee-tsew*.

> To my astonishment I walked up to the place that had been
> described to me and heard it! I had no trouble seeing the bird.
> A full-plumaged male, it sat on an open branch about 20 feet
> up and gave me a perfect view while it sang. It hardly stopped
> singing during the two hours I spent there. Reluctantly, I pulled
> myself away, wondering whether this was an experience I would
> ever repeat. It was not.

As other birders were to testify, the male returned to the same
spot the next two springs. No female ever joined him. The
extraordinary exertions of the Bachman's male were a sign that
he was in prime breeding condition, but he was destined to go
undiscovered by any female of the same species.

> I imagine that each spring a tiny remnant of birds crossed the
> Gulf of Mexico and fanned out into a huge area in the Southeast,
> where they became, so to speak, needles in a haystack. Toward
> the end, it is likely that most of the males in the population,
> like the one at Pohick Creek, were never discovered by females.
> Once this situation developed, there could have been no possible
> salvation for the species in the wild.

In parallel manner, Kirtland's warbler winters in the pine
woodland of two islands in the northern Bahamas, Grand
Bahama and Abaco. Terborgh has written that, however zealously the Kirtland's warbler and its habitat may be protected
in Michigan, its fate probably lies at the mercy of commercial
interests in the Bahamas. Migratory birds as a whole are declining across the United States from the same environmental
malady that afflicts the warblers: wintering grounds are being
demolished by logging and burning. The prospects are especially grim for species that depend on the rapidly shrinking
forests of Mexico, Central America and the West Indies.

Earlier I spoke of specialization, that tender trap of evolutionary opportunism, and how it is affected by natural selection

at the species level. A rich resource appears, and a species adapts to use and hold it against all competitors. To keep the edge, the members of the species surrender their ability to compete for other resources. Driven by natural selection, the advantage gained by those members one generation at a time, the species shrinks inside a smaller range. It is then more vulnerable to environmental change. Individual organisms bearing the specialist genes have triumphed, but in the end the species as a whole will lose the struggle and all its organisms will die. During the Paleozoic era, an entire family of snails, the platycerids, flourished by attaching themselves to the anuses of crinoids, a group of echinoderms called sea lilies. They fed on their hosts' fecal matter, which they were able to appropriate directly and with minimal competition. When the crinoids became extinct, so did all the multitude of ingenious platycerids.

On high, crumbling bluffs above western Florida's Apalachicola River grow the last wild trees of the Florida torreya or stinking cedar (*Torreya taxifolia*), a small understory conifer. Among them are found relics of a cooler climate, when the advance of the last ice sheet forced boreal elements southward to the southeastern United States. When the glacier retreated 10,000 years ago, most of the plant and animal species spread back, eventually to most of their former wide distributions. The torreya could not expand, in part because of its dependence on rich, moist soils of limestone origin. In the late 1950s a fungus disease struck the small Apalachicola population and brought the species close to extinction.

The streams of the Apalachicola system around the dying torreya stands are occupied by small populations of Barbour's map turtle, a handsome species with large sawteeth running the length of the carapace midline and with curlicues decorating the ventral rim of the shell. The female bears the most unusual feature. She is much larger than the male and has a grotesquely enlarged head. Evolved in only this one river system, the species has not spread beyond, and it is now vulnerable to extinction as the freshwater environments of Florida are increasingly disturbed.

Hidden in the muck-bottomed outflows of springheads of the same region live one-toed amphiumas, dwarf members of a genus of giant salamanders. I visited the habitat of this rare

The rarest songbird: Bachman's warbler of the southeastern United States is on the brink of extinction, if not already gone. This drawing of a singing male is based on one of the last photographs taken.

and possibly threatened species on the same day I examined the last stands of the torreyas. Walking with another natural- ist in the broiling sun through a turkey-oak flat, one of the most unpromising environments in the eastern United States, we found the springhead I sought, a small, narrow gorge 20 meters deep. It was like an oasis, its walls covered by thick broad-leaved woodland and its interior mercifully cool. A thin stream meandered across the flat muddy bottom. This is where the shy one-toed amphiumas live. They prey on a breed of equally unusual aquatic worms, also limited to this habitat. We found the worms but did not stay to locate the salamanders, because even in daytime the mosquitoes were so ferocious that the oak flat seemed bearable after all.

A small geographical range like those of the Apalachicola endemics carries an extra risk: a single sweep of disease (called *epizootics* to distinguish them from human epidemics), a forest fire, a deep freeze, or a day's work with chain saws can carry the species away. Specialization is perilous even for widespread species, whose local populations, however numerous and far-flung, are individually more likely to dip to extinction, until all happen to vanish.

The fossil record ascribes to this general principle. Recently I studied ants preserved in amber from the Dominican Republic, early Miocene in age, about 20 million years old. Abundant amber, the fossilized gum of trees, is one of the treasures of this Caribbean country. Columbus acquired pieces by trade there during his second voyage in 1493–94, from a mining region still active near present-day Santiago. Ants are among the most abundant insects in the clear golden matrix, many as exquisitely preserved as if they had been set in tinted glass by a master jeweler. From brokers I acquired a total of 1,254 pieces containing specimens. I cut and polished them until I could examine the ants at several angles under the microscope. I was able to study and illustrate them in fine detail, counting near-invisible hairs on their legs, measuring their head widths to the hundredth of a millimeter, writing out their dental formulas (you can identify species and even individual ants by the shapes and arrangements of their teeth). Comparing the amber species with those that live in tropical America today, I classified some as specialized and relatively rare, because their closest living relatives either take only certain types of prey, such as millipedes or arthropod eggs, or else nest in unusual places. I found that more of the specialized species and their descendants became extinct in the Dominican Republic and elsewhere in the West Indies than was the case for generalized species.

The same trend, of abundance favoring survival, was observed independently by Steven Stanley in mollusk species that lived along the North Pacific rim in Pleistocene times, about two million years ago. The patterns he found suggest that abundance, or total population size, is the most important control on survival.

> One pattern has to do with mode of life for clams, which burrow in the sea floor. Species possessing siphons have enjoyed a much greater rate of survival during the past two million years than

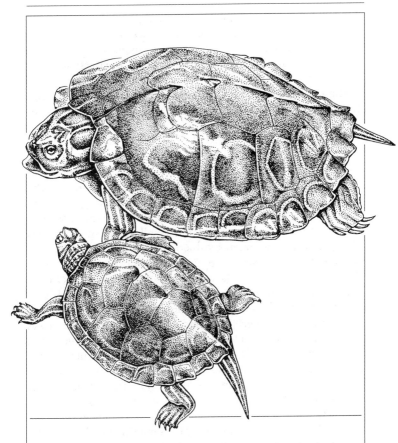

Barbour's map turtle is a threatened species limited to the Apalachicola River system of the Florida panhandle and adjacent portions of Alabama and Georgia. The female of the species is much larger than the male and in addition possesses an outsized head.

species lacking siphons. Siphons are fleshy tubes that channel water to and from the buried animal. Deep burial and a capacity for rapid burrowing make siphonate species less vulnerable to predators than are nonsiphonate species. As a result, most of the highly abundant species of burrowing bivalves are siphonate. Many nonsiphonate species are quite rare. Indeed, the survivorship of species in the Pacific regions (84 percent) has been twice that of the nonsiphonate species (42 percent). This pattern is compatible with the idea that abundance is of first order importance in determining probability of extinction.

The same principle applies widely through the animal kingdom: large body size, like specialization, means smaller populations and earlier extinction. The large mammals of North America and Eurasia were the first to yield to the invasions of human hunters. Wolves, lions, bear, bison, elk, and ibex largely disappeared. Foxes, raccoons, squirrels, rabbits, mice, and voles flourished. In their analysis of the resident landbirds of the British coastal islands, Stuart Pimm and his colleagues found that large-bodied species such as hawks and crows have become locally extinct more frequently than smaller ones such as wrens and sparrows. The greater vulnerability of the larger birds is due in part to their smaller populations, but not entirely. Even when population size is factored out (only populations of the same size are considered, regardless of species), the vulnerability remains. This added weakness of large birds apparently stems from their lower reproductive rates. Hawks and crows raise fewer young than wrens and sparrows. When struck once by high mortality, they are slower to recover; when struck again, they are more likely to fall all the way to extinction. The disadvantage is reversed when populations of both large and small birds are so small as to be on the verge of extinction, when (to be exact) they comprise seven or fewer breeding pairs. Then the greater longevity of the individual larger birds becomes the deciding factor. Hawks live longer than sparrows and are less likely to die off entirely before any one pair can raise offspring to maturity.

As populations decrease to a few individuals, they flirt with extinction through what geneticists term *inbreeding depression*. Imagine the extreme case of a population of birds, say an ill-fated warbler species, whittled down to a single mated pair, brother and sister. They are both heterozygous for a recessive lethal gene. This means that each bird carries a lethal gene on one of its chromosomes and a normal gene at the same site on the corresponding chromosome. The normal gene prevails over the lethal gene, and the birds stay more or less healthy. If they were homozygous for the lethal gene, possessing two instead of one, they would be dead. Brother and sister mate. There is half a chance that any given sperm carries a lethal gene, and half a chance that any given egg carries the same gene: each outcome has the same probability as the flip of a coin coming up tails. The chance that any given offspring gets two lethal genes and dies is the same as throwing two tails with a double flip of a coin: one half (bad sperm) times one half (bad egg) equals one fourth (afflicted offspring). The population gives up a quarter of its reproductive potential for being so small.

Why does inbreeding, as opposed to ordinary haphazard mating with individuals other than close relatives, cause depression of life and reproduction? Siblings, first cousins, and parents and their offspring are so closely related that the lethal recessive genes they carry are likely to be the same. Each human being and each fruit fly, typical organisms in this respect, carries an average of one to several lethal recessive genes. But there are many such genes in the population as a whole, and each one occurs in only one out of hundreds or even thousands of individuals. The chances that two unrelated individuals will carry the same defective gene are very small, even though both carry one kind of defective gene somewhere on their chromosomes. The odds against matching the same gene are so great in human beings that deadly hereditary conditions, such as Tay-Sachs disease and cystic fibrosis, are mercifully rare. The chances that one or the other such syndrome will appear is greatly elevated, however, if the child's parents are closely related, and that juxtaposition is more likely to occur if the population is small and closed.

Such is the basic concept of inbreeding depression. But real populations subscribe to it in idiosyncratic and subtle ways. Only a small fraction of deleterious genes are lethal. Most are "sublethal" or "subvital." To varying degrees they interfere with development, reduce strength, and diminish fertility. These are the genes that shorten life and cause sterility in cheetahs and gazelles kept in zoos, and they are the genes that inflict congenital heart defects on cocker spaniels bred too pure.

Conservation biologists have tried to draw danger lines below which a species is at conspicuously higher risk of extinction from genetic depression. They speak loosely of a 50–500 rule of genetic health in populations. When the effective population size falls below 50 and defective genes are present, inbreeding depression becomes common enough to slow population growth. Breeders of domestic animals generally do not worry about the amount of inbreeding depression encountered in populations with an effective size of 50 individual animals or more. But they consider themselves in trouble when the number falls below 50. When the effective population size is below 500, genetic drift (the chance fluctuation of gene percentages) is strong enough to eliminate some genes and reduce the variability of the population as a whole. At the same time, the mutation rate is not high enough to replenish this loss. So the species steadily loses its ability to adapt to changes in the environment. Inbreeding depression, turning the screw generation by generation, shortens the longevity of species. So does the shrinking of genetic reserves over many generations. To express this as concisely as possible: a population of 50 or more is adequate for the short term only, and one of 500 is needed to keep the species alive and healthy into the distant future.

I have used the phrase "effective population size" to make sense of genetic deterioration. It is a measure of considerable importance in the theory of conservation biology. Think of an actual population such as the tree sparrows on Scotland's Isle of May. It could consist of all males, and the effective size would be zero. Or it could be a thousand adults too old to breed plus five healthy females and five healthy males randomly mated; in that case the effective size would be ten. The effective size of a population refers to an idealized population, with random mating of individuals, possessing the same amount of genetic drift as the actual population. There are, in the last imaginary case, 1,010 tree sparrows but the thousand post-reproductive individuals don't count. All those tree sparrows are genetically the same as a population of ten birds living by themselves. The effective size declines as sterility rises, by aging or any other cause. It also declines to the extent that the adults forsake random mating and turn to relatives in order to breed. The point is that the age, health, and breeding patterns of individuals have an important effect on the genetic trajectory of a population and eventually its very survival. Even if the woods and fields are swarming with plants and animals of a certain kind, the species might be destined for extinction.

Conservation biologists and geneticists understand such matters in a general way. They have built a loose framework of theory clothed by a smattering of laboratory and zoo studies. They have learned that if the depression of fitness emanates from close inbreeding and the rapid juxtaposition of deleterious genes already present at high levels, the population is in immediate peril. But if inbreeding is gradual, the population has a better chance to pass through the bottleneck. Furthermore, as generations pass, the inbreeding depression will moderate, since natural selection purges the deleterious genes from the population. As the most harmful genes attain the homozygous (double-dose) state in individuals, they are eliminated and their frequency drops throughout the population.

The ruling consideration in the death of a species, however, is not in most cases the bite taken from the population by its defective genes. More important is the size of the population and the manner by which it subdivides and spreads across the terrain. It is risky to say, "Get the species up to an effective population size of 500 and it will be safe." If the species has been reduced to one population in one refuge, a single fire could destroy it, even if it contains 5,000 members. The population could be wiped out by one disease; there could be a local killing freeze; the food species on which it depends might become extinct; the crucial pollinator might vanish. Such events are "demographic accidents"—irregular and drastic reductions in population size caused by environmental change—and they are deadly. For species passing through the narrows of small population size, the Scylla of demographic accident is more dangerous than the Charybdis of inbreeding depression.

Only a few species consist of a single vulnerable population in one locality. They include the giant flightless darkling beetle (*Polposipus herculeanus*), which is restricted to dead trees on tiny Frigate Island in the Seychelles. There is the hau kuahiwi tree (*Hibiscadelphus distans*), which consists of exactly ten 6-meter-high trees growing on a dry rocky cliff on the island of Kauai. And perhaps the most intriguing example of all: the Socorro sowbug (*Thermosphaeroma thermophilum*), an aquatic crustacean that has lost its natural habitat and survives in an abandoned bathhouse in New Mexico. Most species are not like this. In some cases the constituent populations are so isolated from one another that they can never exchange individuals, but more typically the species is arrayed as a metapopulation, a population of populations, among which organisms do occasionally migrate.

Watched across long stretches of time, the species as metapopulation can be thought of as a sea of lights winking on and off across a dark terrain. Each light is a living population. Its location represents a habitat capable of supporting the species. When the species is present in that location the light is on, and when it is absent the light is out. As we scan the terrain over many generations, lights go out as local extinction occurs, then come on again as colonists from lighted spots reinvade the same localities. The life and death of species can then be viewed in a way that invites analysis and measurement. If a species manages to turn on as many lights as go out from generation to generation, it can persist indefinitely. When the lights wink out faster than they are turned on, the species sinks to oblivion.

The metapopulation concept of species existence is cause for both optimism and despair. Even when species are locally extirpated, they often come back quickly, provided the vacated habitats are left intact. But if the available habitats are reduced in sufficient number, the entire system can collapse. All the lights go out even if some intact habitats remain. A few jealously guarded reserves may not be enough. When the number of populations capable of populating empty sites becomes too small, they cannot achieve colonization elsewhere before they themselves go extinct. The system spirals downward out of control, and the entire sea of lights turns dark.

One metapopulation in the process of collapsing is the Karner blue butterfly (*Lycaeides melissa samuelis*), which lives in a pine barren in upstate New York called the Albany Pine Bush. A pine barren is an area of relatively sterile, sandy soils and dunes supporting forests and shrublands—dominated in the case of the Albany woods by pitch pine, scrub oak, and dwarf chestnut oak. The vegetation is frequently burned by

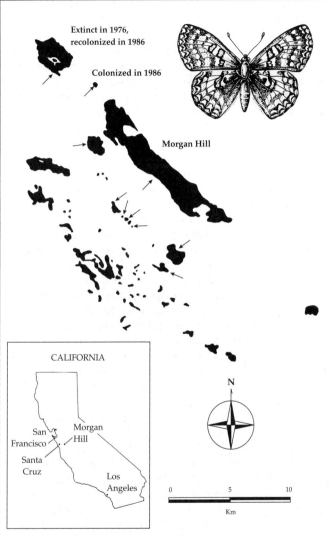

A metapopulation of Bay checkerspot butterflies lives south of San Francisco on serpentine grassland, a habitat represented here in black; the patches occupied in 1987 are indicated by arrows. In metapopulations, the occupancy of suitable environments changes from one year to the next.

lightning-sparked fires. The Karner blue lives within the barren as a scattered network of small local populations, each bound to the wild lupine (*Lupinus perennis*) on which its caterpillars exclusively feed. The lupine itself exists in scattered small patches. It is a fire plant, which means that it grows up after fires burn through the lower vegetation, clearing the ground and allowing more sunlight to reach small herbaceous plants. Fires are simultaneously a curse and a blessing for the Karner blue. They destroy the local population, but they also prepare the way for recolonization and even more vigorous growth by later generations. As each site is burned and the lupine plants proliferate, Karner blue adults fly in from nearby surviving populations. The butterfly, in other words, is what ecologists call a fugitive species, driven from place to place by its evolutionary commitment to an unstable niche.

The Albany Pine Bush once covered 10,000 hectares, enough to allow the Karner blue metapopulation to play this game of gambler's ruin indefinitely. Urban development in the Albany-Schenectady region has reduced it to 1,000 hectares, not enough to sustain the metapopulation. The butterfly will probably perish, as did a similar population on Manhattan Island long ago, unless the remaining habitat is kept intact at its present area and the lupine and butterfly populations are sustained at healthy levels by fires set and carefully controlled by human managers. The Karner blue is at the crossroad toward which thousands of other species are journeying: it must cease to be a truly wild race or die.

Every species makes its own farewell to the human partners who have served it so ill. I started this account with a New Zealand mistletoe, passed on to a frail butterfly ruined by urban development in New York, and will finish with a Brazilian parrot, Spix's macaw. It is the most endangered bird in the world, and one of the most beautiful: totally blue, darkest on top, with a greenish tinge on the belly, black mask around the lemon-yellow eye. The species, *Cyanopsitta spixii*, is so distinctive that it has been placed in its own genus. Never common, it was limited to palm groves and river-edge woodland across southern Pará to Bahia near the center of Brazil. It was driven to extreme rareness by bird fanciers, who near the end, in the mid-1980s, were paying up to $40,000 for a single bird. The

The Karner blue and its host plant, wild lupine.

Brazilians hunting Spix's macaw say that the decline was hastened by imported Africanized bees, whose colonies occupy the tree holes favored by the macaw. Their claim is self-serving but has the ring of truth. It is plausible natural history and therefore too far-fetched for the ordinary imagination. In any case, the collectors and their suppliers were the killing force. By 1987 four birds were left in the wild, and by late 1990 only a single male. This last Spix's macaw, according to Tony Juniper of the International Council for Bird Preservation, is "desperate to breed. It is investigating nesting holes and showing all the signs of breeding behavior." At last report, the male has paired with a female Illiger's macaw (*Ara maracana*). No hybrids are expected.

CHAPTER TWELVE

Biodiversity Threatened

HIDDEN AMONG the western Andean foothills of Ecuador, a few kilometers from Rio Palenque, there is a small ridge called Centinela. Its name deserves to be synonymous with the silent hemorrhaging of biological diversity. When the forest on the ridge was cut a decade ago, a large number of rare species were extinguished. They went just like that, from full healthy populations to nothing, in a few months. Around the world such anonymous extinctions—call them "centinelan extinctions"—are occurring, not open wounds for all to see and rush to stanch but unfelt internal events, leakages from vital tissue out of sight. Only an accident of timing led to an eyewitness account of the events on Centinela.

The eyewitnesses were Alwyn Gentry and Calaway Dodson, working out of the Missouri Botanical Garden, St. Louis. Gentry and Dodson made their discovery because they are born naturalists. By that I mean they are members of a special cadre of field biologists, those who do not practice science in order to be a success but try to succeed in order to practice science—at least this kind of science. Even if they have to pay for the trip themselves, they will go into the field to do biology, to blend sun and rain with the findings of evolution and to give memory thereby to places like Centinela.

When Gentry and Dodson visited the ridge in 1978, they were the first to explore it botanically. Centinela is only one of a vast number of little-known spurs and saddles arrayed on either side of the Andes for 7,200 kilometers from Panama to Tierra del Fuego. At middle to high elevations in the tropical latitudes these mountain buttresses are covered by cloud forests. A traverse reveals that they are ecological islands, closed off above by the treeless paramos, surrounded below by the lowland rain forests, and segregated from one another by deep mountain valleys. Like conventional islands in the ocean, they tend to evolve their own species of plants and animals, which are then the endemics of that place, found nowhere else or at most in a few nearby localities. On Centinela Gentry and

Dodson discovered about 90 such plant species, mostly herbaceous forms growing under the forest canopy, along with
orchids and other epiphytes on the trunks and branches of
trees. Several of the species had black leaves, a highly unusual
trait and still a mystery of plant physiology.

In 1978 farmers from the valley below were moving in along
a newly built private road and were cutting back the ridge
forest. This is standard operating procedure in Ecuador. Fully
96 percent of the forests on the Pacific side have been cleared
for agriculture, with little notice taken by conservationists
outside Ecuador and no constraining policy imposed by local
governments. By 1986 Centinela was completely cleared and
planted in cacao and other crops. A few of the endemic plants
have persisted in the shade of the cacao trees. Several others
hold on in the forest of neighboring ridges, which themselves
are in danger of clear-cutting. I don't know if any black-leaved
plant species survived.

The revelation of Centinela and a growing list of other such
places is that the extinction of species has been much worse
than even field biologists, myself included, previously understood. Any number of rare local species are disappearing just
beyond the edge of our attention. They enter oblivion like the
dead of Gray's *Elegy*, leaving at most a name, a fading echo in
a far corner of the world, their genius unused.

Extinction has been much greater even among larger, more
conspicuous organisms than generally recognized. During
the past ten years, scientists working on fossil birds, especially
Storrs Olson, Helen James, and David Steadman, have uncovered evidence of massive destruction of Pacific Island landbirds
by the first human colonists centuries before the coming of
Europeans. The scientists obtain their data by excavating fossil
and subfossil bones wherever the dead birds dropped or were
thrown, in dunes, limestone sinkholes, lava tubes, crater lake
beds, and archaeological middens. On each of the islands the
deposits were mostly laid down from 8,000 years ago up to
nearly the present, bracketing the arrival of the Polynesians.
They leave little room for doubt that in the outer Pacific in
particular, from Tonga in the west to Hawaii in the east, the
Polynesians extinguished at least half of the endemic species
found upon their arrival.

This vast stretch of Pacific islands was colonized by the
Lapita people, ancestors of the modern Polynesian race. They
emigrated from their homeland somewhere in the fringing
islands of Melanesia or Southeast Asia and spread steadily eastward from archipelago to archipelago. With great daring and
probably heavy mortality, they traveled in single outrigger or
double canoes across hundreds of kilometers of water. Around
3,000 years ago they settled Fiji, Tonga, and Samoa. Stepping
from island to island they finally reached Hawaii, with Easter
the most remote of the habitable Pacific islands, as recently as
300 A.D.

The colonists subsisted on crops and domestic animals
carried in their boats but also, especially in the early days of
settlement, whatever edible animals they encountered. They
ate fish, turtles, and a profusion of bird species that had never
seen a large predator and were easily caught, including doves,
pigeons, crakes, rails, starlings, and others whose remains are
only now coming to light. Many of the species were endemics,
found only on the islands discovered by the Lapita. The voyagers ate their way through the Polynesian fauna. On Eua, in
present-day Tonga, twenty-five species lived in the forests when
the colonists arrived around 1000 B.C., but only eight survive
today. Nearly every island across the Pacific was home to several endemic species of flightless rails before the Polynesian
occupation. Today populations survive only on New Zealand
and on Henderson, an uninhabited coral island 190 kilometers
northeast of Pitcairn. It used to be thought that Henderson
was one of the few virgin habitable islands of any size left in
the world, never occupied by human beings. But recently discovered artifacts reveal that Polynesians colonized Henderson,
then abandoned it, probably because they consumed the birds
to less than sustainable levels. On this and other small islands
lacking arable soil, birds were the most readily available source
of protein. The colonists drove the populations down, erasing
some species in the process, then either starved or sailed on.

Hawaii, last of the Edens of Polynesia, sustained the greatest damage measured by lost evolutionary products. When
European settlers arrived after Captain Cook's visit in 1778,
there were approximately fifty native species of landbirds. In
the following two centuries, one third disappeared. Now we

know from bone deposits that another thirty-five species identified with certainty, and very likely twenty other species less
well documented, had already been extinguished by the native
Hawaiians. Among those identified to date are an eagle similar to the American bald eagle, a flightless ibis, and a strange
parliament of owls with short wings and extremely long legs.
Most remarkable of all were bizarre flightless forms evolved
from ducks but possessing tiny wings, massive legs, and bills
resembling the beaks of tortoises. Helen James and Storrs
Olson record that

> although they were terrestrial and herbivorous, like geese, we
> now know from the presence of a duck-like syringeal bulla that
> these strange birds were derived either from shelducks (Tador-
> nini), or more likely from dabbling ducks (Anatini), quite pos-
> sibly from the genus *Anas*. They may have had an ecological
> role similar to that of the large tortoises of the Galápagos and
> islands of the western Indian Ocean. Because we now recognize
> three genera and four species of these birds, and because they
> are neither phyletically geese nor functionally ducks, we have
> coined a new word, *moa-nalo*, as a more convenient general term
> for all such flightless, goose-like ducks of the Hawaiian Islands.

The surviving native Hawaiian birds are for the most part
inconspicuous relicts, small, elusive species restricted to the
remnant mountain forests. They are a faint shadow of the
eagles, ibises, and moa-nalos that greeted the Polynesian colonists as the Byzantine empire was born and Mayan civilization
reached its zenith.

Centinelan extinctions also occurred on other continents
and islands as human populations spread outward from Africa
and Eurasia. Mankind soon disposed of the large, the slow,
and the tasty. In North America 12,000 years ago, just before
Paleo-Indian hunter-gatherers came from Siberia across the
Bering Strait, the land teemed with large mammals far more
diverse than those in any part of the modern world, including Africa. Twelve millennia back may seem like the Age of
Dinosaurs, but it was just yesterday by geological standards.
Humanity was stirring then, some eight million people alive
and many seeking new land. The manufacture of hooks and
harpoons for fishing was widespread, along with the cultivation

of wild grains and the domestication of dogs. The construction of the first towns, in the Fertile Crescent, lay only a thousand years in the future.

In western North America, just behind the retreating glacial front, the grasslands and copses were an American Serengeti. The vegetation and insects were similar to those alive in the west today—you could have picked the same wildflowers and netted the same butterflies—but the big mammals and birds were spectacularly different. From one spot, say on the edge of riverine forest looking across open terrain, you could have seen herds of horses (the extinct, pre-Spanish kind), long-horned bison, camels, antelopes of several species, and mammoths. There would be glimpses of sabertooth cats, possibly working together in lionish prides, giant dire wolves, and tapirs. Around a dead horse might be gathered the representatives of a full adaptive radiation of scavenging birds: condors, huge condor-like teratorns, carrion storks, eagles, hawks, and vultures, dodging and threatening one another (we know from the species that survived), the smaller birds snatching pieces of meat and waiting for the body to be whittled down enough to be abandoned by their giant competitors.

Some 73 percent of the large mammal genera that lived in the late Pleistocene are extinct. (In South America the number is 80 percent.) A comparable number of genera of the largest birds are also extinct. The collapse of diversity occurred about the same time that the first Paleo-Indian hunters entered the New World, 12,000 to 11,000 years ago, and then spread southward at an average rate of 16 kilometers a year, It was not a casual, up-and-down event. Mammoths had flourished for two million years to that time and were represented at the end by three species—the Columbian, imperial, and woolly. Within a thousand years all were gone. The ground sloths, another ancient race, vanished almost simultaneously. The last known surviving population, foraging out of caves at the western end of the Grand Canyon, disappeared about 10,000 years ago.

If this were a trial, the Paleo-Indians could be convicted on circumstantial evidence alone, since the coincidence in time is so exact. There is also a strong motive: food. The remains of mammoths, bison, and other large mammals exist in association with human bones, charcoal from fires, and stone weapons of the Clovis culture. These earliest Americans were skilled big-game hunters, and they encountered animals totally unprepared by evolutionary experience for predators of this kind. The birds that became extinct were also those most vulnerable to human hunters. They included eagles and a flightless duck. Still other victims were innocent bystanders: condors, teratorns, and vultures dependent on the newly devastated populations of heavy-bodied mammals.

In defense of the Paleo-Indians, their counsel might argue the existence of another culprit. The end of the Pleistocene was a time not only of human invasion of the New World, but also of climatic warming. As the continental glacier retreated across Canada, forests and grasslands shifted rapidly northward. Changes of this magnitude must have exerted a profound effect on the life and death of local populations. Between 1870 and 1970, by way of comparison, Iceland warmed an average 2°C in the winter and somewhat less in the spring and summer. Two Arctic bird species, the long-tailed duck and the lesser auk, declined to near extinction. At the same time, lapwings, tufted ducks, and several other southern species established themselves on the island and began to breed. There are hints of similar responses during the great Pleistocene decline. Mastodons, for example, were apparently specialized for life in coniferous forests. As this belt of vegetation migrated northward, the proboscideans moved with it. In time they became concentrated along the spruce forest zone in the northeast, then disappeared. Their extinction might have stemmed not only from overkill by hunters but also from fragmentation and reduction of the populations forced by a shrinking habitat.

Let the defense now speak even more forcefully: for tens of millions of years before the coming of man, mammal genera were born and died in large numbers, with the extinction of some accompanied by the origin of others to create a rough long-term balance. The changes were accompanied by climatic shifts much like those in evidence 11,000 years ago, and perhaps they were driven by them. During the last 10 million years, David Webb has pointed out, six major extinction episodes leveled the land mammals of North America. Among them the terminating event of the Pleistocene (the Rancholabrean, named after Rancho La Brea, in California) was not the most catastrophic. The greatest, according to available records,

> was the late Hemphillian (nearly five million years ago) when more than sixty genera of land mammals (of which thirty-five were large, weighing more than 5 kg) disappeared from this continent. The late Rancholabrean extinction pulse (about 10,000 years ago) was the next greatest; over forty genera became extinct, of which nearly all were large mammals ... Some evidence shows that these extinction episodes were correlated with terminations of glacial cycles, when climatic extremes and instability are thought to have reached their maxima.

In at least two of the great extinction spasms, the large browsing mammals were destroyed as the climate deteriorated and the broad continental savannas gave way to steppes. At the end of the Hemphillian, even grazing mammals such as horses, rhinos, and pronghorns precipitously declined.

It may seem that the debate between experts who favor overkill by humans and those who favor climatic change resembles a replay, in a different theater, of the debate over the end of the Age of Dinosaurs. The Paleo-Indians have replaced the giant meteorite in this new drama. Circumstantial evidence is countered by other circumstantial evidence, while both sides search for a smoking gun. The dispute is the product of neither ideology nor clashing personalities. It is the way science at its best is done.

That said, I will lay aside impartiality. I think the overkill theorists have the more convincing argument for what happened in America 10,000 years ago. It seems likely that the Clovis people spread through the New World and demolished most of the large mammals during a hunters' blitzkrieg spanning several centuries. Some of the doomed species hung on here and there for as long as 2,000 years, but the effect was the same: swift destruction, on the scale of evolution that measures normal lifespans of genera and species in millions of years.

There is an additional reason for accepting this verdict provisionally. Paul Martin, who revived the idea in the mid-1960s (a similar proposal had been made a century earlier for the Pleistocene mammals of Europe), called attention to this important circumstance: when human colonists arrived, not only in America but also in New Zealand, Madagascar, and Australia, and whether climate was changing or not, a large part of the megafauna—large mammals, birds, and reptiles— disappeared soon afterward. This collateral evidence has been pieced together by researchers of various persuasions over many years, and it points away from climate and toward people.

Before the coming of man around 1000 A.D., New Zealand was home to moas, large flightless birds unique to the islands. These creatures had ellipsoidal bodies, massive legs, and long necks topped by tiny heads. The first Maoris, arriving from their Polynesian homeland to the north, found about thirteen species ranging in size from that of large turkeys to giants weighing 230 kilograms or more, the latter among the largest birds ever evolved. There had in fact been a moa radiation, filling many niches. It was of the kind normally occupied by medium-sized and large mammals, of which there were none on New Zealand. The Maoris proceeded to butcher the birds in large numbers, leaving conspicuous moa-hunting sites all over New Zealand. On South Island, where most of the remains occur, the deposits are piled with moa bones dating from 1100 to 1300. During this brief interlude the colonists must have obtained a substantial portion of their diet from cooked moa. The peak kills began on the northern part of the island, the Maori point of entry, and spread slowly to the southern dis- tricts. Several Europeans claimed to have seen moas in the early 1800s, but the records cannot be verified. Archaeological and public opinion alike hold the Maori hunters responsible, as declared in the popular New Zealand song:

> No moa, no moa,
> In old Ao-tea-roa.
> Can't get 'em.
> They've et 'em;
> They've gone and there aint no moa!

The moa extinction was only part of the New Zealand carnage. A total of twenty other landbirds, including nine additional flightless species, were also wiped out in short order. The tuatara, only living member of the reptilian order

Rhynchocephalia, along with unique frogs and flightless insects, were driven to the edge of extinction. Their demise was partly due to the deforestation and firing of large stretches of land. It was hastened by rats that came ashore with the Maoris and bred in huge numbers, against which the autochthons had few natural defenses. In the 1800s the British settlers came upon a beautiful but already much-damaged archipelago. As elsewhere, they proceeded to reduce its biodiversity still further, with a pernicious ingenuity of their own.

Madagascar, fourth largest island in the world, is a small continent virtually on its own. Fully isolated during a northward drift through the Indian Ocean for 70 million years, it was the theater for a biological tragedy like New Zealand's. Despite the proximity of Africa, the first human colonists came to Madagascar not from that continent but from far-off Indonesia. They arrived around 500 A.D. In the centuries immediately following, the megafauna of the great island vanished. No important climatic change accompanied this event; it appears to have been solely the work of the Malagache pioneers. Six to a dozen elephant birds, large and flightless like the moas, disappeared. They included the heaviest birds of recent geological history, *Aepyornis maximus*, a feathered giant almost 3 meters tall with massive legs. Its eggs, the size of soccer balls, can still be pieced together from fragments piled around Malagache archaeological sites. Also erased were seven of the seventeen genera of lemurs, primates most closely related among living mammals to monkeys, apes and men. The lemuroids had undergone a spectacular adaptive radiation on Madagascar. The forms that disappeared were the largest and most interesting of all. One species ran on all fours like a dog, and another had long arms and probably swung through the trees like a gibbon. A third, as big as a gorilla, climbed trees and resembled an oversized koala. Also erased were an aardvark, a pygmy hippopotamus, and two huge land tortoises.

Essentially the same story of destruction was repeated when aboriginal human populations came to Australia about 30,000 years ago, also by way of Indonesia. A number of large mammals soon vanished, including marsupial lions, gigantic kangaroos 2.5 meters (8 feet) tall, and others separately resembling ground sloths, rhinos, tapirs, woodchucks, or, perhaps more accurately expressed, blends of these more familiar types of World Continent fauna. The case for overkill by the aboriginal Australians, however, is complicated by the remote time of their arrival, the longer period during which the extinctions took place, and the scarcity of fossils and kill sites to document the role of hunting. It is also true that Australia experienced a severe arid period from 15,000 to 26,000 years ago, during which the greatest number of animal extinctions occurred. We know that the Australian aboriginals hunted skillfully and burned large stretches of arid land in their search for prey. They still do. Men must have played a role in extinction, but the evidence does not yet allow us to weigh their influence against the drying out of the continent's interior.

In 1989 Jared Diamond summed up for the prosecution in the case of the extinguished megafaunas. Climate, he said, cannot be the principal culprit. He asked: how could changes in climate and vegetation during the retreat of the last glacier lead to mass extinction in North America but not in Europe and Asia? The differences between the land masses were not climatic but the first-time colonization of America, confronting a megafauna with no previous experience of human hunters. And in North America, why did this hecatomb occur at the end of the last glacial cycle, which closed the Quaternary period, but not at the end of the twenty-two glacial cycles preceding it? Again, the difference was the coming of the Paleo-Indian hunters. How, Diamond pressed, did Australia's reptiles manage to survive the prehistoric human invasions better, as did the smaller mammals and birds? And, finally, why did such large forms as the marsupial wolf and giant kangaroos disappear about the same time from both Australia's arid interior and rain forests, as well as from nearby New Guinea's wet mountain forests?

> Quaternary extinctions were selective in space and time because they appear to have occurred at those places and times where naive animals first encountered humans. It is further argued that they were selective in taxa and in victim size because human hunters concentrate on some species (e.g. large mammals and flightless birds) while ignoring other species (e.g. small rodents).

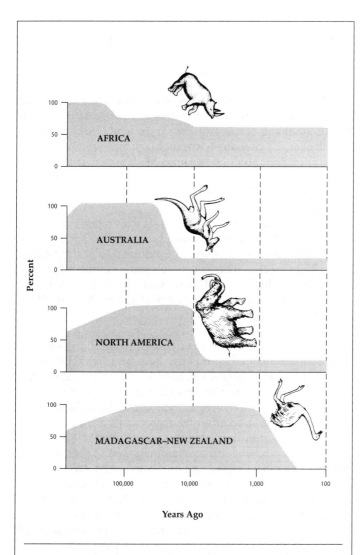

The extinction of large mammals and flightless birds coincided closely with the arrival of humans in North America, Madagascar, and New Zealand, and less decisively earlier in Australia. In Africa, where humans and animals evolved together for millions of years, the damage was less severe.

It is argued that Quaternary extinctions befell species in all habitats because humans hunt in all habitats, and human hunters
help no species except as an incidental consequence of habitat
changes and of removing other species.

"Human hunters help no species." That is a general truth
and the key to the whole melancholy situation. As the human
wave rolled over the last of the virgin lands like a smother-
ing blanket, Paleo-Indians throughout America, Polynesians
across the Pacific, Indonesians into Madagascar, Dutch sailors
ashore on Mauritius (to meet and extirpate the dodo), they
were constrained by neither knowledge of endemicity nor any
ethic of conservation. For them the world must have seemed to
stretch forever beyond the horizon. If fruit pigeons and giant
tortoises disappear from this island, they will surely be found
on the next one. What counts is food today, a healthy family,
and tribute for the chief, victory celebrations, rites of passage,
feasts. As the Mexican truck driver said who shot one of the last
two imperial woodpeckers, largest of all the world's woodpeckers, "It was a great piece of meat."

From prehistory to the present time, the mindless horsemen
of the environmental apocalypse have been overkill, habitat
destruction, introduction of animals such as rats and goats,
and diseases carried by these exotic animals. In prehistory the
paramount agents were overkill and exotic animals. In recent
centuries, and to an accelerating degree during our generation,
habitat destruction is foremost among the lethal forces, followed by the invasion of exotic animals. Each agent strengthens the others in a tightening net of destruction. In the United
States, Canada, and Mexico, 1,033 species of fishes are known
to have lived entirely in fresh water within recent historical
times. Of these, 27 or 3 percent have become extinct within the
past hundred years, and another 265 or 26 percent are liable
to extinction. They fall into one or the other of the categories utilized by the International Union for Conservation of
Nature and Natural Resources (IUCN), which publishes the
Red Data Books: Extinct, Endangered, Vulnerable, and Rare.
The changes that forced them into decline are:

Destruction of physical habitat	73% of species
Displacement by introduced species	68% of species
Alteration of habitat by chemical pollutants	38% of species
Hybridization with other species and subspecies	38% of species
Overharvesting	15% of species

(These figures add up to more than 100 percent because more than one agent impinges on many of the fish populations.) When habitat destruction is defined as both the physical reduction in suitable places to live and the closing of habitats by chemical pollution, then it is found to be an important factor in over 90 percent of the cases. Through a combination of all these factors, the rate of extinction has risen steadily during the past forty years.

In fishes and in all other groups of which we have sufficient knowledge, the depredations were started in prehistory and early historical times and are being pressed with a vengeance by modern generations. Early peoples exterminated most of the big animals on the spot. They also decimated less conspicuous plants and animals on islands and in isolated valleys, lakes, and river systems, where species live in small populations with their backs to the wall. Now it is our turn. Armed with chainsaws and dynamite, we are assaulting the final strongholds of bio-diversity—the continents and, to a lesser but growing extent, the seas.

Will it ever be possible to assess the ongoing loss of biological diversity? I cannot imagine a scientific problem of greater immediate importance for humanity. Biologists find it difficult to come up with even an approximate estimate of the hemorrhaging because we know so little about diversity in the first place. Extinction is the most obscure and local of all biological processes. We don't see the last butterfly of its species snatched from the air by a bird or the last orchid of a certain kind killed by the collapse of its supporting tree in some distant mountain forest. We hear that a certain animal or plant is on the edge, perhaps already gone. We return to the last known locality to search, and when no individuals are encountered there year after year we pronounce the species extinct. But hope lingers on. Someone flying a light plane over Louisiana swamps thinks he sees a few ivory-billed woodpeckers start up and glide back down into the foliage. "I'm pretty sure they were ivorybills, not pileated woodpeckers. Saw the white double stripes on the back and the wing bands plain as day." A Bachman's warbler is heard singing somewhere, maybe. A hunter swears he has seen Tasmanian wolves in the scrub forest of Western Australia, but it is probably all fantasy.

In order to know that a given species is truly extinct, you have to know it well, including its exact distribution and favored habitats. You have to look long and hard without result. But we do not know the vast majority of species of organisms well; we have yet to anoint so many as 90 percent of them with scientific names. So biologists agree that it is not possible to give the exact number of species going extinct; we usually turn palms up and say the number is very large. But we can do better than that. Let me start with a generalization: *in the small minority of groups of plants and animals that are well known, extinction is proceeding at a rapid rate, far above prehuman levels. In many cases the level is calamitous: the entire group is threatened.*

To illustrate this principle, I will present a few anecdotes, out of many available: whenever we can focus clearly, we usually see extinction in progress. Then I will take a more theoretical approach, using models of island biogeography, to arrive at an estimate of extinction rates in tropical rain forests, which contain half or more of the world's species of plants and animals. Here are the examples:

• One fifth of the species of birds worldwide have been eliminated in the past two millennia, principally following human occupation of islands. Thus instead of 9,040 species alive today, there probably would have been about 11,000 species if left alone. According to a recent study by the International Council for Bird Preservation, 11 percent or 1,029 of the surviving species are endangered.

• A total of 164 bird species have been recorded from the Solomon Islands in the southwest Pacific. The *Red Data Book* lists only one as recently extinct. But in fact there have been no records for twelve others since 1953. Most of these are ground nesters especially vulnerable to predators. Solomon Islanders who know the birds best have stated that at least some of the species were exterminated by imported cats.

• From the 1940s to the 1980s, population densities of migratory songbirds in the mid-Atlantic United States dropped 50 percent, and many species became locally extinct. One cause appears to be the accelerating destruction of the forests of the West Indies, Mexico, and Central and South America, the principal wintering grounds of many of the migrants. The fate of Bachman's warbler will probably befall other North American summer residents if the deforestation continues.

• About 20 percent of the world's freshwater fish species are either extinct or in a state of dangerous decline. The situation is approaching the critical stage in some tropical countries. A recent search for the 266 species of exclusively freshwater fishes of lowland peninsular Malaysia turned up only 122. Lake Lanao on the Philippine Island of Mindanao is famous among evolutionary biologists for the adaptive radiation of cyprinid fishes that occurred exclusively within the confines of the lake. As many as 18 endemic species in three genera were previously known; a recent search found only three species, representing one of the genera. The loss has been attributed to overfishing and competition from newly introduced fish species.

• The most catastrophic extinction episode of recent history may be the destruction of the cichlid fishes of Lake Victoria, which I described earlier as a paradigm of adaptive radiation. From a single ancestral species 300 or more species emanated, filling almost all the major ecological niches of freshwater fishes. In 1959 British colonists introduced the Nile perch as a sport fish. This huge predator, which grows to nearly 2 meters in length, has drastically reduced the native fish population and extinguished some of the species. It is projected eventually to eliminate more than half of the endemics. The perch affects not only the fishes but the lake ecosystem as a whole. As the alga-feeding cichlids disappear, plant life blooms and decomposes, depleting oxygen in the deeper water and accelerating the decline of cichlids, crustaceans, and other forms of life. A task force of fish biologists observed in 1985, "Never before has man in a single ill advised step placed so many vertebrate species simultaneously at risk of extinction and also, in doing so, threatened a food resource and traditional way of life of riparian dwellers."

• The United States has the largest freshwater mollusk fauna in the world, especially rich in mussels and gill-breathing snails. These species have long been in a steep decline from the damming of rivers, pollution, and the introduction of alien mollusk and other aquatic animals. At least 12 mussel species are now extinct throughout their ranges, and 20 percent of the remainder are endangered. Even where extinction has not yet occurred, the extirpation of local populations is rampant. Lake Erie and the Ohio River system originally held dense populations of 78 different forms; now 19 are extinct and 29 are rare. Muscle [sic] Shoals, a stretch of the Tennessee River in Alabama, once held a fauna of 68 mussel species. Their shells were specialized for life in riffles or shoals, shallow streams with sandy gravel bottoms and rapid currents. When Wilson Dam was constructed in the early 1920s, impounding and deepening the water, 44 of the species were extinguished. In a parallel development, impoundment and pollution have combined to extinguish two genera and 30 species of gill-breathing snails in the Tennessee and nearby Coosa rivers.

• Freshwater and land mollusks are generally vulnerable to extinction because so many are specialized for life in narrow habitats and unable to move quickly from one place to another. The fate of the tree snails of Tahiti and Moorea illustrates the principle in chilling fashion. Comprising 11 species in the genera *Partula* and *Samoana*, a miniature adaptive radiation in one small place, the snails were recently exterminated by a single species of exotic carnivorous snail. It was folly in the grand manner, a pair of desperate mistakes by people in authority, which unfolded as follows. First, the giant African snail *Achatina fulica* was introduced to the islands as a food animal. Then, when it multiplied enough to become a pest, the carnivorous snail *Euglandina rosea* was introduced to control the *Achatina*. *Euglandina* itself multiplied prodigiously, advancing along a front at 1.2 kilometers a year. It consumed not only the giant African snail but every native tree snail along the way. The last of the wild tree snails became extinct on Moorea in

1987. On nearby Tahiti the same sequence is now unfolding. And in Hawaii the entire endemic tree-snail genus *Achatinella* is endangered by *Euglandina* and habitat destruction. Twenty-two species are extinct and the remaining 19 are endangered.

• A recent survey by the Center for Plant Conservation revealed that between 213 and 228 plant species, out of a total of about 20,000, are known to have become extinct in the United States. Another 680 species and subspecies are in danger of extinction by the year 2000. About three fourths of these forms occur in only five places: California, Florida, Hawaii, Puerto Rico, and Texas. The predicament of the most endangered species is epitomized by *Banara vanderbiltii*. By 1986 this small tree of the moist limestone forests of Puerto Rico was down to two plants growing on a farm near Bayamon. At the eleventh hour, cuttings were obtained and are now successfully growing in the Fairchild Tropical Garden in Miami.

• In western Germany, the former Federal Republic, 34 percent of 10,290 insect and other invertebrate species were classified as threatened or endangered in 1987. In Austria the figure was 22 percent of 9,694 invertebrate species, and in England 17 percent of 13,741 insect species.

• The fungi of western Europe appear to be in the midst of a mass extinction on at least a local scale. Intensive collecting in selected sites in Germany, Austria, and the Netherlands has revealed a 40 to 50 percent loss in species during the past sixty years. The main cause of the decline appears to be air pollution. Many of the vanished species are mycorrhizal fungi, symbiotic forms that enhance the absorption of nutrients by the root systems of plants. Ecologists have long wondered what would happen to land ecosystems if these fungi were removed, and we will soon find out.

For species on the brink, from birds to fungi, the end can come in two ways. Many, like the Moorean tree snails, are taken out by the metaphorical equivalent of a rifle shot—they are erased but the ecosystem from which they are removed is left intact. Others are destroyed by a holocaust, in which the entire ecosystem perishes.

The distinction between rifle shots and holocausts has special merit in considering the case of the spotted owl (*Strix occidentalis*) of the United States, an endangered form that has been the object of intense national controversy since 1988. Each pair of owls requires about 3 to 8 square kilometers of coniferous forest more than 250 years old. Only this habitat can provide the birds with both enough large hollow trees for nesting and an expanse of open understory for the effective hunting of mice and other small mammals. Within the range of the spotted owl in western Oregon and Washington, the suitable habitat is largely confined to twelve national forests. The controversy was engaged first within the U.S. Forest Service and then the public at large. It was ultimately between loggers, who wanted to continue cutting the primeval forest, and environmentalists determined to protect an endangered species. The major local industry around the owl's range was affected, the financial stakes were high, and the confrontation was emotional. Said the loggers: "Are we really expected to sacrifice thousands of jobs for a handful of birds?" Said the environmentalists: "Must we deprive future generations of a race of birds for a few more years of timber yield?"

Overlooked in the clamor was the fate of an entire habitat, the old-growth coniferous forest, with thousands of other species of plants, animals, and microorganisms, the great majority unstudied and unclassified. Among them are three rare amphibian species, the tailed frog and the Del Norte and Olympic salamanders. Also present is the western yew, *Taxus brevifolia*, source of taxol, one of the most potent anticancer substances ever found. The debate should be framed another way: what else awaits discovery in the old-growth forests of the Pacific Northwest?

The cutting of primeval forest and other disasters, fueled by the demands of growing human populations, are the overriding threat to biological diversity everywhere. But even the data that led to this conclusion, coming as they do mainly from vertebrates and plants, understate the case. The large, conspicuous organisms are the ones most susceptible to rifle shots, to overkill and the introduction of competing organisms. They are of the greatest immediate importance to man and receive the greater part of his malign attention. People hunt deer and pigeons rather than sowbugs and spiders. They cut roads into a forest to harvest Douglas fir, not mosses and fungi.

Not many habitats in the world covering a kilometer contain fewer than a thousand species of plants and animals. Patches of rain forest and coral reef harbor tens of thousands of species, even after they have declined to a remnant of the original wilderness. But when the *entire* habitat is destroyed, almost all of the species are destroyed. Not just eagles and pandas disappear but also the smallest, still uncensused invertebrates, algae, and fungi, the invisible players that make up the foundation of the ecosystem. Conservationists now generally recognize the difference between rifle shots and holocausts. They place emphasis on the preservation of entire habitats and not only the charismatic species within them. They are uncomfortably aware that the last surviving herd of Javan rhinoceros cannot be saved if the remnant woodland in which they live is cleared, that harpy eagles require every scrap of rain forest around them that can be spared from the chainsaw. The relationship is reciprocal: when star species like rhinoceros and eagles are protected, they serve as umbrellas for all the life around them.

And so to threatened and endangered species must be added a growing list of entire ecosystems, comprising masses of species. Here are several deserving immediate attention:

Usambara Mountain forests, Tanzania. Varying widely in elevation and rainfall, the Usambaras contain one of the richest biological communities in East Africa. They protect large numbers of plant and animal species found nowhere else, but their forest cover is declining drastically, having already been cut to half, some 450 square kilometers, between 1954 and 1978. Rapid growth of human populations, more extensive logging, and the takeover of land for agriculture are pressing the last remaining reserves and thousands of species toward extinction.

San Bruno Mountain, California. In this small refuge surrounded by the San Francisco metropolis live a number of federally protected vertebrates, plants, and insects. Some of the species are endemics of the San Francisco peninsula, including the San Bruno elfin butterfly and the San Francisco garter snake. The native fauna and flora are threatened by offroad vehicular traffic, expansion of a quarry, and invasion by eucalyptus, gorse, and other alien plant species.

Oases of the Dead Sea Depression, Israel and Jordan. These humid refuges in a quintessentially desert area, called *ghors*, are isolated tropical ecosystems sustained by freshwater springs. They contain true pockets of an ancient African fauna and flora cut off by the dry terrain of the Jordan Rift Valley. Species that flourish thousands of kilometers to the south are joined here by others restricted to the vicinity of ghors or even to single springs. In 1980 I walked most of the length of Ein Gedi, one of these sites, through the lush bankside vegetation, marveling at the crystalline water of the spring-fed brook, with its endemic cichlid fish and emerald algae. I studied large weaver ants that nest in the banks—a little slice of Africa an hour's drive from Jerusalem. Climbing away from the bank trail for a hundred meters, I was back in the desert terrain of the Middle East. The ghors are of exceptional scientific interest because they bring an African fauna and flora into direct contact with a different set of species that together range from Europe across the Middle East to temperate Asia. The oases are threatened by overgrazing, mining, and commercial development. In an exquisitely symbolic reflection of the region's politics, several are used as minefields.

If species vanish en masse when their isolated habitats collapse, they die even more catastrophically when entire systems are obliterated. The logging of a mountain ridge in the Andes may extinguish scores of species, but logging all such ridges will erase hundreds of thousands. Such broad areas were labeled "hot spots" by Norman Myers in 1988. The emergency-care cases of global conservation, they are defined as areas that both contain large numbers of endemic species and are under extreme threat; their major habitats have been reduced to less than 10 percent of the original cover or are destined to fall that low within one to several decades. Myers has listed eighteen hot spots. Although they collectively occupy a tiny amount of space, only half a percent of the earth's land surface, they are the exclusive home of one fifth of the world's plant species. The hot spots comprise a far-flung array of forests and Mediterranean-type scrubland and are represented on every continent except Antarctica. Each deserves special and immediate mention.

California floristic province. This familiar Mediterranean-climate domain, stretching from southern Oregon to Baja California and recognized by botanists as a separate evolutionary center, contains one fourth of all the plant species found in the United States and Canada combined. Half, or 2,140 species, are found nowhere else in the world. Their environment is being rapidly constricted by urban and agricultural development, especially along the central and southern coasts of California.

Central Chile. South America's preeminent Mediterranean vegetation contains 3,000 plant species, slightly over half of the entire Chilean flora, crowded into only 6 percent of the national territory. The surviving cover is only one third that of the original and unfortunately is located in the most densely populated part of the country. It is being pressed especially hard by rural families, who rely on natural vegetation for fuel and livestock fodder.

The Colombian Chocó. The forest of Colombia's coastal plain and low mountains extends the entire length of the country. The Chocó, as the region is called after the state it includes, is drenched with extreme rainfall and blessed with one of the richest but least explored floras in the world. At present, 3,500 plant species are known but as many as 10,000 may grow there, of which one fourth are estimated to be endemic and a smaller but still substantial fraction are new to science. Since the early 1970s, the Chocó has been relentlessly invaded by timber companies and, to a lesser extent, by poor Colombians hungry for land. The forests are already down to about three quarters of their original cover and are being destroyed at an accelerating rate.

Western Ecuador. The wet forests of the lowlands and foothills of Ecuador west of the Andes, including the small portion that formerly clothed the Centinela ridge, once contained about 10,000 plant species. Of these one quarter, as in the closely similar Chocó region to the north, were endemic. The forests, so notable for the richness of their orchids and other

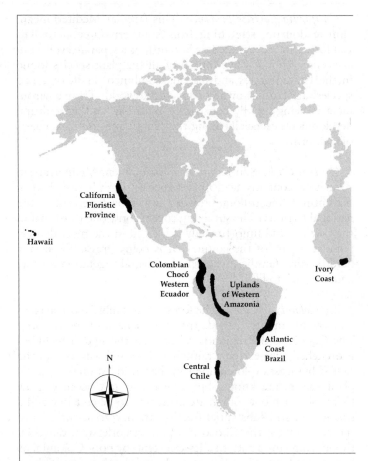

California
Floristic
Province

Hawaii

Colombian
Chocó
Western
Ecuador

Uplands
of Western
Amazonia

Ivory
Coast

N

Atlantic
Coast
Brazil

Central
Chile

Hot spots are habitats with many species found nowhere else and in greatest danger of extinction from human activity. The 18 hot spots identified here are forests and Mediterranean scrubland well enough known to be included with certainty. But the map, based on preliminary study, is far from complete. Other forest types not shown are

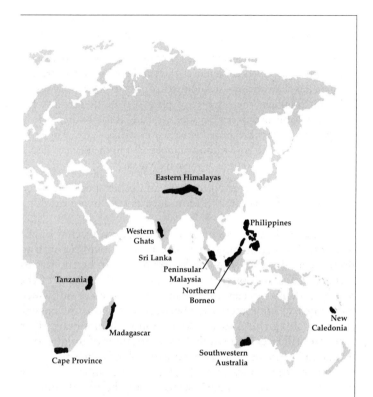

Eastern Himalayas

Philippines

Western
Ghats

Sri Lanka

Peninsular
Malaysia

Northern
Borneo

Tanzania

New
Caledonia

Madagascar

Southwestern
Australia

Cape Province

Forests and Heathland: Hot-Spot Areas

endangered, as well as a large number of lakes, river systems, and coral reefs. The broader areas depicted, such as the coastal forests of Brazil and the Philippines, actually consist of many smaller hot spots scattered across local mountain ridges, valleys, and islands.

epiphytes, have been almost completely wiped out. They are, in Myers' expression, among the hottest of the hot spots.

> An idea of the former biotic diversity may be gained from the Rio Palenque Science Center at the southern tip of the area, where less than one square kilometer of primary forest survives. In this fragment there are 1200 plant species, 25 percent of them endemic to western Ecuador. As many as 100 of these Rio Palenque species have proved to be new to science; 43 are known only from the site, and a good number exist in the form of just a few individuals, some as a single individual.

Daniel Janzen has referred to these and other species reduced to a population too small to reproduce as the "living dead."

Uplands of western Amazonia. The western reaches of the Amazon Basin stretching in an arc from Colombia south to Bolivia contain what some biologists believe to be the largest fauna and flora of any place on earth. And the richest of the rich in endemic species are the uplands of the region, which form a belt 50 kilometers wide between 500 and 1,500 meters' elevation along the Andean slopes. On each mountain ridge there are still largely unstudied concentrations of unique local plants and animals. The Amazonian uplands, like the western side of the Andes in Colombia and Ecuador, are being rapidly settled. In Ecuador's sector alone the population has grown from 45,000 to about 300,000 during the past forty years. About 65 percent of the upland forests have already been cleared or converted into palm-oil plantations. The loss is projected to approach 90 percent by the year 2000.

Atlantic coast of Brazil. A unique rain forest once reached from Recife southward through Rio de Janeiro to Florianópolis, of which the young Charles Darwin once wrote, "twiners entwining twiners—tresses like hair—beautiful lepidoptera—silence—hosanna—silence well exemplified—lofty tree . . . Wonder, astonishment, & sublime devotion, fill & elevate the mind." That was 1832 when, as naturalist on the *Beagle*, Darwin first put ashore in South America and jotted his impressions in a notebook. The Atlantic forests originally covered about a million square kilometers. Geographically isolated from the

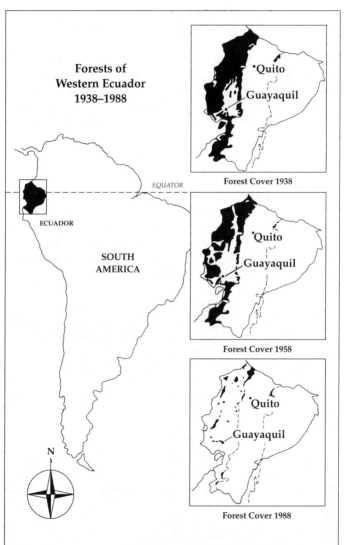

Forests of Western Ecuador 1938–1988

EQUATOR

ECUADOR

SOUTH AMERICA

N

Forest Cover 1938

•Quito

Guayaquil

Forest Cover 1958

•Quito

Guayaquil

Forest Cover 1988

•Quito

Guayaquil

More than 90 percent of the forests of western Ecuador have been destroyed during the past four decades. The loss is estimated to have extinguished or doomed over half of the species of the area's plants and animals. Many other biologically diverse areas of the world are under similar assault.

Amazonian forests to the north and west, they contain one of the most diverse and distinctive biotas in the world. But Brazil's south Atlantic coast is also the agriculturally most productive and densely populated part of the country. The forests have been reduced to less than 5 percent of the original cover, and that part survives mostly in steep mountainous regions. A good part of this remnant is protected as parks and reserves, a last glimpse of Eden for future generations teeming around it.

Southwestern Ivory Coast. The towering rain forest of the Ivory Coast and adjacent areas of Liberia, a distinct botanical province of West Africa, once covered 160,000 square kilometers. Unrestricted logging and slash-and-burn farming have reduced it to 16,000 square kilometers. What remains is being cleared at the rate of up to 2,000 square kilometers a year. Only the Taï National Park, with 3,300 square kilometers, is officially protected, and even this solitary reserve is under pressure from illegal logging and gold prospecting.

Eastern arc forests of Tanzania. The Usambara forest, described earlier, is one of nine sections of montane forests strung across eastern Tanzania. Isolated to some extent since prehuman times, these habitats are the sites of profuse local evolution. They are the native home, for example, of 18 of the known 20 species of African violets and 16 of the species of wild coffee. The forests are down to half their original cover and shrinking fast from the incursions of Tanzania's exploding population.

Cape floristic province of South Africa. At the southern tip of Africa is a specialized heathland called *fynbos*, which is graced with one of the world's most unusual and diverse floras. In the 89,000 square kilometers of the environment still surviving, 8,600 plant species can be found. Of these, 73 percent exist nowhere else in the world. A third of the fynbos has been lost to agriculture, development, and the incursion of exotic plant species. The remainder is being rapidly fragmented and degraded. Most of the native species occur in local areas of a square kilometer or less in extent. At least 26 are known to be extinct and another 1,500 are rare and threatened, a total exceeding the entire flora of the British Isles. Unless swift action is taken, South Africa will lose a large part of its greatest natural heritage.

Madagascar. Madagascar, the most isolated of the great islands of the world, has a fauna and flora independently evolved to a corresponding degree: 30 primates, all lemurs; reptiles and amphibians that are 90 percent endemic, including two thirds of all the chameleons of the world; and 10,000 plant species of which 80 percent are endemic, including a thousand kinds of orchids. The impoverished Malagasy people have relied heavily on slash-and-burn agriculture on poor rain forest soils to sustain their growing populations, causing them to grind their way through and destroy most of the world-class biological environment they inherited. In 1985 the forest remaining intact was down to a third of the cover encountered by the first colonists fifteen centuries ago. The destruction is accelerating along with population growth, with most of the loss having occurred since 1950.

Lower slopes of the Himalayas. A girdle of lush mountain forest encircles the southern and eastern edges of the Himalayas, from Sikkim in northern India across Nepal and Bhutan to the western provinces of China. It comprises a complex mixture of tropical species of southern origin and temperate species from the north. A seemingly endless succession of deep valleys and knife-edge ridges breaks the fauna and flora into large numbers of local assemblages, containing for example about 9,000 plant species of which 39 percent are limited to the region as a whole. The original extent of the forests was roughly 340,000 square kilometers. Occurring in or near some of the most densely populated regions of the world, the forests are down by two thirds and disappearing quickly through unregulated logging and conversion to farmland.

Western Ghats of India. Along the seaward slopes of the Western Ghat mountains, extending the length of peninsular India, is a zone of tropical forest covering about 17,000 square kilometers. It is home to 4,000 known plant species, of which 40 percent are endemic. The pressure from the expanding local populations is intense, and clearing for timber and agriculture has been rapid. About a third of the cover is gone already, and the remainder is disappearing at a rate of 2–3 percent a year.

Sri Lanka. The wet forests of this island off the southern tip of India are relics of an ancient, largely vanished floristic province that once covered all of the Indian peninsula. The Sri Lankan remnant itself contains over a thousand plant species, of which half are endemic. With a population density of 260 persons per square kilometer and timber and agricultural land in heavy demand, the forest cover has been reduced to slightly less than 10 percent of its original area. Much of the primary forest growth is limited to a 56-square-kilometer tract within the Sinharaja forest near the southwestern corner of the island. This sector is also the most densely settled part of the island. To make matters worse, most of the local people depend on shifting cultivation and forest products for their livelihood.

Peninsular Malaysia. Most of the Malay peninsula was once covered by tropical forest. It contained at least 8,500 plant species, of which as many as a third were endemic. By the mid-1980s half the forest was gone. Almost all of the remaining lowland sector, the richest repository of diversity, had been degraded to some degree. About a half of the endemic tree species are now classified as endangered or extinct.

Northwestern Borneo. In earlier times Borneo was equated in lore with the perfect image of vast pristine jungle. That image has mostly faded. The forest is being stripped back swiftly, and many of the resident 11,000 plant and uncounted animal species are under siege. The northern third of the island, where biodiversity is deep and plant endemicity approaches 40 percent, has been extensively cleared by logging. In the state of Sarawak, part of Malaysia, the forest cover has been reduced by nearly a half, and most of the remainder has been consigned to timber companies.

The Philippines. This island nation is at the edge of a full-scale biodiversity collapse. Isolated from the Asian mainland but close enough to Indonesia to receive many plant and animal colonists, fragmented into 7,100 islands in a pattern that promotes species formation, the Philippines had evolved a very large fauna and flora with high levels of endemicity. In the past fifty years, two thirds of the forest has been cleared, including all but 8,000 square kilometers of the original lowland cover. From island to island intensive logging was pursued until it became uneconomical, to be followed in lockstep by full agriculture settlement. The demand for new land by a growing population endangers the remaining upland forest. Preserves are planned for 6,450 square kilometers, 2 percent of the nation's land surface. At best the ultimate losses will be heavy. As I write, the Philippine or monkey-eating eagle, majestic symbol of the nation's fauna, is down to 200 or fewer individuals.

New Caledonia. My favorite island: far enough off the east coast of Australia to spawn a unique fauna and flora; large enough to accommodate large numbers of animals and plants; and close enough to the Melanesian archipelagoes to the north to have received elements from that different biogeographical realm. For the naturalist, New Caledonia is a melting pot and a place of mystery. One of the finest days of my life was spent climbing Mt. Mou and then hiking along the summit ridge in mist-shrouded araucaria forest, where I found a pure native biota, not a single species of which I had ever seen in the wild before. The forests of New Caledonia hold 1,575 species of plants, of which an astonishing 89 percent are endemic. The New Caledonians, including the colonial French, have exploited the environment with abandon, logging, mining, and setting brushfires that push back the edges of the drier woodlands. Less than 1,500 square kilometers of undisturbed forests survive, covering 9 percent of the island. To see New Caledonia as it was, you must climb to mountain slopes too remote or steep for the loggers to clear.

Southwestern Australia. The extensive heathland west of the Nullarbor Plain evolved in a Mediterranean climate and state of isolation similar to those of the South African fynbos. It also resembles the fynbos in physical appearance and rivals it in diversity, harboring 3,630 plant species of which 78 percent are found nowhere else in the world. When I visited it in 1955, the environment was in nearly pristine condition. You could stand in the midst of the waist-high scrub in many places and see an unbroken horizon in all directions. In spring the flowers bloomed in splendid profusion. The cover has been reduced by half since then, mostly through agricultural conversion. It is being degraded further by mining operations, invasion of exotic weeds, and frequent wildfires. One quarter of its species are now classified as rare or threatened.

These are the eighteen hot spots, but the list is not closed. There are other candidates among forested regions, including the remnant rain forests of Mexico, Central America, West Indies, Liberia, Queensland, and Hawaii. To them can be added a large assemblage of entirely different habitats: the Great Lakes of East Africa and their counterpart in Siberia, Lake Baikal; virtually every river drainage system in the world near heavily populated regions, from the Tennessee to the Ganges and even some of the tributaries of the Amazon; the Baltic and Aral seas, the latter dying not just as an ecosystem but as a body of water; and a myriad of isolated tracts of species-rich tropical deciduous forests, grasslands, and deserts.

Then there are the coral reefs. These fortresses of biological diversity in the shallow tropical seas are giving way to a combination of natural and human assaults. The reefs have a permanent look but are highly dynamic in composition. Subject to the vagaries of weather and climate, they have always experienced local advances and retreats. Hurricanes periodically turn portions of the Caribbean reefs into rubble, but they grow back. El Niño events, the warming of water currents in the equatorial eastern Pacific, cause widespread mortality. The 1982–83 phenomenon, strongest recorded during the past two centuries, killed huge quantities of coral along the coasts of Costa Rica, Panama, Colombia, and Ecuador.

In normal circumstances, the reefs recover from natural destruction within a few decades. But now these natural stresses are being augmented by human activity, and the coral banks are being steadily degraded with less chance for regeneration. Reefs of twenty countries are affected around the world, from the Florida Keys and the West Indies to the Gulf of Panama and the Galápagos Islands, from Kenya and the Maldives east across a large swath of tropical Asia and south to Australia's Great Barrier Reef. In some places the reduction of reef area approaches 10 percent. Off Florida's Key Largo it is 30 percent, with most of the damage having occurred since 1970. In no particular order the principal causes are pollution (the oil spill during the Persian Gulf war being a disastrous example), accidental grounding of freighters, dredging, mining for coral rock, and harvesting of the more attractive species for decoration and amateur collections.

The decline of the reefs has been accompanied by coral bleaching. The loss of pigment is due to the failure of the zooxanthellae, the single-celled algae that live in the tissues of the coral animals and share a large fraction of the energy fixed by photosynthesis. The algae either die or lose much of the photosynthetic pigment they hold in their own cells. Like etiolated and dwarfed green plants germinated in the dark, the corals are as sickly as they look, and unless the process is reversed they die. Bleaching is a generalized stress reaction. It results variously from excessive heat or cold, chemical pollution, or dilution by fresh water, all of which are promoted by human activity.

During the 1980s, coral-reef bleaching occurred over a large part of the tropics. Rapid change most often proceeded in places where water temperatures rose conspicuously. It has been estimated that if the shallow tropical seas warm by as little as one or two degrees Celsius over the next century, many coral species would become extinct (three were lost from the eastern Pacific during the El Niño event of 1982–83 alone) and some reefs might disappear altogether. It is therefore possible that the bleaching of the last decade was the first step toward a catastrophe foretold by the rising levels of carbon dioxide in the atmosphere—possible, but still unproved. Coral bleaching in the 1980s occurred in some localities around the world but not in others. It was probably due to a variety of causes of which warming was only one. As they await further developments, marine biologists are inclined to agree that the greatest immediate peril for coral reefs comes from physical damage and pollution, not a worldwide warming trend.

But the long-term danger from climatic change looms in the decades ahead, for most ecosystems. If even the more modest projections of global warming prove correct, the world's fauna and flora will be trapped in a vise. On one side they are being swiftly reduced by deforestation and other forms of direct habitat destruction. On the other side they are threatened by the greenhouse effect. Whereas habitat loss on the land is most destructive to tropical biotas, climatic warming is expected to have a greater impact on the biotas of the cold-temperate and polar regions. A poleward shift of climate at the rate of 100 kilometers or more each century, equal to one meter or more a day, is considered at least a possibility. That rate of progression would soon leave wildlife preserves behind in a warmer regime, and many animal and plant species simply could not depart from the preserves and survive. The fossil record supports this forecast of limited dispersal. As the last continental ice sheet retreated from North America 9,000 years ago, spruce managed to spread at a rate of 200 kilometers a century, but the ranges of most other tree species spread at rates of only 10 to 40 kilometers. This history suggests that unless transplantings of entire ecosystems are undertaken, many thousands of native species are likely to be dislocated. How many will adapt to the changing climate, not having emigrated northward, and how many will become extinct? No one knows the answer.

It seems to follow that the organisms of the tundra and polar seas have no place to go even with a modest amount of global warming; the north and south poles are the end of the line. All the species of the high latitudes, reindeer moss to polar bears, risk extinction.

In another arena, large numbers of species around the world, at all latitudes, are restricted to low-lying coastal areas that will be flooded as the sea rises from the melting of polar ice. Various estimates have bracketed the rise somewhere between half a meter and two meters. In the United States, Florida will be the hardest hit region biologically. More than half of the rare animals and plants specialized for existence on the extreme coastal fringe live there. In the western Pacific many atolls, and even two small island nations, Kiribati and Tuvalu, would be largely covered by the sea.

Human demographic success has brought the world to this crisis of biodiversity. Human beings—mammals of the

50-kilogram weight class and members of a group, the primates, otherwise noted for scarcity—have become a hundred times more numerous than any other land animal of comparable size in the history of life. By every conceivable measure, humanity is ecologically abnormal. Our species appropriates between 20 and 40 percent of the solar energy captured in organic material by land plants. There is no way that we can draw upon the resources of the planet to such a degree without drastically reducing the state of most other species.

An awful symmetry of another kind binds the rise of humanity to the fall of biodiversity: the richest nations preside over the smallest and least interesting biotas, while the poorest nations, burdened by exploding populations and little scientific knowledge, are stewards of the largest. In 1950 the industrialized nations held a third of the world's population. The proportion fell to a quarter by 1985 and is expected to decline further to a sixth by 2025, when the total world population will have risen by 60 percent to 8 billion. One cannot help being struck with an irony, that if nineteenth-century technology had been born midst tropical rain forests instead of temperate-zone oaks and pines, there would be very little biodiversity left for us to save.

But what precisely is the magnitude of the crisis—how many species are disappearing? Biologists cannot tell in absolute terms because we do not know to the nearest order of magnitude how many species exist on earth in the first place. Probably fewer than 10 percent have even been given a scientific name. We cannot estimate the percentage of species going extinct each year around the world in most habitats, including coral reefs, deserts, and alpine meadows, because the requisite studies have not been made.

It is possible, though, to get a handle on the richest environment of all, the tropical rain forests, and to make a rough estimate of the extinction rates of species there. That much is possible because, thanks to the efforts of the Food and Agriculture Organization of the United Nations and a few pioneer researchers, such as Norman Myers, the rate of destruction of the rain forests has been ascertained. From the loss in forest area we can infer the rates at which species are being extinguished or doomed. And since the tropical forests contain more than half the species of plants and animals on earth, estimates pertaining to them allow us to make a rough qualitative assessment of the general severity of the biodiversity crisis.

Before attempting this projection, I am obliged to say something about the regenerative powers of rain forests. Despite their extraordinary richness, despite their reputation for exuberant growth ("the jungle quickly reclaimed the settlement as though nothing had existed there before"), these forests are among the most fragile of all habitats. Many of them grow on "wet deserts"—an unpromising soil base washed by heavy rains. Two thirds of the area of the forest surface worldwide consists of tropical red and yellow earths, which are typically acidic and poor in nutrients. High concentrations of iron and aluminum form insoluble compounds with phosphorus, decreasing the availability of that element to plants. Calcium and potassium are leached from the soil soon after their compounds are dissolved in the rain water. Only a tiny fraction of the nutrients filters deeper than 5 centimeters (2 inches) beneath the soil surface.

During the 150 million years of their existence, rain-forest trees have nevertheless evolved to grow thick and tall. At any given time, most of the carbon and a substantial fraction of the nutrients of the ecosystem are locked up in the tissue and dead wood of the vegetation. So the litter and humus on the ground are, in most cases, as thin as in any forests in the world. Here and there, patches of bare earth show through. At every turn there are signs of rapid decomposition by termites and fungi. When the forest is cut and burned, the ash and decomposing vegetation flush enough nutrients into the soil to support vigorous new herbaceous and shrubby growth for two or three years. Then the nutrients decline to levels too low to support healthy crops and forage. Farmers must add artificial fertilizer or move on to the next patch of rain forest, perpetuating the cycle of slash-and-burn.

The regeneration of rain forests is also limited by the fragility of the seeds of its trees. Those of most species germinate within a few days or weeks. They have little time to be carried by animals or water currents across the stripped land into sites favorable for growth. Most sprout and die in the hot, sterile soil of the clearings. The monitoring of logged sites indicates that regeneration of a mature forest may take centuries. Even though the forest at Angkor, for example, dates back to the abandonment of the Khmer capital in 1431, it is still structurally different from even older forests in the same region. The process of rain-forest regeneration is generally so slow, particularly after agricultural development, that few projections of its progress have been possible. In some areas, where the greatest damage is combined with low soil fertility and no native forest exists nearby to provide seeds, restoration might never occur without human intervention.

The ecology of rain forests stands in sharp contrast to that of northern temperate forests and grasslands. In North America and Eurasia, organic matter is not locked up so completely in the living vegetation. A large portion lies relatively fallow in the deep litter and humus of the soil. Seeds are more resistant to stress and able to lie dormant for long periods of time until the right conditions of temperature and humidity return. That is why it is possible to cut and burn large portions of the forest and grassland, graze cattle or grow crops for years on the land, and then see the vegetation grow back to nearly the original state a century after abandonment. Ohio, in a word, is not the Amazon. On a global scale, the north has been luckier than the south.

In 1979 tropical rain forests were down to about 56 percent of the prehistoric cover. Surveys made by satellite, by low-altitude overflights, and on the ground disclosed that the remainder, along with the much less extensive monsoon forests, was being removed at the rate of approximately 75,000 square kilometers, or 1 percent of the cover a year. Removal means that the forest is completely destroyed, with hardly a tree standing, or else degraded so severely that most of the trees die a short time later. The main causes of deforestation continue to be small-scale farming, especially slash-and-burn cultivation that leads to permanent agricultural settlement; only somewhat less important are commercial logging and cattle ranching.

During the 1980s the rate of deforestation was accelerated everywhere. It soared to tragic proportions in the Brazilian Amazon. There the people recognize three seasons, the dry, the wet, and the *queimadas*, or burnings. During the last brief period, armies of small farmers and peons employed by land barons set fires to clear the land of fallen trees and brush.

About 50,000 square kilometers in four states of the Amazon (Acre, Mato Grosso, Pará, Rondonia) were cleared and burned during four months, July through October, in 1987. A similar amount was destroyed the following year. Deforestation was driven by government-sponsored road building and settlement, sanctioned as government policy. It approached holocaust proportions, with effects spreading outward across larger parts of Brazil. "At night, roaring and red," observed the journalist Marlise Simons, "the forest looks to be at war." According to a report of the Institute for Space Research, "The dense smoke produced by the Amazonian burnings, at the height of the season, spread over millions of square kilometers, bringing health problems to the population, shutting down airports, hampering air traffic, causing various accidents on riverways and on roads, and polluting the earth's atmosphere in general." Global pollution did indeed occur. The Brazilian fires manufactured carbon dioxide containing more than 500 million tons of carbon, 44 million tons of carbon monoxide, over 6 million tons of particles, and a million tons of nitrogen oxides and other pollutants. Much of this material reached the upper atmosphere and traveled in a plume eastward across the Atlantic.

By 1989 the tropical rain forests of the world had been reduced to about 8 million square kilometers, or slightly less than half of the prehistoric cover. They were being destroyed at the rate of 142,000 square kilometers a year, or 1.8 percent of the standing cover, nearly double the 1979 amount. The loss is equal to the area of a football field every second. Put another way, in 1989 the surviving rain forests occupied an area about that of the contiguous forty-eight states of the United States, and they were being reduced by an amount equivalent to the size of Florida every year.

What impact does this destruction have on biodiversity in the tropical forests? In order to set a lower limit above which the species extinction rate can be reasonably placed, I will employ what we know about the relation between the area of habitats and the numbers of species living within them. Models of this kind are used routinely in science when direct measurements cannot be made. They yield first approximations that can be improved stepwise as better models are devised and more data added.

The first model is based on the widely observed area-species curve earlier given, $S = CA^Z$, where S is the number of species, A is the area of the place where the species live, and C and z are constants that vary from one group of organisms to another and from one place to another. For purposes of calculating the rate of species extinction, C can be ignored; z is what counts. In the great majority of cases the value of z falls between 0.15 and 0.35. The exact value depends on the kind of organism being considered and on the habitats in which the organisms are found. When species are able to disperse easily from one place to another, z is low. Birds have a low z value, land snails and orchids a high z value.

The higher the z value, the more the species numbers will eventually fall after the area is reduced. I say "eventually fall": whereas some doomed species may vanish quickly when a forest is trimmed back or a lake partly drained, other species decline slowly and linger a while before disappearing. In more precise language, when an area is reduced, the extinction rate rises and stays above the original background level until the species number has descended from a higher equilibrium to a lower equilibrium. The rule of thumb, to make the result immediately clear, is that when an area is reduced to one tenth of its original size, the number of species eventually drops to one half. This corresponds to a z value of 0.30 and is actually close to the number often encountered in nature.

In 1989 the area of the combined rain forests was declining by 1.8 percent each year, a rate that can be reasonably assumed to have continued into the early 1990s. At the typical z value, 0.30, each year's area reduction can be expected to reduce the number of species by 0.54 percent. Let us try to bracket the extinction rate for most kinds of organisms by estimating the minimal and maximal numbers possible. At the lowest likely z value, 0.15, this extinction rate would be 0.27 percent a year; at the highest likely z value, 0.35, the extinction rate would be 0.63 percent. *Very roughly, then, reduction in the area of tropical rain forest at the current rate can be expected to extinguish or doom to extinction about half a percent of the species in the forests each year.* More precisely, groups with a low z value will be affected the least, those with high z values the most. If most groups of organisms have low z values, the overall extinction rate will be closer to 0.27 percent; if most have high z values, the overall extinction rate will approach 0.63 percent. Not enough data exist to guess where the true overall value falls between these extremes.

If destruction of the rain forest continues at the present rate to the year 2022, half of the remaining rain forest will be gone. The total extinction of species this will cause will lie somewhere between 10 percent (based on a z value of 0.15) and 22 percent (based on a z value of 0.35). The "typical" intermediate z value of 0.30 would lead to a cumulative extinction of 19 percent over that span of time. Roughly, then, if deforestation continues for thirty more years at the present rate, one tenth to one quarter of the rain-forest species will disappear. If the rain forests are as rich in diversity as most biologists think, their reduction alone will eliminate 5 to 10 percent or more—probably considerably more—of all the species on earth in thirty years. When other species-rich but declining habitats are added, including heathland, dry tropical forests, lakes, rivers, and coral reefs, the toll mounts steeply.

The area-species relation accounts for a great deal of extinction, but not for all of it. We need a second model. As the last trees are cut, the last patch turned into a pasture or cornfield, the area-species curve plunges off the extrapolated line down to zero. So long as a small remnant of forest exists somewhere, say on a ridge in western Ecuador, a substantial number of species will hang on, most in tiny populations. Some may be doomed unless heroic efforts are made to culture and transplant them to new sites. But for the moment they hang on. When the last bit of forest or other natural habitat is removed and the area falls from 1 percent to zero, a great many species immediately perish. Such is the condition of legions of Centinelas around the world, the silent extinctions occurring as the last trees are felled. When Cebu in the Philippines was completely logged, nine of the ten bird species unique to the island were extinguished, and the tenth is in danger of joining them. We don't know how to assess global species loss from all these small-scale total extinctions. One thing is certain: because they do occur, the estimation of global rates based purely on the area-species curve must be on the low side. Consider the impact of removing the final few hundred square kilometers of natural reserves: in most cases, more than half of the original species would vanish immediately. If these were the refuges of species found nowhere else, the circumstance for so many rain-forest animals and plants, the loss in diversity would be immense.

The concept of a world peppered by miniature holocausts can be extended. Take the extreme imaginary case in which all species dwelling in the rain forests were local in distribution, limited to a few square kilometers, in the manner of the endemic plant species of Centinela. As the forest is cut back, the percentage loss in species approaches but never quite equals the percentage loss in forest area. In the next thirty years, the world would lose not only half of its forest cover but nearly half of the forest species. Fortunately, this assumption is excessive. Some species of animals and plants dwelling in rain forests have wide geographical distributions. So the rate of species extinction is less than the reduction in area.

It follows that the amount of species loss from halving of the rain-forest area will be greater than 10 percent and less than 50 percent. But note that this range of percentages is the loss expected from the area effect only, and it is still on the low side. A few species in the remnant patches will also be lost by rifle-shot extinction, the hunting out of rare animals and plants in the manner of Spix's macaw and the New Zealand mistletoe. Others will be erased by new diseases, alien weeds, and animals such as rats and feral pigs. That secondary loss will intensify as the patches grow smaller and more open to human intrusion.

No one has any idea of the combined magnitude of these additional destructive forces in all habitats. Only the minimal in the case of tropical rain forests—10 percent extinction with a halving of area—can be drawn with confidence. But because of the generally higher z values prevailing and the additional and still unmeasured extinction factors at work, the real figure might easily reach 20 percent by 2022 and rise as high as 50 percent or more thereafter. A 20 percent extinction in total global diversity, with all habitats incorporated, is a strong possibility if the present rate of environmental destruction continues.

How fast is diversity declining? The firmer numbers I have given are the estimates of species extinctions that will *eventually* occur as rain forests are cut back. How long is "eventually"? When a forest is reduced from, say, 100 square kilometers to 10, some immediate extinction is likely. Yet the new equilibrium described by the equation $S = CA^Z$ will not be reached all at once. Some species will linger on in dangerously reduced populations. Elementary mathematical models predict that the number of species in the 10-kilometer-square plot will decline at a steadily decelerating rate, swiftly at first, then slowing as the new and lower equilibrium is approached. The reasoning is simple: at first there are many species destined for extinction, which therefore vanish at a high overall rate; later only a few are endangered and the rate slows. In ideal form, with species going extinct independently of one another, this course of events is called *exponential decay*.

Employing the exponential-decay model, Jared Diamond and John Terborgh approached the problem in the following way. They took advantage of the fact that rising sea levels at the end of the Ice Age 10,000 years ago cut off small land masses that had once been connected to South America, New Guinea, and the main islands of Indonesia. When the sea flowed around them, these land masses became "land-bridge islands." The islands of Tobago, Margarita, Coiba, and Trinidad were originally part of the South and Central American mainland and shared the rich bird fauna of that continent. In a similar manner, Yapen, Aru, and Misool were connected to New Guinea and shared its fauna before becoming islands fringing its coast. Diamond and Terborgh studied birds, which are good for measuring extinction because they are conspicuous and easily identified. Both investigators arrived at the same conclusion: after submergence of the land bridges, the smaller the land-bridge island, the more rapid the loss. The extinctions were regular enough to justify use of the exponential-decay model. Extending the analysis in the American tropics, Terborgh turned to Barro Colorado Island, which was created by the formation of Gatun Lake during the construction of the Panama Canal. In this case the clock started ticking not 10,000 years ago but fifty years before the study. Applying the land-bridge decay equation to an island of this size, 17 square kilometers, Terborgh predicted an extinction of 17 bird species during the first 50 years. The actual number known to have vanished during that time is 13, or 12 percent of 108 breeding species originally present.

For a process as complex as the decline of biodiversity, the conformity of the Barro Colorado bird data to the same equation based on much larger islands and longer times, even if just within a factor of two, seemed too good to be true. But several other studies of new islands have produced similar results, which are at least consistent with the decay models, and depressing. The islands are patches of forest isolated in cleared agricultural land. When the islands are in the range of 1 to 25 square kilometers, the extinction rate of bird species during the first hundred years is 10 to 50 percent. Also, as predicted by theory, the extinction rate is highest in the smaller patches and rises steeply when the area drops below a square kilometer. Three patches of subtropical forest in Brazil surrounded by agricultural land for about a hundred years varied in area from 0.2 to 14 square kilometers; the resident bird species suffered a 14 to 62 percent extinction, in reverse order. On the other side of the world, a 0.9-square-kilometer forest patch, the Bogor Botanical Garden, was also isolated by clearing. In the first fifty years it lost 20 of its 62 breeding bird species. Still another example in a different environment: comparable rates of local extinction of bird species occurred in the wheat belt of southwestern Australia when 90 percent of the original eucalyptus woodland was removed and the remainder was broken into fragments.

There is no way to measure the absolute amount of biological diversity vanishing year by year in rain forests around the world, as opposed to percentage losses, even in groups as well known as the birds. Nevertheless, to give an idea of the dimension of the hemorrhaging, let me provide the most conservative estimate that can be reasonably based on our current knowledge of the extinction process. I will consider only species being lost by reduction in forest area, taking the lowest z value permissible (0.15). I will not include overharvesting or invasion by alien organisms. I will assume a number of species living in the rain forests, 10 million (on the low side), and I will further suppose that many of the species enjoy wide geographical ranges. Even with these cautious parameters, selected in a biased manner to draw a maximally optimistic conclusion, the number of species doomed each year is 27,000. Each day it is 74, and each hour 3.

If past species have lived on the order of a million years in the absence of human interference, a common figure for some groups documented in the fossil record, it follows that the normal "background" extinction rate is about one species per one million species a year. Human activity has increased extinction between 1,000 and 10,000 times over this level in the rain forest by reduction in area alone. Clearly we are in the midst of one of the great extinction spasms of geological history.

RIGHT: Ed Wilson (left) and his best friend Ellis MacLeod at the Hubbard School, Washington, D.C., in 1940, dressed for duty as student traffic-crossing guards.

BELOW: The three-year-old future zoologist, in 1932, with an early animal acquaintance.

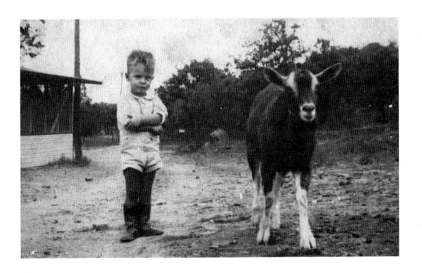

In 1937, when Ed was eight, he lived for several months in an Atlanta boardinghouse with his father, Edward O. Wilson, Sr.

The author, devoted to entomology at the age of thirteen, collects insects in a vacant lot next to his Mobile, Alabama, home in the summer of 1942. *(Photograph by Ellis MacLeod)*

As an Eagle Scout, in Brewton, Alabama, 1944.

ABOVE: The author with police escort near Gemenhen on April 11, 1955, during the long trek through the mountains of New Guinea's Huon Peninsula.

LEFT: With Methuselah, his pet giant Cuban anole, at the Atkins Gardens near Cienfuegos, in July 1953.

RIGHT: William L. Brown, Ed's mentor in his early studies of ants, at the Museum of Comparative Zoology in 1955.

BELOW: Renee and Ed at Holden Green, Harvard University's housing project for married students and young faculty, in November 1956.

Field work in the Florida Keys during the 1966–1968 experiment on island biogeography.

RIGHT: The author, perched on the crown of a red mangrove tree, identifies insects in and around an abandoned osprey nest.

BELOW: His student and collaborator Daniel Simberloff approaches a mangrove islet during the recolonization surveys.

ABOVE: Ed Wilson examines ants at Spring Hill, in Trinidad's North Range, in 1961.

LEFT: Renee Wilson assists with butterfly net at Gulf Shores, Alabama, in 1956.

Wilson receives the National Medal of Science from President Carter at a White House ceremony, November 22, 1977.

ABOVE: Robert Trivers, William Brown, and Hope Hara (Trivers' assistant) on a field trip in Concord, Massachusetts, on May 15, 1975 (the day *Sociobiology* was featured in a front-page story of the *New York Times*).

RIGHT: Ed with his daughter Catherine in 1969.

The author with his mother, Inez Huddleston, at the University of West Florida, Pensacola, in 1979.

ABOVE: Ed, Renee, and Cathy Wilson attending a parents' dinner at Regis College, in September 1982.

BELOW: Kathleen Horton, Ed, and Bert Hölldobler with the newly finished manuscript of *The Ants*, June 29, 1988.

ABOVE: Founders of sociobiology, at the first meeting of the Human Behavior and Evolution Society, Evanston, Illinois, in August 1989. Left to right: Irenäus Eibl-Eibesfeldt, George Williams, E. O. Wilson, Richard Dawkins, and William Hamilton.

BELOW: Wilson and Hölldobler in 1990, at the time of publication of *The Ants*.

Wilson receives the annual gold medal of the Worldwide Fund for Nature from Prince Philip in Sydney, Australia, on November 26, 1990.

ABOVE: The author's home laboratory in Lexington, Massachusetts, in 1944.

BELOW: Paul Ehrlich and Wilson receive the 1990 Crafoord Prize of the Royal Swedish Academy of Sciences from H. M. Carl XVI Gustaf, in Stockholm, September 26. The award was established to recognize fields of science, including ecology, not covered by the Nobel Prize.

ABOVE: Thomas Eisner and Ed Wilson in the Ant Room of the Museum of Comparative Zoology, spring 1991.

BELOW: The author accepting the International Prize for Biology, with Emperor Akihito and Empress Michiko in attendance, Tokyo, November 29, 1993.

E. O. Wilson in 1990, with "Dacie," the metallic sculpture of one of his favorite ant species, *Daceton armigerum*, of South America.

CHAPTER THIRTEEN

Unmined Riches

B IODIVERSITY is our most valuable but least appreciated resource. Its potential is brilliantly illustrated by the maize species *Zea diploperennis*, a wild relative of corn discovered in the 1970s by a Mexican college student in the west central state of Jalisco, south of Guadalajara. The new species is resistant to diseases and unique among living forms of maize in possessing perennial growth. Its genes, if transferred into domestic corn (*Zea mays*), could boost domestic production around the world by billions of dollars. The Jalisco maize was found just in time, however. Occupying no more than 10 hectares (25 acres) of mountain land, it was only a week away from extinction by machete and fire.

It can be safely assumed that a vast array of other beneficent but still unknown species exist. A rare beetle sitting on an orchid in a remote valley of the Andes might secrete a substance that cures pancreatic cancer. A grass down to twenty plants in Somalia could provide green cover and forage for the saline deserts of the world. No way exists to assess this treasure house of the wild except to grant that it is immense and that it faces an uncertain future.

For a start we need to reclassify environmental problems in a way that more accurately reflects reality. There are two major categories, and two only. One is alteration of the physical environment to a state uncongenial to life, the now familiar syndrome of toxic pollution, loss of the ozone layer, climatic warming by the greenhouse effect, and depletion of arable land and aquifers—all accelerated by the continued growth of human populations. These trends can be reversed if we have the will. The physical environment can be guided back and held rock-steady in a state close to optimum for human welfare.

The second category is the loss of biological diversity. Its root cause is the despoliation of the physical environment, but is otherwise radically different in quality. Although the loss cannot be redeemed, its rate can be slowed to the barely perceptible levels of prehistory. If what is left is a lesser biotic world than the one humanity inherited, at least an equilibrium will have been reattained in the birth and death of species. There is in addition a positive side not shared by the reversal of physical deterioration: merely the attempt to solve the biodiversity crisis offers great benefits never before enjoyed, for to save species is to study them closely, and to learn them well is to exploit their characteristics in novel ways.

A revolution in conservation thinking during the past twenty years, a New Environmentalism, has led to this perception of the practical value of wild species. Except in pockets of ignorance and malice, there is no longer an ideological war between conservationists and developers. Both share the perception that health and prosperity decline in a deteriorating environment. They also understand that useful products cannot be harvested from extinct species. If dwindling wildlands are mined for genetic material rather than destroyed for a few more boardfeet of lumber and acreage of farmland, their economic yield will be vastly greater over time. Salvaged species can help to revitalize timbering, agriculture, medicine, and other industries located elsewhere. The wildlands are like a magic well: the more that is drawn from them in knowledge and benefits, the more there will be to draw.

The old approach to the conservation of biodiversity was that of the bunker. Close off the richest wildlands as parks and reserves, post guards. Let the people work out their problems in the unreserved land, and they will come to appreciate the great heritage preserved inside, much as they value their cathedrals and national shrines. Parks and guards are necessary, without doubt. The approach has worked to some extent in the United States and Europe, but it cannot succeed to the desired degree in the developing countries. The reason is that the poorest people with the fastest-growing populations live next to the richest deposits of biological diversity. One Peruvian farmer clearing rain forest to feed his family, progressing from patch to patch as the soil is drained of nutrients, will cut more kinds of trees than are native to all of Europe. If there is no other way for him to make a living, the trees will fall.

Proponents of the New Environmentalism act on this reality. They recognize that only new ways of drawing income from land already cleared, or from intact wildlands themselves, will save biodiversity from the mill of human poverty. The race is on to develop methods, to draw more income from the wildlands without killing them, and so to give the invisible hand of free-market economics a green thumb.

This revolution has been accompanied by another, closely related change in thinking about biodiversity: the primary focus has moved from species to the ecosystems in which they live. Star speries such as pandas and redwoods are no less esteemed than before, but they are also viewed as protective umbrellas over their ecosystems. The ecosystems for their part, containing thousands of less-conspicuous species, are assigned equivalent value, enough to justify a powerful effort to conserve them, with or without the star species. When the last tiger on Bali was shot in 1937, the rest of the island's diversity lost none of its importance.

The humble and ignored are in fact often the real star species. An example of a species lifted from obscurity to fame by its biochemistry is the rosy periwinkle (*Catharanthus roseus*) of Madagascar. An inconspicuous plant with a pink five-petaled flower, it produces two alkaloids, vinblastine and vincristine, that cure most victims of two of the deadliest of cancers, Hodgkin's disease, mostly afflicting young adults, and acute lymphocytic leukemia, which used to be a virtual death sentence for children. The income from the manufacture and sale of these two substances exceeds $180 million a year. And that brings us back to the dilemma of the stewardship of the world's biological riches by the economically poor. Five other species of periwinkles occur on Madagascar. One, *Catharanthus coriaceus*, is approaching extinction as the last of its natural habitat, in the Betsileo region of the central highlands, is cleared for agriculture.

Few are aware of how much we already depend on wild organisms for medicine. Aspirin, the most widely used pharmaceutical in the world, was derived from salicylic acid discovered in meadowsweet (*Filipendula ulmaria*) and later combined with acetic acid to create acetylsalicylic acid, the more effective painkiller. In the United States a quarter of all prescriptions dispensed by pharmacies are substances extracted from plants. Another 13 percent come from microorganisms and 3 percent

more from animals, for a total of over 40 percent that are
organism-derived. Yet these materials are only a tiny fraction of
the multitude available. Fewer than 3 percent of the flowering
plants of the world, about 5,000 of the 220,000 species, have
been examined for alkaloids, and then in limited and haphazard fashion. The anticancer potency of the rosy periwinkle was
discovered by the merest chance, because the species happened
to be widely planted and under investigation for its reputed
effectiveness as an antidiuretic.

The scientific and folkloric record is strewn with additional
examples of plants and animals valued in folk medicine but
still unaddressed in biomedical research. The neem tree (*Aza-
dirachta indica*), a relative of mahogany, is a native of tropical
Asia virtually unknown in the developed world. The people
of India, according to a recent report of the U.S. National
Research Council, treasure the species. "For centuries, millions
have cleaned their teeth with neem twigs, smeared skin disor-
ders with neem-leaf juice, taken neem tea as a tonic, and placed
neem leaves in their beds, books, grain bins, cupboards, and
closets to keep away troublesome bugs. The tree has relieved so
many different pains, fevers, infections, and other complaints
that it has been called the 'village pharmacy.' To those millions in India neem has miraculous powers, and now scientists
around the world are beginning to think they may be right."

One should never dismiss the reports of such powers as superstition or legend. Organisms are superb chemists. In a sense
they are collectively better than all the world's chemists at syn-
thesizing organic molecules of practical use. Through millions
of generations each kind of plant, animal, and microorganism
has experimented with chemical substances to meet its special
needs. Each species has experienced astronomical numbers of
mutations and genetic recombinations affecting its biochemical machinery. The experimental products thus produced have
been tested by the unyielding forces of natural selection, one
generation at a time. The special class of chemicals in which the
species became a wizard is precisely determined by the niche
it occupies. The leech, which is a vampire annelid worm, must
keep the blood of its victims flowing once it has bitten through
the skin. From its saliva comes the anticoagulant called hirudin, which medical researchers have isolated and used to treat

The rosy periwinkle, a Madagascar plant that is the source of two alkaloid chemicals with powerful anticancer activity.

hemorrhoids, rheumatism, thrombosis, and contusions, conditions where clotting blood is sometimes painful or dangerous. Hirudin readily dissolves blood clots that threaten skin transplants. A second substance obtained from the saliva of the vampire bat of Central and South America is being developed to prevent heart attacks. It opens clogged arteries twice as fast as standard pharmaceutical remedies, while restricting its activity to the area of the clot. A third substance called kistrin has been isolated from the venom of the Malayan pit viper.

The discovery of such materials in wild species is but a fraction of the opportunities waiting. Once the active component is identified chemically, it can be synthesized in the laboratory, often at lower cost than by extraction from raw harvested tissue. In the next step, the natural chemical compound provides the prototype from which an entire class of new chemicals can be synthesized and tested. Some of these less-than-natural substances may prove even more efficient on human subjects than the prototype, or cure diseases never confronted with chemicals of their structural class in nature. Cocaine, for example, is used as a local anesthetic, but it has also served as a blueprint for the laboratory synthesis of a large number of specialized anesthetics that are more stable and less toxic and addictive than the natural product. Here is a brief list of pharmaceuticals derived from plants and fungi:

Drug	Plant source	Use
Atropine	Belladonna (*Atropa belladonna*)	Anticholinergic
Bromelain	Pineapple (*Ananas comosus*)	Controls tissue inflammation
Caffeine	Tea (*Camellia sinensis*)	Stimulant, central nervous system
Camphor	Camphor tree (*Cinnamomium camphora*)	Rubefacient
Cocaine	Coca (*Erythroxylon coca*)	Local anesthetic
Codeine	Opium poppy (*Papaver somniferum*)	Analgesic

Drug	Plant source	Use
Colchicine	Autumn crocus (*Colchicum autumnale*)	Anticancer agent
Digitoxin	Common foxglove (*Digitalis purpurea*)	Cardiac stimulant
Diosgenin	Wild yams (*Dioscorea* species)	Source of female contraceptive
L-Dopa	Velvet bean (*Mucuna deeringiana*)	Parkinson's disease suppressant
Erogonovine	Smut-of-rye or ergot (*Claviceps purpurea*)	Control of hemorrhaging and migraine headaches
Glaziovine	*Ocotea glaziovii*	Antidepressant
Gossypol	Cotton (*Gossypium* species)	Male contraceptive
Indicine N-oxide	*Heliotropium indicum*	Anticancer (leukemias)
Menthol	Mint (*Mentha* species)	Rubefacient
Monocrotaline	*Crotalaria sessiliflora*	Anticancer (topical)
Morphine	Opium poppy (*Papaver somniferum*)	Analgesic
Papain	Papaya (*Carica papaya*)	Dissolves excess protein and mucus
Penicillin	Penicillium fungi (esp. *Penicillium chrysogenum*)	General antibiotic
Pilocarpine	*Pilocarpus* species	Treats glaucoma and dry mouth
Quinine	Yellow cinchona (*Cinchana ledgeriana*)	Antimalarial
Reserpine	Indian snakeroot (*Rauvolfia serpentina*)	Reduces high blood pressure
Scopolamine	Thornapple (*Datura metel*)	Sedative

Drug	Plant source	Use
Strychnine	Nux vomica (*Strychnos nuxvomica*)	Stimulant, central nervous system
Taxol	Pacific yew (*Taxus brevifolia*)	Anticancer (esp. ovarian cancer)
Thymol	Common thyme (*Thymus vulgaris*)	Cures fungal infection
D-tubocu-rarine	*Chondrodendron* and *Strychnos* species	Active component of curare; surgical muscle relaxant
Vinblastine, vincristine	Rosy periwinkle (*Catharanthus roseus*)	Anticancer

The same bright prospect exists with wild plants that can serve as food. Very few of the species with potential economic importance actually reach world markets. Perhaps 30,000 species of plants have edible parts, and throughout history a total of 7,000 kinds have been grown or collected as food but, of the latter, 20 species provide 90 percent of the world's food and just three—wheat, maize, and rice—supply more than half. This thin cushion of diversity is biased toward cooler climates, and in most parts of the world it is sown in monocultures sensitive to disease and attacks from insects and nematode worms.

Fruits illustrate the pattern of underutilization. A dozen temperate-zone species—apples, peaches, pears, strawberries, and so on down the familiar roster—dominate the northern markets and are also used heavily in the tropics. In contrast, at least 3,000 other species are available in the tropics, and of these 200 are in actual use. Some, like cherimoyas, papayas, and mangos, have recently joined bananas as important export products, while carambolas, tamarindos, and coquitos are making a promising entry. But most consumers in the north have yet to savor lulos (the "golden fruit of the Andes"), mamones, rambutans, and the near-legendary durians and mangosteens, esteemed by aficionados as the premier fruits of the world. Here are other plant foods that could be developed:

Species	Location	Use
Arracacha (*Arracacia xanthorrhiza*)	Andes	Carrot-like tubers with delicate flavor
Amaranths (3 species of *Amaranthus*)	Tropical and Andean America	Grain and leafy vegetable; livestock feed; rapid growth, drought-resistant
Buffalo gourd (*Curcurbita foetidissima*)	Deserts of Mexico and southwestern United States	Edible tubers, source of edible oil; rapid growth in arid land unusable for conventional crops
Buriti palm (*Mauritia flexuosa*)	Amazon lowlands	"Tree of life" to Amerindians; vitamin-rich fruit; pith as source for bread; palm heart from shoots
Guanabana (*Annmona muricata*)	Tropical America	Fruit with delicious flavor; eaten raw or in soft drinks, yogurt, and ice cream
Lulo (*Solanum quitoense*)	Colombia, Ecuador	Fruit prized for soft drinks
Maca (*Lepidium meyenii*)	High Andes	Cold-resistant root vegetable resembling radish, with distinctive flavor; near extinction
Spirulina (*Spirulina platensis*)	Lake Tchad, Africa	Cyanobacterium producing vegetable supplement; very nutritious; rapid growth in saline waters
Tree tomato (*Cyphomandra betacea*)	South America	Elongated fruit with sweet taste
Ullucu (*Ullucus tuberosus*)	High Andes	Potato-like tubers, leafy part a nutritious vegetable; adapted to cold climates

Species	Location	Use
Uvilla (*Pouroma cecropiaefolia*)	Western Amazon	Fruit eaten raw or made into wine; fast-growing and robust
Wax gourd (*Benincasa hispida*)	Tropical Asia	Melon-like flesh used as vegetable, soup base, and dessert; rapid growth, several crops each year

Our narrow diets are not so much the result of choice as of accident. We still depend on the plant species discovered and cultivated by our neolithic ancestors in the several regions where agriculture began. These cradles of agriculture include the Mediterranean and Near East, Central Asia, the horn of Africa, the rice belt of tropical Asia, the uplands of Mexico and Central America, and middle to high elevations in the Andes. A few favored crops were spread around the world, woven into almost all existing cultures. Had the European settlers of North America not followed the practice, had they stayed resolutely with the cultivated crops native to the new land, citizens of the United States and Canada today would be living on sunflower seeds, Jerusalem artichokes, pecans, blueberries, cranberries, and muscadine grapes. Only these relatively minor foods originated on the continent north of Mexico.

Yet even when stretched to the limit of the neolithic crops, modern agriculture is only a sliver of what it could be. Waiting in the wings are tens of thousands of unused plant species, many demonstrably superior to those in favor. One potential star species that has emerged from among the thousands is the winged bean (*Psophocarpus tetragonolobus*) of New Guinea. It can be called the one-species supermarket. The entire plant is palatable, from spinach-like leaves to young pods usable as green beans, plus young seeds like peas and tubers that, boiled, fried, baked or roasted, are richer in protein than potatoes. The mature seeds resemble soybeans. They can be cooked as they are or ground into flour or liquified into a caffeine-free beverage that tastes like coffee. Moreover, the plant grows at a phenomenal pace, reaching a length of 4 meters in a few weeks. Finally, the winged bean is a legume; it harbors nitrogen-fixing nodules in its roots and has little need for fertilizer. Apart from its

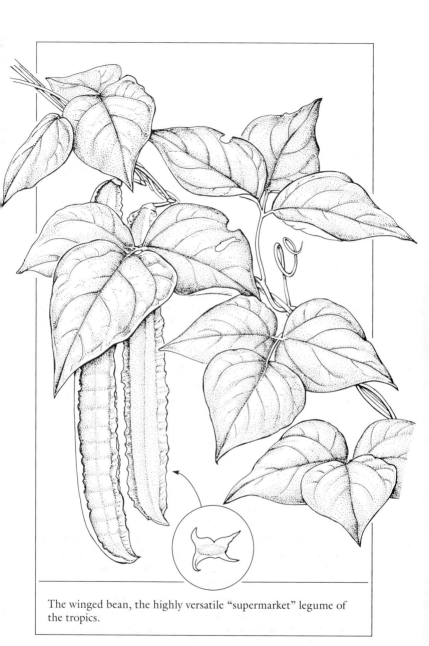

The winged bean, the highly versatile "supermarket" legume of the tropics.

potential as a crop, it can be used to raise soil fertility for other crops. With a small amount of genetic improvement through selective breeding, the winged bean could raise the standard of living of millions of people in the poorest tropical countries.

From the mostly unwritten archives of native peoples has come a wealth of information about wild and semicultivated crops. It is a remarkable fact that with a single exception, the macadamia nut of Australia, every one of the fruits and nuts used in western countries was grown first by indigenous peoples. The Incas were arguably the all-time champions in creating a reservoir of diverse crops. Without the benefit of wheels, money, iron, or written script, these Andean people evolved a sophisticated agriculture based on almost as many plant species as used by all the farmers of Europe and Asia combined. Their abounding crops, tilled on the cool upland slopes and plateaus, proved especially suited for temperate climates. From the Incas have come lima beans, peppers, potatoes, and tomatoes. But many other species and strains, including a hundred varieties of potatoes, are still confined to the Andes. The Spanish conquerors learned to use a few of the potatoes, but they missed many other representatives of a vast array of cultivated tuberous vegetables, including some that are more productive and savory than the favored crops. The names are likely to be unfamiliar: achira, ahipa, arracacha, maca, mashua, mauka, oca, ulloco, and yacon. One, maca, is on the verge of extinction, limited to 10 hectares in the highest plateau region of Peru and Bolivia. Its swollen roots, resembling brown radishes and rich in sugar and starch, have a sweet, tangy flavor and are considered a delicacy by the handful of people still privileged to consume them.

Another premier native crop of the Americas is amaranth. It is only now coming into the markets of the United States, mostly as a cereal supplement. Out of 60 wild species available to them, Indians from Mexico to South America cultivated three species widely during pre-Columbian times. Amaranth seeds yield a nutritious grain, and the young leaves when cooked become a palatable spinach-like green. The plants grow so well in cool, dry climates that they were favored as much as corn in Mexico at the time of the Conquest. Amaranth might have become one of the world's several leading crops after the

A cultivated amaranth, one of the principal food plants of the
Amerindians and a crop of outstanding worldwide potential.

Spanish conquest except for a bizarre historical circumstance, described by Jean Marx:

> Five hundred years ago, amaranth grain was a staple of the Aztec diet and an integral part of their religious rites. The Aztecs made idols out of a paste composed of ground, toasted amaranth seeds mixed with the blood of the human sacrifice victims. During the religious festivals, the idols were broken into pieces that were consumed by the faithful, a practice that the Spanish conquistadors considered a perverse parody of the Catholic Eucharist. When the Spanish subjugated the Aztecs in 1519, they banned the Aztec religion and with it the cultivation of amaranth.

Prejudice and inertia have always slowed the advance of agriculture. The mystery of untapped wild species is illustrated as in a parable by the case of natural sweeteners. A plant has been found in West Africa, the katemfe (*Thaumatococcus daniellii*) that produces proteins 1,600 times sweeter than sucrose. A second West African plant, the serendipity berry (*Dioscoreophyllum cumminsii*), yields a substance 3,000 times sweeter. The parable is the following: where among wild species do such progressions end? Human ingenuity has never been stretched to find the answer in this or any other domain of practical application. Consider a second, equally instructive case. The Amazonian babassu palm (*Orbignya phalerata*), even though still harvested in the wild and semiwild states, gives the world's highest known yield of vegetable oil. A stand of 500 trees produces about 125 barrels a year from huge, 100-kilogram masses of fruit. Different parts of the tree are used by local peoples to make feedcakes for livestock, pulpwood, thatching materials for roots and baskets, and finally charcoal. The babassu has not been bred to bring it to fuller commercial use, nor has it been planted extensively away from the fertile upland soils and alluvial bottomlands on which it originally grew as a wild plant.

Another frontier awaiting capital investment is saline agriculture, using salt-tolerant plants to cultivate land not previously arable. In an experimental farm in Mexico, farmers have begun to use seawater to irrigate salicornia, a native of saltflats. The small, succulent plants produce an oil resembling that of safflower. They yield two tons of oil seeds per hectare annually, leaving a residual straw that can be used to feed livestock. In

Pakistan, kallar grass is grown in soil saturated with saltwater, then harvested as animal fodder. In the forbidding Atacama Desert of northern Chile, where seven years may pass without rain, the tamarugo tree sends roots through a meter of salt to tap brackish water deep within the desert soil. This extraordinary plant can create open woodland and ground vegetation in otherwise sterile wastelands. Sheep reared in tamarugo forests grow about as rapidly as those reared in high-quality pastures elsewhere in the world.

The history of animal husbandry has been just as haphazard as that of agriculture. Like crop plants, the animals of the barnyard and range are mostly limited to those first domesticated by our neolithic ancestors 10,000 years ago in the temperate zones of Europe and Asia. We have been stuck with a narrow range of ungulate mammals, horses, cattle, donkeys, camels, pigs, and goats, ill suited for most habitats of the world and often spectacularly destructive of the natural environment. In many cases these species are locally inferior in yield to wild species that humanity has left unattended.

A good example of wildlife superiority is provided by Amazon river turtles of the genus *Podocnemis*. The seven known species are highly regarded as a protein source by local people. The meat is of excellent quality and the base of a pleasing native cuisine. As the river banks have been more thickly settled, the turtles have been overhunted and several species are now endangered. But they are also easy to cultivate. Each female lays a clutch of up to 150 eggs, and the young grow rapidly. One species, the giant *Podocnemis expansa*, reaches a length of nearly a meter and a weight of 50 kilograms. It can be confined in cement tanks and natural ponds along the broad flood plains while being fed on aquatic vegetation and fruit, all at minimal cost. Under these conditions the turtle produces each year about 25,000 kilograms of meat per hectare (22,000 pounds per acre), more than 400 times the yield of cattle raised in nearby pasture cut from surrounding forests. Since floodplains compose 2 percent of the land surface of the Amazon region, the commercial potential of the species is enormous. It carries far less cost to the environment than the cattle and other exotic animals now being thrust upon the land with disastrous result.

Similar advantages are offered by the green iguana, the "chicken of the trees." A large lizard with light and tasty flesh, it has been favored as a delicacy for centuries by farmers in the humid regions of Central and South America. To be sure the iguana is a lizard, and some may flinch at the idea of eating a reptile. But it is all a matter of cultural perspective. In a phylogenetic sense, chickens and other birds are just hot reptiles with wings, and in any case our cuisines are already filled with creatures far more visually horrifying, from lobsters to thresher sharks.

But I digress. Iguanas have been rendered scarce over most of their range by overhunting and each animal now fetches $25 on the Panamanian black market. Even though they are protected by law in several Latin American countries, the big reptiles are

The giant Amazon river turtle, an easily cultured species that far outstrips cattle as a potential source of meat on the river floodplains.

declining because of the accelerating destruction of their forest habitat. If farmers were to leave more forest standing, there would be more iguanas for the stew pot. "But if you're a farmer with a family to feed, even a family with a taste for iguana meat," Chris Wille and Diane Jukofsky have observed, "you're likely more interested in chopping or burning down the trees on your land to make way for cattle or crops—something you can sell. After all, iguanas make a delectable dinner, but they won't keep the kids in clothes."

Although the result is a downward spiral both for the forests and the farmers, it can be reversed. As Dagmar Werner has shown in a series of impressive field experiments, the iguanas can be made to yield up to ten times the amount of meat as cattle on the same land if managed carefully, while leaving a large part of the forest intact. The trick is to cultivate a breeding stock, incubate the eggs, then protect the hatchlings during their earliest and most vulnerable growth before releasing them into the forest. The iguanas are left to feed on leaves in the tree canopies, perhaps helped along by kitchen scraps, until they are large enough to be harvested. It will also be necessary to cultivate a broader export market, while easing laws protecting the iguana in areas where cultivation is practiced. Here in summary are a few wild animals that could be raised commercially for food products:

Species	Distribution	Uses
Babirusa (*Babyrousa babyrussa*)	Indonesia: Moluccas and Sulawesi	A deep-forest pig, thrives on vegetation high in cellulose and hence less dependent on grain
Capybara (*Hydrochoeris hydrochoeris*)	South America	World's largest rodent; meat esteemed; easily ranched in open habitats near water
Chachalacas (*Ortalis*, many species)	South and Central America	Birds, potentially tropical chickens, thrive in dense populations, adaptable to human habitations, fast-growing

Species	Distribution	Uses
Gaur (*Bos gaurus*)	India to Malay peninsula	Threatened relative of domestic cattle; alternative cattle species
Green iguana (*Iguana iguana*)	American tropics	Chicken of the trees: traditional native food for 7,000 years; rapid growth; low rearing costs
Guanaco (*Lama guanicoe*)	Andes to Patagonia	Threatened species related to llama; excellent source of meat, fur, and hides; can be profitably ranched
Olive ridley sea turtles (*Lepidochelys olivacea*)	Beaches of India and Pacific coast of Mexico and Central America	Turtles emerge from sea to lay eggs; egg harvesting productive when beaches protected
Paca (*Cuniculus paca*)	American tropics	Large rodents, flesh esteemed; usually caught in wild but can be maintained in small packs in forested areas
Pigmy hog (*Sus salvanus*)	Northeastern India	One of most endangered mammal species on earth; potential source of new genes for domestic pig
Sand grouse (*Pterocles*, many species)	Deserts of Africa and Asia	Pigeon-like birds adapted to harshest deserts; domestication a possibility
Vicuna (*Lama vicugna*)	Central Andes	Threatened species related to llama; valuable source of meat, fur, and hides; can be profitably ranched

The goal of all such innovations is to increase productivity and wealth with a minimal disturbance of natural ecosystems and loss of biological diversity. Chosen and managed wisely, the exotic becomes the familiar and favored—and remains environmentally benign.

To river turtles and iguanas in the category of potential elites add the babirusa, a pig-like animal inhabiting the rain forests of Sulawesi, the Sula and Togian Islands, and Buru in eastern Indonesia. The babirusa is a bizarre creature of the kind normally seen only in zoos—slender, gray-skinned and mostly hairless, with males whose upper canines grow upward as tusks, piercing the flesh of the snout, and curving back toward the forehead without ever entering the mouth. The closest known relatives of the babirusa, all extinct, once roamed the forests of Europe. An adult is larger than most men, weighing up to 100 kilograms. Despite its resemblance to a Hindu demon, the species has been tamed by Indonesian forest peoples and serves as an important source of meat. Its most promising commercial feature, however, is its status as a possible ruminant pig. Its stomach is enlarged and chambered like that of sheep, a unique trait that apparently enables it to feed extensively on leaves and other vegetation heavy in cellulose. With luck the babirusa might enter the ranks of domestic pigs elsewhere in the world, sustained on an inexpensive and universally available fodder.

The goals of economic growth and conservation might both be served by cultivating species within their natural ecosystems, in the manner of river turtles, iguanas, and babirusas, or by the transfer of hardy species to marginal lands possessing few endemic species. The greatest potential expansion in production is by aquaculture, the rearing of fish, oysters and other mollusks, and other marine and freshwater organisms in artificial ponds or, in the case of mollusks, on the surfaces of support racks set up in estuaries. More than 90 percent of the fish consumed by human beings worldwide is obtained by the hunting of wild species in fully natural environments. This primitive industry prevails despite the fact that sophisticated aquaculture techniques are available, and fish in particular have been reared in ponds and other enclosed structures for 4,000 years. If pressed aggressively, the production of animal protein by aquaculture could easily be increased many times over within one or two decades. "There are two reasons for this vast potential," Norman Myers has written:

First, water dwelling creatures enjoy a distinct advantage over their terrestrial relatives in that their body density is almost the same as that of the water they inhabit, so they do not have to direct energy into supporting their body weight; this means, in turn, that they can allocate more food energy to the business of growing than is the case for land animals. Second, fishes, as cold-blooded creatures, do not consume large amounts of energy to keep themselves warm. Carp, for instance, can convert one unit of assimilated food into flesh one and a half times as quickly as can pigs or chickens, and twice as rapidly as cattle or sheep. The tiny shrimplike crustaceans called *Daphnia* can, when raised in a nutrient-nourished environment, generate almost 20 metric tons of flesh per hectare in just under five weeks, which is ten times the production rate for soybeans—and at one-tenth the cost per unit of protein produced.

Present-day aquaculture resembles the rearing of conventional crops and livestock in utilizing only a small fraction of available diversity. It depends heavily on those species encountered happenstance for the first time by the cultures that invented the practice. About 300 kinds of fish—finfish, to be exact, as opposed to shellfish—are cultured for food somewhere in the world. But 85 percent of the yield comes from only several carp species, while tilapias contribute a large part of the remainder. There are 18,000 more species known to science, and undoubtedly thousands of others still unknown. In the end only a small minority will prove to be commercially valuable, but even if that figure is only 10 percent, it will vastly increase the utilized diversity.

It is within the power of industry to increase productivity while protecting biological diversity, and to proceed in a way that one leads to the other. The forests of the world, for example, are under pressure from a rising demand for paper pulp. Stands of a thousand species in Borneo and ancient forests in North America are converted to pulp at a rising rate, expected to reach 400 million metric tons of annual manufacture by the end of the century. There are better ways to make newspapers and carton boxes than the conversion of wildlands. Kenaf (*Hibiscus cannabinus*), an East African plant related to cotton and okra, is superior to traditional woody plants in almost every respect. Its stands, resembling those of bamboo but bearing white, hibiscus-like flowers, grow to a mature height of 5 meters in just four to five months. In the southern United States, kenaf yields three to five times more pulp than trees, and only minor chemical treatment is needed to whiten the fibers. The young stands can be cut and gathered with a machine similar to a sugarcane harvester.

Pulp and fiber can also be mass-produced from saplings raised in a remarkable form called "wood grass." In a procedure still in the experimental phase, trees are grown in dense stands and mowed like grass while still young and flexible. Their vegetation is then converted into pulp, fuel, or livestock feed. When the right tree species are chosen, the stands are fast-growing, sprout from the root stocks like grass and do not require reseeding. And when predominantly legumes, they nitrify the soil, reducing the need for fertilizer.

Kenaf and wood-grass plantations are among the latest discoveries in a saga that began with the origins of agriculture. The crucial innovations made here and there five to ten thousand years ago were the cultivation of certain food species already harvested in the wild, followed by selection of the best varieties within the species. Hunter-gatherers must have understood for millennia that plants produce seeds, which then grow into plants. It was but a small step for them to plant the seeds in convenient places. When they also learned to cultivate the plants on prepared ground, selecting the best ones to spawn the next generation, they became farmers, and agriculture was born. A chain of events had been set in motion that took the plants and their descendants, in Erich Hoyt's image, on a strange and wondrous ride into modern history.

Today the old geographic localities of neolithic agriculture support not only the domestic varieties on agricultural land but also those original wild species still surviving in the dwindling natural habitats nearby. The combination of domestic and native strains makes these sites the headquarters of genetic diversity. They are called Vavilov centers, in recognition of the pioneering work of the Russian botanist Nikolai Vavilov, who in the 1920s and 1930s traveled through Afghanistan, Ethiopia, Mexico, Central America, and the far reaches of the Soviet Union to collect plants for agricultural use. Our geographical knowledge of diversity centers has been augmented by other botanists in recent decades. There is nothing mysterious about Vavilov centers. They are for the most part simply the places where agriculture began, and hence fall within the ranges of the plant species chosen by the first farmers. In southwest Asia, for example, lived the grasses that became barley and wheat. In Mexico grew wild corn (maize), squashes, and beans, and in Peru the ancestors of potatoes.

With cultivation comes evolution by artificial selection of the succulent foliage, large tubers, and tender fruits favored by human beings. Specialization of this kind means a reduced ability to persist unattended in the original habitats. No domestic strain I know of has reentered the natural habitat of its ancestors and competed successfully there. Domestic strains are also more vulnerable to diseases and plant-eating insects and other pests. Artificial selection has always been a tradeoff between the genetic creation of traits desired by human beings and an unintended but inevitable genetic weakness in the face of natural enemies.

With the Green Revolution of agrotechnology, the tradeoff became more pronounced. Highly productive strains have been bred and mass-cultivated during the past forty years, and domestic species have become even more specialized and homogeneous than before. In India farmers originally grew as many as 30,000 varieties of rice. That diversity is being whittled down so fast that by the year 2005 three quarters of the ricefields may contain no more than ten varieties.

In a world created by natural selection, homogeneity means vulnerability. Purity of stock lowers resistance to disease, while monocultures spread contiguously over vast areas are an invitation to enemies made newly formidable. The consolidated rice paddies of Asia, rendered even more vulnerable by year-round cropping, have been opened to the swift spread of maladies that can threaten the livelihood of millions. During the 1970s the grassy-stunt virus devastated fields from India to Indonesia. Fortunately, enough wild species and varieties of rice existed to handle the problem. The International Rice Institute assayed 6,273 kinds of rice for resistance to grassy stunt. Of this array only one, the relatively feeble Indian species *Oryza nivara* (which has been known to science only since 1966), had genes with the desired qualities. It was bred with the prevailing cultivated type to create a resistant hybrid, which is now grown across 110,000 square kilometers of ricefields in Asia.

Most of the coffee plantations of Brazil are descended from a single tree that originated in East Africa. Plantings were first made in the West Indies, and some of the progeny reared there were transferred to South America. In 1970 coffee rust, a disease that had already destroyed most of the crops of Sri Lanka, appeared in Brazil and spread to Central America, threatening the economies of several countries. It happens that wild varieties of coffee still grow in the Kaffa region of southwestern Ethiopia, the presumed ancestral land of domestic coffee. Genes resistant to coffee rust were found there and bred into the Brazilian and Central American crops just in time to save the industry.

Crop species owe roughly 50 percent of their increased productivity to selective breeding and hybridization, in other words to agricultural programs that deliberately reshuffle genes among species and varieties. The modern tomato (*Lycopersicon esculentum*) is benefited by genes of many related species and races. At least nine stocks, all native to Central and South America, have either contributed valuable traits to this crop or at least possess genes that can make such a contribution:

Lycopersicon cheesmanii. Endemic of the Galápagos Islands; can be irrigated with seawater.

Lycopersicon chilense. Drought resistance.

Lycopersicon chmielewskii. Color intensity, increased sugar content.

Lycopersicon esculentum cerasiforme. Tolerance of high temperature and humidity.

Lycopersicon hirsutum. High-altitude form, resistant to many diseases and pests.

Lycopersicon parviflorum. Color intensity, increased soluble solids.

Lycopersicon pennellii. Drought resistance, increased vitamin C and sugar content.

Lycopersicon peruvianum. Pest resistance, rich source of vitamin C.

Lycopersicon pimpinellifolium. Wide disease resistance, lower acidity, higher vitamin content.

The creation of today's domestic tomato was a skilled feat of plant breeding, but one that requires many generations to accomplish. A wild species or race bred into the domestic stock also carries with it baggage of less desirable genes that reduce yield and quality. Breeders must delete these traits through repeated backcrossing, mating the hybrids back to the domestic strains, in a way that preserves only the desirable genes of both domestic and wild forms in the breeding stock. Finally, conventional hybridization can be accomplished solely among species and strains similar enough to be bred together, as in the case of the multiple parents of *Lycopersicon esculentum.*

Now, however, traditional selective breeding can be short-circuited. New methods of genetic engineering have made it possible to transfer genes directly, excising them from the chromosomes of one species and placing them into the chromosomes of another species without hybridization of the entire genomes. In other words, sex has been bypassed. Furthermore, the exchange can be accomplished among species of plants and animals so different as to make ordinary hybridization impossible. Thomas Eisner has described the possibilities in striking imagery:

> A biological species, nowadays, must be regarded as more than a unique conglomerate of genes. As a consequence of recent advances in genetic engineering, it must be viewed also as a depository of genes that are potentially transferable. A species is not merely a hard-bound volume of the library of nature. It is also a loose-leaf book, whose individual pages, the genes, might be available for selective transfer and modification of other species.

A species in the tomato genus treated as a loose-leaf notebook might share genes with species outside the genus, say plants in the larger nightshade family, or even beyond, with radically different flowering plants—donating or acquiring disease resistance, larger fruit mass, cold hardiness, the ability to grow all year, and so on through the full gamut of desired biological qualities. The possibilities are there, and they raise the potential importance for humanity of every wild species and race.

I do not mean to suggest that every ecosystem now be viewed as a factory of useful products. Wilderness has virtue unto itself and needs no extraneous justification. But every ecosystem, including those in wilderness reserves, can be the source of species to be cultivated elsewhere for practical purposes or of genes for transfer to domesticated species.

The supreme test of the utilitarian principle will be the rain forests. It is now highly profitable in most tropical countries simply to log all the trees from one tract and move on to the next. Land is cheap enough to turn a profit by the destruction of primal forests, allowing more land to be purchased and the cycle continued until the last of the trees are down. The alternative is to use the rain forests as extractive reserves, for the harvesting of "minor" products such as edible fruits, oils, latex, fibers, and medicines.

The key question from an economic viewpoint is whether the income from minor products is high enough to justify preserving rain forests as extractive reserves. The answer turns out to be yes, at least in some localities, even with the limited knowledge at hand. In 1989 Charles Peters, Alwyn Gentry, and Robert Mendelsohn demonstrated that not only are minor products in the Peruvian Amazon potentially more profitable in the long run, but considerably more so than conventional one-time logging. Among 275 kinds of trees they identified in a 1-hectare plot near the town of Mishana, 72 (26 percent) yielded fruits, vegetables, wild chocolate, and latex that could be sold in Peruvian markets. The annual net yield, after deducting costs for collecting and transport, was estimated to be $422. The Mishana plot contains enough timber to generate a net revenue of $1,000 on delivery to the sawmill if cut once, the usual practice. Within a short time, then, sustained harvesting of fruits and latex can be made more profitable than clear-cutting, and the forest is left intact. Even if trees of high profitability were removed at intervals that allowed maximal yield of their timber, the long-term income would still be less than one tenth that from fruit and latex harvesting. These are the unmined riches of the Mishana rain forest (1 hectare):

Product	Number of plants	Annual production per plant	Value (U.S. $)
Palm fruits			
Aguaje	8	195.8 lbs.	$177.60
Aguajillo	25	66.15 lbs.	75.00
Sinamillo	1	3,000 fruits	22.50
Ungurahui	36	80.48 lbs.	115.92
Other edible fruits			
Charichuelo	2	100 fruits	1.50
Lachehuaya	2	1,060 fruits	70.67
Naranjo podrido	3	150 fruits	112.50
Masaranduba	1	800 fruits	3.75
Tamamuri	3	500 fruits	11.25
Other edible products			
Sacha cacao (wild chocolate)	3	50 fruits	22.50
Shimbillo (legume)	9	200 fruits	27.00
Rubber-tree products			
Shiringa (latex)	24	4.41 lbs.	57.60
Total	117		$697.79
Cost of harvesting and transport			$276.00
Net value			$421.79

The computed yield of extractive products is actually the most conservative for the Mishana plot, since it was based exclusively on the inventory of commercially tested materials and a still poorly developed market. Little effort has been made to perform bioeconomic assays of whole ecosystems, identifying the species that can produce food and pharmaceuticals, as well as serve as agents of pest control and as restorers and enrichers of the soil. Almost all of the potent species are destroyed when the forest is clear-cut for timber and agricultural land. The old ways of using the land, chained to markets handed down by the traditions of conquistadors and the vagaries of foreign markets, extract only a small portion of the wealth while discarding the rest. The same is true, to only slightly lesser degree, in the valuation and full use of forests in the temperate zones. Economists are struggling to fit wilderness and living species into their equations. They have created a new field, ecological economics, devoted to the sustainability of the environment and its long-term productivity. I believe they will succeed in an accurate assessment of the fraction of biodiversity subject to inventory and cost-benefit analyses, as already accomplished for the consumable plant products of the Mishana tract. They will also be able to add the revenues from "ecotourism." More and more people from developed countries are willing to pay to experience, however briefly, the prehuman earth. In 1990 tourism had risen to become the second most important source of outside income in Costa Rica, ahead of bananas and closing fast on coffee. Rain forests used for the purpose have become many times more profitable per hectare than land cleared for pastures and fields. Ecotourism is the third most important source of income in Rwanda, rising fast behind coffee and tea, largely because that tiny, overpopulated East African country is home to the mountain gorilla. As Rwanda protects the gorilla, the gorilla will help to save Rwanda.

Beyond commodity value, the economists fall short. In dimensions other than produce and tourist dollars, their yardsticks are elastic and poorly calibrated. They have no sure way of valuating the ecosystem services that species provide singly and in combination—the soil we plow, the air we breathe, the water we draw. Natural ecosystems regulate the atmospheric gases, which in turn alter temperature, wind patterns, and precipitation. The vast Amazonian rain forests create half their own rainfall. As the forests are cut, the water supply is diminished to corresponding degree. Mathematical models of the cycle of precipitation and evaporation suggest that a critical threshold of green cover exists below which the forests will no longer be able to last, converting much of the great river basin irreversibly into scrubby grassland. The pall might then travel southward to desiccate parts of Brazil's rich agricultural heartland.

When forests are leveled, the elements that composed the wood and tissue are partially converted into greenhouse gases. Then when forests are regrown, an equivalent amount of the elements are recalled into solid matter. The net loss of tropical forest cover worldwide during 1850–1980 contributed between 90 and 120 billion metric tons of carbon dioxide to the earth's atmosphere, not far below the 165 billion metric tons emanating from the burning of coal, oil, and gas. These two processes together have raised the concentration of carbon dioxide in the global atmosphere by more than 25 percent, setting the stage for global warming and a rise in sea level. The second most important greenhouse gas, methane, has doubled at about the same time; and 10 to 15 percent of the increase is thought to be due to tropical deforestation. If 4 million square kilometers of the tropical regions were replanted in forest, in other words an area half the size of Brazil, all of the current buildup of atmospheric carbon dioxide from human agents would be canceled. In addition, the rise in methane and other greenhouse gases would be slowed.

The very soils of the world are created by organisms. Plant roots shatter rocks to form much of the grit and pebbles of the basic substrate. But soils are much more than fragmented rock. They are complex ecosystems with vast arrays of plants, tiny animals, fungi, and microorganisms assembled in delicate balance, circulating nutrients in the form of solutions and tiny particles. A healthy soil literally breathes and moves. Its microscopic equilibrium sustains natural ecosystems and croplands alike.

The mere phrase "ecosystems services" has a mundane ring, rather like waste disposal or water-quality control. But if only a small percentage of the journeyman organisms filling these roles were to disappear, human life would be diminished and strikingly less pleasant. It is a failing of our species that we ignore and even despise the creatures whose lives sustain our own.

What then is biodiversity worth? The traditional econometric approach, weighing market price and tourist dollars, will always underestimate the true value of wild species. None has been totally assayed for all of the commercial profit, scientific knowledge, and aesthetic pleasure it can yield. Furthermore, none exists in the wild all by itself. Every species is part of an ecosystem, an expert specialist of its kind, tested relentlessly as it spreads its influence through the food web. To remove it is to entrain changes in other species, raising the populations of some, reducing or even extinguishing others, risking a downward spiral of the larger assemblage.

Downward by how much? The relation between biodiversity and stability is a gray area of science. From a few key studies of forests we know that diversity enlarges the capacity of the ecosystem to retain and conserve nutrients. With multiple plant species, the leaf area is more evenly and dependably distributed. Then the greater the number of plant species, the broader the array of specialized leaves and roots, and the more nutrients the vegetation as a whole can seize from every nook and cranny at every hour through all seasons. The extreme reach of biodiversity anywhere may be that attained by the orchids and other epiphytes of tropical forests, which harvest soil particles directly from mist and airborne dust otherwise destined to blow away. In short, an ecosystem kept productive by multiple species is an ecosystem less likely to fail.

If species composing a particular ecosystem begin to go extinct, at what point will the whole machine sputter and destabilize? We cannot be sure because the requisite natural history of most kinds of organisms does not exist, and experiments on ecosystem failure have been generally lacking. Yet think of how such an experiment *might* unfold. If we were to dismantle an ecosystem gradually, removing one species after another, the exact consequences at each step would be impossible to predict, but one general result seems certain: at some point the ecosystem would suffer a collapse. Most communities of organisms are held together by redundancies in the system. In many cases two or more ecologically similar species live in the same area, and any one can fill the niches of others extinguished, more or less. But inevitably the resiliency would be sapped, efficiency of the food webs would drop, nutrient flow would decline, and eventually one of the elements deleted would prove to be a keystone species. Its extinction would bring down other species with it, possibly so extensively as to alter the physical structure of the habitat itself. Because ecology is still a primitive science, no one is sure of the identity of most keystone species. We are accustomed to thinking of the organisms in this vital category as being large in size—sea otters, elephants, Douglas firs, coral heads—but they might as easily include any of the tiny invertebrates, algae, and microorganisms that teem in the substratum and that also possess most of its protoplasm and move the mass of nutrients.

Economists speak of the "option value" of a species whose worth is still unmeasured, and no measure in all of economics is more intriguing or more elusive. Its greatest difficulty is that it applies equally to commodity, amenity, and morality, the three standard domains of valuation. "As time passes," Bryan Norton has observed,

> we gain knowledge in all these areas, and new knowledge may lead to new commodity uses for a species or to a new level of aesthetic appreciation, or our moral values may change and some species will, in the future, prove to have moral value that we cannot now recognize. If placing a dollar figure on these option values seems a daunting task, the situation is actually far worse than it first seems. Calculations of option value can only be begun after we identify a species, guess what uses that species might have, place some dollar value on those uses, and estimate the likelihood of discoveries occurring at any future date.

The attempt to valuate species has led to two competing guidelines of conservation. The first is cost-benefit (CB) analysis, which singles out each threatened species in turn, weighs visible and possible future benefits against the costs of keeping it alive, and decides whether to invest enough land and time to preserve it. The second guideline is the safe minimum standard (SMS), which treats each as an irreplaceable resource for humanity, to be preserved for posterity unless the costs are unbearably high.

Surely prudence and a decent concern for posterity demand the safe minimum standard. Cost-benefit studies consistently undervalue the net benefits conferrable by species since it is much easier to measure the costs of conservation than the ultimate gains, even in purely monetary units. The riches are there, fallow in the wildlands and waiting to be employed by our hands, our wit, our spirit. It would be folly to let any species die by the sole use of the criterion of economic return, however potent, simply because the name of that species happens to be written in red ink.

CHAPTER FOURTEEN

Resolution

E VERY COUNTRY has three forms of wealth: material, cultural, and biological. The first two we understand well
because they are the substance of our everyday lives. The
essence of the biodiversity problem is that biological wealth is
taken much less seriously. This is a major strategic error, one
that will be increasingly regretted as time passes. Diversity is a
potential source for immense untapped material wealth in the
form of food, medicine, and amenities. The fauna and flora
are also part of a country's heritage, the product of millions of
years of evolution centered on that time and place and hence
as much a reason for national concern as the particularities of
language and culture.

The biological wealth of the world is passing through a
bottleneck destined to last another fifty years or more. The
human population has moved past 5.4 billion, is projected to
reach 8.5 billion by 2025, and may level off at 10 to 15 billion
by midcentury. With such a phenomenal increase in human
biomass, with material and energy demands of the developing
countries accelerating at an even faster pace, far less room will
be left for most of the species of plants and animals in a short
period of time.

The human juggernaut creates a problem of epic dimensions:
how to pass through the bottleneck and reach midcentury with
the least possible loss of biodiversity and the least possible cost
to humanity. In theory at least, the minimization of extinction
rates and the minimization of economic costs are compatible:
the more that other forms of life are used and saved, the more
productive and secure will our own species be. Future generations will reap the benefit of wise decisions taken on behalf of
biological diversity by our generation.

What is urgently needed is knowledge and a practical ethic
based on a time scale longer than we are accustomed to apply.
An ideal ethic is a set of rules invented to address problems so
complex or stretching so far into the future as to place their

solution beyond ordinary discourse. Environmental problems are innately ethical. They require vision reaching simultaneously into the short and long reaches of time. What is good for individuals and societies at this moment might easily sour ten years hence, and what seems ideal over the next several decades could ruin future generations. To choose what is best for both the near and distant futures is a hard task, often seemingly contradictory and requiring knowledge and ethical codes which for the most part are still unwritten.

If it is granted that biodiversity is at high risk, what is to be done? Even now, with the problem only beginning to come into focus, there is little doubt about what needs to be done. The solution will require cooperation among professions long separated by academic and practical tradition. Biology, anthropology, economics, agriculture, government, and law will have to find a common voice. Their conjunction has already given rise to a new discipline, biodiversity studies, defined as the systematic study of the full array of organic diversity and the origin of that diversity, together with the methods by which it can be maintained and used for the benefit of humanity. The enterprise of biodiversity studies is thus both scientific, a branch of pure biology, and applied, a branch of biotechnology and the social sciences. It draws from biology at the level of whole organisms and populations in the same way that biomedical studies draw from biology at the level of the cell and molecule. Where biomedical studies are concerned with the health of the individual person, biodiversity studies are concerned with the health of the living part of the planet and its suitability for the human species. What follows, then, is an agenda on which I believe most of those who have focused on biodiversity might agree. All the enterprises I will list are directed at the same goal: to save and use in perpetuity as much of earth's diversity as possible.

1. *Survey the world's fauna and flora.* In approaching diversity, biologists are close to traveling blind. They have only the faintest idea of how many species there are on earth or where most occur; the biology of more than 99 percent remain unknown. Systematists are aware of the urgency of the problem but far from agreed on the best way to solve it. Some

have recommended the initiation of a global survey, aimed at the discovery and classification of all species. Others, sensibly noting the shortage of personnel, funds, and time, think the only realistic hope lies in the rapid recognition of the threatened habitats that contain the largest number of endangered endemic species (the hot spots).

In order to move systematics into the larger role demanded by the extinction crisis, its practitioners have to agree on an explicit mission with a timetable and cost estimates. The strategy most likely to work is mixed, aiming at a complete inventory of the world's species, but across fifty years and at several levels, or scales in time and space, from hot-spot identification to global survey, audited and readjusted at ten-year intervals. As each decade comes to a close, progress to that point could be assessed and new directions identified. Emphasis from the outset would be placed on the hottest spots known or suspected.

Three levels can be envisioned. The first is the RAP approach, from the prototypic Rapid Assessment Program created by Conservation International, a Washington-based group devoted to the preservation of global biodiversity. The purpose is to investigate quickly, within several years, poorly known ecosystems that might be local hot spots, in order to make emergency recommendations for further study and action. The area targeted is limited in extent, such as a single valley or isolated mountain. Because so little is known of classification of the vast majority of organisms and so few specialists are available to conduct further studies, it is nearly impossible to catalog the entire fauna and flora of even a small endangered habitat. Instead a RAP team is formed of experts on what can be called the elite focal groups—organisms, such as flowering plants, reptiles, mammals, birds, fishes, and butterflies, that are well enough known to be inventoried immediately and can thereby serve as proxies for the whole biota around them.

The next level of inventory is the BIOTROP approach, from the Neotropical Biological Diversity Program of the University of Kansas and a consortium of other North American universities formed in the late 1980s. Instead of pinpointing brushfires of extinction at selected localities in the RAP manner, BIOTROP explores more systematically across broad areas believed to be major hot spots or at least to contain multiple hot spots. Examples of such regions include the eastern slopes of the Andes and the scattered forests of Guatemala and southern Mexico. Beyond identifying critical localities, the larger goal is to set up research stations across the area that embrace different latitudes and elevations. The work begins with a few focal organisms. It expands to less familiar groups, such as ants, beetles, and fungi, as enough specimens are collected and experts in the groups are recruited to study them. In time, close studies of rainfall, temperature, and other properties of the environment are added to the species inventory. The most important and best equipped of the stations are likely then to evolve into centers of long-term biological research, with leadership roles taken by scientists from the host countries. They can also be used to train scientists from different parts of the world.

We now come to the third and highest stage of the biodiversity survey. From inventories at the RAP and BIOTROP levels in different parts of the world, accompanied by monographic studies of one group of organisms after another, the description of the living world will gradually coalesce to create a fine-grained image of global biodiversity. The growth of knowledge will inevitably accelerate, even given a constant level of effort, by producing its own economies of scale. Costs per species logged into the inventory fall as new methods of collecting and distributing specimens are devised and procedures for accessing information are improved. Costs are not simply additive when nonelite groups of organisms are included, but instead decline on a per-species basis. Botanists, for example, can collect insects living on the plants they study, while identifying these hosts for the entomologists, and entomologists can run the procedure in reverse, gathering plant specimens in company with the insects they collect. Groups such as reptiles, beetles, and spiders can be sampled across entire habitats, then distributed to specialists on each group in turn.

As biodiversity surveys proceed at the several levels, the knowledge gathered becomes an ever more powerful magnet for other kinds of science. Field guides and illustrated treatises open doors to the imagination, and networks of technical information draw geologists, geneticists, biochemists, and others into the enterprise. It will be logical to gather much of the activity into biodiversity centers, where data are gathered and new inquiries planned. The prototype is Costa Rica's National Institute of Biodiversity (Instituto Nacional de Biodiversidad), INBio for short, established on the outskirts of the capital city of San José in 1989. The aim of INBio is nothing less than to account for all the plants and animals of this small Central American country, over half a million species in number, and to use the information to improve Costa Rica's environment and economy. It is perhaps odd that a developing nation should lead the way in such a concerted scientific enterprise, but others will follow. Detailed distribution maps of plants and many kinds of animals have been drawn up in Great Britain, Sweden, Germany, and other European countries under governmental and private auspices. As I write, plans for a national biodiversity center in the United States have been advanced by the Smithsonian Institution and are under wide discussion. Enabling legislation has been placed before Congress but is not yet passed.

The national center of the United States will not have to start from scratch. Many kinds of organisms have been already carefully studied and mapped. Several of the states, including Massachusetts and Minnesota, have undertaken programs to locate endangered species of plants and vertebrate animals within their borders. For fifteen years the Nature Conservancy, one of the premier private American foundations, has conducted a similar effort across all the states. The operation, setting up Natural Heritage Data Centers, has recently been extended to fourteen Latin American and Caribbean countries.

Another key element of biodiversity studies at all levels will be microgeography, the mapping of the structure of the ecosystem in sufficiently fine detail to estimate the populations of individual species and the conditions under which they grow and reproduce. A working technology already exists in the form of Geographic Information Systems, a collection of layers of data on topography, vegetation, soils, hydrology, and species distributions that are registered electronically to a common coordinate system. When applied to biodiversity and endangered species, the cartography is called *gap analysis.* Even though incomplete, gap analysis can reveal the effectiveness of existing parks and reserves. It can be used to help answer the larger questions of conservation practice. Do protected areas in fact embrace the largest possible number of endemic species? Are the surviving habitat fragments large enough to sustain the populations indefinitely? And what is the most cost-effective plan for further land acquisition?

The same information can be used to zone large regions. Parcels of land will have to be set aside as inviolate preserves. Others will be identified as the best sites for extractive reserves, for buffer zones used in part-time agriculture and restricted hunting, and for land convertible totally to human use. In the expanded enterprise, landscape design will play a decisive role. Where environments have been mostly humanized, biological diversity can still be sustained at high levels by the ingenious placement of woodlots, hedgerows, watersheds, reservoirs, and artificial ponds and lakes. Master plans will meld not just economic efficiency and beauty but also the preservation of species and races.

The layered data can further aid in defining "bioregions," areas such as watersheds and forest tracts that unite common ecosystems but often extend across the borders of municipalities, states, or even countries. A river may make economic or military sense in dividing two political units, but it makes no sense at all in organizing land-use management. Bioregionalism has had a long but inconclusive history within the United States. It dates back at least as far as John Muir's successful championing of national parks and the establishment of the national forest system in 1891. Since the 1930s it has received increasing governmental sanctions with variable specific agendas, from the Tennessee Valley Authority, which managed land and created hydroelectric power through a large part of the southeast, to the establishment of the Appalachian National Scenic Trail, federal and state management of the south Florida water system and the Everglades, and the multiple regulatory and promotional activities of the New England River Basins Commission during its tenure from 1967 to 1981.

Other examples of bioregionalism abound in the United States, but it cannot be said that the movement has coalesced around any single philosophy of land management. Nor has the preservation of biodiversity ranked as more than an auxiliary

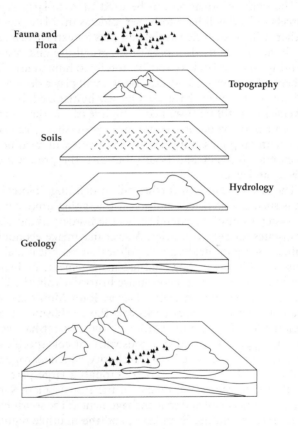

Geographic Information Systems

Fauna and Flora

Topography

Soils

Hydrology

Geology

Geographic Information Systems combine information on physical and biological environments by joining layered data sets. These can be used to manage the landscape in a way that protects endangered species and ecosystems, including the designation of natural reserves.

goal. In fact the great dams built by the Tennessee Valley Authority, while providing cheap electric power to an impoverished part of the nation, inadvertently wiped out a substantial part of the native river fauna. The lower priority given diversity has not been by deliberation but from incomplete knowledge of the faunas and floras of the affected regions.

Systematics, having emerged as a prerequisite for effective long-term zoning and bioregionalism, is a labor-intensive enterprise. Scientists who study the classification of particular organisms, such as centipedes and ferns, are often by default the only authorities on the general biology of those organisms. About 4,000 such specialists in the United States and Canada attempt to manage the classification of the many thousand species of animals, plants, and microorganisms living on the continent. To varying degree they are also responsible for the millions of species occurring elsewhere in the world, since even fewer systematists are active in other countries. Probably a maximum of 1,500 trained professional systematists are competent to deal with tropical organisms, or more than half of the world's biodiversity. A typical case is the shortage of experts on termites, which are premier decomposers of wood, rivals of earthworms as turners of the soil, owners of 10 percent of the animal biomass in the tropics, and among the most destructive of all insect pests. There are exactly three people qualified to deal with termite classification on a worldwide basis. A second revealing case: the oribatid mites, tiny creatures resembling a cross between a spider and a tortoise, are among the most abundant animals of the soil. They are major consumers of humus and fungus spores, and therefore key elements of land ecosystems almost everywhere. In North America only one expert attends to their classification on a full-time basis.

With so few people prepared to launch it, a complete survey of earth's vast reserves of biological diversity may seem beyond reach. But compared with what has been dared and achieved in high-energy physics, molecular genetics, and other branches of big science, the magnitude of its challenge is not all that great. The processing of 10 million species is achievable within fifty years, even with the least efficient, old-fashioned methods. If one systematist proceeded at the cautious pace of ten species per year, including field trips for collecting, analysis of specimens in the laboratory, and publication, taking time out for vacations and family, about one million person-years of work would be required. Given forty years of productive life per scientist, the effort would consume 25,000 professional lifetimes. The number of systematists would still represent less than 10 percent of the current population of scientists active in the United States alone, and it falls well short of the number of enlisted men in the standing armed forces of Mongolia, not to mention the trade and retail personnel of Hinds County, Mississippi. The volumes of published work, one page per species, would fill 12 percent of the shelves of the library of Harvard's Museum of Comparative Zoology, one of the larger institutions devoted to systematics.

I have based these estimates on what is the least efficient procedure imaginable, in order to establish the plausibility of a total inventory of global biodiversity. Systematic work can be speeded up many times over by new techniques now coming into general use. The Statistical Analysis System (SAS), a set of computer programs already running in several thousand institutions worldwide, records taxonomic identifications and localities of individual specimens and automatically integrates data in catalogs and maps. Other computer-aided techniques compare species automatically across large numbers of traits, applying unbiased measures of similarity, the procedure called *phenetics*. Still others assist in deducing the most likely family trees of species, the method called *cladistics*. Scanning-electron microscopy has accelerated the illustration of insects and other small organisms. Computer technology will in time include image scanning that can identify species instantly while flagging specimens that belong to new species. Biologists are also close to electronic publication, which will allow retrieval of descriptions and analyses of particular groups of organisms by desktop personal computers.

Every other form of biological information on species—ecology, physiology, economic uses, status as vectors, parasites, agricultural pests—can be layered in the databases. DNA and RNA sequences and gene maps can be added. GenBank, the genetic-sequence bank, has been chartered to provide a computer database for all known DNA and RNA sequences and related biological information. By 1990 it had accumulated 35 million sequences distributed through 1,200 species of plants, animals, and microorganisms. The rate of data accession is ascending swiftly with the advent of improved sequencing methods.

2. *Create biological wealth.* As species inventories expand, they open the way to bioeconomic analysis, the broad assessment of the economic potential of entire ecosystems. Every community of organisms contains species with potential commodity value—timber and wild-plant products to be harvested on a sustained basis, seeds and cuttings that can be transplanted to grow crops and ornamentals elsewhere, fungi and microorganisms to be cultured as sources of medicinals, organisms of all kinds offering new scientific knowledge that points to still more practical applications. And the wild habitats have recreational value, which will grow as a larger sector of the public travels and learns to enjoy natural history.

The decision to make bioeconomic analysis a routine part of land-management policy will protect ecosystems by assigning them future value. It can buy time against the removal of entire communities of organisms ignorantly assumed to lack such value. When local faunas and floras are better known, the decision can be taken on how to use them optimally—whether to protect them, to extract products from them on a sustainable yield basis, or to destroy their habitat for full human occupation. Destruction is anathema to conservationists, but the fact remains that most people, lacking knowledge, regard it as perfectly acceptable. Somehow knowledge and reason must be made to intrude. I am willing to gamble that familiarity will save ecosystems, because bioeconomic and aesthetic values grow as each constituent species is examined in turn—and so will sentiment in favor of preservation. The wise procedure is for law to delay, science to evaluate, and familiarity to preserve. There is an implicit principle of human behavior important to conservation: *the better an ecosystem is known, the less likely it will be destroyed.* As the Senegalese conservationist Baba Dioum has said, "In the end, we will conserve only what we love, we will love only what we understand, we will understand only what we are taught."

A key enterprise in bioeconomic analysis is what Thomas Eisner has called *chemical prospecting,* the search among wild species for new medicines and other useful chemical products. The logic of prospecting is supported by everything we have learned about organic evolution. Each species has evolved to become a unique chemical factory, producing substances that allow it to survive in an unforgiving world. A newly discovered species of roundworm might produce an antibiotic of extraordinary power, an unnamed moth a substance that blocks viruses in a manner never guessed by molecular biologists. A symbiotic fungus cultured from the rootlets of a nearly extinct tree might yield a novel class of growth promoters for plants. An obscure herb could be the source of a sure-fire blackfly repellent—at last. Millions of years of testing by natural selection have made organisms chemists of superhuman skill, champions at defeating most of the kinds of biological problems that undermine human health.

Because chemical prospecting depends so heavily on classification, it is best conducted in tandem with biodiversity surveys. In order to succeed, investigators must also work in laboratories equipped with advanced facilities, which are usually available only in industrialized countries. In 1991 Merck and Company, the world's largest pharmaceutical firm, agreed to pay Costa Rica's National Institute of Biodiversity $1 million to assist in such a screening effort. The institute will collect and identify the organisms, sending chemical samples from the most promising species to the Merck laboratories for medicinal assay. If natural substances are marketed, the company is committed to pay the Costa Rican government a share of the royalties, which will then be earmarked for conservation programs. Merck has previously marketed four drugs from soil organisms originating from other countries. One, derived from a fungus, is Mevacor, an effective agent for lowering cholesterol levels. In 1990 Merck sold $735 million worth of this substance alone. It follows that a single success in Costa Rica—a commercial product from, say, any one species among the 12,000 plants and 300,000 insects estimated to live in the country—could handsomely repay Merck's entire investment.

There are historical reasons why Merck and other research and commercial organizations are increasingly inclined to take on chemical prospecting. The search for naturally occurring drugs and other chemical products has been cyclical through the years. In the 1960s and 1970s pharmaceutical companies phased out the screening of plants on the grounds that it was too complicated and expensive. With only one in 10,000 species yielding a promising substance (by procedures then in use) and millions of dollars needed to bring a product fully on line, the eventual payoff seemed marginal. The companies turned to new technologies in microbiology and synthetic chemistry, hoping to design the magic bullets of the new medical age with chemicals taken from the shelf. To rely on human ingenuity rather than evolved natural chemistry in distant jungles seemed much more "scientific" and direct, and perhaps less expensive. Yet natural products remained a potential shortcut, a Columbus-like journey west, for those willing to acquire the essential skills. Now the pendulum has begun to swing back, again from advances in technology, because high-volume, robot-controlled biological assays allow larger companies to screen up to 50,000 samples a year using only bits of fresh tissue or extract flown to them from any part of the world.

The path from wild organism to commercial production can sometimes be shortened further by taking clues from the lore and traditional medicine of indigenous peoples. It is a remarkable fact that of the 119 known pure pharmaceutical compounds used somewhere in the world, 88 were discovered through leads from traditional medicine. The knowledge of all the world's indigenous cultures, if gathered and catalogued, would constitute a library of Alexandrian proportions. The Chinese, for example, employ materials from about 6,000 of the 30,000 plant species in their country for medicinal purposes. Among them is artemisinin, a terpene derived from the annual wormwood (*Artemisia annua*), which shows promise as an alternative to quinine in the treatment of malaria. Because the molecular structures of the two substances are entirely different, artemisinin would have been discovered much less quickly if not for its folkloric reputation.

Because the lives of people and the reputations of shamans have depended on it for generations, much of the traditional pharmacopoeia is reliable. Extraction procedures and dosage have been tested by trial and error countless times. But this preliterate knowledge, like so many of the plant and animal species to which it pertains, is disappearing rapidly as tribes move from their homelands onto farms and into cities and villages. When they take up new trades, their languages fall into disuse and the old ways are forgotten. During the 1980s, all but 500 of the 10,000 Penans of Borneo abandoned their centuries-old seminomadic life in the forests and settled in villages. Today their memories are fading quickly. Eugene Linden notes, "Villagers know that their elders used to watch for the appearance of a certain butterfly, which always seemed to herald the arrival of a herd of boar and the promise of good hunting. These days, most of the Penans cannot remember which butterfly to look for." On the other side of the world, 90 of Brazil's 270 Indian tribes have vanished since 1900, and two thirds of those remaining contain populations of less than a thousand. Many have lost their lands and are forgetting their cultures.

Small farms around the world are giving way to the mono-cultures of agrotechnology. The raised garden squares of the Incas have all but vanished; the densely variegated gardens of Mesoamerica and West Africa are threatened. The revitalization of local farming is another aim of biodiversity studies. The goal is to make the practice more economically practical, while conserving the genetic reserves that will contribute to crops of the future. Species and strains of high economic efficiency, from perennial corn to amaranth and iguanas, can be fed through research centers into the local regions best suited to use them. A successful prototype of such enterprises is the Tropical Agricultural Research and Training Center (CATIE) at Turrialba, Costa Rica. Created by the Organization of American States in 1942, CATIE maintains large samples of plant species, including disease-resistant strains of cacao and other tropical crops. Its staff members experiment with propagation methods for crops and timber, design wildland preservation programs, search for new crop species and varieties, and train students in the new methods of agriculture and conservation. Institutions of the future can be profitably built to include not only these activities but also chemical prospecting and molecular techniques of gene transfer from wild to domestic species.

3. *Promote sustainable development.* The rural poor of the Third World are locked onto a downward spiral of poverty and the destruction of diversity. To break free they need work

that provides the basic food, housing, and health care taken for granted by a great majority of people in the industrialized countries. Without it, lacking access to markets, hammered by exploding populations, they turn increasingly to the last of the wild biological resources. They hunt out the animals within walking distance, cut forests that cannot be regrown, put their herds on any land from which they cannot be driven by force. They use domestic crops ill suited to their environment, for too many years, because they know no alternative. Their governments, lacking an adequate tax base and saddled with huge foreign debts, collaborate in the devastation of the environment. Using an accountant's trick, they record the sale of forests and other irreplaceable natural resources as national income without computing the permanent environmental losses as expense.

The poor are denied an adequate education. They cannot all move into the cities; in most countries, and especially those in the tropics, industrialization will be too slow to absorb more than a small fraction into the labor force. Their striving billions will, for the next century at least, have to be accommodated in rural areas. So the issue comes down to this: how can people in developing countries achieve a decent living from the land without destroying it?

The proving ground of sustainable development will be the tropical rain forests. If the forests can be saved in a manner that improves local economies, the biodiversity crisis will be dramatically eased. Within that "if" are folded technical and social difficulties of the most vexing kind. But many paths to the goal have been suggested, and some have successfully tested.

One of the most encouraging advances to date is the demonstration, cited in the last chapter, that the extraction of nontimber products from Peruvian rain forests can yield similar levels of income as logging and farming, even with the limited outlets available in existing local markets. The practice has been regularized by the rubber tappers of Brazil without a bit of theory or cost-benefit analysis. The tappers, or *seringueiros* as they are locally called, are the descendants of immigrants from northeastern Brazil who colonized portions of the Amazon during the late nineteenth century and found a steady living in latex harvesting. Half a million strong, they draw their principal income today not only from rubber but also from Brazil nuts, palm hearts, tonka beans, and other wild products. Each family owns a house in the midst of harvesting pathways shaped like clover leaves. In addition to harvesting natural products, rubber tappers also hunt, fish, and practice small-scale agriculture in forest clearings. Because they depend on biological diversity, the tappers are devoted to the preservation of the forests as stable and productive ecosystems. They are in fact full members of the ecosystems. In 1987 the Brazilian government authorized the establishment of *seringueiro* extractive reserves on state land, with thirty-year renewable leases and a prohibition on the clear-cutting of timber.

Extractive reserves represent a major conceptual advance, but they are not enough to save more than a small portion of the rain forests. In 1980 rubber-tapper households occupied 2.7 percent of the area of the North Region of Brazilian Amazonia, including the states of Amazonas and Acre, while farms and ranches occupied 24 percent. Only a small fraction of the flood of new immigrants now pouring into the region can become extractivists. The rest will seek income wherever they find it, primarily by advancing the agricultural frontier. The key to the future of Amazonia and other forested tropical regions is whether employment made available to them saves or destroys the environment. "The real challenge," John Browder writes, "is not where to designate extractive reserves, but rather, how to integrate sustainable extraction and other natural forest management practices into the production strategies of those existing rural properties, small farms and large ranches alike, that are responsible for most of the devastation being visited upon Amazonian rainforests. Fundamentally, the problem is not where to sequester forests, but how to turn people into better forest managers."

It is possible to harvest timber from the Amazonian wilderness and other great remaining rain forests extensively and profitably with little loss of biodiversity. The method of choice, first suggested by Gary Hartshorn in 1979 and extended by other foresters, is strip logging. While lowland forested basins are not rugged in terrain, most are moderately rolling with well-defined slopes and dense systems of drainage streams. Strip logging imitates the natural fall of trees that create linear gaps through the forest, with the artificial gaps being aligned along the contours. The technique is described by Carl Jordan:

> In this scheme, a strip is harvested on the contour of a slope, parallel to the stream. Along the upper edge of the strip is a road used for hauling out the logs. After harvesting, the area is left for a few years until saplings begin to grow in the cut areas. Then the loggers clear-cut another strip, this time above the road. The advantages of this system are that the nutrients from the freshly cut second strip wash downslope into the rapidly regenerating first strip, where the trees can quickly use the nutrients, and that seeds from the mature forest above the cut area will roll down into the recently cut strip. In contrast, in clear-cutting there are no saplings with well-developed roots capable of retaining nutrients in the system, nor is there a source of seed for regeneration of the forest.

So far so good, but how can governments and local peoples be persuaded to adopt such innovations as extractive reserves and strip logging? The shift to sustainable development will depend as much on education and social change as on science. Around the world modest projects are being advanced with one common result: if procedures tailored to the special case are used, economic development and conservation can both be served. People can be persuaded; they understand their own long-term interest and they can adapt. Here are three successful programs from Latin America.

• By Panama law, the Kuna Indians hold sovereign rights over the San Blas Islands and 300,000 hectares of adjacent mainland forest. The Kuna maintain "spirit sanctuaries," areas of primary forest in which only certain kinds of trees may be cut and no farming is allowed. Local communities depend on the sea for most of their protein, on the forests for wood, game, and medicine, and on limited patches of cleared land for domestic crops. When a spur of the Pan-American Highway was brought to the edge of their land, the Kuna established a forest reserve and guarded it with their own people. Well aware of the outside world, welcoming to visitors, the tribes have nevertheless chosen to discourage immigration and to preserve their own culture within the bountiful natural environment that has sustained them for centuries.

• Most of Central America, unlike the land of the Kuna, is plagued by soil erosion and nutrient loss owing to the excessive cultivation of maize and other crops, leading to the cutting of forests on ever steeper slopes, all driven in turn by overpopulation. As production declines, farmers invade the remaining natural areas in search of more arable land. The process is especially acute in the Güinope region of Honduras. In 1981 two private foundations, one international and one Honduran, commenced a pilot program in some of the Güinope villages under government auspices to raise productivity and restore the land. They introduced drainage ditches, contour furrows, grassy barriers, and intercropping with nitrogen-restoring legumes. The field labor and implementation costs were provided entirely by the farmers. Within several years, yields tripled and emigration nearly ceased. The new agricultural methods began to spread to surrounding areas.

• When a highway, the Carretera Marginal de la Selva, was cut into Peru's Palcazú Valley, 85 percent of the land was still clothed by rain forest. Like most of the eastern tropical slopes of the Andes, the valley is biologically rich, containing for example more than a thousand species of trees. The region also supported about 3,000 Amuesha Indians and an equal number of settlers who had established small landholdings over the previous fifty years. Once opened to outside commerce, the typical fate of a western Amazonian valley is to be clearcut by new immigrants and logging companies, then used for cattle ranches and small farms. The thin, acidic soil soon loses most of its free phosphates and other nutrients, launching the next phase: erosion, poverty, partial abandonment. For this valley, however, an alternative plan was proposed by the U.S. Agency for International Development and approved by the Peruvian government. It is to extract timber by strip cutting, regulated to allow perpetual regeneration of the forest through thirty- to forty-year rotations. The plan permits limited permanent conversion of the most arable land to agriculture and livestock production. But it also calls for the establishment of a watershed reserve in the adjacent San Matias mountain range and the designation of the neighboring Yanachaga range as the Yanachaga-Chemillén National Park. With luck, the Palcazú will support a healthy human population and a slice of Peru's biodiversity into the next century.

Wildlands and biological diversity are legally the properties of nations, but they are ethically part of the global commons. The loss of species anywhere diminishes wealth everywhere. Today the poorest countries are rapidly decapitalizing their natural resources and unintentionally wiping out much of their biodiversity in a scramble to meet foreign debts and raise the standard of living. By perceived necessity they follow environmentally destructive policies that yield the largest short-term profits. The rich debt-holding nations aggravate the practice by encouraging a free market in poor countries while providing subsidies to farmers at home.

Consider the infamous "hamburger connection" between the United States and Central America. By 1983, in response to the excellent U.S. market for beef, Costa Rican landowners had accelerated the creation of new pastures until only 17 percent

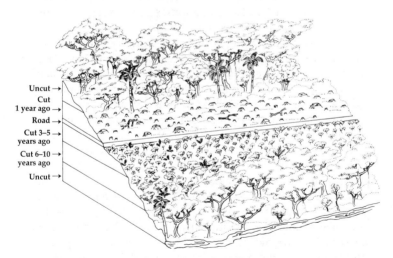

Strip logging allows a sustainable timber yield from forests, including the relatively fragile rain forests. A corridor is cleared along the contours of the land, narrow enough to allow natural regeneration within a few years. Another corridor is then cut above the first, and so on, through a cycle lasting many decades.

of the country's original forest cover was left. For a time it was the world's leading exporter of beef to the United States. When northern tastes changed somewhat and the market fell, Costa Rica was left with a denuded landscape and widespread soil erosion. It had also lost part of its biological diversity.

Developing countries competing in an international free market have a strong incentive to transfer capital into single-money crops such as bananas, sugar cane, and cotton. To that end governments often subsidize the clearing of wildlands and the overuse of pesticides and fertilizers. The rush to maximize export income also concentrates ever more acreage in the hands of a relatively few, politically favored landowners. Small farmers are then forced to seek new land of marginal productivity, including natural habitats. Faced with ruin, they have no choice but to press into nutrient-poor tropical forests, steep hillside watersheds, coastal wetlands, and other final refuges of terrestrial diversity.

This journey to the precipice is hastened by the agricultural support systems of the richest nations. At the present time subsidies to developed-world farmers total $300 billion a year, six times the official foreign aid to Third World countries. When European Community countries recently underwrote a large program of feedlot cattle raising, they created a huge artificial market for cassava. Landowners in Thailand responded by clearing more tropical forest to grow cassava, and in the process displaced large numbers of subsistence farmers into the deep forest and up the eroding hillsides. When the United States tightened import quotas of cane sugar to aid domestic growers, U.S. imports from the Caribbean countries dropped 73 percent in ten years, forcing many of the rural poor out of jobs in the plantations and into marginal habitats for subsistence farming. Japan's extravagant subsidy to its own rice farmers, intended to continue an ancient agricultural tradition (the Japanese written character for rice means "root of life"), has a depressing effect on the rice-growing populations of tropical Asia. Once again, the impact on natural environments is increased.

The richest countries set the rules for international trade. They provide the bulk of loans and direct aid and control technology transfer to the poor nations. It is their responsibility to use this power wisely, in a manner that both strengthens these trading partners and protects the global environment. They themselves will suffer if the wildlands and biological diversity are not entered into the calculus of trade agreements and international aid.

The raging monster upon the land is population growth. In its presence, sustainability is but a fragile theoretical construct. To say, as many do, that the difficulties of nations are not due to people but to poor ideology or land-use management is sophistic. If Bangladesh had 10 million inhabitants instead of 115 million, its impoverished people could live on prosperous farms away from the dangerous floodplains midst a natural and stable upland environment. It is also sophistic to point to the Netherlands and Japan, as many commentators incredibly still do, as models of densely populated but prosperous societies. Both are highly specialized industrial nations dependent on massive imports of natural resources from the rest of the world. If all nations held the same number of people per square kilometer, they would converge in quality of life to Bangladesh rather than to the Netherlands and Japan, and their irreplaceable natural resources would soon join the seven wonders of the world as scattered vestiges of an ancient history.

Every nation has an economic policy and a foreign policy. The time has come to speak more openly of a population policy. By this I mean not just the capping of growth when the population hits the wall, as in China and India, but a policy based on a rational solution of this problem: what, in the judgment of its informed citizenry, is the *optimal* population, taken for each country in turn, placed against the backdrop of global demography? The answer will follow from an assessment of the society's self-image, its natural resources, its geography, and the specialized long-term role it can most effectively play in the international community. It can be implemented by encouragement or relaxation of birth control and the regulation of immigration, aimed at a target density and age distribution of the national population. The goal of an optimal population will require addressing, for the first time, the full range of processes that lock together the economy and the environment, the national interest and the global commons, the welfare of the present generation with that of future generations. The matter should be aired not only in think tanks but in public debate. If humanity then chooses to breed itself and the rest of life into impoverishment, at least it will have done so with open eyes.

4. *Save what remains.* Biodiversity can be saved by a mixture of programs, but not all the programs proposed can work. Consider one often raised in discussions by futurists. Suppose that we lost the race to save the environment, that all natural ecosystems were allowed to vanish. Could new species be created in the laboratory, after genetic engineers have learned how to assemble life from raw organic compounds? It is doubtful. There is no assurance that organisms can be generated artificially, at least not any as complex as flowers or butterflies—or amoebae for that matter. Even this godlike power would solve only half the problem, and the easy one at that. The technicians would be working in ignorance of the history of the extinct life they presumed to simulate. No knowledge exists of the endless mutations and episodes of natural selection that inserted billions of nucleotides into the now-vanished genomes, nor can it be deduced in more than tiny fragments. The neospecies would be creations of the human mind—plastic, neither historical nor adaptive, and unfit for existence apart from man. Ecosystems built from them, like zoos and botanical gardens, would require intensive care. But this is not the time for science-fiction dreams.

On then to the next technical remedy that springs up in scientific conferences and corridor arguments. Can extinct species be resurrected from the DNA still preserved in museum specimens and fossils? Again the answer is no. Fractions of genetic codes have been sequenced from a 2400-year-old Egyptian mummy and magnolia leaves preserved as rock fossils 18 million years ago, but they constitute only the smallest portion of the genetic codes. Even that part is hopelessly scrambled. To clone these organisms or a mammoth or a dodo or any other extinct organism would be, as the molecular biologist Russell Higuchi recently said, like taking a large encyclopedia in an unknown language previously ripped into shreds and trying to reassemble it without the use of your hands.

Consider the next possibility raised with regularity: why not just forget the problem and let natural evolution replace the species that are disappearing? It can be done if our descendants are willing to wait several million years. Following the five great extinction episodes of geological history, full recovery of biodiversity required between 10 and 100 million years. Even if *Homo sapiens* lasts that long, the recovery would require returning a large part of the land to its natural state. By appropriating or otherwise disturbing 90 percent of the land surface, humanity has already closed most of the theaters of natural evolution. And even if we did that much and waited that long, the new biota would be very different from the one we destroyed.

Then why not scoop up tissue samples of all living species and freeze them in liquid nitrogen? They could be cloned later to produce whole organisms. The method works for some microorganisms, including viruses, bacteria, and yeasts, as well as the spores of fungi. The American Type Culture Collection, located at Rockville, Maryland, contains over 50,000 species suspended in the deep sleep of absolute biochemical inactivity, ready for warming and reactivation as needed. The cultures are used in research, primarily in molecular biology and medicine. It is possible that many larger organisms could be similarly preserved in nitrogen sleep, at least as fertilized eggs, to be reared later into mature individuals. Even scraps of undifferentiated tissue might be stimulated into normal growth and development. It has been done for organisms as complex as carrots and frogs.

So let us suppose for argument that all kinds of plants and animals are salvageable by such means, that biologists will perfect the techniques of total inactivation and total recovery. The cryotorium in which they would rest, the new Noah's ark, must house tens of millions of species. The preservation of the content of even one endangered habitat (say a mountain-ridge forest in Ecuador) would be an immense operation enveloping thousands of species, most of which are still unknown to science. Even if completed at the species level, only a small fraction of the genetic variability of each species could be practicably included. Unless the samples numbered into the millions, great arrays of naturally occurring genetic strains would be lost. And when the time comes to return the species to the wild, the physical base of the ecosystem, including its soil, its unique nutrient mix, and its patterns of precipitation, will have been altered so as to make restoration doubtful. Cryopreservation is at best a last-ditch operation that might rescue a few select species and strains certain to die otherwise. It is far from the best way to save ecosystems and could easily fail. The need to put an entire community of organisms in liquid nitrogen would be tragic. Its enactment would be, in a particularly piercing sense of the word, obscene.

I have spoken so far of the maintenance of species and genetic stocks away from their natural habitats. Not all such methods are fantastic or repugnant. One that works for many plants is the maintenance of seed banks: seeds are dried and kept in repositories over long periods. The banks are kept in cool temperatures (about −20°C is typical) but not in the suspended animation of liquid nitrogen. Botanists have proved the technique effective for preserving most strains of crop species. About a hundred countries maintain seed banks and are adding to them steadily by exchanges and new collecting expeditions. Their efforts are aided by the "Green Board," the International Board for Plant Genetic Resources (IBPGR), an autonomous scientific organization located in Rome that composes part of the network of the International Agricultural Research Centers. In 1990 over 2 million sets of seeds were on deposit, representing more than 90 percent of the known local geographic varieties—landraces, as they are called—of many of the basic food crops. Especially well represented are wheat, maize, oats, potatoes, rice, and millet. An effort has begun to include the wild relatives of existing crop species, such as the richly promising perennial maize of Mexico. The method can be extended to wild, noncrop floras of the world.

But there are serious problems with seed banks. Up to 20 percent of plant species, some 50,000 in all, possess "recalcitrant" seeds that cannot be stored by conventional means. Even if seed storage were perfected for all kinds of plants, an unlikely prospect for the immediate future, the task of collecting and maintaining many thousands of endangered species and races would be stupendous. All the efforts of the existing seed banks to date have been barely enough to cover a hundred species, and even those are in many cases poorly recorded and of uncertain survival ability. Another difficulty: if reliance were placed entirely on seed banks, and the species then disappeared in the wild, the bank survivors would be stripped of their insect pollinators, root fungi, and other symbiotic partners, which cannot be put in cold storage. Most of the symbionts would go extinct, preventing the salvaged plant species from being replanted in the wild.

Other *ex situ* methods rely more realistically on captive populations that grow and reproduce. There are about 1,300 botanical gardens and arboretums in the world, many harboring plant species that are endangered or extinct in the wild. As of June 1991, twenty such institutions in the United States that subscribe to the registry of the National Collection of Endangered Plants contained seeds, plants, and cuttings of 372 species native to the United States. Some of the gardens in North America and Europe are more global in their reach. Harvard's Arnold Arboretum, for one, is famous for its collection of Asiatic trees and shrubs. England's magnificent Kew Gardens is engaged in a bold attempt to preserve and cultivate the last remnants of the nearly vanished tree flora of St. Helena.

Animals are vastly more difficult than plants and microorganisms to maintain *ex situ*. Zoos and other animal facilities have attempted the task in heroic fashion. By the late 1980s, those around the world whose stocks are known had gathered breeding populations of 540,000 individuals belonging to more than 3,000 species of mammals, birds, reptiles, and amphibians. The collections include roughly 13 percent of the known land-dwelling species of vertebrate animals. The better-financed zoos, including those in London, Frankfurt, Chicago, New York, San Diego, and Washington, D.C., conduct basic and veterinarian research with results that are applied to both captive and wild populations. The rosters of 223 zoos in Europe and North America are tracked by the international Species Inventory System (ISIS), which uses the data to coordinate preservation and crossbreeding. The ISIS zoos and research institutions aim not only to save endangered animals but to reintroduce species into their native habitats when land is made available. They have been successful with three species, the Arabian oryx, the black-footed ferret, and the golden lion tamarin. Attempts are underway or planned for at least four other species, the California condor, the Bali starling, the Guam rail, and the Przewalski horse, the ancestor of all domestic horses. The ISIS facilities are trying to get ready if the giant panda, the Sumatran rhinoceros, and the Siberian tiger, now on the brink, should go extinct in the wild.

The best efforts by zoos, zooparks, aquariums, and research facilities, however, slow the tide of extinction by a barely perceptible amount. Even the groups of animals most favored by the public cannot be completely served. Conservation biologists estimate that as many as 2,000 species of mammals, birds, and reptiles can only be salvaged if they are bred in captivity, a task beyond reach with the means at hand. William Conway, director of the comprehensive zoo maintained by the New York Zoological Society, believes that existing facilities worldwide can sustain viable populations of no more than 900 species. At best these survivors would contain only a small fraction of their species' original genes. And far worse: no provision at all has been made for the many thousands of species of insects and other invertebrates that are equally at risk.

The dreams of scientists come to this: *ex situ* conservation is not enough and will never be enough. Some of the methods are invaluable as safety nets for the fraction of endangered species that biology best understands and the lay public is willing to support. But even if countries everywhere chose to finance greatly enlarged cryobiological vaults, seed banks, botanical gardens, and zoos, the facilities could not be assembled quickly enough to save a majority of species close to extinction from habitat destruction alone. Biologists are hampered by lack of knowledge of more than 90 percent of the species of fungi, insects, and smaller organisms on earth. They have no way to ensure a reasonable sampling of genetic variation even in the species rescued. They have only the faintest idea of how to reassemble ecosystems from salvaged species, if indeed such a feat is possible. Not least, the entire process would be enormously expensive.

All these considerations converge to the same conclusion: *ex situ* methods will save a few species otherwise beyond hope, but the light and the way for the world's biodiversity is the preservation of natural ecosystems. If that is accepted, we must face two realities squarely. The first is that the habitats are disappearing at an accelerating rate and with them a quarter of the world's biodiversity. The second is that the habitats cannot be saved unless the effort is of immediate economic advantage to the poor people who live in and around them. Eventually idealism and high purpose may prevail around the world. Eventually an economically secure populace will treasure their native biodiversity for its own sake. But at this moment they are not secure and they, and we, have run out of time.

The rescue of biological diversity can only be achieved by a skillful blend of science, capital investment, and government: science to blaze the path by research and development; capital investment to create sustainable markets; and government to promote the marriage of economic growth and conservation.

The primary tactic in conservation must be to locate the world's hot spots and to protect the entire environment they contain. Whole ecosystems are the targets of choice because even the most charismatic species are but the representatives of thousands of lesser-known species that live with them and are also threatened. The most inclusive federal legislation in the United States is the Endangered Species Act of 1973, which throws a protective shield around species of "fish, wildlife, and plants" that are "endangered and threatened" by human activities; as amended in 1978, the act also includes subspecies. A bold and creative advance, the legislation is nevertheless destined to be an arena of rising litigation. As any natural environment is reduced in area, the number of species that can live in it indefinitely is also reduced. In other words, some species are doomed to extinction even if all of the remaining habitat were to be preserved from that time on. One of the principles of ecology, as I have stressed, is that the number of species eventually declines by an amount roughly equal to the sixth to third root of the area already lost. Because the great majority of species of microorganisms, fungi, and insects are not well known, it follows that they have been slipping unnoticed through the cracks in the Endangered Species Act. Conflicts between developers and conservationists over birds, mammals, and fishes are already commonplace. As ecosystems are better explored, less-conspicuous endangered species will come to light and the number of clashes will grow.

There is a way out of the dilemma, other than abandoning legal protection of America's fauna and flora altogether.

As biodiversity surveys are improved, the hot spots will come more sharply into focus. Well-documented examples already include the embattled coral reef of the Florida Keys and the rain forests of Hawaii and Puerto Rico. As other local habitats are pinpointed, they can be assigned the highest priority for conservation. This means, in most cases, that they will be set aside as inviolate reserves. Warm spots, areas less threatened or containing fewer species not found elsewhere, can be zoned for partial development, with core preserves centered on endemic species and races and buffer strips around the preserves kept partly wild. Agricultural landscapes and harvested forest tracts can be better designed to harbor rare species and races.

All these actions together, wisely administered, will be effective. But the Endangered Species Act or an equivalent is also needed to serve as a safety net for threatened forms of life in all environments, whether harbored in reserves or not. Finally, in those rare cases where "the costs are perceived as intolerable by the electorate, a compromise can be sought by means of population management. This means transplantation of the species to suitable habitats nearby, or restoring its environment in places where it was previously extinguished outside the zone of conflict, or—when all else fails—exile to botanical gardens, zoos, or other *ex situ* preserves.

The area-species relation governing biodiversity shows that maintenance of existing parks and reserves will not be enough to save all the species living within them. Only 4.3 percent of the earth's land surface is currently under legal protection, divided among national parks, scientific stations, and other classes of reserves. These fragments represent recently shrunken habitat islands, whose faunas and floras will continue to dwindle until a new, often lower equilibrium is reached. Over 90 percent of the remaining land surface, including most of the surviving high-diversity habitats, has been altered. If the disturbance continues until most of the natural outside reserves are swept away, a majority of the world's terrestrial species will be either extinguished or put at extreme risk. And more: even the existing reserves are in harm's way. Poachers and illegal miners invade them, timber thieves work their margins, developers find ways to convert them in part. During recent civil wars in

Ethiopia, Sudan, Angola, Uganda, and other African countries, many of the national parks were left to ruin.

So we should try to expand reserves from 4.3 percent to 10
percent of the land surface, to include as many of the undisturbed habitats as possible with priority given to the world's
hot spots. One of the more promising means to attain this goal
is by debt-for-nature swaps. As currently practiced, conservation organizations such as Conservation International, the
Nature Conservancy, and the World Wildlife Fund (U.S.) raise
funds to purchase a portion of a country's commercial debt
at a discount, or else they persuade creditor banks to donate
some of it. This first step is easier than it sounds because so
many developing countries are close to default. The debts are
then exchanged in local currency or bonds set at favorable
rates. The enlarged equity is used to promote conservation,
especially by the purchase of land, environmental education,
and the improvement of land management. By early 1992 a
total of twenty such agreements totaling $110 million had been
arranged in nine countries, including Bolivia, Costa Rica,
Dominican Republic, Ecuador, Mexico, Madagascar, Zambia,
the Philippines, and Poland.

In February 1991, to take one example, Conservation International was authorized to buy $4 million in debt from Mexico's
creditors. After discounting on secondary markets, the actual
cost is expected to be as little as $1.8 million. The conservation
organization has agreed to forgive the full amount in return
for the expenditure of $2.6 million by the Mexican government
on a broad range of conservation projects. The most important
initiative will be to preserve the Lacandan tract in the extreme
south of Mexico, the largest rain forest in North America.

The debt of Third World countries has been reduced so far
by only one part in 10,000 through debt-for-nature swaps.
Nor are the arrangements without risk for the receiving country, notably in the crowding out of domestic expenditures and
a sparking of local inflation. But these temporary effects are
offset by the immense gain, dollar for dollar, in the stabiliza-
tion of the environment.

More potent still are unencumbered contributions from
wealthier nations channeled and carefully targeted through

international assistance organizations. The most important enterprise of this kind is the Global Environment Facility (GEF), established in 1990 by the World Bank, the United Nations Environmental Program, and the United Nations Development Program. At this writing, $450 million has been committed to set up national parks, promote sustainable forestry, and establish conservation trust funds in developing countries. Under consideration or already approved are proposals from Bhutan, Indonesia, Papua New Guinea, the Philippines, Vietnam, and the Central African Republic. Two principal difficulties have appeared within the GEF agenda. One is the limited absorptive power of the recipient nations. With limited trained personnel and expert knowledge, national leaders find it difficult to select the best projects and initiate them effectively. Of much greater significance, the brief terms of funding leave little prospect for the proper management and protection of reserves when the money runs out. Fearing loss of employment, the brightest professionals are likely to look to other activities to ensure their futures. The solution to both problems may lie in the establishment of national trust funds, producing income that can be fed into the conservation programs gradually and over a period of many years. One such fund has recently been established for Bhutan with the help of the World Wildlife Fund.

We come then to the design of the reserves themselves. As land is set aside, the primary goal is to place the reserves in the regions of highest diversity and to make them as large as possible. Another goal is to design their shape and spacing for retaining efficiency. In approaching that secondary end, a debate has arisen in conservation circles on the so-called SLOSS problem: whether to invest allotted land into a Single Large reserve Or into Several Small reserves. A single large reserve, to put the matter as simply as possible, possesses larger populations of each species, but they all fit into one basket. A single catastrophic fire or flood could extinguish a large part of the diversity of the region. Breaking the reserve into several pieces reduces that problem, but it also diminishes the size of the constituent populations and hence threatens each with extinction. All might easily decline in the face of widespread stress, such as drought or unseasonable cold.

Some biologists have suggested a compromise solution to the SLOSS problem, which is to create small reserves connected by corridors of natural habitat. For example, several forest patches (say 10 kilometers square each) might be joined by strips of forest 100 meters across. Then if a species vanishes from one of the patches, it can be replaced by colonists immigrating along the forest corridor from another patch. The disadvantage that critics of the compromise have been quick to identify is that disease, predators, and exotic competitors can also use corridors to move through the network. Since populations in the patches are small and vulnerable, all might fall like a row of dominoes. I doubt that any general principle of population dynamics exists that can resolve the SLOSS controversy, at least not in the clean manner suggested by its simple geometric imagery. Instead each ecosystem must be studied in turn to decide the best design, which will depend on the species the system contains and the year-by-year fluctuation of its physical environment. For the time being, conservation biologists will agree on the cardinal rule: to save the most biodiversity, make the reserves as large as possible.

5. *Restore the wildlands.* The grim signature of our time has been the reduction of natural habitats until a substantial portion of the kinds of plants and animals, certainly more than 10 percent, have already vanished or else are consigned to early extinction. The toll of genetic races has never been estimated, but it is almost certainly much higher than that of species. Yet there is still time to save many of the "living dead"—those so close to the brink that they will disappear soon even if merely left alone. The rescue can be accomplished if natural habitats are not only preserved but enlarged, sliding the numbers of survivable species back up the logarithmic curve that connects quantity of biodiversity to amount of area. Here is the means to end the great extinction spasm. The next century will, I believe, be the era of restoration in ecology.

In haphazard manner, largely through the abandonment of small farms, the area of coniferous and hardwood forests in the eastern United States has increased during the past hundred years. Deliberate efforts to enlarge wild areas are also underway. In 1935 a pioneering effort resulted in the planting of

24 hectares of tall-grass prairie at the University of Wisconsin Arboretum. The arboretum has also served as the headquarters of the Center for Restoration Ecology, devoted to research and the collation of information from projects in other parts of the country. Elsewhere in the United States, small restoration projects by the hundreds have been initiated, all devoted to the increase in area of natural habitats and the return of degraded ecosystems to full health. They range broadly in ecosystem types, from the ironwood groves of Santa Catalina Island to the Tobosa grassland of Arizona, the oakland understory of California's Santa Monica Mountains, the magnificent open mountain woodlands of Colorado, and last savanna remnants of Illinois. They include fragments of salt and freshwater wetlands from California to Florida and Massachusetts.

In Costa Rica an audacious effort by the American ecologist Daniel Janzen and local conservation leaders has led to the establishment of Guanacaste National Park, a 50,000-hectare reserve in the northwestern corner of the country. The park will be created—literally created—by the regrowth of dry tropical forest planted on cattle ranches. The Guanacaste dream was born of recognition that in Central America dry forest is even more threatened than rain forest, down to only 2 percent of its original cover. The plan is to use existing patches of the original forest to seed a steadily growing area of ranchland. The conversion will be made easier by the low density of the human population in the area. The regenerating woodland will provide a protected watershed, an income from tourism expected to reach $1 million or more annually, and a net increase in employment of the area's residents. Most important in the long run, it will save a significant part of Costa Rica's natural heritage.

I have spoken of the salvage and regeneration of existing ecosystems. There will come a time when even more is possible with the aid of scientific knowledge. The return to biology's Eden might also include the creation of synthetic faunas and floras, assemblages of species carefully selected from different parts of the world and introduced into impoverished habitats. The idea struck home for me one late afternoon as I sat at the edge of the artificial lake near the center of the University of Miami campus, surrounded by the densely urbanized community of Coral Gables. At least six species of fishes swarmed in the clear brackish water within 2 meters of shore, some as solitary foragers, others in schools. Most were exotics. Their unusual diversity and beauty reminded me of a newly created coral reef. As the sun set and the water darkened, a large predator fish, probably a gar, broke the surface in the middle of the lake. A small alligator glided out from reeds across the way and cruised into open water. Well beyond the far shore, a flock of parrots returned noisily to their palm-top evening roost. They belonged to one of more than twenty exotic species that breed or occur in the Miami area, all originating from individuals that escaped or were deliberately released from captivity. Thus has the parrot family, the Psittacidae, returned to Florida with a vengeance, only decades after the extermination of the Carolina parakeet, last of the endemic North American species. With flashing wings they salute the vanished native.

It is dangerous, I must quickly add, to think too freely of introducing exotics anywhere. They might or might not take to the new environment—between 10 and 50 percent of bird species have succeeded, depending on the part of the world and the number of attempts made to introduce them. Exotics might become economic pests or force out native species. A few, like rabbits, goats, pigs, and the notorious Nile perch are capable not only of extinguishing individual species but of degrading entire habitats. Ecology is still too primitive a science to predict the outcome of the synthesis of predesigned biotas. No responsible person will risk dumping destroyers into the midst of already diminished communities. Nor should we delude ourselves into thinking that synthetic biotas increase global diversity. They only increase local diversity by expanding the ranges and population sizes of selected species.

Yet the search for the safe rules of biotic synthesis is an enterprise of high intellectual daring. If the effort is successful, regions already stripped of their native biotas can be restored to places of diversity and environmental stability. A wilderness of sorts can be reborn in the wasteland. Species already extinct in the wild, those now maintained in zoos and gardens, deserve high priority. Transplanted into impoverished or synthetic biotas, they can endure as orphan species in foster ecosystems. Even though their original home has been closed to them, they will regain security and independence. They will repay us by attaining one criterion of wilderness—that we are allowed to lay down the burden of their care and visit them as equal partners, on our own time. A few species will be prosthetic. As keystone elements, such as a tree able to grow rapidly and shelter many other plant and animal species, they will play a disproportionate role in holding the new communities together.

Finally, the question of central interest is how much of the world's biodiversity we can expect to carry with us out of the bottleneck fifty or a hundred years hence. Let me venture a guess. If the biodiversity crisis remains largely ignored and natural habitats continue to decline, we will lose at least one quarter of the earth's species. If we respond with the knowledge and technology already possessed, we may hold the loss to 10 percent. At first glance the difference may seem bearable. It is not; it amounts to millions of species.

I feel no hesitance in urging the strong hand of protective law and international protocols in the preservation of biological wealth, as opposed to tax incentives and marketable pollution permits. In democratic societies people may think that their government is bound by an ecological version of the Hippocratic oath, to take no action that knowingly endangers biodiversity. But that is not enough. The commitment must be much deeper—to let no species knowingly die, to take all reasonable action to protect every species and race in perpetuity. The government's moral responsibility in the conservation of biodiversity is similar to that in public health and military defense. The preservation of species across generations is beyond the capacity of individuals or even powerful private institutions. Insofar as biodiversity is deemed an irreplaceable public resource, its protection should be bound into the legal canon.

CHAPTER FIFTEEN

The Environmental Ethic

THE SIXTH GREAT extinction spasm of geological time is upon us, grace of mankind. Earth has at last acquired a force that can break the crucible of biodiversity. I sensed it with special poignancy that stormy night at Fazenda Dimona, when lightning flashes revealed the rain forest cut open like a cat's eye for laboratory investigation. An undisturbed forest rarely discloses its internal anatomy with such clarity. Its edge is shielded by thick secondary growth or else, along the river bank, the canopy spills down to ground level. The nighttime vision was a dying artifact, a last glimpse of savage beauty.

A few days later I got ready to leave Fazenda Dimona: gathered my muddied clothes in a bundle, gave my imitation Swiss army knife to the cook as a farewell gift, watched an overflight of Amazonian green parrots one more time, labeled and stored my specimen vials in reinforced boxes, and packed my field notebook next to a dog-eared copy of Ed McBain's police novel *Ice,* which, because I had neglected to bring any other reading matter, was now burned into my memory.

Grinding gears announced the approach of the truck sent to take me and two of the forest workers back to Manaus. In bright sunlight we watched it cross the pastureland, a terrain strewn with fire-blackened stumps and logs, the battlefield my forest had finally lost. On the ride back I tried not to look at the bare fields. Then, abandoning my tourist Portuguese, I turned inward and daydreamed. Four splendid lines of Virgil came to mind, the only ones I ever memorized, where the Sibyl warns Aeneas of the Underworld:

> The way downward is easy from Avernus.
> Black Dis's door stands open night and day.
> But to retrace your steps to heaven's air,
> There is the trouble, there is the toil . . .

For the green prehuman earth is the mystery we were chosen to solve, a guide to the birthplace of our spirit, but it is slipping away. The way back seems harder every year. If there is danger in the human trajectory, it is not so much in the survival of our own species as in the fulfillment of the ultimate irony of organic evolution: that in the instant of achieving self-understanding through the mind of man, life has doomed its most beautiful creations. And thus humanity closes the door to its past.

The creation of that diversity came slow and hard: 3 billion years of evolution to start the profusion of animals that occupy the seas, another 350 million years to assemble the rain forests in which half or more of the species on earth now live. There was a succession of dynasties. Some species split into two or several daughter species, and their daughters split yet again to create swarms of descendants that deployed as plant feeders, carnivores, free swimmers, gliders, sprinters, and burrowers, in countless motley combinations. These ensembles then gave way by partial or total extinction to newer dynasties, and so on to form a gentle upward swell that carried biodiversity to a peak—just before the arrival of humans. Life had stalled on plateaus along the way, and on five occasions it suffered extinction spasms that took 10 million years to repair. But the thrust was upward. Today the diversity of life is greater than it was 100 million years ago—and far greater than 500 million years before that.

Most dynasties contained a few species that expanded disproportionately to create satrapies of lesser rank. Each species and its descendants, a sliver of the whole, lived an average of hundreds of thousands to millions of years. Longevity varied according to taxonomic group. Echinoderm lineages, for example, persisted longer than those of flowering plants, and both endured longer than those of mammals.

Ninety-nine percent of all the species that ever lived are now extinct. The modern fauna and flora are composed of survivors that somehow managed to dodge and weave through all the radiations and extinctions of geological history. Many contemporary world-dominant groups, such as rats, ranid frogs, nymphalid butterflies, and plants of the aster family Compositae, attained their status not long before the Age of Man. Young or old, all living species are direct descendants of the organisms that lived 3.8 billion years ago. They are living genetic libraries, composed of nucleotide sequences, the equivalent of words and sentences, which record evolutionary events all across that immense span of time. Organisms more complex than bacteria—protists, fungi, plants, animals—contain between 1 and 10 billion nucleotide letters, more than enough in pure information to compose an equivalent of the *Encyclopaedia Britannica*. Each species is the product of mutations and recombinations too complex to be grasped by unaided intuition. It was sculpted and burnished by an astronomical number of events in natural selection, which killed off or otherwise blocked from reproduction the vast majority of its member organisms before they completed their lifespans. Viewed from the perspective of evolutionary time, all other species are our distant kin because we share a remote ancestry. We still use a common vocabulary, the nucleic-acid code, even though it has been sorted into radically different hereditary languages.

Such is the ultimate and cryptic truth of every kind of organism, large and small, every bug and weed. The flower in the crannied wall—it is a miracle. If not in the way Tennyson, the Victorian romantic, bespoke the portent of full knowledge (by which "I should know what God and man is"), then certainly a consequence of all we understand from modern biology. Every kind of organism has reached this moment in time by threading one needle after another, throwing up brilliant artifices to survive and reproduce against nearly impossible odds.

Organisms are all the more remarkable in combination. Pull out the flower from its crannied retreat, shake the soil from the roots into the cupped hand, magnify it for close examination. The black earth is alive with a riot of algae, fungi, nematodes, mites, springtails, enchytraeid worms, thousands of species of bacteria. The handful may be only a tiny fragment of one ecosystem, but because of the genetic codes of its residents it holds more order than can be found on the surfaces of all the planets combined. It is a sample of the living force that runs the earth—and will continue to do so with or without us.

We may think that the world has been completely explored. Almost all the mountains and rivers, it is true, have been named, the coast and geodetic surveys completed, the ocean floor mapped to the deepest trenches, the atmosphere transected and chemically analyzed. The planet is now continuously monitored from space by satellites; and, not least, Antarctica, the last virgin continent, has become a research station and expensive tourist stop. The biosphere, however, remains obscure. Even though some 1.4 million species of organisms have been discovered (in the minimal sense of having specimens collected and formal scientific names attached), the total number alive on earth is somewhere between 10 and 100 million. No one can say with confidence which of these figures is the closer. Of the species given scientific names, fewer than 10 percent have been studied at a level deeper than gross anatomy. The revolution in molecular biology and medicine was achieved with a still smaller fraction, including colon bacteria, corn, fruit flies, Norway rats, rhesus monkeys, and human beings, altogether comprising no more than a hundred species.

Enchanted by the continuous emergence of new technologies and supported by generous funding for medical research, biologists have probed deeply along a narrow sector of the front. Now it is time to expand laterally, to get on with the great Linnean enterprise and finish mapping the biosphere. The most compelling reason for the broadening of goals is that, unlike the rest of science, the study of biodiversity has a time limit. Species are disappearing at an accelerating rate through human action, primarily habitat destruction but also pollution and the introduction of exotic species into residual natural environments. I have said that a fifth or more of the species of plants and animals could vanish or be doomed to early extinction by the year 2020 unless better efforts are made to save them. This estimate comes from the known quantitative relation between the area of habitats and the diversity that habitats can sustain. These area-biodiversity curves are supported by the general but not universal principle that when certain groups of organisms are studied closely, such as snails and fishes and flowering plants, extinction is determined to be widespread. And the corollary: among plant and animal remains in archaeological deposits, we usually find extinct species and races. As the last forests are felled in forest strongholds like the Philippines and Ecuador, the decline of species will accelerate even more. In the world as a whole, extinction rates are already hundreds or thousands of times higher than before the coming of man. They cannot be balanced by new evolution in any period of time that has meaning for the human race.

Why should we care? What difference does it make if some species are extinguished, if even half of all the species on earth disappear? Let me count the ways. New sources of scientific information will be lost. Vast potential biological wealth will be destroyed. Still undeveloped medicines, crops, pharmaceuticals, timber, fibers, pulp, soil-restoring vegetation, petroleum substitutes, and other products and amenities will never come to light. It is fashionable in some quarters to wave aside the small and obscure, the bugs and weeds, forgetting that an obscure moth from Latin America saved Australia's pastureland from overgrowth by cactus, that the rosy periwinkle provided the cure for Hodgkin's disease and childhood lymphocytic leukemia, that the bark of the Pacific yew offers hope for victims of ovarian and breast cancer, that a chemical from the saliva of leeches dissolves blood clots during surgery, and so on down a roster already grown long and illustrious despite the limited research addressed to it.

In amnesiac revery it is also easy to overlook the services that ecosystems provide humanity. They enrich the soil and create the very air we breathe. Without these amenities, the remaining tenure of the human race would be nasty and brief. The life-sustaining matrix is built of green plants with legions of microorganisms and mostly small, obscure animals—in other words, weeds and bugs. Such organisms support the world with efficiency because they are so diverse, allowing them to divide labor and swarm over every square meter of the earth's surface. They run the world precisely as we would wish it to be run, because humanity evolved within living communities and our bodily functions are finely adjusted to the idiosyncratic environment already created. Mother Earth, lately called Gaia, is no more than the commonality of organisms and the physical environment they maintain with each passing moment, an environment that will destabilize and turn lethal if the organisms are disturbed too much. A near infinity of other mother planets can be envisioned, each with its own fauna and flora, all producing physical environments uncongenial to human life. To disregard the diversity of life is to risk catapulting ourselves into an alien environment. We will have become like the pilot whales that inexplicably beach themselves on New England shores.

Humanity coevolved with the rest of life on this particular planet; other worlds are not in our genes. Because scientists have yet to put names on most kinds of organisms, and because they entertain only a vague idea of how ecosystems work, it is reckless to suppose that biodiversity can be diminished indefinitely without threatening humanity itself. Field studies show that as biodiversity is reduced, so is the quality of the services provided by ecosystems. Records of stressed ecosystems also demonstrate that the descent can be unpredictably abrupt. As extinction spreads, some of the lost forms prove to be keystone species, whose disappearance brings down other species and triggers a ripple effect through the demographies of the survivors. The loss of a keystone species is like a drill accidentally striking a powerline. It causes lights to go out all over.

These services are important to human welfare. But they cannot form the whole foundation of an enduring environmental ethic. If a price can be put on something, that something can be devalued, sold, and discarded. It is also possible for some to dream that people will go on living comfortably in a biologically impoverished world. They suppose that a prosthetic environment is within the power of technology, that human life can still flourish in a completely humanized world, where medicines would all be synthesized from chemicals off the shelf, food grown from a few dozen domestic crop species, the atmosphere and climate regulated by computer-driven fusion energy, and the earth made over until it becomes a literal spaceship rather than a metaphorical one, with people reading displays and touching buttons on the bridge. Such is the terminus of the philosophy of exemptionalism: do not weep for the past, humanity is a new order of life, let species die if they block progress, scientific and technological genius will find another way. Look up and see the stars awaiting us.

But consider: human advance is determined not by reason alone but by emotions peculiar to our species, aided and tempered by reason. What makes us people and not computers is emotion. We have little grasp of our true nature, of what it is to be human and therefore where our descendants might someday wish we had directed Spaceship Earth. Our troubles, as Vercors said in *You Shall Know Them,* arise from the fact that we do not know what we are and cannot agree on what we want to be.

The primary cause of this intellectual failure is ignorance of our origins. We did not arrive on this planet as aliens. Humanity is part of nature, a species that evolved among other species. The more closely we identify ourselves with the rest of life, the more quickly we will be able to discover the sources of human sensibility and acquire the knowledge on which an enduring ethic, a sense of preferred direction, can be built.

The human heritage does not go back only for the conventionally recognized 8,000 years or so of recorded history, but for at least 2 million years, to the appearance of the first "true" human beings, the earliest species composing the genus *Homo.* Across thousands of generations, the emergence of culture must have been profoundly influenced by simultaneous events in genetic evolution, especially those occurring in the anatomy and physiology of the brain. Conversely, genetic evolution must have been guided forcefully by the kinds of selection rising within culture.

Only in the last moment of human history has the delusion arisen that people can flourish apart from the rest of the living world. Preliterate societies were in intimate contact with a bewildering array of life forms. Their minds could only partly adapt to that challenge. But they struggled to understand the most relevant parts, aware that the right responses gave life and fulfillment, the wrong ones sickness, hunger, and death. The imprint of that effort cannot have been erased in a few generations of urban existence. I suggest that it is to be found among the particularities of human nature, among which are these:

• People acquire phobias, abrupt and intractable aversions, to the objects and circumstances that threaten humanity in natural environments: heights, closed spaces, open spaces, running water, wolves, spiders, snakes. They rarely form phobias to the recently invented contrivances that are far more dangerous, such as guns, knives, automobiles, and electric sockets.

• People are both repelled and fascinated by snakes, even when they have never seen one in nature. In most cultures the serpent is the dominant wild animal of mythical and religious symbolism. Manhattanites dream of them with the same frequency as Zulus. This response appears to be Darwinian in origin. Poisonous snakes have been an important cause of mortality almost everywhere, from Finland to Tasmania, Canada to Patagonia; an untutored alertness in their presence saves lives. We note a kindred response in many primates, including Old World monkeys and chimpanzees: the animals pull back, alert others, watch closely, and follow each potentially dangerous snake until it moves away. For human beings, in a larger metaphorical sense, the mythic, transformed serpent has come to possess both constructive and destructive powers: Ashtoreth of the Canaanites, the demons Fu-Hsi and Nu-kua of the Han Chinese, Mudamma and Manasa of Hindu India, the triple-headed giant Nehebkau of the ancient Egyptians, the serpent of Genesis conferring knowledge and death, and, among the Aztecs, Cihuacoatl, goddess of childbirth and mother of the human race, the rain god Tlaloc, and Quetzalcoatl, the plumed serpent with a human head who reigned as lord of the morning and evening star. Ophidian power spills over into modern life: two serpents entwine the caduceus, first the winged staff of Mercury as messenger of the gods, then the safe-conduct pass of ambassadors and heralds, and today the universal emblem of the medical profession.

• The favored living place of most peoples is a prominence near water from which parkland can be viewed. On such heights are found the abodes of the powerful and rich, tombs of the great, temples, parliaments, and monuments commemorating tribal glory. The location is today an aesthetic choice and, by the implied freedom to settle there, a symbol of status. In ancient, more practical times the topography provided a place to retreat and a sweeping prospect from which to spot the distant approach of storms and enemy forces. Every animal species selects a habitat in which its members gain a favorable mix of security and food. For most of deep history, human beings lived in tropical and subtropical savanna in East Africa, open country sprinkled with streams and lakes, trees and copses. In similar topography modern peoples choose their residences and design their parks and gardens, if given a free choice. They simulate neither dense jungles, toward which gibbons are drawn, nor dry grasslands, preferred by hamadryas baboons. In their gardens they plant trees that resemble the acacias, sterculias, and other native trees of the African savannas. The ideal tree crown sought is consistently wider than tall, with spreading lowermost branches close enough to the ground to touch and climb, clothed with compound or needle-shaped leaves.

• Given the means and sufficient leisure, a large portion of the populace backpacks, hunts, fishes, birdwatches, and gardens. In the United States and Canada more people visit zoos and aquariums than attend all professional athletic events combined. They crowd the national parks to view natural landscapes, looking from the tops of prominences out across rugged terrain for a glimpse of tumbling water and animals living free. They travel long distances to stroll along the sea-shore, for reasons they can't put into words.

These are examples of what I have called *biophilia*, the connections that human beings subconsciously seek with the rest of life. To biophilia can be added the idea of wilderness, all the land and communities of plants and animals still unsullied by human occupation. Into wilderness people travel in search of new life and wonder, and from wilderness they return to the parts of the earth that have been humanized and made physically secure. Wilderness settles peace on the soul because it needs no help; it is beyond human contrivance. Wilderness is a metaphor of unlimited opportunity, rising from the tribal memory of a time when humanity spread across the world, valley to valley, island to island, godstruck, firm in the belief that virgin land went on forever past the horizon.

I cite these common preferences of mind not as proof of an innate human nature but rather to suggest that we think more carefully and turn philosophy to the central questions of human origins in the wild environment. We do not understand ourselves yet and descend farther from heaven's air if we forget how much the natural world means to us. Signals abound that the loss of life's diversity endangers not just the body but the spirit. If that much is true, the changes occurring now will visit harm on all generations to come.

The ethical imperative should therefore be, first of all, prudence. We should judge every scrap of biodiversity as priceless while we learn to use it and come to understand what it means to humanity. We should not knowingly allow any species or race to go extinct. And let us go beyond mere salvage to begin the restoration of natural environments, in order to enlarge

Dreams of Amazonia. The exuberance of rain-forest life is captured in this detail from a work by Roxana Elizabeth Marín, a native student of the Usko-Ayar Amazonian school of painting in Peru. The founder of the school is Pablo César Amaringo, a *vegetalista*, one of the mestizo shamans and healers of the region who derive their knowledge and personal powers from plants. Drugs, such as *ayahuasca* extracted from a local species of vine (*Banisteriopsis caapi*), are used to summon dreams of animals, plants, and spirits, which range from the realistic to the fantastical and from literal natural history to tribal mythology.

wild populations and stanch the hemorrhaging of biological wealth. There can be no purpose more enspiriting than to begin the age of restoration, reweaving the wondrous diversity of life that still surrounds us.

The evidence of swift environmental change calls for an ethic uncoupled from other systems of belief. Those committed by religion to believe that life was put on earth in one divine stroke will recognize that we are destroying the Creation, and those who perceive biodiversity to be the product of

Two visions painted by the Peruvian shaman Pablo Amaringo. Above: "The session of the Chullachaki," in which the forest spirit Chullachaki transmits instructions for the care of wild animals. Below: "The Sachamama," the nearly invisible giant snake of the deep forest.

blind evolution will agree. Across the other great philosophical divide, it does not matter whether species have independent rights or, conversely, that moral reasoning is uniquely a human concern. Defenders of both premises seem destined to gravitate toward the same position on conservation.

The stewardship of environment is a domain on the near side of metaphysics where all reflective persons can surely find common ground. For what, in the final analysis, is morality but the command of conscience seasoned by a rational examination of consequences? And what is a fundamental precept but one that serves all generations? An enduring environmental ethic will aim to preserve not only the health and freedom of our species, but access to the world in which the human spirit was born.

Glossary

Included in this list of terms is biographical information for those scientists and other principal contributors to biodiversity studies mentioned in the text.

abyssal benthos The community of organisms living on or close to the floor of the deep sea.

adaptation A particular part of the anatomy (such as color), a physiological process (such as respiration rate), or behavior pattern (such as a mating dance) that improves an organism's chances to survive and reproduce. Also the evolution that creates such a trait.

adaptive radiation The evolution of a single species into many species that occupy diverse ways of life within the same geographical range. Example: the origin of kangaroos, koalas, and other present-day Australian marsupials from a single distant ancestor.

allele A particular form of a gene, where multiple such forms occur. Sickle-cell anemia is caused by one such variant of a gene; another variant of the same gene contributes to normal hemoglobin.

allometry The condition in which one part of the body grows faster relative to another part, so that the larger the organism, the greater the disproportion; the large males of many kinds of beetles and deer, for example, develop horns that are enormous in comparison with the rest of the body.

allopatric Occupying different geographical ranges.

allopatric speciation The same as geographic speciation: the splitting of a population into two or more subpopulations by a geographic barrier, followed by the evolutionary divergence of the population until it attains the status of full species.

alternation of generations The alternation of haploid organisms (possessing one chromosome of each kind per cell) in one generation with diploid organisms (two chromosomes of each kind per cell) in the next generation, then back to haploid organisms, and so on. In most species the haploid generation consists of only the eggs and sperm, which fuse to create the diploid generation, which in time produces more eggs and sperm, the next haploid generation.

Alvarez, Luis W. (1911–1988)   Particle physicist at the University of California, Berkeley; led the team that discovered a high level of iridium at the Cretaceous-Tertiary boundary and interpreted it as the result of massive meteorite strikes.

amphibian   A member of the vertebrate class Amphibia, such as a frog or salamander.

analogy   In biology, a resemblance in appearance and function between structures that occur in two kinds of organisms but not because of common ancestry. The wings of birds and insects are analogues, they are analogous to each other, and the similarity constitutes an analogy; they were not evolved from the same organ in a common ancestor. Cf. *homology.*

angiosperm   A flowering plant, member of the plant phylum that dominates land vegetation, characterized by seeds from fruits.

annelid   A worm of the phylum Annelida, such as an earthworm, leech, or nereid.

area-species curve   The relationship between the area of an island or some other discrete geographic region and the number of species living there. Approximated by the equation $S = CA^Z$, where A is the area, S is the number of species, and C and z are constants that depend on the place and group of organisms (such as birds or trees) being considered. Also called the species-area curve.

arthropod   A member of the phylum Arthropoda, such as an insect, spider, or crustacean, bearing an articulated, external skeleton.

asexual species   Populations of organisms that are different enough to be conveniently distinguished as species, even though they do not reproduce sexually and the criterion of reproductive isolation cannot be applied to them.

assembly rules   The combinations of species that can live together in a community of animals and plants and the sequences in which they can invade and persist in the community.

autochthon   A species that originated in a certain place, such as New Zealand or Lake Victoria, and is found only there. Cf. *endemic.*

bacteria   Microscopic single-celled organisms that are prokaryotic, or lacking nuclear membranes around the genes.

base pair   A pair of organic bases constituting a letter of the genetic code; usually adenine (A) paired with thymine (T), or cytosine (C) paired with guanine (G). Each base is sited on one strand of the DNA double helix and opposes the other base at the same position on the second strand. The code is then read off as a sequence of four possible letters on the double helix, AT, TA, CG, and GC. Versions of the same gene differ by the sequence of these four letters.

biodiversity   The variety of organisms considered at all levels, from genetic variants belonging to the same species through arrays of species to arrays of genera, families, and still higher taxonomic levels; includes the variety of ecosystems, which comprise both the communities of organisms within particular habitats and the physical conditions under which they live.

biodiversity studies   The systematic examination of the full array of different kinds of organisms, together with a consideration of the technology by which the diversity can be maintained and used for the benefit of humanity.

bioeconomic analysis   The assessment of the potential economic value of all the organisms in an ecosystem, from their natural products to their use in ecotourism.

biogeography   The scientific study of the geographic distribution of organisms.

biological diversity   See *biodiversity.*

biomass   The total weight (usually, dry weight) of a designated group of organisms in a particular area, as of all the birds living in a woodlot or all the algae in a pond or all the organisms in the world.

biome   A major category of habitat in a particular region of the world, such as the tundra of northern Canada or the rain forest of the Amazon basin.

bioregion   A continuous natural area, such as a river system or mountain range, large enough to extend beyond political boundaries.

biota   The combined flora, fauna, and microorganisms of a given region. Microorganisms are often referred to as fauna or flora depending on the group to which they belong, such as the bacterial flora.

biotic   Biological, especially referring to the characteristics of faunas, floras, and ecosystems.

Bush, Guy L. (1929–)   Entomologist and evolutionary biologist at Michigan State University; the foremost researcher on host races and their role in the formation of species.

Cambrian   Referring to the earliest period of the Paleozoic era, extending 550–500 million years ago, when larger marine animals vastly increased in both numbers and diversity—the Cambrian explosion of animal evolution.

centimeter   One-hundredth of a meter; equal to 0.39 inch (2.5 centimeters is approximately 1 inch).

centinelan extinctions   An expression (proposed in this book) to designate extinctions of species unknown before their demise and hence unrecorded.

character   A trait that varies so as to be useful in classification, such as a flower part that varies among plants or a dental formula that varies among mammals. The differences from one species to the next are called character states.

character displacement   The process by which two species evolve away from one another, acquiring greater differences, as a result of competition or the risk of lowered survival and fertility caused by hybridization.

chemical prospecting   The screening of wild species of plants, animals, and microorganisms for natural substances of practical use, especially in medicine.

chromosome   A structure, visible under the light microscope and usually rod-shaped, that carries genes. Chromosomes are made up of DNA, which compose the genes, and a supporting matrix of proteins.

chronospecies   A population that evolves so much that it is regarded as a different species, even though the population has not divided into multiple coexisting species; its division into two species across time is based on subjective judgment about the degree of change.

clade   A group of species all descended from a common ancestor. Cats of the genus *Felis*, living and extinct, are a clade descended from a single ancestor that lived in the geological past. The clade includes the ancestor.

class In classification, a group of species of common ancestry ranked below the phylum and above the order; hence one or more orders.

coevolution The evolution of two or more species due to mutual influence; for example, many species of flowering plants and their insect pollinators have coevolved in a way that makes the relationship more effective.

Cohen, Joel E. (1944–) Professor of population biology at Rockefeller University; a major contributor to the interpretation of food webs in ecosystems.

commensalism A form of symbiosis (intimate coexistence) in which one species profits from the association without harming or benefiting the other.

community All the organisms—plants, animals, and micro-organisms—that live in a particular habitat and affect one another as part of the food web or through their various influences on the physical environment.

competitive exclusion The extinction of one species by another in a habitat through competition.

conservation biology The relatively new discipline that treats the content of biodiversity, the natural processes that produce it, and the techniques used to sustain it in the face of human-caused environmental disturbance.

continental drift The gradual breaking up of the continents that has occurred steadily over the past 200 million years.

convergence In evolutionary biology, the same as convergent evolution, the increasing similarity during evolution of two or more unrelated species. Example: the placental wolf of the northern hemisphere and its remarkable look-alike, the marsupial "wolf" of Australia.

Conway Morris, Simon (1951–) A paleontologist at the University of Cambridge; a leading scholar on the Cambrian explosion of invertebrate animals and early evolution of arthropods.

cryopreservation The storage of organisms and tissue samples at extremely cold temperatures, usually in liquid nitrogen.

cyanobacteria Formerly called blue-green algae, these organisms are not true algae but single-celled prokaryotes resembling bacteria. They were dominant elements during the early history of life and are still ecologically prominent.

Darwin, Charles Robert (1809–1882)   Originator with Alfred Russel Wallace of the theory of evolution by natural selection, author of *On the Origin of Species,* and hence founder of the mode of evolutionary thinking that pervades biology today.

Darwinism   Evolution by natural selection, originally proposed by Charles Darwin. The modern interpretation of the process is called neo-Darwinism; it incorporates all we know about evolution from genetics, ecology, and other disciplines.

debt-for-nature swap   The purchasing or forgiving of portions of the debt of poorer countries in exchange for local conservation projects, especially the purchase of land.

deme   A population of organisms in which breeding is completely random—an important idealized concept used as a standard to calculate the degrees of inbreeding and genetic drift.

demography   The study of birth rates, death rates, age distributions, sex ratios, and size of populations—a fundamental discipline within the larger field of ecology. Also the properties themselves, as in the demography (demographic traits) of a particular population.

density dependence   The increasing severity by which factors in the environment slow down growth of a population as the organisms become more numerous and hence densely concentrated. Density-dependent factors include competition, food shortage, disease, predation, and emigration.

DeVries, Philip J. (1952–)   Tropical field biologist and author of the widely admired field guide *The Butterflies of Costa Rica* (1987).

Diamond, Jared M. (1937–)   Professor at the University of California, Los Angeles, School of Medicine; explorer of New Guinea bird fauna, inventor of the concept of assembly rules in community organization, and an influential scholar of the extinction process.

diploid   Having two copies of the same complement of chromosomes in each cell. The diploid condition usually arises from fertilization, during which one set of chromosomes from the male is joined with a second set from the female. Cf. *haploid*.

disharmony   In biodiversity studies, the gross overrepresentation of some groups of organisms and underrepresentation or absence of others on an island or continent owing to accidents of dispersal. Example: there are no native woodpeckers or ants on Hawaii but a great variety of honeycreepers and wasps.

diversity   See *biodiversity.*

DNA   Deoxyribonucleic acid. The fundamental hereditary material of all living organisms; the polymer composing the genes.

dominance   In genetics, the expression of one form of a gene over another form of the same gene when both occur on different chromosomes in the same organism; the gene for normal blood clotting, for example, is dominant over the one for hemophilia (failure to clot) in human beings. In ecology, the abundance and ecological influence of one species or group of species over others: pines are dominant plants and beetles are dominant animals. In animal behavior, the control of one individual over another in social groupings.

echinoderm   A member of the phylum Echinodermata, such as a starfish or a sea urchin.

ecological economics   A new interdisciplinary field devoted to protection of the environment and to attainment of sustained economic production.

ecology   The scientific study of the interaction of organisms with their environment, including the physical environment and the other organisms living in it.

ecosystem   The organisms living in a particular environment, such as a lake or a forest (or, in increasing scale, an ocean or the whole planet), and the physical part of the environment that impinges on them. The organisms alone are called the community.

ecosystem services   The role played by organisms in creating a healthful environment for human beings, from production of oxygen to soil genesis and water detoxification.

ecotourism   Tourism focused on attractive and interesting features of the environment, including the fauna and flora.

Ehrlich, Paul R. (1932–)   Professor at Stanford University; a leading researcher on population dynamics and the extinction process; in addition his many books and articles with Anne H. Ehrlich have brought environmental problems to the attention of a worldwide audience.

Eisner, Thomas (1929–2011)   Professor at Cornell University; a foremost entomologist and founder of chemical ecology; developed the concept of chemical prospecting.

Eldredge, Niles (1943–)   Curator of fossil invertebrates at the American Museum of Natural History; a leading authority on trilobites and originator, with Stephen Jay Gould, of the thesis of punctuated equilibrium.

electrophoresis   A method by which substances, especially proteins, are separated from one another on the basis of their electric charge and molecular weights. Used in the study of diversity among both species and organisms belonging to the same species.

endangered   Near extinction. Referring to a species or ecosystem so reduced or fragile that it is doomed or at least fatally vulnerable.

endemic   A species or race native to a particular place and found only there. If it originated in the same place by evolution, it is also called an autochthon.

environment   The surroundings of an organism or a species: the ecosystem in which it lives, including both the physical environment and the other organisms with which it comes in contact.

eon   The major division of geological time. The most recent such division, the Phanerozoic eon, spans the past 550 million years.

epiphyll   A plant that grows on the leaves of other kinds of plants, hence a specialized form of epiphyte.

epiphyte   A plant specialized to grow on other kinds of plants in a neutral or beneficial manner, not as a parasite. Examples: most species of orchids and bromeliads.

epoch   The division of geological time just below the period in rank. We live in the Recent epoch, which began 10,000 years ago with the end of the Pleistocene epoch, or Ice Age.

equilibrium   See *species equilibrium.*

era   A major subdivision of geological time, just below the eon in rank. The Phanerozoic eon, for example, is divided into three eras: the Paleozoic (oldest), Mesozoic, and Cenozoic (most recent).

Erwin, Terry L. (1940–2020)   Curator in entomology at the United States National Museum; a leading expert on beetles, best known for his estimation of the diversity of insects and other arthropods in rain forests.

ethnobotany   The study of plant biology as understood by other cultures, especially those of preliterate peoples, and the practical uses of plants made by those cultures.

eukaryote   An organism whose DNA is enclosed in nuclear membranes. The vast majority of kinds of organisms are eukaryotic; only bacteria and a few other microscopic forms lack such a nuclear envelope.

evolution   In biology, any change in the genetic constitution of a population of organisms. Evolution can vary in degree from small shifts in the frequencies of minor genes to the origins of complexes of new species. Changes of lesser magnitude are called microevolution, and changes at or near the upper extreme are called macroevolution.

evolutionary agent (or force)   Any factor in the external environment or in the bodies of the organisms themselves that induces shifts in the frequencies of genes within populations, hence evolution.

evolutionary biology   An umbrella term for a broad array of disciplines that have in common their focus on the evolutionary process and hence the creation of biodiversity. Evolutionary biology includes the study of molecular evolution, ecology, population biology, systematics, biogeography, and comparative aspects of anatomy, physiology, and animal behavior.

extinction   The termination of any lineage of organisms, from subspecies to species and higher taxonomic categories from genera to phyla. Extinction can be local, in which one or more populations of a species or other unit vanish but others survive elsewhere, or total (global), in which all the populations vanish. When biologists speak of the extinction of a particular species without further qualification, they mean total extinction.

extractive reserve   A wild habitat from which timber, latex, and other natural products are taken on a sustained yield basis with minimal environmental damage and, ideally, without the extinction of native species.

family   In the hierarchical classification of organisms, a group of species of common descent higher than the genus and lower than the order; hence a group of genera. Examples: Felidae (cats) and Fagaceae (beeches and oaks).

fauna   All the animals found in a particular place.

flora   All the plants found in a particular place.

food chain   Part of the food web of a particular community of organisms, consisting of predators and their prey, predators that feed on the predators, and so on all the way from the photosynthesizing plants to top predators (such as eagles and cats) and decomposers that consume the remains of dead organisms.

food web   The complete array of food links in a particular habitat, represented in diagrams by the direction in which energy and nutrients flow from consumed to consumer.

fossil   Any remains left by an organism, such as a track or mineralized bone, that has been preserved through geological time, by which is usually meant a timespan of 10,000 years or longer.

gene   The basic unit of heredity.

gene frequency   For the population as a whole, the percentage of genes at a particular locus that are of one form (allele) as opposed to another, such as the allele for sickle-cell hemoglobin that can be distinguished from the allele for normal hemoglobin.

gene pool   All the genes in all the organisms belonging to a population.

genera   The plural of genus; groups of closely related, similar species.

genetic drift   Evolution in the genetic constitution of a population by chance processes alone.

genome   All the genes of a particular organism or species.

genotype   The genetic constitution of an organism, either prescribing a single trait (such as eye color) or a set of traits (eye color, blood type, etc.).

Gentry, Alwyn H. (1945–1993)   Tropical botanist of the Missouri Botanical Garden; a principal modern explorer of South American plants.

genus   A group of similar species of common descent. Examples: *Canis*, comprising the wolf, domestic dog, and similar species; and *Quercus*, the oaks.

geographic speciation   Also called allopatric speciation. The divergence to species level by populations that originally belonged to the same species but were isolated by a physical barrier such as a sea strait, river valley, or mountain range.

Goksøyr, Jostein (1922–2000) Professor of microbiology at the University of Bergen, Norway; a pioneer in estimation techniques of bacterial diversity.

Gould, Stephen Jay (1941–2002) Professor of geology and curator of fossil invertebrates at Harvard University; the most influential modern popularizer and commentator on evolutionary biology; originator with Niles Eldredge of the thesis of punctuated equilibrium.

Grant, Peter R. (1936–) Professor of zoology at Princeton University; vertebrate ecologist and a leader in the study of the ecology and microevolution of Darwin's finches.

Great American Interchange The migration of North American mammals south and South American mammals north when the Panamanian land bridge came into existence 2.5 million years ago. The process continues to the present day. Attention has been focused on mammals because of their excellent fossil record, but plants and other animals participated as well.

guild A group of species found in the same place that share the same food resource. Examples: the insects of a Rhode Island field that feed on goldenrod pollen; the hawks of a Bolivian rain forest that prey on songbirds.

habitat An environment of a particular kind, such as lake shores or tall-grass prairie; also a particular environment in one place, such as the mountain forest of Tahiti.

habitat island A patch of habitat separated from other patches of the same habitat, such as a glade separated by forest or a lake separated by dry land. Habitat islands are subject to the same ecological and evolutionary processes as "real" islands.

haploid Possessing a chromosome set composed of only one chromosome of each kind, usually encountered in eggs and sperm and characterizing the haploid generation in the alternation of generations.

hectare Metric unit of area; equal to 2.47 acres.

heterozygous Possessing two gene forms (alleles) at the same chromosome position but on different chromosomes. A person carrying an allele for sickle-cell hemoglobin on one chromosome and an allele for normal hemoglobin on the other chromosome is said to be heterozygous for those traits. Cf. *homozygous.*

homology   In biology, a similarity in structure, physiology, or behavior in two species due to inheritance from a common ancestor, whether or not the function is the same. Example: human arms and bat wings. Also two chromosomes of the same type found in the same individual are spoken of as homologous. Cf. *analogy.*

homozygous   Possessing the same gene form (allele) on both chromosomes. A person carrying alleles for sickle-cell hemoglobin on both chromosomes is said to be homozygous for that condition. Cf. *heterozygous.*

host race   A genetically distinct population of organisms that feeds on one kind of plant and lives in the midst of other populations of the same species feeding on other kinds of plants; thought to be an intermediate stage in the formation of full species.

hot spot   A region of the world, such as the island of Madagascar, that is both rich in endemic species and environmentally threatened.

Hubbell, Stephen P. (1942–)   Professor of biology at Princeton University; a leading tropical ecologist and originator of the long-term study of tree diversity on Barro Colorado Island, Panama.

hybrid   The offspring of parents that are genetically dissimilar, especially of parents that belong to different species.

intrinsic isolating mechanism   Any hereditary difference between species that prevents them from interbreeding freely under natural conditions. Examples: different breeding seasons, courtship behavior, or local habitats.

invertebrate   Any animal lacking a backbone of bony segments that enclose the central nerve cord. Most animals are invertebrates, from sea anemones to earthworms, spiders, and butterflies.

island biogeographic theory   The concepts and mathematical models that account for the number of species of organisms found on islands and fragments of habitats. A central idea in the theory is the equilibrium in the species numbers attained when new species arrive and old residents go extinct at the same rate.

Janzen, Daniel H. (1939–)   Professor at the University of Pennsylvania; a leading tropical biologist widely known for his program to regenerate the threatened deciduous forest of Central America.

keystone species   A species, such as the sea otter, that affects the survival and abundance of many other species in the community in which it lives. Its removal or addition results in a relatively significant shift in the composition of the community and sometimes even in the physical structure of the environment.

kilometer 1,000 meters; equal to 0.62 mile.

kingdom The highest category used in classification. Five kingdoms are commonly recognized: Plantae (plants), Animalia (animals), Fungi (mushrooms and other fungi), Protista (or Protoctista, algae and single-celled "animals"), and Monera (bacteria and close relatives).

Knoll, Andrew H. (1951–) Professor of paleobotany at Harvard University; a principal scholar of the history of life, from the earliest Precambrian microorganisms to modern flowering plants.

latitudinal diversity gradient The trend, widespread but not universal among plants and animals, toward greater diversity when passing from polar regions toward the equator.

lichen A compound organism composed of a fungus that harbors either cyanobacteria or single-celled algae. The symbiosis of the two kinds of organisms is mutually beneficial.

life cycle The entire lifespan of an organism from the moment it is conceived (usually at fertilization) to the time it reproduces.

Lovejoy, Thomas E. (1941–) Assistant secretary for external affairs, Smithsonian Institution; expert on South American birds, originator of the giant Forest Fragment Project in the Brazilian Amazon.

MacArthur, Robert H. (1930–1972) Professor at the University of Pennsylvania and Princeton University; a brilliant theoretician of ecology and originator of the theory of island biogeography.

macroevolution Large-scale evolution, entailing major alterations in anatomy or other biological traits, sometimes accompanied by adaptive radiation. Cf. *microevolution*.

mammal An animal of the class Mammalia, characterized by production of milk in the female mammary gland and possession of a body covering of hair.

marsupial An animal, such as an opossum or a kangaroo, characterized (in most species) by a pouch, the marsupium, containing milk glands and serving as a receptacle for the young.

Martin, Paul S. (1928–2010) Paleontologist and professor at the University of Arizona; chief architect of the hypothesis of mass extinctions of the megafauna by prehistoric humans.

May, Robert M. (1936–2020)   Professor of ecology at Oxford University; a foremost theoretical population biologist and an important scholar of the natural processes underlying biodiversity.

Mayr, Ernst (1904–2005)   Professor emeritus at Harvard University; doyen of evolutionary biology; architect of neo-Darwinism and the biological-species concept.

megafauna   The largest animals, weighing over 10 kilograms, such as deer, big cats, elephants, and ostriches.

meiosis   Cell division leading to reduction of the number of chromosomes from two sets to one, which in most kinds of higher organisms leads directly to the production of sex cells. Cf. *mitosis*.

Mesozoic era   Age of Reptiles or Age of Dinosaurs, extending from 245 million years to 66 million years ago.

metamorphosis   A radical change in body form, physiology, and behavior during the growth and development of an organism.

metapopulation   A set of partially isolated populations belonging to the same species. The populations are able to exchange individuals and recolonize sites in which the species has recently become extinct.

meter   Basic metric unit of length; equal to 39.37 inches.

microbial mat   A thin layer of bacteria and cyanobacteria ("blue-green algae") that forms on bare surfaces, sometimes secreting a carbonate base called a stromatolite; one of the earliest of ecosystems, still persisting in some modern environments such as shallow intertidal waters.

microbiology   The scientific study of microscopic organisms, especially bacteria.

microevolution   Evolutionary change of minor degree, such as an increase in size or body part, usually controlled by a relatively small number of genes. Cf. *macroevolution*.

millimeter   One-thousandth of a meter; equal to 0.04 inch (25 millimeters is approximately 1 inch).

mitosis   Cell division in which the chromosomes are exactly duplicated without loss in numbers. Cf. *meiosis*.

mollusk   An animal belonging to the phylum Mollusca, such as a snail or clam.

mutation   Broadly defined as any genetic change in an organism, either from an alteration of DNA composing individual genes or from a shift in the structure or number of chromosomes. Mutations form the new material for evolution.

mutualism   Symbiosis in which both of the partner species benefit.

mycorrhiza   A symbiotic association between fungi and plant roots.

Myers, Norman (1934–2019)   British botanist and conservation biologist; identifier of hot spots; a major scholar in biodiversity studies.

natural selection   The differential contribution of offspring to the next generation by various genetic types belonging to the same population; the mechanism of evolution proposed by Darwin. Distinguished from artificial selection, the same process but carried out with human guidance.

neo-Darwinism   The modern study of the evolutionary process that assigns a central role to natural selection, the idea originally suggested by Darwin and now informed by substantial new knowledge from genetics, ecology, and other modern disciplines of biology.

niche   A vague but useful term in ecology, meaning the place occupied by a species in its ecosystem—where it lives, what it eats, its foraging route, the season of its activity, and so on. In a more abstract sense, a niche is a potential place or role within a given ecosystem into which species may or may not have evolved.

nucleus   In biology, the dense central body of the cell, surrounded by a double nuclear membrane and containing the chromosomes and genes.

Olson, Storrs L. (1944–)   Curator in paleontology at the United States National Museum; authority on fossil birds and pioneer in the study of extinction of island bird faunas, especially those on Hawaii (conducted with Helen F. James).

paleontology   The scientific study of fossils and all aspects of extinct life.

period   A division of geological time just below the era. The Mesozoic era (Age of Reptiles), for example, is divided into three periods: Triassic, Jurassic, and Cretaceous.

Permian period   Last period of the Paleozoic era, extending from 290 million years to 245 million years ago and closing with the greatest extinction spasm of all time.

Phanerozoic eon   The major division of geological time during which most biodiversity has evolved and existed, 550 million years ago to the present.

phenotype   The observed traits of an organism, created by an interaction of the organism's genotype (hereditary material) and the environment in which it developed.

phylogeny   The evolutionary history of a particular group of organisms, such as antelopes or orchids, with special reference to the family tree of the species composing the group.

phylum   The highest level of classification below the kingdom. Examples: phylum Mollusca (snails, clams, octopuses) and phylum Pterophyta (ferns).

phytoplankton   The plant part of the plankton, as opposed to the zooplankton, the animal part.

Pimm, Stuart L. (1949–)   A theoretical population biologist and ecologist at the University of Tennessee; important scholar of the extinction process.

placental   Pertaining to a group of mammals characterized by use of a placenta to nurture the unborn young; comprises the great majority of living mammalian species. Cf. *marsupial.*

plankton   Organisms that float passively in the sea and air, comprising mostly microorganisms and small plants and animals.

Pleistocene epoch   The span of geological time preceding the Recent epoch, during which continental glaciers advanced and retreated and the human species evolved. The epoch began about 2.5 million years ago and closed with the end of the Ice Age 10,000 years ago.

polyploidy   The condition in a cell or an organism in which the number of complete sets of chromosomes (per cell) is greater than two. Polyploidy is a common means by which species multiply, especially in plants.

population   In biology, any group of organisms belonging to the same species at the same time and place.

prokaryote   An organism whose DNA is not enclosed in nuclear membranes; hence the cells of a prokaryotic organism do not contain a well-defined nucleus. Most prokaryotes are bacteria. Cf. *eukaryote*.

protistan   A member of the kingdom Protista (or Protoctista), comprising the protozoans, algae, and related forms.

protozoan   A member of a group of single-celled organisms, including amoebas and cilitates, usually placed in the kingdom Protista.

Quaternary period   The second and final period of the Cenozoic era, following the Tertiary period and including the Pleistocene and Recent epochs, thus extending from about 2.5 million years ago to the present.

rain forest   See *tropical rain forest*.

Raup, David M. (1933–2015)   Professor of paleontology at the University of Chicago; a foremost contributor to the analysis of diversification and extinction.

Raven, Peter H. (1936–)   Director of the Missouri Botanical Garden, St. Louis; authority on tropical botany, initiator of studies of plant diversity around the world.

restoration ecology   The study of the structure and regeneration of plant and animal communities, aimed at the enlargement or restitution of threatened ecosystems.

seed bank   A central facility for the storage of seeds representing a diversity of species and genetic strains, especially of domestic plants and their wild relatives.

selection   See *natural selection*.

Sepkoski, J. John, Jr. (1948–1999)   Professor of paleontology at the University of Chicago; with his collaborator David Raup a leading scholar of diversification and extinction.

sibling species   Species so similar to each other as to be difficult to distinguish, at least by human observers.

sickle cell   A hereditary condition prescribed by a change in a single gene that causes a warping of the red blood cells; induces anemia when present in double dose.

Simberloff, Daniel S. (1942–)   Professor of ecology at Florida State University; a pioneer of island biogeography.

Soulé, Michael E. (1936–2020)   Professor of environmental studies at the University of California, Santa Cruz; a founder of conservation biology.

source-sink model The hypothesis that species diversity, especially in tropical forests, builds up when restricted localities favorable to certain species allow them to produce a surplus of emigrants, hence to be a source of new individuals dispersing to less favorable sites nearby, the sinks.

Southwood, T. R. E. (1931–2005) Sir Richard Southwood, vice chancellor of Oxford University; a major contributor to the theory and measurement of diversity.

speciation The process of species formation: the full sequence of events leading to the splitting of one population of organisms into two or more populations reproductively isolated from one another.

species The basic unit of classification, consisting of a population or series of populations of closely related and similar organisms. In sexually reproducing organisms, the species is more narrowly defined by the biological-species concept: a population or series of populations of organisms that freely interbreed with one another in natural conditions but not with members of other species.

species-area curve See *area-species curve.*

species equilibrium The steady-state number of species, or biodiversity, found on an island or isolated patch of habitat due to a balance between the immigration of new species and the extinction of old residents. See also *island biogeographic theory.*

species selection The differential multiplication and extinction of species as a result of differences in certain traits possessed by the organisms belonging to the various species, and causing a spread of the favoring traits through the fauna or flora as a whole.

Stanley, Steven M. (1941–) Professor of paleobiology at Johns Hopkins University; authority on fossil invertebrates and a developer of the theory of species selection.

Steadman, David W. (1951–) Senior scientist in zoology at the New York State Museum; with Storrs Olson a leading researcher on the fossil history and extinction history of island birds, especially in Polynesia.

strip logging Removal of timber in narrow strips along contours, allowing rapid regrowth, sustained yield, and protection of the native fauna and flora.

subspecies   Subdivision of a species. Usually defined narrowly as a geographical race: a population or series of populations occupying a discrete range and differing genetically from other geographical races of the same species. Cf. *host race*.

sustainable development   The use of land and water to sustain production indefinitely without environmental deterioration, ideally without loss of native biodiversity.

symbiosis   The living together of two or more species in a prolonged and intimate ecological relationship, such as the incorporation of algae and cyanobacteria within fungi to create lichens.

sympatric   Occurring in the same place, as in the case of two species sharing parts of their geographic ranges.

sympatric speciation   The splitting of an ancestral species into two daughter species without an intervening geographic barrier that first breaks the ancestral population into isolated populations.

systematics   The scientific study of the diversity of life. Sometimes used synonymously with taxonomy to mean the procedures of pure classification and reconstruction of phylogeny (relationship among species); on other occasions it is used more broadly to cover all aspects of the origins and content of biodiversity.

taxonomy   The science (and art) of the classification of organisms. See also *systematics*.

tephra   Fragmented rock and ash ejected during a volcanic eruption.

Terborgh, John (1936–)   Professor of biology at Duke University; botanist and zoologist best known for long-term studies of bird and mammal ecology in rain forests.

Tertiary period   The first period of the Cenozoic era, beginning with the end of the Mesozoic era (Age of Reptiles) 66 million years ago and closing with the start of the Pleistocene epoch about 2.5 million years ago; succeeded by the Quaternary period (Pleistocene plus Recent epochs).

Thornton, Ian W. B. (1926–2002)   Professor of zoology at La Trobe University, Australia; leader of modern expeditions to Krakatau.

triploid   A cell or organism possessing three complete sets of chromosomes.

trophic level   A group of organisms that obtain their energy from the same part of the food web of a biological community. Examples: the primary producers, which are mostly plants, and herbivores, the animals that consume plants.

tropical rain forest   Also known more technically as *tropical closed moist forest:* a forest with 200 centimeters of annual rainfall spread evenly enough through the year to support broad-leaved evergreen trees, typically arrayed in several irregular canopy layers dense enough to capture over 90 percent of the sunlight before it reaches the ground.

Vavilov center   A region containing crop plants in both wild and cultivated states, hence a center of unusual genetic diversity in the species. Named after the Russian botanist Nikolai Vavilov.

vertebrate   Any animal that possesses a backbone of bony segments enclosing the central nerve cord. There are five major groups of living vertebrates: fishes, amphibians (frogs, salamanders, and caecilians), reptiles, birds, and mammals.

Vrba, Elisabeth S. (1942–)   Professor of geology at Yale University; authority on the fossil history of African mammals and a leading contributor to species-selection theory.

Webb, S. David (1936–2019)   Curator of vertebrate zoology at the Florida State Museum; major contributor and theorist of mammalian evolution in the New World.

World Continent fauna   The dominant fauna that evolved in Africa, Europe, Asia, and North America (the World Continent) during the Cenozoic era (the past 66 million years). The several constituent land masses have been connected closely enough to allow periodic exchanges of species, illustrated especially by the mammals.

zooplankton   The animal portion of the plankton, as opposed to the phytoplankton, the plant part.

Acknowledgments

In a sense, the preparation of this book began when I was a student at the University of Alabama in the late 1940s, working my way around red-clay gullies and toxic creeks in search of remnants of the natural environment. Though often discouraged, I always hoped that the world was in better shape elsewhere. My intellectual journey gathered momentum in 1953 during a field trip to Cuba's Sierra Trinidad, as I toiled up muddy roads in search of rain forest, past logging trucks on their way to Cienfuegos with the final fragments of the trees. I repeated that experience many times in other countries in later years. The world, I discovered, was *not* in better shape elsewhere. The book took solid form in my mind in September 1986 during the National Forum on Biodiversity, held in Washington, D.C., under the auspices of the National Academy of Sciences and the Smithsonian Institution. There I joined sixty biologists, economists, agricultural experts, and related professionals to consider, at long last comprehensively and with unaccustomed attention from the media, the full scale of global diversity as a central issue of the environment.

The study of biological diversity as it relates to contemporary human affairs is an eclectic subject just beginning to coalesce. In attempting this synthesis I have benefited from the advice and encouragement of colleagues in a wide range of disciplines. It is a pleasure to list them here, while exonerating them from the errors and omissions that, like hidden tripwires, may survive as the book goes into production (February 1992).

Larry D. Agenbroad (Quaternary megafauna extinction)
Peter S. Ashton (tropical floras)
Richard O. Bierregaard Jr. (rain-forest diversity)
Elizabeth Boo (ecotourism)
Kenneth J. Boss (mollusks)
William H. Bossert (area-species modeling)
Bryan C. Clarke (mollusks)
Rita R. Colwell (microbiology)
Simon Conway Morris (Cambrian diversity)
Jared M. Diamond (extinction)

Eric Dinerstein (bioreserve analysis)
Victoria C. Drake (debt-for-nature)
Donald A. Falk (U.S. plants)
Richard T. T. Forman (landscape, policy analysis)
Charles H. W. Foster (bioregionalism)
David G. Furth (beetle diversity)
Douglas J. Futuyma (evolutionary theory)
Alwyn H. Gentry (tropical floras)
Thomas J. Givnish (butterflies, metapopulations)
Jostein Goksøyr (bacterial diversity)
Jerry Harrison (world natural reserves)
Karsten E. Hartel (ichthyology)
Gary S. Hartshorn (forestry, public policy)
Michael Huben (mites)
Helen F. James (Hawaiian birds, extinction)
David P. Janos (mycorrhizal fungi)
Robert E. Jenkins (biodiversity inventories)
Carl F. Jordan (tropical forestry)
Laurent Keller (entomology)
Andrew H. Knoll (geological history of life)
Russell Lande (genetic diversity)
Robert J. Lavenberg (sharks)
Karel F. Liem (ichthyology)
Hans Löhrl (European birds)
Jane Lubchenco (marine ecosystems)
Ariel E. Lugo (tropical forests, extinction)
Denis H. Lynn (protozoan diversity)
David R. Maddison (genetics, systematics)
Michael A. Mares (extinction)
Ernst Mayr (species formation)
Kenton R. Miller (conservation and public policy)
Russell A. Mittermeier (conservation biology)
Gary Morgan (Cenozoic mammals)
Norman Myers (deforestation, extinction)
Storrs L. Olson (Hawaiian birds, extinction)
Michael O'Neal (plant extinctions)

Raymond A. Paynter Jr. (ornithology)

Tila M. Pérez (mites)

David Pilbeam (human evolution)

Mark J. Plotkin (economic botany, ethnobotany)

James F. Quinn (mammalian extinctions)

Katherine Ralls (cetacean diversity)

David M. Raup (paleontology, extinction)

Peter H. Raven (plant diversity, ethnobotany)

Jamie Resor (economics, foreign aid)

Michael H. Robinson (zoological parks)

Gustavo A. Romero (orchids)

José P. O. Rosado (reptiles)

Cristián Samper K. (South American forests)

G. Allan Samuelson (beetles)

J. William Schopf (geological history of life)

Richard E. Schultes (ethnobotany)

Raymond Siever (Cenozoic era)

Daniel S. Simberloff (extinction)

Tom Simkin (Krakatau)

Otto T. Solbrig (plant evolution)

Andrew Spielman (mosquitoes)

Steven M. Stanley (geological history of life, evolutionary theory)

David W. Steadman (Pacific birds, extinction)

Martin H. Steinberg (sickle-cell anemia)

Peter F. Stevens (plant diversity)

Roger D. Stone (conservation and policy analysis)

Nigel E. Stork (arthropod diversity)

Jorgen B. Thomsen (parrots)

Ian W. B. Thornton (Krakatau)

Barry D. Valentine (beetle diversity)

Noel D. Vietmeyer (economic botany)

Elisabeth S. Vrba (evolutionary theory, mammalian evolution)

S. David Webb (mammalian evolution)

T. C. Whitmore (tropical forestry, extinction)

Delbert Wiens (plant evolution)

Irene K. Wilson (editorial process)

As I have for all my books and articles back to 1966, I acknowledge the meticulous and invaluable work of Kathleen M. Horton in bibliographical research and preparation of the manuscript. It has also been a pleasure to work with Sarah Landry, George Ward, and Amy Bartlett Wright as they prepared the illustrations, and with Mark Moffett and Darlyne Murawski in selecting photographs from their fine natural-history collections.

Several of the line drawings were derived from previously published work by other authors. **Chapter 3:** The time curve of the five great extinction spasms, as evidenced in families of marine animals, is derived from David M. Raup and J. John Sepkoski Jr., "Mass Extinction in the Marine Fossil Record," *Science,* 215:1501–1503 (1982). **Chapter 8:** The structure of leaf litter and humus in deciduous forest, together with data on the distribution of arthropods living in these strata, is based loosely on figures and data in Gerhard Eisenbeis and Wilfried Wichard, *Atlas zur Biologie der Bodenarthropoden* (Stuttgart: Gustav Fischer, 1985). The speciesscape, representing the amount of species diversity in each group by the size of the representative organism, was introduced by Quentin D. Wheeler, "Insect Diversity and Cladistic Constraints," *Annals of the Entomological Society of America,* 83(6):1031–1047 (1990). The figure used here is an interpretation by Amy Bartlett Wright. **Chapter 9:** The idea of representing ecosystem assembly rules as a jigsaw puzzle originated with James A. Drake, "Communities as Assembled Structures: Do Rules Govern Pattern?," *Trends in Ecology and Evolution,* 5(5):159–164 (1990). **Chapter 9:** The depiction of a driver-ant swarm was prepared by Katherine Brown-Wing and published in my book *Success and Dominance in Ecosystems: The Case of the Social Insects* (Oldendorf/Luhe, Germany: Ecology Institute, 1990). **Chapter 10:** The cross section of a microbial mat is derived from David J. Des Marais, "Microbial Mats and the Early Evolution of Life," *Trends in Ecology and Evolution,* 5(5):140–144 (1990). The figure of the rise of local plant diversity is based on a diagram by Andrew H. Knoll, "Patterns of Change in Plant Communities through Geological Time," in Jared M. Diamond and Ted J. Case, eds., *Community Ecology* (New York: Harper and Row, 1986), pp. 126–141. **Chapter 11:** The map of the California checkerspot metapopulation is modified from a map presented by Susan Harrison, Dennis D. Murphy, and Paul R. Ehrlich, "Distribution of the Bay Checkerspot Butterfly, *Euphydryas editha bayensis:* Evidence for a Metapopulation Model," *American Naturalist,* 132(3):360–382 (1988). **Chapter 12:** The diagram of mass extinction of mammalian megafaunas during the past 100,000 years is modified from one by Paul S. Martin, "Prehistoric Overkill: The Global Model," in P. S. Martin and Richard G. Klein, eds., *Quaternary Extinctions: A Prehistoric Revolution* (Tucson: University of Arizona Press, 1984), pp. 354–403. The hot-spot maps are based on the publications of Norman Myers, as cited in my notes. The historical map of deforestation in Ecuador is derived from an illustration provided by Calaway H. Dodson and Alwyn H. Gentry, "Biological Extinction in Western Ecuador," *Annals of the Missouri Botanical Garden*, 78(2):273–295 (1991). **Chapter 14:** The map used in geographical systems analysis is modified from one presented by J. Michael Scott, as cited in my notes. The diagram of strip logging is modified slightly from a figure by Carl F. Jordan, "Amazon Rain Forests," *American Scientist*, 70:394–401 (1982), used with the permission of the author and the publisher, Sigma Xi.

Credits

Mark W. Moffett

associate of entomology at Harvard's Museum of Comparative Zoology, has studied and photographed tropical fauna around the world. He has contributed both articles and photographs to many publications, particularly the *National Geographic*.
Photographs following page 424.

Darlyne Murawski

studies tropical trees and butterflies, working in Central America and Sri Lanka. Her findings have appeared in numerous scientific journals, and currently she is writing and photographing a story for the *National Geographic*.
Photographs following page 424.

Amy Bartlett Wright

who studied scientific illustration at the Smithsonian Institution and Rhode Island School of Design, has contributed to several books on insects and wildlife, among them *The Social Biology of Wasps*. Her work in progress includes writing and illustrating *The Peterson First Guide to Caterpillars*.
Figures on pages 151, 177, 192–93, 197, 211, 219, 233, 243, 250–51, 260–61, 274–75, 284–85, 321, 347, 355, 371, 373, 381, 429, 435, 437, 440.

George Ward prepared the maps, charts, and graphs. The illustration of driver ants on page 304 is by Katherine Brown-Wing. The paintings in the last chapter, by Pablo César Amaringo and Roxana Elizabeth Marín, are reproduced courtesy of Luis Eduardo Luna and the North Atlantic Press, publisher of *Ayahuasca Visions*.

Note: Those wishing to reproduce illustrations from this book must obtain written permission from the publisher and must carry a credit line for the artists and photographers.

NATURALIST

For Renee and Cathy

Contents

Prelude

I HAVE BEEN, as the physicist Victor Weisskopf once said of himself, a happy man in a terrible century. My preoccupation was not, however, with nuclear swords and breathtaking technological advances, but of a wholly different kind: I have served as a close witness to fundamental changes in Nature.

Nature, with a capital *N*, the concept: for me it holds two meanings. When the century began, people could still easily think of themselves as transcendent beings, dark angels confined to Earth awaiting redemption by either soul or intellect. Now most or all of the relevant evidence from science points in the opposite direction: that, having been born into the natural world and evolved there step by step across millions of years, we are bound to the rest of life in our ecology, our physiology, and even our spirit. In this sense, the way in which we view the natural world, Nature has changed fundamentally.

When the century began, people still thought of the planet as infinite in its bounty. The highest mountains were still unclimbed, the ocean depths never visited, and vast wildernesses stretched across the equatorial continents. Now we have all but finished mapping the physical world, and we have taken the measure of our dwindling resources. In one lifetime exploding human populations have reduced wildernesses to threatened nature reserves. Ecosystems and species are vanishing at the fastest rate in 65 million years. Troubled by what we have wrought, we have begun to turn in our role from local conquerer to global steward. Nature in this second sense, our perception of the natural world as something distinct from human existence, has thus also changed fundamentally.

Because my temperament and profession predispose me, I have followed these changes closely. As a younger scientist and naturalist, my own worldview shifted in concert with the advances of evolutionary biology and the decline that practitioners of this science perceived to be occurring in the natural environment. From childhood into middle age, my ontogeny repeated the larger phylogeny. Nature metamorphosed into something new.

My childhood was blessed. I grew up in the Old South, in a beautiful environment, mostly insulated from its social problems. I became determined at an early age to be a scientist so that I might stay close to the natural world. That boyhood enchantment remains undiminished, but it exists in a Heraclitean stream in which everything else has changed, all that I first thought about how the world works and all that I believed of humanity's place in the world. I have written this account to learn more fully why I now think the way I do, to clarify the elements at the core of my beliefs to you and to myself, and perhaps to persuade.

PART I
DAYBREAK IN ALABAMA

When I get to be a composer
I'm gonna write me some music about
Daybreak in Alabama
And I'm gonna put the purtiest songs in it
Rising out of the ground like swamp mist
And falling out of heaven like soft dew.

LANGSTON HUGHES

CHAPTER ONE

Paradise Beach

WHAT HAPPENED, what we *think* happened in distant memory, is built around a small collection of dominating images. In one of my own from the age of seven, I stand in the shallows off Paradise Beach, staring down at a huge jellyfish in water so still and clear that its every detail is revealed as though it were trapped in glass. The creature is astonishing. It existed outside my previous imagination. I study it from every angle I can manage from above the water's surface. Its opalescent pink bell is divided by thin red lines that radiate from center to circular edge. A wall of tentacles falls from the rim to surround and partially veil a feeding tube and other organs, which fold in and out like the fabric of a drawn curtain. I can see only a little way into this lower tissue mass. I want to know more but am afraid to wade in deeper and look more closely into the heart of the creature.

The jellyfish, I know now, was a sea nettle, formal scientific name *Chrysaora quinquecirrha*, a scyphozoan, a medusa, a member of the pelagic fauna that drifted in from the Gulf of Mexico and paused in the place I found it. I had no idea then of these names from the lexicon of zoology. The only word I had heard was *jellyfish*. But what a spectacle my animal was, and how inadequate, how demeaning, the bastard word used to label it. I should have been able to whisper its true name: *scyph-o-zo-an*! Think of it! I have found a scyphozoan. The name would have been a more fitting monument to this discovery.

The creature hung there motionless for hours. As evening approached and the time came for me to leave, its tangled undermass appeared to stretch deeper into the darkening water. Was this, I wondered, an animal or a collection of animals? Today I can say that it was a single animal. And that another outwardly similar animal found in the same waters, the Portuguese man-of-war, is a colony of animals so tightly joined as to form one smoothly functioning superorganism. Such are the general facts I recite easily now, but this sea nettle was special. It came into my world abruptly, from I knew not

539

where, radiating what I cannot put into words except—*alien purpose and dark happenings in the kingdom of deep water.* The scyphozoan still embodies, when I summon its image, all the mystery and tensed malignity of the sea.

The next morning the sea nettle was gone. I never saw another during that summer of 1936. The place, Paradise Beach, which I have revisited in recent years, is a small settlement on the east shore of Florida's Perdido Bay, not far from Pensacola and in sight of Alabama across the water.

There was trouble at home in this season of fantasy. My parents were ending their marriage that year. Existence was difficult for them, but not for me, their only child, at least not yet. I had been placed in the care of a family that boarded one or two boys during the months of the summer vacation. Paradise Beach was paradise truly named for a little boy. Each morning after breakfast I left the small shorefront house to wander alone in search of treasures along the strand. I waded in and out of the dependably warm surf and scrounged for anything I could find in the drift. Sometimes I just sat on a rise to scan the open water. Back in time for lunch, out again, back for dinner, out once again, and, finally, off to bed to relive my continuing adventure briefly before falling asleep.

I have no remembrance of the names of the family I stayed with, what they looked like, their ages, or even how many there were. Most likely they were a married couple and, I am willing to suppose, caring and warmhearted people. They have passed out of my memory, and I have no need to learn their identity. It was the animals of that place that cast a lasting spell. I was seven years old, and every species, large and small, was a wonder to be examined, thought about, and, if possible, captured and examined again.

There were needlefish, foot-long green torpedoes with slender beaks, cruising the water just beneath the surface. Nervous in temperament, they kept you in sight and never let you come close enough to reach out a hand and catch them. I wondered where they went at night, but never found out. Blue crabs with skin-piercing claws scuttled close to shore at dusk. Easily caught in long-handled nets, they were boiled and cracked open and eaten straight or added to gumbo, the spicy seafood stew of the Gulf coast. Sea trout and other fish worked deeper water out to the nearby eelgrass flats and perhaps beyond; if you had a boat you could cast for them with bait and spinners. Stingrays, carrying threatening lances of bone flat along their muscular tails, buried themselves in the bottom sand of hip-deep water in the daytime and moved close to the surf as darkness fell.

One late afternoon a young man walked past me along the beach dangling a revolver in his hand, and I fell in behind him for a while. He said he was hunting stingrays. Many young men, my father among them, often took guns on such haphazard excursions into the countryside, mostly .22 pistols and rifles but also heavier handguns and shotguns, recreationally shooting any living thing they fancied except domestic animals and people. I thought of the stingray hunter as a kind of colleague as I trailed along, a fellow adventurer, and hoped he would find some exciting kind of animal I had not seen, maybe something big. When he had gone around a bend of the littoral and out of sight I heard the gun pop twice in quick succession. Could a bullet from a light handgun penetrate water deep enough to hit a stingray? I think so but never tried it. And I never saw the young marksman again to ask him.

How I longed to discover animals each larger than the last, until finally I caught a glimpse of some true giant! I knew there were large animals out there in deep water. Occasionally a school of bottlenose porpoises passed offshore less than a stone's throw from where I stood. In pairs, trios, and quartets they cut the surface with their backs and dorsal fins, arced down and out of sight, and broke the water again ten or twenty yards farther on. Their repetitions were so rhythmic that I could pick the spot where they would appear next. On calm days I sometimes scanned the glassy surface of Perdido Bay for hours at a time in the hope of spotting something huge and monstrous as it rose to the surface. I wanted at least to see a shark, to watch the fabled dorsal fin thrust proud out of the water, knowing it would look a lot like a porpoise at a distance but would surface and sound at irregular intervals. I also hoped for more than sharks, what exactly I could not say: something to enchant the rest of my life.

Almost all that came in sight were clearly porpoises, but I was not completely disappointed. Before I tell you about the one exception, let me say something about the psychology of monster hunting. Giants exist as a state of the mind. They are defined not as an absolute measurement but as a proportionality. I estimate that when I was seven years old I saw animals at about twice the size I see them now. The bell of a sea nettle averages ten inches across, I know that now; but the one I found seemed two feet across—a grown man's two feet. So giants can be real, even if adults don't choose to classify them as such. I was destined to meet such a creature at last. But it would not appear as a swirl on the surface of the open water.

It came close in at dusk, suddenly, as I sat on the dock leading away from shore to the family boathouse raised on pilings in shallow water. In the failing light I could barely see to the bottom, but I stayed perched on the dock anyway, looking for any creature large or small that might be moving. Without warning a gigantic ray, many times larger than the stingrays of common experience, glided silently out of the darkness, beneath my dangling feet, and away into the depths on the other side. It was gone in seconds, a circular shadow, seeming to blanket the whole bottom. I was thunderstruck. And immediately seized with a need to see this behemoth again, to capture it if I could, and to examine it close up. Perhaps, I thought, it lived nearby and cruised around the dock every night.

Late the next afternoon I anchored a line on the dock, skewered a live pinfish on the biggest hook I could find in the house, and let the bait sit in six feet of water overnight. The following morning I rushed out and pulled in the line. The bait was gone; the hook was bare. I repeated the procedure for a week without result, always losing the pinfish. I might have had better luck in snagging a ray if I had used shrimp or crab for bait, but no one gave me this beginner's advice. One morning I pulled in a Gulf toadfish, an omnivorous bottom-dweller with a huge mouth, bulging eyes, and slimy skin. Locals consider the species a trash fish and one of the ugliest of all sea creatures. I thought it was wonderful. I kept my toadfish in a bottle for a day, then let it go. After a while I stopped putting the line out for the great ray. I never again saw it pass beneath the dock.

Why do I tell you this little boy's story of medusas, rays, and sea monsters, nearly sixty years after the fact? Because it illustrates, I think, how a naturalist is created. A child comes to the edge of deep water with a mind prepared for wonder. He is like a primitive adult of long ago, an acquisitive early *Homo* arriving at the shore of Lake Malawi, say, or the Mozambique Channel. The experience must have been repeated countless times over thousands of generations, and it was richly rewarded. The sea, the lakes, and the broad rivers served as sources of food and barriers against enemies. No petty boundaries could split their flat expanse. They could not be burned or eroded into sterile gullies. They were impervious, it seemed, to change of any kind. The waterland was always there, timeless, invulnerable, mostly beyond reach, and inexhaustible. The child is ready to grasp this archetype, to explore and learn, but he has few words to describe his guiding emotions. Instead he is given a compelling image that will serve in later life as a talisman, transmitting a powerful energy that directs the growth of experience and knowledge. He will add complicated details and context from his culture as he grows older. But the core image stays intact. When an adult he will find it curious, if he is at all reflective, that he has the urge to travel all day to fish or to watch sunsets on the ocean horizon.

Hands-on experience at the critical time, not systematic knowledge, is what counts in the making of a naturalist. Better to be an untutored savage for a while, not to know the names or anatomical detail. Better to spend long stretches of time just searching and dreaming. Rachel Carson, who understood this principle well, used different words to the same effect in *The Sense of Wonder* in 1965: "If facts are the seeds that later produce knowledge and wisdom, then the emotions and the impressions of the senses are the fertile soil in which the seeds must grow. The years of childhood are the time to prepare the soil." She wisely took children to the edge of the sea.

The summer at Paradise Beach was for me not an educational exercise planned by adults, but an accident in a haphazard life. I was parked there in what my parents trusted would be a safe and carefree environment. During that brief time, however, a second accident occurred that determined what kind of naturalist I would eventually become. I was fishing on the dock with minnow hooks and rod, jerking pinfish out of the water as soon as they struck the bait. The species, *Lagodon rhomboides*, is small, perchlike, and voracious. It carries ten needlelike spines that stick straight up in the membrane of the dorsal fin when it is threatened. I carelessly yanked too hard when one of the fish pulled on my line. It flew out of the water and into my face. One of its spines pierced the pupil of my right eye.

The pain was excruciating, and I suffered for hours. But being anxious to stay outdoors, I didn't complain very much. I continued fishing. Later, the host family, if they understood the problem at all (I can't remember), did not take me in for medical treatment. The next day the pain had subsided into mild discomfort, and then it disappeared. Several months later, after I had returned home to Pensacola, the pupil of the eye began to cloud over with a traumatic cataract. As soon as my parents noticed the change, they took me to a doctor, who shortly afterward admitted me to the old Pensacola Hospital to have the lens removed. The surgery was a terrifying nineteenth-century ordeal. Someone held me down while the anesthesiologist, a woman named Pearl Murphy, placed a gauze nose cone over my nose and mouth and dripped ether into it. Her fee for this standard service, I learned many years later, was five dollars. As I lost consciousness I dreamed I was all alone in a large auditorium. I was tied to a chair, unable to move, and screaming. Possibly I was screaming in reality before I went under. In any case the experience was almost as bad as the cataract. For years afterward I became nauseous at the smell of ether. Today I suffer from just one phobia: being trapped in a closed space with my arms immobilized and my face covered with an obstruction. The aversion is not an ordinary claustrophobia. I can enter closets and elevators and crawl beneath houses and automobiles with aplomb. In my teens and twenties I explored caves and underwater recesses around wharves without fear, just so long as my arms and face were free.

I was left with full sight in the left eye only. Fortunately, that vision proved to be more acute at close range than average—20/10 on the ophthalmologist's chart—and has remained so all my life. I lost stereoscopy but can make out fine print and the hairs on the bodies of small insects. In adolescence I also lost, possibly as the result of a hereditary defect, most of my hearing in the uppermost registers. Without a hearing aid, I cannot make out the calls of many bird and frog species. So when I set out later as a teenager with Roger Tory Peterson's *Field Guide to the Birds* and binoculars in hand, as all true naturalists in America must at one time or other, I proved to be a wretched birdwatcher. I couldn't hear birds; I couldn't locate them unless they obligingly fluttered past in clear view; even one bird singing in a tree close by was invisible unless someone pointed a finger straight at it. The same was true of frogs. On rainy spring nights my college companions could walk to the mating grounds of frogs guided only by the high-pitched calls of the males. I managed a few, such as the deep-voiced barking tree frog, which sounds like someone thumping a tub, and the eastern spadefoot toad, which wails like a soul on its way to perdition; but from most species all I detected was a vague buzzing in the ears.

In one important respect the turning wheel of my life came to a halt at this very early age. I was destined to become an entomologist, committed to minute crawling and flying insects, not by any touch of idiosyncratic genius, not by foresight, but by a fortuitous constriction of physiological ability. I had to have one kind of animal if not another, because the fire had been lit and I took what I could get. The attention of my surviving eye turned to the ground. I would thereafter celebrate the little things of the world, the animals that can be picked up between thumb and forefinger and brought close for inspection.

CHAPTER TWO

Send Us the Boy

WHO CAN SAY what events formed his own character? Too many occur in the twilight of early childhood. The mind lives in half-remembered experiences of uncertain valence, where self-deception twists memory further from truth with every passing year. But of one event I can be completely sure. It began in the winter of 1937, when my parents, Edward and Inez Freeman Wilson, separated and began divorce proceedings. Divorce was still unusual at that time and in that part of the country, and there must have been a great deal of gossiping and head-shaking among other family members. While my parents untangled their lives, they looked for a place that could offer a guarantee of security to a seven-year-old. They chose the Gulf Coast Military Academy, a private school located on the shore road four miles east of Gulfport, Mississippi.

So one January morning I traveled with my mother to Gulf-port on a westbound bus out of Pensacola by way of Mobile and Pascagoula. We arrived at GCMA in the afternoon. I looked around and judged the landscape, which was classic leisured-Gulf-Coast, instantly inviting. Brick buildings with verandas and borders crowded by ornamental shrubs were dispersed over beautifully tended grass lawns. Old live oaks (I grant that all live oaks *look* old) and towering loblolly pines offered generous shade. U.S. 90, then a quiet two-lane road, bordered the campus on the south. A few dozen feet beyond, at the bottom of a seawall, peaceful waves rolled in from the Gulf of Mexico. I brightened at this ocean view. Paradise Beach again? It was not to be. We entered the Junior Dormitory to meet the housemother and some of the other grammar-school cadets. I looked at my military-style cot, the kind you can bounce a coin on when properly made. I listened to an outline of the daily regimen. I examined my uniform, patterned after that at West Point. I shook hands with my roommate, who was inordinately stiff and polite for a seven-year-old. All dreams of languor and boyhood adventure vanished.

GCMA was a carefully planned nightmare engineered for the betterment of the untutored and undisciplined. It was a military academy of the original mold, all gray-wool clothed and ramrod-straight. The school prospectus guaranteed—it did not "offer" or "make available"; it *guaranteed*—a solid traditional education. Some of its graduates went on to civilian colleges and universities across the country. But at heart GCMA was a preparatory school for West Point, Annapolis, and private equivalents such as the Virginia Military Academy whose central purpose was to train America's officer corps.

All of this was consistent with white middle-class culture in the South of 1937. Young men could aspire to no higher calling than officer rank in the military. The South continued her antebellum dream of the officer and gentleman, honorable, brave, unswerving in service to God and country. He comes to our mind, the newly graduated second lieutenant, clad in dress white, escorting his bride, pretty and sweet, out of the church beneath the raised crossed sabers of his classmates, as his proud family watches. His conduct will henceforth affirm the generally understood historical truth that we lost the War Between the States for lack of arms and the exhaustion of battle-depleted troops. Our men, and especially our officers, were nonetheless individually the finest soldiers in the world at that time. They were southerners, men not to be trifled with.

Now you understand why commanding officers interviewed on television at Vietnam firebases so often spoke with southern accents. They had thin lips and highway eyes, and they didn't joke around. Medicine, law, and engineering made admirable careers for a southern man, and business and the ministry were all right of course. Golf champions and quarterbacks who came from Alabama were heroes, and we were all real proud when one of our relatives (his nickname was Skeeter or T. C. or something like that, in any case your third cousin Hank's oldest boy) was elected to Congress. But military command was the profession that bore the cachet of strength and honor.

The Gulf Coast Military Academy was classed each year without fail as an Honor School by the United States War Department. In other words, it was a boot camp. Its regimen was designed to abrade away all the bad qualities inhering in the adolescent male, while building the kind of character that does not flinch at a whiff of grapeshot. "Send Us the Boy and We Will Return the Man" was its motto. The 1937 yearbook, from which my childish face stonily gazes, explains the formula with pitiless clarity:

— The daily work is a systematic routine in which every duty has its place in the day and, therefore, will not be overlooked.

— By association with other cadets, each cadet begins to recognize himself as an integral part of a body and, with this in view, he assumes the correct attitude toward the rights of others.

— By being thrown on his own resources, a boy develops initiative and self-dependence and grows away from the helpless, dependent spirit into which many boys have been coddled.

The systematic routine the author had in mind (and was he, I wonder, square-faced Major Charles W. Chalker, Professor of Military Science and Tactics, whose photograph gazes out at me from the yearbook?) emulated those of the adult service academies. It could be used today, if softened a bit, at the Marine training camp on Parris Island. For seven days a week real bugles, played by cadets proud of their job, led us lockstep through the Schedule. First Call 6:00, Reveille 6:05, Assembly 6:10, Sick Call 6:30, Police Inspection 6:40, Waiter's Warning 6:45, Assembly and March to Mess 7:00, School Call 7:40. Then, without bugles, came calls to change class, Chapel Assembly 10:20, Intermission 4 minutes, Warning Call, Return to Class. And so tramp forward through the day, finally to dinner. The bugles resumed with Call to Quarters 6:50, Study (no radios!) 7:00, Tattoo 9:15, and Taps 9:30. No talking afterward, or you go on delinquency report.

On Saturday the schedule was similar but lighter, with time off for leisure, athletics, and delinquency reports. On Sunday we really snapped to life: shined our shoes, polished our buttons and belt buckles (uniforms mandatory at all times, formal gray and white on Sunday), and attended church. Then we prepared for Battalion Parade, which kicked off at 3:30. We marched out in formation, to be watched and graded by unit and individual, past officer-instructors, visiting parents, and a few curious, respectful townspeople. The youngest boys, of whom I was one, brought up the rear.

The curriculum was laid before the student in resonant single words: arithmetic, algebra, geometry, physics, chemistry, history, English, foreign language. No art, nature field trips, and certainly no enterprises with wimpy titles like "introduction to chemistry" or "the American experience." Some electives were allowed, but only in cheerless subjects such as Latin, commercial geography, and business ethics. There was an implication that if you could not cut the mustard in the military, there was always commerce. Older cadets were trained in rifle marksmanship, mortar and machine-gun fire, surveying, and military strategy. Horsemanship was encouraged. We grammar-school students looked forward to someday enjoying these manly activities.

The school's coat of arms was an Eagle Volant grasping crossed sabers and rifles with bayonets and lances; the shafts of the lances were hung with matched dexter and sinister forty-eight-star American flags. The Navy was represented by a triangular escutcheon enclosing a three-masted barkentine.

All boys at GCMA, from first through twelfth grades, followed the same daily routine and worked their way up the vertically stacked curriculum. We junior cadets, boys in the first six grades, were given a few concessions. There was a dormitory mother, Mrs. R. P. Linfield, whose first name I never knew and whose stiffly composed face in the catalog photographs makes her look exactly like what she was, dormitory mother in a military school. We did not carry a rifle on parade, nor were we trained in weapons and horsemanship. The occasional dances held with young ladies from nearby Gulf Park College were of course irrelevant. In the interests of preserving discipline, parents were urged not to coddle their sons by the dispatching of inappropriate gifts: "Do not send him eatables that will upset his digestion. Send fruit."

Disputes among cadets that could not be talked through were expected to be settled manfully, under adult supervision and in a boxing ring formed by standing cadets. Occasionally fistfights were quietly arranged behind buildings with no instructors or student officers present, but in general all aggression was effectively channeled according to regulations.

Misbehavior of any kind brought time in the bull ring, an activity not mentioned in the brochures. Regular cadets marched with rifles at shoulder arms around a circular track for one to several hours, the length of time depending on the seriousness of the charge. Longer terms were broken up and spread over a succession of days. Junior cadets "marched"— actually, most of the time we just strolled—without rifles. It was a good time to get away from the others and daydream. I was a frequent rule breaker, and spent what seems in retrospect to have been an unconscionable amount of my time at GCMA traveling in circles. As I recall, most of my sins involved talking with other cadets during class. If so, the lesson did not take. Now, as a university professor, I spend almost all of my time talking in class.

In my heart I know that I was a reasonably good kid. I was neither laggard nor rebellious, and time in the bull ring usually came as a surprise. Little or nothing was said to us junior cadets directly about discipline and punishment. We learned mostly by example and word of mouth. Infractions and sentences were posted each Saturday afternoon at 1:50 on the bulletin board next to the mail window, under "Delinquencies." We ran there each time to see who would play and who would march. No further recreation was allowed until all bull-ring time was completed. We heard rumors of legendary sentences imposed on older boys for unspeakable violations.

Wednesday afternoons were for fun, in the GCMA way of thinking. From 1:30 to 5:30, all cadets free of punishment went on leave. Buses conveyed us the four miles west to Gulfport for milkshakes, movies, and just walking around.

This dollop of frivolity was all well and good, but I pined for my beloved Gulf of Mexico, always in full sight from the front lawn of the Academy. I could not go down to the water; cadets were sensibly forbidden to cross the two-lane highway that separated the school grounds from the seawall and beach. On several occasions toward the end of the term, I joined a group of other boys with the housemother for a supervised swim in the surf. A photograph in the catalog shows us filing down in our regulation swimsuits, complete with shoulder straps. Its caption reads, "Boys going out on the beach under supervision where in the warm sunshine they can frolic on the clean white sand and bathe in the sparkling briny water of the Gulf." No fishing, no time during the frolic to wander dreaming along the strand, no chance to see stingrays or other monsters rising from the deep.

The most notable event during my stay at the Gulf Coast Military Academy was a visit by President Franklin Delano Roosevelt. Fresh into his second term of office, he came to Mississippi and Louisiana to press the flesh and extend thanks to his constituents in the most solidly Democratic of all the states. Along the coast schools were let out and businesses closed. Storefronts were painted and streets cleaned. Even the "negro kiddies turned out in their best attire," as the *Biloxi-Gulfport Daily Herald* unselfconsciously reported. Upward of 100,000 people converged along the route to be followed by the President and his entourage. In those days chief executives were rarely seen in person, and the New Deal had made Roosevelt a demigod in the Deep South. He brought relief to what was then in many respects a Third World country.

Arriving in Biloxi by train from Washington on the morning of April 29, the President and his staff were escorted to a motorcade of twenty-four cars already packed with local politicians, military officers, and journalists. The party visited Biloxi's high spots, including a lighthouse painted black after Lincoln's assassination, the Veterans Administration Hospital, and Beauvoir, Jefferson Davis' Mississippi home, where a band struck up "Dixie" and the last eight surviving Confederate veterans of the city greeted the President with rebel yells. At frequent intervals Roosevelt lifted his fedora and flashed his famous grin. Finally, the motorcade headed west on U.S. 90 toward Gulfport, passing the Gulf Coast Military Academy at ten o'clock. The entire cadet corps stood at attention in dress gray and white, forming a single line shoulder to shoulder at roadside. Roosevelt at first believed that he was to inspect the corps, and so instructed his military officers to don the formal gold braid worn by presidential aides. On learning that the schedule was too tight to allow a pause of this length, he directed the motorcade instead to slow down as it passed the Academy. We all saluted as the long line of automobiles rolled

by. Somehow I failed to distinguish the President among all
the passing faces, but I like to think he saw me, standing at the
end of the line, one of the two smallest cadets.

To all this strange new life I adjusted reasonably well. For
the first few days after arrival I was seized with confusion and
black loneliness, crying myself to sleep when the lights were
out—quietly, however, so no one would hear. But after a while
I came to feel that I belonged, that GCMA was a family of
sorts, one moved by benevolent intention. I hated the place
then but came to love it later, savoring it ever more in memory
as the years passed and recollections of my distress faded. I
stayed just long enough to be transformed in certain qualities of mind. I still summon easily the images of a perfect
orderliness and lofty purpose. In one of those least faded, a
cadet officer approaches on a Sunday morning as we gather for
parade, a teenager mounted on a horse, resplendent in boots,
Sam Browne belt, sheathed saber, and a cap cloth-covered in
spotless white. He is poised to move in intricate maneuvers.
He works his mount slowly through a tight circle, turning,
turning, as he speaks to a group of other cadets on foot. He
has fallen silent now in my mind, but his visual presence still
shines with grace, decent ambition, and high achievement. I
ask, What achievement? I cannot say, but no matter; the very
ambiguity of his image preserves the power.

I left at the end of the spring term, carrying an inoculum of
the military culture. Up to college age I retained the south-
erner's reflexive deference to elders. Adult males were "sir" and
ladies "ma'am," regardless of their station. These salutations I
gave with pleasure. I instinctively respect authority and believe
emotionally if not intellectually that it should be perturbed
only for conspicuous cause. At my core I am a social conservative, a loyalist. I cherish traditional institutions, the more
venerable and ritual-laden the better.

All my life I have placed great store in civility and good
manners, practices I find scarce among the often hard-edged,
badly socialized scientists with whom I associate. Tone of voice
means a great deal to me in the course of debate. I try to
remember to say "With all due respect" or its equivalent at the
start of a rebuttal, and mean it. I despise the arrogance and
doting self-regard so frequently found among the very bright.

I have a special regard for altruism and devotion to duty, believing them virtues that exist independent of approval and validation. I am stirred by accounts of soldiers, policemen, and firemen who have died in the line of duty. I can be brought to tears with embarrassing quickness by the solemn ceremonies honoring these heroes. The sight of the Iwo Jima and Vietnam Memorials pierces me for the witness they bear of men who gave so much, and who expected so little in life, and the strength ordinary people possess that held civilization together in dangerous times.

I have always feared I lack their kind of courage. They kept on, took the risk, stayed the course. In my heart I admit I never wanted it; I dreaded the social machine that can grind a young man up, and somehow, irrationally, I still feel that I dropped out. I have tried to compensate to erase this odd feeling. When I was young I made it a habit to test myself physically during field excursions, pushing myself just enough in difficult or dangerous terrain to gain assurance. Later, when I had ideas deemed provocative, I paraded them like a subaltern riding the regimental colors along the enemy line. I asked myself fretfully over and again: Could I have measured up if ever I had been put to a *real* test, with my life at stake, at Château-Thierry or Iwo Jima? Such are the fantasies imposed by egoism and guilt.

I have spent a good deal of time during my career as a scientist thinking about the origins of self-sacrifice and heroism, and cannot say I understand them fully in human terms. The Congressional Medal of Honor I find to be more mysterious and exalting than the Nobel Prize. In leisure reading I browse through the stories of those who won it. I am pleased to know one of them personally, James Stockdale, winner of the Medal of Honor and among this country's most celebrated military heroes. A copy of my book *On Human Nature*, which addresses the biological foundations of altruism and leadership and the possible meaning of the Medal of Honor, rests by his arm in his formal portrait as vice admiral.

Jim Stockdale endured eight years in a North Vietnamese prisoner-of-war camp. During much of that time he was tortured by his captors, who wanted information about his air missions at the time of the Gulf of Tonkin incident. He never broke. Once, feeling that his will might not hold, he slashed his wrists with broken window glass to stop the questioning. The tactic worked; he was given better treatment afterward. All the while, as the senior ranking American officer, he organized his fellow prisoners with secret "tap codes" passed from cell to cell, and pulled them together in semblance of a wartime unit. It would have been as easy to rationalize, to say, I have done my part; I am just a little cog and a forgotten player; why risk my life?

I do not doubt that the steel in Stockdale's spine was put there in part by the self-discipline and sense of honor the best of the military academies aim to instill. True, the qualities of military heroism can easily be hardened into blind obedience and suborned. But they remain in my eyes the codification of certain virtues necessary for civilization. Their acceptance sacralizes unfurled colors, serried ranks, and ribbons of honor.

You will understand, then, that the people I find it easiest to admire are those who concentrate all the courage and self-discipline they possess toward a single worthy goal: explorers, mountain climbers, ultramarathoners, military heroes, and a very few scientists. Science is modern civilization's highest achievement, but it has few heroes. Most is the felicitous result of bright minds at play. Tricksters of the arcane, devising clever experiments in the laboratory when in the mood, chroniclers of the elegant insight, travelers to seminars in Palo Alto and Heidelberg. For it is given unto you to be bright, and play is one of the most pleasurable of human activities, and all that is well and good; but for my own quite possibly perverse reasons I prefer those scientists who drive toward daunting goals with nerves steeled against failure and a readiness to accept pain, as much to test their own character as to participate in the scientific culture.

One such was Philip Jackson Darlington, for many years Curator in Entomology at Harvard University's Museum of Comparative Zoology. In the spring of 1953, when I was twenty-three and preparing to leave on a collecting trip to Cuba and Mexico, my first into the tropics, I called upon Darlington seeking advice. We met at his cluttered bench in the far corner of the Coleoptera Room.

Phil was held in deep respect by young entomologists. A private and single-minded man, he chose, as his wife, Libbie, once put it, to live an unfragmented life. He devoted his career to the study of beetles and the geographic distribution of animals. He conducted research around the world in an era when foreign travel was difficult and expensive. Darlington was a collector of legendary prowess, unmatched in the field. He would zero in on just the right habitat, then toil hour after hour, day after day and sometimes into the evenings, to bottle hundreds of specimens, many belonging to species rare or new to science. If his scholarly interests seem recondite, it should be recalled that Charles Darwin was also an impassioned beetle collector with a special interest in the geographic distribution of animals.

Darlington was pleased to see me but not inclined to waste words. His manner was professional and reserved, eased by a frequent ironic smile and pursed lips—the scholar's look. He had thick, black eyebrows that sheltered his eyes, pulled attention slightly upward, and made it easier to hold contact with his face as he spoke. He came quickly to the point.

"Ed, don't stay on the trails when you collect insects. Most people take it too easy when they go in the field. They follow the trails and work a short distance into the woods. You'll get only some of the species that way. You should walk in a straight line through the forest. Try to go over any barrier you meet. It's hard, but that's the best way to collect."

He then mentioned some good collecting spots he had worked himself, described the best way to drink Cuban coffee, and the interview was over.

It was exactly what I wanted to hear. How to do it the right way, the *hard* way. Words from a master to chosen disciple: grit, work, determination, some pain, and new species—success—await the tough-minded. No admonitions to watch my health or to have a good time in Havana. Just collect in straight lines and get it right. Bring back the good stuff to the M.C.Z.

As a young man Darlington lived that advice. He climbed the Sierra Nevada de Santa Marta of northern Colombia, collecting all along the 6,000 meters of elevation. He toiled up in an almost straight line to the summit of Pico Turquino, Cuba's highest mountain, in the Sierra Maestra, the backcountry made famous by Fidel Castro's guerrilla campaign of the 1950s. Darlington repeated the feat on La Hotte, Haiti's highest peak, in the remote massif of the Tiburon Peninsula. He covered the last 1,000 meters of elevation alone, cutting his way straight through the undergrowth of virgin cloud forest, twisting his body through narrow breaks between tree trunks. At the summit he was disappointed to find that a team of Danish surveyors had already ascended the opposite slope, hacked out a clearing, and set up cloth targets for their transects. He had supposed that this would be the wildest spot in Haiti, free and safe for a while from the teeming human population below. He hoped in any case that rare endangered native mammals, possibly new to science, might still live on the summit. That night he searched for them with a jacklight and found none. All that appeared was one black rat, an Old World species accidentally introduced to the West Indies in early European cargo and the scourge of the native fauna. But the trip was rewarding just the same. Darlington was indeed the first biologist to visit the top. While working across La Hotte's upper slopes he collected many new kinds of insects and other animals, including a new genus of snakes, later named *Darlingtonia* in his honor.

His adventures continued. Shortly after Pearl Harbor, Darlington enlisted in the Army Sanitary Corps Malaria Survey as a first lieutenant. He served in the Sixth Army during the campaigns on New Guinea, Bismarck Archipelago, central Philippines, and Luzon, retiring as a major in 1944. Before leaving New Guinea, he managed to collect great numbers of ground beetles and other insects in several regions of the country, including the summit of Mount Wilhelm, the highest peak in the Bismarck Range.

One Darlington exploit from that adventure has become a standard of zoology lore. Alone in the jungle looking for specimens, he ventured out on a submerged log to sample water from the middle of a stagnant jungle pool, when a giant crocodile rose from the depths and swam toward him. As he edged back gingerly toward shore, he slipped from the slimy log into the water. The crocodile rushed him, mouth gaping, huge teeth bared. He tried to grasp its jaws, got one grip, then lost it.

"I can't describe the horror of that instant," he told an Army reporter at the time, "but I was scared and I kept thinking: What a hell of a predicament for a naturalist to be in."

The thirty-nine-year-old Darlington was six feet two inches tall and weighed 190 pounds; the crocodile, weighing several hundred pounds, was in its element. It spun him over and over, finally carrying him to the bottom.

"Those few seconds seemed hours," he said. "I kicked, but it was like trying to kick in a sea of molasses. My legs seemed heavy as lead, and it was hard to force my muscles to respond." Whether because of a well-placed kick or for some other reason, the animal suddenly opened its jaws, and Darlington swam free. Flailing his legs and torn arms, he made the shore, scrambled frantically up the bank, and tried to keep going, knowing that crocodiles sometimes pursue prey onto land. At the last moment he slipped in the mud and rolled back again toward the water. The predator closed once again.

"It was a nightmare. That's the first time I've ever hollered for help," he said. "But there was no one to hear me." He scrambled up the bank again and this time made it into the shelter of the jungle. Only then did he become aware of the pain in his arms and his weakness from loss of blood. "That hike to the hospital, which I knew was nearby, was the longest I've ever made." The muscles and ligaments of both arms were lacerated, and bones in his right arm were crushed. The crocodile's teeth had also pierced both his hands. Only his left arm and hand were marginally functional.

I grant that to fight off a crocodile is an act of survival, not proof of character. But to go where crocodiles live is, and especially to do what Darlington did next. He was in a cast for several months, convalescing at Dobadura, Papua. But nothing could stop this driven man. He perfected a left-handed technique for collecting insects. Have someone tie a vial to the end of a stick. Walk out into the forest, jam the stick into the ground, pull the cork out with the left hand, drop the specimens into the vial, replace the cork. Over several months he eventually regained full use of his hands and arms, pursuing his collecting and research all the while. He was able during that time to assemble a world-class collection. After returning to Harvard he continued to work year after year, expanding our knowledge of the insects of New Guinea and other parts of the world.

The standards I use for my heroes were first implanted by

a tough military academy, whose instructors believed that little boys should be treated essentially like big boys. It was an accident of timing with an odd result. I thought the academy presaged real life. For the rest of my childhood and through adolescence I assumed, despite mounting evidence to the contrary, that hard work and punishingly high standards are demanded of all grown men, that life is tough and unforgiving, that slipups and disgrace are irreparable.

This ethic stirs faint and deep within me even now, although I know it is not entirely reasonable. Nothing will change. There are certain experiences in childhood that surge up through the limbic system to preempt the thinking brain and hold fast for a lifetime to shape value and motivation. For better or for worse, they are what we call character.

I have told you of two such early formative experiences, the embracing of Nature and military discipline in turn, very different from each other in quality and strangely juxtaposed. There were three such episodes during my early childhood. I now come to the last, which in the genesis of a scientist may seem the most peculiar of them all.

CHAPTER THREE

A Light in the Corner

ON A SUNDAY MORNING in January 1944, I sat alone in a back pew of Pensacola's First Baptist Church. The service was almost over. The pastor, Dr. Wallace Roland Rogers, walked from the pulpit to the center aisle, raised his arms with palms turned up, and began to intone the Invitation:

Won't you come? Jesus is calling; Jesus is our friend; let us
    weep with Him;
let us rejoice with Him in knowledge of everlasting life.

As he repeated the cantorial phrases with variations, a familiar hymn rose from the organ behind him in muted reinforcement. The congregation did not need to sing the words. They are graven in the hearts of every born-again Baptist. Loved like scripture, the text teaches the suffering and redemption of evangelistic Christianity.

On a hill far away stood an old rugged cross,
the emblem of suffering and shame.
And I love that old cross where the dearest and best
For a world of lost sinners was slain.

Rogers as I beheld him was a dignified, friendly man in his mid-forties, with the broad, open face, the steel-rimmed glasses, the quick smile of a Rotary Club president. A leader of the Pensacola community, he was also much respected at the nearby naval air base for his religious leadership during the early months of the war. He was a builder of his church, a warm but disciplinarian friend of youth, a crusader against alcohol and legalized gambling, and a surprising progressive in a racist city in racist times. He represented northwest Florida actively in the national affairs of the Southern Baptist Conference. His sermons and lectures were intelligent and well crafted.

This morning as always the service began at eleven, and as always the choir and congregation rose to sing the Doxology: *Praise God from whom all blessings flow.* Rogers offered a prayer, and we sang another hymn. The congregation resumed their seats. Then came announcements from the pastor, of parish news, special church events, and the names of ill members to be included in the prayers of the congregation. A second hymn followed, and a welcome to visitors. Ushers proceeded down the aisle to collect the offering, with organ music in the background. Next the choir sang; then a soloist rendered "Amazing Grace," in achingly pure soprano. The hymn captured the central theme of redemption: *How precious did that Grace appear, The hour I first believed.*

Finally Rogers rose to deliver the sermon, starting, in traditional form, with a reading from the Bible.

Sorrowful, yet always rejoicing; poor, yet making many rich; having nothing, yet possessing everything.

Second Corinthians, chapter 6, verse 10. After the reading he told a story, lightly humorous in tone as usual, this one about two young farm boys who had come to the big city in order to enlist. Having risen as was their habit just before daybreak, they were wandering somewhat bewildered through the still-empty city streets between tall buildings, lost, unable to find anyone for directions. One turned to the other and said, "Where do you reckon they all go in the morning?"

The congregation chuckled in appreciation. The sermon had been personalized, brought to earth. We were now relaxed, in friendly contact. Rogers paused, grew serious. He began the homily, leading off from the quoted scripture and his story. We Americans, like these boys, may be simple people, he intoned; we may be innocent, but we are winning the war, because this country was founded on faith in God and Christian values, on the pioneers' courage in adversity, and on a willingness to sacrifice for others. If the cause is just, if Jesus is truly in our hearts, no one can stop us. That is the way it was with the disciples, simple men who left all behind them and abandoned all their material desires to suffer unto death if need be in the name of Jesus Christ. They were not the mighty rulers of Rome or rich Sadducees of Jerusalem. They were not men of steel and power. But they changed the world! They came as little children in their hearts to serve the Lord.

And Jesus said, Whosoever shall not receive the Kingdom of God as a little child, he shall not enter therein. And Jesus said, Except a man be born again, he cannot see the Kingdom of God. And Jesus said, over and over, to each of us, Hold fast till I come.

Now came the Invitation, the call to those not yet saved and those who wished to reaffirm their fellowship in Jesus Christ. Several rose in response and walked forward down the aisle. They shook the pastor's hand, took his embrace, and turned to pray with the congregation. There were tears in their eyes. I was one of them, fourteen years old, able and ready to make this important decision on my own. The choir sang "In the Garden," sweetly, adagio,

> And He walks with me, and He talks with me,
> And He tells me I am His own,
> And the joy we share as we tarry there,
> None other has ever known.

Evangelical Protestantism does not waste time with philosophy. It speaks straight to the heart. The message draws power from the simplicity of the story of Christ expressed as a mythic progression. From the pain and humiliation of earthly existence, the soul can be redeemed by union with the sacred fellowship and thereby enter eternal life in heaven.

Instruction and ritual are minimal, belief all. Jesus is with you always, in the spirit; He waits to comfort you. He will return in the flesh on the appointed day, which will be soon, perhaps in our lifetime. Our Lord is the incarnation of the fellowship and the perfect eternal patriarch. Suffer the little children to come unto me, He said, and hinder them not. Of the Trinity, He is the personal God. Each Christian must find Him individually, out of free choice, and be guided thereafter by readings of the Bible and in communion with others who have also found Grace. The Southern Baptists have no bishops. The pastor can do no more than advise and lead. Members of the congregation form the priesthood of the believer. They learn to speak in a scripture-laced discourse called the language of Zion.

The service codifies the morals of the believers. It makes explicit what the consensus holds to be decent and right. But it is much more, and those who do not grasp that added dimension will always underestimate religious faith. It catches the *power*. It is the parabolic mirror that bends the rays of tribalism to one white-hot focus: Saved by Jesus Christ, united in the fellowship of the Lord. Born again!

There was no question I would be raised in this faith, which is today shared by 15 million people in the United States and ranks second in numbers only to Roman Catholicism. Virtually all my forebears on both sides of the family as far back as the mid-1800s, close to the founding of the denomination in 1845, were Southern Baptists. All lived in Alabama and Georgia, the fundamentalist Bible Belt. Theirs was an activist religion. How well I remember sitting as a six-year-old in the Sunday School class, held at the Pensacola First Baptist Church just before service, learning to sing the heart-thumping refrain,

> Onward, Christian soldiers, marching as to war,
> With the cross of Jesus going on before.

When I was sprung from the Gulf Coast Military Academy in the summer of 1937, my religious training acquired a new dimension. I was now eight years old. My parents had separated and divorced. My relationship with my mother, who retained custody, remained loving and close, but the divorce had shorn her of support. Pensacola was gripped by the Depression, and times were hard. My mother found work as a secretary, but the wages were marginal, and it was several years before she acquired the additional training and experience to move to a better job. During that first year she placed me in the care of a trusted family friend.

Belle Raub, Mother Raub as I quickly came to call her, lived on East Lee Street in Pensacola with her husband, E. J., a retired carpenter. She was a heavyset, bosomy woman in her late fifties. She eschewed makeup and favored long, floral-print dresses. She wore a cougar-claw pendant that I found fascinating. ("Where did you get it? Where are cougars found? What do they do?" Monsters of the land.)

Mother Raub was in fact the perfect grandmother. She was forever cheerily working in and around the house, from before I woke in the morning to after I fell asleep, gardening, cleaning, cooking, and crocheting spokewheel-patterned bedspreads that she gave to friends and neighbors. She was attentive to my every need and listened carefully to every story of my life, which I considered to have been both long and filled with meaning. I gave her no problems with my manners or discipline; the Gulf Coast Military Academy had taken care of that.

Mother Raub had edged the porch with a small botanic garden of ornamental plants. I set out to learn the many kinds as best she could teach me. I found the living environment fascinating: elephant-ear arums the size of kitchen tables sprang from the soil along the front of the porch; a persimmon tree next to the street yielded winter fruit; and in the vacant lot behind the Raub property, second-growth turkey oaks formed a miniature forest. All this and the surrounding neighborhood I enthusiastically explored, freed from the twenty-four-hour discipline of the military school.

I acquired a black cat and planted a small garden of my own in the backyard. In a soft sandy spot nearby I began to dig a hole to China or to wherever else the shaft might lead, a project never completed. I learned the joy of fried grits at breakfast time: Mother Raub indulged me in this and in all other things. E. J., who had a drinking problem and was periodically scolded by his wife, was kindly in a gruff, distracted way.

In the fall I entered the nearby Agnes McReynolds Elementary School, located beyond the vacant lot and one block west, each day carrying my lunch of a sandwich and banana in a tin pail and finding, miraculously, that by noon the banana was always squashed and oozing out of its skin onto the bottom.

Soon judged by my teacher at Agnes McReynolds to be too advanced for the third grade, I was given a written test and skipped to the fourth. This was a serious misdetermination for me socially. I was already small for my age, and growing more shy and introverted all the time. For the rest of my school years I was destined to be the runt of my class. (Upon entering Murphy High School in Mobile four years later, I was the only boy still wearing short pants. I soon switched to the brown corduroy knickers then in vogue, the ones whose legs rubbed together and squeaked when you walked.)

Mother Raub was a Methodist, not a Baptist. This meant that the services she and E. J. attended weekly, after dropping me off at the First Baptist Church, were a bit more sedate and less evangelical. She was nothing, however, if not a fierce moralist in the stricter Methodist tradition. Smoking, drinking, and gambling were in her eyes among the gravest of sins. She was undoubtedly aware that my father zealously over-indulged in all these vices. She asked me to swear that I would never in my whole life give in to the same temptations, and I gladly agreed. It was easy: eight-year-old boys then were not prone to vices beyond occasionally betting marbles on games of mumblety-peg. In the nearly sixty years since I have kept my promise, except for the odd glass of wine or beer with meals—not because of piety, but rather because I don't much care for the taste of the stuff, and, at a deeper and probably more cogent level, because of the later downward spiral of my father's life as a result of alcoholism, which I witnessed with helpless despair.

Mother Raub was a woman with a steadfast heart and a mystic soul. Holiness was for her a state to be ardently sought. She told me a story about a very religious friend who wished to unite with Jesus through prayer. One day this good woman looked up from her devotions and saw a strange light in the room. It was a sign from God.

"Where in the room?" I interrupted.

"Well, in the corner."

"Where in the corner?"

"Well . . . in the upper part of the corner, next to the ceiling."

My mind raced. Her friend had seen God! Or, at least, she had received a Sign. Therefore, she must have been a chosen person. Maybe the light gives you the answer to everything, whatever that is. It was the *Grail*! The leap was possible if you prayed in some special way.

So I prayed long and hard many evenings after that, glancing around occasionally to see if the light had arrived or if any other change had occurred in the room. Nothing happened. I decided I just wasn't up to bringing God into my life, at least not yet. I would have to wait, maybe grow a little more.

At the end of that school year I left Mother Raub, this time to rejoin my father. My interest in the mysterious light faded. Perhaps (I cannot remember exactly) I stopped believing altogether in the existence of the light. But I never lost faith in the immanence of the Lord. He would come soon as a light unto the world.

In the fall of 1943, when I was fourteen, I came back to spend another year with Belle Raub. I was old enough to be baptized and born again by my own free will. No one counseled me to take this step; I could have waited for years before the weekly altar call struck home. One evening it just happened. Mother Raub and I had walked over to the McReynolds School to attend a recital of gospel hymns sung a cappella by a traveling tenor soloist. I have forgotten his repertoire as a whole. But one song, delivered in measured, somber tones, deeply moved me. It was a dissonant piece that gripped the listener in Pentecostal embrace:

> Were you there when they crucified my Lord?
> Were you there when they nailed Him to the cross?
> Sometimes it makes me to tremble, tremble.
> Were you there when they crucified my Lord?

An otherwise restless and free-spirited adolescent, I wept freely in response to the tragic evocation. I wanted to do something decisive. I felt emotion as though from the loss of a father, but one retrievable by redemption through the mystic union with Christ—that is, if you believed, if you really believed; and I did so really believe, and it was time for me to be baptized.

Dressed in my Sunday clothes and accompanied by Mother Raub, I called on the Reverend Wallace Rogers at the First Baptist Church to announce my decision and to select a time for baptism. For a teenager to meet the pastor of a large congregation was an exceptional event. I was tense and nervous as we walked into Rogers' office. He rose from his desk to greet us.

He was dressed in sports clothes and *smoking a cigar*. A cigar! In his friendly, casual way he congratulated me on my decision, and together we chose a date for the baptism. I filled out the application form as he watched and drew on his cigar. Mother Raub said nothing, then or later, about his transgression. But I knew what was on her mind!

One Sunday evening in February 1944, I stood in the line of the newly converted in a room behind the pulpit. While the congregation watched, we came out one by one to join the pastor in a large tank of chest-deep water in the choir loft at the front of the church. I was dressed in a light gown over my undershorts. When my turn came Rogers recited the baptismal dedication and bent me over once like a ballroom dancer, backward and downward, until my entire body and head dipped beneath the surface.

Later, as I dried off and rejoined the congregation, I reflected on how totally physical, how somehow common, the rite of passage had been, like putting on swimming trunks and jumping off the tower at the Pensacola Bay bathhouse the way it was done in 1943, letting your toes squish in the bottom mud for a moment before you kicked back up to the surface. I had felt embarrassed and uncomfortable during the baptism itself. Was the whole world completely physical, after all? I worried over Dr. Rogers' comfortable clothing and cigar. And something small somewhere cracked. I had been holding an exquisite, perfect spherical jewel in my hand, and now, turning it over in a certain light, I discovered a ruinous fracture.

The still faithful might say I never truly knew grace, never had it; but they would be wrong. The truth is that I found it and abandoned it. In the years following I drifted away from the church, and my attendance became desultory. My heart continued to believe in the light and the way, but increasingly in the abstract, and I looked for grace in some other setting. By the time I entered college at the age of seventeen, I was absorbed in natural history almost to the exclusion of everything else. I was enchanted with science as a means of explaining the physical world, which increasingly seemed to me to be the complete world. In essence, I still longed for grace, but rooted solidly on Earth.

My fictional heroes in late adolescence were the protagonists of *Arrowsmith*, *The Sea Wolf*, and *Martin Eden*, the Nietzschean loners and seekers. I read Trofim D. Lysenko's *Heredity and Its Variability*, a theory officially sanctioned by Stalin as sound Marxist-Leninist doctrine, and wrote an excited essay about it for my high school science class. Imagine, I scribbled, if Lysenko was right (and he must be, because otherwise why would traditional geneticists be up in arms against him?), biologists could change heredity in any direction they wished! It was rank pseudoscience, of course, but I didn't know it at the time. And I didn't care by then; I had tasted the sweet fruit of intellectual rebellion.

I was exhilarated by the power and mystery of nuclear energy. Robert Oppenheimer was another far-removed science hero. I was especially impressed by a *Life* magazine photograph of him in a porkpie hat taken as he spoke with General Leslie Groves at ground zero following the first nuclear explosion. Here was Promethean intellect triumphant. Oppenheimer was a slight man, as I was a slight boy. He was vulnerable in appearance like me, but smilingly at ease in the company of a general; and the two stood there together because the physicist was master of arcane knowledge that had tamed for human use the most powerful force in nature.

Shortly afterward, during my first year at college, someone lent me a book that was creating a sensation among biologists, Erwin Schrödinger's *What Is Life?* The great scientist argued not only that life was entirely a physical process, but that biology could be explained by the principles of physics and chemistry. Imagine: biology transformed by the same mental effort that split the atom! I fantasized being Schrödinger's student and joining the great enterprise. Then, as an eighteen-year-old sophomore, I read Ernst Mayr's *Systematics and the Origin of Species*. It was a cornerstone of the Modern Synthesis of evolutionary theory, one of the books that combined genetics with Darwin's theory of evolution by natural selection. Mayr's writing reinforced in my mind the philosophy implicit in Schrödinger. He showed that variety among plants and animals is created through stages that can be traced by the study of ordinary nature around us. Mayr's text told me that I could conduct scientific research of a high order with the creatures I already knew and loved. I didn't need to journey to a faraway place and sit at the feet of a Schrödinger or a Mayr in order to enter the temple of science.

Science became the new light and the way. But what of religion? What of the Grail, and the revelation of purest ray serene that gives wholeness and meaning to life? There must be a scientific explanation for religion, moral precepts, the rites of passage, and the craving for immortality. Religion, I knew from personal experience, is a perpetual fountainhead of human emotion. It cannot be dismissed as superstition. It cannot be compartmentalized as the manifestation of some separate world. From the beginning I never could accept that science and religion are separate domains, with fundamentally different questions and answers. Religion had to be explained as a material process, from the bottom up, atoms to genes to the human spirit. It had to be embraced by the single grand naturalistic image of man.

That conviction still grips me, impelled and troubled as I am by emotions I confess I do not even now fully understand. There was one instructive moment when the subterranean feelings surfaced without warning. The occasion was the visit of Martin Luther King Sr. to Harvard in January 1984. He came under the auspices of a foundation devoted to the improvement of race relations at the university. Its director, Allen Counter, an old friend with a similar Southern Baptist background, invited me to attend a service conducted by the father of the martyred civil rights leader, and to join a small group at a reception afterward.

It was the first Protestant service I had sat through in forty years. It was held in Harvard's Memorial Church. Reverend King gave a quiet hortatory address organized around scripture and moral principle. He omitted the altar call—this was, after all, Harvard. But at the end a choir of black Harvard students surprised me by singing a medley of old-time gospel hymns, with a professionalism equaling anything I ever heard in the churches of my youth. To my even greater surprise, I wept quietly as I listened. My people, I thought. My people. And what else lay hidden deep within my soul?

CHAPTER FOUR

A Magic Kingdom

COME BACK WITH me to October 1935, to Pensacola, for a walk up Palafox Street. Let's start by peeking over the seawall that closes the south end of the street. Down there on the rocks, kept wet and alga-covered by the softly lapping water of the bay, sits a congregation of grapsid crabs. They resemble large black spiders, with crustaceous skins, carapaces the size of silver dollars, and clawlike legs that spread straight out from the sides of the body. They rest on needle-tipped feet, alert and ready to sprint forward, backward, or sideways at the slightest disturbance. Drop a pebble among them, and those closest to impact scurry for cover.

Let's turn and stroll north along the street, just looking around. On the right is the Childs Restaurant, where the courthouse crowd gathers for lunch. Stop for a moment and pass your hand through the light beam across the entrance. The door swings open, a miracle of modern technology. Do it again; this time let the couple waiting behind you walk on through. People don't seem to mind kids' playing with the beam. A little farther along is the Saenger Theater, Pensacola's premier palace of pleasure, in summer "cooled by iced air." This Saturday's matinee bill is an episode of the Flash Gordon serial (Flash escapes from the lair of the fire dragon), followed by Errol Flynn in *Captain Blood*. You've seen the feature; there's a scary scene near the end in which Flynn skewers Basil Rathbone in a duel on the treasure island, and the treacherous French pirate falls dying in the surf.

This section of lower Palafox, west to Reus and east to Adams, is the busy part of town. Model A Fords are on the street, a pretty good crowd of shoppers on the sidewalks. Be careful when you cross Romana Street up ahead; a kid on a bicycle got run over there last year; that's what they told me.

It's hot, as always. North Florida is still a tropical place in the early fall. Afternoon thunderheads are gathering to the south and west across the bay. No breeze has kicked in yet; the air hangs heavy and moist and laced with engine fumes. Let's cross Palafox over to the courthouse, on your left. On the lawn next to the sidewalk a fire ant colony is swarming. The ants are pouring out of a mound nest, here no more than an irregular pile of dirt partly flattened by the last pass of a lawnmower. Winged queens and males are taking off on their nuptial flight, protected by angry-looking workers that run up and down the grass blades and out onto the blistering-hot concrete of the sidewalk. The species is unmistakably *Solenopsis geminata*, the native fire ant, I can tell you now. Another fifteen years will pass before the infamous *Solenopsis invicta*, imported from South America, will spread this far east from its point of introduction, Mobile, Alabama. I'll be here as a college student to watch that happen.

Walk a few more blocks, on past the old San Carlos Hotel (they've torn it down since), and cut left on West Gregory Street. My parents' apartment, one of two on the second floor of a stucco Spanish-style building, is several blocks farther down. There is a large live oak in the side yard where bluejays land and shriek at one another. Their call is like a fire engine's siren, always announcing some emergency or other.

On the sidewalk (please turn your gaze downward occasionally and look with me for insects on the ground) lion ants of the genus *Dorymyrmex* run like whirligigs on the stove-hot surface. Crush one, and the unmistakable smell of a dolichoderine ant hits your nose. I can tell you now that the odor of this species comes from a mix of heptanone and methylheptenone, secretions from the pygidial glands the workers use to defend the colony against enemies and alarm nestmates to approaching danger.

Forty years later I will return to within a few feet of this exact spot. I will get down on my knees (an elderly black man passing by will ask me if I need help) and look again for the lion ants. The dirt and cracked concrete slabs will look the same, but this time the ants running around will be *Pheidole dentata*, which lack a strong odor. Same thing fifteen years later. I'll keep coming here from time to time whenever I visit Pensacola, to see if the *Dorymyrmex* have returned to this special square yard or two of space. So far, my surveillance has lasted for nearly sixty years; if I am fortunate, it will last for eighty. Meanwhile, I can tell you that the ant present in 1935 was a *Dorymyrmex*.

You will think this a strange journey and a stranger obsession, but not I. Consider how long-term memory works. With each changing moment, the mind scans a vast landscape of jumbled schemata, searching for the one or two decisive details upon which rational action will be based. The mind with a search image is like a barracuda. The large predatory fish pays scant attention to the rocks, pilings, and vast array of organisms living among them. It waits instead for a glint of silver that betrays the twisting body of a smaller fish. It locks on this signal, rushes forward, and seizes the prey in its powerful jaws. Its singlemindedness is why swimmers are advised not to wear shiny bracelets or wrist watches in barracuda waters.

The human mind moving in a sea of detail is compelled like a questing animal to orient by a relatively few decisive configurations. There is an optimum number of such signals. Too few, and the person becomes compulsive-obsessive; too many, and he turns schizophrenic. Configurations with the greatest emotional impact are stored first and persist longer. Those that give the greatest pleasure are sought on later occasions. The process is strongest in children, and to some extent it programs the trajectory of their lives. Eventually they will weave the decisive images into a narrative by which they explain to themselves and others the meaning of what has happened to them. As the Talmud says, we see things not as they are, but as we are.

Our remembered images are reinforced like pictures improved by one overlay upon the next, each adding finer detail. In the process edges are sharpened, content refined, emotional colors nuanced. In this way, for me Pensacola on a hot autumn day in 1935 has evolved into a network of vividly remembered small animals. There is a backdrop of people, streets, a theater marquee, and houses; but although these parts of my world were important then, they have faded since.

I was a normal boy, within reason. I had friends, played rough-and-tumble games in the yard of nearby P. K. Yonge Elementary School, was ashamed and tearful when kept after class for misbehavior, had a wonderful Christmas that year, obeyed my parents but had to be forced to eat asparagus, and in the gentle Gulf winter, when leaves had fallen and a scum of paper-thin ice coated the gutter puddles, I searched the ground with other boys for pecans and chinquapins. But sixty years have drained these memories of most of their importance, and the fine details and emotional force have largely eroded away to nothing.

They have done so, and natural history has been reinforced, because at an early age I resolved to become a naturalist and a scientist. And I took that course in part, if there must be an explanation, because I was an only child who lived something of a gypsy's existence. My mother had legal custody, and we were to remain thereafter very close, but she allowed my father, who had a better job, to care for me on a provisional basis. He remarried in 1938. I acquired a devoted stepmother named Pearl, and the temporary arrangement was extended. My father, a government accountant, for some reason preferred road assignments. He began an odyssey across the southeastern United States, changing his job and the location of his home every year or two. My itinerary from the fourth grade to high school graduation circled round and about the South as follows: Pensacola, Mobile, Orlando, Atlanta, the District of Columbia, Evergreen (Alabama), back to Mobile, back to Pensacola, and finally Brewton and Decatur in Alabama, with intervening summer sojourns in Boy Scout camps and homes of friends in Alabama, Florida, Virginia, and Maryland. Over eleven years I attended fourteen different public schools. In the summer before I started college we finally came back to Mobile, my father's birthplace, which he had said earlier he hated and had "taken off the map"; but now he changed his mind, and it was there he was to remain until he died, in 1951.

A nomadic existence made Nature my companion of choice, because the outdoors was the one part of my world I perceived to hold rock steady. Animals and plants I could count on; human relationships were more difficult. With each move I had to insert myself into a new group of peers, mostly boys. At first, before my father remarried, we lived in boardinghouses, from which I ventured cautiously. In Orlando, our first port of call, I avoided schoolmates for a few weeks, out of fear. I conversed silently with myself, creating three boys in my head: I, me, and myself. I rescued bits of Spanish moss that had fallen to the ground and replaced them on the low branches of the schoolyard oaks. They were my friends, I thought; but the emotion I felt was self-pity.

I studied plants and insects around the streets of Orlando, a beautiful little city in 1938. I kept harvester ants in a jar of sand under my bed and watched them excavate. I discovered fairy tales in the school library, and took to reading every one I could find. I was transfixed by the magical choices between grisly death or eternal happiness. I did well in class, and came close to winning the school spelling bee ("Indain" instead of "Indian" was my undoing, still burned on my brain). No one paid any attention to my eccentricities. There were no programs for gifted or disturbed children in those days.

Nor was there any educational theory in 1938 to suggest that loneliness in a beautiful environment might be a good if risky way to create a scientist, at least a field biologist. After a few weeks in Orlando I discovered to my joy that one of the city lakes was within walking distance. I started fishing there for minnows and bream, taking care with the hook I baited and the fins of the fish I caught to protect the one eye I had left after the accident at Paradise Beach. I spent hours admiring a large alligator gar held captive in a cement pool in the small waterfront park. I went back and forth to the lake alone, keeping my own company and organizing my thoughts. Lantana hedges, laden grapefruit trees, and men in Panama hats still float through my surviving memories.

In towns and cities we settled in later I learned how to adapt more quickly in my perpetual role as new kid on the block. Once, early on in Mobile, I pretended to be deaf and dumb until I had a small, fascinated group of boys and girls following me in an attempt to read my ersatz sign language. When I admitted the hoax, all was well. They were relieved, and still fascinated enough to make me a popular new member of the gang. Usually I approached the problem straight on, by working myself into baseball games as an extra, or talking with other boys I saw standing alone at the edges of the schoolyard or lunchroom.

My worst difficulties came from the fistfights. They were merciless and brutal. I suspect that most adults, especially those reared in middle-class suburbs, cannot bring themselves to acknowledge the innate savagery of preadolescent boys. From the ages of nine to fourteen they are naturally predisposed to set up blockwide territories, run in gangs, and bully to gain acceptance, to swagger, boast, dare, and call back and forth to one another in the loud honking voices of emerging male adolescence. Strange kids in the neighborhood, especially those without brothers or parents in view, are fair game. In the South of the 1930s and 1940s—here I have enough experience to speak like a sociologist—there was a certain protocol to the combat. One boy, usually the local bully or the "champion" of a group, challenged another boy, usually the newcomer. The fight was held after school in some secluded spot in the neighborhood where both lived. It was stand-up combat with fists, fought the way we saw Joe Louis do it in the newsreels. Except there were no rounds, and no conclusion until one boy gave up or an adult broke in. Word got around in advance, and a circle of boys gathered to hoot the pugilists on.

"One of 'em's scared and the other's glad of it!"

"Hey, that was just a love tap; wait'll he gets goin'."

Most of the fights I watched, I was in, maybe a dozen up to the age of fourteen. After that time boys of my age found other outlets for aggression, most commonly football or hunting. I can recall these battles like a sports historian ticking off rounds in the Dempsey-Tunney fight. I was deeply anxious about challenges, especially in a new neighborhood, and tried to steer clear of the more aggressive-seeming boys, always without success. My father told me never to back down, and the Gulf Coast Military Academy ethos forbade it. It was unmanly to refuse a fight. I did decline, however, twice, because the boys were too big to beat, and with a gang and from different schools anyway, and I knew I wouldn't see them again. I retreated before their taunts, to my everlasting shame. It is ridiculous, of course, but I still burn a little when I think about my cowardice. I never picked a fight. But once started I never quit, even when losing, until the other boy gave up or an adult mercifully pulled us apart.

"Hey, hey, he's had enough!"

"Okay, okay, let's stop; I wanna stop."

"This is a lot of shit, anyhow. I got to get home."

I couldn't stop. Somehow I felt that, having invested this much of my limited store of courage to take on a challenge, I must never throw it all away and suffer the added shame of losing. My face was sometimes a bloody mess; I still carry old lip and brow split scars, like a used-up club fighter. Even my father, proud that I was acting "like a little man," seemed taken aback. But later I savored the memories of my combat, and especially the victories. There is no finer sight on green Earth than a defeated bully.

My childhood was nevertheless relatively serene. Most of the time I simply found a best friend in new neighborhoods, a boy the same age and physical size who enjoyed riding bikes and exploring the nearest woods for snakes and insects. I was drawn to conspicuous introverts, and they to me. We stayed away from social activities at the school and the clubs and from the roving gangs of boys. Throughout, I was just as happy to be entirely alone. I turned with growing concentration to Nature as a sanctuary and a realm of boundless adventure; the fewer people in it, the better. Wilderness became a dream of privacy, safety, control, and freedom. Its essence is captured for me by its Latin name, *solitudo.*

So inevitably, and given that I was looking at the world with only one visually acute eye, I came to be an entomologist, a scientist who specializes in insects. To put the matter as simply as possible: most children have a bug period, and I never grew out of mine. But as in the lives of scientists generally, there is more to the story. Every child wants to visit a magic kingdom. Mine was given to me at the age of ten, when my father moved Pearl and me to Washington, D.C. We took up residence in a basement apartment on Fairmont Street near Fourteenth Street, within walking distance of the National Zoo and a five-cent streetcar ride to the National Museum of Natural History. A year later (possibly not wanting to risk putting down roots), my father moved us again, to a second apartment six blocks away, on Monroe Street. For me the central-city location, in what is now an all-black neighborhood, was an extraordinary stroke of good luck.

Here I was in 1939, a little kid, nine years old, tuned to any new experience so long as it had something to do with natural history, with a world-class zoo on one side and a world-class museum on the other, both free of charge and open seven days a week. Unaffected by the drabness of our working-class neighborhood, I entered a fantasy world made weirdly palpable by federal largesse. I spent hours at a time wandering through the halls of the National Museum, absorbed by the unending variety of plants and animals on display there, pulling out trays of butterflies and other insects, lost in dreams of distant jungles and savannas. A new vision of scientific professionalism took form. I knew that behind closed doors along the circling balcony, their privacy protected by uniformed guards, labored the curators, shamans of my new world. I never met one of these important personages; perhaps a few passed me unrecognized in the exhibition halls. But just the awareness of their existence—experts of such high order going about the business of the government in splendid surroundings—fixed in me the conception of science as a desirable life goal. I could not imagine any activity more elevating than to acquire their kind of knowledge, to be a steward of animals and plants, and to put the expertise to public service.

The National Zoo, the second focus of my life, was a living museum of equal potency with the National Museum of Natural History. It was and is administered as part of the same umbrella organization, the Smithsonian Institution. Here I spent happy days following every trail, exploring every cage and glass-walled enclosure, staring at the charismatic big animals: Siberian tigers, rhinoceros, cassowaries, king cobras, reticulated pythons, and crocodiles big enough to consume a boy in two bites. There were also smaller animals that eventually became equally fascinating. I developed a liking for lizards, marmosets, parrots, and Philippine tree rats.

Close to the zoo was Rock Creek Park, a wooded urban retreat, into which I ventured on "expeditions." In those confines, within earshot of passing automobiles and the conversations of strollers, I found neither elephants to photograph nor tigers to drop-net. But insects were everywhere present in great abundance. Rock Creek Park became Uganda and Sumatra writ small, and the collection of insects I began to accumulate at home a simulacrum of the national museum. During excursions with a new best friend, Ellis MacLeod (who was later to become a professor of entomology at the University of Illinois), I acquired a passion for butterflies. Using homemade nets made of broomsticks, coat hangers, and cheesecloth bags, we captured our first red admirals and great spangled fritillaries and sought the elusive mourning cloak along the shaded trails of Rock Creek. We were inspired by Frank Lutz's *Field Guide to the Insects* and W. J. Holland's *Butterfly Book*. Poring over R. E. Snodgrass' *Principles of Insect Morphology*, which we could barely begin to understand but revered because it was *real* science, we decided we would devote our lives to entomology.

The course of my life had been set. While sorting through dusty files I recently discovered a letter to my parents written by my fifth-grade teacher at the Hubbard School on February 2, 1940, when I was ten years old: "Ed has genuine writing ability, and when he combines this with his great knowledge of insects, he produces fine results."

About this time I also became fascinated with ants. One day as Ellis and I clambered over a steep wooded slope in the park, I pulled away the bark of a rotting tree stump and discovered a seething mass of citronella ants underneath. These insects, members of the genus *Acanthomyops*, are exclusively subterranean and can be found only in the soil or in fallen pieces of decaying wood. The worker ants I found were short, fat, brilliant yellow, and emitted a strong lemony odor. The smell was the chemical citronellal, which thirty years later (in my laboratory at Harvard) I discovered is secreted by glands attached to the mandibles of the ants and, like the pygidial substances of the Pensacola *Dorymyrmex*, is used to attack enemies and spread alarm through the colony. That day the little army quickly thinned and vanished into the dark interior of the stump heartwood. But it left a vivid and lasting impression on me. What netherworld had I briefly glimpsed? What strange events were happening deep in the soil?

I devoured an article titled "Stalking Ants, Savage and Civilized," by William M. Mann, in the August 1934 issue of the *National Geographic*. In what was to be one of the more remarkable coincidences of my life, Mann was at that time director of the National Zoo. Like the still anonymous keepers of the museum, he became my hero from afar. To run a great zoo while writing about his adventures around the world with ants—what a role model! In 1957, when I was a beginning assistant professor at Harvard and Mann was in the last year of his directorship, he gave me his large library on ants (an important source for my later research) and escorted me and my wife, Renee, on a special tour of the zoo. In 1987 I was awarded the silver medal of the National Zoological Park for my work on ants and other animals; at the ceremony I came home in a deeply satisfying way.

Always prone to closing and repeating circles in my life, I have often returned to the National Museum of Natural History. The denizens of that Olympus, all a new generation since 1940, have acquired names and faces and become friends and colleagues. The great collections they attend behind the closed doors are familiar ground.

There is today a quickening of purpose, a sense of rising importance and responsibility, at both of the institutions that influenced me fifty years ago. Michael Robinson, the director of the National Zoo as I write, in 1994, prefers to speak of his domain as a biopark, where animals will be released from the isolation of cages and terraria and placed in natural settings of plants and animals from their place of origin. The public can then view them not as caged curiosities but as parts of ecosystems, on which biological diversity—and the health of the planet itself—ultimately depend.

A short distance away, on the Mall, the curators of the National Museum of Natural History continue building one of the world's largest collections of plants and animals. They too must feel the future in their bones. Recent studies indicate that between 10 and 100 million species of plants, animals, and microorganisms exist on Earth, but only about 1.4 million have been studied well enough to receive scientific names. Many of these species are vanishing or being placed in imminent danger of extinction by the reduction of habitat and other human activities. The loss in tropical rain forests in particular, thought to contain a majority of the species on Earth, may exceed half a percent a year.

So there is a lot for those who study the diversity of life to do, a new respectability, and a great responsibility. But that is not the reason I am wedded to the subject. The boy who experienced the magic of the zoo and museum is still strong inside me. He is the puppet master of the man. I would have followed the same path regardless of what happened in the rest of the world.

CHAPTER FIVE

To Do My Duty

IN THE SPRING of 1941 my grandmother Mary Emma Joyner Wilson, known to her family as May, died in Mobile of a heart attack in the house where she had been born in 1868, married, attended a private school run by her mother, raised four sons, and stayed the remainder of her life. Since 1916, when her husband died, she had lived in the company of her bachelor son, Herbert. During all those seventy-three years she had seldom journeyed beyond the edge of the city.

My father brought Pearl and me to live in the large rambling structure that he and Herbert inherited from my grandmother. The house had a long history, at least for the young state of Alabama. Built by May's grandfather in 1838, it was for a few years the only house on Charleston Street, though located only a dozen blocks from Bienville Square and the commercial heart of the old city. Here then, if anywhere, were the roots of my peregrine family.

Alabama's seaport was a small town in the early 1800s when my father's forebears arrived, a junior version of New Orleans complete with muddy streets, balcony grillwork, creole cooking, and epidemics of yellow fever. In 1815, two years after American troops took it from the Spanish on orders from President Madison, Mobile was nothing more than fourteen city blocks grouped in a large square north of Fort Charlotte. By the 1830s and 1840s the town was growing rapidly, but many of the streets, including Charleston, still led down to what an early map labeled "low and miry land"—mud banks—lining the Mobile River estuary. The Hawkinses, Joyners, and Wilsons could ride there by carriage in a few minutes and walk over long wharves to reach the ferry slips. Often, no doubt, they just went to fish and net blue crabs lured with soup bones. The wildlands south of the city still existed in a remnant condition. Large stretches of hardwood and pine forest extended south all the way to Cedar Point, the southernmost tip of mainland Alabama on the west side of Mobile Bay. Beyond that, across Mississippi Sound, a mostly uninhabited Dauphin Island formed a line along the horizon.

When my father was a teenage boy, just before the First World War, he was able, he told me, to step out the front door of the Charleston Street house, stroll down the road a mile or two with a .22 rifle under his arm to the wooded terrain now occupied by Brookley Airport, and hunt quail, rabbits, or whatever else took his fancy. When I was the same age in the 1940s, I often rode my bicycle around Brookley to reach uninhabited woodland and pitcher plant and pine savanna along the Dog and Fowl rivers. I sometimes paused to eat sandwiches and drink Royal Crown Colas on the two-lane wooden bridges spanning these two streams. Around midday an hour or more might pass without the approach of a single automobile. I leaned on the wooden rails in reverie, looking deeply into the slow-moving and limpid water for glimpses of gars and soft-shelled turtles. Today this land is thickly settled, and heavy traffic rumbles all the way down to a bridge running from Cedar Point to Dauphin Island.

My father was proud of his family history. The Hawkinses and Joyners had emigrated from New England to the Mobile Bay area not long after it became American territory; one, my great-great-grandmother Mary Ann Hawkins, was born there in 1826. They prospered as marine engineers, pilots, and ship-owners. My great-grandfather James Eli Joyner, who married Mary Ann's daughter Anna Amelia, operated a ferry that ser-viced the Baldwin County shore out of Mobile. One November day in 1870 his ship caught fire and sank close to Mobile, and he drowned while attempting to swim ashore. His young wife was holding my grandmother May in her arms on the porch of the Charleston Street house as she gazed at the distant plume of smoke, not realizing that it meant she would be a widow. To make ends meet thereafter, she opened a private school in the house, the first in Mobile. I own her pendant containing a portrait of her mother, as well as the heavy gold watch chain with dolphin catch taken from her husband's body.

In the War Between the States virtually every able-bodied male on both my father's and mother's sides fought for the Confederacy. Of the two paternal great-grandfathers, James Joyner served for the duration of the war as artilleryman and teamster; the other was a special case, the undoubted star of all my forebears as far back as I have been able to trace them: William Christopher Wilson.

Black Bill, as his friends called him, was a man whose blood I like to imagine coursing through my veins, even though after three generations I carry only one-eighth of his genes. He was born William Christopher O'Conner in 1816 in a family of Dublin printers, whose customers I was told included the Bank of England. He must have been a rebel of considerable fire. His parents wanted him to train for the Episcopal ministry, but he yearned for a life at sea. So he left home as a teenager, took a job as a cabin boy on a ship bound for Baltimore, and changed his name to Wilson en route when a passenger by that name died.

In Baltimore he proceeded to take a Jewish bride named Maria Louise Myers, daughter of Jacob Myers and Sarah Solomon Myers, late of Germany. The newlyweds soon moved to Mobile to seek their fortune. Black Bill—his name came later from the color of his long beard and not from a Black Irish complexion—found employment as a bar pilot. He advanced to master's status and eventually acquired his own boat, with which he guided merchant ships through the treacherous shallows between Fort Morgan and Fort Gaines. In the early 1840s he became a founding member of the Mobile Bar Pilots Association, a guild still in operation today. He moved his family to Navy Cove, on Fort Morgan Peninsula, where the sails of approaching merchantmen could first be sighted on the Alabama coast as they approached across the open Gulf.

In 1863, when Admiral Farragut blockaded Mobile Bay, Black Bill and his fellow pilots used their fast ships to run supplies in from Havana. Often pursued, Wilson was finally cornered on a small island outside the harbor. Instead of being simply thrown in irons, he was brought before Farragut and his staff, who made an offer: if Wilson would lead the fleet into Mobile Bay so that it could move swiftly past the guns of Fort Morgan and Fort Gaines without running aground on the shoals, he would receive a large monetary reward and be resettled with his family somewhere in the North. He refused, crying, according to his own account, "I'd see the whole Yankee fleet damned in hell before I'd betray my country!" Not exactly the famously historical flourish of Farragut's authenticated "Damn the tor-pedoes! Go ahead!" that followed soon afterward (or it might have been "Damn the torpedoes, full speed ahead!" or, most likely and least euphoniously, "Damn the torpedoes, Jouett, full speed!"), but good enough for a southern family in a city in which, even into this century, a sign of respect to an older man was to call him "Cap'n." Oddly, William Christopher Wilson was still an Irish citizen at the time of his capture and remained so. And he never legally changed his name to Wilson. Had I known that as a young man, I might have changed my own back to O'Conner, the sound of which has a nice swing around the apostrophe and a pleasing consonantal bark at the hard C, in contrast to the whispery syllables of Wil-son.

Black Bill was sent off to a succession of federal prisons in New York and Maryland for the remaining two years of the war, and Farragut and his men soon got what they wanted anyway. They captured another Mobile bar pilot who was at that time fishing for a living off the coast at nearby Pasca-goula, the legitimate pilot business having been pretty well closed down. His name was Martin Freeman (no relation to my mother's people of the same surname, who were then living in northern Alabama). He and other fishermen were armed and prepared to resist a Yankee invasion, but one salvo from the Union guns offshore changed their minds. Freeman agreed to pilot the fleet, and on August 5, 1864, when a double column of monitors and wooden frigates charged into the bay, he was coolly riding the main top of the flagship *Hartford*. Members of Black Bill's family watched from their Navy Cove house as Federal shells burst on nearby Fort Morgan. Among Freeman's rewards after Mobile had been captured was the Congressional Medal of Honor, making him—I hope I do not put too heavy a spin on it after 130 years—the only traitor ever to receive America's highest military honor.

When we arrived in Mobile in 1941, the old house was dilapi-dated and the surrounding neighborhood in decay. Most of the men in the Wilson clan had either died or dispersed, leaving behind widows and spinster daughters sprinkled about the city. We addressed them all as either Aunt or Cousin, depending on their blood ties and age. These survivors were of surprising interest to my father, who at this point turned out to be a family historian seized by nostalgia and a yearning for reflected glory. On Sunday afternoons we visited these living monuments of the treasured past—Aunt Nellie, the younger Cousin Nellie, Aunt Vivian, and Cousin Mollie—in their respective parlors. Obedient to instructions, cleaned up and dressed in my visiting clothes, I kissed each on the cheek and sat on a chair to one side until I could slip away without notice. The reminiscing droned on, a recycling of the stories and sketches of the late Grandma May, Aunt Hope, Aunt Georgia, Aunt Sarah, and all their stalwart departed husbands and brothers and sons, and what happened in the lamented War Between the States, and the many things families did in Old Mobile. Occasionally we visited Magnolia Cemetery, where our forebears and their multitudinous relations and friends lay at rest. Pearl and I stood patiently by as my father located graves, checked dates, and reconstructed lives and genealogies.

I had no interest in this world of ghosts. I considered my father a bore and my great-aunts and cousins an ordeal. For me Mobile was a place of vibrant life—not of spirits, however, nor of people, and certainly not of relatives, but of *butterflies*. At twelve years of age, I had arrived with a burning desire to collect and study butterflies. I was keenly aware that the city is on the edge of the subtropics and home to many species not found in Washington, D.C.

At every opportunity I charged out on my balloon-tired, single-gear Schwinn bicycle, pumping my way down Charleston Street to the rubble-strewn weedlots of the riverfront, west to the scattered pine-and-hardwood copses of Spring Hill, south on the Cedar Point road as far as Fowl River, and east across the Mobile-Tensaw delta on old U.S. 90 to Spanish Fort in Baldwin County. I greeted the sight of each new species of butterfly with joy, and when I caught my first specimen I thought myself a big-game hunter with net. The zebra and golden-winged julia, northernmost representatives of a group that abounds in tropical forests; the goatweed butterfly, bright orange-red with a swift erratic flight, hard to net; the little fairy sulfur, average-sized dog face sulfur, outsized cloudless sulfur, all tropical-looking with the flamboyant flashing of their yellow wings; giant swallowtail (and what a thrill to see how different it looked in life from the common tiger swallowtail of the

North); zebra swallowtail in the shadowed woods; great purple hair-streak, a stunning iridescent gem I first spotted resting on a weed in a vacant lot; and the large Brazilian skipper, which I reared from translucent gray-green caterpillars feeding on canna lilies in our backyard—all these I added to my butterfly life list.

During the next two years, before we hit the road again, as it seemed inevitably we must, my interest in natural history soared. I went looking for pileated woodpeckers rumored to nest at Spanish Fort, and on the way saw my first wild alligators, in the marshes of the Tensaw estuary. I scoured the riverine hardwood forests for holly trees and orchids. I built a secret outdoor shelter partly from the stems of poison oak and paid for it with an agonizing rash over a large part of my body (afterward I could identify *Rhus quercifolia* at a hundred paces). I hunted reptiles: stunned and captured five-lined skinks with a slingshot, and learned the correct maneuver for catching Carolina anole lizards (approach, let them scuttle to the other side of the tree trunk and out of sight, peek to see where they are sitting, then take them by grabbing blind with one hand around the trunk). One late afternoon I brought home a coachwhip snake nearly as long as I was tall and walked into the house with it wrapped around my neck. Pearl sent me back out with instructions to release it as far from the house as could be traveled round-trip during the remaining daylight hours. I owned a machete and used it to chop my way through tangled undergrowth, imagining myself to be in the jungles of South America. One day I misjudged the downward stroke and slashed my left index finger to the bone. Blood streamed down my arm on the long bicycle ride home. Pearl let me keep the knife, nonetheless, figuring I'd learned the hard way to be more careful.

When America entered the war in December 1941, the tempo of life in Mobile picked up sharply. Tanker traffic in and out of the harbor increased, and overflights of B-17 bombers and other warplanes became commonplace. Poor rural whites—peapickers we derisively called them—and blacks poured into the city looking for work. Jobs were plentiful and labor was short. One anecdote making the rounds at the time involved a white woman who stopped a local Negro woman (to use the idiom of 1942) near her house, saying that she was looking for domestic help. The other responded: "Why, so am I." If you were white you were supposed to gasp with amused surprise. Change was in the air.

I was sanguine about the war, knowing that since Franklin Delano Roosevelt had already fixed just about everything else in the country, and since the Democratic Party and Joe Louis, both of whom enjoyed my allegiance, had always won for as far back as I could remember, this new crisis would also work out all right. In my insouciance I retrieved an identity card discarded by my father, doctored it with swastikas and pseudo-German phrases, and dropped it on the sidewalk in front of our house. Someone found it and took it to the district office of the Federal Bureau of Investigation. My father was called in and questioned by agents. All quickly agreed on the explanation, and my father, to his credit, found the incident hilarious. Indeed, he dined out on it for a while.

My friends and I were indignant about the Japanese attack on Pearl Harbor, of course, and we knew the Nazis to be evil incarnate. As cartoonist for the Barton Academy junior high school newspaper, I depicted a bestial Japanese soldier stabbing Uncle Sam in the back. In school assembly we sang "The White Cliffs of Dover" and other songs in solidarity with the British war effort. But mostly my mind was elsewhere. I was absorbed in my own interests and never tried to follow the course of the war.

In June 1942 Ellis MacLeod came down from Washington to stay with me for the summer. We visited my favorite haunts, shared again our old fantasies, and renewed our intention to become entomologists. That fall after he returned home, I set out to collect and study all the ants in a vacant lot next to the Charleston Street house. I still remember the species I found, in vivid detail, enhanced by the knowledge acquired in later studies: a colony of the trap-jawed *Odontomachus insularis*, whose vicious stings drove me away from their nest at the foot of a fig tree; a colony of a small yellowish-brown *Pheidole*, possibly *Pheidole floridanus*, found nesting beneath an amber-colored whiskey bottle in midwinter and which I kept for a while in a vertical observation nest of sand between two glass plates. And colonies of imported fire ants, unmistakably *Solenopsis invicta*, were there. The vacant lot discovery was the earliest record of the species in the United States, and I was later to publish it as a datum in a technical article, my first scientific observation.

My energies and confidence were gathering. By the fall of 1942, at the age of thirteen, I had become in effect a child workaholic. I took a job with backbreaking hours of my own free will, without adult coercion or even encouragement. Soon after the start of the war there was a shortage of carriers for the city newspaper, the *Mobile Press Register*. Young men seventeen and over were departing for the service, and boys aged fifteen to sixteen were moving up part-time into the various jobs they vacated. On the lowest rung of unskilled labor, many paper routes came open as the fifteen- and sixteen-year-olds moved up. Somehow, for reasons I do not recall, an adult delivery supervisor let me take over a monster route: 420 papers in the central city area.

For most of that school year I rose each morning at three, slipped away in the darkness, delivered the papers, each to a separate residence, and returned home for breakfast around seven-thirty. I departed a half-hour later for school, returned home again at three-thirty, and studied. On Monday nights from seven to nine I attended the meeting of my Boy Scout troop at the United Methodist Church, on Government and Broad streets. On Sunday mornings I went to service at the First Baptist Church. On Sunday evenings I stayed up through Fibber McGee and Molly on the radio. On other nights I set the alarm soon after supper, went to bed, and fell asleep.

Four hundred and twenty papers delivered each morning! It seems almost impossible to me now. But there is no mistake; the number is etched in my memory. The arithmetic also fits: I made two trips to the delivery dock at the back of the *Press Register* building, each time filling two large canvas satchels. When stacked vertically on the bicycle front fender and strapped to the handlebars, the bags reached almost to my head and were close to the maximum bulk and weight I could handle. The residences receiving the papers were not widely spaced suburban houses but city dwellings, apartment buildings with two or three stories. It took perhaps a maximum of one hour to travel back and forth to the *Press Register* dock, load the papers twice, and make two round trips in and out; the delivery area was only a few minutes' ride away. That leaves three and a half hours for actual on-the-scene work, or an average of two papers a minute—during which I reached down, pulled out the paper, dropped it, or threw it rolled up for a short distance, and passed on, moving faster and more easily after one satchel was emptied.

The supervisor collected the week's subscription money from the customers on Saturday, twenty-five cents apiece, so I didn't have to work extra hours that day and had time to continue my field excursions. I made thirteen dollars a week, from which I bought my Boy Scout paraphernalia, parts for my bike, and whatever candy, soft drinks, and movie tickets I wanted.

At the time it did not occur to me that my round-the-clock schedule was unusual. I felt fortunate to have a job and to be able to earn money. It was the kind of regimen I had learned to expect as normal from my brief experience at the Gulf Coast Military Academy. I still assumed, without any real evidence, that the same level of effort would be required of me as an adult. And what of my father and Pearl, asleep in their bed as I headed out in the predawn hours in all kinds of weather? Pearl, who came from a hardscrabble life in rural North Carolina, seemed well pleased that I showed the kind of spunk it takes to survive. And the feelings of my father, who never worked that hard in his life—who can say?

But the labor over long hours did not really matter; I had discovered the Boy Scouts of America. All that I had become by the age of twelve, all the biases and preconceptions I had acquired, all the dreams I had garnered and savored, fitted me like a finely milled ball into the socket of its machine when I discovered this wonderful organization. The Boy Scouts of America seemed invented just for me.

The 1940 *Handbook for Boys*, which I purchased for half a dollar, became my most cherished possession. Fifty years later, I still read my original annotated copy with remembered pleasure. Richly illustrated, with a cover by Norman Rockwell, it was packed with useful information on the subjects I liked the most. It stressed outdoor life and natural history: camping, hiking, swimming, hygiene, semaphore signaling, first aid, mapmaking, and, above all, zoology and botany, page after page of animals and plants wonderfully well illustrated, explaining where to find them, how to identify them. The public schools and church had offered nothing like this. The Boy Scouts legitimated Nature as the center of my life.

There were rules, uniforms, and a crystal-clear set of practical ethics to live by. If I jog my memory today by raising my right hand with the middle three fingers up, thumb and little finger down and crossed, I can still recite the Scout Oath:

> On my honor I will do my best:
> To do my duty to God and my country,
> and to obey the Scout law;
> To help other people at all times;
> To keep myself physically strong,
> mentally awake and morally straight.

And the Scout Law: A Scout is trustworthy, loyal, helpful, friendly, courteous, kind, obedient, cheerful, thrifty, brave, clean, reverent. Finally, there was the Scout Motto, *Be Prepared*.

I drank in and accepted every word. Still do, as ridiculous as that may seem to my colleagues in the intellectual trade, to whom I can only reply, Let's see you do better in fifty-four words or less.

The work ethic was celebrated from cover to cover. There was a clearly marked Boy Scout of America route to success through virtue and exceptional effort. In the chapter titled "Finding One's Life Work," I read: "A Scout looks ahead. He prepares for things before they happen. He therefore meets them easily." Never be satisfied, the instructions warned. Just to wait and hope and accept whatever comes is the road to failure. Reach high, strive long and hard toward honorable goals, and keep ever in mind Longfellow's invocation:

> The heights by great men reached and kept, were not
> attained by sudden flight; but they, while their compan-
> ions slept, were toiling upward in the night.

I found something else the public schools never offered, a ladder of education to be taken at your own pace, better fast than slow, with each new step successively harder. I saw the whole challenge of scouting as a competition I would enjoy and surely win. The Scout program was my equivalent of the Bronx High School of Science.

I plunged into the new regimen. In three years I advanced to Eagle Scout with palm clusters, the highest rank, and was made junior assistant scoutmaster of my troop. I earned forty-six merit badges, almost half of those available in the organization. I happily crunched through the programs for subjects as diverse as Bird Study, Farm Records and Book-Keeping, Life Saving, Journalism, and Public Health. I pored over the requirements of all the badges at night to see which one I could best do next. My heart sang when I first read the prescription for Insect Life, beginning: "To obtain this Merit Badge, a Scout must: 1. Go into the country with the Examiner and show to him the natural surroundings in which certain specified insects live, and find and demonstrate living specimens of the insects, telling of their habits or of the nature of their fitness for life in their particular surroundings."

All the time I attended to my schoolwork in an adequate but desultory manner. The subjects were relatively easy, and I maintained passing grades. But most of the curriculum seemed dull and pointless. My most memorable accomplishment in my freshman year at Murphy High School in Mobile was to capture twenty houseflies during one hour of class, a personal record, and lay them in rows for the next student to find. The teacher found these trophies instead, and had the grace to compliment me on my feat next day in front of the class. I had developed a new technique for catching flies, and I now pass it on to you. Let the fly alight, preferably on a level and unobstructed surface, such as a restaurant table or book cover. Move your open hand carefully until it rests twelve to eighteen inches in front of the sitting fly's head. Bring the hand very slowly forward, in a straight line, taking care not to waggle it sideways; flies are very sensitive to lateral movement. When your hand is about nine inches away, sweep it toward the fly so that the edge of the palm passes approximately one or two inches above the spot where the fly is resting. Your target will dart upward at about the right trajectory to hit the middle of the palm, and as you close your fingers you will feel the satisfying buzz of the insect trapped inside your fist. Now, how to kill the fly? Clap your hands together—discreetly, if you are in a restaurant or lecture hall.

Scouting also proved to be the ideal socializing environment for an undersized and introverted only child. Our gangs were the Scout patrols, groups about the size of an army squad, several of which made up the larger troop. We automatically became members of one when we joined the Scouts, and we were esteemed or criticized on our own merits according to Scout rules. I never met a bully in the Scouts, and relatively few braggarts. The questions before each boy were: Can you walk twenty miles, tie a tourniquet, save a swimmer in Red Cross lifeguard exercises, build a sturdy sapling bridge with nothing but ax and rope? For me the answers were yes, yes, *yes*!

Scouting added another dimension to my expanding niche. I became a teacher. In the summer of 1943 I was asked to be the nature counselor at Camp Pushmataha, the Boy Scout summer camp near Citronelle used by the Mobile Area District. At fourteen, I was the youngest counselor, with no experience at instruction, but I quickly figured out what interested other boys, what would get them talking about natural history and make them respect the subject: snakes. Several volunteers and I built cages and searched the surrounding woods for as many different kinds of snakes as we could find. Somehow in the process I learned how to capture a poisonous snake. Pin its body with a staff as close to the head as you can manage, then roll the staff forward until the head is pressed firmly to the ground and the neck clear, grasp the neck closely behind the posterior jaw angles, and lift the whole body up. Few boys would touch a snake of any kind, so when one was discovered the word was brought to me by yelling messengers: "Snake! Snake!" And off I would go to perform my derring-do, which I followed with a brief lecture on the species discovered. In a short time we had a row of cages filled with a partial representation of the rich fauna that inhabits the Gulf states. I worked like a zoo director, talking to visitors about the diversity of species. I could then segue into discourses about the insects and plants of Greater Pushmataha. I had become a successful natural history instructor.

But before long my inexperience and reckless pride did me in. It happened one afternoon when I was cleaning a cage containing several pygmy rattlesnakes, my star attractions. The adults of this cryptically colored species (*Sistrurus miliarius*) grow to no more than fifty centimeters in length. They are less deadly than their larger cousins found in the same region, the diamondback and canebrake rattlesnakes, but they are still poisonous and moderately dangerous. In a moment of carelessness I moved my left hand too close to one of the coiled rattlers. Like a quarrel sprung from a crossbow, it uncoiled and struck the tip of my index finger. The two fang punctures felt like a bee sting. I knew I was in trouble. Off I went with an adult counselor to a nearby doctor in town, who administered the old-fashioned first-aid treatment as quickly as he could: deep X-shaped scalpel incisions centered on each of the fang punctures, followed by suction of the blood with a rubber cup. I knew the drill; I had learned it when I earned the merit badge for Reptile Life. I didn't cry during the operation, which was performed without anesthesia. I held my hand steady and cursed loudly nonstop with four-letter words, at myself for my stupidity and not at the innocent doctor or the snake, in order to keep my mind off the procedure. I knew a great deal of off-color language at fourteen and must have surprised the adults helping me. The next morning I was sent home for convalescence. I lay gloomily on a couch for a week, holding my swollen left arm as still as possible.

It was a bad time for herpetology at Camp Pushmataha. When I returned to resume my duties, I found that the camp supervisor had wisely disposed of the pygmy rattlesnakes. I was forbidden to touch any more poisonous species, and nothing more was said to me about the matter.

Contrary to the impression this account may have created, the Boy Scouts of America in southern Alabama in the early 1940s was not an ideal organization in all respects. It retreated helplessly from the Gorgons of sex and race.

Sex education was not on the agenda of the Boy Scouts of America, or of any school or other youth organization for that matter. The 1940 *Handbook for Boys* went no further than to caution that boys of a certain age have nocturnal seminal emissions once or twice a week. Scouts were not, it said, to worry about these episodes, which were normal. They were not to "excite" themselves to produce the emissions; the practice was "a bad habit." If urges became too troublesome to handle, one should try a cool hip bath, 55° to 60°F. If more help on this or related matters was required, "Seek advice from wise, clean, strong men." No warnings were given about pederasts. They surely lurked somewhere in the ranks of the adult leaders. I heard rumors of one, but never met him personally.

Then where did I learn all my filthy language? From other boys, who spiced their conversation whenever they were beyond adult earshot. Because sex was taboo and bizarre enough to be conceptually exciting, boys of Scout age talked about it all the time. We approached the subject obliquely, with Rabelaisian humor. A substantial fraction of campfire and trailside conversation consisted of raucous jokes that dwelled on every imaginable sexual perversion and grotesquerie: homosexuals trying to have babies, necrophiliac undertakers carrying the severed genitalia of female corpses as trophies, sex with animals, impossibly large sexual equipment on both men and women, insatiable appetites and marathon infidelities, and so on through a fantastical *Psychopathia Sexualis*. Every teenage Alabamian male, it seemed, was a budding Krafft-Ebing. But of normal heterosexual relations scant was known and nothing spoken. We could not cross the barrier to what we guessed our parents and married sisters did each night, or what we ourselves hoped to experience with girls. To discuss such matters would be a shocking invasion of privacy. So we left orthodoxy alone, and sketched the vague outline of acceptable behavior by circumscribing it with everything that was explicitly forbidden. Normalcy was like the image in a photograph created in silhouette by developing its chromatically sensitive background.

Race was also emphatically not on the official agenda. All the boys in the *Handbook for Boys* were white, and so were all the Scouts I knew. So were the students in the school I attended and the people in our church. I grew up mostly unconcerned about segregation and its dehumanizing effects. The impact of discrimination entered my mind only secondhand. But not entirely. In 1944 I was invited by a senior counselor to put on my Eagle Scout regalia and visit a troop of Boy Scouts just starting up in a black rural area near Brewton, Alabama. Standing in the front of the church meeting room I gave a short talk on the many advantages of Scouting. When we left I did not feel pride in the example I was supposed to have set; I felt shame. I was depressed for days. I knew in my heart that those boys, mostly two or three years younger than I, would have few real advantages no matter how gifted or how hard they tried. The doors open to me were shut to them.

Then I gradually forgot about the matter. What could I do? My mind was on other things. I was filled with ambition and anxiety and did not have a strong social conscience. Twenty years later the Old South came to an end. The civil rights activists who risked their lives to break segregation were heroes to my liking: singlemindedly true to a moral code, physically courageous, enduring. That was enough to make me look again at this part of my social heritage. And by then I had left Alabama. The world changed; I changed. But I cannot claim to have been a liberal as a boy and young man, certainly not one with any foresight or courage. The trajectory that took me into science would have been the same regardless; and I will not now be so presumptuous or hypocritical as to offer an extraneous apology for the proud and tortured culture through which I passed to the naturalist's calling.

CHAPTER SIX

Alabama Dreaming

IN AUGUST 1944 I weighed 112 pounds. I know this to be true because in that month I reported with my best friend, Philip Bradley, for football practice at Brewton High School, and we were put on the scales in the locker room. At fifteen years, I was probably the youngest, and certainly the smallest, of the players. Bradley was a bit heavier, at 116 pounds, every ounce of which I envied, while the largest member of the team came in at a hulking 160 pounds. I was allowed to strap on my ridiculously oversized uniform because the team needed every man (well, every boy) it could get. And I was there despite my obvious lack of qualifications because this was Alabama. In small towns across the state, football was what young males between the ages of fifteen and nineteen aspired to do when not in class or occupied with part-time jobs. At the other end of the statistical curve of athletic promise from me, boys with heavy shoulders and quick hands could hope for college athletic scholarships. There happened to be, however, none in our school well enough endowed for college play that year.

Brewton was and still is a town of about five thousand on the Alabama side of the border with Florida, forty miles north of Pensacola. It has changed very little since 1944. I have returned twice in middle age while on my way by automobile across the state, to drift like a phantom through the grid of residential streets down to the main commercial section that runs parallel to the railroad tracks, and to pause at the grounds of the high school, where I summoned the memory of boys hitting the worn tackle bag, grunting, and joking back and forth in reasonably close imitations of grown men. Once I stopped to ask a young fireman for directions, and when I mentioned that I had attended the high school in 1944, he said, "Boy, that was a l-o-o-ng time ago!" I replied that it didn't seem very long to me, not in a pleasant little town that had conceded so little to the rush of the twentieth century. And not when I could close my eyes and summon uniforms caked with dried mud and turned aromatic by stale sweat.

There were twenty-three on the football squad that year, composing the first and second teams of eleven each, each member playing both offense and defense, plus me, the third-string left end and, by accident of numbers, the entire third string. I couldn't catch the football half the time, I couldn't even see a pass coming with my one good eye, and I was too light to block. About all I could manage was a shoestring tackle. If I dived to the ground and threw my arms around both ankles of the onrushing ball carrier, I could trip him, hoping he wouldn't fall onto me too hard. Somehow, perhaps because the opposing teams were even punier than our own, we managed to defeat every one of the other ten high schools we played except archrival Greenville. I was allowed on the field just once all season, in the fourth quarter of the final game, played at home, and this once because it was toward the end of the fourth quarter and the enemy had been crushed beyond all hope of recovery. How warmly I remember and cherish the command, "Wilson, take left end!" It was an act of charity on the part of the coach, whose name I have forgotten but toward whom I will always feel gratitude. Because of him I was thereafter authorized to say in that part of Alabama, "I played football for Brewton," in the same way a New York corporate executive in a Century Club dining room says, "I rowed for Yale."

Most of the players had nicknames such as Bubba (it was not a joke then; Bubbas were future good old boys and managers of Chevrolet dealerships; they were big, heavyset, and good-natured), J. C., Buddy, Skeeter, Scooter, and Shoe. Mine was Snake, not because of my body shape, which would have been apropos, and certainly not because I could weave magically through crowds of tacklers head down with the ball tucked hard on my waist, as in my dreams, but because I had maintained my enthusiasm for real snakes. After our sojourn in Mobile, my father had left me with Belle Raub in Pensacola and gone on the road with Pearl, to a destination I never knew. The three of us reunited in a small house in Brewton in the early spring of 1944. That summer I served as Boy Scout nature counselor at Camp Bigheart, on the shore of Pensacola Bay. Once again I relied on snakes to enliven accounts of natural history.

By this time reptiles and amphibians had become my central interest. The fauna of the region would excite passion in a herpetologist of any age. Forty species of snakes, one of the richest assemblages in the world, are native to the western Florida panhandle and adjacent border counties of Alabama. Over a period of a year I managed to capture most of them. And a majority of those I could not take alive I either saw at a distance, such as the marsh-dwelling flat-tailed water snake (*Natrix compressicauda*), or else were brought to me dead, most memorably a large diamondback rattlesnake killed by a group of men not far from our house.

On the western edge of Brewton, next to a dense swamp, was a goldfish hatchery run by an affable sixty-year-old Englishman named Mr. Perry. I never learned his first name; polite southern youth did not address their elders in such familiar terms. Nor did I ask him how he came to such an unusual occupation in a backwater southern town. But we became good friends and spent hours talking freely on many subjects. He was always glad to see me when I rode my bicycle up to the edge of the property. He never had other visitors that I saw, lived quietly with his wife in a small house on the property, and always worked alone. His water came from artesian wells that have since dried up, and he fed his goldfish cornmeal mixed with pig blood received weekly from a local slaughterhouse. The goldfish were sold for bait, both locally and out of town. His canisters of young fish, some monochromatic gold, others gold marbled with white, departed at regular intervals by rail from the Brewton station.

Perry had excavated the ponds, each twenty to thirty feet square, in an irregular double row along the edge of the swamp. Thick weeds choked their borders, and tall trees walled them in on the swamp side. A six-foot-wide stream of artesian water flowed into the swamp from each end of the hatchery. The whole ensemble was a textbook diagram from an ecology textbook made literal: the rich nutrients pumped in continuously gave birth to an exuberance of algae, aquatic plants, and fish. The net produce of biomass fed swarms of insects and thence of frogs, snakes, herons, and other larger predators; and all the excess food and all the waste draining into the exit streams fructified the biota of a deep swamp that stretched east for an indeterminate distance.

Into this paradise I threw myself with abandon. The hours I spent there were among the happiest of my life. At every opportunity I came down to the hatchery ponds. After talking with Mr. Perry for a while, mostly about his pisciculture and my own explorations, I donned calf-length rubber boots from the row of pairs he kept in his equipment shed and walked into my private world. At home I politely ignored the nagging of my stepmother, who seemed almost distraught at my failure to find a job after school. I in turn grew increasingly abashed and resentful at her singleminded efforts to prepare me for the grim Depression-era life she had experienced. I had already worked longer hours than she, I had proved myself, and now I needed space. Pearl saw little value in my swamp expeditions, and, looking back, I cannot blame her.

Adults forget the depths of languor into which the adolescent mind descends with ease. They are prone to undervalue the mental growth that occurs during daydreaming and aimless wandering. When I focused on the ponds and swamp lying before me, I abandoned all sense of time. Net in hand, khaki collecting satchel hung by a strap from my shoulder, I surveilled the edges of the ponds, poked shrubs and grass clumps, and occasionally waded out into shallow stretches of open water to stir the muddy bottom. Often I just sat for long periods scanning the pond edges and vegetation for the hint of a scaly coil, a telltale ripple on the water's surface, the sound of an out-of-sight splash. Then, sooner on hot days than otherwise, I worked my way down for a half-mile or so along one of the effluent streams into the deep shade of the swamp, crossed through the forest to the parallel stream, and headed back up it to the hatchery. Sometimes I cut away to explore pools and mudflats hidden in the Piranesian gloom beneath the high closed canopy. In the swamp I was a wanderer in a miniature wilderness. I never encountered another person there, never heard a distant voice, or automobile, or airplane. The only tracks in the mud I saw were those of wild animals. No one else cared about this domain, not even Mr. Perry. Although I held no title, the terrain and its treasures belonged entirely to me in every sense that mattered.

Water snakes abounded at abnormally high densities around the ponds and along the outflow streams, feeding on schools of blood-gorged fish and armies of frogs. Mr. Perry made no attempt to control them. They were, he said, no more than a minor source of goldfish mortality. Although neither of us had the vocabulary to express such things, we shared the concept of a balanced ecosystem, one in which man could add and take out energy but otherwise leave alone without ill consequence. Mr. Perry was a natural-born environmentalist. He trod lightly upon the land in his care.

A swamp filled with snakes may be a nightmare to most, but for me it was a ceaselessly rotating lattice of wonders. I had the same interest in the diversity of snakes that other fifteen-year-old boys seemed automatically to develop in the years and makes of automobiles. And knowing them well, I had no fear. On each visit I found something new. I captured live specimens, brought them home to cages I had constructed of wood and wire mesh, and fed them frogs and minnows I collected at the hatchery.

My favorites included the eastern ribbon snakes, graceful reptiles decorated with green and brown longitudinal stripes, which spent their time draped in communal bunches on tree limbs overhanging the pond waters. With their bulging, lidless eyes they could see at a considerable distance and were wary. I stalked them to within a few feet by wading in the shallow water of the pond edges and seized one or two at a time as they plunged into the water and tried to swim away. They grew tame in captivity and fed readily on small frogs. Green water snakes were memorable in another way. Found lying half-concealed in vegetation at the edge of the ponds, they were big, up to four feet in length, and heavy-bodied. Catching one was an unpleasant experience unless I could take them quickly back of the head. Most larger snakes try to bite when first handled, and many can break the skin to leave a horseshoe row of needle pricks; but green water snakes have an especially violent response, and their sharp teeth can slash the skin and make blood run freely. They were also difficult to maintain in captivity. Once I found a mud snake, a species that uses the hardened tip of its tail to help hold giant amphiuma salamanders while subduing and swallowing them. The tip can prick human skin; hence the species' alternate name of stinging snake.

One species, the glossy watersnake *Natrix rigida*, became a special target just because it was so elusive. The small adults lay on the bottom of shallow ponds well away from the shore and pointed their heads out of the alga-green water in order to breathe and scan the surface in all directions. I waded out to them very slowly, avoiding the abrupt lateral movements to which reptiles are most sensitive. I needed to get within three or four feet in order to dive and grab them by the body, but before I could close the distance they always pulled their heads under and slipped quietly away into the deeper, opaque waters. I finally solved the problem with the aid of the town's leading slingshot artist, a taciturn loner my age who liked me because I praised his skills as a hunter. He aimed pebbles at the heads of the snakes with surprising accuracy, stunning several long enough for me to seize them underwater. After they recovered, I kept the captives for a while in the homemade cages, where they thrived on live minnows offered in dishes of water.

The tigers and lords of this place were the poisonous cot-tonmouth moccasins, large semiaquatic pit vipers with thick bodies and triangular heads. Young individuals, measuring eighteen inches or so, are brightly patterned with reddish-brown crossbands. The adults are more nearly solid brown, with the bands mostly faded and confined to the lower sides of the body. When cornered, moccasins throw open their jaws, sheathed fangs projecting forward, to reveal a conspicuous white mouth lining, the source of their name. Peterson's *A Field Guide to Reptiles and Amphibians of Eastern and Central North America*, written by the herpetologist Roger Conant, warns, "Don't ever handle a live one!" I did so all the time, with the fifteen-year-old's naive confidence that I would never make a mistake.

Immature cottonmouths were never a problem, but one day I met an outsized adult that might easily have killed me. As I waded down one of the hatchery outflow streams, a very large snake crashed through the vegetation close to my legs and plunged into the water. I was especially startled by the movement because I had grown accustomed through the day to modestly proportioned frogs, snakes, and turtles quietly tensed on mudbanks and logs. This snake was more nearly my size as well as violent and noisy—a colleague, so to speak. It sped with wide body undulations to the center of the shallow watercourse and came to rest on a sandy riffle. It was the largest snake I had ever seen in the wild, more than five feet long with a body as thick as my arm and a head the size of my fist, only a bit under the published size record for the species. I was thrilled at the sight, and the snake looked as though it could be captured. It now lay quietly in the shallow clear water completely open to view, its body stretched along the fringing weeds, its head pointed back at an oblique angle to watch my approach. Cottonmouths are like that, even the young ones. They don't always undulate away until they are out of sight, in the manner of ordinary watersnakes. Although no emotion can be read in the frozen half-smile and staring yellow eyes, their reactions and postures give them an insolent air, as if they see their power reflected in the caution of human beings and other sizable enemies.

I moved into the snake handler's routine: pinned the body back of the head, grasped the neck behind the swelling masseteric muscles, and lifted the snake clear of the water. The big cottonmouth, so calm to that moment, reacted with stunning violence. Throwing its heavy body into convulsions, it twisted its head and neck slightly forward through my tightened fingers and stretched its mouth wide open to unfold inch-long fangs. A fetid musk from its anal glands filled the air. In the few seconds we were locked together the morning heat became more noticeable, reality crashed through, and at last I awoke from my dream and wondered why I was in that place alone. If I were bitten, who would find me? The snake began to turn its head far enough to clamp its jaws on my hand. I was not strong even for a boy of my slight size, and I was losing control. Reacting as by reflex, I heaved the giant out into the brush, and it thrashed frantically away, this time until it was out of sight and we were rid of each other.

This narrow escape was the most adrenaline-charged moment of my year's adventures at the hatchery. Since then I have cast back, trying to retrieve my emotions to understand why I explored swamps and hunted snakes with such dedication and recklessness. The activities gave me little or no heightened status among my peers; I never told anyone most of what I did. Pearl and my father were tolerant but not especially interested or encouraging; in any case I didn't say much to them either, for fear they would make me stay closer to home. My reasons were mixed. They were partly exhilaration at my entry into a beautiful and complex new world. And partly possessiveness; I had a place that no one else knew. And vanity; I believed that no one, anywhere, was better at exploring woods and finding snakes. And ambition; I dreamed I was training myself someday to be a professional field biologist. And finally, an undeciphered residue, a yearning remaining deep within me that I have never understood, nor wish to, for fear that if named it might vanish.

Too quickly the enchanted interlude came to an end. In the late spring of 1945, a few weeks after sirens blew across the little town to celebrate the surrender of Germany, we moved again, to the city of Decatur, in north central Alabama. This time I yielded to the pertinacity of my stepmother and found work. In the ensuing year I held a series of jobs: paperboy, lunch-counter attendant and short-order cook at a downtown drugstore, stock clerk at a five-and-ten department store, and finally, in the summer of 1946, just before leaving for college, office boy in a nearby steel manufacturing plant. My income rose steadily with each step, to about twenty-five dollars a week. All this was good for my soul—maybe. I know it made Pearl happy; but, more important, it persuaded me to strive thereafter to my limit in order to go any distance, master any subject, take any risk to become a professional scientist and thereby avoid having to do such dull and dispiriting labor ever, ever again.

I managed to continue my relationship with Nature on a part-time basis that summer and fall. On warm days when I could get away from school and work I wandered the banks and tributary streams of the Tennessee River to the north and east of Decatur. Surrounded by one of the richest variegations of aquatic environment in North America, I took an interest in freshwater ecology. I discovered and studied sponges and the odd larvae of the spongillaflies that live in them. Soon after my arrival I learned, to my delight, that a local research station of the Tennessee Valley Authority had a complete collection of local freshwater fishes (Alabama has more kinds than any other state). After ingratiating myself with the personnel, I set out to learn this fauna species by species. The Tennessee Valley is also riddled with limestone caverns. I heard of one cave close enough to reach by bicycle and began exploring it in search of bats and blind subterranean insects. I shed most of my immediate interest in snakes, those in the Tennessee Valley being less diverse and harder to find than the ones in southern Alabama.

To my relief, there was no hope of playing football; the high school in Decatur was much larger than the one in Brewton and well peopled with natural athletes. There was no point in even showing up for practice; most male students did not. Thus I was spared the humiliation of my physical inadequacy.

Suddenly, in the fall of 1945, having reached sixteen years of age and with college only a year away, I recognized that I must get serious about my career as an entomologist. The time had come to select a group of insects on which I could become a world authority. Butterflies were out; they were too well known and were being studied by a great many obviously capable scientists. Flies looked much more promising. They occur everywhere in dazzling variety, and they have environmental importance. I liked their clean looks, acrobatics, and insouciant manner. Although houseflies and dung flies, not to mention mosquitoes, have given the dipterous clan a bad name, most species are little jewels in nature's clockwork, fastidious, unobtrusive, and efficient at what they do, which is scavenging, pollinating flowers, or preying on other insects. I was especially taken by long-legged flies of the family Dolichopodidae, many of which are metallic green and blue and skitter about on leaves in the sunshine like animated gemstones. More than a thousand species in North America were known at that time, and hundreds more were undoubtedly waiting to be discovered. I set out to order the equipment I needed to collect these insects: killing jar, Schmitt specimen boxes, and the special long black insect pins made chiefly in Czechoslovakia. But it was 1945; Czechoslovakia had recently been a war zone and was soon to fall under Soviet occupation. No pins were available.

Without pause I cast about for another group of insects in which to invest my energies, one that could be preserved in small bottles of alcohol obtained locally. I quickly hit upon ants. Of course, ants: my old acquaintances, the source of some of my earliest passions. From a local drugstore I purchased dozens of five-dram prescription bottles, the old-fashioned glass ones with metal screwtops, and filled them with rubbing alcohol. I ordered a copy of William Morton Wheeler's 1910 classic *Ants: Their Structure, Development, and Behavior* from a Decatur bookstore, built glass observation nests to the author's specifications, and prepared to launch my career as a myrmecologist. I rode my bicycle into the woods and fields all around Decatur, building a sizable collection of species and annotating the habitat and nests of each. Such museum series have lasting value. To this day, nearly fifty years later, I still occasionally consult my early Alabama specimens and notes on questions of classification and ecology. I have studied ants in European museums that were collected as early as 1832. They are all beautifully preserved, their exoskeletons as complete and finely sculptured as in life.

About this time I learned of a myrmecologist named Marion R. Smith who worked at the National Museum of Natural History. I knew that he was a middle-aged gentleman who had grown up in Mississippi and devoted his early research to the ants of that state. In a laboriously typed letter I announced my intention to conduct a survey of the ants of Alabama. Without a pause Smith wrote back to say, *Good idea!* He himself, he informed me, had surveyed the ants of Mississippi, and he enclosed a copy of a binary taxonomic key he had written to identify the species known from that state. In such keys you follow the specimen through a succession of two-way choices until you arrive at one that tells you the name of the species. Here, for example, is the beginning of the key to the ant genus *Monomorium* from William S. Creighton's classic 1950 monograph on the ants of North America. I have changed some words to make the language less technical:

1. The three terminal joints of the antenna ("feeler") thicken successively toward the tip of the antenna; workers in a given colony are all about the same size . . . . . . .Go to **2**

OR

The first two of these joints are about equal in size; workers in each colony are of two sizes. . . *Monomorium destructor*

2. The head is densely covered with small punctures, which make its surface dull; a common house ant in the United States ("Pharaoh's ant"). . . . . *Monomorium pharaonis*

OR

The head has only scattered punctures, its entire surface shiny . . . . Go to 3

And so on until all the known species are covered from a particular geographic area, say Mississippi or all of North America or even the whole world. I got busy, put names on the specimens I had collected, and sent them to Smith for verification. He responded quickly: You got half of them right. You are off to a *good start*! He didn't say, You got half of them wrong. And he did not say, Why don't you study a few years more and see me then? He said, Keep up the good work and write me soon. With each passing year I cherish yet more warmly the memory of Dr. M. R. Smith, myrmecologist of the National Museum of Natural History.

I redoubled my efforts and began to discover unusual and interesting species. One day I found a marching column of army ants in my backyard—not the famous voracious hordes of South American rain forests, but miniature army ants of the genus *Neivamyrmex*, whose colonies of 10,000 to 100,000 workers search for prey through grass clumps around human habitations and across leaf-carpeted forests in the southern United States. At first glimpse a *Neivamyrmex* raiding group resembles nothing more than a large column of slender, dark-brown workers of some other species, running back and forth between nest and a dead animal or spilled sugar. A close examination, however, reveals them to be armies on the march, invading the nests of other kinds of ants, often changing their own nest site from one day to the next. I tracked the *Neivamyrmex* colony for several days until finally, on a rain-soaked afternoon, they marched across the street and out of sight into the tangled weeds of a neighbor's yard. In future years I would encounter and study *Neivamyrmex* colonies many times, in many places, from the Carolinas to the Amazon. I would write on army ants from all around the world.

During my senior year in high school this late-adolescent idyll was invaded by a rising anxiety: to be a scientist, one must go to college, and no member of my family on either side had ever progressed that far. They had been successful business-men, farmers, shipowners, even engineers in an era when a high school diploma sufficed for such occupations. College was still thought of as a costly luxury, and the ordinary middle-class life trajectory up to that time was to pass directly from high school graduation to gainful employment. To further my ambitions I had to enter uncharted waters.

Unfortunately, my father's health was failing. A thin, frail-looking man, his 130 pounds stretched over a five-foot, nine-inch frame, he had been sickly for years, worn down by bouts of alcoholism and bronchitis. The latter was made chronic and severe by chain smoking, two to three packs a day. Now, in the winter of 1945, he was stricken with a bleeding duode-nal ulcer. He checked into the Naval Hospital in Charleston, South Carolina, where treatment was free to him as a veteran of the First World War. The operation, during which a large section of his small intestine was removed, was nearly fatal. He returned home for a long convalescence, never complaining to me, never expressing anything but optimism about our future; but I knew better.

Although I loved my father, my concern at this point was mostly selfish. I realized I could not depend on him for further support and feared that I might have to postpone college and take work to assist him and Pearl (who never took a job of her own). Later I learned that my mother, now married to a suc-cessful businessman and herself a civilian employee of the Army Quartermaster Corps, would have been more than willing to cover all my expenses. She was soon to supply partial support in any case. But I was a proud, closemouthed kid, frankly igno-rant in such matters, and did not tell her of my father's troubles or my own anxieties.

How, then, to get to college? Grades. For the first time I focused on my course work and began to receive straight *A*s. Financial aid. I competed for a scholarship from Vanderbilt University, a respected private institution in nearby Nashville,

Tennessee. The application consisted of a written test, transcripts, and letters from teachers. As a newcomer at Decatur High School with a spotty previous academic record, I must have seemed easily dismissible to the Vanderbilt scholarship committee. There was no way to convey my passion and special expertise in natural history, nor did I think these qualities should weigh much in comparison with formal classroom performance. Probably I was right. In any case I was turned down.

The GI Bill of Rights offered a way to college. If I enlisted in the Army immediately after my seventeenth birthday, I would be technically a veteran of the Second World War and eligible for veterans' benefits, including financial support for later college attendance. Three years in the service, four years of college, graduate at the age of twenty-four. My father and Pearl enthusiastically approved. So in June 1946 I rode a Greyhound bus to the induction center at Fort McClellan near Anniston, Alabama, where I intended to enlist. I hoped to train and qualify as a medical technician, to learn all the biology possible during my period of servitude, perhaps to travel, and to spend all my spare time improving my skills in entomology.

At the end of the physical examination the attending physician and a recruitment officer took me aside. They informed me that I could not be accepted into the Army because I was blind in my right eye. Physical standards, they said, had tightened with the end of the shooting war. Once again the little pinfish of Paradise Beach, whose dorsal spine had pierced my eye, changed the course of my life. I stood on the veranda of the administration building, my hands on the railing, enviously watching successful recruits drill on the field below, as I waited for transportation back to Anniston. Bitterly disappointed by this unfair outcome, I wept. I vowed that although I had failed here, I would go on, make it through college and succeed some other way, work on the side as needed, live in basements or attics if I had to, keep trying for scholarships, accept whatever help my parents could give, but regardless of what happened, let nothing stop me. In a blaze of adolescent defiance against the fates, I swore I would not only graduate from college but someday become an important scientist.

CHAPTER SEVEN

The Hunters

THE UNIVERSITY OF Alabama saved me. It was open to all graduates of Alabama high schools, by which I mean all qualified *white* graduates, an exclusion that was to endure two more decades. The expense was minimal: $42 a quarter in tuition and fees, $168 for the full year of four quarters, including summer; room rent $7 a month; laundry costs negligible; textbooks $2 to $10 apiece, less if you got them secondhand. Travel back and forth from home, by either hitchhiking or Greyhound bus, cost less than $20. I found a boardinghouse that offered three meals a day, heavy on eggs, flapjacks, grits, turnip greens, corn bread, and fried chicken necks and wings, for $30 a month. My total expenses for attending the University of Alabama in the 1946–47 academic year plus an extra summer term were about $700. By finishing in three years through an accelerated program, I earned my bachelor of science degree with an expenditure of a little more than $2,000, somewhat less than the annual salary of a government clerk or schoolteacher at the time.

None of it came from loans and scholarships. All of it came from my parents. My luck was holding as I started classes in September 1946. My father's health had improved somewhat. He moved with Pearl yet once again, this time back to Mobile, where they settled in half of a duplex house owned by one of my aging aunts. My father found a job as an accountant at Brookley Air Field and was able to defray part of my expenses. My mother, alerted by this time to our precarious financial state, gave me the balance. As the only child of four parents, I was blessed, proceeding on safer ground than I had expected. Nevertheless, the generous admission standards and low cost of the University of Alabama were important preconditions of upward mobility for me, as they have been for thousands of others even less well situated. Faithful alumnus I have been ever since. My journey came full circle in 1980, when I was invited to give the spring commencement address. There before me,

to my relief, sat black graduates among white, the doors of
opportunity by then having been opened to all.

When my father and I rode into Tuscaloosa that first September afternoon in his new Hudson Commodore sedan, the
campus was verging on chaos. Veterans were pouring in to
the university to use the educational benefits of the GI Bill
of Rights. All the physical facilities were overcrowded, traffic
around the campus was snarled, and teachers, administrators,
and counselors were forced to work overtime to cope with the
greatest crisis since that sorry day in 1865 when the teenage
Corps of Cadets had marched out to engage an advancing
column of Union cavalry, lost, and watched as the Federals
burned the university down.

I entered college in the company of men as much as ten years
my senior, many of whom had undergone harrowing combat
only a year or two before. One, Hugh Rawls, a biology student
with whom I became good friends, had seen just ten minutes
of action. He had gone ashore at Saipan as commander of an
amphibious tank; on the beach Japanese shells fell first left,
then right, then dead center on his tank. Only he and the
gunner were able to crawl out. As he staggered back to the
water's edge, seven sniper bullets struck and permanently dis-
abled him. Another good friend, Herbert Boschung, survived
three plane crashes during combat missions over Germany. My
companions seldom spoke of these events. They had begun a
new life.

Many of the men came from outside Alabama, having found
colleges and universities closer to their homes too crowded to
admit them. I had no problem in adjusting to their company.
They were used to mingling good-naturedly with seventeen-
year-old recruits. College life was in any case as strange to
them as it was to me, and I found reassurance in their shared
bewilderment.

The university solved many of its problems by acquiring and
converting part of a military hospital two miles away on the
outskirts of Tuscaloosa. Thus was created the Northington
Campus, where I lived and at first attended most of my classes
in Quonset huts and recreation rooms. Because the hospital
had been constructed during the war and was large, many
of us were assigned private rooms. Mine was a padded cell

in the former mental ward. In 1978, thirty-two years later, I watched the tall smokestack crumple, felled by an explosive charge, and the surrounding buildings destroyed. I viewed the scene on film, at the climax of the motion picture *Hooper*, starring Burt Reynolds and Sally Field. Thus Northington Campus ended its existence in the service of a Warner Brothers Gotterdämmerung.

It was in the university, padded cell notwithstanding, that I found my natural home. Shortly after classes started, I climbed the balustraded steps to the main entrance of Nott Hall, built in the 1920s but antebellum in design. I had come to call on Professor J. Henry Walker, head of the Department of Biology, to introduce myself and to discuss my career plans. I was moved to this bold maneuver not by any sense of self-importance—I was still a timid boy, and hubris was only later to fester in my soul—but by the mistaken belief that college students normally chose their careers immediately, and should therefore at an early point consult the faculty for guidance on research and special study. I was reinforced in my presumption by the manly talk I heard among the returned veterans, most of whom had firm career plans of their own.

Walker was a slenderized replica of Warren Harding, a handsome, middle-aged man with blue eyes, prematurely white hair, and meticulous grooming befitting a gentleman of the Deep South. He communicated with soft accent and precise hand gestures. He was careful in all things, I later learned: he kept the department's postage stamps in his office safe. He nodded encouragingly as he peered into my Schmitt box of specimens and listened to my disquisition on the ants of Alabama. He murmured reinforcement as though it were entirely routine for freshman students to launch entomological careers in his office: "Yes, yes, very interesting, fella, very interesting, you've done very well." (All younger males, it turned out, were called fella.) He then made a telephone call and escorted me one flight up to the office of Bert Williams, a young professor of botany newly arrived from Indiana University.

Williams, a tall, gangling man in his thirties with a slight stoop and Lincolnesque face, greeted me warmly without hesitation, as though I were a fellow academic on sabbatical leave. After we talked ants, natural history, and botany for a while, he took me to a table space in his laboratory where, he suggested, I might wish to conduct my research. His largess knew no bounds thereafter. He lent me a dissecting micro-scope, glassware, and alcohol. He offered to take me along on future field trips. Later in the year he gave me a part-time research assistantship, tracing radioactive phosphorus through the roots of plants. Perhaps because Williams had no other research students at that time, and certainly in part because he was by nature a modest and caring man, he treated me as though I were a graduate student or postdoctoral fellow. I even came to feel as though I had joined his wife and infant daugh-ter as part of the family, like a favored nephew. I have known no kinder or more effective mentor. Forty-seven years later, in 1993, I had the great pleasure of welcoming his granddaughter to her freshman year at Harvard University and offering her my assistance.

I received less personal but equally cordial treatment from the other half-dozen members of the biology faculty. They were used to devoting their time to large classes of premedical students, whose strictly defined needs in anatomy, physiology, histology, and parasitology called for formal lectures and by-the-book laboratory exercises. Undergraduate students who followed in their own footsteps, who were bent on careers in pure science, were relatively rare. I flourished under the guid-ance of these multiple elders. In addition to training, they gave me the most priceless gifts an apprentice can receive: they let me know that they did not understand everything, that I might acquire information they did not have, and that my efforts were valued.

I set up an aquarium just inside the lower entrance of the biology building and exhibited a giant amphiuma salamander I had captured on one of our field trips. Fascinated students watched as it slithered back and forth crunching live crayfish. I captured entire colonies of *Neivamyrmex* army ants, seething masses of thousands of workers, housed them in artificial nests I built in Williams' laboratory, and studied the parasitic beetles and flies living with them. One of these guests, a near-micro-scopic beetle in the genus *Paralimulodes*, rides on the backs of the worker ants like a flea and lives by licking oily secretions from their bodies. My observations later became the basis of one of my early scientific papers. The biology faculty let me know with passing smiles and fragments of corridor conversation that they considered all these efforts useful and important.

To much of the rest of the country the University of Alabama means football, the Rose Bowl in the golden 1930s, the Sugar Bowl in the 1970s and 1980s, the blood feud with Auburn University—Harvard versus Yale with 280-pound tackles—and the legend of Paul William "Bear" Bryant. But those are only the most visible aspects of an excellent public university. The University of Alabama was and is the home of first-rate scholars and teachers, and of abounding opportunity for students who come there, as I did in 1946, to learn about the world, to enter a profession, and, if you will permit an old-fashioned expression, to make something of themselves. I found it as good a place for undergraduate training in my field of science as I would later judge Harvard, Princeton, and Cambridge to be, among other universities I have come to know reasonably well. The personal attention and encouragement I received could not have been surpassed.

What counts heavily in the shaping of a scientist is the accessibility and approval of the faculty. What is truly decisive, however, is the desire and ability of the student. Otherwise, failure awaits regardless of the learning environment, and no excuse can be made for it. If you are a lousy hunter, the woods are always empty.

Unencumbered by the need to hold a job on the side, I devised a time budget that was optimal for my progress through the university; I paid just enough attention to formal courses to get mostly As. I spent the rest of my time doing research, reading, and talking with faculty and other students, usually about evolutionary biology but ranging widely into subjects as diverse as geography, philosophy, and the techniques of creative writing.

I never joined one of the fraternities that dominated social life on the campus, for the simple reason that I was never invited. At the end of my senior year I was inducted into Phi Beta Kappa, the national honor society, as a reward for my overall high grade average. At commencement I hitched a ride to the campus from a middle-aged couple from Tuscaloosa. As they let me off near the president's house, the woman told me her son belonged to Sigma Epsilon Alpha and asked which fraternity I belonged to. "Phi Beta Kappa," I replied. "Why, I never heard of *that* one," she said. Too bad, I thought.

During my first two years I was a part-time cadet in the Reserve Officers Training Corps (ROTC), which was compulsory for all male students at the University of Alabama. I was by then in my late-teens radical period and anxious to see the world rise to meet my own empyrean and wholly untested moral standards. I now held much of American culture in contempt. My guidebooks to radicalism were Philip Wylie's *Generation of Vipers* and *An Essay on Morals*, wonderfully humorous jeremiads against organized religion, Babbitry, Mom worship, and sundry other national foibles. If radical left students had existed and been active then, I might have linked arms on behalf of each week's nonnegotiable demand. At ROTC drill one day, I explained to our sergeant, a regular Army lifer waiting out his retirement in this remote outpost, that marching and rifle practice had been made obsolete by the atom bomb. What we were doing on the parade ground, I declared, was a useless exercise to commemorate the past, like dancing around maypoles. Without changing expression, he growled something inaudible that might have been an expletive.

My feelings about the military were decidedly mixed by this time. On Governor's Day in my sophomore year, His Excellency James Folsom Senior (Junior was also to become governor in the 1990s) traveled from his capital office in Montgomery to review the ROTC cadet corps. A great populist and pro-education governor, referred to fondly as Big Jim because of his towering height and hefty body, and Kissing Jim for reasons ambiguously reported by the press, Folsom was already a legend in the Camellia State. I waited out in front of formation in a special line of cadets to be honored for scholarship, together with others to be recognized for rifle marksmanship. Folsom arrived in a gaggle of state troopers, military officers, and school officials. He was that day conspicuously under the influence of alcoholic refreshments, a common condition for him on public occasions after eight in the morning, and he wove a bit on his feet as he moved from honoree to honoree, speaking to each before handing him a medal. When he came to me he said, "Wheah you from, boy? Mobile? That's a mighty fine place, mighty fine." He reached into a box held by a staff aide and handed me a marksmanship medal. I was delighted to own this unearned award, even for a short while. I much preferred it to the wimpy scholarship medal—it did not seem right for a soldier to be decorated for doing well in English literature. The next day I reluctantly returned my prize to ROTC headquarters.

Leftist radicalism and an uneven passage through ROTC were minor deflections from my chosen trajectory. My resolve to be a biologist was reinforced when I discovered the ideal social environment for developing a scientist—or at least one of several possible ideals. It is the same as for a political revolutionary. Start with a circle of ambitious students who talk and work together and conspire against their elders in order to make their way into a particular discipline. They can be as few as two or as many as five; more than five makes the unit unstable. Give them an exciting new idea that can transform the discipline and with which they can advance their ambitions: let them believe that they own a central truth shared by few others and therefore a piece of the future. Add a distant authority figure, in this case a scientist who has written a revolutionary text, or at least a circle of older revolutionaries who have generated the accepted canon. The farther away these icons are from their acolytes, the better. At midcentury, Europe was best of all. French and German pedants, especially if their texts are hard to translate (and therefore require exegesis by English-speaking disciples), are especially potent. Bring on a local role model, an older man or woman who promotes The Idea and embodies in his character and working habits the ideals of the youthful discipline.

The circle I joined in my sophomore year, though all older than I by two to seven years, were also novices, committed naturalists, and ambitious. They included future successful academics: George Ball from Detroit, later to become professor of entomology at the University of Alberta; Herbert Boschung, a fellow native Alabamian who was to remain at the university first as professor and subsequently as director of the Alabama

State Natural History Museum; Hugh Rawls, whose love of mollusks led to a professorship in Illinois; and Barry Valentine, a New Yorker who became a professor of zoology and entomology at Ohio State University.

Our mentor in this formulaic mix was Ralph Chermock, newly arrived from Cornell University as an assistant professor. A relative of Erich Tschermak von Seysenegg, one of the three rediscoverers of the Mendelian laws of heredity, he was a highly competent specialist in butterfly classification and deeply committed to research on evolutionary biology. At thirty, Chermock was physically impressive, an amateur boxer with a compact gymnast's body and thick arms, who occasionally performed one-arm pushups on his office floor to intimidate his followers, but also a tense man who chain-smoked and often snorted and giggled when he laughed. He had the disconcerting habit of listening intently to everything you had to tell him, head cocked and wearing an inviting but quizzical smile, like a psychiatrist or a skeptical job interviewer.

Perhaps I overinterpret Chermock's demeanor from his particular reaction to me. On arriving at the university in 1947 he immediately spotted me as a youngster turning spoiled and overconfident from too much praise. An adjustment was in order. He scandalized me by giving me an *A–* instead of *A* in a course on evolutionary theory when I was convinced I had done brilliantly—at least until thirty years later, when I reread my final examination paper. In any case he took every opportunity to grind my ego down to size. When I completed a careful laboratory study of prey selection in the trapjawed ant *Strumigenys louisianae*, using a "cafeteria" technique I had invented myself, and showed him the article I had written on my findings, his response was muted. He gravely informed me that I could never publish the article until I had confirmed the laboratory data by going back into the field and actually finding the same prey captured and dead in undisturbed *Strumigenys* nests. I knew it would be like searching for a needle in a haystack, but out I went, day after day, locating these elusive little ants and carefully opening their nests until finally I discovered one with freshly caught prey that were still intact enough to be identified before the voracious larvae had eaten them; and

Chermock relented. The several best teachers of my life, including Chermock, have been those who told me that my very best was not yet good enough.

Ball and Valentine had come to Alabama explicitly to work with Ralph Chermock. With him they brought the Cornell mystique, the reputation of an entomology department whose history extended back to the great nineteenth-century pioneer John Henry Comstock, and whose reputation for total dedication to insect research at the highest professional level was and remains internationally respected. Awed by the legends, I felt myself to be in the best of company.

The prophets of the Chermock circle were the architects of the Modern Synthesis of evolutionary theory. All were, in 1947, men of middle age who worked in prestigious places like Columbia, the University of Chicago, and New York's American Museum of Natural History. The sacred text of the Chermock circle was Ernst Mayr's 1942 work *Systematics and the Origin of Species.* Mayr was the curator of birds at the American Museum, but his training had been in Germany, a source of added cachet. The revolution in systematics and biogeography that Mayr promulgated was spreading worldwide, but especially in England and the United States, the national strongholds of Darwinian evolutionary theory.

Bear with me while I explain the reason for the extraordinary impact of the new Darwinian movement. By 1920, just a quarter-century before I encountered it as a student, evolutionary biology had dissolved into a jumble of natural history observations, with its best theory consisting of a few rules and geographic trends mounted upon statistical correlations. The principle of natural selection, the core of the Darwinian theory, was itself in doubt. Geneticists thought that evolution might proceed not so much by incremental episodes of natural selection (acting upon continuously varying traits such as size, instinct, and digestion) as by mutations that change heredity in discontinuous steps. In retrospect it seems obvious that both propositions had to be true. Variation, we now understand very well, arises by mutations and also by recombinations of mutations during sexual reproduction; the changes can be large or small in effect; and natural selection—differential survival and reproduction—determines which mutations and combinations survive and reproduce by virtue of the traits they prescribe in such properties as size, instinct, and digestion.

This synthetic view is essentially the Darwinian theory of natural selection with mutating genes added. The close connection to Darwinism is why the modern theory came to be called Neo-Darwinism or, just as often, the Modern Synthesis. In the 1920s and early 1930s a group of population geneticists, most prominently Sergei Chetverikov of Russia, Sewall Wright of the United States, and J. B. S. Haldane and Ronald A. Fisher of England, used mathematical models to demonstrate that one gene form created by mutation can replace another throughout a population even if its advantage in survival and reproduction is quite small, say 1 or 2 percent. In theory at least, the substitution can occur rapidly, with most of it completed in as few as ten generations. Such microevolution, entailing one or a few genes at a time, can accumulate to become macroevolution, producing whole new structures such as eyes and wings. It can also cause the splitting of species into two or more daughter species, a process that is the fount of higher-level biodiversity.

The Modern Synthesis reconciled the originally differing worldviews of the geneticists and naturalists. It empowered scientists in both disciplines to examine the entire evolutionary cavalcade as an extension of Mendelian heredity and, later, to add the refinements of genetics brought by molecular biology.

The natural history phase of the Modern Synthesis followed genetics and natural selection theory. If there was any single moment of birth, it was the publication in 1937 of Theodosius Dobzhansky's landmark *Genetics and the Origin of Species*. For the first time, new data from the field and laboratory defined the differences among species and races with precision, illuminating the nature of variation within populations in chromosomes and genes, and the steps of microevolution. Evolution seemed firmly grounded in genetics, at least to the following extent: nothing the geneticists could say by the late 1940s, when I came along as a student, seemed likely to overturn the Modern Synthesis. Only a complete surprise, something major and out of the blue, could accomplish that. To this day nothing so radical has occurred, although many an ambitious biologist has tried to play the role of revolutionary.

The naturalists were given a hunting license, and for the Chermock circle Mayr's *Systematics and the Origin of Species*, following upon Dobzhansky's book, was the hunter's vade mecum. From Mayr we learned how to define species as biological units. With the help of his written word we pondered the exceptions to be expected and the processes by which races evolved into species. We acquired a clearer, more logical way to think about classification by using the phylogenetic method. This system measures differences between species by the amount of evolution that has occurred since they split apart.

Also in our armamentarium was George Gaylord Simpson's *Tempo and Mode in Evolution*, published in 1944. The great paleontologist argued that the fossil record is consistent with the evidences of ongoing evolution seen in living species. And finally, in 1950, botany entered the mainstream with the publication of Ledyard Stebbins' *Variation and Evolution in Plants*.

We thus were equipped with the texts of radical authority. We also had field guides and our own previously acquired expertise: fishes, amphibians, and reptiles for Boschung; mollusks for Rawls; beetles for Ball and Valentine; and ants for me. And providence shone bright on all of us together: Valentine had an automobile. We were scientifically licensed hunters, with the means to roam an ecologically diverse state that had to that time been only partly explored by naturalists.

Chermock encouraged us to collect not only our favored organisms but also amphibians and reptiles for the University of Alabama collection. On weekends and holidays we struck out across the state, to the farthest corners and back and forth. We pulled the car over to roadsides and clambered down into baygum swamps, hiked along muddy stream banks, and worked in and out of remote hillside forests. On rainy spring nights we drove along deserted rural back roads, falling silent to listen for choruses of frogs. Sometimes I sat on the front fender of the car as Rawls or Valentine drove slowly. Perched that way, with my left arm curled around a headlight and a collecting jar held in my right hand, I watched for frogs and snakes spotlighted by the high beams of the car. When one was sighted the driver stopped the car, and I dashed ahead to bottle the specimen. On other nights we walked the streets of Tuscaloosa, observing and collecting insects attracted to the lights of storefronts and service stations. During these expeditions I soaked up new information on dryinids, perlids, limulodids, entomobryomorphs, plethodontids, lithobiids, sphingids, libelludids, and so on and on deep into the heart of biodiversity. Chermock was unimpressed by our growing expertise. He told us, half seriously, that we could not call ourselves biologists until we knew the names of 10,000 kinds of organisms. I doubt that he could have passed the test himself, but it didn't matter. Hyperbole from the chief kept our juices flowing.

By the age of eighteen I had been converted to scientific professionalism. Barely out of my Boy Scout years, I was back on the trail of merit badges, this time through research, discovery, and publication. I came to understand that science is a social activity. Previously I had spent most of my time in natural history to learn about wild creatures and to enjoy personal adventure. I didn't care much what others thought of my activity. Now, as Alfred North Whitehead once said of scientists generally, I did not discover in order to learn; I learned in order to discover. My private pleasure was now tinged with social value. I came routinely to ask: What have I acquired in my studies that is new not just for me but for science as a whole?

The poorly explored Alabama environment offered the Chermock circle boundless opportunity for discovery even with minimal training. One night we drove slowly from the central part of the state into the Florida panhandle, stopping the car frequently to listen to the songs of chorus frogs mating in the accumulated rainwater of roadside ditches. (For a close approximation of a chorus frog call, run the edge of your fingernail along the fine teeth of a pocket comb.) We were searching for the zone where the northern race *Pseudacris nigrita triseriata*, which sings with one trill pattern, meets and interbreeds with the southern race, *Pseudacris nigrita nigrita*, which sings with a different pattern. Near dawn we encountered the changeover close to the Florida border, and then it proved to be very abrupt. We reasoned that the two forms are actually reproductively isolated species, not interbreeding races, and deserved their formal distinction as *Pseudacris triseriata* and *Pseudacris nigrita*. Research by later specialists proved us right.

At another time, wading far up the underground stream of a cave in northern Alabama, we discovered a new kind of blind white shrimp. And again: in mixed hardwood and pine forests, Barry Valentine and I collected the first Alabama specimens of the rare insect order Zoraptera, and soon afterward published our records in an entomological journal. Occasionally I worked alone, an old habit. While digging into soil on the fringes of a swamp near Tuscaloosa, I discovered a new species of a pretty little ant with dark-brown body and yellow legs and described it as *Leptothorax tuscaloosae*.

Scientific discovery at this elementary level was all so easy, all such fun. I could not understand why most of the other students at the university did not also aspire to be biologists.

Meanwhile I developed a strong new research interest in the imported fire ant, which I had first observed in Mobile in 1942. The notorious pest species was beginning to spread out of the city and into the fields and woodlands of the rural countryside. In 1948 Bill Ziebach, the "Outdoors" editor of the *Mobile Press Register*, began a series of articles on the threat by the ant to crops and wildlife. He consulted me on the species and quoted me in the paper. As a result, in early 1949 the Alabama Department of Conservation asked me to conduct a study of the ant and evaluate its impact on the environment. I took leave from the university for the spring term to begin, at the age of nineteen, a four-month stint as entomologist, my first position as a professional scientist. I was joined by James Eads, another biologist, like my other companions a war veteran in his mid-twenties and, most crucially again, owner of a car. Jim and I crisscrossed southwestern Alabama and the western counties of the Florida panhandle, mapping the expanding semicircular range of the ant. We dug up colonies to analyze nest structure, explored fields for crop damage, and interviewed farmers. In July we submitted a fifty-three-page analysis to the Department of Conservation office in Montgomery titled "A Report on the Imported Fire Ant *Solenopsis saevissima* var. *richteri* Forel in Alabama." It contained original findings on the ant still in use today, including the rate of spread (five miles a year along all borders), the partial elimination of native fire ant species, and documentation of moderate crop damage caused by direct consumption of seeds and seedlings.

How this notorious insect got its common name is itself a story worth telling. Up to the time of our first meeting with state officials in Montgomery, the species was called the Argentine fire ant, in recognition of its presumed native origins (it is now known to occur widely through northern Argentina as far as the Paraguayan border). Someone in the Department of Conservation suggested that the name might prove offensive to Argentinians; we already had too many German cockroaches, English sparrows, and the like. We should change it, he said, while we had time. Someone else, I can't remember who, suggested the imported fire ant. That name was used in our report and subsequently by the media and in scientific literature.

In the year following, while working on my master's degree at the University of Alabama, I intensified my studies of the imported fire ant. Eads and I, along with Marion Smith at the National Museum, had observed that workers of the species belonging to different colonies vary in color from dark brown to light reddish brown. I noticed further that the light workers were smaller, and that their colonies appeared to be displacing those of the dark workers. By 1949 the dark form was limited mostly to peripheral areas in Alabama and Mississippi. It had disappeared entirely from Mobile, its point of origin. I set out to test experimentally whether the two forms were genetically distinct. One method I invented was to introduce light queens into dark colonies and observe the color of their offspring reared in a socially altered environment. The color remained true to their mother queen, providing evidence— but not definitive proof—that the difference between light and dark was hereditary.

In the course of my switching experiments I discovered that when more than one queen was introduced into a new colony at the same time, the workers executed all but one by stinging and dismembering them. They never made the mistake of eliminating the final queen, which would have destroyed the colony's ability to produce more workers. This result foreshadowed the discovery, by other entomologists thirty years later, that the workers are able to discriminate among many queens and select the healthiest and most fecund.

In a history of the imported fire ant I published later, in 1951, I considered the color forms to be varieties of the same species. In 1972 William Buren, after an exhaustive new study, confirmed my general findings but elevated the light form to full species rank. He gave it the name *Solenopsis invicta*, meaning the "unconquered" *Solenopsis*. In 1972 the ant was spreading throughout the southern United States in the teeth of intense efforts and the expenditure of over $100 million to stop it. In a widely quoted interview at the time I summed up the futility of the enterprise in a phrase: the fire ant eradication program, I said, is the Vietnam of entomology.

I was exhilarated by the successes of my early fire ant research. I found that the vagrant learning of my boyhood could be focused in a way that was of interest and practical use to the public. The self-confidence I acquired helped to carry me through the critical years of intellectual growth and testing ahead.

Meanwhile, a second obsession intruded briefly into my college training. I was transfixed by the legend of the four-minute mile, the supposedly unbreakable barrier of track and field. In 1945, when Gunder Hägg brought the record down to 4:01.4, there was much talk about whether the great Swedish miler had reached the limit of human endurance. Such speculation was entirely misdirected, of course: an inspection of the history of the event shows that the mile record had been descending in a nearly straight line for eighty years; the curve showed no sign of bottoming out when Hägg led the world, and a simple extrapolation in the late 1940s would have indicated that the four-minute mile could be expected at any time. That moment came on May 6, 1954, when England's Roger Bannister ran the distance in 3:59.4. Thereafter hundreds of athletes repeated the feat, pushing the finishing time steadily downward. As I write, the record stands at 3:46.31.

But in 1948, while athletes around the world prepared for the first postwar Olympic Games, distance running was still in its romantic period. The four-minute mile was the Everest of track and field. In the July 10 issue of the *Saturday Evening Post* I came upon an article declaiming that European athletes would "run Americans ragged" in the distance events. They trained longer, the author said, were willing to endure more discipline and pain than the soft Americans, and would sweep the medal rounds. Gunder Hägg was pictured cruising along a track in six-foot strides, long dark hair flying. I became enchanted by the idea of breaking records by will and discipline. If you were not large in body, I thought, perhaps you could triumph by being large in spirit. It was my kind of activity: do it alone, avoid the drag of teams, have no one witness your trials and failures, until you can accomplish some exceptional feat.

So I bought a pair of surplus Army boots to add weight to my feet and endurance to my body, and started running through the back streets of Mobile, into the countryside, and, back in Tuscaloosa, round and round Northington Campus, which I treated as a giant track. I trained in solitude, mostly at night, all through the late summer and into the winter of 1948. I scaled the chain-link fence surrounding the University of Alabama athletic grounds in order to run on the cinder track when the regular athletes had left, to get the feel of a quarter-mile. I ran for an hour or two hours at a time. I had neither coach nor training schedule, and spoke to no one about my effort. I just ran in the heavy shoes that I thought would lend wings to my feet when I later switched to lighter gear.

In February I tried out for the track team. I simply reported to the locker room, put on spiked shoes for the first time in my life, walked out onto the track, and ran a trial mile while the coach timed me with a stopwatch. I came in at "a little over five minutes." The coach mercifully didn't tell me the exact time, and I didn't want to hear it. I was bitterly disappointed and humiliated. Not just my body but my philosophy had failed. But—surely if I tried harder I could do better! The coach was kindly disposed. He suggested that I practice for the two-mile race. There were no longer distances, such as 10,000-meter races and marathons, in the Southern Conference programs of 1949. So I started coming to practice for two-mile runs every afternoon, adding speed sprints to my endurance training. But it was too late, and obviously hopeless. At nineteen, I was already a senior, and I must have seemed to the coach his poorest prospect. We were both saved further embarrassment when shortly afterward I was offered the temporary position to survey fire ants in Alabama. I told the coach I was dropping out, and handed back my spiked shoes. He did not burst into tears.

My failure galled me for years afterward. What, I sometimes mused, if I had started at sixteen or seventeen, with proper coaching? Might I have at least made the team? Would Gunder Hägg have found an American rival? In 1970, at the age of forty-one, I started jogging again, then running wind sprints, this time to lose weight and safeguard my health. These goals attained, I felt the old fire rekindling and crazy hopes rising: maybe I could compete in races at the master's level, for men over forty. Obviously no four-minute miles were in the cards, but perhaps a five-minute mile? As my times dropped in solitary runs, I consulted the world records for different age groups, from childhood to old age. They are kept for all distances, based on times registered from all over the world. I found that even though most of the records from one age to the next, say twenty-nine to thirty to thirty-one years, were made by different individuals at various meets in widely separated parts of the world, they formed a tight line of points for each event. The curves peaked in the early twenties for the hundred-meter dash and in the late twenties for the marathon. This statistical evidence suggested that the best in the world, whoever they were, wherever and whenever they ran, turned in a record time that is precisely predictable once age is known. Age alone accounts for almost all the variation in world record times.

This result impressed me deeply. It seemed to show that heredity is destiny, at least in one important sense: taken to the limit of human capacity, performance follows a predetermined trajectory. No athlete can break away, not even an iron-willed distance runner. I applied the results to my own capacity. I took the ratio of my mile in 1949, "a little over five minutes," to the world record set in that period, a tick over four minutes. Multiplying it by the world record held in 1970 by men in their early forties, I arrived at my own likely best personal time, about six minutes.

Pathetic! In a sport where a tenth of one percent can mean victory or defeat, I was carrying a 25 percent hereditary deficit. Then I felt a last adolescent surge. I would break the apparent genetic bond and wipe away the stain of 1949! This was the period just before the jogging craze of the mid-1970s. I ran the streets of my hometown, Lexington, Massachusetts, in tennis shoes, almost never encountering another jogger. Dogs chased me, neighbors stared, teenage boys hurled taunts. I ran quarter-mile wind sprints on the high school track. I entered races and did time trials. My three best times were 6:01, 6:01, and 6:04. Returning to my track tables, I estimated that my fastest time for two miles would be about 13 minutes. One day I ran my personal best, 12:58. Heredity was destiny after all.

Meanwhile, I witnessed one triumph after another by my friend Bernd Heinrich, a distinguished entomologist and champion master's distance runner. He won the over-forty laurels in the 1980 Boston Marathon, and variously set national or world records in the 50-mile, 100-kilometer, and 24-hour endurance runs, the latter by covering 158 miles in nonstop running. I went out with him one day for a 4-mile practice run, during which he patiently held back as I padded alongside. "Ed," he said, "you could go faster if you ran on the balls of your feet." He might as well have said, you could fly if you flapped your wings. He seemed made of aluminum tubes and wires. His lungs were leather-lined. He was Mozart to my envious Salieri.

The experience has often made me think more objectively about my own limitations and more generally about those of the species to which I belong. For the obsessed and ambitious, the only strategy is to probe in all directions and learn where one's abilities are exceptional, where mediocre, where poor, then fashion tactics and prostheses to achieve the best possible result. And never give up hope that the fates will allow some unexpected breakthroughs.

I am blind in one eye and cannot hear high-frequency sounds; therefore I am an entomologist. I cannot memorize lines, have trouble visualizing words spelled out to me letter by letter, and am often unable to get digits in the right order while reading and copying numbers. So I contrived ways of expressing ideas that others can recite with quotations and formulas. This compensation is aided by an unusual ability to make comparisons of disparate objects, thus to produce syntheses of previously unconnected information. I write smoothly, in part I believe because my memory is less encumbered by the phrasing and nuances of others. I pushed these strengths and skirted the weaknesses.

I am a poor mathematician. At Harvard as a tenured professor in my early thirties I sat through two years of formal courses in mathematics to remedy my deficiency, but with little progress. It was distance running all over again. I remain mathematically semiliterate. When walked through step by step, I have been able to solve partial differential equations and grasp the elements of quantum mechanics, although I soon forget most of what I have learned. I have no taste for the subject. I have succeeded to some extent in theoretical model building by collaborating with mathematical theoreticians of the first class. They include, in successive periods of my research, William Bossert, Robert MacArthur, George Oster, and Charles Lumsden. My role was to suggest problems to be addressed, to combine my intuition with theirs, and to lay out empirical evidence unknown to them. They were my intellectual prosthesis and I theirs. Like my fellow field biologists who waded with me into swamps and climbed forested hillsides, we were civilized hunters searching for something new that might be captured, something valuable enough to take back home and display at the tribal campfire.

I have evolved a rule that has proved useful for myself and might be for others not born with championship potential: for every level of mathematical ability there exists a field of science poorly enough developed to support original theory. The advice I give to students in science is to move laterally and up and down and peer all around. If you have the will, there is a discipline in which you can succeed. Look for the ones still thinly populated, where fine differences in raw ability matter less. Be a hunter and explorer, not a problem solver. Perhaps the strategy can never work for track, with one distance and one clock. But it serves wonderfully well at the shifting frontiers of science.

CHAPTER EIGHT

Good-Bye to the South

W HEN I GRADUATED from the University of Alabama, in 1949, my father's health had begun to decline steeply. Chronic bronchitis, worsened by two packs of cigarettes a day, racked his body far into the night. As a member of a class and generation of men who took pride in fingers stained yellow by tar, he had no inclination to quit the habit. His alcoholism was also severe, and that addiction he took seriously. He feared becoming, as he put it, "like a Bowery bum." Already a member of Alcoholics Anonymous, he checked himself in at intervals to a rehabilitation center for detoxification and yet another stab at recovery. Nothing worked for long; the problem seemed insoluble. Given that he was already seeking professional help, there was nothing left for Pearl and me to offer but sympathy and attempts at persuasion. I hid the frustration and anger I felt: a son does not easily instruct his father on right behavior and self-control.

In early 1951 my father grew noticeably depressed and his behavior erratic. I was not able to read the signs, and was in any case away from home most of the time. I did not suspect what was coming. Early in the morning of March 26, he wrote a calm note of apology to his family, drove his car to an empty section of Bloodgood Street near the Mobile River, seated himself by the side of the road, put his favorite target pistol to his right temple, and ended his pain. He was forty-eight years old when he died.

He was given a military funeral at Magnolia Cemetery, graced with rifle volleys and the folding of the American flag from atop his coffin. The painful disorder of his life made this strictly prescribed rite of passage deeply comforting to me. My father was laid to rest close to the last of his three brothers, Herbert, dead from heart failure only a year before.

After a few days the shock of grief was infiltrated by feelings of relief, for my father now released, for Pearl whose desperate siege had been broken, and for myself—the filial obligation I had feared might tie me to a crumbling family was now forgiven. The impending tragedy finally took form, and happened, and was over. I could now concentrate entirely on my new life. As the years passed, sorrow and guilt-tinged relief were replaced by admiration for my father's courage. It is easy to say that the greater courage would have been to try again, to pull himself back and struggle toward a normal life. I am reasonably certain, however, that he had considered the matter very carefully and decided otherwise.

No son knows his father well enough to matter until it is too late; then understanding comes in fragments. I can say of him that he was an intelligent man who cheated himself of his own potential. Before he finished high school he ran away from home to go to sea in the boiler room of a cargo ship, made one round trip to Montevideo, and joined the Army. In the Quartermaster Corps he learned his trade as an accountant, which carried him through a long succession of jobs in private business and, in his last twelve years, the federal government. He was by nature loyal, warm, and sympathetic. He was quick-tongued in mixed company, given to frequent bouts of nostalgic and embroidered tales of personal adventure and to little, short-lived flashes of anger. He loved poetry but, like me, could not memorize enough lines to recite it competently. The youngest of four brothers, he lost his father when he was thirteen, and in his remaining time at home was spoiled by a mother whose permissiveness had become a family legend. His lifelong self-indulgence was made worse by a restiveness that never found ease because, I suspect, he had no destination in mind. His dream of retirement was to pilot his own houseboat back and forth on the Intracoastal Waterway of the Atlantic and Gulf Coasts, with no place chosen as home port.

My father's reading was limited to magazines and newspapers. He paid scant attention to music or to history other than that of his family, and he had little interest in current affairs. He loved hunting and fishing but did not take the time to develop his skills. He turned instead to the more quickly satisfying recreation of target practice with his collection of guns. From him I learned how to blow cans and bottles off fence

posts with pistols and shotguns at twenty paces, how to fire a
U.S. Army Colt .45 with both hands to keep it from bucking
too far out of line.

He drew strength from his conception of southern white
male honor. Never lie, he told me, never break your word,
be always respectful of others and protective of women, and
never back down if honor is at stake. He rested his dicta upon
the remembered traditions of his family, which he rewove and
annotated endlessly. He meant every word of this credo, and
he was a physically courageous man. I think he would have
died rather than accept humiliation or disgrace as defined by
his lights. In truth, he did just that in the end. But otherwise
the world in which he chose to live was too confining, too
ambiguous, and too nearly obsolete to test his code of honor
in any decisive way.

I sometimes reflect on the fact that my father and the old
house on Charleston Street are not just physically gone but
absolutely gone, except for a handful of photographs, official
records, and now this brief memoir. The neighborhood of old
trees and sagging Victorian homes has been scraped away and
replaced with cinderblock public housing. When I and a few
other older family members die, the man and our family home
will vanish almost as though they never existed. This observation on the human condition is one that I find both altogether
banal and eternally astonishing. When my cousin Jack Wilson,
son of my father's oldest brother and a lifelong resident of
Mobile, died in 1993, a large section of the cerebral memory
of his and my father's generation was erased. I have felt a small
pleasure from this, certainly not from Jack's death but from the
fact that I am now the sole inheritor of my father's existence. I
have been freed to recreate my father not just from his scantily
remembered actions but also from what I can reconstruct of
his character. Some of that I will keep private and let go to
oblivion, when I die.

Strong father, weak son; weak father, strong son; either way,
pain drives the son up or down in life. I do not dare to take the
full measure of my father's influence on me. But I would say to
him if I could that his self-image was a worthy template, and I
tried to bring it to fruition.

My mother, Inez Linnette Freeman, had achieved a better life after the divorce, and she encouraged and assisted me to do the same. She had come from a background similar to my father's in many respects, her roots reaching far back in Alabama. Her forebears, all of English descent, had come from the Mississippi Delta and Georgia to settle in the northern half of the state. Several of them helped to found the little towns of Bremen, Falkville, and Holly Pond during the early and mid-1800s. Most were farmers and merchants. One, my great-grandfather Robert Freeman, Jr., was both a farmer and a renowned (I hesitate to use the word notorious) horse trader. His wife, Isabel "Izzie" Freeman, practiced as a country doctor, which I interpret to mean the equivalent of practical nurse and midwife in a rural region where M.D.s were scarce. Being freeholders with property well north of the cotton belt and main river ports, these people by and large held tepid opinions about the Confederacy and the Civil War. When captured by Federal forces, Private Robert Freeman readily forswore further military service in order to return to his family and farm near Falkville.

In 1938 my mother married Harold Huddleston, a native of Stevenson, Alabama, near the Tennessee line. He was a successful businessman, and later advanced by the time of his retirement to a vice presidency of the Citizens Fidelity Bank and Trust, one of the several largest banks in the Southeast. Each September from my early teens until my graduation from college, I lived with my mother and Harold in their home first in Louisville, Kentucky, and then in the adjacent town of Jeffersonville, across the Ohio River in Indiana. They were supportive of my plans to attend college and train to be a biologist. Harold himself had attended the University of Alabama. Upward mobility into the professional class was something both he and my mother embraced as a fundamental ethic. They frequently took me to local parks where I could collect butterflies and ants. In what was a brave expedition on her part, my mother accompanied me when I was fourteen to Mammoth Cave in Kentucky. The cavern system, one of the largest in the world, had recently been set aside as a national park. As we descended into the gloom, I held back from the tour group to search (illegally) along the walkways for blind yellow ground beetles, cave crickets, and any other cave-dwelling insects I could find, giving the specimens to my mother to hold. She lost them somewhere near the cave exit, and I sulked all the way back to Louisville.

My mother provided not just encouragement but also financial help when I entered college. Later, as I prepared to enter the Ph.D. program at Harvard, she offered to pay my way through medical school. She wanted to be sure, as she put it, that I had not excluded dreams of a medical career for lack of money. But this more traditional profession held no interest for me. An entomologist I would be, and I was confident that I could make it the rest of the way on my own.

I did support myself from the time of my master of science candidacy at the University of Alabama until I completed the Ph.D. at Harvard five years later. I relied on scholarships and teaching assistantships throughout and never incurred debt. Long-term student loans were scarce to absent at the time. In any case the possibility simply never occurred to me.

In 1950 I transferred to the University of Tennessee, in Knoxville, to begin work on the Ph.D., mainly because of the presence there of Arthur Cole, a professor of entomology who specialized in the classification of ants. That year I searched the nearby Chilhowee and Great Smoky mountains for my favored insects, building my personal collection, while studying Cole's collection from the United States, the Philippines, and India. I finished a comprehensive review of the history and genetic change in the imported fire ant and sent it to the journal *Evolution*. While serving as Cole's laboratory teaching assistant, I honed my skills in the anatomy and classification of insects.

The academic challenge was not great at the University of Tennessee, and I grew restive. Out of boredom I also became a bit reckless. I was intrigued by the fact that a statute was still on the books forbidding the teaching of evolution in the state. In 1925 the Tennessee legislature had declared unlawful any doctrine that questioned the divine origin of man. A young high school teacher, John T. Scopes, was brought to trial that same year for presenting the theory of evolution to his biology class. In one of the most celebrated legal proceedings of American history, William Jennings Bryan led the prosecution and Clarence Darrow the defense. Since Scopes was undeniably guilty, he was convicted and fined $100, but not before expert testimony from scientists in favor of evolution and Darrow's scarifying courtroom examination of Bryan on the Bible sent shock waves through the ranks of the Christian fundamentalists. The state supreme court later acquitted Scopes, but only on the ground that the fine was excessive. The law stayed in place and was still untested in the higher courts when I came to Knoxville.

In the fall of that year, while teaching laboratory sessions in the general biology course at the University of Tennessee, I learned about the extraordinary discovery of the first of the South African man-apes. These erect, small-brained hominids seemed to place the origin of humanity one to two million years ago on the African continent. They were the key missing links between remote apelike ancestors and the most primitive true humans of the genus *Homo* known at the time of the Scopes trial, the so-called Javan Man and Peking Man—both of which are now placed in the single species *Homo erectus.*

Here, I thought, was one of the most important scientific discoveries of the century: Eden revealed in Africa by the lights of Darwin! I was intrigued by the prospect of a complete human phylogeny, with its deep significance for the self-image of our species. I also had a mischievous itch to shake things up just to see what would happen. I might get into the same trouble as Scopes, but I would spring out of it immediately—I guessed—because the evidence was so much more solid—I felt sure—and the faculty would support me—I hoped. In any case I could not resist spreading the word about the amazing South African man-apes.

I was granted permission to give a lecture on the subject to the elementary biology class. I told them the matter was settled: we *did* descend from apes, or a close approximation thereof, and scientists knew when these distant ancestors had lived and even something about how they had lived—they were carnivores, and Eden was no garden. The students, some my own age, were mostly Protestants, and many had been raised in fundamentalist families. Some, I am sure, had been taught that Darwin was the devil's parson, the spokesman of evil heresy. They scribbled notes; some glanced at the clock as time wore on. Finally the hour ended, and I waited for a reaction. The students filed out, talking among themselves about this and that but not, so far as I could overhear, about evolution, until only one remained, a large blond boy who looked me in the eye and asked, "Will this be on the final exam?" I told him no, please don't worry. He seemed relieved; one less thing to memorize. Nothing more was heard of my lecture. It was as though I had declaimed for an hour on the life cycle of the fruit fly.

The state legislature, yielding to reason or perhaps just resigned to the inevitable, repealed the anti-evolution law in 1967. The religious movement against the theory of evolution has sputtered on in a few other states, unsuccessfully promoting laws to force the teaching of the biblical account of creationism as an alternative theory. Either way, I learned a lesson of my own in Tennessee: the greater problems of history are not solved; they are merely forgotten.

By early 1951 I had decided to move on to Harvard University. It was my destiny. The largest collection of ants in the world was there, and the tradition of the study of these insects built around the collection was long and deep. To this end I had the support of Aaron J. Sharp, a distinguished botanist and professor at the University of Tennessee, who quietly advised me to apply to Harvard and nominated me for a fellowship there. A second supporter was William L. Brown, then a graduate student in Harvard's Department of Biology. I had first contacted Brown in 1948, when I was still an undergraduate at the University of Alabama, because I had heard from Marion Smith of his interest in the biology of ants. He turned out to be a fellow fanatic on the subject. Brown also was, and is, one of the warmest and most generous people I have ever known. He fueled my already considerable enthusiasm with a stream of advice and urgings. Equally important, he treated me from the start as an adult and a fellow professional. His attention was focused on the good of the discipline. He rallied others to the cause and urged them to take up significant research topics. What you must do, he wrote me in so many words, is to broaden the scope of your studies. Never mind a survey of the Alabama ants; start on a monograph of an important ant group. Make it continentwide, or even global if circumstances warrant. You and I and others who join us must get myrmecology on the move. Right now, you have the advantage of living in the Deep South, where there are a great many dacetine ants. These are extraordinarily interesting insects, and we still don't know much about them. There is an opportunity to do some really original research. See what you can come up with, and keep me posted.

I plunged into the dacetine project at once, tracking down species one after another, turning over rocks and tearing apart decaying stumps and logs, dissecting nests, and capturing colonies to be cultured in the laboratory. The dacetines are slender, ornately sculptured little ants with long, thin mandibles. Their body hairs are modified into little clubs, scales, and sinuous whips. In many species a white or yellow spongy collar surrounds their waists. Clean and decorative, they are under the microscope among the most aesthetically pleasing of all insects. The workers hunt springtails and other soft-bodied, elusive insects by approaching them with extremely cautious movement, legs lifting and swinging forward as though in slow motion. They open their jaws wide during the stalk, in some species by more than 180 degrees, to reveal rows of needle-sharp teeth. When the huntresses draw very close, they are able to touch the prey with the tips of paired slender hairs that project forward from their mouths. The instant contact is made, they snap the jaws shut like a bear trap, impaling the prey on their teeth. Each dacetine species uses a variation on this technique, mostly in the speed and degree of caution of the stalking approach, and each hunts a particular range of prey species.

Largely because I enjoyed grubbing in dirt and rotting wood on hands and knees, I was very successful in my pursuit of dacetine ants. In two articles, published in 1950 and 1953, I presented detailed accounts of the comparative behavior of the dacetine ants found in the southern states. In 1959 Brown and I combined our data to prepare a synthesis of dacetine biology. We correlated the food habits of large numbers of species from around the world with their social organization. We discovered that the anatomically most primitive species, which are native to South America and Australia, forage above ground for larger and more various insect prey, such as flies, grasshoppers, and caterpillars. They form large colonies and often have well-differentiated castes, including large-headed "majors" or "soldiers" and small-headed "minors," each playing a different role in the colony. The majors, for example, are adept at defending the colony against invaders, while the minors are prone to serve as brood nurses and attendants of the queen. As evolution proceeded, its trends evidenced by living species that are anatomically more advanced, the ants came to specialize more on springtails and other very small insects. The body size of the workers correspondingly decreased and became more nearly uniform within each colony, and the division of labor was diminished. Colony populations became smaller, and the nests more subterranean and inconspicuous.

Ours was a novel approach to the study of behavior. So far as I am aware, the dacetine study was the first of its kind on the evolution of social ecology in animals. It preceded the work of John Crook and others in the 1960s on primate socioecology; and in some respects it was more definitive, principally because we had more species and could use experiments on food choice. But despite the fact that we published our findings in the *Quarterly Review of Biology*, a premier journal with a worldwide distribution, our summary article was cited only rarely thereafter, and principally by fellow entomologists. It had little effect on the later development of behavioral ecology and sociobiology. Part of the reason is that monkeys, birds, and other vertebrates are more nearly human-sized and more familiar than ants, and therefore textbooks and popular accounts treat them as more "important."

William Brown, Uncle Bill as he was to become affectionately known by younger entomologists in later years, urged me to visit Harvard. I did so in late June 1950, traveling three days and nights on a Greyhound bus from Mobile to Boston. We seemed to stop at every city and town of greater than 50,000 population along the way, and I was exhausted by the time I reached the ant room of the Museum of Comparative Zoology. Bill and his wife, Doris, were gracious hosts. They put me up in their Cambridge apartment, where I slept on a sofa next to the crib of their two-year-old daughter Allison. Early the next morning I watched apprehensively as Allison reached through the crib bars for pages of Bill's newly completed Ph.D. thesis.

During the next several days, as Bill made final preparations with Doris to leave for field work in Australia, he took time to guide me through the ant collection. Once again he encouraged me to select large, important projects and to aim for publishable results. Your dacetine and fire ant studies are very promising, he said; but now you should come to Harvard in order to work more effectively with projects of even greater scope. Take a global view; don't sell yourself short with local studies and limited goals. He introduced me to Frank M. Carpenter, the professor of entomology and great authority on insect fossils and evolution, who was later to serve as my doctoral supervisor. Both men urged me to apply to the Ph.D. program at Harvard. I did so, even though I had already enrolled for the coming academic year of 1950–51 at the University of Tennessee.

The following spring I was admitted to Harvard for the coming fall semester with a scholarship and teaching assistantship that covered all expenses. In late August 1951 I sold my only suit to a secondhand store in Knoxville for ten dollars, packed all my belongings, including my research notebooks, into a single suitcase, and traveled by bus to visit my mother and Harold in Louisville. After taking one look at me, dressed— how shall I say it—in Salvation Army grunge, Harold escorted me to a men's clothing store and bought me a wardrobe befitting a 1951 Harvard student. I walked out in an Irish tweed jacket, Oxford button-down white shirt, narrow knit tie, chino slacks, and white duck shoes and socks. With a fresh crew cut added, I was ready to pass into a new life.

I arrived in Boston by bus and took the subway to Harvard Square, crossed over to the Harvard Yard entrance next to Massachusetts Hall, and asked the first person I met for directions across campus. He was evidently a student, and he spoke in a cultured English voice. So this, I said to myself, is the famous Harvard accent. Several weeks later the same student, who was indeed a Harvard sophomore, turned up in a laboratory section of beginning biology I was teaching. I learned then that his name was John Harvard Baker; that he was British, having only recently arrived in this country; and that he was a descendant of the uncle of the legendary John Harvard, whose donation founded the university in 1636 (Harvard himself had no children). Our meeting was, I think, a fittingly symbolic introduction to the university where I was to spend the rest of my professional life.

I walked on that day in September to Richards Hall, one of the graduate student dormitories in Harkness Commons, picked up my keys at the manager's office, and proceeded to my assigned room, number 101. My roommate had already arrived and posted his name on the door: Hezekiah Oluwasanmi. I thought, what kind of a name is that? Polynesian, maybe Samoan? He was Nigerian, and another Ph.D. candidate. We soon became good friends, continuing on through his eventual tenure as vice-chancellor of the University of Ife. Through the fall of 1951, as I sat reading at my desk, I half listened to Hezekiah and his friends, some of whom wore tribal scars on their cheeks, discuss the coming liberation of Nigeria. They were among the first intellectuals to plot such a movement in British Africa. I wondered if perhaps I was getting involved in something illegal just by being in the same room. I could see the headline in the *Mobile Press Register*: "Alabamian arrested with African revolutionaries as FBI closes in." It was all exhilarating, a proper introduction to the expanding and infinitely interesting world I had entered.

CHAPTER NINE

Orizaba

A LMOST ALL MY LIFE I have dreamed of the tropics. My boyhood fantasies drifted far beyond the benign temperate zone of Thoreau and Muir. Nor did I have any interest in arctic glaciers or the high Himalayas. I hungered instead for the frontiers of Frank Buck and Ivan Sanderson, hunters of tropical exotic animals, and William Beebe, naturalist-explorer of the Venezuelan jungle. My favorite novel was Arthur Conan Doyle's *Lost World*, which hinted that dinosaurs might yet be found on the flat summit of some unclimbed South American *tepui*. I was besotted with *National Geographic* articles on tortoise beetles and butterflies, winged jewels that entomologists—of the kind I hoped to become when I grew up—netted during journeys to remote places with unpronounceable names. The tropics I nurtured in my heart were the untamed centers of Creation.

When I was a boy most of the tropical forests and savannas were indeed still wildernesses in a nineteenth-century sense. They covered vast stretches of land waiting to be explored on foot, and sprinkled through them were unrafted rivers and mysterious mountains. In the farthest reaches of the Amazon-Orinoco basins and New Guinea highlands lived Stone Age peoples never seen by white men. But more compelling than all these wonders, more than white water, talking drums, arrows quivering in tent poles, and virgin peaks awaiting the flags of explorers' clubs, the fauna and flora of the tropics called to me. They were the gravitational center of my hopes, a vertiginous world of beauty and complexity I longed to enter. When I grew impatient during my late teens, I looked around for some passable equivalent nearer home. The Alabama bay-gum swamps and riverine hardwood forests, I realized, were somewhat like tropical forests writ small. After I entered college I explored the edges of the Mobile-Tensaw delta floodplain with that comparison in mind. I was attracted by the dense shrubby vegetation and meanders of unnavigable shallow mud-bottom creeks.

It was a place no field biologist had visited—and was seldom entered by anyone for any reason—and I wondered if it might contain undiscovered species of ants and other insects living in ecological niches new to science. I decided I would conduct a one-man expedition into the interior and thus inaugurate my career as a tropical explorer, at least in spirit.

I never made it into the delta. I was too occupied with the demands of college life at the University of Alabama and my ongoing studies of fire ants and other research projects across the state. Then successive transfers to the University of Tennessee and Harvard to continue graduate studies removed me from the region altogether.

In my first year at Harvard I was delayed further. I settled on a sensible thesis project that could be reliably finished in three or four years. *Then*, I figured, I could go to the tropics. My research would be on the ant genus *Lasius*, one of the most abundant but poorly understood assemblages of the north temperate insect fauna. The forty or more species are distributed through the cooler habitats of Europe, Asia, and North America. Their colonies excavate a large percentage of the little crater nests that dot cornfields, lawns, golf courses, and sidewalk cracks across the United States and Canada. If you go out and look for small brown chunky ants along the streets of cities such as Philadelphia, Toronto, and Boise, the first ones you are likely to see are foraging workers of a species of *Lasius*.

My project required a great deal of museum and laboratory work, but my explorer's urge destined me for the open air. I made it back decisively into the field in the summer of 1952 when I teamed up with Thomas Eisner. He was, like me, a first-year graduate student at Harvard. We found we had a great deal of scientific interests in common and soon became best friends. In one sense he was the perfect Harvard intellectual: multicultural and driven. His father, Hans, was a chemist and German Jew who, with his wife, Margarete, three-year-old Tom, and Tom's older sister Beatrice, left in 1933 when Hitler rose to power. They settled in Barcelona, only to witness the outbreak of the Spanish Civil War and the expansion of fascism. In 1936 Tom, then seven years old, heard the sound of dive bombers attacking the city, as the family prepared to flee to Marseilles and then to Paris. In 1937 the elder Eisner took his family to Montevideo, Uruguay. In this neutral country Tom spent the rest of his childhood in relative peace. The war had been left largely beyond the horizon, but Tom was kept aware of its progress. He was one of the spectators who watched the distant smoke plume rising from the pocket battleship *Admiral Graf Spee* as it was scuttled in the River Plate outside Montevideo after being chased there by British cruisers.

In Uruguay Eisner kindled a lifelong interest in butterflies and other insects. As he approached college age, his family moved to New York. Tom came to Harvard fluent in German, Spanish, French, and English, with a smattering of Italian. He was a virtuoso at the concert piano and, most important to me, a committed entomologist. We were kindred spirits in that one central pursuit. On a grander scale he had repeated the pattern of my own childhood, having been towed from one locality to another, anxious and insecure, turning to natural history as a solace.

Eisner was, and is (he has changed remarkably little over the years), a slender man with wispy hair and a tense and energy-charged manner, spinning in perpetual motion from one research scheme to another. He is a great biologist by virtue not just of extraordinary lifelong dedication but also of a masterly application of what I like to call the pointillist technique, which works wonderfully well in evolutionary biology. Eisner completes one meticulous study after another, usually a pinpoint analysis of some aspect of the way insects and other arthropods use chemical secretions to communicate and defend themselves. Taken separately, any one of the individual contributions may seem to apply to only a few species and hence to be of limited interest. Taken together and viewed from a distance, however, they form a novel evolutionary pattern of biology.

When I met him in the fall of 1951, Eisner was, like me, on the threshold of the serious part of his career. We had the good fortune to fall in together with other students destined for achievement and whose influence on us was immediate and considerable. They included Donald Kennedy, who became president of Stanford University; Howard Schneiderman, in later years vice president for research at Monsanto Company;

and Sheldon Wolff, destined for a distinguished career in cytology and medical research.

Tom and I decided to spend the summer of 1952 in search of insects across North America, traveling fast and free. In late June we took off in his 1942 Chevrolet, which he had named Charrúa II after the old Amerindian warrior tribe of Uruguay. We went north from Massachusetts to Ontario, then proceeded across the Great Plains states to Montana and Idaho, from there to California, Nevada, Arizona, New Mexico, through the Gulf states, and, finally, northward home in late August. We were naturalist hobos. We lived on the margin of society. Each night we slept on the ground, sometimes in the feeless camping areas of state parks, more often on the edge of open fields and woodlots off the side of the road. We ate canned food and washed our clothes under campground faucets, putting most of our negligible funds into the care and fueling of Charrúa II. The car required a quart of oil every hundred miles and frequent repairs of its frazzled tires. While I collected and studied ants, Eisner collected ants for his own future thesis research on anatomy, along with dustywings, snakeflies, and other insects of the order Neuroptera.

It was a time when national parks were uncrowded and many of the nation's major highways were still winding two-lane roads. We wandered almost aimlessly through cypress swamps, alpine meadows, and searing deserts, observing and collecting insects. On one ovenlike July night we made a swift traverse of Death Valley, cooled only by wet handkerchiefs tied around our heads. We saw most of the major ecosystems of North America close up, and all we learned in that remarkable summer cemented our lifelong passion for field biology.

A few months later, in the spring of 1953, I was handed the opportunity of a lifetime: election as a Junior Fellow in Harvard University's Society of Fellows. The Society, patterned after the prize Fellows of Trinity College in Cambridge University, gave three years of unrestricted financial support to young men (and, in later years, young women) who demonstrated exceptional scholarship potential. Junior Fellows were encouraged to study any subject, conduct any form of research, go anywhere in the world their interests directed them. The Society was made up of two dozen Junior Fellows and nine

Senior Fellows, the latter being distinguished Harvard professors who served as mentors and dinner companions to the
younger men. Each year the Senior Fellows chose eight new
members to replace the third-year, graduating class. In 1953, as
one of the fortunate few, I found myself lodged free in Lowell
House with a generous stipend, a book allowance, and travel
funds available upon application.

At the first dinner of the fall term we new Fellows stood as
the Society chairman, the historian Crane Brinton, read the
statement written by Abbott Lawrence Lowell, who as president of Harvard in 1932 had given a substantial portion of his
fortune to found the Society:

> You have been selected as a member of this Society for your
> personal prospect of achievement in your chosen field, and
> your promise of notable contribution to knowledge and
> thought. That promise you must redeem with your whole
> intellectual and moral force . . .
>
> You will seek not a near, but a distant, objective, and
> you will not be satisfied with what you have done. All that
> you may achieve or discover you will regard as a fragment
> of a larger pattern, which from his separate approach every
> true scholar is striving to descry.

Fair enough. On that first evening I savored the expertly
selected wine, the rare roast beef, the postprandial cigar, the
self-conscious scholar's talk. Like Thackeray's Barry Lyndon,
I was a happy indigent admitted to the company of lords. The
Society proceeded to transform my self-image and my career.
Its greatest immediate impact was a sharp rise in my expectations. I had been examined by first-rate scholars in diverse fields
and judged capable of exceptional research across an expanding
terrain. I thought, I have three years to justify the confidence
placed in me, the same amount of time it took to make Eagle
Scout. No problem. The Society's alumni and Senior Fellows
were outstanding achievers; they included Nobel and Pulitzer
winners. I thought, That's a reasonable standard to shoot for.

The second gift from the Society of Fellows was to place
me in the weekly company of other young men, all in their
twenties, who had begun to excel in widely diverse fields of

learning. My new companions included Noam Chomsky, with whom I discussed the instinctive behavior of animals; the poet Donald Hall; and Henry Rosovsky, economic historian and future dean of the Faculty of Arts and Sciences at Harvard. Among the many notable dinner guests I met during my three years as a Junior Fellow were Bernard DeVoto, T. S. Eliot, Robert Oppenheimer, and Isidor I. Rabi. I spent an especially memorable evening arguing with Rabi about the evolutionary consequences of atomic bomb tests; he defended the position that the explosions were good because radiation increases the rate of mutation, which can speed evolution. And that is a good thing, is it not? Was he serious? I was not completely sure, but the conversation was information-packed and exciting either way.

The final gift of the Society of Fellows was to launch me into the tropics at last. As quickly as I could arrange it, in mid-June, I departed for Cuba. On the flight from Miami to Havana the pilot invited the younger passengers into the cockpit, where I watched the Cuban coast come into view and my dream become reality.

In Havana I joined a small group of other Harvard graduate students to commence a course in tropical botany. We first traveled by car to the western province of Pinar del Río to visit patches of forest on the *mogotes*, outcroppings of limestone too rugged to convert into fields of sugarcane. The rest of the land had been cleared almost everywhere down to dirt and grass, left dotted for the most part only by towering royal palms. Several days later we proceeded to the Atkins Gardens, a Harvard-owned property at Soledad, near Cienfuegos on the southwestern coast forty miles east of the Bay of Pigs.

From the Gardens I traveled with three botanists, Robert Dressier, Quentin Jones, and the course instructor, Grady Webster, to search for remnants of the Cuban forest in Las Villas Province. The very difficulty of this quest bore shocking witness to the ecological destruction of the island. For centuries Cuban landowners had relentlessly cleared the forests with no concession to the native fauna and flora. To find the last refuge we had to go beyond the reach of bulldozers and chain saws, mostly up onto the slopes of steeper mountains and down the banks of river gorges. Traveling across the west central part of the island in 1953, I began to undergo a fundamental change in my view of the tropics.

On one memorable morning, we climbed into a Jeep to visit Blanco's Woods, a locally famous woodlot left uncut because its wealthy absentee owner had for some reason neglected to "develop" it. Blanco's Woods was one of the few parcels of relatively undisturbed lowland forest remaining in all of Las Villas Province, and probably all of Cuba. We drove for miles along rutted dirt roads through sugarcane fields and cattle pastures, crossing small fordable streams lined with corridors of weeds and second-growth woody vegetation. Occasionally we had to stop to open cattle gates and shut them behind us. We found it next to useless to probe the baked clay ground along the way for native plants and insects, of which there were few to none. It was equally pointless to look for endemic Cuban birds and other vertebrates. Among the few to be seen in the vicinity were the abundant brown anole lizards on the fence posts and, once in a great while, a giant Cuban anole in the crown of a royal palm.

When at last we came to Blanco's Woods it seemed unprepossessing, not a rain forest of popular expectation, but a stand of small to medium-sized trees, mostly torchwood, undergirded by dense shrubby undergrowth. If we had not been able to identify the trees, we might have imagined ourselves to be on the edge of an Iowa woodlot. Still, the little forest proved to be rich in representatives of the original Cuban fauna and flora, and we reveled in the discovery of one native species after another. While mosquitoes feasted on my sweating face and arms, I turned up two treasures of the ant world: the Cuban species *Thaumatomyrmex cochlearis*, its proportionately huge pitchforklike jaws wrapping all the way around the head so that the longest tines stick out beyond the rear border; and *Dorisidris nitens*, one of the rarest ants in the world, a shiny black species of a genus and species also found only in Cuba—and then known from only one previous collection. These specimens and others gathered in a few hours made valuable additions to the Museum of Comparative Zoology ant collection.

We decided to visit next the nearby Trinidad Mountains to study residues of forest still surviving there. Our car trip was even more difficult than the one to Blanco's Woods. We drove southeast along the uncertain two-lane, mostly unpaved highway from Cienfuegos to the town of Trinidad, and we were held up for an hour at a ford on the Rio Arimao jammed with trucks and cars. We had heard that a new road was being cut up the east slopes of the massif, and now we resolved, in order to make up lost time, to take it as a short cut to the vicinity of San Blas, where we knew forest was most likely to be found. The route proved to be a muddy nightmare. We toiled up it, occasionally pushing our four-wheel-drive vehicle out of the deeper ruts. We passed treaded earth-moving equipment and trucks loaded with newly cut logs (our forest!) on the way down. At the top, where we stopped at last to rest and to collect specimens, groups of people came from their houses to offer congratulations: ours was the first vehicle to make it up the new road.

Across the island on this day, young Fidel Castro was preparing to storm the Moncada Barracks in Santiago de Cuba, which was defended by 1,000 of Batista's troops. His near-suicidal attack would be launched a week later. Seven years later the Harvard station would be appropriated and Cuba largely closed to American naturalists.

The groves of trees we encountered in the Trinidad Mountains were mostly *cafetals*, small family-owned coffee plantations. I found and duly sampled a few native Cuban ants and other insects living there. We then hiked off the road to higher slopes, working our way along the edges of bluffs and around spurs of dogtooth limestone. The land was either too steep or too rugged to support agriculture, yet fertile enough to shelter patches of native rain forest and other vegetation. If it were not for mountains and limestone, I reflected, all of Cuba would be a sugarcane field. At Mina Carlota we found ourselves at last in the midst of an abundance of the old fauna and flora of the Cuban mountains. Forty years earlier William Mann, then a Harvard graduate student studying ants and now, in 1953, director of the National Zoo, had traveled to this exact spot. After a few hours of random collecting, he stumbled upon a new species of ants, which he later named *Macromischa wheeleri* in honor of his sponsoring professor, William Morton Wheeler. In 1934 he recounted his discovery thus in the *National Geographic*:

I remember one Christmas Day at the Mina Carlota, in the Sierra de Trinidad of Cuba. When I attempted to turn over a large rock to see what was living underneath, the rock split in the middle, and there, in the very center, was a half teaspoonful of brilliant green metallic ants glistening in the sunshine. They proved to be an unknown species.

Ever since reading that passage as a ten-year-old, I had been enchanted by the idea of prospecting in a faraway place for ants that resembled living emeralds. Now here I was at the very same place, climbing the steep forested hillside of Mina Carlota. Searching for ants, I turned over one limestone rock after another, perhaps a few of the very ones that Mann had handled. Some cracked; some crumbled; most stayed intact. Then one rock broke in half, exposing a cavity from which poured a teaspoonful of the beautiful metallescent *Macromischa wheeleri*. I took a special satisfaction in repeating Mann's discovery in exact detail after such a long interval of time. It was a reassurance of the continuity of both the natural world and the human mind.

As my companions and I proceeded across the Trinidad massif on our way to Mayarí, I encountered another ant, *Macromischa squamifer*, whose workers glistened golden in the sunlight. The color resembled the scintillations of tortoise beetles found in many parts of the world. This striking and unlikely color is most likely produced by microscopic ridges on the body that refract strong light. Bright colors are a widespread trait among West Indian species of *Macromischa* (the genus has since been reclassified as a group with the genus *Leptothorax*), and it is a fair guess that the ants use their raiment to warn predators of strong stings at the tips of their abdomens or poisonous chemicals held within the glands of their bodies. In the natural world, beautiful usually means deadly. Beautiful plus a casual demeanor *always* means deadly.

On this special day the old Cuban animals and plants continued to reveal themselves, like surviving spirits in a sacred ruin. On a tree fern near Naranjo, at 1,000 meters altitude, I found a species of anole lizard new to science, light brown with a greenish tinge, overlaid by cream rectangles along the back. While trying to escape me, it hopped like a frog instead of running like most of the other members of its genus.

The botanists brought me another kind of anole, a giant by the standards of its group of lizards, nearly a foot in length. Its eyelids were partly fused, giving it a permanently sleepy look, and it bore a strange crescent-shaped ridge along the back of its skull. The creature was very slow-moving for an anole and had the unique ability to rotate its eyes in different directions independently. I later found that my little monster was a known species, *Chamaeleolis chamaeleontides*, the sole member of an endemic Cuban genus. As the nineteenth-century zoologist who named it was aware, the species resembles the true chameleons of Africa and Madagascar in the traits I have just cited. Its superficially similar anatomy is not due to kinship with these lizards, however: it did not descend from an African species that rafted from Africa across the Atlantic to the West Indies. Rather, its peculiarities are the product of convergent evolution, a true all-Cuban creation.

I named the lizard Methuselah for its craggy features and gray wrinkled skin, and kept it as a pet for the rest of my summer's travels. I was fond of Methuselah but also recognized an unusual opportunity for original research. No one had previously studied a live *Chamaeleolis*. Was the species convergent with the true chameleons in behavior as well as in anatomy? In the fall I brought Methuselah back to Harvard, continued to study it daily, and found that its behavior was indeed convergent with that of the chameleons, as I had guessed. It stalked flies and other insects with extremely slow movements of its body, following its targets by a covert rotation of its eyes and then seizing them with startling speed by a forward lash of tongue and snap of jaws. Methuselah's manner was strikingly different from that of other anoles, which dash forward from their resting place to catch prey, then back again, like flycatchers. It diverged even though the species is related to them by common ancestry. I thus held in my possession an important bit of Cuban natural history never before reported. I subsequently published an article on my findings. Only later did I come to appreciate that *Chamaeleolis chamaeleontides* is probably a threatened species; as a consequence, having removed an individual even for scientific study is not something of which I can be proud.

In late July, accompanied by Robert Dressler, Quentin Jones, and Methuselah, I flew from Havana to Mérida, on Mexico's Yucatán Peninsula. We departed immediately for a week's collecting in the thorn forest along the Progreso–Campeche Road, with a side trip to the ruins at Uxmal. We found the great temples and courtyards of the Mayan city only partly cleared of vegetation. No tourists or guides were present, and we enjoyed a free run of the grounds. Ants abounded on and among the crumbling edifices, as no doubt they had done 1,400 years previously when the first stones were laid. I climbed the stairs of the Temple of the Magician to a fig tree growing on its apex, and from the branches of the tree collected workers of *Cephalotes atratus*, a large, shiny black ant with compound spines. Resting briefly by the tree, I reflected on this triumph of the ever-abounding life of insects over the works of man.

We next flew out of Mérida to Mexico City, where I left Dressler and Jones and began a solitary all-entomology expedition. I took a bus eastward, through the pine-dotted uplands of the Mexican Plateau and down the winding road that drops thousands of feet to the coastal plain and city of Veracruz. I arrived for the first time in what I like to call the *serious* tropics: not the island habitats of the West Indies, with their eccentric and interesting but limited fauna and flora; not the mangrove fringes of the Florida Keys and Caribbean coasts, however verdant; but the inland continental lowland tropics, the true Neotropics, with its vast biota deployed in endless combinations of species from Tampico in Mexico through Central and South America to Misiones Province in northern Argentina. Here in almost any patch of moist forest I could find more species of ants in an hour than would be possible in a month's travel through Cuba.

I searched for residues of the vanishing rain forest along the coast, finding them in the vicinity of El Palmar, Pueblo Nuevo, and San Andrés Tuxtla. All were under heavy siege, already cut back along the edges and high-graded in the interior. Off the highway other such refugia could be seen on distant hilltops and the slopes of steep ravines. Such was and remains the pattern of access left to visitors everywhere in the tropical world. It can be expressed in the form of a standard route: leave the road, climb through a barbed-wire fence, hike across a pasture, and slide down a slope to the edge of a stream. Cross the stream—if it is shallow enough—and start up the other side to the edge of the forest. Cut through fringing second growth until you reach the shade of trees. At this point you have arrived at your destination but are likely to be on an incline so steep that it is necessary to hold on to the trunks and exposed roots of bushes and small trees to avoid tumbling head over heels back down to the bottom.

How much longer will these precarious refuges last until they too are cut away? It was frustrating and heartbreaking to travel in Mexico with such thoughts in mind. When at last I made it into the rain forests of Veracruz State I operated like a vacuum cleaner, taking samples of every kind of ant I could find. At night I identified species, labeled my vials, and wrote natural history notes. I had remarkably quick success by entomological standards: I captured colonies of two genera, *Belonopelta* and *Hylomyrma*, that had never been studied before, and recorded my observations on their social organization and predatory behavior for later publication.

As I prepared to leave the Veracruz coast two weeks later, my attention was drawn to Pico de Orizaba, the great volcanic mountain just north of the city of Orizaba. Its beautiful symmetric cone rises 5,747 meters—18,855 feet—above sea level to a permanently snow-covered peak. Orizaba is not just a prominence atop an already towering mountain range or plateau like Popocatepetl and mighty Aconcagua, but a mountain of more solitary and mystic qualities, a lone giant born of Mexico's ring of fire, standing sentinel over the southern approaches of the central plateau.

I was drawn not just by the amazing sight but also by the very concept of Orizaba. I thought of the mountain as an island. It was isolated from the plateau, yet I believed that a lone climber could travel in one relatively short straight pass from tropical forest to cold temperate forest and finally into the treeless arctic scree just below the summit. The cooler habitats constituted the island. The surrounding tropical and subtropical lowlands were the sea. Orizaba's uplands were close enough to the plateau to receive immigrant plants and animals adapted to the middle and upper slopes, yet isolated enough for unique races and species to have evolved and dwell only there.

So what might I hope to discover if I climbed Orizaba? No one had toiled up the slopes of the mountain to study ants, generally the most abundant of small terrestrial animals, with the possible exception of the much smaller mites and spring-tails. For every bird there might be a hundred thousand or million ants, and I could reasonably expect to sample the species effectively during a single fast traverse. I knew that the change in fauna and flora from tropical to temperate was likely to be dramatic. The southeastern face of the Mexican escarpment, where Orizaba sits, is the site of the most abrupt changeover of biogeographic realms found anywhere in the world, except perhaps in the Indian and Bhutanese Himalayas. On the plateau live large numbers of plants and animals typical of the Nearctic Region, a realm extending northward to encompass all of North America. While descending earlier the tortuous road from Puebla over the plateau to the Veracruz plain, I had left this world of beech, oak, sweetgum, and pine and entered the Neotropical Region, where aroids and orchids cling in masses to arrow-straight tree boles and lianas hang like ropes from the lofty horizontal branches.

I expected to find all this and more if I climbed Pico de Orizaba. Let me put it more strongly: I was foreordained to try it. I would start at La Perla, at 3,000 feet, and follow a donkey trail I had heard about to the hamlet of Rancho Somecla, at 11,000 feet. I would simply ask for the hospitality of the people there, who were rumored to be friendly to strangers, and proceed the next day on up to the snow line, at about 16,000 feet. I would collect ants and make notes on the environment all along the way.

I was a fool of course, traveling alone on foot up a high mountain without a map and no more than phrasebook Spanish. But I did make it most of the way. Early in the morning of a beautiful late August day, I took a bus from the city of Orizaba to La Perla and started walking. The mountain's south slope was mostly uninhabited; I encountered no one on the trail until I reached Rancho Somecla, my destination, late that afternoon.

My journey began in subtropical vegetation. At 5,500 feet I entered a forest dominated by hornbeam and sweetgum, both temperate-zone trees, with tree ferns abundant in the understory. Scattered through the habitat at lower elevations were dense, wet patches of tropical hardwoods. The ants in this transition belt, which composed nothing less than the passage from the Neotropical to the Nearctic regions, were a mix of tropical and temperate species: army ants and fire ants mingled with species of the typical north temperate genus *Formica*. Two of the *Formica* species later proved new to science. At 8,000 feet I found a mixture of pines, making their first appearance along the ridges, and broadleafed trees dominated by hornbeam on the slopes. The woodland was tessellated by pastures and stump-filled glades recently cleared by woodcutters.

When I arrived at Rancho Somecla, which turned out to be a collection of about a dozen houses, I was close to exhaustion. To the people who came out to meet me, I explained as best as I could why I was there. I doubt that they really understood my words or gestures, but one family promptly offered me lodging. I rested while they prepared a chicken dinner. Then, as the light failed, I headed out for one more try at ant collecting in the surrounding pine forest, this time accompanied by several young men who listened gravely as I explained why I was tearing up the bark of rotting logs and putting insects in bottles. One of my companions agreed to guide me to the snow line the next day.

That night I slept not at all. My bed was a table, and the single blanket given me offered little warmth when the temperature fell into the forties Fahrenheit. Occasionally I rose to look through the door at the brilliant full moon in a cloudless sky. It would be a wonderful place to live, I thought, if you brought a lot of blankets.

At dawn the next morning, after pressing some pesos on my hosts, I headed on up the mountain with my guide. When we reached an elevation of between 12,000 and 13,000 feet, we entered open cloud forest, where the pine trunks were gnarled and the branches draped with epiphytes. My excitement was growing, but I could go no further. The air was too thin for someone who had been living at sea level, and I was gasping for breath. I estimate that I had come to within 400 feet of timberline and perhaps 3,000 to 4,000 feet below the snowcap. Of course I was at my physical limit. I had been entirely naive to suppose that anyone could walk three miles from the lowlands straight up into the air in thirty-six hours and keep on going.

In any case ants had become very scarce, even in the clearings warmed by the morning sun. I searched for an hour before finding one colony nesting beneath a wood chip. Then I turned around and started walking back down. At Rancho Somecla I shook hands with my guide and headed alone down the trail to La Perla, moving rapidly now, then to my hotel in Orizaba, where, the sated adventurer, I slept for twelve hours.

PART 2
STORYTELLER

If you're a storyteller, find a good story and tell it.

HOWARD HAWKS, FILMMAKER

CHAPTER TEN

The South Pacific

O N A COLD MARCH day in 1954, the season when Cambridge is its least lovable, Philip Darlington called me to his office. How would you like, he said, to go to New Guinea? The Society of Fellows and Museum of Comparative Zoology had agreed to cover my expenses for an extended visit. No specialist had collected ants in that fabulously rich and still mostly unexplored fauna. Other islands such as New Caledonia might be visited en route. I could work hands on in the very arena where a hundred years before young Alfred Russel Wallace had begun to turn zoogeography, the study of animal distribution, into a scientific discipline. Who knows how the experience might transform my own thinking as a zoogeographer? And if I picked up some ground beetles, Darlington's favorite group, that would be all right too.

Here was the brass ring for a young field biologist. Many years were to pass before another generation of well-funded researchers were to descend in teams on New Guinea and other South Pacific islands and set up field stations. I could be a pioneer. Darlington said, Go, while you're still footloose and fancy-free.

I was not footloose and fancy-free. I was in love. The previous fall I had met a beautiful young woman, Renee Kelley, from Boston's Back Bay, and we were engaged to be married. She was a fellow introvert, pleasured by long hours of quiet conversation, a budding poet, deeply interested in literature, a scholar by temperament, and thus, though not a scientist, able to understand my dreams of pursuits in faraway places. Our marriage was to be happy and enduring.

We were young then, in 1954, and a parting seemed almost unbearable at the start of our engagement. But we agreed I should go to New Guinea. I would be away ten months, much of the time in remote areas. No jetliners existed then to shuttle me back and forth, and the great distance and high costs of transportation by other means made interim visits improbable.

Telephone calls were difficult and expensive, to be used only for emergencies.

On the morning of November 24 the Eastern Airlines carrier to San Francisco taxied from the gate at Boston's Logan Airport and out onto the runway. I could see Renee pressed to the visitors' observation window, her right hand waving slowly. Around her neck she wore a long woolen scarf striped in Harvard maroon and white, its tasseled ends nearly touching the floor. We were both weeping. Divided by two passions, the tropics and romantic love, I was a young seafarer venturing into another age on a long and uncertain voyage. Until I came home we would write to each other daily and at length, accumulating a total of some six hundred diarylike letters.

I had decided on a tour of the outer Melanesian archipelagoes, then Australia, and finally New Guinea. From San Francisco I took a propeller-driven Pan American Super Constellation across the Pacific. It touched down for refueling at Honolulu and Canton Island—a dry, cheerless atoll in the Phoenix group—before proceeding to Fiji. As it descended toward Nandi Airport in Viti Levu the next morning, I looked down upon a passage of white and green atolls in a turquoise sea. Never before or afterward in my life have I felt such a surge of high expectation—of pure exhilaration—as in those few minutes. I know now that it was an era in biology closing out, a time when a young scientist could travel to a distant part of the world and be an explorer entirely on his own. No team of specialists accompanied me and none waited at my destination, whatever I decided that was to be. Which was exactly as I wished it. I carried no high-technology instruments, only a hand lens, forceps, specimen vials, notebooks, quinine, sulfanilamide, youth, desire, and unbounded hope.

The South Pacific is a galaxy of thousands of islands, spread out in configurations that have served many of the key advances in evolutionary biology. Darwin conceived of evolution by natural selection from what he learned about birds in the Galápagos Islands, and Wallace had the same idea after studying butterflies and other organisms in the old Malay Archipelago, now modern Malaysia, Brunei, and Indonesia.

A true, biogeographic island, I knew as I stepped off the plane at Viti Levu and looked around, is a world that holds most of its organisms tight within its borders. It is the ideal unit for the study of evolution. Enough immigrants fly, swim, or drift ashore to colonize the island, yet not so many in each generation to form the commanding elements of its populations. If the island is large and old and distant enough, the descendants of the immigrants evolve into new races peculiar to the new home. Given enough time the races diverge still further from their sister populations on the continent and on neighboring islands to deserve the taxonomic rank of species. We speak of such local races and species as endemic: they are native to the island and nowhere else in the world. The Hawaiian hawk is a good example of an endemic, as well as the Jamaican giant swallowtail and the Norfolk Island pine. By factoring in the age of the island and the origin of the immigrants, biologists can reconstruct the evolution of the plants and animals there more easily than on continents. The simplicity of islands makes them the best of all natural laboratories.

The experiments are conducted in opposite manner from those in conventional laboratories. They are retroactive rather than anticipatory. Whereas most biologists vary a few factors under controlled conditions and observe the effects of each deviation, the evolutionary biologist observes the results already obtained, as learned from studies of natural history, and tries to infer the factors that operated in the past. Where the experimental biologist predicts the outcome of experiments, the evolutionary biologist retrodicts the experiment already performed by Nature; he teases science out of history. And because so many factors may have operated in the guidance of evolution and the creation of wild species, the best result of the retrodictive method can be obtained if the ecosystems are relatively small and simple. Hence islands.

Unlike experimental biologists, evolutionary biologists well versed in natural history already have an abundance of answers from which to pick and choose. What they most need are the right questions. The most important evolutionary biologists are those who invent the most important questions. They look for the best stories Nature has to tell us, because they are above all storytellers. If they are also naturalists—and a great majority of the best evolutionary biologists are naturalists—they go into the field with open eyes and minds, complete opportunists looking in all directions for the big questions, for the main chance.

To go this far the naturalist must know one or two groups of plants or animals well enough to identify specimens to genus or species. These favored organisms are actors in the theater of his vision. The naturalist lacking such information will find himself lost in a green fog, unable to tell one organism from another, handicapped by his inability to distinguish new phenomena from those already well known. But if well equipped, he can gather information swiftly while continuously thinking, every working hour, What patterns do the data form? What is the meaning of the patterns? What is the question they answer? What is the story I can tell?

This is the strategy I brought with me to the ant fauna of the Pacific Islands. I would collect samples of every species I found and write notes on all the aspects of ecology and behavior I observed, all the while watching for patterns in the form of geographic trends and adaptation of species to the environment. I was well aware of existing theory and the conventional wisdom of my discipline, but I would hold my mind open to any phenomena congenial enough to enter it.

Nadala, Viti Levu, December 1954. Fiji in one dreadful sense was Cuba and Mexico all over again: the native biota had already been driven back to scattered and nearly inaccessible enclaves. At Nandi I hired a driver and traveled along the north coast road of Viti Levu through villages, domestic groves, and pastureland. Virtually no natural forest survived along this thoroughfare, crowded as it was by the dense settlements of immigrant East Indians. We turned south at Tavua toward the central hills to search for patches of native forest, all on the land of the aboriginal Fijian population. One elderly man I met remembered another ant collector who had visited nearby Nandarivatu forty years back. He could not recall the name, but I knew it was William Mann, also my predecessor in Cuba, sent by Harvard to the islands in 1915–16 to collect for the Museum of Comparative Zoology. The forest I could reach in one day was similar to what he had known except that it had been disturbed by high-grade lumbering and perforated by slash-and-burn clearings. At Nadala I crawled up a steep slope over scattered pumice rocks to a well-shaded pocket of native trees, densely hung with lianas, where I found elements of the endemic ant fauna. One of them shot adrenaline into my veins: *Poecilomyrma*, a genus known solely from Fiji and collected only one time previously—of course by William Mann.

The next day, working south off the coastal road near Korovou, I learned another melancholy fact about conservation. In a small patch of what appeared to be natural forest I found only exotic ant species. I realized that on islands harboring native species of limited diversity, the ecosystems are vulnerable to invasion by aliens even if left physically intact. Much of the Pacific fauna has gone under in the path of pigs, goats, rats, Argentine ants, beard grass, and other highly competitive forms introduced by human commerce. Strangers have savaged the islands of the world.

I did not linger in Fiji. The ant fauna was already reasonably well known, thanks to Mann's lengthy residence. The next day I caught the Qantas flying boat from Suva to Noumea, the French colonial center of New Caledonia.

Mount Mou, New Caledonia, December 1954. With my arrival in New Caledonia, I had reached what I would thereafter consider my favorite island, a large, pencil-shaped land 1,200 kilometers off the east coast of Australia, the southernmost reach of Melanesia. The very name of the place meant, and still means, "alien" and "distant" to me. I knew from the work of previous naturalists that plants and animals had come to its shores during millions of years, for the most part eastward from Australia and southward from the Solomon Islands through the New Hebrides. They had mingled and evolved to form unique ecosystems. Among the native species were archaic trees and other plants, a few of ultimate Gondwanaland origin, whose ancestors had lived as far afield as Antarctica, when warmer climates prevailed. Also present were stocks of animals and plants that have evolved into extreme forms found nowhere else, including the famous kagu, representing an entire family of birds, the Rhynchochetidae. This flightless endemic, its shrill call piercing the night, had been reduced to near extinction since the French colonized the island in the 1860s. Early records indicated that the same broad biogeographic pattern of mixed origin and endemicity, with extreme rarity for some species, held for ants. I intended to find out.

On one hot day in this austral midsummer I caught the northbound bus from Noumea and got off at the little village of Païta. I hiked six kilometers up a small dirt road to the estate of a family named Bourdinet and dropped my gear at the gazebo, where I set up camp. The Bourdinets were not home that week and could offer no other accommodation. No matter; I was happy to concentrate entirely on my work. I walked another kilometer to the house of the Pentecost family, their nearest neighbors, climbing in the process to 300 meters elevation. My goal was the summit ridge forest of Mount Mou rising beyond, to 1,220 meters. To get there I pushed my way through a broad thicket of dense, dry bracken. When I reached the western crest of the ridge, I found myself still in the midst of bracken. But at least I had broken out onto the summit trail. The going grew easier, and the forested mountaintop was in full view only a kilometer away.

Because I was completely alone and had seen no one since Païta, it occurred to me that if I were crippled in an accident it would be three or four days before my contacts in Noumea became aware that anything was wrong. I stepped more carefully for the rest of the climb. As newly gathering mountaintop mist closed around me, I entered the forest. There I encountered first low shrubs and scattered trees, then continuous stands of *Araucaria* and *Podocarpus* conifers, their trunks and branches laden with mosses and other epiphytes. A little farther on, near the summit, I entered the true cloud forest. Here the trees were gnarled and stunted, and their canopies closed overhead at only ten meters. Their trunks and the surface of the soil in which the trees grew were coated with an unbroken blanket of wet moss.

I had arrived on an island within an island, a world of my own. The warm proprietary feeling of my boyhood flooded back. My imagination drifted back across epochs. The conifers there were ancient members of the Antarctic realm, still distributed across southern Australia, New Zealand, temperate South America, and here, the uplands of New Caledonia. Some of the species of plants and animals dated back to the Mesozoic Era, when they were surely browsed by dinosaurs, and when parts of the Antarctic continent itself were still habitable by all. As I began hunting ants, a little green parrot with a red cap landed on a branch close by and stayed there. At intervals he squawked at me in some mysterious psittacine language. We were perfect companions in the mossy forest, native and exotic joined in momentary harmony. I would do no harm, I told the parrot, and leave soon, but this place would live forever in my memory.

Not just the ants but everything I saw, every species of plant and animal, was new to me. These creatures were a fully alien biota, and it is time to confess: I am a neophile, an inordinate lover of the new, of diversity for its own sake. In such a place everything is a surprise, and I could make a discovery of scientific value anytime I wished. My archetypal dream came clear:

Take me, Lord, to an unexplored planet teeming with new life forms. Put me at the edge of virgin swampland dotted with hummocks of high ground, let me saunter at my own pace across it and up the nearest mountain ridge, in due course to cross over to the far slope in search of more distant swamps, grasslands, and ranges. Let me be the Carolus Linnaeus of this world, bearing no more than specimen boxes, botanical canister, hand lens, notebooks, but allowed not years but centuries of time. And should I somehow tire of the land, let me embark on the sea in search of new islands and archipelagoes. Let me go alone, at least for a while, and I will report to You and loved ones at intervals and I will publish reports on my discoveries for colleagues. For if it was You who gave me this spirit, then devise the appropriate reward for its virtuous use.

Ciu, near Mount Canala, New Caledonia, December 1954. I had to go to what seemed the edge of nowhere in order to sample ants on the northern coast, in moist lowland and foothills forest. The insects were likely to belong to different species from those around Noumea and might include at least two rare endemic genera taken by earlier collectors. I had risen at 3:45 in the morning to catch the daily bus to Canala. The antique vehicle followed a route that wound 170 kilometers across the central massif of the island. The driver made countless detours and stops to pick up and drop off native New Caledonians. We arrived at Canala at 10:30 in a driving rain that continued for the rest of the day. I lunched at the Hôtel de Canala, tumbled into bed, and dreamed of blue skies as I fell asleep.

Canala in 1954 was a collection of twenty rundown houses, the hotel, and a Catholic mission. The marquee social activity of the village was cricket played by men and women on the same teams, supported by cheerleaders who beat bamboo sticks together. The glamor ended there. The Hôtel de Canala contained a kitchen, a dining room, and a row of six square cubicles, each three meters on a side and furnished with a bed, table, and water basin. Lodging was U.S. $4.80 a night. The cubicle next to mine was the entertainment parlor of a prostitute who practiced her trade very noisily. All the guests used the same shower and evil-smelling outhouse. Meals were of uncertain provenance and often inexplicably cold, but I didn't care. Dinner was only $1.60 with wine, and all I wanted anyway was enough nourishment to fuel me to the nearby forests and back free of dysentery.

The next morning, packing a sandwich and a bottle of diluted red wine, I walked the one-lane dirt road seven kilometers south to Ciu, an aggregate of farms on the edge of the inland forest. For part of this distance the road passed through a marsh, from which clouds of striped *Aedes* mosquitoes poured through the hot sunlight like sniper bullets. In the manner of *Aedes* in other parts of the world, they commenced biting as soon as they landed on a patch of bare skin. The repellant I splashed on myself meant little to them. I named this stretch of the road Mosquito Alley, and started to jog-walk its length, head down and arms folded like a man running a gauntlet.

My destination was the Fèré farm, bordered by a small river of the size we call a creek in Alabama. I followed once again the universal formula for gaining access to tropical forest: crawl through a barbed-wire fence, walk across a cow pasture, wade a shallow part of the river (in this case adorned by an upstream waterfall), and climb a hillside into the forest. The effort proved worth the trouble. I soon entered the shade of native timber, a prehistoric New Caledonian world. I had passed no one on the Canala–Ciu road, and as I worked back into the forest I could see no sign of recent human disturbance. The solitude, as usual, felt right. Human beings mean comfort, but they also mean a loss of time for a field biologist, a break in concentration and, for strangers in an unknown land, always a certain amount of personal risk.

The Fèré tract was not a true rain forest in the familiar, Amazonian sense. It comprised only two stories of trees, with the upper canopy twenty meters high and broken in enough places to let sunshine fall in large, radiant patches on the forest floor. The habitat was ideal for ants. It abounded with pure New Caledonian species, many new to science. I was struck by the prevalence of red-and-black coloration among the workers foraging above ground. At Chapeau Gendarme near Noumea the same species were predominantly yellow. What was the meaning of this local color code? Perhaps it was just coincidence. But I suspected mimicry. My guess was that one to several of the species were poisonous, as I had supposed to be true of the metallescent ants of Cuba. The bright and distinctive color says to potential visual predators such as birds and lizards: Don't even try to eat me, you'll be sorry. In theory it pays for all the local poisonous species to evolve the same color, forming a consortium among the advertisers. It also pays for harmless, tasty species to acquire the same appearance and enjoy a free ride on the repellant forms they imitate. I had neither the means nor the time, however, to test either hypothesis.

My attention soon shifted to a phenomenon that was more tractable to immediate study and would later prove important in tracing certain aspects of ant evolution. Near Noumea I had collected the first ants of the genera *Cerapachys* and *Sphinctomyrmex* and hence of the entire ant tribe Cerapachyini ever recorded from New Caledonia. Here at Ciu they were so abundant I could observe them within the first hours of my arrival. I discovered that the peculiarly cylindrical, hard-bodied workers feed on other ants. To overcome their formidable prey, they hunt in packs much like the army ants of the mainland tropics. Their sorties, I saw, are much smaller and less well organized yet also effective in breaking down the defenses of the target colonies. "Real" army ants, the kind that march in thick columns in Asia and Australia, never managed to cross the Coral Sea and colonize New Caledonia. The less spectacular cerapachyines somehow succeeded, and even though they are less formidable huntresses they have the army ant niche to themselves. That is why, I conjectured, they are so abundant on New Caledonia and so rare on most continents. The concept was ill-formed in my mind then. I recorded the habits of the cerapachyine ants only because they are interesting in their own right. But three years later I would use the field notes as a key piece in my reconstruction of the evolutionary origin of army ants.

Ratard Plantation, Luganville, Espiritu Santo, New Hebrides, January 1955. Curiosity and opportunism brought me to this most remote and least-known large island of all the South Pacific. Still mostly covered by undisturbed rain forest, the northern New Hebrides had never been collected for ants, so every record I put in my notebook would be new. Even a brief glimpse of the fauna as a whole might allow me to place the New Hebrides (now the republic of Vanuatu) in the larger biogeographic picture. The archipelago is a potential stepping-stone to the more distant islands of the western Pacific. It can receive both Asian elements from the fully tropical Solomon Islands, to the north, and Australian elements from subtropical New Caledonia, to the south.

On this day, however, my exploration had been cut short. I was in bed with a high fever; the opening strains of *Swan Lake* ran inexplicably through my head, over and over, scrambling my thoughts into chaos. To add to my distress, I was occasionally bounced by aftershocks of an earthquake that had struck from an epicenter near Malekula three days earlier. Several large circular bruises were spaced at regular intervals across my chest. They had been inflicted by a doctor—he *claimed* he was a doctor—in Luganville in an attempt to draw the fever out with powerful suction cups. Surely I was one of the last patients in the Western world to endure this archaic and useless remedy.

My hosts were Aubert Ratard, his wife, Suzanne, and their two teenage sons. The Ratards were among the wealthiest of the two hundred French families who owned copra plantations on Espiritu Santo. Down the road from their coastline property were an airstrip and Quonset huts, remnants of an American base from the Second World War. The American forces and the people of the New Hebrides were the inspiration for James Michener's *Tales of the South Pacific.* Michener had also been a houseguest of the Ratards a decade earlier, and Ratard himself was the inspiration for the French planter in the book and musical. At dinner Aubert told me about the real Bloody Mary, who still lived in the central administrative town of Vila, on the island of Éfaté. From the shore of his property he pointed to Bali-ha'i across the Segond Channel, in real life the island of Malo.

Literary history was forgotten when I turned my attention to the wilderness that surrounded us. Soon after arriving and before falling ill, I walked into lush rain forest that reached all the way to the sandy beach, a rarity in the overpopulated tropics. It was home to undisturbed flocks of parrots and crowing jungle fowl, the wild ancestral species of the domestic chicken. Flying foxes, giant fruit-eating bats, flapped leisurely above the treetops. I soon fixed the affinities of the ant species I found there: Melanesian, as expected, Solomon Islands most likely, hence ultimately Asian. I made a general observation on the ecology of these insects that would find a place in my later synthesis of island evolution. It is as follows. Relatively few species of ants inhabit Espiritu Santo; the island is just too distant and geologically young to have received many immigrants. Freed from heavy competition, some of the colonists have dramatically increased their niche; they occur in dense populations across a wide range of local environments and nest sites. I would later call this phenomenon "ecological release," and help to establish it as an important early step in the proliferation of biodiversity.

Esperance to Mount Ragged, Western Australia, January–February 1955. I disliked leaving Espiritu Santo just when I had begun to study its fauna, but now I had to go to Australia for a potentially even more important excursion scheduled some months earlier. I took the weekly Qantas flying boat back to Noumea, then to Sydney and, after a brief stay in the city and a collecting excursion into the surrounding countryside, flew on to Kalgoorlie. From this inland center of Western Australia's sheep country, I proceeded south by rail to Norseman for a round of ant collecting. At a local bar I fell in with a group of construction workers, who invited me to collect ants out at their workplace in the nearby eucalyptus scrub. A full day in the bush completely dehydrated me; two months in the humid tropics had rendered my system unable to handle evaporation in such a hot, semidesert environment. When we arrived back at the bar late that afternoon, I chugalugged four beers in a row. My hosts, themselves heavy consumers in a country known for Olympic-class beer drinking, were impressed. So was I: I am normally a one-beer-maximum occasional drinker.

I then went farther south to Esperance, an isolated coastal town just west of the Great Australian Bight. Here I was joined by Caryl Haskins, a fellow entomologist and the newly appointed president of the Carnegie Institution of Washington. This outpost was our point of departure on a quest for the grail of ant studies. A hundred kilometers to the east, out across the sandplain heath, lived the grail: *Nothomyrmecia macrops*, the most primitive known ant, a lost species since its discovery twenty-three years before, and quite possibly the key to the origin of social life in ants. We meant to rediscover the species and be the first to study it in life.

Before departing, we decided to look for ants around Esperance. We walked out of the small town to the top of nearby Telegraph Hill, a low granitic rise covered by woody shrubs and patches of open gravelly soil ideal for ant nests. We stood quietly for a while admiring the long sweep of bush-covered land down to Esperance Bay, where huge combers thundered in from Antarctica. Out on the horizon were scattered islands of the uninhabited Recherche Archipelago. We had been told they were home to dense populations of poisonous elapine snakes. Great white sharks were known to be common in the blue-black waters. We were a long way from home, as far as it was possible to be from Boston, and Renee, and remain on land.

Telegraph Hill and its environs were strange and beautiful to behold, but not comfortable. January is the hottest month of the year. Just four days previously the temperature in Esperance had reached 41° Celsius (106°F). On the day of our excursion, the sun beat down from a nearly cloudless sky, and a hard dry wind blew in from the mainland semidesert behind us. Bush flies, aggressive relatives of house flies, swarmed around our heads, ran about on our ears and faces, and attempted to feed on the moisture of our eyes, nostrils, and mouths. We responded by continuously performing the "Australian salute," a wave of the hand around the head to chase bush flies away.

Caryl set out at once to collect colonies of bulldog ants, his favorite insects. This was no casual undertaking. The workers, measuring up to three centimeters in length, possess large bulging eyes with excellent vision, long saw-toothed mandibles, and painful stings. They are among the most belligerent insects in the world. Imagine a crater nest one to two meters across, with an opening in the center several centimeters wide, from which come and go dozens of surly red-and-black ants the size of hornets. Disturb them in the slightest and they charge you fearlessly. A few will follow your retreat for as much as ten meters from the nest. These ants, in short, are not the furtive picnic and kitchen raiders of America.

Caryl showed me how to gather entire bulldog ant colonies without risking one's life in the process. One needs a bit of courage and a willingness to endure pain. He went straight at the nest, snatching up each attacking ant closest to him and popping it into a large bottle—quickly, before it had a chance to curl its abdomen around and sting him. Usually the method worked, but occasionally a charging worker climbed onto his ankle or forearm and nailed him before he could brush it away. When all the outside guards had been cleared off, he started to dig into the entrance hole. More angry workers poured out, only to join their nestmates in the bottle. Caryl continued until he had excavated a shaft a meter or so below the surface where, in every case, he found the mother queen hiding in one of the deepest nest chambers. He walked away with each colony clean and healthy, ready for transport back to the United States and laboratory study.

The next day our thoughts turned entirely to *Nothomyr-mecia macrops*. The idea of the "missing link" ant is about as romantic a concept as is possible for an entomologist. The whole story began on December 7, 1931, when a holiday party set out by truck and horseback from Balladonia, a sheep ranch and beer stop on the cross-Australia highway northeast of Esperance. They traveled leisurely for 175 kilometers south-ward across the vast, uninhabited eucalyptus scrub forest and sandplain heath. In this first leg they passed close to Mount Ragged, a forbidding treeless granitic hill. Then they stopped for a few days at the abandoned Thomas River station on the coast before turning west to Esperance, where they took rail and automobile transportation back home. The habitat they traversed is botanically one of the richest in the world, har-boring large numbers of shrubs and herbaceous plants found nowhere else. A naturalist and artist resident at Balladonia, Mrs. A. E. Crocker, had asked members of the party to collect insects along the way. These they placed in jars of alcohol tied to the saddles of their horses. The specimens, including two large, oddly shaped yellow ants, were turned over to the National Museum of Victoria in Melbourne. There the ants were described by the entomologist John Clark in 1934 as a new genus and species, *Nothomyrmecia macrops.*

Our hopes were high as we left Esperance the next day, retracing the 1931 party's route in reverse. We were accompanied by the Australian naturalist Vincent Serventy and Bob Douglas, an Esperance native who served as camp manager and cook. We rode on the flatbed of a huge hand-cranked truck that had seen service on the Burma Road during the war. On the nearly invisible rutted dirt road to the Thomas River farm, we encountered not a single person. The sun bore down from the blue summer sky, from which bush flies descended in relentless swarms. When we stopped the only sound we heard was the wind whispering through sandheath shrubs.

We found the Thomas River to be a dry bed—an arroyo—in a basin depressed twenty-five to thirty meters below the level of the sandplain. Its floor had once been shaded by tall yate trees and carpeted by grass. Not long after their arrival in the 1890s, the first settlers had thinned the yate forest, and their flocks of sheep had destroyed the grass. Now, a half-century later, the groves were composed of a mix of yate, paperbark, and wattle, and the forage had been replaced by patches of succulent salt-tolerant herbs. Huge nests of meat ants, five to ten meters across and seething with hundreds of thousands of big red-and-black workers, dominated the more disturbed swaths of open terrain.

Nothomyrmecia could have been anywhere in such a varied environment. I was excited and tense, knowing that we might find scientific gold with a single glance to the ground. Haskins and I set to work immediately, each hoping to be the lucky discoverer. We searched back and forth through the basin grove, turning logs, scanning the tree trunks, inspecting every moving light-colored ant remotely resembling a *Nothomyrmecia*, but found nothing. We hiked up onto the sandplain heath and swept the low bushes back and forth with a net to capture foraging ants, again without success. That night, armed with flashlights and net, we walked back out onto the sandplain, and this time lost our way. Rather than risk wandering farther from camp in a dangerous desert-like environment, we settled down to wait for daybreak. To my surprise Caryl found a football-sized rock, pulled and rocked it as though positioning a pillow, lay on his back on the ground, and fell asleep. I was too keyed up to attempt the same feat and spent the rest of the night searching for the ant in the immediate vicinity. How marvelous it would be, I thought, if I could hand Caryl a specimen when he awoke!

But again, no luck. The four days we spent at the Thomas River station, broken by a side trip north to Mount Ragged, were a textbook introduction to wild Australia. Dingoes, the feral dogs of Australia, whined unseen around our camp at night. Kangaroos and emus could be seen moving across the sandplain at a distance during the day. One morning, while absorbed in close inspection of insects on the sandplain, we were startled by the sound of an animal snorting behind us. We turned to find a white stallion standing ten meters away, gazing placidly at us as though waiting to be saddled up. In a few moments he turned and trotted away. Returning to our work, we looked up now and then to locate him again until he passed out of sight in the distant gray-green heath.

Research progress was rapid and satisfying around the Thomas River, at least by ordinary standards of field biology. We discovered new species, in the course of which we also defined an entire ecological guild of sandplain ants specialized for foraging on the low vegetation at night. Large-eyed and light-colored, they represent members of the genera *Camponotus*, *Colobostruma*, and *Iridomyrmex* that have evidently converged in evolution to fill this arid niche. Because *Nothomyrmecia* is also large-eyed and pale, we reasoned that it was a member of the guild, and so we concentrated our efforts on the sandplain.

We never found *Nothomyrmecia*, but we made it famous. In the years to follow, other teams of Americans and Australians scoured the area with equal lack of success. The ant acquired a near-legendary status in natural history circles. The break finally came in 1977 when Robert Taylor, a former Ph.D.

student of mine at Harvard and at that time chief curator of the Australian National Insect Collection, stumbled upon *Nothomyrmecia* in eucalyptus scrub forest near the little town of Poochera, in South Australia, a full thousand miles east of the Thomas River. It was a totally unanticipated discovery. Taylor came running into camp shouting (his exact words) in pure Australianese, "The bloody bastard's here! I've got the *Notho*-bloody-*myrmecia*!"

A small industry then grew up among ant specialists, who studied every aspect of the ant's biology. Many visited the Poochera site. The details supported a theory, originally promoted by William Morton Wheeler while professor of entomology at Harvard and furthered by Haskins, that social life in ants began when subordinate daughters remained in the nest to assist their mother in rearing more sisters. At that point in distant geological time, according to this now strengthened scenario, solitary wasps evolved into ants.

Brown River Camp, Papua, March 1955. After taking the train back to Kalgoorlie I flew to Perth, then to Sydney, and on to New Guinea. The great island was to be the "ultimate" tropics and the climax of my odyssey. Immediately after my arrival in Port Moresby in a Qantas DC-3, I contacted Joseph Szent-Ivany, a Hungarian refugee from the postwar communist takeover and the Mandated Territory's only resident entomologist. We called on G. A. V. Stanley, a longtime resident and planter known as "bush pig" by the natives for his exploits as a civilian scout for the Allied forces during the war. Both men were expert field travelers in New Guinea. After filling me with advice and a couple of good dinners, they accompanied me while I set up a camp near the Brown River, a tributary of the Laloki near Port Moresby. Thanks to their selfless assistance, I was able within five days to commence work in primary rain forest. My little expedition was supported by a native cook, a driver, and a personal assistant. I had no trouble supporting this crew, despite my own impecunious state as a mere postdoctoral fellow far from home. Each man's salary was thirty-three cents U.S. a day plus rations; this rate was the going standard, and Szent-Ivany and Stanley cautioned me not to exceed it.

Our tents stood in a small clearing ringed by giant trees whose finlike plank buttresses gave their trunks the appearance of rockets poised for flight. More than thirty meters above us, the dense canopy, festooned with lianas and epiphytes, closed out most of the sunlight. Only a few shafts cut to the floor through scattered breaks caused by naturally falling trees and branches.

I was immersed in a pandemonium of life. The racket made by parrots and other birds, frogs, and singing insects beat incessantly upon my ears—a cacophony but one of great eloquence so long as I listened to the separate instruments and not the orchestra all at once. Less congenial were the mosquitoes, gnats, and stingless bees that hummed around my head in merciless attendance. The body fluids paid these pests are the expected tariff for tropical field work. But I was where I most wanted to be in all the world; I had no complaint.

Tree trunks, lianas, and rotting logs teemed with thousands of species of insects. I scurried about continuously through the days and on into the nights, closely followed by my assistant, who rapidly turned into an enthusiastic amateur entomologist. Together we gathered more than fifty species of ants, many of them new to science. During all that time my eyes were fixed on the ground and lower vegetation. Rarely did I look upward, and then only to watch giant birdwing butterflies churning the air, or flocks of parrots rushing back and forth through the canopy, a riot of different species, first one brown in plumage, then another green, then another yellow. I heard a bird of paradise calling once, but stared skyward too late to find it. I never saw one during my four-month visit to New Guinea, although I must have passed close to a variety of species many times. My eyes were locked to the ground, head bent and shoulders hunched in the lifetime posture I had already acquired by my late teens. At dusk we dined on wallaby, wild pigeon, and, for appetizers, nut-flavored grubs of long-horned beetles dug from nearby decaying logs and roasted like marshmallows on sticks over the fire.

In New Guinea I felt like a real explorer. I *was* an explorer, at least in the world of entomology. Shortly after returning from the Brown River and then a second, week-long trip to the foothills rain forest of the Sogeri Plateau (where I discovered an extraordinary new kind of ant living as a social parasite in the nest of another species), I drove with Joe Szent-Ivany to the Port Moresby airport to meet Linsley Gressitt. The great entomologist was soon to turn Hawaii's Bishop Museum into a world center for research on Pacific insects. He was rightly to be known as the pioneering expert on insect biodiversity in this region. He arrived on that day for his first visit to New Guinea. I had beaten him there by two weeks, and I think back now with pride on the advice I gave him on local collecting.

The Huon Peninsula, Northeast New Guinea, April 1955. The Huon Peninsula is the horn of New Guinea, a mountainous extension of the northeastern corner of the island that projects into the Solomon Sea. Its spine is the Sarawaget Range, which runs most of the length of the peninsula before dividing eastward into the Rawlinson and Cromwell Mountains. At its tip, below the foothills of these satellite spurs, is the little coastal town of Finschhafen, where I arrived on April 3 to begin the greatest physical adventure of my life.

I had been invited there by Bob Curtis, an Australian patrol officer, to accompany him on a government-sponsored tour into the mountainous interior. His mission was an official visit to villages in the Hube country as far west as the Sarawaget highlands. He would consult with the village chiefs, settle disputes within and between villages, offer advice on agriculture, and if possible capture and return two suspected murderers for trial in Finschhafen. He anticipated no special risks, the murderers notwithstanding, but a great many unknowns did lie ahead. Although messages were carried regularly by natives back and forth between the coast and mountains, the villages of the region were visited by patrol officers at intervals of only one to two years. Some had not been contacted since 1952.

The most remarkable aspect of the patrol may have been Bob Curtis' age. He was twenty-three, and would turn twenty-four on April 19, near the end of the trip. As he reviewed the itinerary with me, he seemed as self-possessed and competent as a man twice his age. Curtis was blond, powerfully muscled, and possessed of movie-star good looks. I was reminded that a quarter-century earlier, Errol Flynn, up from Tasmania, had also started his career as a New Guinea patrol officer. Prior to his employment on the Huon Peninsula Bob had played semiprofessional rugby in Australia, the distinctive Australian rough-and-tumble kind, and had lost his upper front teeth in the process. He now wore a bridge. Any concern he may have felt about the patrol he kept to himself.

He seemed above all delighted to have my company. I was even happier. I had nothing to do but tag along, collecting and studying ants in a remote mountainous region that had never been visited by an entomologist. Curtis proposed as a bonus that we also climb to 3,600 meters on the central range, where the summits and passes rise in near-freezing cold to a treeless grassland. Natives had died there trying to cross from one side to another, and rumor had it that Japanese soldiers had also perished in the cold after being driven inland from Finschhafen by Australian forces in 1944. Were there mummified bodies still here, we wondered, preserved like the fabled leopard atop Mount Kilimanjaro? I didn't expect to find any ants so high, but looked forward to the climb with enthusiasm. It would be a difficult and possibly risky trek. Looking back later, I wondered whether Bob and I were bold nineteenth-century adventurers or just a pair of excited kids having fun. Almost certainly, both.

We left Finschhafen at the head of forty-seven bearers, three camp assistants, and a uniformed police boy with rifle. Boy: that is what a native policeman was still called in the days of waning colonial rule. Other men were assigned their various tasks as cook boy, hunt boy, and so on. Aside from Curtis' platoon-sergeant manner, however, we all treated one another with courtesy, and once on the trail we were effectively close to being equals. Perhaps it would have been dangerous to act otherwise. These people still belonged to a culture whose recent hallmark was war and the collecting of blood debts.

The carriers changed at successive villages along the way, and each man was paid between twenty-five and fifty cents U.S. for a day's work. On one leg of the trip, when too few adult males were available, Curtis recruited women and children to fill the gap.

The land that we entered, as we pushed north across the Mape River and then westward into Hube country, was sparsely populated. Most of the terrain was covered by rain forest, with little evidence of disturbance except for cultivated fields close to the settlements. The villages were four to seven hours' walk apart. Each settlement contained several hundred men, women, and children.

Because the villages were invariably set on mountain ridges, we almost never walked on level ground. On a typical day we set out at around nine in the morning on narrow trails that descended tortuously through as much as 1,000 meters of elevation. At the bottom we crossed white-water rivers on a suspension bridge, in some cases no wider than the span of a hand with a bamboo railing added above for balance. From there we zigzagged uphill for a similar distance to the next village. The paths were muddy and slick, often widening into almost impassable traps of deep black mire resembling pig wallows. Their steep, slippery banks were surmounted by dense undergrowth that made off-trail travel nearly impossible. Most of the time we were forced to walk in single file.

These treks were a great deal harder on the bearers than on Curtis and me, or on the native policeman, who was responsible only for his rifle and a few personal belongings. Each man handled twenty kilograms or more of cargo. Most of the neatly packed pieces were supported by a headband and slung over the back or else suspended from a bamboo pole carried in tandem by two of the porters. But the men were almost invariably cheerful. Hard climbing with heavy loads is a part of daily life for mountain Papuans. Even though I was in excellent physical condition at this time, they were far tougher, especially at the higher elevations. I came to suspect that they were also genetically better adapted to the harsh physical conditions in which they lived. I wondered: Would they make good marathoners?

The men were also serenely indifferent to the land leeches that attacked their bare feet and legs during the long walks. At intervals they stopped to pull the engorged worms off with about the same casualness one reties loosened shoe laces. After passing through heavily infested woodland their skin was often streaked with drying rivulets of blood. I never saw evidence that any suffered illness from these attacks.

In researching the trip for this memoir, I pored over the most detailed and updated maps I could find of the region, principally official topographic charts drawn from patrol reports and 1973 stereoscopic aerial photographs. I was able to locate only about half of the villages we visited in 1955, whose names I recorded in my notebooks: Mararuo, Boingbongen, Nanduo (or Nganduo), Yunzain, Homohang, Joangen, Zinzingu,

Buru, Gemeheng, Zengaru, Tumnang, Ebabaang, Wamuki, Sambeang, and Butala. Had some of the settlements already been abandoned? In 1955 I took photographs of some of them for future archives.

The deeper we penetrated into the Hube country, the more excited and pleased the people were to see us. At Zinzingu we were honored by a sing-sing, an intricately choreographed program of dances and songs, lasting a large part of the day. At Gemeheng the *luluai* (chief) organized an archery contest, with meter-long bamboo arrows fired from black palm bows at banana-stalk targets ten paces away. I took one shot that passed a meter to the side, and braced myself for laughter. Instead, I heard the man next to me grunt, "Him all right," which probably meant, Not bad for someone who doesn't have to work for a living.

To the Hube people, Curtis and I were curiosities of the first order. In several of the remotest villages women and children ran and hid until we had settled in the thatched guest house, then quietly filtered back home. Throughout the day crowds of people stood outside our lodging, watching our every move with open curiosity. Curtis remained at ease with all of them, conversing smoothly in pidgin English. Once, at Joangen, we put on the brief equivalent of a magic show, mostly for the children. Standing in front of the house entrance, Curtis pulled out the bridge of his front teeth and held them up. I turned around, put my glasses on the back of my head and pointed backward at the crowd as though I could see them from behind. We were both met with gasps of amazement. One child broke out crying, and I decided not to try the trick again. I felt more like a fraud than an entertainer. And who knows what taboos we broke, or ghosts summoned?

At every opportunity I collected ants. I would rush ahead a few hundred meters, work for a few minutes, and fall in again as the others came by, or linger behind the bearers to work promising patches of forests and clearings, then walk fast or jog to catch up. I ventured from the villages at trail's end to work nearby habitats in the fading afternoon light. When we stayed in one place for two or three days, as at Gemeheng and Ebabaang, I left on more distant excursions. My worst problem was the weather: on most days afternoon mist and sporadic light rain settled throughout the mountain country. Most of the time, day and night, the temperature swung between a chilly 10° and barely comfortable 20°C. Only near midday in open sunny ground did it rise to the upper twenties, and it was under these conditions that ants were most numerous and active.

I was also hampered by the oppressively curious and helpful Hube people. Because I spent so much time in the forest, they called me "bush man." In the early days I was accompanied by mobs of boys of all ages and a few men, who pressed in so closely to see what I was doing that I could barely work. When they understood my purpose, they began to hunt for me, with the result that all the nearby logs and stumps were quickly torn to pieces with little result. I started telling them, as politely as I could, to stay away. But on several occasions I asked my retinue to collect spiders, frogs, and lizards and to ignore the ants and other smaller insects I was picking up myself. They scattered in all directions and in short order filled all the spare bottles I carried. I remember especially one boy, about twelve years old, who ran up to me clutching a giant silk spider in one hand, its fangs gnawing at a callused pad on his thumb. Grinning widely, he held it up for me to take. I am a bit of an arachnophobe, and for an instant I panicked. Then I grinned widely myself and held out my open lunch bag to receive the monster.

To my relief, we never caught the murderers. We also failed to make it to the high Sarawaget. At Buru a runner from Finschhafen brought Curtis a message: he was to report to Port Moresby as soon as possible to be interviewed for another, better-paying position. The next day we turned southward along the Bulum River, away from the Sarawaget and toward Butala on the coast, where a truck had been reserved to carry us the rest of the way to Finschhafen. As I walked homeward one day during this final leg I experienced an insight about the diversity of tropical ants. All through my stay in New Guinea I had studied long stretches of relatively undisturbed rain forest, first at the Brown River, then at the Busu River near Lae, and now partly along the length of the Huon Peninsula. I had paid close attention to the identity and relative abundance of all the ants I could find, writing notes on every colony. I noticed that although the forest seemed to change very little in outward appearance from one kilometer to the next at the same elevation, the composition of the ant fauna usually shifted in a striking manner. It was possible to find, say, fifty species in one hectare, and another fifty species in a second hectare a short distance away, but at most only thirty or forty of the same species occurred in both places. Some of this variation was due to a local change in the physical properties of the habitat: the second hectare, for example, might contain a small sago palm slough or a clearing caused by a falling tree. That is the kind of easily understood change ecologists now call beta diversity—variety in species based on local differences in habitat. But much of the shift could not be so easily explained. It represented what is today called gamma diversity, the changeover in species with growing geographic distance.

The pattern I observed was very different from that in temperate forests, where gamma diversity occurs to the same extent only over tens or hundreds of kilometers. I had discovered something new about the structure of ant faunas in the tropics, and perhaps about the origin of the fabulous diversity of rain forest faunas and floras as well. In 1958 I concluded a formal article on the subject as follows:

> *In any appraisal of comparative ecology, the New Guinea ant fauna is first of all to be characterized by the exceptional richness of its species and the great size of its biomass . . . In addition to sheer size, an additional factor adds greatly to the total faunal complexity. This is the discordant patchy distribution of individual species . . . As a result of discordant patchiness, no two localities harbor exactly the same fauna. Considering that several hundreds of species are thus involved, it is clear that the spatiotemporal structure of the entire New Guinea fauna must present the appearance of a great kaleidoscope. The effects of such a structure on the evolution of individual species of ants, as well as of other kinds of animals, must be considerable. It very possibly hastens the genetic divergence of local populations and plays an important role in the "exuberance" and amplitude that characterize evolution in the tropics.**

---

*"Patchy Distributions of Ant Species in New Guinea Rain Forests," *Psyche* 65(1) (1958): 26–38.

I later learned that André Aubreville and Reginald Ernest Moreau had earlier and independently noted similar patchiness in African rain forest trees and birds. So patchiness is a general phenomenon, as I had hoped in 1955. My own insight was to be an early step in the development of my theory of the taxon cycle and, later, in collaboration with Robert MacArthur, the theory of island biogeography. Most important, it fixed my attention on biological diversity as a subject worthy of study in its own right.

I felt gratified—indeed, exuberant—that I had discerned what appeared to be a broad ecological pattern from my undisciplined collections and journals. But this was the way it is supposed to be. Nature first, then theory. Or, better, Nature and theory closely intertwined while you throw all your intellectual capital at the subject. Love the organisms for themselves first, then strain for general explanations, and, with good fortune, discoveries will follow. If they don't, the love and the pleasure will have been enough.

This insight came to me at Wamuki, which faces southward on the mountain ridge separating the Bulum Valley on the west from the Mongi Valley on the east. A day's walk farther south, the two rivers converge to form the greater Mongi, which runs on to the sea at Butala. As I strolled back at dusk one day at the end of one of my final excursions, I watched the clouds clear over the entire Bulum Valley below me. I could then see unbroken forest rolling down to the river and beyond for fifteen kilometers to the lower slopes of the Rawlinson Range. All that domain was bathed in an aquamarine haze, whose filtered light turned the valley into what seemed to be a vast ocean pool. At the river's edge 300 meters below, a flock of sulfur-crested cockatoos circled in lazy flight over the treetops like brilliant white fish following bottom currents. Their cries and the faint roar of the distant river were the only sounds I could hear. My tenuous thoughts on evolution, about which I had felt such enthusiasm, were diminished in the presence of sublimity. I could remember the command on the fourth day of Creation, "Let the waters teem with countless living creatures, and let birds fly above the earth across the vault of heaven."

Central summit ridge, Sarawaget Mountains, May 1955. When I returned to Lae, the administrative center of northeastern New Guinea, at the coastal base of the Huon Peninsula, my mind stayed on the Sarawaget Mountains. Standing in the main street on a clear morning, I could look north all the way to the blue-gray ridge of the center of the range. I learned that no European had ever climbed to the top of this main part of the mountain crest, which thrust up in such plain sight. Eight men, including the ornithologist Ernst Mayr, had independently reached the eastern end of the Sarawaget out of Finschhafen along the route partly traveled by Bob Curtis and me, but none had pushed on west toward the center ridge. In 1955 this lack of exploration did not seem very surprising: the Lae area had been settled only in the 1920s, and when I was there the population of planters, lumbermen, and colonial officials was still small; clearly, they had other things to do.

I hungered to get to the top of the Sarawaget. I was excited by the possibility of being the first white to visit the center of the crest. The problem was how to get there. Hearing that there was a Lutheran mission at Boana, halfway up the Bunbok Valley between Lae and the summit ridge, I inquired about Boana at the Agricultural Service office and soon received word that the resident minister, Reverend G. Bergmann, would be glad to have me as a guest and provide native help if I decided to attempt the climb. Bergmann was one of the eight Europeans to have climbed to the eastern end. He believed that it would not be too difficult to reach the middle as well.

On May 3 I walked into the office of Crowley Airways, the main transportation link to the Bunbok Valley. The president and sole full-time employee, Mr. L. Crowley, was seated behind a battered wooden desk. He rose, we shook hands, he shuffled some papers, and I paid the round-trip fare to Boana, four pounds and ten shillings Australian. A few days later Crowley and I walked across the tarmac to his 1929 biplane for the weekly flight to Boana. He stepped into the forward cockpit and I into the rear passenger cockpit, and we took off for the Bunbok Valley. I enjoyed the open-air, low-altitude view while peering over the edge of the cockpit. I also noticed that the double wings on each side waved up and down slightly throughout the flight. I supposed—hoped may be the better word—that the birdlike movement was a normal part of the airplane's aerodynamics.

The approach to the Boana was tricky. The mission was perched on a mountain spur to the east of the valley, and the airstrip could be reached only from the tributary valley by a southward approach. So we flew north along the river, mountains rising on both sides, turned right at the tributary, then immediately turned right again, now pointed south, and there, coming up fast, were the mountain ridge and, plastered on its side, the airstrip. As we glided in for a landing on the grassy surface, I saw the second aircraft of the Crowley fleet, also a 1929 biplane, this one recently crashed and still nose down on the edge of the airstrip.

In the late morning the Sarawaget crest, blue-gray against a lowering cloud bank, seemed close to Boana, a day's walk perhaps, and I ached to get started. It was in fact five days' walk away. I departed two days later, accompanied by a half-dozen young Papuan men hired as guides and carriers. The trip proved to be the physically most demanding of my life. It beggared even the hardest parts of the just-finished Sarawaget patrol. We reached the village of Bandang the first day, then proceeded up into virtually untracked country in daily marches lasting five to seven hours. Seldom were we able to travel for more than a hundred meters in a straight line on level ground. We wove, stumbled, waded, climbed, and sometimes just crawled our way, following stream banks, tracing animal trails up and along the crests of ridges, down into stream valleys and then up again. To my dismay my guides occasionally became lost, and we had to wait while one or two left to reconnoiter higher terrain and reestablish our position.

Our little party was almost constantly wet from intermittent rain, which set in predictably by early afternoon and continued into the early evening. Our clothing was stained by the soaked and puddled moss-grown earth over which we struggled. Beyond 2,100 meters, the temperature dropped to below 15°C at night and never rose as high as 20° during the day. Land leeches were everywhere—big, aggressive, and black, quickly swelling to half the size of a thumb with fresh blood—and periodically we stopped to take them off our legs and feet. To sit on a muddy stream bank in near exhaustion, pull off boots and peel down socks, and burn free a half-dozen engorged leeches, then watch blood trickle down from the bite wounds: that is an experience best savored after a few years have passed.

I was afraid, at times, of a crippling accident, of the possible unreliability of my assistants, with whom I could communicate very little, but most of all of the inexpressible unknown. Would I fail from physical incapacity or lack of will? Would I have to turn back as I did in Mexico, short of the Orizaba snowfield? Why had I come here anyway, except to be able to say I was the first white man to climb the central Sarawaget? That prideful goal was part of the truth, but I was looking for something more. I wanted the unique experience of being the first naturalist to walk on the alpine savanna of this part of the Sarawaget crest and collect animals there. And I wanted release from the mountain-peak compulsion that gripped me. I decided to keep going, even if I crawled or had to be carried.

Slogging on, we began to pass from the midmountain rain forest into mossy forest at about 2,000 meters. Here ferns, orchids, and other epiphytes thickly encrusted the branches and trunks of low gnarled trees. By 3,000 meters moss flowed like a continuous carpet from the trunks onto the forest floor. The canopy was also low here, only five meters above the ground, so in places progress was restricted to wide tree-tunnels with mossy floors, walls, and roofs. Then at 3,200 meters, as we worked up at last onto the central ridge, the moss forest began to give way to scattered patches of *Eugenia* shrubs and alpine grasses.

Early on the fifth day, before the inevitable clouds descended and cold rain began to fall, we walked the final two hours to the summit. Here, at 3,600 meters, we were in a savanna composed of tall grass sprinkled with cycads, squat gymnospermous plants that resemble palm trees and date from the Mesozoic Era; a very similar scene might have existed here 100 million years before. The ground was mostly a mountaintop bog difficult to navigate. I made it on up to the nearest high point, sat down, wrote my name and the date on a scrap of journal paper, put the memento in a tightly capped bottle, then buried the bottle beneath a cairn of small rocks. From that position I could look south all the way down to the grasslands of the

Markham Valley and the more distant Herzog Mountains, and north to the Bismarck Sea. Zigzagging back down the savanna, I collected every insect I could find, as well as small frogs that turned out to be a new species, while my companions hunted alpine wallabies with bows and arrows and dogs.

Then we turned back and began the two-day return trip to Boana, a far easier trek than the climb up. As we fast-walked and slid our way down the ridges and into the upper Bunbok Valley, something resistant and troubling finally broke and receded inside me. The Sarawaget, cold and daunting, had proved a test of my will severe enough to be satisfying. I had reached the edge of the world I wanted, and knew myself better as a result. By passing from the sea to its peaks I had finally encompassed the serious tropics of my dreams, and I could go home.

CHAPTER ELEVEN

The Forms of Things Unknown

F ROM NEW GUINEA I continued my journey westward on around the world. After pausing for a week of field work in the rain forests of Queensland, I boarded an Italian liner in Sydney that carried me south around the Victoria coast and then west along the Bight to Perth. From this most distant city on Earth from my home, the ship churned slowly north across the Indian Ocean toward its eventual destination of Europe. I got off at Ceylon, now the republic of Sri Lanka—in any case the island "Pearl of Asia" that hangs like a teardrop from the tip of India.

I traveled inland from the port of Colombo to search for one of the rarest ants in the world. *Aneuretus simoni* is the apparent evolutionary link between two of the great worldwide groups of ants, the Myrmicinae and Dolichoderinae. Fifty million years ago the group to which it belongs, the Aneuretinae, abounded throughout the northern half of the world. Now only one species remains, the endangered *Aneuretus simoni*. I began my quest in the Botanic Gardens at Peradeniya, where the only specimens in museums had been collected around 1890.

I hoped to find *Aneuretus* at the original find site, near the center of the island. No luck; the place had been shorn of native vegetation. For three more days I worked without result in the forest of the neighboring Udawaddatekele Sanctuary, close to Dalada Maligava, the temple holding a giant tooth said to be that of Buddha. Unrewarded by His blessed aura, I took a bus south to the gem center of Ratnapura. There, somewhere in the remnants of rain forest scattered along the road to Adam's Peak, Sri Lanka's highest elevation, the prize might be found. I checked into the government rest house, eager to get started. Dropped my Army-issue duffel bag in my room. Swung my stained canvas collector's bag over my shoulder. Walked down the stairs and out the back entrance a hundred meters to a line of trees fringing the town reservoir. Looked around. Picked up a dead twig lying on the ground, broke it open, and stared as a colony of small, yellow ants ran out over my hand. *Aneuretus!*

I could not have been happier if I had discovered a priceless Ratnapura sapphire lying unclaimed on the ground. Settled back in my room, I turned the vial of specimens slowly over and over in my hand, looking at the first queen, larvae, and soldiers of the living aneuretine ever seen (the 1890 specimens had all belonged to the worker caste). This was one of the great thrills of my life. Dinner tasted fit for a gourmet that evening, and afterward, sleep came easily.

In the days that followed, I ventured into forest closer to Adam's Peak. Though sometimes delayed for hours by monsoon downpours, the kind called gully-washers or frog-stranglers back home, I easily secured more colonies, finding the *Aneuretus* in one locality to be among the most common ants. In a short while I was able to put together a picture of the social life of the last surviving aneuretine. Twenty years later one of my undergraduate students, Anula Jayasuriya, a native Sri Lankan, found the species rare or absent in the same localities. I recommended placement of *Aneuretus simoni* in the Red Data Book of the International Union for Conservation of Nature and Natural Resources, and in time it became one of the first of several ants to be officially classified as a threatened or endangered species.

My field adventure was now finished. I continued on by a second Italian liner (cheap fare) to Genoa, where I worked on the ant collection of Carlo Emery at the Museo Civico di Storia Naturale. Then I proceeded by train across Switzerland and France and finally to London, visiting other ant collections in museums along the way. That was my European tour as a young academic: the great ant collections of the world. When others at Harvard spoke of their experiences at Hagia Sophia and the Prado, I reminisced about the wondrous ants I examined in Geneva and Paris.

On September 5, 1955, I flew to New York. The last four hours up to Boston by train, eased only slightly by an anesthetic reading and rereading of a copy of that week's *Life* magazine, were the longest of my life. Finally, clad in khaki and heavy boots, crew-cut, twenty pounds underweight, and tinted faint yellow from the antimalarial drug quinacrine, I fell into Renee's arms. I came home innocent of hula hoops, Davy Crockett, Tommy Manville's ninth divorce, and other Western Hemispheric events of 1955. I had not heard Vice President Richard Nixon say, "Sincerity is the quality that comes through on television." I was ignorant of the latest in men's leisure clothing, which, should you wish to learn or be reminded, consisted of cotton jersey pullovers, bateau necklines, and moccasin slip-ons worn in chaste combinations to create a European effect. American culture had begun to fade for me around the edges in the ten months of my absence. I began quickly to recover, however, with the help of ten-cent hamburgers at the White Tower and hours of serious television watching.

Six weeks later Renee and I were married in Boston's St. Cecilia Church. We moved to Holden Green, university housing on the Cambridge-Somerville line that seemed to be the starting point of most young Harvard couples. There were young Holden Greeners on their way up and out and older Holden Greeners who apparently wanted to stay there and at Harvard forever.

The following winter I was offered an assistant professorship in Harvard's Department of Biology. My main function would be to assist William H. ("Cap") Weston, an elderly professor and fungus expert, in the creation of a new beginning biology course for nonscience students. The department chairman, Frank Carpenter, who had previously served as my doctoral adviser, cautioned me that the appointment would be for five years only. After that I had a distant chance of further employment, but the position was not tenure track. Would I be interested just the same? I had already sent letters of inquiry to thirty other prospective employers and gleaned offers from the University of Florida and the University of Michigan, both tenure track. Then, as now for most colleges and universities, this meant that all I had to do was perform well for the first few years to be guaranteed a lifetime position.

The impermanence of the Harvard position did not faze me. I was young, only twenty-six, and wanted the added time there to settle in a world-class collection and library and to expand my research program. I accepted Harvard's offer and set out to plan my first lectures.

Partway into the first year, however, my nerve began to fail. Like all assistant professors at the great university, I felt disposable. And obviously, I *was* disposable. With Renee's assistance,

I made plans to find a new position long before the end of my five-year term put me on the street. Then providentially, in the spring of 1958, with more than three years to go, Stanford University offered me an associate professorship with lifetime tenure as part of the package. The invitation came out of the blue in a letter from Victor Twitty, chairman of the Department of Biology, flat and definitive, prefaced by no tentative inquiry, and conditional upon no visit and seminar. Twitty said in effect: Here it is; will you join us?

Soon afterward Frederick Terman, dean of the Stanford faculty, came to visit me in my office in the Biological Laboratories. He was accompanied by an older gentleman whom he introduced as Wallace Sterling.

When the two were seated, I turned to Sterling. "Are you with Stanford, too?"

Terman answered for him: "Yes, he's the president."

Leaving me time to regain my composure, the two men went on smoothly to explain that the biology faculty hoped that I would come to Palo Alto to build a new program in entomology. The incumbent professor, a specialist on scale insects, was retiring. I listened with intense excitement. Stanford was the Harvard of the West, and California was the golden state of the 1950s. Come to this land of opportunity, they said, and help us to grow. I knew that others were responding to the call. Earlier that year *Time* had reported a westward surge of scientists and other academics from the older eastern universities.

Renee and I were thrilled by Stanford's commitment. I was really wanted for my special skills in entomology and for my myrmecophilia, the love of ants! The salary was also good for the times, $7,500 a year, and Stanford would assist us in buying a house, a policy unheard of at Harvard. The next morning I told Carpenter that I was going to Stanford. Thank you, I said, for all you have done. He said, Wait a few weeks before making a firm commitment. Let's see what Harvard can do. Over the next two months the biology faculty and McGeorge Bundy, dean of the Faculty of Arts and Sciences, moved to review my status and decide whether to match the offer from Stanford. As I waited I tossed the matter around in my mind. To the present day this is how Harvard makes most offers of tenure to its own young faculty. It reacts to outside threat, and even then turns down a majority of the hopefuls. The process seems more ponderous than anything west of the Vatican, usually taking a year or more to complete. But I was favored by an accelerated schedule. I received the offer from Mac Bundy, and I decided.

Nowadays on freezing January mornings in Cambridge, as I pick my way across Kandinsky landscapes of snow painted with automobile smudge and dog urine, I remind myself that the New England winter is a hard but fair price to pay to work closely with the best collection of ants in the world. Thirty-five years after accepting Harvard's tenure offer, having reached the same age as Cap Weston when he walked with me into the lecture room of Allston Burr Hall that first September morning, I still teach a large class in beginning biology for nonmajors. My contentment with this repetitive schedule does not rise from the comfortable stagnation of the tenured academic. Rather, it exists because I find Harvard undergraduates wonderfully talented, and because each year's contact with them renews me. Most share my own restlessness and optimistic rationalism. We work each other up into enthusiastic conversation. The nonscientists in particular are a prime investment. I know they will carry with them into great ventures, in law, government, business, and art, the commitments they first acquire in the university's hothouse environment, and a few (it has happened) will convert to biology. I speak to these students as intellectual equals, keeping in the back of my mind their prospective if not immediate attainment of that status. In 1992 the Committee on Undergraduate Education, consisting of undergraduate students, awarded me the Levenson Prize, as outstanding tenured teacher in the college. But I have another, more selfish reason for lecturing on biology to nonscientists. The bourgeois life of the college teacher, if one's schedule is not too crowded, frees the mind for creative work.

By 1958 I had temporarily forsaken field biology to press research in the laboratory and museum. My central aim was the classification and analysis of the ants of New Guinea and the surrounding regions of tropical Asia, Australia, and the South Pacific. I had embarked on a bread-and-butter task, of a largely descriptive nature. It was time-consuming, tedious, fact-centered, and, for this combination of reasons, virtuous in my own mind.

So let me digress for a moment to explain the special satisfaction of taxonomy. It is a craft and a body of knowledge that builds in the head of a biologist only through years of monkish labor. The taxonomist enjoys the status of mechanic and engineer among biologists. He knows that without the expert knowledge accumulated through his brand of specialized study, much of biological research would soon come to a halt. Only a specialist expert enough to recognize the species chosen for study ("Ah, that is a carabid beetle of the genus *Scarites*") can unlock all that is already known about it in the literature. From journal pages and museum specimens he is able to go promptly from the already discovered to the exhilarating unknown. If a biologist does not have the name of the species, he is lost. As the Chinese say, the first step to wisdom is getting the right name.

There is much more. A skilled taxonomist is not just a museum labeler. He is a world authority, often *the* world authority since there are so few taxonomists, on the group he has chosen. He is steward and spokesman for a hundred, or a thousand, species. Other scientists come to him to seek entry to his taxon—sharks, rotifers, soldier flies, weevils, conifers, dinoflagellates, cyanobacteria, and so on down the long roster comprising over a million species. He knows not only the classification but also the anatomy, physiology, behavior, biogeography, and evolutionary history of the group, in fine detail both published and unpublished. In conversation he will speak as follows: "Come to think of it, there *is* an enchytraeid I ran into in Honduras with a reddish color, and that just might be the invertebrate hemoglobin you're looking for." Or, "No, no, the main center of that particular moth family is the temperate forests of southern Chile. Those species haven't been worked on yet, but there is a big collection in the National Museum made by the Hensley expedition in 1923. Let's check it out." No CD-ROM, no encyclopedia can replace the taxonomic expert. Once, after receiving an award in Japan for such studies, I had the additional honor of spending an evening in conversation with Emperor Akihito, a noted specialist on the taxonomy of gobiid fishes. Soon I fell into a comfortable routine of listening to him speak about gobies and Japan's endangered fish species, while he and his family asked me questions about ants.

It was like a Harvard seminar. At times I almost (but never completely) forgot with whom I was speaking.

In 1958, sitting in my office on the first floor of the Biological Laboratories, occasionally glancing out the window past the monumental bronze statues of the Indian white rhinoceros, I felt on temporarily safe professional ground returning to this kind of enterprise. It guaranteed a stream of tangible results, the kind for which grants-in-aid and other professional emoluments are awarded. At the age of twenty-nine, I had fifty-five technical articles published or in press. Being thoroughly professional in attitude by that time, I knew that every young scientist needs such proof of productivity. Otherwise the National Science Foundation and J. S. Guggenheim Selection Committee will wave his grant applications aside. But if truly creative he does not always hug the coast. He gambles repeatedly on risky projects, stays alert and aggressive, ready to move whenever a long shot shows a hint of promise.

What, then, were my gambles? Highly diverse in nature, they came to me as unplanned products of pedestrian daily research. During my accumulation of facts about ant biology, vaporous notions—constructs, definitions, inchoate patterns (the perfect phrase escapes me)—drifted in and out of my mind like Celtic fog. My daydreams were mostly about the origins of biological diversity. Most took coherent form, only to prove marginal or unattractive, then to fade and disappear. A few went on to gain robust life in the course of my daily reverie. They then turned into narratives, which I began to repeat to myself like stories. I prepared to speak about the matter to others. I imagined how the narrative would look in print, how it might sound in a lecture before a skeptical audience. I rehearsed, edited, and performed in silence. I was a storyteller, sorting and arranging pieces of nonfiction, dreaming in order to fill in the gaps. Then I tried the performance before a real audience.

One of my first constructions was a critique of subspecies, the formal category of race used universally in biological classifications. My coauthor in this endeavor was William L. Brown, seven years my senior, who had enticed me to come to Harvard as a graduate student. During 1952 we met almost daily at lunch to gossip and mull over issues in evolutionary biology. Brown, I soon learned from his pungent remarks, was

a scientific curmudgeon. He seemed happiest when he could
sprinkle doubt on the reputation of a reigning academic pooh-
bah. He tended to divide all scientific ideas into two piles:
those he embraced passionately and those he ridiculed. Passionate he was (and remains), but also thoroughly professional.
And proletarian in spirit, a hater of pomposity and pretense.
Grinning impishly, he would hold up an imaginary "phony
meter" when certain faculty notables walked nearby, and take
a reading: Red zone! Off the scale! In other lives he would
have been the first sergeant who wisecracks about the foibles
of the company commander, or the engineer, oil-streaked in
the bowels of the plant, who makes up for and grumbles about
the incompetence of management. He enjoyed having a beer,
or two or three, with working-class stiffs at a nearby bar after a
day in the Museum of Comparative Zoology. He was annoyed
with me for not joining him. "Can't completely trust someone
who doesn't like beer." It seems never to have occurred to
him that he was himself a member of the ruling class. But no
matter; his animadversions were usually on target. Management *was* incompetent, a lot of the time. This year, as we began
collaborating, he had been roused by the subspecies.

It was a subject deserving close inspection. Everywhere taxonomists were treating the subspecies as an objective category
and one of the key steps of evolution. Consider their logic:
species are divided into subspecies, which we must assume to
be real and objective because given enough time they evolve
into species, which are real and objective. Subspecies were (and
still are) given formal latinized names by taxonomists. The bald
eagle *Haliaeetus leucocephalus*, for example, is a species divided
by taxonomists into two such races, the southern bald eagle
Haliaeetus leucocephalus leucocephalus and the northern bald
eagle *Haliaeetus leucocephalus washingtoniensis*.

For reasons not immediately clear to Brown and me, subspecies seemed insubstantial and arbitrary. We set out to conduct
a critical review of the premises behind their recognition, by
looking at real cases. The foundations proved even weaker than
we had imagined. We discovered that the geographic limits of
subspecies are often hard or impossible to draw, because the
traits used to define them vary in a discordant pattern. The
nature of the discordance can be most immediately understood

with an imaginary but typical example: color in a butterfly species varies east to west, size decreases from north to south, and an extra band appears on the hind wing in a few localities near the center. And so on for any number of traits the taxonomist might choose from an almost endless list available for classification. It follows that the identity of the subspecies into which the butterfly species is divided depends on the traits chosen to define them. Pick color, and you have two east–west races. Pick color plus size, and four races in a quadrant come into existence. Add the hind-wing band, and the number of races can double again. Hence the subspecies are arbitrary. In 1953 we published a report recommending that subspecies not be given formal names.* We argued that the geographic variation is real all right but should instead be analyzed trait by trait. It is more informative to focus on the traits and not on the subspecies that might be concocted from them.

Our critique of the subspecies triggered a tempest of controversy in the journals of systematic biology. When the debate subsided several years later, opinion had shifted to our side. Fewer triple-name races were formally described thereafter, and emphasis was increasingly placed on the properties of independently varying traits. Nevertheless, I realize now that Brown and I overstated our case in 1953. Some populations can be defined clearly with sets of genetic traits that do change in a concordant, not a discordant manner. Furthermore, the subspecies category is often a convenient shorthand for alluding to important populations even when their genetic status is ambiguous. What, for example, is the Florida panther? It is a subspecies, a nearly vanished remnant of a series of populations once widespread across the United States, now further altered by hybridization with panthers of South American origin released to the wild in south Florida. Biologists rightly speak of the Florida population in a way that calls attention to its genetic distinctness, using one sharp phrase: the Florida subspecies (or race, meaning the same thing) of *Felis concolor*.

Soon afterward Brown and I made a second conceptual discovery, this one unfettered by controversy. We found a new

---

*E. O. Wilson and W. L. Brown, "The Subspecies Concept and Its Taxonomic Application," *Systematic Zoology* 2(3) (1953): 97–111.

phenomenon in biodiversity, which we came to call character displacement. The process is the exact opposite of hybridization. In hybridization, two species exchange genes where they meet, and as a result become overall more similar. In character displacement two species spring apart where they meet, like particles with the same charge. I first encountered the mysterious effect in the ant genus *Lasius*, which I had chosen as the subject of my Ph.D. dissertation. During our lunchtime dialogues Brown and I explored the possible causes and searched the literature for a similar pattern in other kinds of organisms.

We learned that the British ornithologist David Lack had already delineated character displacement in his 1947 study of Darwin's finches on the Galápagos Islands. In the 1970s and 1980s Peter Grant of Princeton University, his wife, Rosemary, and their students were to work out displacement in exquisite detail through a lengthy field study of the same finches in the Galápagos. Thus this little group of birds has been favored by three of the best field studies in the history of evolutionary biology: those by Darwin, Lack, and the Grants, respectively. The principal contribution that Brown and I made in our 1956 report was to show that the repellant effect is widespread among animals and is caused, according to the species pair considered, by either competition or the active avoidance of hybridization.*
We brought character displacement to prominence in biology and gave it the name now in general use.

Character displacement, we also realized, is one means by which species can be packed together more tightly in ecosystems. The evolution of greater differences between species reduces the chance that one of them will erase the other through competition or hybridization. The better the mutual adjustment that avoids competition and hybridization, the more species that can live together indefinitely; hence the richer will be the biodiversity, as an outcome of evolution in the community as a whole. In 1959 Evelyn Hutchinson of Yale University, the doyen of ecology, used our presentation of character displacement as a key point in his highly influential article, "Homage to Santa Rosalia, or Why Are There So Many

---

*W. L. Brown and E. O. Wilson, "Character Displacement," *Systematic Zoology* 5(2) (1956): 49–64.

Kinds of Animals?" The question he posed in this title became the entrée for ecologists who later tried to analyze the basis of biodiversity in more quantitative terms. They asked, Why are there a certain number of butterfly species in Florida, and not some other? of snakes in Trinidad, of marsupials in Australia? Just posing such questions presaged the effort to understand more deeply the causes of species formation and extinction, which by the 1950s was to become a prominent social issue in biology.

I had grown more interested in theory by the late 1950s, but down deep I was still possessed by an elemental self-image: hunter in the magical forest, searching not just for animals now but also for ideas to bring home as trophies. A naturalist, real and then more metaphorical, a civilized hunter, I was destined to be more of an opportunist than a problem solver. The boy inside still made my career decisions: I just wanted to be the first to find something, anything, the more important the better, but something as often as possible, to own it a little while before relinquishing it to others. I confess that to the degree I was insecure, I was also ambitious. I hungered for the recognition and support that discovery in science brings. To make this admission does not embarrass me now as it would have when I was young. All the scientists I know share a desire for fair recognition of their work. Acknowledgment is their silver and gold, and why they are usually very careful to grant deserved priority to others while so jealously guarding their own. New knowledge is not science until it is made social. The scientific culture can be defined as new verifiable knowledge secured and distributed with fair credit meticulously given.

Scientists, I believe, are divided into two categories: those who do science in order to be a success in life, and those who become a success in life in order to do science. It is the latter who stay active in research for a lifetime. I am one of them, and I suspect that all my fellow devotees in this category are also driven by a childhood dream of one kind or another, possibly closer to my own than I have guessed, for evolutionary biology has become the last refuge of the explorer naturalist.

So I hunted this time for abstract principles in one of the more poorly mapped domains of evolutionary biology. As a result of my experience in zoogeography, I was drawn by the

idea of the fountainheads of evolution, places in which dominant groups rise and spread to the rest of the world. Serious
discussion of the phenomenon had begun in 1915 with William
Diller Matthew, curator in invertebrate paleontology at the
American Museum of Natural History and later professor of
paleontology at the University of California at Berkeley. In his
monograph *Climate and Evolution* he constructed a picture of
dominance in mammals and other vertebrate animals. Matthew
instructed the reader to look at a north polar projection of the
globe. Europe, Asia, and North America are so close together
as to form a single supercontinent. Citing evidence from his
own and previous studies of fossil mammals, he audaciously
posited that the dominant groups arise on this supercontinent
and spread outward, displacing formerly dominant groups
southward to the peripheral regions of tropical Asia, Africa,
and South America. In our own time, he noted, the winners
include deer, camels, pigs, and the most familiar of the rats
and mice, members of the family Muridae. Among the losers
retreating to the edges of the world are horses, tapirs, and
rhinoceroses. Like a true Aryan biologist, Matthew suggested
that competitive superiority is the result of the adaptation of
the dominant groups to the harsh, constantly changing environments of the northern latitudes.

Then along came Philip Darlington with a new twist, presented first in a 1948 article in the *Quarterly Review of Biology* and later in his 1957 full-dress treatise *Zoogeography: The
Geographical Distribution of Animals*. Matthew, Darlington
wrote, was only half right. The fossil remains he had studied
were biased toward the Northern Hemisphere, where most collecting had been concentrated during the early years of paleontology. Thirty years after *Climate and Evolution*, Darlington
argued, we have more data on fossils to examine, and these
come from all over the world, including Matthew's "peripheral" areas. In addition we must examine more carefully the
evidence from the distribution of living groups, especially
fishes, frogs, and other cold-blooded vertebrates, animals for
which most of the new evidence is available. When we put all
the pieces together, we see that the evolutionary crucible is not
the north temperate land mass, but the Old World tropics. For
the past 50 million years or so, groups of vertebrates have arisen

in the vast greenhouse region that comprises southern Asia, sub-Saharan Africa, and, until recent geological times, much of the Middle East. The most dominant of the animal groups pressed on northward to Europe and Siberia, across the Bering Sea, a barrier periodically breached by the rise of isthmuses, and into the New World. People living in North America or Europe today need only look around to see the current hegemonic groups: cervid deer, canid dogs, felid cats, murid rats and mice, ranid frogs, bufonid toads, and other pioneer groups familiar to every child. Their species are extending their ranges into, not out of, the harsher climates.

I was enchanted by the idea of dominant animals and the succession of dynasties. There was a main center of evolution on the land, and Darlington seemed closer than Matthew to pinning it down. Either way, there remained a second question of major importance: What is the *biological* nature of dominance? More precisely, What hereditary traits cause an assemblage of species to spread into new lands and overwhelm the old endemics? The surrender of any group is all the more puzzling because endemics have had thousands or millions of years to become adapted to the habitats they occupy before the invaders appear.

This problem of the biological cause of dominance was not clearly in my mind as I began my own biogeographic study, a monograph on the ants of New Guinea and surrounding regions. But Matthew and Darlington, who never asked the question directly themselves, had primed me to formulate it. All I needed, I realize in retrospect today, was a small set of data to fall into place in order for the question to form somewhere in my subconscious. Then, driven by the power of the mythic conqueror archetype evoked by Matthew and Darlington, I would put together a tentative scenario, a story, and a phrase to capture it all, in the manner of Shakespeare's muse,

> And as imagination bodies forth
> The forms of things unknown, the poet's pen
> Turns them into shapes, and gives to airy nothing
> A local habitation and a name.

I asked the question and got a persuasive and verifiable answer because in the course of my taxonomic drudge-work

I had sketched out on paper the geographic ranges of the ant species one at a time and possessed a large body of quantitative information. I knew what I was talking about. Matthew and Darlington had developed their images in a coarser resolution, at the level of entire genera and families of animal species. I understood the ants of the western Pacific, not all larger groups of land vertebrates of the world, but in more detail than they had available. As part of my omnibus approach, I had collected a great many data on the places the ant species lived, their nest sites, their colony size, what they ate, and anything else that might find use—somehow, I hoped, somewhere, someday. Because I considered all the information valuable in its own right, I had swept up specimens like a vacuum cleaner in the field and continued close work on them in the museum. My ultimate aim was to find interesting patterns of evolution, but I would have continued on to the end of my descriptive work even if I had expected nothing of superordinate value.

A pattern did emerge, however. Evolutionary biology always yields patterns if you look hard enough, because there are a hundred parameters and a thousand patterns awaiting examination. It came clear this time as I mapped ranges of one species after another. I saw that some of the ants were in the early stages of invading New Guinea and the eastern Melanesian archipelagoes. Other species, apparently survivors from older invasions, were splitting off as forms limited to one island or another. Some had fragmented into many such endemic residents. And still other ensembles of species were clearly in retreat, their populations now scattered here and there in pockets of island terrain. Finally, a small percentage had begun to expand again, this time from New Guinea. It dawned on me that the whole cycle of evolution, from expansion and invasion to evolution into endemic status and finally into either retreat or renewed expansion, was a microcosm of the worldwide cycle envisioned by Matthew and Darlington.

To find the same biogeographic pattern in miniature was a surprise then, although in retrospect it seems almost self-evident. But for some reason I just didn't anticipate that particular sequence at the time. It came within a few minutes one January morning in 1959 as I sat in my first-floor office next to the entrance of the Biological Laboratories, sorting my newly sketched maps into different possible sequences—early evolution to late evolution. Which came first, which came late? I occasionally glanced up at the giant metallic rhinoceros outside the window and the intermittent stream of students and faculty walking into and out of the building. My mind drifted round and about, home, museum, field trips, lectures. I looked back down to the maps, and up again, and at some point the pattern became obvious, the only one possible.

Discovery of the cycle of advance and retreat was followed immediately by recognition of another ecological cycle. As I reflected on the expanding and retreating species, I drew on my memories of the long walks in New Guinea. I saw that the expanding, hence dominant, species are adapted for ecologically marginal habitats, in which relatively small numbers of ant species occur. Such places include the savannas, the monsoon forests, the sunny margins of lowland rain forest, and the salt-lashed beaches. They are marginal not just in having smaller numbers of ant species than the inland rain forests, but also in a purely geographic sense. Located near river banks and sea coast, they are staging areas from which it is easiest to disperse by wind and by floating vegetation from one island to another. The marginal species, I also realized, are most flexible in terms of the places in which they live. Because they face only a small number of competitors, they have been ecologically "released," able to live in more habitats and in denser populations than would otherwise be possible. It seemed likely that these ants not only could move more easily but also would tend to press older native species back into the inner rain forests, reducing their dispersal power and shattering their populations into fragments prone to evolve into endemic species.

I knew I had a candidate for a new principle of biogeography. Though far from definitive, and attributable to only one animal group—ants—the concept is at least rested on solid data. I gathered my maps, stepped next door to the office of my old companion in Cuba, Grady Webster, spread the papers out, and recited my scenario. What did he think? Reasonable, he said, looks good. Congratulations! (What did he really think? It didn't matter. I was too pleased with myself to worry.) Over the next few months I presented a full-dress review to the evolutionary biologists I considered most knowledgeable in zoogeography: Bill Brown, Phil Darlington, the geneticist Theodosius Dobzhansky, Ernst Mayr, and the senior entomologists Alfred Emerson, Carl Lindroth, and Elwood Zimmerman. This is the way it is done on the path to publication, especially by young scientists. And these luminaries all wrote back: Okay, they said, no obvious flaws that stand out.

I named the phenomenon the taxon cycle. Let me explain here that a taxon is any subspecies, species, or group of species, such as a genus, recognized in taxonomic classifications as being similar by virtue of common descent and labeled as such in taxonomic classifications. The grizzly bear, *Ursus horribilis*, being a species, is a taxon; so is the genus *Ursus*, containing the grizzly and all the other species of bear, including the black bear and brown bear, close enough to each other to be reasonably considered to share a recent common ancestor. I conjectured that if the principle held for species, it would hold for other taxa as well. In two articles, I refined my analysis.* The expanding species, I reported, have certain characteristics associated with life in the marginal habitats. The colonies are more populous and tend to nest in the soil rather than in decaying logs and tree limbs on the ground. The workers possess more spines on the body, an armament used against enemies in the open spaces of the marginal habitats. They orient more frequently by odor trails laid by scouts over the ground.

These traits are not the source of the dominance, however. They are only adaptations to life in the marginal habitats. I had no basis to infer the existence of special "dominance genes," a powerful ichor flowing in the blood of warrior ants. All that mattered in the history of the fauna was a happenstance: the dominant species had become adapted to the marginal habitats, which served as potent dispersal centers. Like the people of some island civilizations, a few ant species achieve dominance simply by their ability to cross the sea.

The taxon cycle led me to reconsider a very old concept, that of the balance of Nature: when one species is established, eventually another species has to go. But the replacement is rarely

---

*"Adaptive Shift and Dispersal in a Tropical Ant Fauna," *Evolution* 13(1) (1959): 122–144; "The Nature of the Taxon Cycle in the Melanesian Ant Fauna," *American Naturalist* 95 (1961): 169–193.

so precise; in fact, nothing in evolution ever is. The principle is more correctly defined as a statistical generalization. If a hundred species invade a certain ecological guild, say night-flying fruit eaters or orchid-pollinating bees, roughly a hundred comparable species will disappear, with many exceptions accruing to special places and times. The rule was reinforced in my mind by the discovery of a simple relation between the area of each of the Melanesian islands and the number of ant species found in it. The greater the area, the larger the number of species. When I plotted the logarithms, the points formed an approximately straight line. I expressed the area-species curve simply as follows: $S = CA^z$, where S is the number of species found on the island, A is the area of the island, and C and z are fitted constants. In 1957 Darlington had expressed the same relation in the reptiles and amphibians of the West Indies not as an equation but as the following general rule: with each tenfold increase in island area, the number of species on the island doubles. There are, for example, approximately forty species of reptiles and amphibians on Jamaica, and eighty-five on the nearby island of Cuba, which has about ten times the area of Jamaica. His expression is more readily understood in the many cases in which it applies, but the logarithmic equation is the more precise and flexible expression and therefore more generally true.

I did not grasp its significance just then, but the area-species relation that Darlington and I had defined would soon lead to a deeper understanding of the balance of species diversity. In order to explain clearly and congenially how that next step was taken, however, I need first to describe certain developments that were unfolding in biology as a whole and in Harvard biology in particular during the 1950s and 1960s.

CHAPTER TWELVE

The Molecular Wars

WITHOUT A TRACE of irony I can say I have been blessed with brilliant enemies. They made me suffer (after all, they were enemies), but I owe them a great debt, because they redoubled my energies and drove me in new directions. We need such people in our creative lives. As John Stuart Mill once put it, both teachers and learners fall asleep at their posts when there is no enemy in the field.

James Dewey Watson, the codiscoverer of the structure of DNA, served as one such adverse hero for me. When he was a young man, in the 1950s and 1960s, I found him the most unpleasant human being I had ever met. He came to Harvard as an assistant professor in 1956, also my first year at the same rank. At twenty-eight, he was only a year older. He arrived with a conviction that biology must be transformed into a science directed at molecules and cells and rewritten in the language of physics and chemistry. What had gone before, "traditional" biology—*my* biology—was infested by stamp collectors who lacked the wit to transform their subject into a modern science. He treated most of the other twenty-four members of the Department of Biology with a revolutionary's fervent disrespect.

At department meetings Watson radiated contempt in all directions. He shunned ordinary courtesy and polite conversation, evidently in the belief that they would only encourage the traditionalists to stay around. His bad manners were tolerated because of the greatness of the discovery he had made, and because of its gathering aftermath. In the 1950s and 1960s the molecular revolution had begun to run through biology like a flash flood. Watson, having risen to historic fame at an early age, became the Caligula of biology. He was given license to say anything that came to his mind and expect to be taken seriously. And unfortunately, he did so, with a casual and brutal offhandedness. In his own mind apparently he was *Honest Jim*, as he later called himself in the manuscript title of his memoir of the discovery—before changing it to *The Double Helix*. Few dared call him openly to account.

Watson's attitude was particularly painful for me. One day at a department meeting I naively chose to argue that the department needed more young evolutionary biologists, for balance. At least we should double the number from one (me) to two. I informed the listening professors that Frederick Smith, an innovative and promising population ecologist, had recently been recruited from the University of Michigan by Harvard's Graduate School of Design. I outlined Smith's merits and stressed the importance of teaching environmental biology. I proposed, following standard departmental procedure, that Smith be offered joint membership in the Department of Biology.

Watson said softly, "Are they out of their minds?"

"What do you mean?" I was genuinely puzzled.

"Anyone who would hire an ecologist is out of his mind," responded the avatar of molecular biology.

For a few moments the room was silent. No one spoke to defend the nomination, but no one echoed Watson either. Then Paul Levine, the department chairman, jumped in to close the subject. This proposal, he said, is not one we are prepared to consider at this time. With documentation, we might examine the nomination at some future date. We never did, of course. Smith was elected a member only after the molecular biologists split off to form a department of their own.

After this meeting I walked across the Biological Laboratories quad on my way to the Museum of Comparative Zoology. Elso Barghoorn hurried to catch up with me. A senior professor of evolutionary biology, he was one of the world's foremost paleobotanists, the discoverer of Pre-Cambrian microscopic fossils, and an honest man. "Ed," he said, "I don't think we should use 'ecology' as an expression anymore. It's become a dirty word." And sure enough, for most of the following decade we largely stopped using the word "ecology." Only later did I sense the anthropological significance of the incident. When one culture sets out to erase another, the first thing its rulers banish is the official use of the native tongue.

The molecular wars were on. Watson was joined to varying degrees in attitude and philosophy by a small cadre of other

biochemists and molecular biologists already in the department. They were George Wald, soon to receive a Nobel Prize
for his work on the biochemical basis of vision; John Edsall,
a pioneering protein chemist and a youngish elder statesman
who smiled and nodded a lot but was hard to understand; Matthew Meselson, a brilliant young biophysicist newly recruited
from the California Institute of Technology; and Paul Levine,
the only other assistant professor besides Watson and myself
promoted to tenure during the 1950s. Levine soon deserted
population biology and began to promote the new doctrine
aggressively on his own. Zeal of the convert, I thought
to myself.

At faculty meetings we sat together in edgy formality, like
Bedouin chieftains gathered around a disputed water well. We
addressed one another in the old style: "As Professor Wetmore
has just reminded us . . ." We used Robert's Rules of Order.
Prestige, professorial appointments, and laboratory space were
on the line. We all sensed that our disputes were not ordinary,
of the academic kind that Robert Maynard Hutchins once
said are so bitter because so little is at stake. Dizzying change
and shifts of power were in the air throughout biology, and
we were a microcosm. The traditionalists at Harvard at first
supported the revolution. We agreed that more molecular and
cellular biology was needed in the curriculum. The president
and several successive deans of the Faculty of Arts and Sciences were also soon persuaded that a major shift in faculty
representation was needed. The ranks of molecular and cellular
biologists swelled rapidly. In one long drive, they secured seven
of eight professorial appointments made. No one could doubt
that their success was, at least in the abstract, deserved. The
problem was that no one knew how to stop them from dominating the Department of Biology to the eventual extinction
of other disciplines.

My own position was made more uncomfortable by the
location of my office and laboratory in the Biological Laboratories, the bridgehead from physics and chemistry into which
the richly funded molecular biologists were now pouring. I
found the atmosphere there depressingly tense. Watson did not
acknowledge my presence as we passed in the hall, even when
no one else was near. I was undecided whether to respond in

kind by pretending to be unaware of his own existence (impossible) or to humiliate myself by persisting with southern politesse (also impossible). I settled on a mumbled salutation. The demeanor of Watson's allies ranged from indifferent to chilly, except for George Wald, who acquired an Olympian attitude. He was friendly indeed, but supremely self-possessed and theatrically condescending. On the few occasions we spoke, I could not escape the feeling that he was actually addressing an audience of hundreds seated behind me. He would in fact adopt political and moral oratory before large audiences as a second calling during the late 1960s. At the height of the campus turmoil at Harvard and elsewhere, Wald was the speaker of choice before cheering crowds of student activists. He was the kind of elegant, unworldly intellectual who fires up the revolution and is the first to receive its executioner's bullet. And on the future of our science he agreed completely with Watson. There is only one biology, he once declared, and it is molecular biology.

My standing among the molecularists was not improved by my having been granted tenure several months before Watson, in 1958. Although it was an accident of timing—I had received an unsolicited offer from Stanford and Harvard counteroffered—and in any event I considered him to be far more deserving, I can imagine how Watson must have taken the news. Badly.

Actually, I cannot honestly say I knew Jim Watson at all. The skirmish over Smith's appointment was only one of a half-dozen times he and I spoke directly to each other during his twelve years at Harvard and in the period immediately following. On one occasion, in October 1962, I offered him my hand and said, "Congratulations, Jim, on the Nobel Prize. It's a wonderful event for the whole department." He replied, "Thank you." End of conversation. On another occasion, in May 1969, he extended his hand and said, "Congratulations, Ed, on your election to the National Academy of Sciences." I replied, "Thank you very much, Jim." I was delighted by this act of courtesy.

At least there was no guile in the man. Watson evidently felt, at one level, that he was working for the good of science, and a blunt tool was needed. Have to crack eggs to make an omelet, and so forth. What he dreamed at a deeper level I never knew. I am only sure that had his discovery been of lesser magnitude he would have been treated at Harvard as just one more gifted eccentric, and much of his honesty would have been publicly dismissed as poor judgment. But people listened carefully, and a few younger colleagues aped his manners, for the compelling reason that the deciphering of the DNA molecule with Francis Crick towered over all that the rest of us had achieved and could ever hope to achieve. It came like a lightning flash, like knowledge from the gods. The Prometheans of the drama were Jim Watson and Francis Crick, and not just by a stroke of good luck either. Watson-Crick possessed extraordinary brilliance and initiative. It is further a singular commentary on the conduct of science that (according to Watson in a later interview) no other qualified person was interested in devoting full time to the problem.

For those not studying biology at the time in the early 1950s, it is hard to imagine the impact the discovery of the structure of DNA had on our perception of how the world works. Reaching beyond the transformation of genetics, it injected into all of biology a new faith in reductionism. The most complex of processes, the discovery implied, might be simpler than we had thought. It whispered ambition and boldness to young biologists and counseled them: Try now; strike fast and deep at the secrets of life. When I arrived at Harvard as a graduate student in 1951, most outside the biochemical cognoscenti believed the gene to be an intractable assembly of proteins. Its chemical structure and the means by which it directs enzyme assembly would not, we assumed, be deciphered until well into the next century. The evidence nevertheless had grown strong that the hereditary substance is DNA, a far less complex macromolecule than most proteins. In 1953 Watson and Crick showed that pairing in the double helix exists and is consistent with Mendelian heredity. ("It has not escaped our notice," they wrote teasingly at the end of their 1953 letter to *Nature*, "that the specific pairing we have postulated immediately suggests a possible copying mechanism for the genetic material.") Soon it was learned that the nucleotide pairs form a code so simple that it can be read off by a child. The implication of these and other revelations rippled into organismic and evolutionary biology, at least among the younger and more entrepreneurial researchers. If heredity can be reduced to a chain of four molecular letters—granted, billions of such letters to prescribe a whole organism—would it not also be possible to reduce and accelerate the analysis of ecosystems and complex animal behavior? I was among the Harvard graduate students most excited by the early advances of molecular biology. Watson was a boy's hero of the natural sciences, the fast young gun who rode into town.

More's the pity that Watson himself and his fellow molecularists had no such foresights about the sector of biology in which I had comfortably settled. All I could sift from their pronouncements was the revolutionary's credo: Wipe the slate clean of this old-fashioned thinking and see what new order will emerge.

I was of course disappointed at this lack of vision. When Watson became director of the Cold Spring Harbor Laboratory in 1968 (he kept his Harvard professorship by joint appointment until 1976) I commented sourly to friends that I wouldn't put him in charge of a lemonade stand. He proved me wrong. In ten years he raised that noted institution to even greater heights by inspiration, fund-raising skills, and the ability to choose and attract the most gifted researchers.

A new Watson gradually emerged in my mind. In October 1982, at a reception celebrating the fiftieth anniversary of Harvard's Biological Laboratories, he pushed his way across a crowded room to compliment me on a throwaway remark I had made during a lecture earlier that afternoon. "The history of philosophy," I had said, "consists largely of failed models of the brain." Afterward I realized that my phrasing was the kind of preemptive dismissal he would have made twenty years earlier. Had I been corrupted in the meantime? Yes, a little perhaps. I had never been able to suppress my admiration for the man. He had pulled off his achievement with courage and panache. He and other molecular biologists conveyed to his generation a new faith in the reductionist method of the natural sciences. A triumph of naturalism, it was part of the motivation for my own attempt in the 1970s to bring biology into the social sciences through a systematization of the new discipline of sociobiology.

The conflict set in motion another and ultimately positive effect of the molecular revolution. By the late 1950s the atmosphere in the department had become too stifling for members to plan the future of Harvard biology in ordinary meetings. So the professors in organismic and evolutionary biology prepared to exit. We formed a caucus and met informally to chart our own course. We began to think as never before about our future position in the biological sciences. I am reminded of another anthropological principle by this development. When savage tribes reach a certain size and density they split, and one group emigrates to a new territory. Among the Yanomamö of Brazil and Venezuela the moment of fission can be judged to be close at hand when there is a sharp increase in ax fighting. By the fall of 1960 our caucus had hardened to become the new Committee on Macrobiology.

Odd name that: macrobiology. In 1960 we realized that zoology, botany, entomology, and other disciplines centered on groups of organisms no longer reflected the reality of biology. The science was now being sliced crosswise, according to levels of biological organization, that is, oriented to the molecule, cell, organism, population, and ecosystem respectively. Biology spun through a ninety-degree rotation in its approaches to life. Specialists became less concerned with knowing everything about birds or nematode worms or fungi, including their diversity. They focused more on the search for general principles at one or two of the organizational levels. To do so many contracted their efforts to a small number of species. Colleges and universities throughout the country accordingly reconfigured their research and teaching programs into departments of molecular biology, cell biology, developmental biology, and population biology, or rough equivalents of these divisions.

During this transitional period, which continued throughout the 1960s and into the 1970s, the expression "evolutionary biology" gained wide currency. It was meant to combine the higher strata of biological organization with multilevel approaches to the environment, animal behavior, and evolution. Conceding a spotty memory and not having undertaken archival research to improve upon it, I nevertheless believe that "evolutionary biology" was launched from Harvard and probably originated there. I know that in the spring of 1958 I concocted the term on my own and entered it in the Harvard catalog as a course title for the following year. It was then spread at Harvard as follows.

One fall day in 1961, after teaching the subject for three years, I was seated in the main seminar room of Harvard's new herbarium building across the table from George Gaylord Simpson, waiting for other members of the Committee on Macrobiology to arrive for one of our regular meetings. Simpson, considered the greatest paleontologist of the day, was then in the last years of his professorship at Harvard. I struck up a conversation, a necessity if we were not to sit looking at each other in silence: G. G., as we called him, almost never spoke first. He was shy, self-disciplined to an extreme, and totally absorbed in his work. I suspect that he prized every minute saved from talking with other people, which could then be invested in the writing of articles and books. He avoided committee work with stony resolution, refused to take graduate students, and gave lectures sparingly even by the cavalier standards of the general Harvard faculty. That day I approached him with a challenge. I was fretting about the proper name for our embattled end of biology. Macrobiology, we agreed, was a terrible word. Classical biology was out; that was what our molecular adversaries were calling it. Just "plain biology"? What about *real* biology? No and no. Population biology? Accurate but too restrictive. Well then, I said, what about evolutionary biology? That would cover the ground nicely. Given that evolution is the central organizing idea of biology outside the application of physics and chemistry, its use as part of the disciplinary name might serve as the talisman of intellectual independence. I tried the expression on others, and it was received very well. By the fall of 1962 we had a formal Committee on Evolutionary Biology.

As the time for a complete departmental split approached, our conflict with the molecular faction centered with increasing heat on new faculty appointments, taken up case by painful case. The Harvard faculty is a well-known pressure cooker in the sciences, in most subjects most of the time. Peer pressure among the tenured professors is superintended by vigilant deans and presidents determined to keep quality high. That combination of intent is responsible in large part for Harvard's lofty reputation. The explicit goal of all concerned is to select the best in the world in every discipline represented, or at least a workaholic journeyman tolling at the forefront. The probing questions invariably asked by both faculty and administration are, What has he discovered that is important? Does Harvard need someone in his discipline? Is he the best in that discipline? More than half the assistant professors either fail to make tenure or go elsewhere before being put to the test. Such was intensively the case in the Department of Biology in the late 1950s and early 1960s. Every appointment recommended by one of the two camps was scrutinized with open suspicion by the other.

The rising tension was due not just to the clash of megafaunistic egos. The fissure ran deeper, into the very definition of biology. The molecularists were confident that the future belonged to them. If evolutionary biology was to survive at all, they thought, it would have to be changed into something very different. They or their students would do it, working upward from the molecule through the cell to the organism. The message was clear: Let the stamp collectors return to their museums.

The evolutionary biologists were not about to step aside for a group of test-tube jockeys who could not tell a red-eyed vireo from a mole cricket. It was foolish, we argued, to ignore principles and methodologies distinctive to the organism, population, and ecosystem, while waiting for a still formless and unproved molecular future.

We were forced by the threat to rethink our intellectual legitimacy as never before. In corridor conversations and caucus meetings, we tried to reach agreement on an agenda of future research and teaching that would soar and present the best of organismic and evolutionary biology to the world. But in these first years of molecular triumphalism our position was weak. We were moreover sharply divided in our individual interests and aspirations. Most of the caucus members were too specialized, too fixed in their ways, or too weak to resist. They sat through department meetings numbly, preferring to seek common ground by dwelling on subjects of lesser import: Who will teach the elementary course? What is the status of the Arnold Arboretum? Shall we be active partners in the new Organization for Tropical Studies? For their part the molecular

biologists made little effort to articulate a philosophy of biological research. To them the future had already been made
clear by the heady pace of their own progress. Unspoken but
heavily implied was the taunt: Count our Nobel Prizes. Ernst
Mayr and George Simpson, giants of the Modern Synthesis,
heroes of my youth, and incidentally denied Nobel Prizes
because none are given in evolutionary biology, seemed oddly
reluctant to broach these central issues openly in the meetings. Part of the reason, I suspect, was the narrow spectrum
of attitudes shown by the molecularists, ranging from indifference to contempt. Why rile them, and make an unpleasant
situation worse?

    In the absence of strong statesmanship in evolutionary biol-
ogy, our potential allies were falling away. One of the two most
distinguished organismic biologists of the time, Donald Griffin, discoverer of animal sonar, was early on persuaded by the
molecularist philosophy. We are all evolutionary biologists, he
declaimed at one meeting, are we not? Doesn't what we learn at
every level contribute to the understanding of evolution? The
eminent insect physiologist Carroll Williams remained ami-
ably neutral. A courtly Virginian who had spent his adult life
at Harvard with tidewater accent intact, he insisted on maintaining the manners that had prevailed in the old department.
More important than personality, however, was the plain fact
that the evolutionary biologists could point to no recent great
advances comparable to those in molecular and cellular biology
swelling the pages of *Nature*, *Science*, and the *Proceedings of the
National Academy of Sciences*.

    The mood of the era is caught in a personal letter I wrote
to Lawrence Slobodkin, a young evolutionary biologist and
newly acquired friend, on November 20, 1962. He had bravely
journeyed from the University of Michigan to give a lecture
on ecology.

> *You will be glad to know that the students, both under-
> graduate and graduate, are nearly unanimous in their
> praise. They found the subject matter and your particular
> style exciting. . . . The faculty were less impressed. While
> quick to state that you were disturbingly original, your
> argumentation and data were not convincing. The reasons*

for this feeling are complex. My impression is that they arise in large part from the ancient prejudice against ecology whose stereotype includes that it is not "solid" or rigorously experimental. Had a well-known biochemist, speaking on a more "solid" subject, given an exactly analogous lecture, he would have been cheered for his limber imagination and boldness.

There is a final principle of social behavior to help keep these many developments in perspective. When oppressed peoples have no other remedy they resort to humor. In 1967 I composed a "Glossary of Phrases in Molecular Biology" that was soon distributed in departments of biology throughout the country and praised—by evolutionary biologists—for capturing the strut of the conquerors. My samizdat included the following expressions, which I have changed here from alphabetical order to create a logical progression of the concepts:

Classical Biology. That part of biology not yet explained in terms of physics and chemistry. Classical Biologists are fond of claiming that there is a great deal of Classical Biology that individual Molecular Biologists do not know about; but that is all right because it is probably mostly not worth knowing about anyway, we think. In any case, it doesn't matter, because eventually it will all be explained in terms of physics and chemistry; then it will be Molecular Biology and worth knowing about.

Brilliant Discovery. A publishable result in the Mainstream of Biology.

Mainstream of Biology. The set of all projects being worked on by me and my friends. Also known as Modern Biology and Twenty-first Century Biology.

Exceptional Young Man. A beginning Molecular Biologist who has made a Brilliant Discovery (*q.v.*).

First-rate. Pertaining to biologists working on projects in the Mainstream of Biology.

Molecular Biology. That part of biochemistry which has supplanted part of Classical Biology. A great deal of Molecular

Biology is being conducted by First-rate Scientists who make Brilliant Discoveries.

Third-rate. Pertaining to Classical Biologists.

First-rate, Brilliant, Wave of the Future . . . believe me, this was the phrasing actually used. Today those once oft-heard mantras clink with antique brittleness. The passage of thirty years has done much to close the divide between molecular and evolutionary biology. As I write, systematists, the solitary experts on groups of organisms, have unfortunately been largely eliminated from academic departments by the encroachment of the new fields. That is the worst single damage caused by the molecular revolution. Ecologists, pushed to the margin for years, have begun a resurgence through the widespread recognition of the global environment crisis. Molecular biologists, as they promised, have taken up evolutionary studies, making important contributions whenever they can find systematists to tell them the names of organisms. The surviving evolutionary biologists routinely use molecular data to pursue their Darwinian agenda. The two sides sometimes speak warmly to each other. Indeed, teams from both domains increasingly collaborate to conduct First-rate Work in what may now safely and fairly be called part of the Mainstream of Biology. The corridor language one overhears from molecular biologists has grown more chaste and subtle. Only hard-shelled fundamentalists among them think that higher levels of biological organization, populations to ecosystems, can be explained by molecular biology.

I did not foresee this accommodation in the 1960s, caught as I was in the upheaval. Worse, I was physically trapped in the Biological Laboratories among the molecular and cellular biologists, who seemed to be multiplying like the *E. coli* and other microorganisms on which their finest work had come to be based. In buildings a hundred feet and a world of ideas away were the principalities and margravates of the senior evolutionary biologists. They were mostly curators and professors in charge of Harvard's "Associated Institutions," comprising the Museum of Comparative Zoology, the University herbaria, the Botanical Museum, the Arnold Arboretum, and the Harvard

Forest. I envied them mightily. They could retreat to their
collections and libraries and continue to be supported by venerable endowments bearing the names of nineteenth-century
Anglo-Saxons.

What I desired most was to emigrate across the street to
the Museum of Comparative Zoology, to become a curator
of insects, to surround myself with students and like-minded
colleagues in an environment congenial to evolutionary biology, and never have to pass another molecular biologist in the
corridor. But I held off requesting such a move for ten years,
while Ernst Mayr was director. Perhaps I was overly timid,
but the great man seemed forbiddingly stiff and cool toward
me personally. There was also the twenty-five-year difference
in age, and the fact that I had felt filial awe ever since adopting his book *Systematics and the Origin of Species* as my bible
when I was eighteen. We have since become good friends, and
I speak to him frankly on all—well, most—matters (he is still
fully active in his ninetieth year as I write), but at that time
I felt it would be altogether too brash to ask for haven in his
building. My self-esteem was fragile then to a degree that now
seems beyond reason. I felt certain that Mayr thought little
of me. I dared not risk the humiliation of a refusal. I figured
the odds at no better than fifty-fifty he would give it. When
a new director, A. W. ("Fuzz") Crompton, was installed and
proved as approachable in personality as the nickname implies,
I asked him for entry. Fuzz promptly invited me to the newly
erected laboratory wing of the Museum ("You've made my
day, Ed") and soon afterward had me appointed Curator in
Entomology. I do not doubt that the molecular biologists were
also pleased to see me leave. One day near the end, while I sat
at my desk, Mark Ptashne, one of the younger shock troopers
of this amazing group, walked into my quarters unannounced
with a construction supervisor and began to measure it for
installation of equipment.

By this time I had been radicalized in my views about the
future of biology. I wanted more than just sanctuary across
the street, complete with green eyeshades, Cornell drawers of
pinned specimens, and round-trip air tickets for field work in
Panama. I wanted a revolution in the ranks of the young evolutionary biologists. I felt driven to go beyond the old guard

of Modern Synthesizers and help to start something new. That might be accomplished, I thought, by the best effort of men my age (men, I say, because women were still rare in the discipline) who were as able and ambitious as the best molecular biologists. I did not know how such an enterprise might be started, but clearly the first requirement was a fresh vision from the young and ambitious. I began to pay close attention to those in other universities who seemed like-minded.

A loose cadre in fact did form. In January 1960 I was approached by an editorial consultant of Holt, Rinehart and Winston, a leading publisher of scientific texts, who asked me to referee the manuscript of a short book by Larry Slobodkin. The title was *Growth and Regulation in Animal Populations.* As I flipped through the manuscript pages I was excited by Slobodkin's crisp style and deductive approach to ecology. He advanced simple mathematical models to describe the essential features of population dynamics, then expanded on the premises and terms of the equations to ask new questions. He argued that such complex phenomena as growth, age structure, and competition could be broken apart with minimalist reasoning, leading to experiments devised in the postulational-deductive method of traditional science. He went further: the hypotheses and experimental results could be greatly enriched by explanations from evolution by natural selection.

Slobodkin was not the first scientist to advance this prospectus for the invigoration of ecology, but the clarity of his style and the authority implied by a textbook format rendered the ideas persuasive. It dawned on me that ecology had never before been incorporated into evolutionary theory; now Slobodkin was showing a way to do it. He also posed, or so I read into his text, the means by which ecology could be linked to genetics and biogeography. Genetics, I say, because evolution is a change in the heredity of populations. And biogeography, because the geographic ranges of genetically adapted populations determine the coexistence of species. Communities of species are assembled by genetic change and the environmentally mediated interaction of the species. Genetic change and interaction determine which species will survive and which will disappear. In order to understand evolution, then, it is necessary to include the dynamics of populations.

With this conception in mind, and my hopes kindled that Slobodkin would emerge as a leader in evolutionary biology, I wrote an enthusiastic report to the editor. A short time later I approached Slobodkin himself, suggesting that the time had come to produce a more comprehensive textbook on population biology. Would he be interested in writing one with me? In such a collaboration, he might introduce population dynamics and community ecology, while I added genetics, biogeography, and social behavior. The material would serve as an intermediate-level textbook. It would also promote a new approach to evolutionary biology founded on ecology and mathematical modeling.

Slobodkin said he was interested. He would talk the matter over with me. Soon afterward we met in Cambridge to outline our prospective work. We went so far as to draw up individual assignments in the form of chapter headings.

Slobodkin was then an assistant professor at the University of Michigan. A rising star in the admittedly still depauperate field of American ecology, he was later to move to the Stony Brook campus of the State University of New York, where he founded a new program in evolutionary biology. His reputation as a researcher has been securely based on a series of eclectic studies conducted before I met him and in the years immediately following. He studied the red tide phenomenon, the periodic population bloom of toxic dinoflagellate protozoans that poison fish and other marine life. He pioneered the measurement of energy transfer across trophic levels in ecosystems by the use of the bomb calorimeter. In the realm of theory, he elaborated the concept of a balancing relationship between the "prudent predator" and "efficient prey."

During the years to follow, I never failed to find Slobodkin's physical appearance arresting: red-haired, alternately clean-shaven and dramatically mustachioed, an ursine body relaxed in scholar's informality. Not given to easy laughter, he preferred ironic maxims over funny stories. His conversational tone was preoccupied and self-protective, and to a degree unusual in a young man tended toward generalizations about science and the human condition. It was leavened in the company of friends with discursive sentences and fragments of crude humor, seemingly contrived to throw the listener off balance, especially when combined with Delphic remarks of the kind philosophers use to stop conversations. These latter asides implied: There is more to the subject of our banter, much more; see if you can figure it out. Slobodkin in fact was a philosopher. I came to think of him as progressing through a scientific career to a destiny somewhere in the philosophy of science, where he would become a guru, a rabbi, and an interpreter of the scripture of natural history. Some of our friends complained that his persona was a pose, and perhaps it was to some degree, but I enjoyed Slobodkin's subtle and penetrating mind, and his company. Not least, we were opposites in cultural origin, which made him all the more interesting to me. He was a New York intellectual, a Jew, as far in every dimension of temperament and style as it is possible to get from the sweat-soaked field entomologist I still fancied myself to be, then, in the early 1960s.

Slobodkin was heavily influenced by his Ph.D. adviser at Yale, G. Evelyn Hutchinson, himself as different from Slobodkin and me as Larry and I were from each other: our relationship formed an equilateral triangle. Born in 1903, the son of Arthur Hutchinson, the Master of Pembroke College at Cambridge University, Evelyn—"Hutch" to those who dared call him an intimate—was a creation of British high-table science. True to the Oxbridge prize Fellow tradition, he never bothered to earn the doctorate, but instead trained himself into a polymath of formidable powers. He was a free spirit, an eclecticist who proved brilliant at fitting pieces together into large concepts. He never seemed to have met a fact he didn't like or couldn't use, somewhere, to start an essay or at least place in a footnote. He began his career as field entomologist studying aquatic "true bugs" as experts call them—members of the order Hemiptera—and especially notonectid backswimmers. He worked as far from home as Tibet and South Africa. Then he turned to pioneering research on algae and other phytoplankton of lakes and ponds. He broadened his scope to include the cycles and stratification of nutrients on which life in these bodies of water depends. He was among the first students of biogeochemistry, a complex discipline combining analyses of land, water, and life. Still later, after becoming professor of zoology at Yale in 1945, he turned to the evolution of population dynamics, which also became Slobodkin's forte.

Hutchinson's insights were deep and original, and, notwithstanding that such tropes have been worn to banality through overuse, he deserves to be called the father of evolutionary ecology. Among his notions that proved most influential was the "Hutchinsonian niche." Like most successful ideas in science, it is also a simple one: the life of a species can be usefully described as the range of temperatures in which it is able to live and reproduce, the range of prey items it consumes, the season in which it is active, the hours of the day during which it feeds, and so on down a list as long as the biologist wishes to make it. The species is viewed as living within a space defined by the limits of these biological qualities each placed in turn on a separate scale. The niche, in short, is an n-dimensional hyperspace.

Hutchinson's independence was such that he remained unperturbed by molecular triumphalism; at least I never heard of his protesting in the manner of his colleagues in Harvard's overheated department. In his later years he metamorphosed gracefully from field biologist to guru, seated in his office with wispy white hair and basset eyes. Beside him presided a stuffed specimen of the giant Galapagos tortoise. In a teaching career spanning nearly three decades, he trained forty of the best ecologists and population biologists in the world to the doctoral level. They included Edward Deevey, Thomas Edmondson, Peter Klopfer, Egbert Leigh, Thomas Lovejoy, Robert MacArthur, Howard Odum, and, of course, Larry Slobodkin. They all seemed to admire and love the man, and to have drawn strength and momentum from his example. Fanning out across the country to represent the many growing fields of ecology, they exerted a crucial influence in American biology.

I asked several after they became my friends what "Hutch" did to inspire such enterprise in his disciples. The answer was always the same: nothing. He did nothing, except welcome into his office every graduate student who wished to see him, praise everything they did, and with insight and marginal scholarly digressions, find at least some merit in the most inchoate of research proposals. He soared above us sometimes, and at others he wandered alone in a distant terrain, lover of the surprising metaphor and the esoteric example. He resisted successfully the indignity of being completely understood.

He encouraged his acolytes to launch their own voyages. It was pleasant, on the several occasions I lectured at Yale before Hutchinson's death in 1991, to encounter him and receive his benediction. Head bobbing slightly between hunched shoulders, a wise human Galápagos tortoise, he would murmur, Wonderful, Wilson, well done, very interesting. It would have been pleasant to stay near him, the kindly academic father I never knew. I came to realize that the overgenerous praise did not weaken the fiber of our character. Hutchinson's students criticized one another, and me as well, and that was enough to spare us from major folly most of the time.

Hutchinson and Slobodkin were then what today are called evolutionary ecologists. In my formative years they caused me to try to become one as well. Through them I came to appreciate how environmental science might be better meshed with biogeography and the study of evolution, and I gained more confidence in the intellectual independence of evolutionary biology. I was encouraged to draw closer to the central problem of the balance of species, which was to be my main preoccupation during the 1960s, as the molecular wars subsided to their ambiguous conclusion.

CHAPTER THIRTEEN

Islands Are the Key

D URING A SEMINAR break at the 1961 meeting of the
American Association for the Advancement of Science,
held in New York's Biltmore Hotel, Larry Slobodkin told me
that there was someone I had to meet, someone we should ask
to be another coauthor of the book on population biology.
Two months had passed since we had first met to discuss our
joint venture. "It's Robert MacArthur, and he's a real theoretician—very bright. I think we need someone else closer to pure
theory, with a better mathematical background, to help with
the book."

MacArthur, I learned, was a thirty-year-old assistant professor at the University of Pennsylvania. He had received his
Ph.D. under Evelyn Hutchinson in 1957. After a year spent at
Oxford studying with the British ornithologist David Lack,
he had begun to move swiftly into a brilliant career. But neither Larry nor I could have guessed on that day we waited
for him just how brilliant, that in one decade he would come
to rival Hutchinson in influence. MacArthur was to bring
population and community ecology closer to genetics. By
reformulating some of the key parameters of ecology, biogeography, and genetics into a common framework of fundamental theory he set the stage, more than any other person
during the decisive decade of the 1960s, for the unification
of population biology. Then he was cut down by a fatal renal
cancer, and became a legend. Today one of the most coveted
honors for midcareer researchers in evolutionary biology is to
be invited to give the MacArthur Lecture of the Ecological
Society of America.

He joined Slobodkin and me, a thin, diffident young man
who spoke with an American accent but in the British style
of cautious understatement, perhaps acquired at Oxford. The
book is an attractive idea, he said. We should explore it further.
He had a headache. He wanted to go home. We shook hands,
and he left.

Nothing more happened for nearly a year. That was my fault. I simply put the book project completely aside in order to return to field work. The tropics had reasserted their pull. A dream stirred deep within me that El Dorado was still there unattained. I had to go. In February 1961 Renee and I traveled to Trinidad, where we stayed as guests of an Icelandic native and widow named Asa Wright. Her property, Spring Hill Estate, was perched near the head of Arima Valley, in the North Range. It had become a popular stopover for naturalists and serious birders from around the world. Broken rain forest ran down the valley to Simla, the research station founded by William Beebe. The great naturalist was in the last year of his life, and I was grateful for the opportunity to meet him. Renee and I dined occasionally with him and his capable assistant Jocelyn Crane at Simla, admiring the silver candlesticks given to Beebe by his friend Rudyard Kipling and talking tropical natural history here in the place where so much of the best research on it had been conceived.

The tropics in those days nourished a strange collection of intellectuals. At Spring Hill we sat on the screened veranda listening to stories by another famous visitor, Colonel Richard Meinertzhagen, who had first served as an officer under Queen Victoria and fought with T. E. Lawrence in the Middle East during the First World War. I looked him up later in Lawrence's *Seven Pillars of Wisdom*, and sure enough, there was Meinertzhagen, reported in the same episodes he had described to Renee and me. Meinertzhagen was at Spring Hill to visit oil birds in a nearby cave and to collect the fruits of native forest trees. Given all this, and the aging Asa Wright's retrograde colonial attitude toward the Trinidadians of color, we felt we had been propelled backward in time fifty years.

There were adventures to savor, this time shared with Renee. Once a pet donkey wandered across the Spring Hill veranda into the open dining room, its hooves clopping loudly on the hardwood floor, and consumed a chocolate cake set out for afternoon tea. The maid quickly chased him back out. Soon afterward, as Renee sat in a corner of the veranda, waiting for my return from the field, she could not help overhearing Asa's reaction to the news: "Oh my god, Eutrice, do the Wilsons know?"

"No ma'am," Eutrice lied.

The donkey was tethered to a veranda post at night, and sometime during the evening it was usually visited by vampire bats that flew in from the neighboring forest. In the morning one or two dried rivulets of blood streaked some part or other of its flanks or legs. Such bloodletting was a common problem for livestock in the area, and the bats carried rabies. Seated on the veranda, armed with flashlights, Meinertzhagen and I watched eagerly for the arrival of the vampires late into the night. We never saw one. That is the talent of vampires, evading detection.

Two months after arriving Renee and I departed for Suriname, to add field work on the South American mainland. We proceeded by freighter out of Port of Spain to the bauxite mining town of Moengo, then back to the capital of Paramaribo. We lived in a pension there while I explored the forests south as far as Zanderij. We then returned to Spring Hill for a while before proceeding to Tobago for the final three months, June through most of August, of our tour.

I felt completely at home again in the heat and smell of rotting vegetation, although Renee did not, especially when she learned about the vampire bats. Discoveries came easily, as always for me in tropical forests. I acquired a colony of the giant, primitive dacetine ant *Daceton armigerum* from a nest high in a tree in Suriname and made the first study of its social organization. I rediscovered the apparently "true" cave ant *Spelaeomyrmex urichi* in a central Trinidad cavern, and proved that the species also lives in the open forest of Suriname—and thus is not an obligatory cave ant. I puttered here and there, in the opportunistic spirit that had always guided me.

But early on this field trip, while beginning work on Trinidad, I found the tropics less than paradise. To my dismay I slipped into a depression for the first time in my life. I began to worry again about the broader canvases of ecology and evolution, and the need to get on with the agenda of my young evolutionists' conceptual revolution. I hated the corresponding diminution of my naturalist's ardor. I was anxious about my own inadequacy in mathematics. I felt certain that the future principles of evolutionary biology would be written in equations, with the deepest insights expressed by quantitative models. I set out to remedy my deficiency by teaching myself calculus, probability theory, and statistics from textbooks I read on verandas and beach cupolas in Trinidad and Tobago. Progress was slow; I was not gifted; I worried even more. Here I was, thirty-two years old, time and the main chance about to slip away—or so it seemed. Would I miss out on the real action coming?

Soon after our return home in late August, Renee and I bought our first house, a small two-story cape in the suburban town of Lexington, ten miles west of Cambridge. It cost $19,000, about twice my gross annual salary at the time. By scrimping on expenses during our sabbatical trip, we had just managed to save the minimum $3,000 down payment for a first mortgage. Now five years into our marriage, we at last felt rooted and secure. I felt more confident in my work and in the knowledge that I would probably stay at Harvard for the remainder of my career. My math anxiety faded.

Soon afterward MacArthur and Slobodkin joined me at Harvard for a one-day meeting to resume planning our book on population biology. We drew up an outline, divided chapter assignments, and went our separate ways. As much as I admired Slobodkin, I felt a stronger personal attraction to MacArthur. In subsequent correspondence and visits we discovered a surprising range of common interests, among which was a passion for biogeography—the geographic distribution of plants and animals. The traditional discipline, in which I had been steeped throughout my career, was in chaos. Grand chaos, in fact, since the subject matter is the largest in physical scale of all biology, and it spans the entire history of life.

In 1961, when MacArthur and I focused on it, biogeography was still largely descriptive. Its most interesting theory was the Matthew-Darlington cycle of dominance and replacement. Otherwise its main substance comprised such topics as the origin of the fauna and flora of the West Indies—whether by immigration across dry land bridges that once connected the islands to the mainland or by the chance arrival of organisms borne on water and in the air. Biogeography seemed ripe for the new thinking that was emerging in population biology. I showed MacArthur some of the curves in my files linking the area of individual islands to the numbers of resident species of ants and other organisms. I told him about my conception of the taxon cycle and the balance of species.

MacArthur's interest in these and related subjects grew rapidly. As our discussions deepened, and spread to include gossip and personal anecdotes, we became close friends. Our backgrounds proved similar in several respects that matter most in scientific collaboration. Although he had majored in mathematics at Marlboro College, and had a conspicuous talent for it, his heart was in the study of birds. He was a naturalist by calling, and seemed happiest when searching for patterns discovered directly in Nature with the aid of binoculars and field guides. It was his calling to scan the tangled bank of Nature and skeletonize it in his own and others' minds to its essential abstract features. As a mathematician-naturalist he was unique, approached only by his mentor, Evelyn Hutchinson. He was not as expansive in his interests as Hutchinson, but quicker and more deeply penetrating at strategic points. He shared the conviction of the great mathematician G. H. Hardy, whom he resembled in temperament and philosophy, "that a mathematician was a maker of patterns of ideas, and that beauty and seriousness were the criteria by which his patterns should be judged." He wished above all to discover beautiful true-life patterns.

In conversation, MacArthur would say that the best science comes to a great extent from the invention of new classifications of natural phenomena, the ones that suggest hypotheses and new rounds of data gathering. "Art"—he enjoyed quoting Picasso—"is the lie that helps us see the truth." His methodology bore testimony of the strength of an inherent naturalist: he knew what he was talking about, and he was concerned more with the tapestry of Nature and his power to see it independently than with what others thought of it, or of him.

MacArthur watched birds with the patience and skill of a professional ornithologist. He visited the tropics as often as he could, and delighted in relating endless facts of natural history. The store of random information thus accumulated and the play of its intersecting patterns were the inspiration of his theoretical work, by which he described the process of the origin of biological diversity.

When I first met him he was an assistant professor at the University of Pennsylvania, soon to be promoted to associate and then to full professor. He later moved to Princeton, where in a short time he was named Henry Fairfield Osborn Professor of Biology. His demeanor was subdued and pleasing. Of medium height, with a handsomely rectangular face, he met you with a disarming smile and widening of eyes. He spoke with a thin baritone voice in complete sentences and paragraphs, signaling his more important utterances by tilting his head slightly upward and swallowing. He had a calm, understated manner, which in intellectuals suggests tightly reined power. In contrast to the excessive loquacity of most professional academics, MacArthur's restraint gave his words an authority rarely intended. In fact he was basically shy and loathed being caught in a careless error. He was nevertheless conscious of his status among colleagues and felt secure about it. Although he was generous by instinct and capable of lavish, almost Hutchinsonian praise during private conversations for work he thought important, he did not hesitate to describe the foibles and weaknesses of others with pitiless accuracy. But he harbored no malice I could detect, only a taxonomic interest in other scientists and a frequent disappointment that tempered his enthusiasms.

He joined superior talent with an unusual creative drive and decent ambition. He placed his family, Betsy and the four children, above all else. After that came the natural world, birds, and science, in that order. One day as we strolled along a road in the Florida Keys, I told him of the effort I was making with several others to conserve Lignumvitae Key, one of the last islands of Florida with a relatively undisturbed Caribbean forest. He reacted with a warmth that surprised me—I had not even thought of mentioning it earlier. He declared that he would rather save an endangered habitat than create an important scientific theory.

MacArthur launched his scientific career with two articles that revealed his unusual powers. The first, in 1955, suggested a way to predict stability in a community of plants and animals by the use of information theory. It formalized a concept that until then could be expressed only through verbal description.

Soon afterward, in 1957, came the famous "broken stick" model of relative abundance of bird species. To capture the essence of his approach, imagine that the combined numbers of a certain guild of birds, say warblers, found in a particular forest is represented by the length of a stick. Make the stick one meter long, representing 100,000 warblers, so that each bird is represented by a fraction of a millimeter somewhere on the stick. Have the guild consist of ten warbler species. Break the stick into ten pieces at random, with their lengths randomly distributed. Let the length of each piece represent the number of individuals of a particular species. One species, let us say, gets 200 millimeters, or 20 percent of the stick; it is therefore assigned 20,000 individual birds. Another species gets 5 millimeters, or 5,000 birds. And so on for all ten pieces and the individual birds their lengths represent. Because the pieces and thence the species are not allowed to overlap, the array of numbers for the whole ensemble of ten warbler species will be the same as if real warblers divided up the resources of the forest among themselves competitively, so there is no sharing of resources, and the fraction each species has acquired is a random variable. The niche of each species is also unique. If real warblers were found to fit a numbers array like this array (more technically put, if its "species abundance distribution" fits the broken-stick model), we would be justified in supposing that the warblers are really segregated by competition for resources. At least we must keep that possibility open, subject to confirming studies of other kinds. What would be the alternative to the exclusion model? One proposed by MacArthur was that the species receive pieces of the stick with lengths being determined randomly, but that the pieces can be overlapping; in other words, the bird species do not exclude one another by competition. Because the exclusion model turned out to fit one set of bird data and MacArthur's first disposal more closely than did the alternatives he conceived, he concluded that competition is likely to be important in determining the abundance of birds.

The specific hypothesis of competition captured by the broken-stick distribution was later disputed by others, and MacArthur himself eventually dismissed his methodological approach—prematurely, I thought—as obsolete. Even while fading, however, the conception represented a breakthrough in ecological theory. In three pages, MacArthur confronted a central problem of community ecology with competing hypotheses expressed as numbers, in contrast to previous theorists, who had formulated the same general idea more vaguely by words. He characterized the issue in such a way as to allow logically possible alternatives to be tested and a choice made. By working out this example, he showed that the deepest remaining mysteries of natural history might be solved by leaps of the imagination, so long as such efforts are disciplined by clear postulation tested by data taken from the field.

The method of multiple working hypotheses was thereby introduced to the branch of ecology concerned with whole communities of species. MacArthur's 1957 article set the tone of all his later work. Inevitably, his entire approach, not just the broken-stick model, was correctly criticized by some ecologists as oversimplification. That defect matters little in the long haul of history. It was a step in the right direction. Right or wrong in particular applications, it energized a generation of young population biologists and transformed a large part of ecology. It helped us to think clearly.

As MacArthur and I extended our conversations, I expressed three convictions of my own. First, that islands are the key to rapid progress in biogeography. The communities they contain are discrete units that are isolated by the sea and can be studied in multiples. Second, that all biogeography, including even the histories of faunas and floras, can be made a branch of population biology. And finally, that species on islands are somehow in balance in a way that can be modeled quantitatively. MacArthur quickly agreed, and began to apply his powers of abstraction to the data sets I showed him. In the following exchange I have telescoped our conversations and letters on the subject in order to convey the crucial steps in the origin of species-equilibrium theory.*

Wilson: I think biogeography can be made into a science. There are striking regularities no one has explained. For example, the larger the island, the more the species of birds

---

*This account was first presented in my essay collection *Biophilia* (Cambridge, Mass.: Harvard University Press, 1984).

or ants that live on it. Look at what happens when you go from little islands, such as Bali and Lombok, to big ones like Borneo and Sumatra. With every tenfold increase in area, there is roughly a doubling of the number of species found on the island. That appears to be true for most other kinds of animals and plants for which we have good data. Here's another piece in the puzzle. I've found that as new ant species spread out from Asia and Australia onto the islands between them, such as New Guinea and Fiji, they eliminate other ones that settled there earlier. At the level of the species this pattern fits in pretty well with the views of Philip Darlington and George Simpson. They proved that in the past major groups of mammals, such as all the deer or all the pigs taken together, have tended to replace other major groups in South America and Asia, filling the same niches. So there seems to be a balance of Nature down to the level of the species, with waves of replacement spreading around the world.

MacArthur: Yes, a species equilibrium. It looks as though each island can hold just so many species, so if one species colonizes the island, an older resident has to go extinct. Let's treat the whole thing as if it were a physical process. Think of the island as filling up with species from an empty state up to the limit. That's just a metaphor, but it might get us somewhere. As more species establish themselves, the rate at which they go extinct will rise. Let me put it another way: The probability that any given species will go extinct increases as more species crowd onto the island. Now look at the species arriving. A few colonists of each are making it each year on the wind or on floating logs or, like birds, flying in on their own power. The more species that settle on the island, the fewer *new* ones that will be arriving each year, simply because there are fewer that aren't already there. Here's how a physicist or economist would represent the situation. As the island fills up, the rate of extinction goes up and the rate of immigration goes down, until the two processes reach the same level. So by definition you have a dynamic equilibrium. When extinction equals immigration, the *number* of species stays the same, even though there may be a steady change in the particular species making up the fauna.

Look what happens when you play around a little with the rising and falling curves. Let the islands get smaller. The extinction rates have to go up, since the populations are smaller and more liable to extinction. If there are only ten birds of a kind sitting in the trees, they are more likely to go to zero in a given year than if there are a hundred. But the rate at which new species are arriving won't be affected very much, because islands well away from the mainland can vary a lot in size without changing much in the amount of horizon they present to organisms traveling toward them. As a result, smaller islands will reach equilibrium sooner and end up with a smaller number of species at equilibrium. Now look at pure distance as a factor. The farther the island is from the source areas, say the way Hawaii is farther from Asia than New Guinea, the fewer new species that will be arriving each year. But the rate of extinction stays the same because, once a species of plant or animal is settled on an island, it doesn't matter whether the island is close or far. So you expect the number of species found on distant islands to be fewer. The whole thing is just a matter of geometry.

Weeks pass. We are sitting next to the fireplace in MacArthur's living room, with notes and graphs spread out on a coffee table.

Wilson: So far so good. The numbers of bird and ant species do go down as islands get smaller and farther from the mainland. We'll label the two trends the *area effect* and the *distance effect.* Let's take them both as given for the moment. How do we know that they prove the equilibrium model? I mean, other people are almost certainly going to come up with a rival theory to explain the area and distance effects. If we claim that the results prove the model that predicted them, we will commit what logicians call the Fallacy of Affirming the Consequent. The only way we can avoid that impasse is to get results that are uniquely predicted by our model and no one else's.

MacArthur: All right, we've gone this far with pure abstraction—let's go on. Try the following: line up the extinction and immigration curves so that where they cross and create the equilibrium, they are straight lines and tilted at approximately the same angle. As an exercise in elementary differential

calculus, you can show that the number of years an island takes
to fill up to 90 percent of its potential should just about equal
the number of species at equilibrium divided by the number
going extinct every year.

Wilson: Let's look at Krakatau.

Krakatau is the small island between Sumatra and Java that
had been wiped clean of all life in the great volcanic explosion
of August 27, 1883. Scientists from several nations, principally
the Netherlands and Indonesia, then a Dutch colony, began
to visit the reduced remnant of Krakatau within a year of the
event. They managed to keep a spotty but serviceable record
of the return of birds, plants, and a few other organisms to the
bare volcanic slopes. The basic equilibrium model we developed predicted that the birds in particular, for which the best
data of all were available, should reach equilibrium at about
thirty species. Upon approaching that level, the fauna should
be losing one established species by local extinction each year
while acquiring one new species by immigration. The data
gathered by the early researchers indicated that the bird fauna
did indeed appear to be leveling off at approximately thirty
species. But the turnover recorded was one species every five
years, not one each year.

Was the model really off fivefold, or was the discrepancy due
to sampling errors? There was no way to tell. At this point we
saw the need of replicate data sets in order to advance equilibrium theory in a serious way. By 1965 I set out to devise such
an experimental system in the Florida Keys, using the insects
and other arthropods of the smallest islands. That is another
story, an unusual, rather bizarre, adventure of field biology, to
which I will return in the next chapter.

As MacArthur and I progressed on the island biogeography
project, the loose confederation of young population biologists continued to grow. In late July 1964, five of us met at
MacArthur's lakeside home at Marlboro, Vermont, to discuss
our personal research agendas and how they might contribute
to the future of population biology. Joining MacArthur and
me were Egbert Leigh, a young mathematician with a special
interest in the structure of plant and animal communities, later
to join the Smithsonian Tropical Research Institute as research

scientist; Richard Levins, a theoretical population biologist of contemporary renown who later joined the faculty of Harvard's School of Public Health; and Richard C. Lewontin, the rising star of theoretical and experimental genetics, who was to come to Harvard as Agassiz Professor of Zoology in 1973. In close touch with several of us but not present at the lakeside retreat were Slobodkin and Leigh Van Valen, a paleontologist and general evolutionary biologist at the University of Chicago.

For two days between walks in the quiet northern woodland, we expanded upon our common ambition to pull evolutionary biology onto a more solid base of theoretical population biology. Each in turn described his particular ongoing research. Then we talked together about the ways in which that subject might be extended toward the central theory and aligned with it. Besides island biogeography, in which MacArthur and I were now well advanced, I saw myself as adding the study of ants and other social animals to the enterprise. An animal society is a population, I argued, and it should be possible to analyze its structure and evolution as part of population biology. My student Stuart Altmann and I had already, early in 1956, discussed the idea of finding common principles to explain primates and insect societies. We had even used the term "sociobiology" to describe the effort. But we had had little intuition on how to proceed, and our collaboration had advanced no further. I hoped that the combined thinking of this new group, the "Marlboro Circle" as I have come to call it, would provide me with clues. The others were encouraging in their remarks, but few clues were forthcoming. William Hamilton's article on kin selection and altruism, which was to be a keystone of sociobiology, was published that year, but neither I nor the others had yet seen it.

How to proceed with the sociobiological and similar overlapping agendas? There emerged from our freewheeling talk the notion of pooling our work. We would produce a series of essays under the single pseudonym "George Maximin," in imitation of the French mathematicians who have been publishing since the 1930s under the name Nicolas Bourbaki. Maximin was named not in honor of the Roman soldier-emperor but after the point of greatest minimum in optimization theory; George was an arbitrary first name added. With Maximin we thought we could achieve the twin goals of anonymity, with its freedom from ego and authorial jealousy, while acquiring license to be as audacious and speculative as the group decided.

Maximin died an early death. He was an ill-conceived Frankenstein monster. By mid-August MacArthur was expressing serious doubts in letters to me. He argued that we should each take credit and responsibility for his own ideas. Slobodkin disliked the concept from the start. Maximin, he said, would look to others too much like a cabal. I had to admit that down deep I shared these misgivings. Personal idiosyncrasies doomed Maximin. MacArthur was particularly confident of his own powers and inclined to work unimpeded. He seemed to believe that he could generate ideas singly or in groups whenever the spirit moved him. Slobodkin for his part was turning against the idea of unifying theories and heavy dependence on mathematical modeling. I myself was temperamentally ill suited to Maximin, preferring to work alone or at most with a single partner. So the program faded, and for the most part the conspirators went their separate ways. We never met as a group again. But a lot was gained from Maximin's ghostly spirit. I cannot speak for the others, but I believe we all carried away a new confidence in the future of evolutionary biology, and in ourselves.

By the end of the year MacArthur and Slobodkin were growing apart. Slobodkin, Robert wrote me in a letter, "is in an antitheoretical mood." Nature defeats theory, Slobodkin was widely quoted as saying at the time. In August MacArthur pulled out of the biology textbook project we had planned three years earlier. Slobodkin by this time had produced little, and I had not done much better, having become distracted in the meantime by a half-dozen other projects. As a result the book soon followed Maximin into oblivion. We just stopped mentioning it. In 1966, when MacArthur published a short introductory text for freshman courses with Joseph Connell, Slobodkin condemned it with a slashing review. He opposed the very philosophy of science it represented. MacArthur in turn bridled at what he considered gratuitous hostility. He believed he had been misunderstood by retrograde thinkers. "I think I can tell why there are potatoes in the field and where they lie," he mused to me, "but these people say no good, they want to know the size and shape of the potatoes."

None of this had any effect on my collaboration with MacArthur. I believed deeply in the power of reduction-ism, followed by a reconstitution of detail by synthesis. In December 1964, I suggested that we write a full-scale book on island biogeography, with the aim of creating new models and extending our mode of reasoning into as many domains of ecology as we could manage. Robert agreed at once. He was enamored of the subject by this time, and had begun to call himself a biogeographer instead of an ecologist. In this domain existed in most readily definable state the patterns he wanted to discover. When he later brought out a book under his own sole authorship, in 1972, he titled it *Geographical Ecology*.

Off and on in the two years following the Marlboro meet-ing, MacArthur and I assembled the pieces of an expanded theory of island biogeography. We explored the implications of the balance of species in the colonization of islands, lakes, and other isolated habitats. From published data we traced the course of the recolonization of Krakatau and other devastated islands. We examined the general qualities of the niche and the forms of evolution by which species adapt to dispersal and competition. We considered from the bottom up, species by species, the means by which animals and plants are most effi-ciently packed together to create diverse communities.

When our book, *The Theory of Island Biogeography*, was published in 1967, it met with almost unanimous approval in the scientific journals. Some of the reviewers declared it a major advance in biology. A quarter-century later, as I write, it remains one of the most frequendy cited works of evolutionary biology. *The Theory of Island Biogeography* has also become influential in conservation biology, for the following practical reason. Around the world wild lands are being increasingly shattered by human action, the pieces steadily reduced in size and isolated from one another. Nature reserves are by definition islands. The theory serves as a useful tool in conceptualizing the impact of their size and isolation on the biodiversity they contain. Some parts of the formulation made in 1967 have been discarded by later authors, justifiably, and other parts greatly modified. Later researchers have added powerful new insights and definitive data sets unavailable to us at the time. I do not think it an exaggeration to say, however, that MacArthur and

I accomplished most of what we set out to do. We unified, or at least began to unify, biogeography and ecology upon an internally consistent base of population biology.

In the 1960s and 1970s a new wave of population biologists trained in both ecology and mathematics passed through Ph.D. programs in the United States, Canada, and England. They gained the respect of the molecular and cellular biologists, and they were well funded for a while, before the academic recession of the late 1970s and 1980s. They shared the ambition and optimism of their immediate predecessors in the Marlboro Circle. I was able to play a role in this next step, more as a result of my residence at Harvard than of any special talent. My course "Evolutionary Biology," begun in 1958, was relabeled "Population Biology" in 1963 and focused more on basic theory. At first I thought that I had failed by pushing model construction too far at the expense of natural history. An undergraduate complained in the *Crimson Confidential Guide,* the uncensored and often scathing student review, that the course was a dull exercise in numerology. So it might have seemed to some, but I came to realize later that many of the students were greatly influenced by my presentation, and a few were drawn into population biology as a career. They include some of the current leaders in the field: William Bossert, Joel Cohen, Ross Kiester, Jonathan Roughgarden, Daniel Simberloff, and Thomas Schoener. In 1971 Bossert and I collaborated on a short, self-teaching textbook, *A Primer of Population Biology,* which remains popular more than twenty years later.

In the spring of 1971 Robert MacArthur experienced abdominal pains during a field trip to Arizona. Returning home to Princeton, he learned that he had renal cancer. The affected kidney was promptly removed, and he was placed on chemotherapy. Too late: the surgeon told him that he had only months or at most one or two years to live. Robert thereafter conducted his life with even greater intensity than before. He completed his final book, *Geographical Ecology.* He journeyed to Arizona, Hawaii, and Panama for more field work, and while at the university he continued to guide his students. He began a new round of theoretical research, this time with Robert M. May, a brilliant Australian physicist who soon thereafter joined the Princeton faculty. Under MacArthur's influence May converted to biology and developed into one of the world's most influential ecologists. He subsequently moved to Oxford University as a Royal Society Professor.

MacArthur was still reasonably strong as the fall term began at Princeton in 1972. He was coughing frequently as the cancer spread into his lungs, but he was still able to come to his office for short periods to talk to students and friends. In early October his health declined rapidly. By this time I had joined several senior American evolutionary biologists—James Crow, Darlington, Hutchinson, and Eugene Odum—to nominate him for the National Medal of Science. With news that he had only a very short time to live, we redoubled our efforts. Robert sent word through Hutchinson that the nomination was welcome, and he was "pleased that my friends think well of me." *Geographical Ecology* had also just been published, and he awaited the first reviews.

On a Monday afternoon, October 30, John Tyler Bonner, chairman of Princeton's Department of Biology, dropped by my office while visiting Harvard. He told me that MacArthur's condition had deteriorated badly and the end could come in hours or in weeks. The matters most on Robert's mind at this point beside his family were, he reported, the National Medal and the reviews of his book. I dropped everything and inquired about both matters. No progress in the committee office at the National Science Foundation on the medal. But two back-to-back reviews of *Geographical Ecology* had just appeared in *Science*, one by Thomas Schoener and the other by Scott Boorman, both important young population biologists. I called Katherine Livingston, the reviews editor, who said she would send copies directly to Robert.

They arrived too late. The next morning I tried to telephone Robert at his home. A nurse with an unidentifiable foreign accent said he was sleeping. I called again at two in the afternoon, and this time he came on. His voice was thin but level. He coughed frequently, and twice he had to stop to get his position changed in order to continue. I was relieved to find his mind clear and composed. I asked, Had he seen the *Science* reviews? Not yet. I fished out the manuscript of the one by Boorman (who was at that time studying under my direction) and read it. The text was long, detailed, and laudatory. Robert

was fascinated, and stopped me several times to discuss technical points raised. Boorman is clearly bright, he said. Was the
review by Schoener as good? I assured him it was. I'd seen the
manuscript, which after exploring the general methodology of
model building declared Robert's book to be the key synthesis
in the field. He said, Good, it's better than I got from Slobodkin for my elementary biology textbook.

Had I heard more about the National Medal of Science? I
had not, except that eighteen people had been nominated and
the awards would be announced sometime after the November
7 presidential election. Robert was disappointed. I sensed that
he was worried about his place in biology. We then moved on
to gossip and miscellaneous news. Our conversation remained
normal in content and tone, with no serious digression into
his physical condition. We talked as though he had years to
live. He grew tired and quiet. I began to do most of the talking, afraid to let him go. I nattered on about the arrival next
term of my fellow entomologist Bert Hölldobler to assume a
professorship at Harvard; the opening of the new laboratory
wing of the Museum of Comparative Zoology; and Lewontin's
political demonstration at the Chicago meetings of the American Association for the Advancement of Science and his widely
publicized resignation from the National Academy of Sciences.
We drifted on to a recent proposal to exterminate the kiskadee
as a pest bird species on Bermuda. Robert mumbled assent as
I went along.

At last Robert said we had talked enough and should
stop now. We agreed to stay in touch. At dinner, Betsy later
reported, he was calm and happy. He spoke with particular
pleasure about the favorable *Science* reviews. In the early hours
of the next morning he died without distress, in his sleep.
Today I can imagine no more inspiriting intellect or steeper
creative trajectory cut so short with such a loss to others. I wish
he might have known in those final days that his place in the
history of ecology was secure. I owe him an incalculable debt,
that for at least once in my life I was permitted to participate
in science of the first rank.

CHAPTER FOURTEEN

The Florida Keys Experiment

WHERE COULD WE find more Krakataus?

That question dominated my thought for months after MacArthur and I published our first article on island biogeography in 1963. We had conjured a plausible image of the dynamic equilibrium of species, with new colonists balancing the old residents that become extinct, but we could offer very little direct evidence. There are few places in the world where biologists can study the approach to equilibrium on a large scale. Krakatau-sized events, the sterilization of islands the size of Manhattan or larger by volcanic explosions, occur at most once a century. Another hundred years might then be needed, once the smoking tephra cooled down, to observe the full course of recolonization. How might we get data more quickly, say within ten years?

I brooded over the problem, imagined scenarios of many kinds, and finally came up with the solution: a *laboratory* of island biogeography. We needed an archipelago where little Krakataus could be created at will and their recolonization watched at leisure.

My dream embraced more than the search for new experiments in biogeography. I was driven by a more general need to return to the field, to enjoy once again the hands-on kinesthetic pleasures of my youth. I wanted to remain an opportunist, moving among, seeing, and touching a myriad of plants and animals. I needed a place to which I could return for the rest of my life and possess as a naturalist and scientist.

It would have to be a different location and context from those previously enjoyed. I couldn't return to New Guinea to launch my endeavor. Work there would take me for months at a time away from Cambridge, where my duties at Harvard held me tightly. I had also begun experimental work on the social behavior of ants that required a well-equipped laboratory. They were proving too successful to abandon. Not least, I had a family now, Renee and our new daughter, Catherine.

How in the world could I explore an island wilderness while staying close to home? And if I found such a place, how could I turn it into a laboratory? There was only one way to solve the problem: *miniaturize* the system! Instead of relying on conventional islands the size of Krakatau, which are hundreds of square kilometers in area and usually have people living on them, why not use tiny ones, at most a few hundred square meters? Of course such places do not support resident populations of mammals, birds, or any other land vertebrates above the size of small lizards. Vertebrate biologists would not call them islands at all, even in a limited ecological sense. Yet they sustain large breeding populations of insects, spiders, and other arthropods. To an ant or spider one-millionth the size of a deer, a single tree is like a whole forest. The lifetime of such a creature can be spent in a microterritory the size of a dinner plate. Once I revised my scale of vision downward this way, I realized that there are thousands of such miniature islands in the United States, sprinkled along the coasts as well as inland in the midst of lakes and streams.

I thought I had the perfect solution. By exploring such places I would satisfy my emotional and intellectual needs. Working with insects, the organisms I knew best, I could conduct biogeographic research on an accelerated schedule. Succeed or fail, I would stay close to Harvard and my family.

In choosing the site of my laboratory, I preferred marine waters over lakes and rivers—strictly an aesthetic choice. I pored over maps of fringing islands all along the Atlantic and Gulf coasts, from Quoddy Head State Park in down-Easternmost Maine to the Padre Island National Seashore in southernmost Texas. I also studied charts of the small islands around Puerto Rico, still a relatively quick jet flight away. A decisive winner quickly emerged: the Florida Keys, if combined with the nearby northern islands of Florida Bay and the southwest mainland coast, seemed ideal. I turned to more detailed navigational charts and photographs for a closer look. The islands came in all sizes, from single trees to sizable expanses up to a square kilometer or more. They varied in degrees of isolation from a few meters to hundreds of meters from the nearest neighbor. The forests on them were simple, consisting in most cases entirely of red mangrove trees. And they were available in vast numbers. One sprawling miniature archipelago west of the Everglades bore the suggestive name Ten Thousand Islands. Almost all of them could be reached in a single day, if you started with an early four-hour flight from Boston to Miami, drove a rental car down U.S. 1 to the Keys, and finally took a short boat trip out to the island of choice.

In June 1965 I flew to Miami to enter my new island world. I was accompanied by Renee and Cathy—now twenty months old, walking, talking, and pulling down every movable object. For ten weeks I explored the small bayside mangrove keys from along Stock Island and Sugarloaf north to Key Largo. My spirits soared. I was back where I was meant to be! Each morning I pushed away from a marine dock in a rented fourteen-foot boat with outboard motor and moved out along the channels that had been cut through the mangrove swamps to the open waters of Florida Bay. I visited one islet after another, passing over turtle grass flats in water sometimes clear and sometimes, especially on windy days, milky white from the churned-up bottom marl. Once or twice a day I saw a distant fisherman or a powerboat moving to deeper water, but into the swampy archipelagoes of my choice few other people ventured. Less than a mile away U.S. 1, which runs the length of the keys to their southernmost point at Key West, was choked by traffic. It was lined by a noisome thicket of motels, trailer parks, amusement parks, marinas, fishing tackle shops, and fast-food restaurants. But beyond hearing range of the rumble and whine of traffic, the swamps and islets were pristine, a virgin wilderness. Mangrove wood has little commercial value. No one but a naturalist or escaped convict would choose to traverse the gluelike mud flats and climb through the tangled prop roots and trunks of the mangrove trees. So I had it all to myself: one more time, a world I knew so well, more complex and beautiful than anything contrived by human enterprise.

I pushed into the interiors of the islets to examine the arthropod inhabitants. Sometimes the little forests opened at the center into a slightly raised glade carpeted with aerial roots and algal mats. Sometimes I found myself beneath the massed nests of clamoring herons, egrets, and white-crowned pigeons. I drifted along from landfall to landfall, collecting specimens, studying charts, filling my notebook with impressions. Mine was anything but a world-class voyage, but I was as content as Darwin on the voyage of H.M.S. *Beagle*. I ate lunch in the boat while peering over the side at rich marine life along the edge of the islets. Just beneath the reach of low tide, the mangrove prop roots were covered by masses of barnacles, sea squirts, anemones, clams, and green and red algae. Schools of mangrove snappers and young barracuda prowled in and out of the root interstices and alga-slimed cavities of the mudbanks. Should I have become a marine biologist? Too late to think about that now. I was at peace. The only sounds I heard were the call of birds and the slap of waves against the hull of my boat. An occasional jet droned high above, to remind me, you'll come back, dreamer, your life depends on those artifacts you've tried to escape.

I found what I had come for in the mangrove islets. The trees swarmed with scores of species of small creatures: ants, spiders, mites, centipedes, bark lice, crickets, moth caterpillars, and other arthropods. Many flourished in breeding populations, prerequisites for the establishment of an experimental biogeography. And from one mangrove clump to the next, the species changed. For ants the pattern was consistent with competitive exclusion. Below a certain island size, the colonization of some species appeared to preclude the establishment of others. I saw an opportunity in the study of these telescoped patterns. Instead of traveling great distances from one Pacific Island to another to study the distribution of birds, an effort requiring months or years, I could, by guiding a fourteen-footer among the islets, analyze the distribution of arthropods in a period of days or weeks.

How, then, might these mangrove dots be turned into little Krakatus? I saw no easy way, and cast about for some alternative approach. I made the following decision: continue with the mangrove studies, but in addition select other islands lacking trees in order to make sterilization easier. I had learned that treeless sandy islands in the nearby Dry Tortugas are occasionally flooded and swept clean of their low scrubby plant growth by hurricanes. If I could monitor them before and after a big storm, I might observe the recolonization process and establish whether it created an equilibrium. Let the Caribbean's stormy weather be the volcano. At least it was worth a try.

I called on William Robertson, official naturalist of the Everglades National Park, to explain my idea. Bill frequently visited the Dry Tortugas to study sooty terns, a far-ranging species that nests on this remotest of Florida's archipelagoes. He agreed that the procedure might work, and invited me to join his research party on the next boat trip out from the docks at the Everglades town of Flamingo to survey the area. Once settled in dungeonlike rooms at Fort Jefferson, the old Federal stronghold and prison on Garden Key, we took a smaller boat out to the other, smaller islands of the Dry Tortugas. I leaped into the surf and scrambled onto each of the little sandy keys in turn, making a record of the sparse vegetation and arthropods. My notebook was soon complete. All I had to do now was wait for a serious hurricane to pass over in order to begin a study of recolonization.

Providentially, from a biologist's possibly perverse point of view, two hurricanes struck the Dry Tortugas during the next ten months. Betsy, on September 8, 1965, threw gusts up to 125 miles per hour at Fort Jefferson. The milder Alma attained gale force winds on June 8, 1966. Between them they wiped the vegetation off the smallest sandy islands, as I had hoped. By that time, however, I had changed my plan and advanced to a bolder scheme. Why be confined to the haphazard distribution of a few remote keys? And why depend on the passing of hurricanes, which normally strike the Dry Tortugas only once or twice every ten years? The method was in any case not fully experimental. It could not be controlled. Instead, I thought, why not select ideally located mangrove keys from among the hundreds near U.S. 1, then fumigate them with pesticides? It should be possible to kill off all the insects and other arthropods. These islets could be chosen to represent different sizes as well as various distances away from the mainland. Other islets, left unfumigated but otherwise studied in identical fashion, might serve as controls.

At this point, the fall of 1965, Daniel Simberloff joined me as a collaborator. The added vision and inspired effort of this second-year graduate student made it possible to turn the mangrove keys into a laboratory. Dan was primed for an effort of this kind. While an undergraduate at Harvard he had majored in mathematics, graduating magna cum laude. He could have moved on easily to a successful career in mathematics or the physical sciences. But after taking Natural Sciences 5, the famous nonmajors course in biology given by George Wald, he decided this branch of science was more to his liking. During his senior year he interviewed Bill Bossert and me and asked: Is graduate study in biology feasible if one has a stout heart but thin undergraduate training in that subject? Indeed it was, we both responded, especially for a mathematician. If you enter population biology now, the new discipline will reward skills in model building and quantitative analysis. All you need to do is add an all-out effort in biological training.

Simberloff began his Ph.D. study under my sponsorship in the fall of 1964. I hesitate to use the usual expression "studied under me," because in the years to follow I learned as much from him as he did from me. We soon became partners.

Dan at least looked as though he could manage field biology. With somewhat hawkish features, a solid muscular body carried in a relaxed slouch, he might have passed for the kind of Ivy League quarterback who studies calculus or Chinese history too conscientiously to be an athletic star. Like many bright students of the day he was also a leftist radical, of the thinker rather than activist subspecies, suspicious of all authority and fierce enough to be a supporter of Eldridge Cleaver for President. This was quite all right with me. In 1965 the civil rights movement still meant idealism and courage tested on the dangerous back roads of Mississippi. The mere mention of Cuba, recently the site of history's only nuclear confrontation, chilled us both; and the war in Vietnam was slowly gathering momentum. The Florida Keys were bracketed by bases at Homestead and Key West, and the whole area hummed with military activity. That summer I saw my first Green Berets, a platoon riding through the streets of Key West in a troop carrier. My admiration for the military and my vaguely centrist political beliefs were yielding somewhat to uneasiness over the direction the country was headed. Soon Dan and I began to share acerbic jokes about Lyndon Johnson. We watched in resentment as helicopters flew overhead, bearing commanding officers from ships to their homes ashore. We perched on the branches of mangrove trees, collecting spiders and crickets, on a nearly invisible budget, trying to learn how ecosystems are assembled. A dozen helicopter rides would have paid for our entire project. But not one citizen in a hundred would have understood what we were trying to do. It was a time of massive imbalance in favor of military security over environmental security. We had no idea how or when the differential might be redressed, nor did we expect ever to see ecology given national priority as a science. We were just thankful for the opportunity provided us by modest funding from the National Science Foundation. And thankful just to be there, in this beautiful natural environment.

By joining the project, Dan took a career risk. Our endeavor had an uncertain future, because no one had previously tried or even conceived anything like it. If we were unable to eliminate the arthropods completely from the islets, we would be in trouble. If we failed to put scientific names on the myriad of species we found on the islands, our data would be far less valuable. If the colonization of the sterilized islands took ten or twenty years or longer to progress significantly, Dan would have to find other work to complete his Ph.D. thesis. Graduate students were expected to finish their degree requirements, including a complete and reasonably well polished research thesis, in no more than six or seven years. Most accepted low-risk projects, those new enough to generate significant results but close enough to preexisting knowledge and proven techniques to be practicable. Simberloff had none of these assurances. In September 1965 he nonetheless departed for the Florida Keys, with the initial task of selecting the experimental islets.

In the months that followed we divided the labor further. While Dan grew lean and acquired a deep tan laboring on the open waters of Florida Bay, I attended to the administration of the project. The details of my own role ranged from the unusual to the bizarre. For an effort of this kind we had first of all to engage the services of a professional exterminator. Fortunately, there was an abundance of companies in Miami. The executives of the first two I called answered with rich southern accents and clearly thought I was either joking or crazy. On the third try I got Steven Tendrich, vice president of National Exterminators, Inc. He had a northern accent, which gave me hope. Could he manage, I asked carefully, to spray clumps of mangrove in the Florida Bay with short-lived insecticides that would remove all the insects? We would ourselves eliminate by hand the tree snails and other larger animals that might be resistant to the chemical. Tendrich did not hesitate. He said yes, maybe he could do a job like that. Sure, give him some time to study the logistics. But even if it looked promising, he warned, he could not manage much in the field until the fall, when the heavy business of summertime Miami slacked off.

Progress in this sector having been achieved, I went with Simberloff to visit Jack Watson, the resident ranger of the National Park Service, to ask his permission to exterminate the whole faunas of islets. Most of the candidate islets were within the boundaries of the Everglades National Park and Great White Heron National Wildlife Refuge, over which he had partial jurisdiction. Obtaining permission to wipe out animal populations on federally protected land may sound like an impossible dream, but it proved relatively easy. Watson gave it without hesitation, asking only that we keep him briefed. Bill Robertson, our principal contact in the Park Service, was also in sympathy with the rationale and plan of the project. He knew that the targeted islets were no more than clumps of red mangrove among hundreds scattered through Florida Bay. They harbored species or races no different from those abounding elsewhere. We assured Watson and Robertson of our intent to protect the vegetation, and our expectation that the trees would be fully recolonized with insects and other arthropods following the "defaunation," as we now called it. The experiment, Simberloff and I argued, might provide information that would help guide future park management policy. Our earnestness proved persuasive, and we never faced opposition from government officials or the public.

Finally, I set out to contact specialists who could identify the species of insects and other arthropods living on the mangrove islets before fumigation and while the recolonization proceeded. This proved the most difficult task of all. There were at most several hundred entomologists in the United States able to identify insects from the Florida Keys. Their study would be complicated by the fact that many of the creatures we expected to find are immigrants from the West Indies, especially nearby Cuba and the Bahamas. Among our discoveries were to be the first specimens of the tropical spider family Hersiliidae recorded in the eastern United States and several large and striking long-horn beetles previously known only from the Bahamas. In the end we were able to persuade fifty-four specialists to assist us in the classification of our specimens. Most pitched in with enthusiasm. An expert on spiders, Joseph Beatty, went so far as to visit Simberloff in the field to assist with the on-site identification of the colonists.

During the spring of 1966 Simberloff reported in with his recommendation of islets that seemed well placed either for defaunation or to serve as controls. We began surveys prior to spraying by inspecting every square millimeter of trunk and leaf surface, digging into every crevice, prying beneath flakes of dead bark and into hollow twigs and decaying branches. We collected every species of arthropod we found. Later, after the defaunation, Dan took over the heavy duty of regular monitoring. To disturb the colonists as little as possible, he relied on photographs and his own growing familiarity with the mangrove fauna. It was hard and uncomfortable work, demanding the combined skills of insect systematist, roofer, and restaurant health inspector. Simberloff, the city-bred mathematician, did well. He endured the insect bites and lonely hours in the hot sun I had promised him. Once, after his outboard motor failed, he spent the night on one of the islets, managing to escape only when he hailed a passing fisherman the following morning. Exasperated with the gluelike mud through which we had to wade to reach several of our islands, he built a pair of plywood footpads shaped like snowshoes and drilled holes in them to reduce suction when they were lifted. When he tried them out he sank to his knees and had to be pulled out by me and another companion. I called the invention "Simberloffs" afterward. Dan was not noticeably amused.

I joined him at intervals to give assistance. On one memorable occasion—June 7, 1966—Dan met me at the Miami International Airport just as Hurricane Alma was churning up the central Caribbean in the general direction of Florida. A storm watch had been posted for Miami and the keys. When we awoke the next morning the sky had clouded over, wind was picking up from the south, and a light rain had begun to fall. The eye of the storm was expected to pass up the west Florida coast and sideswipe Miami. Here, I thought, was a rare opportunity to watch a hurricane disperse animals out of the mangrove swamps and across the water. Travel in high winds seemed a likely means of colonizing the little islands. I suggested that we stand inside a nearby mangrove swamp during the storm and watch for animals blown along by high winds. For some reason that escapes me now, I didn't think much about danger to ourselves. Simberloff agreed without hesitation. All right, he said, something interesting might happen. Good enough.

We were both a little crazy in those days. As stronger gusts of wind and rain blew in, and the streets began to empty of traffic, we drove to Key Biscayne and hiked into a patch of red mangrove swamp along the bay shore facing Miami. The eye of the hurricane was now passing up the west coast on its way toward landfall in northwest Florida. The gusts on Key Biscayne reached sixty miles an hour, gale but not hurricane force. I was disappointed. The wind was not strong enough to tear insects and other small creatures from the trees. They all stayed hunkered down safely on the branches and leaves as the rain-soaked winds roared through. We saw not a single animal blow by. Nor could we find animals struggling in the water at the edge of the swamp. I said, Well, let's see what would happen if an animal *were* blown free. Would the storm-tossed waves carry it out toward a distant shore? I caught an anole lizard and tossed it ten feet or so out into the water. To my dismay, it popped to the surface, swam expertly back to the shelter of the trees, and climbed up a mangrove trunk. Well, I continued, suppose a full hurricane blew an anole so far away on open water it couldn't get back. Our little experiment shows that it could swim to the nearest islet if it were not too far away. Dan, rainwater streaming from his hat, allowed that the notion was plausible. Our excursion was not a complete loss, but in later years we agreed we were lucky that Alma only brushed Miami. Otherwise we ourselves might have been washed to a distant shore, proving our own hypothesis *in extremis.*

A month later, I joined Steve Tendrich and a crew from National Exterminators on a trip into Florida Bay to spray the first two islets, "Experimental 1" and "Experimental 2," E1 and E2 for short. Simberloff was busy at another location preparing additional islets. We loaded a rented barge with equipment and set forth from a marina on Sugarloaf Key. Halfway out we came upon a stalled sports-fishing cruiser. Observing the law of the sea even in this relatively safe stretch of water, we took the captain and his two fisherman guests on board and back to Sugarloaf. Then we headed forth again. This time we reached E1 and sprayed the little island with parathion. The next morning we proceeded to E2. Here we spotted several nurse sharks, one nearly four feet in length, cruising the shallow waters around the islet. Trouble! The workmen refused to get off the barge. But I knew that nurse sharks never attack people unless hooked or seized by the tail and hauled from the water by the occasional reckless fisherman. They live on a diet of shellfish, crustaceans, and other small bottom-dwelling animals. So I volunteered to stand guard waist deep and drive the sharks away with an oar. Impressed by my specious bravery and with their male pride challenged, the crew got into the water and sprayed E2.

Several days later, after I had returned to Cambridge, Simberloff called with mixed news about E1 and E2. He had made a close inspection of the islets and found that the kill of the arthropods living on the surface of the vegetation had been total. But some beetle larvae living deep in the wood of dead branches survived. We realized we had no way of knowing what other creatures might still live in these deeper spaces. So we quickly agreed that spraying with parathion or some other short-lived insecticide was not enough. In order to run a proper experiment, we had to start with islets scourged of all animal life, with no exception. It would be necessary to fumigate the islets with a poisonous gas, one that penetrates every crack and crevice.

I called Steve Tendrich: could National Exterminators fumigate an island? The ever-resourceful Tendrich responded in his usual positive manner: why not? It was common practice in Miami, he said, to cover entire houses with a rubberized nylon tent and fumigate the interior in order to remove all termites and other insect pests, no matter how deeply hidden in the woodwork. To transfer the method to a large object surrounded by water would be tricky, of course. The crew would need to erect a scaffolding around the islet as a frame for the tent. We couldn't just lay the cover on top of the fragile branches. And something else: the dosage of the gas must be set just right, high enough to kill all the animals but low enough to leave the mangrove trees undamaged. To study a ghost island of dead wood and fallen leaves would have no meaning, I agreed. Not least, I had promised the National Park Service that we would preserve the live vegetation.

So it was to be poison gas. But what kind? We considered and quickly discarded hydrogen cyanide. It was too dangerous for the crew to use under these uncertain conditions, over water with possible stiff winds. Even if we could apply it safely, hydrogen cyanide is water-soluble and would probably kill the marine communities around the mangrove prop roots, an unacceptable side effect. Methyl bromide, Tendrich ventured, might fill the bill, if he could get the dosage just right. Tendrich immediately set up trials, using small mangrove trees in the swamps near Miami. Meanwhile Simberloff collected cockroach egg cases from mangrove swamps for Tendrich to test with various dosages. If these highly resistant insect life stages could be killed without damage to the vegetation, methyl bromide might work.

The window between insect kill and tree kill using methyl bromide was narrow, but Tendrich found it. On October 11, 1966, we all gathered for the first trial on an islet in the shallow waters of Harnes Sound, on the mainland side of Key Largo and a relatively short ride down U.S. 1 from Miami. As the men loaded the gear, we saw ospreys and pelicans flying nearby and herons spearing fish in the shade of the mangrove fringes along walls of barnacles and green algal mats stranded by a dropping tide. Somewhere close by, we had been told, was a nest of bald eagles. The men got the scaffolding up and closed the tent around it without mishap. They pumped the prescribed dose of methyl bromide through a flap-covered opening in the side, in the same manner used to fumigate a small house, then pulled the tent away, allowing the gas to dissipate quickly to harmless levels.

The next day we searched the islet thoroughly and found no trace of animal life. Even the deep-boring insects had been killed. Our colonization experiment was under way at last.

Tendrich, however, was not entirely satisfied with the procedure. It had worked all right at Harnes Sound a hundred yards from the highway, but the metal rods used to create the frame were heavy and clumsy and might be very difficult to transport to the more remote and less accessible mud-flat sites. He began to search for alternative techniques of scaffolding. One day as he drove through Miami he spotted a steeplejack working on a tower atop a hotel, and inspiration struck. Steve stopped the car, took an elevator to the hotel roof, and waited for the man to come down. He asked the steeplejack, whose name was Ralph Nevins, whether it might be possible to erect a small tower like that in the middle of a mangrove swamp, then drape a tent over the guy wires. Sure, Nevins replied—another optimist—why not? Would it be very difficult? Don't think so. Tendrich hired him on the spot. And so it was done thereafter. The rest of our islets were fumigated beneath tents wrapped around the guy wires of a tower raised by Ralph Nevins.

Simberloff continued to carry the main burden of monitoring. He was tied for months to a physically demanding routine of travel, search, and identification. When I found time I came down from Cambridge, and we worked together. Within weeks it was apparent that the project was going to be a success. The recolonization by arthropod species was already well under way. Moths, bark lice, and other flying insects appeared early, at first in small numbers but accumulating and reproducing as time passed. Winged ant queens, newly inseminated during their nuptial flights, landed, shed their wings, and started colonies. Spiders came early in abundance; some were wolf spiders the size of silver dollars. How were they crossing the water? Since there had been no major storms, we guessed that they used ballooning. Many kinds of spiders, when crowded or short of food, prepare to emigrate by standing in exposed places on leaves and twigs and letting out threads of silk into the wind. As the strands lengthen, the drag increases, until the spiders have difficulty holding themselves in place. Finally they let go, allowing the wind on the strands to pull them up and away. With luck they come down again on land, and best of all in some place like a distant mangrove island with few other spiders and an abundance of prey. Those that hit water instead soon become fish food.

Toward the end of the year following the defaunations, a pattern in the colonization began to emerge. With so much of our time invested, we now began to worry that a hurricane might strike, perturbing the new faunas and ruining the continuity of the experimental run. Fortunately, none came close to Florida. In fact no major storm struck the area again until Andrew devastated South Miami and the northern keys in 1992. After a while, we relaxed a bit and broadened our attention to include other aspects of local ecology.

Our first major project was to launch a survey of the arthropods of all the mangrove swamps, in order to gain a picture of the pool of all possible emigrants to the experimental islets. I hired Robert Silberglied, a graduate student working in entomology under my direction, to commence a general survey of the surrounding keys. Bob was a gifted naturalist and a polymath taxonomist who could on sight identify species from a wide array of animal groups. The challenge of a complete arthropod survey was made to order for his talents. He worked tirelessly from island to island, building a large reference collection of insects and other arthropods. His impressive potential was destined never to flower into a full career, however. On January 13, 1982, he died with others in the Potomac River crash of an Air Florida airliner on the outskirts of Washington, D.C. A winter storm was in progress, and the accident was later blamed on wing icing. The flight was to have been the first leg of a flight to Panama, where Bob had planned to continue research on tropical ecology.

Our interest in the Florida Keys, as our research moved onward into 1967, also extended to include conservation. Silberglied and Simberloff heard rumors that Lignumvitae Key, a 280-acre island on the bay side close to Lower Matecumbe Key and its transecting segment of U.S. 1, was an unspoiled paradise covered by large hardwood trees. Undisturbed forest other than mangrove was a rarity in the Keys, and worth investigation. Few people had set foot on Lignumvitae to that time. One was Konrad Lorenz, who later opened his influential book *On Aggression* with a description of the coral reefs there and on nearby Key Largo.

When Silberglied and Simberloff put ashore, they were met by the caretakers, Russell and Charlotte Niedhauk, an elderly and reclusive couple who lived on the island alone. The Niedhauks were suspicious of all visitors, and rudely chased most away. But when Bob and Dan revealed that they were biologists interested in conservation of the island, they were given a warm welcome. As they walked inland from the caretakers' house, they confirmed the rumor: almost all the land was covered by a mature tropical hardwood forest. They were thrilled to find themselves in a near-primeval habitat that once predominated in the high islands of the Keys but had been almost completely obliterated by the 1960s. Huge mahogany and gumbo-limbo, including the largest individual of the latter species in the United States, towered over wild lime, torchwood, Jamaica dogwood, boxleaf stopper, strangler fig, and the only large stand of holywood lignum vitae in Florida. Sixty-five species of trees and woody shrubs, all tropical and subtropical, composed the woody flora. The fauna was also a remnant of the old Keys. Candy-striped tree snails hung like grapes from the trunks and branches. Large butterflies, including showy dagger wings, purple wings, and swallowtails, darted and floated back and forth above the shaded trails. Bald eagles, at that time nearly extinct in the eastern United States, were occasional visitors, and Bahama bananaquits were seen from time to time. Later, after he had visited the island, Archie Carr, the great expert on Caribbean natural history, reminded me that the Lignumvitae forest was a tropical West Indian lowland forest of a quality no longer found in the West Indies themselves. The chances of finding stands of old mahogany and lignum vitae on the islands were close to zero.

The Niedhauks were almost paranoid about the future of Lignumvitae Key. It was owned, they explained, by a private consortium of wealthy Floridians who were planning to convert it into a community of expensive vacation residences. All the owners cared about, they said and I later confirmed, was a financial killing. Could the visitors help find a way to preserve the island in its natural state? Bob and Dan conveyed this information to me as soon as they returned to their base. Soon afterward, I visited the island and was similarly enchanted, and fearful. I invited Thomas Eisner, my old friend on the Cornell faculty, to join me on a second visit. Together we prepared an article for *Natural History* about Lignumvitae and its plight.

While our effort was under way, I spoke at a meeting of the Florida Audubon Society in Miami on the subject, and to my delight an elderly couple living in Coral Gables pledged $100,000 toward the purchase of the island. It was a big first step toward saving Lignumvitae. But we needed more; the owners had set the tentative price at over $2 million. Their spokesman, a septuagenarian Miami dentist, was gleeful that conservationists had entered the bidding. He made it clear that the final price would be raised as high as the owners could make it. He would love to see his beautiful island saved in its natural state, he claimed, but if we did not act soon the land would go to developers. The Lignumvitae ecosystem, in short, had been placed in ransom.

I contacted Thomas Richards, president of The Nature Conservancy, in hopes of pressing the campaign to a successful conclusion. TNC was, as it remains today, famous for its policy of purchasing environmentally important land for preservation in the public domain. After a visit of his own, Richards committed his organization to the effort. He then approached Nathaniel Reed, an influential administrator in Florida's park system, for further assistance. In the end, after long negotiation, a reasonable price was agreed upon. The island was purchased with funds from The Nature Conservancy and the State of Florida, and Lignumvitae Key was turned into a fully protected State Botanical Site. Today visitors walk along trails where tree snails still decorate the gnarled old lignum vitae trees and dagger wings alight among their delicate blue flowers and petard-shaped yellow fruits. The public can in perpetuity, I trust, witness the Florida Keys as they were in prehistory.

Meanwhile, our experimental project continued to move swiftly forward. By the fall of 1967, a year after we fumigated the islets, the results were all but conclusive. In a formal article published two years later, Simberloff and I summarized the events of recolonization and the reattainment of equilibria:

By 250 days after defaunation, the faunas of all the islands except the distant one ("E1") had regained species numbers and composition similar to those of untreated islands even though population densities were abnormally low

> . . . The colonization curves plus static observation on untreated islands indicate strongly that a dynamic equilibrium number of species exists for any island.*

At least the cruder predictions of the theory of island biogeography had been met. The closest island, as expected, had the largest number of species before fumigation, forty-three to be exact, and it regained approximately that number within the year. The most distant island, E1, had the smallest number, twenty-six, and climbed back close to that after defaunation. The other islands, at intermediate distances, had intermediate numbers before fumigation and also returned to their original levels afterward. Two years later, in 1968, these various levels still held.† The turnover in species was very rapid, also as expected from island biogeographic theory applied to small, swiftly occupied islands. In the course of our studies we added many observations on the dispersal and early colonization of various groups of arthropods, including spiders, mites, ants, earwigs, bark lice, crickets, and many others. Dan completed his Ph.D. thesis in the spring of 1968. It had taken us only three years to create miniature Krakataus, in replicate with controls, and follow their histories to an early form of equilibrium.

In 1971 Simberloff and I received the Mercer Award of the Ecological Society of America for our research, a welcome recognition. We had risked a new approach to biogeography, a subject still considered outside the mainstream of ecology, and succeeded. From many employment opportunities open to him, Dan accepted an assistant professorship at Florida State University, in order to be within easy distance of field sites. In time he became an ecologist of international stature. He conducted additional experiments with mangrove islets, varying their size and shape. He expanded his activities to include field studies on other ecosystems, and used his mathematical skills to conduct critiques of ecological theory and to develop

---

*"Experimental Zoogeography of Islands: The Colonization of Empty Islands," *Ecology* 50(2) (1969): 278–295.
†"Experimental Zoogeography of Islands: A Two-Year Record of Colonization," *Ecology* 51(5) (1970): 934–937.

new approaches in quantitative modeling. In time his university appointed him to the Robert O. Lawton Distinguished Professorship.

I did not return to the Florida Keys, and my dream of converting them into a natural laboratory languished. A new possibility—a different opening to the future—had seized my imagination. I wanted to make sociobiology into a single science, one that ranged from ants to chimpanzees.

CHAPTER FIFTEEN

Ants

THEY ARE EVERYWHERE, dark and ruddy specks that zigzag across the ground and down holes, milligram-weight inhabitants of an alien civilization who hide their daily rounds from our eyes. For over 50 million years ants have been overwhelmingly dominant insects everywhere on the land outside the polar and alpine ice fields. By my estimate, between 1 and 10 million billion individuals are alive at any moment, all of them together weighing, to the nearest order of magnitude, as much as the totality of human beings. But a vital difference is concealed in this equivalence. While ants exist in just the right numbers for the rest of the living world, humans have become too numerous. If we were to vanish today, the land environment would return to the fertile balance that existed before the human population explosion. Only a dozen or so species, among which are the crab louse and a mite that lives in the oil glands of our foreheads, depend on us entirely. But if ants were to disappear, tens of thousands of other plant and animal species would perish also, simplifying and weakening land ecosystems almost everywhere.

They are intertwined in our world, too, as illustrated by an incident that occurred in Harvard's Biological Laboratories in the late 1960s. At the risk of melodrama I will call it the Revenge of the Ants. The serious trouble began when an assistant in Mark Ptashne's laboratory, a humming center of research on gene repression, began the routine pipetting of sugar solution for the culture of bacteria. She could not draw the liquid through. Looking more closely, she saw that the narrow pipette channel was plugged with small yellow ants. Other, more subtle signs of a strange invasion had been noted in the building. Here and there yellow ants quickly covered food left out after lunch or afternoon tea. Portions of breeding colonies, with queens and immature stages surrounded by workers, appeared miraculously beneath glass vessels, in letter files, and between the pages of notebooks. But most alarming, researchers found the ants tracking faint traces of radioactive materials from culture dishes across the floors and walls. An inspection revealed that a giant unified colony was spreading in all directions through spaces in the walls of the large building.

I had reason to be concerned with the invasion. It had started by accident in my own quarters. The species was Pharaoh's ant, known to specialists by its formal name *Monomorium pharaonis*, a notorious pest of East Indian origin that infests buildings around the world. When a supercolony occupies hospitals, its workers feed on surgical waste and the wounds of immobilized patients, in the course of which they sometimes spread disease organisms. Portions of colonies transport themselves by moving into luggage, books, clothing, and any other objects with one or two inches of space. Arriving at a new accidental destination, which might equally well be an apartment house in Oslo, a florist shop in St. Louis, or a vacant lot in Caracas, they move out and commence to breed.

Harvard's propagule, as we reconstructed its history later, took passage in the airport at the Brazilian port city of Belém. Portions of a supercolony entered two wooden crates belonging to Robert Jeanne, a Ph.D. candidate studying under my direction. Now a professor of entomology at the University of Wisconsin, Jeanne in 1969 was homeward bound after a lengthy period of field research in the Amazon rain forest. By the time he opened the crates in the Biological Laboratories and discovered the hitchhikers, the ants had established themselves in the walls and were metastasizing.

To eliminate a large population of Pharaoh's ants by conventional means can be expensive and disruptive. An ingenious alternative solution was devised by Gary Alpert, a graduate student in entomology with a special interest in pest control. He was counseled and aided by Carroll Williams, Harvard's professor of insect physiology. Williams provided a chemical compound that mimics the action of the juvenile hormone of insects by sterilizing queens and preventing the full development of the larvae into the adult stage. By mixing this compound with peanut butter, Alpert fashioned baits that he hoped the foraging ants would carry back into the nests and thus spread its stultifying effect. The method was then in its earliest experimental stage, but it worked. Over a period of months the ant population steadily declined. After two years, it disappeared.

The saga of the Pharaoh's ants was, however, not quite over; it was to end on the pages of science fiction. The incident inspired the plot of the 1983 novel *Spirals*, by William Patrick, then the editor for biology and medicine at Harvard University Press. His imaginary ants were suspected of carrying around the laboratory a form of engineered DNA that induced progeria, a disease that fatally accelerates the process of aging. The daughter of a key bioengineer died from the condition, turning into a physiological old lady before she got past her childhood years. In the end the ants were exonerated, when the researcher himself proved at fault: he had cloned the daughter from cells of his dead wife, and her development had gone awry as a result.

One does not need to make ants protagonists of a novel to bring them deserved attention. I placed them at the center of my professional life, the focus of a near obsession, and I think I chose wisely. Yet I also confess that at the time their main appeal was not their environmental importance or the drama of their social evolution. It came from the discoveries they generously offered me. I built my career from easy revelation. The most important topic I addressed was their means of communication, which led me into a long period of productive research on animal behavior and organic chemistry.

My interest in chemical communication began in the fall of 1953, when Niko Tinbergen and Konrad Lorenz visited Harvard University to lecture on the new science of ethology. Twenty years later they shared the Nobel Prize for physiology or medicine, with Karl von Frisch as a third corecipient, for the years of work chronicled during their American tour. Tinbergen, a precise, carefully spoken Dutchman, arrived first. He gave an account of ethology that struck me with the resonance of important discovery. Because I was absorbed in systematics and biogeography, however, subjects remote from behavior, I took only a few notes and otherwise paid little attention. Then Lorenz came. He recounted his work begun in the 1930s, which he now was continuing at the Max Planck Institute in Buldern. He was a prophet of the dais, passionate, angry, and importunate. He hammered us with phrases soon to become famous in the behavioral sciences: imprinting, ritualization, aggressive drive, overflow; and the names of animals: graylag goose, jackdaw, stickle-back. He had come to proclaim a new approach to the study of behavior. Instinct has been reinstated, he said; the role of learning was grossly overestimated by B. F. Skinner and other behaviorists; we must now press on in a new direction.

He had my complete attention. Still young and very impressionable, I was quick to answer his call to arms. Lorenz was challenging the comparative psychology establishment. He was telling us that most animal behavior is preordained. It is composed of fixed-action patterns, sequences of movements programmed in the brain by heredity, which unfold through the life of an animal in response to particular signals in the natural environment. When triggered at the right place and time, they lead the animal through a sequence of correct steps to find food, to avoid predators, and to reproduce. The animal does not require previous experience in order to survive. It has only to obey.

Obedience to instinct: that formula has the ring of an old and tiresome story. Operant conditioning sounds so much more modern. But Lorenz strengthened his case with the logic of evolutionary biology, which secured my allegiance. Each species has its own repertoire of fixed-action patterns. In the case of a particular bird species, for example, the individual spreads its plumes in a certain way to attract mates of its own species; it bonds at a certain time of the year; it builds a nest of the right shape at the right location. Fixed-action patterns are biological events; they are not "psychological." Having a genetic basis, they can be isolated and studied in the same manner as anatomical parts or biochemical reactions, species by species. They are prescribed by particular genes on particular chromosomes. They come into existence and change as one species evolves into another. They serve, no less than anatomy and physiology, as a basis for classification and the reconstruction of trees of evolutionary descent, which clarify the true relationships among species. Instinct, the great ethologist made clear to me, belongs in the Modern Synthesis of evolutionary biology. And that means you can take ethology out into the field and do something with it.

Lorenz's lecture and my supplementary reading in later months drew me in a new direction. The ethologists were giving shape to something I had tried to do earlier with the dacetine ants but for which I had lacked a theory and vocabulary. My thoughts now raced. *Lorenz has returned animal behavior to natural history. My domain. Naturalists, not psychologists with their oversimple white rats and mazes, are the best persons to study animal behavior.*

The fixed-action patterns are what count, I realized. They can be understood only as part of the adaptation of individual species to a particular part of the natural environment. One kind of bird compared to another. One kind of ant against another. If you watch a chimpanzee in a cage, even if you test all its supposed learning ability, you will never see more than a small part of the behavior with which the animal is programmed, and you will miss the full significance of even that part.

What made ethology even more beguiling was the principle that although fixed-action patterns are complex, the signals triggering them are simple. Take the European robin, an early subject of ethological analysis by the British ornithologist David Lack. The male, primed by springtime hormones, uses song and displays to chase other males out of his territory. If these warnings fail, he attacks the intruders with fluttering wings and stabbing beak. His aggression is not provoked by the whole image of a male robin as we see it. He vents his fury instead against a red breast on a tree limb. A stuffed immature male with an olive breast meets no response, but a simple tuft of red feathers mounted on a wire coil evokes the full response.

Lorenz ticked off other examples of the triggering stimuli, or releasers as ethologists call them. The great majority of case studies accumulated by 1953 were of birds and fishes, and he concentrated on them. But the choice of these animals imposes a great bias: their communication is mediated primarily by sight or sound. It occurred to me immediately that the fixed-action patterns of ants and other social insects are triggered by chemicals instead, substances these creatures can smell or taste.

Earlier generations of entomologists had already suggested
something along this line; after all, such creatures cannot see
in the darkness of their nests, and little evidence existed that
they could hear airborne sounds. Some earlier writers had also
believed that ants communicate by tapping one another with
their antennae and forelegs, using a kind of Morse code of
the blind. But in 1953 we knew nothing about the anatomical
source of the chemicals that evoke the smells and tastes, with
one exception—a hindgut trail substance, passed through the
anus, found by the British biologist J. D. Carthy in 1951. Still,
no one had located the ultimate glandular source of the molecules or identified their chemical structure. The idea of fixed-
action patterns and releasers suggested to me a way to enter this
unexplored world of ant communication. The method should,
I reasoned, consist of a set of straightforward steps: break ant
social behavior into fixed-action patterns; then by trial and
error determine which secretions contain the releasers; finally,
separate and identify the active chemicals in the secretions.

   As far as I knew I was the only person thinking along these
lines. So I felt in no hurry to get started. In any case I had first
to finish my Ph.D. thesis, a laborious exercise in anatomy and
taxonomy of the ant genus *Lasius*. With that completed in the
fall of 1954, I left for the South Pacific to launch my studies on
ant ecology and island biogeography. Finally, four years later,
back in Cambridge with a well-equipped laboratory, I began
the search for the chemical releasers of ant communication.
Even then the idea evidently still eluded others; I had plenty
of sea room. It was to be a year before Adolf Butenandt, Peter
Karlson, and Martin Lüscher introduced the word "pheromone" to replace "ectohormone" in the literature of animal
behavior. They used the term "hormone" to designate a chemical messenger inside the body of the organism, "pheromone"
for a chemical messenger passed between organisms.

   I started with the imported fire ant, my favorite ant species
since my college years and one of the easiest social insects to
culture in the laboratory. I devised a new kind of artificial nest
consisting of clear Plexiglas chambers and galleries resting on
broad glass platforms. The arrangement kept the entire colony
in continuous view, allowing me to run experiments and record
the responses of all the ants any time I chose. The ultrasimple

environment did not distress the workers. After a while they habituated to the light and carried on their daily rounds in what appeared to be a normal manner. They flourished in an ant's equivalent of a fishbowl.

The most conspicuous form of communication in fire ants is the laying of odor trails to food. Scouts leave the nest singly to search outward in paths forming irregular loops. When they encounter a particle of food too big or awkward to carry home in one trip, most commonly a dead insect or a sprinkling of aphid honeydew, they head back to the nest in a more or less direct line while laying an odor trail. Some of their nestmates then follow this invisible path back to the food. As I watched from the side while the ants were foraging, I noticed that the returning scout touched the tip of her abdomen (the rearmost part of the body) to the ground and extruded and dragged her sting for short intervals along the surface. The chemical releaser apparently was being paid out through the sting like ink from a pen.

Now I had to locate the source of the chemical, which I presumed to be somewhere inside the abdomen of the worker ant. To take this next step I needed to identify the organ making the chemical and use it to lay artificial trails of my own; I needed to steal the ants' signal and use it to speak to them myself. The abdomen of a worker is the size of a grain of salt and packed with organs barely visible to the naked eye. Making the task more difficult was the fact that the anatomy of the fire ant had not yet been studied. I had to use diagrams drawn of other kinds of ants and add a bit of guesswork.

After placing the severed abdomens of fire ants under a dissecting microscope, I used the tips of fine needles and sharpened watchmaker's forceps to open them up and take out their internal organs one by one. I found myself close to the lower size limit of unaided dissection. Had the organs been only a fraction smaller, I would have been forced to use a micromanipulator, a difficult and expensive piece of equipment I hoped to bypass. If you buy instruments like that, and the experiment fails, you lose a lot of money. Although my hands were steady, I discovered that their natural muscle tremor, barely visible to the naked eye, was enlarged to a palsy under the microscope. Magnified twenty or thirty times, the tips of the needles and forceps spasmed uncontrollably as I brought them close to the abdomens. Then I found the solution: simply make the tremors part of the dissecting technique. Turn the needles and forceps into little jackhammers. Use the muscle spasms to tear open the abdomen and to push the separate organs out of the body cavity.

This much accomplished, I washed each organ in Ringer's solution, which is synthetic insect plasma with concentrations of various salts matching those in insects. Then I made artificial trails in the simplest, most direct way I could conceive, as follows. First I placed drops of sugar water on the glass foraging plate near the nest entrance and let mobs of feeding workers gather around them. With my experimental subjects in place, I crushed each organ in turn on the tip of a sharpened birchwood applicator stick. Then I pressed the tip down on the surface of the glass and smeared the microscopic fleck of semiliquid matter in a line from the assembled workers outward in a direction away from the nest.

First I tried the hindgut, the poison gland, and the fat body, which together fill most of the abdominal cavity. Nothing happened. In the end I came to Dufour's gland, a tiny finger-shaped structure about which almost nothing was known. It empties into a channel at the base of the sting, the conduit known to carry venom to the outside. Might it contain the trail pheromone? Indeed it did. The response of the ants was explosive. I had expected a few workers to saunter away from the sugar-drop crowd to see what might lie at the end of the new trail. What I got was a rush of dozens of excited ants. They tumbled over one another in their haste to follow the path I had blazed for them. As they ran along they swept their antennae from side to side, sampling the molecules evaporating and diffusing through the air. At the end of the trail they milled about in confusion, searching for the reward not there.

That night I could not sleep. After a delay of five years my idea had paid off with only a few hours' work: I had identified the first gland that contributes to ant communication. More than that, I had discovered what seemed to be a new phenomenon in chemical communication. The pheromone in the gland is not just a guidepost for workers who choose to search for food, but the signal itself—both the command and the instruction during the search for food. The chemical was everything. And the bioassay instantly became that much easier. It wasn't necessary, I realized happily, to arrange delicate social settings with a multiplicity of other stimuli to get the desired result. Biologists and their chemist partners should be able to proceed directly to the pheromone's molecular structure, provided they had an effective and easily measured behavioral test. If other pheromones—say, those inducing alarm and assembly—acted the same way as the trail substance, we might decipher a large part of the ant's chemical vocabulary within a short time.

Over the next few days I confirmed the efficiency of the trail pheromone assay over and over. In science there is nothing more pleasant than repeating an experiment that works. When I led my trails all the way back to the entrance of the nest, out poured the ants, even when they had been offered no food to stimulate them first. And when I let a concentrated vapor made from many ants waft down onto the nest, a large percentage of the worker force emerged and spread out in apparent search of food.

Next I enlisted a friend, the Harvard biochemist John Law, in an attempt to identify the structure of the trail substance molecule. We were joined by a gifted undergraduate student, Christopher Walsh, who in later years was to become a leading molecular biologist and president of the Dana Farber Cancer Institute. We were a capable team, but we encountered a technical snag: we learned that each ant carries less than a billionth of a gram of the critical substance in its Dufour's gland at any one time. The problem, however, was not insoluble. The late 1950s and early 1960s were the dawn of coupled gas chromatography and mass spectrometry, which allows the identification of organic substances down to millionths of a gram. That meant we needed tens of thousands or hundreds of thousands of ants, each with its vanishing trace of pheromone, to produce the minimum amount required for analysis.

How to gather such huge quantities? From my field experience, I knew of a relatively easy way. When fire ant nests are flooded in nature by rising stream waters, the workers float to the surface in tightly packed masses. Their bodies form a living raft within which the queen and brood are safely tucked. The colony floats downstream until it reaches solid ground, and there the workers proceed to excavate a new nest. After I had explained this phenomenon to Law and Walsh, we traveled to Jacksonville, Florida, one of the southern cities closest to Boston where fire ants are abundant. We took a rental car to the farm country west of the city, where we found two-foot-high fire ant mounds dotting pastures and grassy strips along the roads. There were as many as fifty to an acre, and within each mound lived 100,000 or more ants. Pulling the car over to the verge of the interstate highway, we shoveled entire nests into the water of a slow-moving stream passing through one of the culverts. The soil settled to the bottom, and large portions of each colony rose to the surface. We scooped up seething masses of ants in kitchen strainers and plopped them into bottles of solvent. Law and Walsh soon learned the source of the ants' common name: the sting of a worker feels like heat from a match brought too close to the skin. And every ant in the nest tries to sting you, ten times or more in succession if you don't squash it first. We took scores of stings on our hands, arms, and ankles, each of which produced an itching red welt. A day or two later, many of these sites erupted into white-tipped pustules. I suspect that my distinguished colleagues resolved then and there to stay with laboratory biology. Having paid the price, we returned home with enough material to proceed with the analysis of the trail pheromone.

Even with enough raw material, however, the structure of the molecule proved elusive. As Law and Walsh closed in on the active part of the spectrographic array, the peak most likely to be the pheromone diminished to levels too low to analyze further. Was the substance unstable during the separation procedure? Possibly, but now we were running out of extract. In the end the two chemists deduced that the material is a farnesene, a terpenoid with fifteen carbon atoms arranged in a basic structure previously found most commonly among the natural products of plants. They fell short of determining the exact structure, in which the location of every double bond is specified. That difficult feat was accomplished twenty years later by Robert Vander Meer and a team of researchers at the U.S. Department of Agriculture Laboratory in Gainesville, Florida. They discovered that the fire ant trail pheromone is actually a mixture of farnesenes, one of which is Z,E-α-farnesene, augmented by at least two other similar compounds. One gallon of the mixture would be sufficient, in theory at least, to summon forth the inhabitants of 10 million colonies.

For several years following my identification of the glandular source of the trail substance I pursued my goal of deciphering as much of the ants' chemical language as possible. As I looked more closely at the fire ant trail, I stumbled on a second phenomenon of social behavior, mass communication. The amount of food or the size of an enemy force cannot, I noticed, be transmitted by signals from a single scout. Such information can be conveyed only by groups of workers signaling to other groups. By laying trails on top of one another during a short interval of time, multiple workers, say a group of ten, can signal the existence of a larger target than one identifiable by only a single worker. A hundred workers acting together can raise the range of the smell volume still more. When the food site becomes crowded or the enemy subdued, fewer workers in the group lay trails, so that excess pheromone evaporates and the signal diminishes, and a smaller number of nestmates thus respond.

The information contained in the combined action of masses of individuals coming and going to a target is surprisingly precise. Later writers pointed to a parallel action in masses of brain cells, and the similarity that exists between the brain, the organ of thought, and the insect colony, the superorganism. The first to make the abstract comparison, I believe, was Douglas Hofstadter in *Gödel, Escher, Bach: An Eternal Golden Braid*, an ingenious disquisition on the nature of organization and creativity. The question then arose and has since been asked many times: Does the resemblance mean that an ant colony can somehow "think"? I believe not. There are too few ants, and those are too loosely organized to form a brain.

I moved on to pheromones that attract and alarm ants. The simplest such substance I found, almost certainly the most elementary pheromone ever discovered, was carbon dioxide. Fire ants use it to hunt subterranean prey and to locate one another in the soil. The most bizarre pheromone, if the generic term can even be used in this case, is the signal of the dead— the means by which a corpse "announces" its new status to nestmates. When an ant dies, and if it has not been crushed or torn apart, it simply crumples up and lies still. Although its posture and inactivity are abnormal, nestmates continue to walk by it as though nothing has happened. Two or three days pass before recognition dawns, and then it is through the smell of decomposition. Responding to the odor, a nestmate picks the corpse up, carries it out of the nest, and dumps it on a nearby refuse pile.

I thought: maybe with the right chemicals I could create an artificial corpse. It should be possible to transfer the odor from one object to another. When I soaked bits of paper with an extract of well-seasoned corpses, the ants carried them to the refuse piles. Thinking back to the basic idea of the chemical releaser I asked, will any decomposition substance trigger the removal instinct, or will the ants respond to just one or two? I found out that a quick answer was possible, because biochemists had already identified a large roster of compounds found in rotting insects. Don't ask me why such research had been conducted. The scientific literature is filled with such information, and however arcane it often proves useful in unexpected ways. Such was the case in my own (also arcane) study. With two newly recruited assistants I gathered an array of the putrid substances and offered them to my ants on bits of paper, one by one. They included skatole, a component of feces; trimethylamine, one of the essences of rotting fish; and several of the more pungent fatty acids that contribute to rancid human body odor. For weeks my laboratory smelled like the combined essences of sewer, garbage dump, and locker room. In contrast to the responses of my human nose and brain, however, the ants' responses to the chemicals were consistently narrow. They removed only the paper scraps treated with oleic acid or its ester.

The experiments proved that the ants are neither aesthetic nor meticulously clean in any human sense. They are programmed to react to narrow cues that reliably identify a decaying body. By removing the source they unconsciously safeguard colony hygiene. To test this conclusion about the simplicity of ant behavior I asked, finally, what would happen if a corpse came to life? To find out, I daubed oleic acid on live workers. Their nestmates promptly picked them up, even though they were struggling to get free, and carried them to the refuse pile.

There the "living dead" cleaned themselves for a few minutes, rubbing their legs against their body and washing the legs and antennae with their mouthparts, before venturing back to the nest. Some were hauled out again, and a few then yet again, until they became clean enough to be certifiably alive.

A new sensory world was opening to biologists. We came fully to appreciate the simple fact that most kinds of organisms communicate by taste and smell, not by sight and sound. Animals, plants, and microorganisms employ among their millions of species an astonishing diversity of devices for transmitting the chemicals. The pheromones are usually sparse enough in the bodies of the organisms to make detection difficult for human beings. Animals are unfailingly ingenious in the methods by which they manufacture and deploy these substances. In the late 1950s I was one of no more than a dozen researchers who studied them in ants and other social insects. It was a bonanza that lay before us. We discovered new forms of chemical messages everywhere we looked, and with minimal effort.

In 1961 I invited William Bossert, a Harvard graduate student in applied mathematics, to join me in a project to synthesize all existing knowledge about chemical communication within a single evolutionary framework. Bill possessed in consummate degree the mathematical skills I so conspicuously lacked. At that time he was also pioneering the use of computers in the modeling of evolutionary change. One day he took me into the computer room in Harvard's Aiken Computation Laboratory, pointed to the spinning tape disks and futuristic control panels, and instructed me that here was housed the future of theoretical biology. Now was the time, he urged, to come aboard and master the powerful new technology. He failed to recruit this naturalist, however. I was just too overwhelmed by the alien culture, easing my way about like an eighteenth-century Pacific islander invited to inspect the armory and rigging of H.M.S. *Endeavour*. In the years thereafter, as hardware with the computing capacity of the Aiken room shrank to the size of a suitcase, Bossert continued his efforts, but I was never motivated to join him. I had no desire to struggle for years in a field in which I could never hope to become more than mediocre.

Instead, on this occasion, I gave Bossert everything I knew or could find about the chemistry and function of the known pheromones and let him devise the models of their dispersal and detection. He incorporated evaporation and diffusion rates of the known or likely candidate molecules, with estimates of the numbers disseminated and the densities required for animals to recognize them. Together we conceived a series of different forms of expanding gases and theorized on active spaces—the zones within which the molecular densities are high enough to trigger a response. Active spaces are hemispheric in shape when the pheromone is released from one spot in still air, and half-ellipsoidal when the material is released into a steady wind or streaked along the ground into still air. We factored in the size of the molecule as it affects the rates of evaporation and diffusion. We showed that the potential variety of signals dramatically increases—it goes up exponentially—as the size of the pheromone molecule is enlarged within a homologous series. We observed that the substance can either evoke an immediate response or else change the physiology of the animal and its propensity to respond over relatively long periods. When the theory was finally stitched together and all the evidence weighed, we concluded that animals have selected chemicals during their evolution that are well suited to particular meanings. For example, the molecules used as alarm pheromones are smaller in size and have higher response concentrations than those used for sexual attraction, allowing their active spaces to flash on and off more quickly. As a rule, the pheromones chosen are among the ones conceivably most effective in transmitting a particular message.*

Even though this theoretical study of the most general properties of pheromones was progressing well, I stayed close to ants and pushed my laboratory research. In time I estimated that the workers and queens of each colony use somewhere between ten and twenty kinds of pheromones to regulate their social organization. The number undoubtedly varies according to species. But this spread, ten to twenty, is only an educated guess, and remains no more than a guess as I write, thirty

---

*W. H. Bossert and E. O. Wilson, "The Analysis of Olfactory Communication among Animals," *Journal of Theoretical Biology* 5 (1963): 443–69; E. O. Wilson and W. H. Bossert, "Chemical Communication among Animals," *Recent Progress in Hormone Research* 19 (1963): 673–716.

years later. The reason is that beyond a few of the most obvious classes, such as the trail and alarm pheromones, the bioassays and chemical analyses turned out to be increasingly difficult. I soon realized that to stay ahead in the field I would have to devote all my time to it and acquire advanced technical training in histology and chemistry. In the late 1960s, ten years after I performed my first crude experiments, the field of pheromone studies was being flooded by a small army of gifted researchers prepared to make this commitment. So I pulled out, an outclassed elder at thirty-five, returning to experiments on chemical communication only when I saw the possibility of a quick result with low technology.

Now we have come to 1969. For some it is the easily remembered year when the Pharaoh's ants began to steal culture media from the molecular biologists, for others as the year student radicals stalked the campus and Harvard Square with raised fists and revolutionary slogans stamped on their T-shirts. For me it marked a significant change of another kind. My interest in pheromones and island biogeography, my two passions of the previous ten years, had begun to wane. But in September a young scientist with whom I had been corresponding, Bert Hölldobler, knocked at the door of my office in the Biological Laboratories. A lecturer in zoology from the University of Frankfurt, he had come to spend a year as a visiting scholar under my sponsorship. I was about to enter the most sustained and productive collaboration of my research career, built upon a close friendship and a common lifelong commitment to the study of ants.

Although we made no such lofty analysis at the time, we met as representatives of two national cultures in behavioral biology, whose melding would soon lead to a better understanding of ant colonies and other complex societies. One of the contributing disciplines was ethology, European in origin and Hölldobler's forte, the study of whole patterns of behavior under natural conditions. Though the product of many minds across two generations, in 1969 ethology had become associated in popular tradition with the leadership of Lorenz, Tinbergen, and von Frisch and was well on its way to cosmopolitan status. The other foundation discipline, the one in which I had been more intensively educated, was population biology.

Mostly of American and British origin, it was radically different from ethology in its approach to behavior. It addressed entire ensembles of individuals, how they grow, how they spread over the landscape, and, inevitably, how they retreat and vanish. Modern population biology, now also cosmopolitan, attempts to span wide stretches of space and time, and consequently it relies as much on the disciplined imagination invested in mathematical models as it does on studies of live organisms. Its techniques are closely allied to those of demography, in the sense that it combines the births, deaths, and migrations of individual members to construct a statistical picture of the whole society.

The key to the second, higher-level approach is the perception that an insect colony is a population. Some colonies, like the queen and 20 million worker force of the African driver ant, have more inhabitants than entire countries. Like human populations, the only way to understand such ensembles fully is to trace the lives and deaths of their separate members. Both the population-level and the individual-level bodies of information, however, require ethology to create a complete science. This discipline alone addresses in concrete terms the heart of social organization, from communication and nest construction to caste and the division of labor. The final element in the mix is evolution. The behavioral descriptions and population analysis are the historical products of natural selection. Put together, population biology, ethology, and evolutionary theory form the content of the new discipline of sociobiology, which I was to define in 1975 as the systematic study of the biological basis of social behavior and of the organization of complex societies.

Bert Hölldobler and I were edging toward sociobiology. We were, however, first and foremost entomologists, committed to the study of insects. At the time we met he was thirty-three, seven years my junior, but had independently arrived at the conviction that ants are worthy of scientific study no matter how it is done, by ethology, sociobiology, or any other biological discipline. We nevertheless also foresaw, in our early occasional conversations, that ethology and population biology are complementary approaches to the study of social behavior, and potent in combination.

It could easily have ended there, as a declaration of common interests. At the end of his second year Hölldobler returned to Frankfurt to resume what he foresaw as a lifetime academic career in Germany. At just this time, however, John Dunlop, dean of Harvard's School of Arts and Sciences, decided to increase faculty representation in behavioral biology. He authorized the appointment of three new professors and placed me in charge of the search committee. In time, after sifting through many letters and evaluations from consultants, we identified this same Dr. Hölldobler as the most promising young scientist in the world working on the behavior of invertebrate animals. He was accordingly invited to come to Harvard as a full professor. He accepted, returning to Cambridge in 1972.

Thereafter we shared the fourth floor of the newly constructed laboratory wing of the Museum of Comparative Zoology. Our contact was close, and we collaborated with increasing frequency in projects in teaching and research. But it was not to be a lifetime arrangement. Sixteen years later, in 1989, Bert returned to Germany, this time to the University of Würzburg in Bavaria, where he had been asked to create a special department devoted to social insects in the newly founded Theodor Boveri Institute of Biological Science. By that time he had come to be greatly appreciated in his native country. Germany, like most other European countries, had a growing interest but weak representation in ecology and related subjects. Bert's hybrid experience in behavior and population biology uniquely qualified him for national leadership—and continues to do so as I write. In 1991 he received the Leibniz Prize, Germany's highest award in science.

Nearly two decades of residence in America had turned Bert Hölldobler into a lover of the Arizona mountains, where he spent summers with his family, and of country-western music. Underneath the new American, however, remained the Bavarian—practical, solid, warm-natured and humorous, flexible, altogether the antithesis of the stereotypical Prussian, a difference he was quick to point out whenever the German national character became the subject of conversation. The bluegrass songs of Doc Watson, he once noted in passing, reminded him of Bavarian folk music. Bert above all was rooted to the earth, a naturalist, perhaps, by hereditary predisposition. Fluent English came slowly during his stay in America, and he never lost a marked accent. But it was an asset at Harvard University, where students assumed, correctly, that they were receiving German science and philosophy straight from the source. They consistently gave his courses the highest ratings.

By strength of character alone no scientist more deserved recognition. Hölldobler was—and remains—the most honest scientist I have ever known. As we filled the hours around tedious replicate experiments with conversation ("Okay, that's—ah—sixty-three seconds, the forager just entered the nest, got it? Now, I want to go back and say just one more thing about Hennig and the original idea of cladism . . ."), he endeavored to make every datum in his notebook, every nuance of expression in his published reports, as straight and transparent as he could. If he had a fault worth mentioning it was one which I shared and which made us the more compatible, an obsessiveness of habit in work, expressed as a sometimes unreasonable need to bring one subject to closure before going on to the next.

In science, obsessiveness under psychological control can be a virtue. To a degree I have not encountered elsewhere Hölldobler extended this urge to the design of experiments and a weighing of evidence. Many successful researchers stop with a single well-conducted procedure, which they repeat often enough for the overall result to be statistically persuasive. Then they are prepared to say in print, "I think it likely that such and such is the case." Others hold back and ask, "What different experiment can I perform, using new kinds of measurements, that will test the conclusion more rigorously?" If they then perform the second procedure and in the second result consistent with the first, they conclude, "That pretty well proves it. Let's move on." Hölldobler is a member of the second group. But sometimes during our collaboration he would pause yet again and ask—to my consternation—"Is there a third way?" He did occasionally press on with yet another method. He was the only third-way researcher I have ever known.

He was a scientist's scientist. He simply loved science as a way of knowing. I believe he would have practiced it without an audience or financial reward. He played no political games.

If new data did not fit, he quickly shifted to a new position. He was one of the few scientists I have known actually willing to abandon a hypothesis. He was meticulous about crediting others, quick to praise research when it was original and solid, harsh in his rejection when it was slovenly. The tone of his conversation was explicitly and uncompromisingly ethical, a posture born neither of arrogance nor of self-regard, but from the conviction of his humanistic philosophy that without self-imposed high standards, life loses its meaning.

But a somber picture of Bert Hölldobler would be misleading. He was fun to be with, the younger brother I never had. In periods of relaxation we confided in all things, both scientific and personal. His manner and even his physical appearance were reassuring. Bearded as he approached middle age, he had a burgher's pleasant countenance fitted upon the short muscular build of a gymnast, the latter a residue of his favorite sport as a youngster. He was deeply devoted to his family, somehow finding time to participate with his wife, Friederike, in every step of the rearing of their three sons. Science was not everything for Hölldobler. A gifted painter and photographer, a good musician, he enjoyed the arts as I never could, locked as I was into my unyielding workaholic's momentum. In darker moments I envied him that.

Though considerably the younger man, he made me a better scientist. During our work in the laboratory and field, I found myself anxious to meet his standards and to let him know I was trying. I am by nature a synthesizer of scientific knowledge, much better than Hölldobler at this activity, but I confess that in my effort to make sense of everything, to fit every piece into my schemes somewhere no matter how procrustean the result, I often overlooked detail. Hölldobler did not. By temperament and training, he belonged to the Karl von Frisch tradition, expressed to me succinctly one day by Martin Lindauer, Frisch's student and Hölldobler's mentor, while I was visiting Würzburg. Lindauer said, grinning as he typically did when speaking of serious matters, "Look for the little things."

That adjuration Bert and I followed many times during the Harvard years. In 1985 we made our first field trip together to Costa Rica. We drove north from San José to La Selva, the field station of the Organization for Tropical Studies. As we entered the rain forest, I used my more general knowledge of ants to find and identify colonies that might be of exceptional interest in behavioral work. I was looking for a quick and exciting payoff. One candidate was the primitive genus *Prionopelta*, which I found nesting in rotting logs. No colonies had ever previously been studied in life. I was eager to record the key facts of the social behavior of this ant, the kind of basic data that go comfortably into syntheses and evolutionary constructions. I plunged into the work with Bert's assistance. We took notes on colony size, the number of queens, division of labor, and the kinds of insects and other small animals captured by the workers. We found, for example, that they preferred silver-fishlike creatures called campodeid diplurans. In the course of our work, Bert's attention fastened on fragments of old cocoon silk plastered on the walls of passageways of the *Prionopelta* nests. He asked, as much to himself as me, What does this mean? Nothing, just trash, I answered. When the new adults emerge from the cocoons, their nestmates throw out the silk fragments, and they don't bother to stack them in separate garbage dumps. No, no, he said, look: the pieces are lined up as a smooth layer on the gallery walls. He went on, with close study of his own and the aid of a scanning electron microscope back at Harvard, to show that the cocoon silk is employed as wallpaper. It keeps the chambers of the moist walls drier than would otherwise be the case, and thereby protects the growing brood. Wallpapering was a technique of climate control previously unknown in ants.

Hölldobler again said, Look, some of the foraging workers appear to be moving more slowly while they drag their hind legs. Again I was unimpressed. Individual ants, I responded, often move slowly or erratically for no particular good reason. Nor is there any cause to believe that these primitive ants lay odor trails in any case. But Hölldobler persevered. He found that not only do the workers lay odor trails—by which they recruit nestmates to new nest sites—but the attractive substance comes from a previously unsuspected gland located in the hind legs. The pheromones are smeared in a line as the ants drag their hind legs over the ground. The existence of the gland provided an important clue to the evolutionary relationships of *Prionopelta*.

From two weeks of data gathered in the La Selva forest we wrote five scientific articles. During our years of collaboration we made uncounted other discoveries while grubbing around and talking back and forth. First one took the lead, then the other. Our partnership in most respects ended when Bert, after being approached by several institutions in Europe, was offered the professorship at Würzburg. He needed the sophisticated new equipment and skilled assistants promised him at the Theodor Boveri Institute. He had a desire, almost an obsession, to get inside the muscles, glands, and brains of ants to learn how these organs mediate social behavior and organization. He wanted to understand a thousand "little things" to make a great whole. That kind of enterprise is expensive, too much so, it appeared, for the National Science Foundation and private U.S. organizations. Most of their support was either inadequate or unstable or both, contingent on appropriations made in three- to five-year cycles of renewal. Although Bert's applications to the NSF were consistently given the highest ratings and funded, the amounts provided fell short of sustaining the effort he envisioned.

One day, as Hölldobler grew more serious about leaving, we decided to write a book recounting everything we knew about ants. And while we were at it, we asked ourselves, why not try for a book that has everything *everybody* ever knew about ants, throughout history? Such a project would take a great deal of effort and time, and it might fall short of the goal we set. But what a worthy conceit! Try for the impossible, as Floyd Patterson, the undersized heavyweight boxing champion of the world once said, in order to accomplish the unusual. The result was *The Ants*, published by Harvard University Press in 1990. It contained 732 double-columned pages, hundreds of textbook figures and color plates, and a bibliography of 3,000 entries. It weighed 7.5 pounds, fulfilling my criterion of a magnum opus—a book which when dropped from a three-story building is big enough to kill a man.

On Tuesday afternoon, the following April 9, the faculty and deans of the College of Arts and Sciences gathered for their monthly meeting in the portrait-encircled main room of University Hall. Just as the meeting was about to be called to order, a secretary entered and handed President Derek Bok a message.

Bok announced its content: *The Ants* had been awarded the 1991 Pulitzer Prize in General Nonfiction. I stood and basked in the applause of the Harvard faculty. Bless my soul, the Harvard faculty. Where could I go from here but down?

I later learned that our book was only the fifth on science ever to receive a Pulitzer Prize, and it was the first with a primarily scientific content, written by specialists for fellow professionals. Soon after I left University Hall that day I called Bert and asked him how it felt to win America's most famous literary award. Note, I reminded him, not scientific, *literary*. Wonderful, he replied. They would celebrate in Würzburg. The accent was still there, of course. It made the occasion more special.

CHAPTER SIXTEEN

Attaining Sociobiology

O N AUGUST 1, 1977, sociobiology was on the cover of *Time*. On November 22 I received the National Medal of Science from President Carter for my contributions to the new discipline. Two months later, at the annual meeting of the American Association for the Advancement of Science, held in Washington, demonstrators seized the stage as I was about to give a lecture, dumped a pitcher of ice water on my head, and chanted, "Wilson, you're all wet!" The ice-water episode may be the only occasion in recent American history on which a scientist was physically attacked, however mildly, simply for the expression of an idea. How could an entomologist with a penchant for solitude provoke a tumult of this proportion? Let me explain.

My interest in sociobiology was not the product of a revolutionary's dream. It began innocently as a specialized zoology project one January morning in 1956 when I visited Cayo Santiago, a small island off the east coast of Puerto Rico, to look at monkeys. I was accompanied by Stuart Altmann, who had just signed up as my first graduate student. Stuart was an academic anomaly, so much so that upon admittance to the Ph.D. program at Harvard the previous fall, he at first found himself without a sponsor. His problem was not his abilities, which were outstanding, but his proposed thesis research, which was too unusual. He had set his sight on the social behavior of free-living rhesus macaques, and particularly those maintained by the National Institutes of Health on Cayo Santiago, newly dubbed "monkey island." He came well prepared. He had recently worked on howler monkeys in the Panama rain forest. His command of the relevant literature was complete.

Unfortunately, no one at Harvard knew what he was talking about. The behavior of primates under natural conditions remained virtually unknown in 1955. C. Ray Carpenter, an American psychologist, had laid a foundation in the 1930s with field observations of howlers, rhesus, and gibbons. His published work was well respected by a small circle of biologists and

775

anthropologists but had not spawned a school of research. It was not easy to journey to where wild primates live. Jane Goodall still lived in England, her first visit to the chimpanzees of the Gombe Reserve four years away. Several Japanese researchers, at the time Altmann began his own work, were observing macaques on Mount Takasoki, on the island of Kyushu, but they published reports in their native language, which was virtually unknown to American and European scientists.

No senior members of the Harvard biology faculty considered primate field studies to lie within their province. Some doubted that the subject even belonged in biology. Thus Stuart came to me. In the late fall of 1955 I had been offered an assistant professorship in biology, effective July 1 of the following year. Frank Carpenter, chairman of the department, asked me whether, given my interest in the social behavior of ants, I would sponsor Stuart even before my faculty term began. I accepted happily. I was hardly more than a graduate student myself, just a year older than Altmann, eager to learn the strange new subject he had chosen.

I had decided wisely. The two days Stuart and I lived among the rhesus monkeys of Cayo Santiago were a stunning revelation and an intellectual turning point. When I first stepped ashore I knew almost nothing about macaque societies. I had read Ray Carpenter but was unprepared for the spectacle unfolding. As Altmann guided me on walking tours through the rhesus troops, I was riveted by the sophisticated and often brutal world of dominance orders, alliances, kinship bonds, territorial disputes, threats and displays, and unnerving intrigues. I learned how to read the rank of a male from the way he walked, how to gauge magnitudes of fear, submission, and hostility from facial expression and body posture.

Altmann issued a warning: "Two things. Don't move too suddenly near an infant, as though you mean to harm it. You might be attacked by a male. And if you do happen to be threatened, don't look the male in the face. A stare is a threat and might provoke an attack. Just hang your head down and look away." Sure enough, in a careless moment on the second day, I twisted my body around suddenly while standing next to a very young monkey, and it let out a shriek. At once the number two male ran up to me and gave me a hard stare, with his mouth gaping—the rhesus elevated-threat expression. I froze, genuinely afraid. Before Cayo Santiago I had thought of macaques as harmless little monkeys. This individual, with his tensed, massive body rearing up before me, looked for the moment like a small gorilla. I needed no reminder. I lowered my head and looked away in my most studiously contrite manner, frantically signaling the message "Sorry, didn't mean anything, sorry." After a few minutes my challenger left.

In the evenings Altmann talked primates and I talked ants, and we came to muse over the possibility of a synthesis of all the available information on social animals. A general theory, we agreed, might take form under the name of sociobiology. Stuart was already using that word to describe his studies; he had picked it up from the Section on Animal Behavior and Sociobiology, a working subgroup of the Ecological Society of America. A belief floated among zoologists even then that animal societies require a different kind of analysis, that they are properly the subject of a separate, minor discipline. But none could say what the general principles of this sociobiology might be, or how they would relate to the rest of biology. Under the guidance of senior zoologists such as Warder Clyde Allee, Alfred Emerson, and John P. Scott, sociobiology was taking form as a discipline but still consisted largely of descriptions of different kinds of social behavior. As Altmann and I talked over the subject during the pleasant Puerto Rico evenings, we could do no better. Primate troops and social insect colonies seemed to have almost nothing in common. Rhesus monkeys are organized strongly by dominance orders based on individual recognition. That much is also true of primitively social wasps but not of the rest of the social insects, whose colonies are composed of hundreds or thousands of anonymous and short-lived siblings living in harmony. Primates communicate by voice and visual gesture, social insects by chemical secretions. Primates fill temporary roles based on personal relationships; social insects have castes and a relatively rigid, lifelong division of labor.

We knew that no science worthy of the name is built wholly from a checklist of similarities and differences with an overlay of phenomena such as dominance and group action. In 1956 there existed no theory to explain diversity—why various traits have arisen in some groups and not others. Altmann had one good idea. He intended to devise probability transition matrices of behavioral acts, to provide a compendium of the following kinds of information: if a rhesus performs act *a*, then there is a certain probability that it will perform *a* again, another probability it will perform *b*, and so on. I agreed on the concept: a great deal of behavior and social interaction can be packed into transition matrices. It should then be possible to use the numbers to compare one kind of society more precisely with another. To quantify social interactions is an important step, but where would it lead? The result would still be a description, offering no explanation of how or why a particular species of monkey or ant arrived at one pattern in the course of its evolution as opposed to another. Neither Altmann nor I had the conceptual tools in 1956 to advance sociobiology further, and we let the subject rest. Stuart pressed on to complete his thesis research.

A congenital synthesizer, I held on to the dream of a unifying theory. By the early 1960s I began to see promise in population biology as a possible foundation discipline for sociobiology. I had entered population biology not to serve sociobiology but to help fashion a counterweight to molecular biology. I believed that populations follow at least some laws different from those operating at the molecular level, laws that cannot be constructed by any logical progression upward from molecular biology. This view of the biological sciences motivated me to collaborate with Lawrence Slobodkin, an alliance that later led to the development of the theory of island biogeography with Robert MacArthur.

By the early 1960s population biology was gaining substantial independent strength, and my confidence in its canonical relation to sociobiology rose. In late July of 1964, when I met with the "Marlboro Circle" in Vermont—Egbert Leigh, Richard Levins, Richard Lewontin, and MacArthur were the others—I represented the idea of sociobiology as a possible derivative of population biology. Societies are populations, I argued, and amenable to the same modes of analysis.

I saw that the quickest way to make the point was to use population biology in a solid account of caste systems and division of labor in the social insects. I was already well prepared for the task. In 1953 I had traced the evolution of caste in ants in a more descriptive manner, using measurements of scores of species from around the world. I showed how the divergence in anatomy among queens, soldiers, and ordinary ("minor") workers is the consequence of changes in allometry, the differential growth of different organs. Allometry, just by increasing or diminishing one dimension of the body relative to another, can produce larger or smaller heads, full-blown or shriveled ovaries, and other divergent products in any part of the final adult form. The idea was not new. It had been advanced earlier by Julian Huxley in his 1932 book, *Problems of Relative Growth*; he in turn had been inspired by D'Arcy Thompson's analysis of the evolution of morphological gradients, published in the 1917 classic *On Growth and Form*. I took the ants from there, following in plausible sequence the evolution of castes from one basic type in small steps all the way to multiple forms differing among themselves radically. Then I gave the subject a new twist. To allometry I added demography, the relative numbers of individuals of different castes within each colony. When allometry and demography are joined closely, the probable evolution of caste becomes much clearer. The anatomy of a particular caste member obviously determines the efficiency of its labor role; a soldier, for example, functions best with large, sharp mandibles and powerful muscles to close them. But the number of soldiers, I pointed out, is also crucial. If there are too few fighting specialists, the colony will be overwhelmed by enemies. If there are too many, on the other hand, the colony cannot gather enough food to care for the young. It follows that colonies must regulate the birth and death rates of the various caste members created by allometry. In later studies I came to call the phenomenon "adaptive demography." I interpreted it as a population-level trait of an advanced society.

Julian Huxley was intrigued by my employment of allometry and demography. When he visited Harvard in 1954, he asked to see me. My faculty advisers were impressed by the request, and I was thrilled to meet the great evolutionary scholar and humanist. Our common interest, we agreed, was a classic topic of general biology. The problem of ant castes had attracted the attention of Charles Darwin, who saw it as a threat to the theory of natural selection. Although Darwin had construed the idea of relative growth intuitively, Huxley and I knew that the ideas and data of our own studies had produced the first full and quantitative evolutionary explanation.

In 1968 I refined the idea of adaptive demography and developed several new principles of caste evolution with the aid of models in linear programming. In 1977 I was joined in a further, year-long study by George Oster, an exceptionally gifted and resourceful applied mathematician from the University of California, Berkeley. This time we explored the theory of caste evolution throughout the social insects. We were able to add other concepts from population biology to my earlier formulation. Oster led the modeling effort. His range of analytic techniques was awesome, affirming his generally held reputation as the mathematically most competent of all theoretical biologists. He often played with novel approaches, and he then had to lead me through the steps before we were able to continue the conversation. My role was the same as the one I had adopted with Bill Bossert in the synthesis of chemical communication fifteen years previously. At the beginning of each new avenue of exploration, I fed in all that I knew about caste and division of labor, information that often consisted of no more than doubtfully related fragments, along with the best intuitive conclusions I could draw. Oster then built formal models with what we could see—or guess—of the empirical relationships and trends, extending our reach in space and time. I responded with new evidence and guesses, he reasoned and modeled again, I responded, he modeled, I responded.* During breaks we gossiped and explored our other common interests. A magician of professional grade, he once dazzled me with sleights of hand I could not fathom even when he repeated them a foot or two from my concentrated gaze. I found my incapacity deeply disturbing. I was a proud scientific materialist, but I had to ask, How much else seemingly real in the world is an illusion? I learned a principle that others have established, often from painful experience: never trust a scientist to evaluate

---

*E. O. Wilson, "The Ergonomics of Caste in the Social Insects," *American Naturalist* 102 (1968): 41–66; George F. Oster and E. O. Wilson, *Caste and Ecology in the Social Insects* (Princeton: Princeton University Press, 1978).

"evidence" of telekinesis and other feats of the paranormal; go instead to an honest magician.

Through the 1960s I searched for other ideas to add to the sociobiology armamentarium. One that I fashioned from population biology was the evolutionary origin of aggression. In his early writings, and again in his famous book *On Aggression* in 1966, Konrad Lorenz postulated aggression to be a widespread instinct that cannot be suppressed. It wells up in organisms and, like a crowded liquid, seeks release in one form or another. In human beings, Lorenz suggested, it is better released in organized sports than in war. In 1968, in the first of two Man and Beast symposia sponsored by the Smithsonian Institution, I showed that a more precise explanation consistent with the growing body of evidence from field studies is the role of aggressive behavior as a specialized density-dependent response.* As populations increase in density, those of many species are constrained by a growing resistance from one or more factors. Among these density-dependent responses are the rise in per capita mortality from predation and disease, the loss of fertility, a greater propensity to emigrate, and—aggression. Whether aggressive behavior originates at all during evolution depends on whether other density-dependent factors reliably intervene to control population growth. Even then the form it takes can vary, emerging as territorial defense, dominance hierarchies, or all-out physical attack and even cannibalism, depending on the circumstances in which population limits are attained. Thus aggression is a specialized response that evolves in some species and not others. Its occurrence can in principle be predicted from a knowledge of the environment and natural history of the species.

The elements of sociobiological theory came from many sources. But when the most important idea of all came along, I at first resisted it with all my ability. In 1964 William Hamilton published his seminal theory of kin selection in the *Journal of Theoretical Biology*, in a two-part article titled "The Genetical Evolution of Social Behaviour." In the decades since, a sizable

---

*"Competitive and Aggressive Behavior," in J. F. Eisenberg and W. Dillon, eds., *Man and Beast: Comparative Social Behavior* (Washington, D.C.: Smithsonian Institution Press, 1971), pp. 183–217.

research industry has been built upon this single paper. Some of Hamilton's reasoning and conclusions have been challenged, then defended by enthusiasts, only to be challenged, and defended, again. The core of the theory has stood up well. Its essence, like that of all great ideas, is simple, of the kind that evokes the response, "Obviously that is true (but why didn't I think of it?)." Conventional Darwinism envisages natural selection as an event occurring directly between generations, from parent to offspring. Different lineages carry different genes, most of which prescribe traits that affect survival and reproduction. How an organism grows in body form, how it searches for food, how it avoids predators: these are among the traits affected by genes. The genes therefore determine survival and reproduction. Because by definition lineages that survive and reproduce better create more offspring in each generation, their hereditary material comes to predominate in the population over many generations. The increase of one set of genes at the expense of another is (again by definition) evolution by natural selection. The history of life has been guided by the appearance of new genes and the rearrangement of chromosomes bearing the genes through random mutations. These ensembles are winnowed by natural selection, which is the increase or decrease of particular combinations of genes and chromosomes through the differential survival and reproduction of the organisms carrying them.

In one important respect this traditional process of natural selection can be called just one type of kin selection. Parents and offspring are, after all, close kin. But Hamilton observed that brothers, sisters, uncles, aunts, cousins, and so forth are also kin; and he thought about what this truism means for evolution. The other kin share genes by common descent no less than parents and offspring. So if there is any interaction among them that is influenced by genes, say, a hereditary tendency toward altruism, or cooperation, or sibling rivalry, the interaction will result in a change in survival and reproduction and should equally well cause evolution by natural selection. Perhaps the ancillary forms of kin selection drive most forms of social evolution.

What made Hamilton's idea immediately attractive was that it helped to resolve the classic problem in evolutionary theory of how self-sacrifice can become a genetically fixed trait. It might seem on first thought—without considering kin selection—that selfishness must reign complete in the living world, and that cooperation can never appear except to enhance selfish ends. But no, if an altruistic act helps relatives, it increases the survival of genes that are identical with those of the altruist, just as the case in parents and offspring. The genes are identical because the altruist and its relative share a common ancestor. True, the corporeal self may die because of a selfless action, but the shared genes, including those that prescribe altruism, are actually benefited. The body may die, but the genes will flourish. In the enduring phrase of Richard Dawkins, social behavior rides upon the "selfish gene."

Hamilton had traveled that high road of science once described by the great biochemist Albert Szent-Györgyi, "to see what everyone has seen and think what no one has thought." But I am reasonably sure that had Hamilton expressed kin selection in merely abstract terms, the response to his formulation would have been tepid. Other biologists upon reading it would have said, "Yes, of course, and Darwin had a somewhat similar idea, did he not?" And, "Correct me if I'm wrong, but haven't notions of this kind been discussed off and on for a long time?" Yet Hamilton did succeed dramatically (although few learned about the theory until I highlighted it in the 1970s). He did so because he went on to tell us something new about the real world in concrete, measurable terms. He provided the tools for real, empirical advances in sociobiology. As Hamilton told me later, he was able to pull off his feat for three loosely related reasons. First, he was "bothered" by the problem of altruism; was the Darwinian explanation completely sound, or was it not? Second, he had a working knowledge of social insects, to which the altruism problem eminently applied. And third, he was intrigued by the mathematics of kinship, into which—impelled by the first two concerns—he had been guided through reading the work of the geneticist Sewall Wright. The closer the kinship, of course, the larger the fraction of genes shared as a result of common descent. Wright had devised an ingenious way of expressing the exact fraction shared, by a measure he called the coefficient of relationship. Working problems with it is an interesting mental game not unlike calculating the odds in gambling. What, for example, is the fraction of genes shared with a second cousin, or a half-sister's full niece? This number, the degree of kinship, Hamilton saw to be crucial in the evolution of altruism. Even this tributary idea is intuitively straightforward. You may be willing to risk your life for a brother, for example, but the most you are likely to give a third cousin is a piece of advice.

With these points in mind, Hamilton now joined the natural history of wasps and other social insects with the calculus of kin selection. At this point he was aware of two more important pieces of relevant information affected by kinship, this time from entomology. One is that most social insects, including the ants, bees, and wasps, are members of the insect order Hymenoptera. The only exceptions are the termites, composing the order Isoptera. The other important fact is that the Hymenoptera have an unusual sex-determining mechanism called haplodiploidy, in which fertilized eggs, with two sets of chromosomes, produce females, and unfertilized eggs, with only one set of chromosomes, produce males. Turning to the coefficient of relationship (or the "concept of relatedness," as he later named it), Hamilton saw that because of haplodiploidy sisters are more closely related to each other—have more genes in common—than are mothers and daughters. At the same time, they are much less related to their brothers. From the occurrence of haplodiploidy alone, he concluded that all of the following should be true if social behavior has evolved in the insects by natural selection.

- The Hymenoptera should have given rise to many more groups of social species than other orders, very few of which are also haplodiploid.
- The worker caste of these species should always be female.
- In contrast, the males should be drones, contributing little or no labor to the colony and receiving little attention from their sisters.

All these inferences are in fact true, and they admit of no easy explanation except kin selection biased by haplodiploidy.

I first read Hamilton's article during a train trip from Boston to Miami in the spring of 1965. This mode of travel was habitual for me during these years, the result of a promise to Renee that I would avoid trips by air as much as possible until our daughter, Catherine, reached high school age. I found an advantage in the restriction. It gave me, in the case of the Miami run, eighteen hours in a private roomette, trapped by my pledge like a Cistercian monk with little to do but read, think, and write. It was on such journeys that I composed a large part of *The Theory of Island Biogeography*. On this day in 1965 I picked Hamilton's paper out of my briefcase somewhere north of New Haven and riffled through it impatiently. I was anxious to get the gist of the argument and move on to something else, something more familiar and congenial. The prose was convoluted and the full-dress mathematical treatment difficult, but I understood his main point about haplodiploidy and colonial life quickly enough. My first response was negative. Impossible, I thought; this can't be right. Too simple. He must not know much about social insects. But the idea kept gnawing at me early that afternoon, as I changed over to the Silver Meteor in New York's Pennsylvania Station. As we departed southward across the New Jersey marshes, I went through the article again, more carefully this time, looking for the fatal flaw I believed must be there. At intervals I closed my eyes and tried to conceive of alternative, more convincing explanations of the prevalence of hymenopteran social life and the all-female worker force. Surely I knew enough to come up with something. I had done this kind of critique before and succeeded. But nothing presented itself now. By dinnertime, as the train rumbled on into Virginia, I was growing frustrated and angry. Hamilton, whoever he was, could not have cut the Gordian knot. Anyway, there was no Gordian knot in the first place, was there? I had thought there was probably just a lot of accidental evolution and wonderful natural history. And because I modestly thought of myself as the world authority on social insects, I also thought it unlikely that anyone else could explain their origin, certainly not in one clean stroke. The next morning, as we rolled on past Waycross and Jacksonville, I thrashed about some more. By the time we reached Miami, in the early afternoon, I gave up. I was a convert, and put myself in Hamilton's hands. I had undergone what historians of science call a paradigm shift.

That fall I attended a meeting of the Royal Entomological Society of London (crossing on the *Queen Mary*) to give an invited lecture on the social behavior of insects. The day before my session I looked up Bill Hamilton. Still a graduate student, he was in some respects the typical British academic of the 1950s—thin, shock-haired, soft-voiced, and a bit unworldly in his throttled-down discursive speech. I found that he lacked the terminal digits of one hand, lost during the Second World War, when as a child he tried to make a bomb in the basement laboratory of his father, an engineer with experience in rock-blasting who invented bombs for the British Home Guard—for use in case of a German invasion. As we walked about the streets of London, rambling on about many subjects of common interest, he told me he had experienced trouble getting approval for his Ph.D. thesis on kin selection. I thought I understood why. His sponsors had not yet suffered through their paradigm shift.

The next day I devoted a third of my hour-long presentation to Hamilton's formulation. I expected opposition, and, having run through the gamut of protests and responses in my own mind, I had a very good idea of what the objections would be. I was not disappointed. Several of the leading figures of British entomology were in the audience, including J. S. Kennedy, O. W. Richards, and Vincent Wigglesworth. As soon as I finished, they launched into some of the arguments I knew so well. It was a pleasure to answer them with simple prepared explanations. When once or twice I felt uncertain I threw the question to young Hamilton, who was seated in the audience. Together we carried the day.

The time was approaching to write a synthesis of knowledge about the social insects. I dreamed of spinning crystal-clear summaries of their classification, anatomy, life cycles, behavior, and social organization. I would celebrate their existence in a single well-illustrated volume. A work of this magnitude had not been attempted in thirty-five years, the last being Franz Maidl's rather opaque *Die Lebensgewohnheiten und Instinkte der staatenbildenden Insekten*, and was badly needed. The literature was scattered through hundreds of journals and books, in a dozen languages, and it varied enormously in quality. The study of social insects had been balkanized for a hundred years: experts on ants seldom spoke to those on termites, honeybee researchers lived in a world of their own, and students of halictine bees and social wasps cultivated their subjects to one side as minor arcane specialties. I wanted to create a showcase for sociobiology using insects and, in so doing, demonstrate the organizing power of population biology. That much, I believe, my book accomplished. *The Insect Societies*, published in 1971, conveyed my vision of the social insects and, in the final paragraph, I looked to the future:

> *The optimistic prospect for sociobiology can be summarized briefly as follows. In spite of the phylogenetic remoteness of vertebrates and insects and the basic distinction between their respective personal and impersonal systems of communication, these two groups of animals have evolved social behaviors that are similar in degree of complexity and convergent in many important details. This fact conveys a special promise that sociobiology can eventually be derived from the first principles of population and behavioral biology and developed into a single, mature science. The discipline can then be expected to increase our understanding of the unique qualities of social behavior in animals as opposed to those of man.**

Where might I go next? Originally I had no intention of extending my studies beyond the social insects. If honeybees are excluded for the moment—apiculture was in 1975 a major applied discipline unto itself, with hundreds of practitioners— the vertebrate animals, comprising fishes, amphibians, reptiles, and mammals, had at least ten times more zoologists attending to their behavior than was the case for insects. The mainstream journals of evolutionary biology tilted toward the natural history of the biggest animals, and vertebrates prevailed among the textbook case studies of ethology. Vertebrate behavior seemed too formidable a subject to enter from the direction

---

*The Insect Societies (Cambridge, Mass.: Harvard University Press, 1971), p. 460.

of entomology. But I found out I was wrong. After probing a bit, talking with specialists, I had a revelation. Vertebrates weren't difficult at all. Very few zoologists appeared to be aiming toward an integrated sociobiology of these animals, at least not with an emphasis on population biology or with the speed and directness that Hamilton and I and a few others had achieved for the social insects. With my inquiry expanding, I saw that entomology is a technically more difficult subject than vertebrate zoology, partly because insects are so much more diverse—750,000 known species versus 43,000 verte-brates—and partly because they seem so alien to *Homo sapiens*, the giant bipedal vertebrates who can see them clearly only through microscopes. They receive little attention in college curricula, and few students turn to them for a career. Not least, advanced insect societies are more complicated and variable than those of the nonhuman vertebrates. So I reasoned that it should be easier for an entomologist to learn about vertebrates than for a vertebrate zoologist to learn about insects.

Once again I was roused by the amphetamine of ambition. Go ahead, I told myself, pull out all the stops. Organize *all* of sociobiology on the principles of population biology. I knew I was sentencing myself to a great deal more hard work. *The Insect Societies* had just consumed eighteen months. When added to my responsibilities at Harvard and ongoing research program in ant biology, the writing had pushed my work load up to eighty-hour weeks. Now I invested two more years, 1972 to 1974, in the equally punishing and still more massive new book, *Sociobiology: The New Synthesis.* Knowing where my capabilities lay, I chose the second of the two routes to success in science: breakthroughs for the extremely bright, syntheses for the driven.

In fact the years spent writing the two syntheses were among the happiest of my life. In 1969 Larry Slobodkin invited me to join him in a summer ecology course at the Marine Biological Laboratory in Woods Hole, Massachusetts. In late June Renee, Cathy, and I journeyed to the small coastal village and moved into one of the cottages maintained by the MBL at Devil's Lane. One mile away, at the end of a winding country road, sat the spectacular lighthouse on Nobska Point, and beyond the lighthouse hill Little Harbor, and yet farther out, across the sail-dotted sound, the vacation island of Martha's Vineyard. Cathy, just then entering kindergarten age, fell in with a gang of other faculty youngsters. She and I also spent hours gazing at butterflies, birds, and, in the swamp behind our cottage, a colony of muskrats. In the late afternoons and evenings the three of us explored the southern reaches of Cape Cod by car. After lunch, outside class time, I took long runs over the Quissitt Hills along the coastal road to Falmouth. The rest of my free time I wrote, and read, and wrote. We continued to return to Woods Hole for another eighteen summers, through Cathy's college years. It was a balanced life during that long period, deeply fulfilling.

In the preparation of the vertebrate sections of *Sociobiology*, I was boosted by an exceptional quality of support resulting, I am inclined to think, from sheer good luck. Decisive parts of the bibliographic search and manuscript editing were conducted by Kathleen Horton, who had joined me in 1965 and acquired a high level of expertise in the difficult and sometimes arcane disciplines that feed into sociobiology. Nearly thirty years later, she continues this vital role across a broad range of biological subjects.

Sarah Landry, then as now one of America's best wildlife illustrators, was miraculously available in the early part of her career as I started work on my big books. She depicted animal societies with composites of animals in behavioral acts that could never be brought together in a single photograph. With a passion for accuracy, she went beyond the effort required for an ordinary book on animal behavior, traveling to zoos and aquaria to sketch captive animals and visiting herbaria to render in detail the plant species found in the natural habitats of the animal societies. To Sarah, the bushes among which a mountain gorilla foraged meant as much as the gorilla itself.

My uneasiness about vertebrate zoologists subsided when I found that they were going to treat me as an ally rather than an intellectual poacher. I sold myself, honestly so, as their chronicler and friendly critic. Literally all with whom I communicated encouraged me to go forward. Many showered me with books, articles, and evaluations of the large literature.

Nineteen seventy-four was one of the earliest years in which a critical mass of sociobiological theory could be assembled.

Studies of important species such as the Florida scrub jay and whiptail wallaby were in their final stages. And new elements of theory continued to pour in. One of the new theoretical concepts destined to be most influential was the natural selection of parent-offspring conflict, originated by Robert Trivers of Harvard. Like Hamilton, Trivers attained the key concept as a graduate student; I had just finished serving on his Ph.D. review committee. Trivers both benefited and suffered from a case of manic-depressive syndrome (now cured). When he was up he was dazzling; when he was down he was terrifying. We came into contact only during the peaks. He would stride through my office door and sit down, oblivious or uncaring of the old Harvard custom of making appointments. Thereupon I figuratively fastened my seat belt and prepared for swift and rocky travel to some unknown destination. Then would come a flood of ideas, new information, and challenges, delivered in irony and merriment. Trivers and I were always on the verge of laughter, and we broke down continually as we switched from concept to gossip to joke and back to concept. Our science was advanced by hilarity. My own pleasure in these exchanges was tinged with a sense of psychological risk, as though testing a mind-altering and possibly dangerous drug. Nor could I just sit and listen to Trivers, and let his mental productions wash over me. It is my nature, my conceit if you wish, to try to match any person with whom I converse fact for fact, idea for idea, and never quit. This is the reason I get killed in the company of my friends Murray Gell-Mann and Steven Weinberg, Nobel laureates in physics, egocentric, supremely self-confident, and said to be competing for the title of World's Smartest Human. Two or three hours with Trivers left me exhausted for the day.

For four spectacular years, 1971 through 1974, Trivers blazed new paths in sociobiological theory. He generated a model of reciprocal altruism, by which humans and more intelligent animals evolve contract rules that reach beyond self-sacrifice based on kin selection. What is undoubtedly his most important contribution, the theory of the family, and especially its undergirding models of parent-offspring conflict, set the foundation for today's substantial research enterprise on these subjects within behavioral biology. The selection pressures that bear on the evolution of nurturing, he pointed out, are different and sometimes opposite for parent and offspring. Shifting in direction and intensity as the young mature, these pressures account for youthful rebellion and family tensions better than the more proximate, conventional explanations of personal maladjustment and stress. At the least, Trivers provided a plausible argument for the ultimate causation of conflict, which persists regardless of the day-to-day proximate stressing events that trigger it.

It was Trivers who finally found the flaw in Hamilton's argument. Then he fixed it in a way that lent kin selection even greater credence. The flaw is the following. So long as social hymenopterans—ants, bees, and wasps—raise an equal number of males and queens among the brood destined to start the next generation of colonies, there is (contrary to Hamilton) no advantage for sisters to behave toward one another with any unusual degree of altruism. As a result of haplodiploidy they share three-fourths of their genes with sisters and only one-fourth with brothers, instead of one-half with both sexes, which is the case in animals using ordinary modes of sex determination. The imbalance in the Hymenoptera would seem to favor the formation of female colonies: more of a worker's genes will go into the next generation if she raises sisters instead of daughters. But, Trivers noted, if the haplodiploid ants, bees, and wasps rear equal numbers of sisters and brothers, they end up with an average relationship for all the offspring of one-half, canceling the apparent advantage. Mathematically, we can express this conclusion as follows:

$$\frac{1}{2} \text{ (fraction of females)} \times \frac{3}{4} \text{ (genes shared)}$$
$$+ \frac{1}{2} \text{ (fraction of males)} \times \frac{1}{4} \text{ (genes shared)} = \frac{1}{2}$$

One-half of the genes shared on average is the same payoff as from the ordinary production of sons and daughters without haplodiploidy. Only if the workers can raise a higher percentage of sisters in the royal brood can they reap the larger rewards of altruism twisted by haplodiploidy. The best possible overall degree of relationship in the Hymenoptera is five-eighths, reached by investing three-fourths of the resources in sisters:

$$\frac{3}{4} \text{ (fraction of females)} \times \frac{3}{4} \text{ (genes shared)}$$
$$+ \frac{1}{4} \text{ (fraction of males)} \times \frac{1}{4} \text{ (genes shared)} = \frac{5}{8}$$

Five-eighths beats one-half and gives the advantage to colonial existence, if all other conditions are equal. Subsequent studies showed that this is indeed approximately the ratio reared by ants. Somehow ant workers manage to obey the expectations of kin selection worked out in the heads of two zoologists.

I meant *Sociobiology: The New Synthesis* to serve as a network of such theory, as a vade mecum, and, not least, as an encyclopedia. I covered all organisms that could even remotely be called social, from colonial bacteria and amoebae to troops of monkeys and other primates. I recognized four "pinnacles" of social evolution, groups of species whose societies were, first, independently derived in evolution, and second, complex or sophisticated in organization, and, finally, possessed of genetic structures and organizations differing radically from those of the others. The pinnacles are respectively the corals, siphonophores, and other invertebrates; social insects; the social vertebrates (especially the great apes and other Old World primates); and man. Yes, *man*; that is the word I used in 1975, before it became unacceptably sexist and still meant generic humanity, while it still exercised the same resonant monosyllabic authority as earth, moon, and sun.

Perhaps I should have stopped at chimpanzees when I wrote the book. Many biologists wish I had. Even several of the critics said that *Sociobiology* would have been a great book if I had not added the final chapter, the one on human beings. Claude Lévi-Strauss, I was later reminded by his friend the historian Emmanuel Ladurie, judged the book to be 90 percent correct, which I took to mean true through the chimpanzees but not a line further.

Still I did not hesitate to include *Homo sapiens*, because not to have done so would have been to omit a major part of biology. By reverse extension, I believed that biology must someday serve as part of the foundation of the social sciences. I saw nothing wrong with the nineteenth-century conception of the chain of disciplines, in which chemistry is obedient to but not totally subsumed by physics, biology is linked in the same way to chemistry and physics, and there is a final, similar connection between the social sciences and biology. *Homo sapiens* is after all a biological species. History did not begin 10,000 years ago in the villages of Anatolia and Jordan. It spans the 2 million years of the life of the genus *Homo*. Deep history—by which I mean biological history—made us what we are, no less than culture. Our basic anatomy and physiology and many of our elementary social behaviors are shared with the Old World nonhuman primates. Even our unique qualities, the tool-using hand with its bizarre opposing thumb and the capacity for swift language acquisition, have a genetic prescription and presumably a history of evolution by natural selection. It felt appropriate to use provocative language as I opened the final chapter of *Sociobiology*:

> *Let us now consider man in the free spirit of natural history, as though we were zoologists from another planet completing a catalog of social species on Earth. In this macroscopic view the humanities and social sciences shrink to specialized branches of biology; history, biography, and fiction are the research protocols of human ethology; and anthropology and sociology together constitute the sociobiology of a single primate species.*

CHAPTER SEVENTEEN

The Sociobiology Controversy

THE SPATE OF reviews that followed the publication of *Sociobiology* in the summer of 1975 whipsawed it with alternating praise and condemnation. Biologists, who as a rule had little stake in the human implications, were almost unanimously favorable. They included Lewis Thomas and C. H. Waddington, elder statesmen of the day. Researchers closest to sociobiology were especially supportive, and they grew more so as time passed. In a 1989 poll the officers and fellows of the international Animal Behavior Society rated *Sociobiology* the most important book on animal behavior of all time, edging out even Darwin's 1872 classic, *The Expression of the Emotions in Man and Animals.*

Social scientists already engaged in biology-accented research also leaned in favor. They included Napoleon Chagnon, ethnographer of the "Fierce People," the Yanomamö of Brazil and Venezuela; and the sociologists Pierre van den Berghe and Joseph Shepher, who sought biological explanations of incest avoidance, marriage customs, and other key aspects of human behavior. Paul Samuelson, Nobel laureate economist turned public philosopher, favored the approach in one of his *Newsweek* columns but said *beware*—this subject is an intellectual and doctrinal minefield.

Samuelson was right. A wave of opposition soon rose among social scientists. Marshall Sahlins, a cultural anthropologist, made a strong attempt to exempt human behavior from the tenets of sociobiology in his 1976 book, *The Use and Abuse of Biology.* In November of that year the members of the American Anthropological Association, gathering in Washington for their annual meeting, considered a motion to censure sociobiology formally and to ban two symposia on the subject scheduled earlier. The arguments of the proposers were mostly moral and political. During the debate on the matter Margaret Mead rose indignantly, great walking stick in hand, to challenge the very idea of adjudicating a theory. She condemned the motion as a "book-burning proposal." Soon afterward the motion was defeated—but not by an impressive margin.

Because such events were widely publicized, with some journalists calling the controversy the academic debate of the 1970s, it is easy to exaggerate the depth of the opposition. The serious literature was in fact always strongly disposed toward human sociobiology. In the nearly twenty years since 1975, more than 200 books have been published on human sociobiology and closely related topics. Those more or less in agreement outnumber those against by a ratio of twenty to one. The basic ideas of sociobiology have expanded (their critics might say metastasized) into fields such as psychiatry, aesthetics, and legal theory. Four new journals were created in the late 1970s to accommodate a rising number of research and opinion articles.

Regardless of its real strength, much of the controversy might have been avoided, and for that I must bear the responsibility. I had written *Sociobiology* as two different books in one. The first twenty-six chapters, composing 94 percent of the text, was an encyclopedic review of social microorganisms and animals, with the information organized according to the principles of evolutionary theory. The second, the twenty-nine double-columned pages of Chapter 27 ("Man: From Sociobiology to Sociology"), consisted mostly of facts from the social sciences interpreted by hypotheses on the biological foundations of human behavior. The differences in substance and tone between books one and two give rise to the dual sociobiologies of popular perception. The first is sociobiology as I intended to portray it: a discipline, the systematic study of the biological basis of social behavior and advanced societies. And then there is the evil twin as perceived by Marshall Sahlins and some members of the American Anthropological Association, the scientific-ideological doctrine that human social behavior is determined by genes.

Genetic determinism, the central objection raised against book two, is the bugbear of the social sciences. So what I said that can indeed be called genetic determinism needs saying again here. My argument ran essentially as follows. Human beings inherit a propensity to acquire behavior and social structures, a propensity that is shared by enough people to be called human nature. The defining traits include division of labor between the sexes, bonding between parents and children, heightened altruism toward closest kin, incest avoidance, other forms of ethical behavior, suspicion of strangers, tribalism, dominance orders within groups, male dominance overall, and territorial aggression over limiting resources. Although people have free will and the choice to turn in many directions, the channels of their psychological development are nevertheless— however much we might wish otherwise—cut more deeply by the genes in certain directions than in others. So while cultures vary greatly, they inevitably converge toward these traits. The Manhattanite and New Guinea highlander have been separated by 50,000 years of history but still understand each other, for the elementary reason that their common humanity is preserved in the genes they share from their common ancestry.

It was the commonality of human nature and not cultural differences on which I focused in *Sociobiology*. At this level what I said could by no stretch be considered original; many others had advanced a similar thesis for decades. Darwin, who seems to have anticipated almost every other important idea in evolutionary biology, cautiously advanced theories of genetic change in aggression and intelligence. But no scientist before me had employed the reasoning of population biology so consistently to account for the evolution of human behavior by natural selection. The human genome is there in the first place, I argued, because it enhanced survival and reproduction during human evolution. The brain, sensory organs, and endocrine systems are prescribed in a way that predisposes individuals to acquire the favored general traits of social behavior.

In order to use models of population genetics as a more effective mode of elementary analysis, I conjectured that there might be single, still unidentified genes affecting aggression, altruism, and other behaviors. I was well aware that such traits are usually controlled by multiple genes, often scattered across many chromosomes, and that environment plays a major role in creating variation among individuals and societies. Yet whatever the exact nature of the genetic controls, I contended, the important point is that heredity interacts with environment to create a gravitational pull toward a fixed mean. It gathers people in all societies into the narrow statistical circle that we define as human nature.

Mine was an exceptionally strong hereditarian position for the 1970s. It helped to revive the long-standing nature–nurture debate at a time when nurture had seemingly won. The social sciences were being built upon that victory. But I hoped that even if sociobiology was dismissed by some of the more established scholars, evolutionary biology, including models of population genetics, would prove attractive to a younger generation of researchers in the social sciences, who might then connect their field to the natural sciences.

That expectation was desperately naive. The sociocultural view favored by most social theorists, that human nature is built wholly from experience, was not just another hypothesis up for testing. In the 1970s it was a deeply rooted philosophy. American scholars in particular were attracted to the idea that human behavior is determined by environment and therefore almost infinitely flexible.

If in fact genes did surrender their control sometime back during human evolution, and if the brain simply resembles an all-purpose computer, biology can play no contributory role in the social sciences. The appropriate domain of sociology would then be variation within cultures, interpreted as the product of environment. And cultural anthropology should concentrate on the internal detailed study of alien societies accepted on their own terms, with minimal reference to extraneous Westernized schemes, including those from biology. There were also important political implications. If human nature is mostly acquired, and no significant part of it is inherited, then it is easier to conclude, as relativists do with passion, that different cultures must be accorded moral equivalency. Differences among them in ethical precepts and ideology deserve respect, for what is thought good and true has been determined more by power than by intrinsic validity. The cultures of oppressed peoples are to be specially valued, because the histories of cultural conflict were written by the victors.

The hypothesis that human nature has a genetic foundation called all these assumptions into question. Many critics saw this challenge from the natural sciences as not just intellectually flawed but morally wrong. If human nature is rooted in heredity, they suggested, then some forms of social behavior are probably intractable or at least can be declared intractable by ruling elites. Tribalism and gender differences might then be judged unavoidable, and class differences and war in some manner "natural." And that would be just the beginning. Because people unquestionably vary in hereditary physical traits, they might also differ irreversibly in personal ability and emotional attributes. Some people could have inborn mathematical genius, others a bent toward criminal behavior.

In the 1970s a great many ordinary people believed these hereditarian propositions to be more or less true. But anyone who advanced such ideas in colleges and universities risked the scalding charges of racism and sexism. In contrast, those who attacked the hereditarian position were praised as defenders of truth and virtue. The psychobiologist Jerre Levy parodied the politically correct formula as follows: "Even without supporting evidence, the sociocultural hypothesis is assumed to be true unless proved false beyond any possible doubt. In contrast, the biological hypothesis is assumed to be false unless evidence is completely unassailable in its support."*

Understandably, then, American scholars, in a society grown hypersensitive to its internal divisions, shrank from the word "sociobiology." When American researchers formed a professional association on human sociobiology in 1989, they named it the Human Behavior and Evolution Society, and they used the word "sociobiology" only sparingly thereafter at their annual meetings.

The Europeans were less chary. One circle of researchers formed the European Sociobiological Society, headquartered in Amsterdam. Another established the Sociobiology Group at King's College, Cambridge University. A third began the Laboratory of Ethology and Sociobiology at the University of Paris–Nord. The word "sociobiology" and the ideas behind it were freely used in China, the Soviet Union, and other socialist countries, with articles written both for and against it in a scattering of journals.

What made *Sociobiology* notorious then was its hybrid nature. Had the two parts of the book been published separately, the biological core would have been well received by specialists in animal behavior and ecology, while the writings on human

---

*Jerre Levy, "Sex and the Brain," *The Sciences* 21, no. 3 (1981): 20–23, 28.

behavior might easily have been dismissed or ignored. Placed between the same two covers, however, the whole was greater than the sum of its parts. The human chapters were rendered creditable by the massive animal documentation, while the biology gained added significance from the human implications. The conjunction created a syllogism that proved unpalatable to many: Sociobiology is part of biology; biology is reliable; therefore, human sociobiology is reliable.

Some of the critics, assuming that I must have a political motive, suggested that the main purpose of the animal chapters was to lend credence to the human chapter. The exact opposite was true. I had no interest in ideology. My purpose was to celebrate diversity and to demonstrate the intellectual power of evolutionary biology. Being an inveterate encyclopedist, I felt an additional obligation to include the human species. As I proceeded, I recognized an opportunity: the animal chapters would gain intellectual weight from their relevance to human behavior. At some point I turned the relationship around: I came to believe that evolutionary biology should serve as the foundation of the social sciences.

Hence my conception of human sociobiology did not spring from any grand Comtean scheme of the relation between the natural and social sciences. I simply expanded the range of the subjects that interested me, starting with ants and proceeding to social insects, then to animals and finally to man. Believing the time ripe for the melding of biology and the social sciences, I used strong, provocative language to start the process. The last chapter of *Sociobiology* was meant to be a catalyst dropped among reagents already present and ready to combine.

Then everything spun out of control. In my calculations I had not counted on the ferocity of the response at my own university. During the McCarthy era, Harvard had been a celebrated—if imperfect—sanctuary for academics accused of being members of the Communist Party. It was supposed to be a forum in which people could exchange ideas with civility, protected from defamation by political ideologues. Yet the fact that it was well populated by leftist ideologues put that genteel goal at risk. Shortly after the publication of *Sociobiology*, fifteen scientists, teachers, and students in the Boston area came together to form the Sociobiology Study Group. Soon afterward the new committee affiliated itself with Science for the People, a nationwide organization of radical activists begun in the 1960s to expose the misdeeds of scientists and technologists, including politically dangerous thinking. The Sociobiology Study Group was dominated by Marxist and New Left scholars from Harvard. Two of the most prominent, Stephen Jay Gould and Richard Lewontin, were my close colleagues and fellow residents of the Museum of Comparative Zoology. Three others, Jonathan Beckwith, Ruth Hubbard, and Richard Levins, held faculty posts in other parts of the university.

Although the unofficial headquarters of the Sociobiology Study Group was Lewontin's office, located directly below my own, I was completely unaware of its deliberations. After meeting for three months, the group arrived at its foreordained verdict. In a letter published in the *New York Review of Books* on November 13, 1975, the members declared that human sociobiology was not only unsupported by evidence but also politically dangerous. All hypotheses attempting to establish a biological basis of social behavior "tend to provide a genetic justification of the *status quo* and of existing privileges for certain groups according to class, race, or sex. Historically, powerful countries or ruling groups within them have drawn support for the maintenance or extension of their power from these products of the scientific community . . . [Such] theories provided an important basis for the enactment of sterilization laws and restrictive immigration laws by the United States between 1910 and 1930 and also for the eugenics policies which led to the establishment of gas chambers in Nazi Germany."

I learned of the letter when it reached the newsstands on November 3. An editor at Harvard University Press called me to say that word about it was spreading fast and might prove a sensation. For a group of scientists to declare so publicly that a colleague has made a technical error is serious enough. To link him with racist eugenics and Nazi policies was, in the overheated academic atmosphere of the 1970s, far worse. But the self-proclaimed position of the Sociobiology Study Group was ethical, and therefore implicitly beyond challenge. And the purpose of the letter was not so much to correct alleged technical errors as to destroy credibility.

In the liberal dovecotes of Harvard University, a reactionary professor is like an atheist in a monastery. As the weeks passed and winter snows began to fall, I received little support from the Harvard faculty. Several friends spoke up in interviews and public radio forums to oppose Science for the People. They included Ernst Mayr, Bernard Davis, Ralph Mitchell, and my close friend and collaborator Bert Hölldobler. But mostly what I got was silence, even when the internal Harvard dispute became national news. I know now after many private conversations that the majority of my fellow natural scientists on the Harvard faculty were sympathetic to my biological approach to human behavior but confused by the motives and political aims of the Science for the People study group. They may also have thought that where there is smoke, there is fire. So they stuck to their work and kept a safe distance.

I had been blindsided by the attack. Having expected some frontal fire from social scientists on primarily evidential grounds, I had received instead a political enfilade from the flank. A few observers were surprised that I was surprised. John Maynard Smith, a senior British evolutionary biologist and former Marxist, said that he disliked the last chapter of *Sociobiology* himself and "it was also absolutely obvious to me—I cannot believe Wilson didn't know—that this was going to provoke great hostility from American Marxists, and Marxists everywhere."* But it was true. I was unprepared perhaps because (as Maynard Smith further observed) I am an American rather than a European. In 1975 I was a political naïf: I knew almost nothing about Marxism as either a political belief or a mode of analysis, I had paid little attention to the dynamism of the activist left, and I had never heard of Science for the People. I was not even an intellectual in the European or New York–Cambridge sense.

Because of my respect for the members of the Sociobiology Study Group I knew personally, I was at first struck by self-doubt. Had I taken a fatal intellectual misstep by crossing

---

*Quoted in Ullica Segerstråle, "Whose Truth Shall Prevail? Moral and Scientific Interests in the Sociobiology Controversy" (Ph.D. diss., Department of Sociology, Harvard University, 1983).

the line into human behavior? The indignant response of the Sociobiology Study Group stood in shocking contrast to the near silence of the other biologists in my department, who failed to offer even casual encouragement during corridor talk. My morale was not helped by the fact that Dick Lewontin, the most outspoken of the critics, was also chairman of the department. I faced the risk, I thought, of becoming a pariah—viewed as a poor scientist and a social blunderer to boot.

Then I rethought my own evidence and logic. What I had said was defensible as science. The attack on it was political, not evidential. The Sociobiology Study Group had no interest in the subject beyond discrediting it. They appeared to understand very little of its real substance.

As my mind settled on the details, anger replaced anxiety. I penned an indignant rebuttal to the *New York Review of Books*. In a few more weeks anger in turn subsided and my old confidence returned, then a fresh surge of ambition. There was an enemy in the field. An important enemy. And a new subject—which, for me, meant opportunity.

I set out to learn the elements of Marxism. I was encouraged in my amateur's effort by Daniel Bell, the distinguished sociologist, and Eugene Genovese, a leading Marxist philosopher. Neither of them cared very much for sociobiology, but they disliked even more the aggressive tactics of Science for the People. I expanded my reading into the social sciences and humanities. I acquired a taste for the history and philosophy of science. Two years after the Sociobiology Study Group published their letter, I wrote *On Human Nature*, which won the 1979 Pulitzer Prize for General Nonfiction (granted, a literary award and not scientific validation). The following year I began an all-out attempt to build a stronger theory to explain the interaction between genetic and cultural evolution.

The sociobiology controversy, I came to realize, ran deeper than ordinary scholarly discourse. The signatories of the Science for the People letter had come to the subject with a different agenda from my own. They viewed science not as separate objective knowledge but as part of culture, a social process compounded with political history and class struggle.

The spirit of their exertions was most clearly embodied, I believe, in the person of Richard C. Lewontin. He was later overshadowed by the scientific and literary celebrity of Stephen Jay Gould, but in 1975 the two men were equally well known and of mostly common political opinion. Gould shared Lewontin's Marxist approach to evolutionary biology, and he afterward maintained a drumfire of criticism in his monthly *Natural History* column and essays published elsewhere. But it was Lewontin who explored more deeply and thoroughly than anyone else every level of the implications of human sociobiology. He was the principal author of the letter in the *New York Review of Books.* Afterward he gave the greatest number of lectures opposing sociobiology, drawing on his extensive knowledge of genetics and the philosophy of science. He devoted the greatest amount of time to rallying opposition among potential converts, and his vigilance never slipped. If there is a truly fatal flaw in the sociobiology argument, he will have explicated it somewhere.

Without Lewontin the controversy would not have been so intense or attracted such widespread attention. He was the kind of adversary most to be cherished, in retrospect, after time has drained away emotion to leave the hard inner matrix of intellect. Brilliant, passionate, and complex, he was stage-cast for the role of contrarian. He possessed a deep ambivalence that kept both friend and foe off balance: intimate in outward manner, private inside; aggressive and demanding constant attention, but keenly sensitive, anxious to humble and to please listeners at the same time; intimidating yet easily set back on his heels by a strong response, revealing a fleeting angry confusion that made one—almost—wish to console him. Robert MacArthur told me, when we three were young men, that Lewontin was the only person who could make him sweat.

Unafflicted by shyness, at committee meetings he almost always seated himself near the head or center of the conference table, speaking up more frequently than others present, questioning and annotating every subject raised. He was the boy prodigy you surely encountered at least once in school, the first to raise his hand, the first reaching the blackboard to crack the algebra problem. His youthful demeanor was preserved into middle age by a round face, easy grin, and knowing stare, a shock of unruly dark hair, and a tieless shirt, always blue, said by amused friends to advertise his solidarity with the working class. Journalists referred to his countenance as owlish, but that was true only in freezeframe. Lewontin was too nervous and active in real life for the strigid image to fit.

He would pivot from one role to another, first the thoughtful and cautious dean, now the lecturer expanding a philosophical idea, then the hearty joking companion, and abruptly, on occasion, the angry radical. To accentuate a point, he would raise his hands above his head with fingers opened, and as his voice evened out and the argument unfolded, slide them back to the table top palms down, at first placed side by side and then eased apart, the mood having turned reflective, then quickly up again to chest level and windmilled one around the other, the subject grown more complex and the listener thereby commanded to pay close attention. He spoke in complete sentences and paragraphs. The stream of words was punctuated at intervals by a slowing delivery, sometimes almost a slurring, to reinforce a key phrase and, finally, the approach of the concluding argument. While he spoke he turned about to make eye contact with each listener within range, flashing the grin, signaling a confidence in his choice of words, revealing an attention to technique as well as to substance.

His self-confidence and style were potent in the academy of the 1960s and 1970s. It was the era when students clamorously asserted their independence and at the same time searched desperately for leaders. Lewontin's lectures at Harvard and abroad were enthusiastically received. His antiestablishment barbs, delivered with the panache of a stand-up comedian, were marvelously witty, even when you happened to be the target; they drew dependable laughter. Here was a scientist, the students knew, and a thinker, drawing from a deep revolutionary wellspring. He impressed journalists, too, who commonly referred to him as "the brilliant population geneticist." Lewontin was an intellectual who preached social change from the temple of hard science.

His scientific credentials were beyond challenge. His genetic research was of the highest caliber. In the mid-1960s, while at the University of Chicago, he collaborated with J. L. Hubby to make the first estimates of gene diversity within populations by means of the electrophoretic separation of closely similar proteins. Their technique soon became standard and inaugurated a new era of quantitative studies in evolutionary biology. He was also one of the first to use computers to study the role of chance in microevolution. Striking out from the same base of expertise, he explored the border area between genetics and ecology by linking the evolution of demography to changes in the rate of population growth.

Very early, at the age of thirty-nine, Richard Lewontin was elected to the National Academy of Sciences, one of the highest honors in American science. Then the contrarian side of his nature emerged. In 1971, amid verbal fireworks, he resigned in protest over the Academy's sponsorship of classified research projects for the Department of Defense. He was one of only twelve members out of the thousands elected during the 130-year history of the organization to quit it for any reason. He had placed himself in distinguished company; the others included Benjamin Peirce, William James, and Richard Feynman.

In the early spring of 1972 a Harvard committee, of which I was a member, recommended to the Department of Biology that Richard Lewontin be offered a full professorship. He was at that time considered the best population geneticist of his generation in the world. Under ordinary circumstances the appointment would have received quick approval and been passed on to the dean and president; but circumstances were no longer normal. Dick by that time was more than just a leading scientist. He had also become a political activist targeting other scientists. At the 1970 annual meeting of the American Association for the Advancement of Science he had been one of a small group who disrupted a session on a politically sensitive topic.

Several of the senior professors, alarmed by what they saw as a trend in his personality, were prepared to vote against Lewontin's candidacy. Wouldn't he be disruptive in his own department, they asked, if brought to Harvard? At the critical meeting of the tenured professors Ernst Mayr and I defended him. We argued (rather stuffily it seems in retrospect) that political beliefs should not influence faculty appointments. Some of the members remained unpersuaded: beliefs are one thing, they said, but what about personal attacks and disruption? I badly wanted Lewontin to come to Harvard. I said, let me call a friend in his department at the University of Chicago and ask if Dick has attacked his own colleagues there on ideological grounds. The proposal was accepted, and the decision postponed. In the interim George Kistiakowsky, one of Harvard's most respected senior professors and wise adviser to the university administration, got wind of the proceedings and telephoned me from the Department of Chemistry. He said in effect, you're going to be sorry if Lewontin comes. I was committed; I made my own call and was assured that Lewontin had not created problems at the University of Chicago. At the next meeting we voted unanimously to recommend him for a professorship. President Derek Bok approved his appointment on November 8, 1972, and the following year he came to Harvard.

Once he was installed, and increasingly after the sociobiology controversy began, I realized that we were opposites in our views of the proper conduct of science. Lewontin was the philosopher-scientist, tightly self-constrained, critical at every step, a stern guardian of standards who opposed—indeed, would have banned, if given the opportunity—plausibility arguments and speculation. I was the naturalist-scientist, in agreement on the need for strict logic and experimental testing but expansive in spirit and far less prone to be critical of hypotheses in the early stages of investigation. A collector and pragmatist by lifelong experience, I believed that every scrap of information and reasonable hypothesis should be put on record, then kept or discarded as knowledge grows. My notebooks were an indiscriminate hodgepodge. To be restrictive in the early stages, to make a moral issue of plausibility arguments, was in my view antithetical to the spirit of science. I wanted to move evolutionary biology into every potentially congenial subject, roughshod if need be, and as quickly as possible. Lewontin did not.

By adopting a narrow criterion of publishable research, Lewontin freed himself to pursue a political agenda unencumbered by science. He adopted the relativist view that accepted truth, unless based upon ineluctable fact, is no more than a reflection of dominant ideology and political power. After his turn to activism he worked to promote his own accepted truth: the Marxian view of holism, a mental universe within which social systems ebb and flow in response to the forces of economics and class struggle. He disputed the idea of reductionism in evolutionary biology, even though it was and is the virtually unchallenged linchpin of the natural sciences. And most particularly, he rejected it for human social behavior. "By reductionism," he wrote in 1991, "we mean the belief that the world is broken up into tiny bits and pieces, each of which has its own properties and which combine together to make larger things. The individual makes society, for example, and society is nothing but the manifestation of the properties of individual human beings. Individual properties are the causes and the properties of the social whole are the effects of those causes."*

This reductionism, as Lewontin expressed and rejected it, is precisely my view of how the world works. It forms the basis of human sociobiology as I construed it. But it is not science, Lewontin insisted. And according to his own political beliefs, expressed over many years, it could not possibly be true. "This individualistic view of the biological world is simply a reflection of the ideologies of the bourgeois revolutions of the eighteenth century that placed the individual at the center of everything."† Lewontin sought instead laws that were transcendent, beyond the reach of natural science. "There is nothing in Marx, Lenin, or Mao," he wrote in collaboration with Richard Levins, "that is or can be in contradiction with the particular physical facts and processes of a particular set of phenomena in the objective world."‡ Only antireductionist, nonbourgeois science would help humanity attain the ultimate, highest goal, a socialist world.

That a distinguished scientist could advocate an approach to science guided by a radically sociocultural version of Marxism in the service of world socialism may seem odd today, and perhaps most of all in the former republics of the Soviet Union. But it helps to explain the distinctive flavor of the controversy at Harvard in the 1970s. In the standard leftward frameshift of academia prevailing then, Lewontin and members of Science for the People were classified as progressives, admittedly a bit

---

*Richard C. Lewontin, *Biology as Ideology: The Doctrine of DNA* (New York: HarperPerennial, 1991), p. 107.
†Ibid.
‡R. C. Lewontin and R. Levins, "The Problem of Lysenkoism," in Hilary Rose and Steven Rose, eds., *The Radicalisation of Science* (London: Macmillan, 1976), pp. 34, 59.

extreme in their methods, while I—Roosevelt liberal turned pragmatic centrist—was cast well to the right.

After the Sociobiology Study Group exposed me as a counterrevolutionary adventurist, and as a result of it, other radical activists in the Boston area conducted a campaign of leaflets and teach-ins to oppose human sociobiology. As this activity intensified through the winter and spring of 1975–76, I grew fearful that it might reach a level embarrassing to my family and the university. I briefly considered offers of professorships from three universities—in case, their representatives said, I wished to leave the physical center of the controversy. But it all came to very little. For a few days a protester in Harvard Square used a bullhorn to call for my dismissal. Two students from the University of Michigan invaded my class on evolutionary biology one day to shout slogans and deliver anti-sociobiology monologues. When it became apparent that they had not read *Sociobiology* and were more interested in using it as a stick to beat the Harvard ruling class, they were heckled by my own students. I received almost no hate mail, and never a death threat.

The most dramatic episode was the water dousing in Washington in 1978. On February 15 I arrived at the Sheraton Park Hotel to speak at a symposium on sociobiology planned as part of the annual meeting of the American Association for the Advancement of Science. The largest organization of scientists in the world, the AAAS was and remains especially concerned with the relation of science to education and public policy. A large crowd was expected at the symposium, which featured a half-dozen of the principal researchers on human sociobiology, as well as one of its most articulate critics, Stephen Jay Gould.

The moderator was to be Margaret Mead, and I looked forward to meeting her for the second time. A year before, at a conference on human behavior in Virginia, she had invited me to have dinner with her to discuss sociobiology. I was nervous then, expecting America's mother figure to scold me about the dangers of genetic determinism. I had nothing to fear. She wanted to stress that she, too, had published ideas on the biological basis of social behavior. One was that each society contains an array of people genetically predisposed toward different tasks, say artist or soldier, and this differentiation creates a more efficient division of labor. Over roast beef and red wine (I was too mesmerized by her presence to taste either) she recommended several of her own writings that she thought I might want to read.

Sadly, I was not to see her again. Shortly before the AAAS meeting, she was stricken with the cancer that would soon take her life.

As the time approached for the symposium to begin, the atmosphere in and around the meeting hall grew tense. I was told that some kind of demonstration was planned by the International Committee Against Racism (INCAR), a group known for violent action. Its leaders, on learning that a session on human sociobiology was scheduled and that I would be present, had alerted members throughout the country. On hearing this news I walked by the INCAR booth to collect the literature they were distributing and to pick up a lapel button. As the crowd of several hundred began to settle in the nearby lecture hall, two INCAR members moved about distributing copies of a protest leaflet. I reached for one, but the young woman offering it recognized me and snatched it away.

Nothing happened as the substitute moderator, Alexander Alland, Jr., an anthropologist from Columbia University, opened the session and several other speakers presented their papers. When my turn came I chose to stay in my seat rather than stand at the lectern; my right leg was in a cast from an ankle fracture incurred while jogging over black ice two weeks previously. As soon as I was introduced, about eight men and women—I never managed an exact count—sprang from their seats in the audience, rushed onto the stage, and lined up behind the row of speakers. Several held up anti-sociobiology placards, on at least one of which was painted a swastika. A young man walked to the lectern to take the microphone away from Alland. AAAS officials had earlier issued instructions to session chairs to surrender their microphones if demonstrators demanded them, to avoid physical scuffling, and then to inform the protestors that if the microphones were not returned within two minutes, hotel security would be called. Alland announced that he was following the AAAS official procedure and turned over the microphone. Meanwhile, some of the members of the audience, fearing a riot, began to move out of their seats and away from the stage. They made little progress, however, because all the seats were filled and the aisles were crowded. Napoleon Chagnon, seated in a middle row, struggled to move the other way, determined to reach the stage and eject the protestors, but his way was also blocked. With several other audience members he shouted back at Alland and the protestors: the surrender of the microphone was wrong; no group should be allowed to take over a session by force. But this was the era of parity and equivalence, and every form of expression was considered free speech. The crowd began to settle down.

Then, as the INCAR leader harangued the audience, a young woman behind me picked up a pitcher of water and dumped the contents on my head. The demonstrators chanted, "Wilson, you're all wet!" In a little over two minutes they left the stage and took their seats. No one asked them to leave the premises, no police were called, and no action was taken against them later. After the symposium, several stayed behind to chat with members of the audience.

As I dried myself off with my handkerchief and a paper towel someone handed me, Alland, in possession of the microphone again, expressed his regret to me for the incident. The audience then gave me a prolonged standing ovation. Of course they did, I thought. What else could they do? They might be next. Before I could proceed with my brief lecture, other members of the panel rose to condemn the INCAR action. Steve Gould seemed to be speaking to the demonstrators when he quoted Lenin on the inappropriateness of violence for mere radical posturing, as opposed to the attainment of worthy political goals. Gould referred to the AAAS incident, using Lenin's words, as an "infantile disorder" of socialism. In that he was correct. It was the grown-up intellectuals I knew I had to worry about.

How did I feel during the incident? Calm—dare I say icy cold, as I let the protestors' anger wash over me? That evening I joined Napoleon Chagnon for dinner and then debated Marvin Harris on human sociobiology at the Smithsonian Institution, with another large audience in attendance—no takeover by radicals this time. Afterward I taxied to Union Station to catch the Night Owl sleeper to Boston. There I ran into the physicist Freeman Dyson, who was on his way home to Princeton. Well, I said, I've had quite a day. I had water dumped on me by protestors at the AAAS sociobiology symposium. Well, he said, I've also had quite a day. I was just in a train wreck. The engine had derailed a few miles north of Washington and the passengers had been ferried back to the station to await a later northbound train.

By this time it was obvious to me that human sociobiology would remain in trouble, both intellectually and politically, until it incorporated culture into its analyses. Otherwise the critics could always cogently argue that since semantically based mind and culture are the defining traits of the human species, explanations of human social behavior without them are useless. This shortcoming was on my mind when Charles Lumsden, a young theoretical physicist from the University of Toronto, arrived in early 1979 to work with me as a postdoctoral research fellow. His interests had lately turned to biology, and he saw great opportunity in the analysis of social behavior. We talked at first about a collaboration on social insects, but soon our conversation gravitated to the subject of heredity and culture. I said, the possible payoff justifies the high risk of failure; let's give it a try. So two or three times a week for eighteen months we sat together and framed the subject piece by piece.

We reasoned as follows. Everyone knows that human social behavior is transmitted by culture, but culture is a product of the brain. The brain in turn is a highly structured organ and a product of genetic evolution. It possesses a host of biases programmed through sensory reception and the propensity to learn certain things and not others. These biases guide culture to a still unknown degree. In the reverse direction, the genetic evolution of the most distinctive properties of the brain occurred in an environment dominated by culture. Changes in culture therefore must have affected those properties. So the problem can be more clearly cast in these terms: how have genetic evolution and cultural evolution interacted to create the development of the human mind?

No doubt we went out of our depth in embarking upon this subject. But so was everyone else, and no one can be sure of anything until the attempt is made. Undaunted then, we sifted through a small mountain of literature in cognitive psychology, ethnography, and brain science. We built models in population genetics that incorporated culture as units of learned information. We studied the properties of semantic thought to make our premises as consistent as possible with current linguistic theory.

We were looking for the basic process that directed the evolution of the human mind. We concluded that it is a particular form of interaction between genes and culture. This "gene-culture coevolution," as we called it, is an eternal circle of change in heredity and culture. Over the course of a lifetime, the mind of the individual person creates itself by picking among countless fragments of information, value judgments, and available courses of action within the context of a particular culture. More concretely, the individual comes to select certain marital customs, creation myths, ethical precepts, modes of analysis, and so forth, from among those available. We called these competing behaviors and mental abstractions "culturgens." They are close to what our fellow reductionist Richard Dawkins conceived as "memes."

Each time an individual modifies his memories or makes decisions, he entrains intricate sequences of physiological events that run first from the perception of visual images, sounds, and other stimuli, then to the storage and recall of information from long-term memory, and finally to the emotional assessment of perceived objects and ideas. Not all culturgens are treated equally; cognition has not evolved as a wholly neutral filter. The mind incorporates and uses some far more readily than others. Examples of heredity-bound culture that Lumsden and I found from the research literature include the peculiarities of color vision, phoneme formation, odor perception, preferred visual designs, and facial expressions used to denote emotions. All are diagnostic of the human species, all part of what must reasonably be called human nature.

Such physiologically based preferences, called "epigenetic rules," channel cultural transmission in one direction instead of another. By this means they influence the outcome of cultural evolution. It is here, through the physical events of cognition, that the genes act to shape mental development and culture.

The full cycle of gene-culture coevolution as we conceived it is the following. Some choices confer greater survival and reproductive rates. As a consequence, certain epigenetic rules, those that predispose the mind toward the selection of successful culturgens, are favored during the course of genetic evolution. Over many generations, the human population as a whole has moved toward one particular "human nature" out of a vast number of natures possible. It has fashioned certain patterns of cultural diversity from an even greater number of patterns possible.

Lumsden and I presented our scheme in several technical articles and two books.* The reviews were mixed; some were enthusiastic, but those in several key journals were unfavorable: Edmund Leach was enraged in *Nature*; Peter Medawar was contemptuous in the *New York Review of Books*; Richard Lewontin, by his own later description, was nasty in *The Sciences.* The subject of gene-culture coevolution simply languished, mostly ignored by biologists and social scientists alike. I was worried, and puzzled. The critics really hadn't said much of substance. Had we nevertheless failed at some deep level they saw but we failed to grasp? During the 1980s a handful of other researchers investigated the subject along conceptual pathways of their own devising. Gifted scientists with diverse expertise from genetics and anthropology, they included Kenichi Aoki, Robert Boyd, Luigi Cavalli-Sforza, William Durham, Marcus Feldman, Motoo Kimura, and Peter Richerson. They too met with only limited success, at least as measured by the spread and advance of the total research enterprise. Kimura, Japan's foremost geneticist, told me that he had received almost no requests for his article on the subject.

It is possible that gene-culture coevolution will lie dormant as a subject for many more years, awaiting the slow accretion of knowledge persuasive enough to attract scholars. I remain in any case convinced that its true nature is the central problem of the social sciences, and moreover one of the great unexplored domains of science generally; and I do not doubt for an instant that its time will come.

---

*C. J. Lumsden and E. O. Wilson, *Genes, Mind, and Culture* (Cambridge, Mass.: Harvard University Press, 1981) and *Promethean Fire* (Cambridge, Mass.: Harvard University Press, 1983). The summary of the theory of gene-culture coevolution presented here is drawn, with minor changes, from our article "Genes, Mind, and Ideology," *The Sciences* 21, no. 9 (1981): 6–8.

CHAPTER EIGHTEEN

Biodiversity, Biophilia

I N 1980 THE editors of *Harvard Magazine* asked seven Harvard professors to identify what they considered to be the most important problem facing the world in the coming decade. Four cited poverty arising from, variously, overpopulation, the influx of rural masses into cities, and capitalism. Another, focusing on the United States, cited the welfare state and excessive governmental control. The sixth chose the global nuclear threat.

None of these scholars mentioned the environment. None gave more than fleeting attention to the impact that problems of the 1980s might have on future generations. As the only natural scientist I chose a radically different subject, and a broader time scale: species are going extinct in growing numbers, I wrote; the biosphere is imperiled; humanity is depleting the ancient storehouses of biological diversity. I was thinking like an evolutionary biologist, in evolutionary time. "The worst thing that can happen, *will* happen," I said, "is not energy depletion, economic collapse, limited nuclear war, or conquest by a totalitarian government. As terrible as these catastrophes would be for us, they can be repaired within a few generations. The one process ongoing in the 1980s that will take millions of years to correct is the loss of genetic and species diversity by the destruction of natural habitats. This is the folly our descendants are least likely to forgive us."*

This article marked my debut as an environmental activist. I was, I will confess now, unforgivably late in arriving. Biodiversity destruction had troubled my mind for decades, but I had made little overt response. In the 1950s, as I worked my way around bare red-clay gullies in Alabama and sought the vanishing rain forests of Cuba, I knew something was terribly wrong. My apprehension grew as I pored over the list of extinct

---

*"Resolutions for the 80s," *Harvard Magazine*, January–February 1980, pp. 22–26.

and endangered animal species in the Red Data Books of the International Union for Conservation of Nature and Natural Resources. In the 1960s the picture darkened further when Robert MacArthur and I found that a reduction of habitat is inexorably followed by a loss of animal and plant species. Very roughly, we learned, a 90 percent reduction of forest cover—or prairie, or river course—eventually halves the number of species living there.

Adding to my concern was the Dream. It was literally an anxiety dream, and one that I occasionally experience to this day. I am on an island near an airport or in a town. I recognize the place immediately: from one night to the next, either Futuna or New Caledonia, both in the South Pacific. I've been there alone for weeks, and now, as my surroundings take increasingly detailed form, I remember that the hour of my departure is approaching. I realize that I have not examined the fauna and flora of the island, nor have I made any attempt to collect the ants, most of whose species remain unknown to science. I begin a frantic search for native forest. In the distance I see what looks like the edge of a copse and run to it, only to find a row of exotic trees planted as a windbreak, with more houses and fields stretching beyond. Now I am in an automobile. I speed down a country road; nothing but houses and fields appear on either side. There are mountains far to the north—in every dream always to the north. Perhaps some forest remains in the mountains. I fumble with a map and locate the access road, but I cannot go; my time has run out. The dream ends, and I awaken knotted with anxiety and regret.

Knowledge and dreams notwithstanding, I hesitated, confining myself in the waking world almost entirely to research and writing on other subjects. As the 1970s passed I wondered, at what point should scientists become activists? I knew from hard experience that the ground between science and political engagement is treacherous. I was gun-shy from the sociobiology controversy. Speak too forcefully, I thought, and other scientists regard you as an ideologue; speak too softly, and you duck a moral responsibility. I hesitated on the side of caution, taking some relief from knowledge that nonacademic organizations were already active in the conservation of biological diversity. They included the World Wildlife Fund and the

International Union for the Conservation of Nature, global in their outlook and highly competent and respected. There was also the Organization for Tropical Studies, a consortium of universities and other institutions dedicated to the training of young biologists, which I had helped found in 1963. Many of these new professionals, I knew, were going into conservation science. I thought, let the next generation do it. Still, the movement needed the voices of senior biologists.

The decisive impetus for me came when, in 1979, the British ecologist Norman Myers published the first estimates of the rate of destruction of tropical rain forests. After adding up data country by country, he calculated the global loss of cover to be a little under one percent per year. This piece of bad news immediately caught the attention of conservationists around the world. The rain forests were and are of crucial importance as reservoirs of diversity. They teem with the greatest variety of plants and animals of all the world's ecosystems, yet at the time of Myers' report they occupied only 7 percent of the world's land surface. Their area was thus about the same as the contiguous forty-eight United States, and the amount of cover removed each year was about equal to half the area of the state of Florida. The reduction in area translated, in terms of the general relation between habitat area and diversity worked out in other ecosystems, to roughly one-quarter of a percent of species extinguished or doomed to early extinction each year. The cutting and burning appeared to be accelerating as a result of incursions by land-hungry rural populations and the increasing global demand for timber products.

Primed by Myers' report, I was finally tipped into active engagement by the example of my friend Peter Raven. A distinguished scientist and director of the Missouri Botanical Garden, increasingly a public figure, Peter was determined and fearless. He had no qualms about activism. By the late 1970s he was writing, lecturing, and debating those still skeptical about the evidences of mass extinction. In 1980 he chaired a National Research Council study of research priorities in tropical biology, putting stress on the urgent problems of deforestation and biological diversity. More than anyone else Raven made it clear that scientists in universities and other research-oriented institutions must get involved; the conservation professionals could not be expected to carry the burden alone. One day on impulse I crossed the line. I picked up the telephone and said, "Peter, I want you to know that I'm joining you in this effort. I'm going to do everything in my power to help." By this time a loose confederation of senior biologists that I jokingly called the "rain forest mafia" had formed. It included, besides Raven and myself, Jared Diamond, Paul Ehrlich, Thomas Eisner, Daniel Janzen, Thomas Lovejoy, and Norman Myers. We were to remain in frequent communication from then on.

A short time later I joined the Board of Directors of the World Wildlife Fund–U.S. and became their key external science adviser. I encouraged the staff to strengthen further their programs in scientific research while broadening the organization's coverage to include entire ecosystems and not just individual star species such as the giant panda and bald eagle. I joined in promoting the "new environmentalism" being formulated within WWF. This more pragmatic approach combines conservation projects with economic advice and assistance to local populations affected by efforts to salvage biological diversity. Nature reserves, we knew and argued, cannot be protected indefinitely from impoverished people who see no advantage in them. Conversely, the long-term economic prospects of these same people will be imperiled to the degree that their natural environment is destroyed.

I lectured and wrote widely on the problems of ecosystem destruction and species extinction, and on possible socioeconomic solutions. In 1985 I published an article in the policy journal of the National Academy of Sciences titled "The Biological Diversity Crisis: A Challenge to Science," which received widespread attention.* The following year I gave one of several keynote addresses at the National Forum on BioDiversity, held in Washington under the combined auspices of the National Academy of Sciences and the Smithsonian Institution. I then served as the editor of the proceedings volume, *BioDiversity*, which became one of the best-selling books in the history of the National Academy Press. The forum was the first occasion on which the word "biodiversity" was used, and after the publication of the book it spread with astonishing speed around the

---

*In *Issues in Science and Technology* 2(1) (Fall 1985): 20–29.

world; by 1987 it was one of the most frequently used terms in conservation literature. It became a favorite subject of museum exhibitions and college seminars. By June 1992, when more than a hundred heads of state met at the Earth Summit in Rio de Janeiro to debate and ratify global protocols on the environment, "biodiversity" approached the status of a household word. President Bush's refusal to sign the Convention on Biological Diversity on behalf of the United States brought the subject into the political mainstream. Finally, the continuing controversies over the Endangered Species Act and the northern spotted owl made it part of American culture.

Biodiversity, the concept, has become the talisman of conservation, embracing every kind of living creature. So what exactly does it mean? The definition soon agreed upon by biologists and conservationists is the totality of hereditary variation in life forms, across all levels of biological organization, from genes and chromosomes within individual species to the array of species themselves and finally, at the highest level, the living communities of ecosystems such as forests and lakes. One slice of biodiversity among the near infinitude possible would be the variety of chromosomes and genes within one species of freshwater fish found in Cuba. Another would be all the freshwater fish species of Cuba, and still another would be the fishes and all other forms of life living in each river in Cuba studied in turn.

Because I edited the volume *BioDiversity* in 1988, it is widely thought that I also coined the term. I deserve no credit at all. The expression was put into play by Walter Rosen, the administrative officer of the National Academy of Sciences who organized the 1986 Washington forum. When Rosen and other NAS staff members approached me to serve as editor of the proceedings, I argued for "biological diversity," the term I and others had favored to that time. Biodiversity, I said, is too catchy; it lacks dignity. But Rosen and his colleagues persisted. Biodiversity is simpler and more distinctive, they insisted, so the public will remember it more easily. The subject surely needs all the attention we can attract to it, and as quickly as possible. I relented.

I am not sure now just why I resisted the word at all, in view of the quickness with which it acquired both dignity and influence. After all, in 1979 I had invented a very similar term, "biophilia," for use in a *New York Times* article on conservation.* Later, in 1984, I employed it as the title and pivotal idea of my book *Biophilia*. It means the inborn affinity human beings have for other forms of life, an affiliation evoked, according to circumstance, by pleasure, or a sense of security, or awe, or even fascination blended with revulsion.

One basic manifestation of what I called biophilia is a preference for certain natural environments as places for habitation. In a pioneering study of the subject, Gordon Orians, a zoologist at the University of Washington, diagnosed the "ideal" habitat most people choose if given a free choice: they wish their home to perch atop a prominence, placed close to a lake, ocean, or other body of water, and surrounded by a parklike terrain. The trees they most want to see from their homes have spreading crowns, with numerous branches projecting from the trunk close to and horizontal with the ground, and furnished profusely with small or finely divided leaves. It happens that this archetype fits a tropical savanna of the kind prevailing in Africa, where humanity evolved for several millions of years. Primitive people living there are thought to have been most secure in open terrain, where the wide vista allowed them to search for food while watching for enemies. Possessing relatively frail bodies, early humans also needed cover for retreat, with trees to climb if pursued.

Is it just a coincidence, this similarity between the ancient home of human beings and their modern habitat preference? Animals of all kinds, including the primates closest in ancestry to *Homo sapiens*, possess an inborn habitat selection on which their survival depends. It would seem strange if our ancestors were an exception, or if humanity's brief existence in agricultural and urban surroundings had erased the propensity from our genes. Consider a New York multimillionaire who, provided by wealth with a free choice of habitation, selects a penthouse overlooking Central Park, in sight of the lake if possible, and rims its terrace with potted shrubs. In a deeper sense than he perhaps understands, he is returning to his roots.

---

*"The Column: Harvard University Press," *New York Times Book Review*, January 14, 1979, p. 43.

Balaji Mundkur, an anthropologist and art historian at the University of Connecticut, has suggested a parallel explanation for another peculiarity of human taste: our fascination with snakes. These reptiles are among the features of mankind's ancient environment for which people can easily acquire phobias. Other strong phobia inducers are spiders, wolves, heights, closed spaces, and running water. Just one frightening experience with snakes—as mild as a scary story—is enough to instill the aversion in a child. The fear experienced thereafter is marked by the onset of panic, nausea, and cold sweat, reactions of the autonomic nervous system beyond ordinary rational control. The responses are quickly acquired, yet strangely difficult to eradicate.

The highly directed reaction against snakes appears to have a genetic foundation. In evidence is the remarkable fact that people rarely acquire phobias toward the objects of modern life that are truly dangerous, such as guns, knives, electric sockets, and speeding automobiles. Our species has not been exposed to these lethal agents long enough in evolutionary time to have acquired the predisposing genes that ensure automatic avoidance.

People everywhere are not just repelled by snakes. They are fascinated by them, and if they can do so safely, they draw close to inspect them. Snakes are the wild animals that appear most often in dreams, and, designated as mystical serpents, in religious symbolism. Variously hybridized with humans or other animals, plumed, twinned, grown gigantic and swift and all-seeing, the dream-mutants are gods who both avenge and transmit wisdom according to the vagaries of mood and circumstance. The caduceus, the staff entwined by a pair of serpents and carried by Mercury as messenger of the gods, now serves as the emblem of the medical profession.

The ultimate source of our attention to snakes may be the same as that of the fear and fascination they excite in other primates: their deadly nature. Poisonous species occur throughout the world, in the Northern Hemisphere as far north as Canada and Finland, and are an important source of mortality in most places where people live close to natural environments. The chain of biophilic evolution, as I interpreted it in 1984 from Mundkur's evidence, runs as follows. The deadliness of some kinds of snakes resulted through evolutionary time in an innate aversion and fascination among human beings. Hence they regularly disturb our dreams with ambiguous symbolism. Shamans and prophets report their own dreams as divine revelation and install the imagery in mythology and religion. From these sacred redoubts the glittering transformed serpent has invaded story and art.

By the ordinary standards of natural science, the evidence for biophilia remains thin, and most of the underlying theory of its genetic origin is highly speculative. Still, the logic leading to the idea is sound, and the subject is too important to neglect. In 1992 a conference of biologists, psychologists, and other scholars meeting at Woods Hole, Massachusetts, reported and evaluated a great deal of ongoing research. Some of it was experimental, consistent with earlier data, and persuasive.*

In my opinion, the most important implication of an innate biophilia is the foundation it lays for an enduring conservation ethic. If a concern for the rest of life is part of human nature, if part of our culture flows from wild nature, then on that basis alone it is fundamentally wrong to extinguish other life forms. Nature is part of us, as we are part of Nature.

Biophilia is the most recent of my syntheses, joining the ideas that have been most consistently attractive to me for most of my life. My truths, three in number, are the following: first, humanity is ultimately the product of biological evolution; second, the diversity of life is the cradle and greatest natural heritage of the human species; and third, philosophy and religion make little sense without taking into account these first two conceptions.

In this memoir I have described, for myself and for you, how I arrived at this naturalistic view of the world. Although the tributary sources extend far back in memory, they still grip my imagination, as I write, in my sixty-sixth year. I am reluctant to throw away these precious images of my childhood and young manhood. I guard them carefully as the wellsprings of my creative life, refining and overlaying their productions

---

*The proceedings of the conference were published as *The Biophilia Hypothesis*, ed. Stephen R. Kellert and E. O. Wilson (Washington, D.C.: Island Press, 1993).

constantly. When obedient to the rules of replicable evidence, the knowledge obtained is what I have called science.

The images created a gravitational force that pulled my career round and round through epicycles of research. They still define me as a scientist. In my heart I will be an explorer naturalist until I die. I do not think that conception overly romantic or unrealistic. Perhaps the wildernesses of popular imagination no longer exist. Perhaps very soon every square kilometer of the land will have been traversed by someone on foot. I know that the Amazon headwaters, New Guinea Highlands, and Antarctica have become tourist stops. But there is nonetheless real substance in my fantasy of an endless new world. The great majority of species of organisms—possibly in excess of 90 percent—remain unknown to science. They live out there somewhere, still untouched, lacking even a name, waiting for their Linnaeus, their Darwin, their Pasteur. The greatest numbers are in remote parts of the tropics, but many also exist close to the cities of industrialized countries. Earth, in the dazzling variety of its life, is still a little-known planet.

The key to taking the measure of biodiversity lies in a downward adjustment of scale. The smaller the organism, the broader the frontier and the deeper the unmapped terrain. Conventional wildernesses of the overland trek may indeed be gone. Most of Earth's largest species—mammals, birds, and trees—have been seen and documented. But microwildernesses exist in a handful of soil or aqueous silt collected almost anywhere in the world. They at least are close to a pristine state and still unvisited. Bacteria, protistans, nematodes, mites, and other minute creatures swarm around us, an animate matrix that binds Earth's surface. They are objects of potentially endless study and admiration, if we are willing to sweep our vision down from the world lined by the horizon to include the world an arm's length away. A lifetime can be spent in a Magellanic voyage around the trunk of a single tree.

If I could do it all over again, and relive my vision in the twenty-first century, I would be a microbial ecologist. Ten billion bacteria live in a gram of ordinary soil, a mere pinch held between thumb and forefinger. They represent thousands of species, almost none of which are known to science. Into that world I would go with the aid of modern microscopy and molecular analysis. I would cut my way through clonal forests sprawled across grains of sand, travel in an imagined submarine through drops of water proportionately the size of lakes, and track predators and prey in order to discover new life ways and alien food webs. All this, and I need venture no farther than ten paces outside my laboratory building. The jaguars, ants, and orchids would still occupy distant forests in all their splendor, but now they would be joined by an even stranger and vastly more complex living world virtually without end. For one more turn around I would keep alive the little boy of Paradise Beach who found wonder in a scyphozoan jellyfish and a barely glimpsed monster of the deep.

The author searches for insects on an epiphyte in the rain forest of Barro Colorado Island, Panama. Photo: Mark W. Moffett

Studies in tropical diversity. On Adam's Peak, Sri Lanka (TOP), live several species of the rare tree genus *Stemonoporus*. Illustrated BELOW is one of the many species of the genus limited to Sri Lanka and in danger of extinction, *S. oblongifolius*. Photos: Darlyne Murawski

TOP: Two canopy researchers, Jack Longino and Nalini Nadkarni, pause for lunch in a fig tree in the mountain forest of Monteverde, Costa Rica. The trunk and branches are so crowded with epiphytes, vines, and parasitic mistletoes that it is difficult to determine which leaves belong to the fig tree itself. Photo: Mark W. Moffett

BOTTOM: Like an arboreal plant garden, this mass of tentacled sea anemones, corals, jug-shaped sea squirts, clams, and other invertebrate animals covers the submerged root of a red mangrove in the Bastimentos National Park of Panama. Photo: Darlyne Murawski

"Microwildernesses" can be found everywhere in tiny spaces, in this case an acorn lying on a forest floor in Massachusetts. Many of the acorns are killed by long-snouted weevils (LEFT). A variety of insects and other small creatures then invade their interiors, creating a miniature community of herbivores, decomposers, predators, and parasites. In the tableau opposite, a millipede feeds on detritus near abandoned insect galls and a snail shell, while at the other end a green fungus spreads across the decaying acorn meat. In closer view (LOWER LEFT), an oribatid mite crawls over the snail shell, probably in search of fungus spores. Over several years a succession of such creatures flourishes and disappears until the remnant of the woody shell crumbles into the humus. Photos: Mark W. Moffett

In the Peruvian rain forest, Terry Erwin fogs a tree with a biodegradable insecticide. Later he will collect hundreds of species of insects, spiders, and other arthropods that fall on the plastic sheets spread below. Photo: Mark W. Moffett

TOP: A square fragment of rain forest has been left standing on land cleared for pasture north of Manaus, Brazil. The plot is one of those monitored before and after the clearing to determine the effect of the size of reserves on the survival of species. Photo: Mark W. Moffett

BOTTOM: Natural storm damage is usually quickly repaired by the secondary growth of native vegetation, a process already begun in this Puerto Rican rain forest six months after the great 1989 hurricane. Photo: Darlyne Murawski

Flowers

Flowers in the rain forests of Panama. In many such habitats of the New World tropics, several hundred plant species can be found within a plot the size of a football field. Photos: Darlyne Murawski

Fruits

The fruits of rain-forest plants are often as showy as the flowers. Their distinctive colors and shapes attract animals or catch the wind to scatter seeds and start a new generation. The species shown here are from Costa Rica and Panama. Photos: Darlyne Murawski

Fungi

About 1.5 million kinds of fungi are believed to exist, of which only 5 percent have been given a scientific name. Among those found in the Panamanian forests are wire-like *Cordyceps* that parasitize ants and other insects, and *Leucocoprinus gongylophorus*, whose giant mushrooms occasionally sprout from the nests of leaf-cutter ants. Photos: Darlyne Murawski

Frogs

Frogs living in the Panamanian rain forest include ground-dwelling species whose color patterns blend with those of fallen leaves, tree frogs easily overlooked in green vegetation, and poison-arrow frogs whose bright colors advertise their toxic nature to would-be predators. Photos: Darlyne Murawski

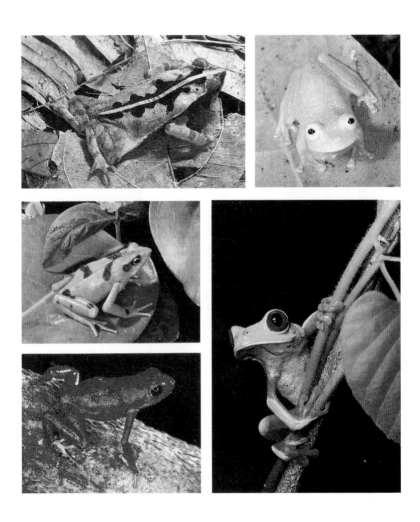

Butterflies

Butterflies, like many other groups of plants and animals, reach maximum diversity in the tropics, especially in the rain forests. More than a thousand species can be found in some localities of the New World tropics. This array represents a small fraction of the huge fauna of Central America. Photos: Darlyne Murawski

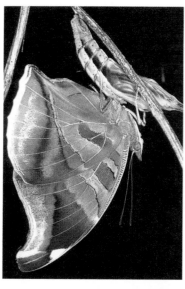

Beetles

Beetles are the most diverse and among the oldest of all land animals. The 290,000 species known worldwide are only a small fraction of those thought to exist. Local diversity is extreme: more than a thousand different kinds, ranging from carnivores to wood borers and leaf eaters, have been found on single trees in the forests of Central and South America. The adults shown here are all from Costa Rica and Panama. Photos: Darlyne Murawski

Spiders

Jumping spiders, composing the worldwide family Salticidae, hunt prey away from webs and use distinctive colors to assist in mating displays. These species are from Sri Lanka. Photos: Mark W. Moffett

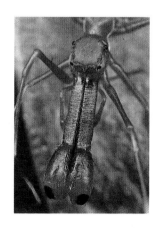

Crabs

Crustaceans are the marine equivalent of insects in the diversity of form and the array of niches they fill. These crabs from Costa Rica and Panama include swift-running beach and tidal-rock species (TOP and LOWER LEFT), a hermit crab inhabiting an abandoned snail shell, and several sluggish but heavily armored dwellers in coastal waters. Photos: Darlyne Murawski

Acknowledgments

I am indebted to a number of persons for important assistance in reconstructing the events of my early childhood. They were, in Biloxi and Gulfport, Mississippi, William B. Carlin II; Edward B. Kitchens, Brigadier General (Ret.); and Murella Powell; in Pensacola, Florida, Frank Hardy, Sr., Barbara McVoy, and Patricia Shoemaker; and in Washington, D.C., Ellis G. MacLeod. I obtained some details of my ancestry from Elizabeth Wilson Covan, our family's genealogist; my mother, Inez Linnette Huddleston; and William M. P. Dunne, professor at the State University of New York, Stony Brook, and an expert on the history of Gulf Coast pilotage. Life at the University of Alabama and Harvard University during my student years was reconstructed with the aid of information from my friends William L. Brown and Thomas Eisner; from Joyce Lamont, librarian at the University of Alabama; and from Aaron J. Sharp, my mentor at the University of Tennessee, who helped me gain admission to Harvard.

I am also grateful to the following friends and colleagues for reading portions of the manuscript and generously providing help and advice: Alexander Alland, Jr., Gary D. Alpert, Stuart Altmann, George E. Ball, George W. Barlow, Herbert T. Boschung, Napoleon Chagnon, Franklin L. Ford, Stephen Jay Gould, William D. Hamilton, Bert Hölldobler, Robert L. Jeanne, Ernst Mayr, Basil G. Nafpaktitis, William Patrick, Reed Rollins, Ullica Segerstråle, Daniel Simberloff, Lawrence B. Slobodkin, Frederick E. Smith, Kenneth Thimann, Robert L. Trivers, Barry D. Valentine, and James D. Watson. My wife Irene (Renee) discussed the work in progress and provided help and encouragement throughout. John P. Scott sent background materials on the earliest days of sociobiology, while Michael Ruse provided wise counsel and advice over the years that enriched my perception of the sociobiology controversy. None of these consultants, of course, is in any way responsible for errors of fact that may have survived, or for my interpretations.

The service at Pensacola's First Baptist Church in 1943 described in Chapter 3 is a composite pieced together, respectfully and I trust without distortion, from my fifty-year-old memories, from conversations with my fellow member (still active) Barbara McVoy, and from *On the Bay—On the Hill*, a history of the Pensacola church by Toni Moore Clevenger and a 1986 publication of the First Baptist Church, Pensacola.

The portion of Langston Hughes's poem "Daybreak in Alabama" that opens Part I is from *Selected Poems of Langston Hughes* (New York: Alfred A. Knopf, 1959) and is reprinted by permission of the publisher. My account of the capture of the cottonmouth moccasin (Chapter 6), together with the reconstruction of early conversations on island biogeography with Robert MacArthur and the description of MacArthur's personality (Chapter 13), is taken with slight modification from *Biophilia* (Cambridge, Mass.: Harvard University Press, 1984). The summary of Konrad Lorenz's 1953 lecture at Harvard (Chapter 15) is based upon an imperfect memory. I may have combined my recollections with some details from reading and discussion conducted soon afterward, but the spirit and main themes I believe to be accurate.

As for all my books in the past, back to *The Theory of Island Biogeography* with Robert MacArthur in 1967, I am grateful to Kathleen M. Horton for her invaluable editorial assistance and advice.

CHRONOLOGY

NOTE ON THE TEXTS

NOTES

Chronology

1929–35   Born Edward Osborne Wilson Jr. in Birmingham, Alabama, on June 10, 1929, the only child of Inez Linnette Freeman, twenty-two, and Edward Osborne Wilson Sr., twenty-nine. His parents, both Alabama natives from families long established in the state, married in Birmingham on June 5, 1926. Father works as an accountant, a trade he learned in the Army. Moving with family to Pensacola, takes Sunday school classes at the First Baptist Church; attends P. K. Yonge Elementary School.

1936   Is sent for the summer to live with a family in Paradise Beach, Florida, where he injures an eye in a fishing accident. Months later, left untreated, develops a traumatic cataract and undergoes surgery at Pensacola Hospital, partially losing his sight.

1937   Parents separate and file for divorce. Attends Gulf Coast Military Academy near Gulfport, Mississippi, beginning in January, remaining until the summer. Mother, who is given custody, puts him in the care of family friend Belle Raub and her husband for a year, while she works as a secretary. Attends Agnes McReynolds Elementary School in Pensacola, skipping the third grade.

1938   Moves briefly to Orlando, Florida, to live with father and new stepmother, Pearl. Attends more than a dozen different schools in cities and towns across the South before his high school graduation, father frequently changing jobs.

1939   Relocates to Washington, D.C., where father works as an audit clerk and stepmother as a saleslady. In November, mother is remarried to businessman Harold Huddleston; beginning in his teens and through the end of his college years, spends each September with them, at first in Louisville, Kentucky, and then in Jeffersonville, Indiana.

1940   In February, a teacher writes: "Ed has genuine writing ability, and when he combines this with his great

829

knowledge of insects, he produces fine results." Spends
time at the National Zoo and the National Museum of
Natural History; explores Rock Creek Park collecting
insects, especially butterflies.

1941   Paternal grandmother Mary Joyner Wilson dies on
March 13, leaving her house in Mobile to father and
uncle; family moves in. Draws cartoons for school
newspaper at Barton Academy junior high; expands
his butterfly collection.

1942   Studies ants in the vacant lot next to his house, determined to pursue a career in entomology; his observations of the invasive fire ant *Solenopsis invicta* are among
the first in the United States. Takes on a paper route in
the fall, delivering for the *Mobile Press Register*. Joins
Boy Scouts of America. Attends First Baptist Church.

1943   Over the summer, works as a nature counselor at Camp
Pushmataha, a Boy Scout camp near Citronelle, Alabama; is bitten by a pygmy rattlesnake, one of many
snakes he collects and raises, forcing a hospital visit.
Visits Mammoth Cave in Kentucky with mother; she
encourages his college aspirations and his entomological interests. Beginning in the fall, spends another year
with Belle Raub.

1944   Is baptized at the First Baptist Church of Pensacola.
Moves to Brewton, Alabama, in the spring; plays football at Brewton High School. Befriends the owner of
a local goldfish hatchery, who allows him to hunt for
snakes and other creatures on his property.

1945   Moves with father and stepmother to Decatur, Alabama. Takes part-time jobs in restaurants and retail
stores. Becomes an Eagle Scout. Continues ant collecting, preserving specimens in alcohol and building glass observation cases; writes to myrmecologist
Marion R. Smith at the National Museum of Natural
History, proposing a survey of the ants of Alabama, and
receives advice and encouragement. Anxious to attend
college, improves his grades; applies unsuccessfully for
a scholarship to Vanderbilt University.

1946   Attempts to enlist in the Army to obtain benefits of
the GI Bill, but is denied because of impaired eye-
sight. Takes job as an office clerk in a steel factory. In

September, his parents agreeing to share the cost of tuition, enrolls at University of Alabama.

1947   Takes part-time job as research assistant to Bert Williams, professor of botany; Williams offers space in his lab. Adds reptile and amphibian specimens to university collections. Reads Erwin Schrödinger's *What Is Life?*

1948   Reads *Systematics and the Origin of Species* by Ernst Mayr. Begins a correspondence with William L. Brown, a Harvard graduate student specializing in ant biology who becomes a mentor and collaborator. Trains with the goal of running a four-minute mile.

1949   Tries out for the track team. Takes job with the Alabama Department of Conservation over the spring term, producing "A Report on the Imported Fire Ant *Solenopsis saevissima* var. *richteri* Forel in Alabama." Graduates from the University of Alabama after three years, with a bachelor's degree in biology; receives Phi Beta Kappa honors. Father, a longtime member of Alcoholics Anonymous, checks himself into a rehabilitation facility.

1950   Receives an MS degree from the University of Alabama. Transfers to the University of Tennessee in Knoxville to work with ant specialist Arthur Cole, serving as Cole's assistant. Continues to build private ant collection, exploring the Great Smoky Mountains for specimens. Visits William Brown at Harvard in June; Brown introduces him to the ant room at the Museum of Comparative Zoology. Gives lecture on hominid evolution in Cole's general biology class, testing Tennessee state law prohibiting the teaching of evolution.

1951   On March 26, father commits suicide in Mobile at age forty-eight; he is buried in Magnolia Cemetery after a military funeral. Travels to Cambridge, Massachusetts, in the fall to pursue graduate studies at Harvard. Begins laboratory and museum investigations of ants, specializing in the genus *Lasius*.

1952   Joins fellow graduate student Thomas Eisner on a cross-country camping and entomological field trip over the summer. Collaborates with William Brown on paper "The Subspecies Concept and Its Taxonomic

Application," published in *Systematic Zoology*; it questions the stability of the subspecies as a taxonomic category.

1953   In the spring is elected to Harvard's Society of Fellows, which provides him with three years' financial support, free housing, and social access to distinguished faculty. Travels to Cuba in June for a field course in tropical botany; collects specimens for the Museum of Comparative Zoology. In late July proceeds to Mérida, touring the Yucatan peninsula with fellow students. The following month, by himself, attempts to climb Pico de Orizaba, Mexico's highest mountain, nearly reaching the summit. Back in Cambridge, meets Irene (Renee) Kelley, a native of Boston's Back Bay, and they are engaged.

1954   In November, with funds from the Society of Fellows and Museum of Comparative Zoology, leaves Massachusetts for a ten-month entomological expedition in Melanesia and Australia. Writes Renee daily. Spends December doing fieldwork in Fiji and New Caledonia.

1955   Travels to the New Hebrides (now Vanuatu) in January and in February across Western Australia. Beginning in March, treks through mountainous and remote areas of Papua New Guinea for more than three months. Returning home, stops in Queensland and Ceylon (now Sri Lanka) for more fieldwork; visits the entomological collections of museums in Genoa, Geneva, Paris, and London. Is reunited with Renee in September; they marry at St. Cecilia Church in Boston on October 30, and move into a university housing complex near campus. Is awarded a PhD for his thesis, "A Monographic Revision of the Ant Genus *Lasius*."

1956   Visits Cayo Santiago, Puerto Rico, in January with Stuart Altmann, who is researching social behavior in rhesus macaques and who will become his first graduate student. Beginning in July, joins the Harvard faculty as assistant professor of biology, a five-year appointment; declines tenure-track jobs at other institutions. Collaborates with William Brown on the concept of "character displacement," by which closely related species evolve to avoid hybridization and competition, publishing paper in *Systematic Zoology*.

1957   Joins William Brown and Herbert Levi, a curator of
                arachnology at the Museum of Comparative Zool-
                ogy, in a letter to U.S. Agriculture Secretary Ezra Taft
                Benson, warning against widespread use of pesticides
                to combat the fire ant problem: spraying "will affect
                not only wildlife, but the native insect fauna, and could
                produce deep-set disturbances of the prevailing system
                of ecological checks and balances," they write.

1958   In the wake of a job offer from Stanford University,
                is awarded tenure at Harvard as an associate professor
                of zoology and decides to remain. Focuses on studies
                of chemical communication in ants, identifying the
                gland with which fire ants produce food-trail phero-
                mones and attempting with colleagues to determine
                these pheromones' molecular structure. Subsequently
                investigates other ant pheromones. Receives first of
                many research grants from the National Science
                Foundation.

1959–62   Introduces concept of taxon cycles in 1959 paper "Adap-
                tive Shift and Dispersal in a Tropical Ant Fauna," pub-
                lished in *Evolution*, and elaborates in 1961 with "The
                Nature of the Taxon Cycle in a Melanesian Ant Fauna,"
                in *American Naturalist*. In 1960, as competition for
                resources within Harvard's biology department increas-
                ingly favors those pursuing cellular and molecular
                studies, joins with organismal and ecosystem-oriented
                faculty to form a "Committee on Macrobiology," sub-
                sequently renamed the "Committee on Evolutionary
                Biology." Continues pheromone research with gradu-
                ate student William Bossert. In February 1961, visits
                the Spring Hill Estate (now the Asa Wright Nature
                Center) in Trinidad, where he meets naturalist and
                writer William Beebe. Travels to Suriname and Tobago
                for fieldwork. Returning to the United States, buys his
                first house, in Lexington, Massachusetts. Meets Robert
                MacArthur at a meeting of the American Association
                for the Advancement of Science, and they begin a col-
                laborative friendship.

1963   Catherine, a daughter, is born in November. Helps to
                establish Organization for Tropical Studies.

1964   In July, meets with a small group of population biolo-
                gists at Robert MacArthur's Marlboro, Vermont, home;

they call themselves the "Marlboro Circle." Receives promotion to full professorship.

1965   Travels to the Florida Keys with wife and daughter over the summer, seeking field sites suitable for island biogeography experiments. In the fall is joined by graduate student Daniel Simberloff. Meets with National Park Service officials and professional exterminators, hoping to completely remove the fauna from islands of varying sizes in order to study reintroduction rates. Lectures on the social behavior of insects at a meeting of the Royal Entomological Society in London.

1966–70   After a successful island fumigation trial in October 1966, island biogeography experiments continue; commutes between Boston and field sites while Simberloff remains in Florida. Visits Lignumvitae Key, whose undisturbed forests are threatened by real estate developers, and works with Thomas Eisner and others to preserve the island (now a State Botanical Site). In 1967, with Robert H. MacArthur, publishes *The Theory of Island Biogeography*. Circulates a satirical "Glossary of Phrases in Molecular Biology," continuing his efforts to defend taxonomic, evolutionary, and ecological approaches to the discipline. Publishes first results of Florida Keys studies with Simberloff in 1969, in *Ecology* ("Experimental Zoogeography of Islands: The Colonization of Empty Islands"), following up in subsequent years. In May 1969 is elected to the National Academy of Sciences. Spends the summer with wife and daughter at the Marine Biological Laboratory in Woods Hole, Massachusetts; returns to Woods Hole during the summer for the next eighteen years, serving on MBL's board of trustees and helping to establish its Ecosystems Center. Meets Bert Hölldobler, a German ant specialist and visiting scholar at Harvard with whom he goes on to collaborate extensively. Takes up jogging.

1971   Publishes *The Insect Societies*, a comprehensive survey of the social insects, and (with William H. Bossert) *A Primer of Population Biology*. Wins Mercer Award of the Ecological Society of America for island zoogeography research, with Daniel Simberloff.

1972   *The Insect Societies* is named a National Book Award finalist. Serves as Hitchcock Visiting Professor at

University of California, Berkeley. Robert MacArthur dies of renal cancer in November.

1973–75   Is named curator in entomology at Harvard's Museum of Comparative Zoology, a position he holds along with his professorship beginning in 1973. Also serves as president of the Society for the Study of Evolution. In June 1975, *Sociobiology: The New Synthesis* is published. Although initial reviews are laudatory, its final chapter, "Man: From Sociobiology to Sociology," draws criticism and protest. Harvard colleagues Richard C. Lewontin and Stephen Jay Gould join with others to form the "Sociobiology Study Group"; in November, they publish a letter "Against 'Sociobiology'" in the *New York Review of Books*, arguing that "Wilson joins the long parade of biological determinists whose work has served to buttress the institutions of their society by exonerating them from responsibility for social problems."

1976   *Sociobiology: The New Synthesis* is a finalist for the National Book Award. Gives Messenger Lectures at Cornell University and Bartram Lecture at Florida State; lectures widely in subsequent years.

1977–78   Is awarded the National Medal of Science by President Jimmy Carter on November 22, 1977. *Time* magazine features sociobiology ("A New Theory of Behavior") on its cover. The following February, at a meeting of the American Association for the Advancement of Science, demonstrators from the International Committee Against Racism carry placards and dump ice water on Wilson's head in an anti-sociobiology protest ("Wilson, you're all wet!"). Agrees to serve on scientific advisory committee of the World Wildlife Fund. Receives honorary degrees from Duke University and Grinnell College, the first of dozens. Publishes *Caste and Ecology in the Social Insects* (with George F. Oster), and, in October 1978, *On Human Nature*.

1979   Receives Pulitzer Prize in general nonfiction for *On Human Nature*.

1980   Asked to name the greatest challenge of the coming decade for *Harvard Magazine* in January, proposes "the loss of genetic and species diversity by the

	destruction of natural habitats." Gives spring commencement address at University of Alabama.
1981–83	*Genes, Mind, and Culture: The Coevolutionary Process*, written with C. J. Lumsden, is published by Harvard University Press in May to mixed reviews; it explores the interrelations between genetic diversity and cultural history. They continue these explorations two years later in *Promethean Fire: Reflections on the Origins of Mind*.
1984–85	*Biophilia*, an essay collection, appears in August. Joins board of directors of the World Wildlife Fund, serving for a decade. Travels to Costa Rica with Bert Hölldobler, spending time at La Selva, a field station of the Organization for Tropical Studies. Publishes "The Biological Diversity Crisis: A Challenge to Science" in *Issues in Science and Technology*, the policy journal of the National Academy of Sciences.
1986–87	Stepfather Harold Huddleston dies on November 4, 1986. Is awarded the National Zoological Park medal in zoology and conservation.
1988–91	Daughter Catherine is engaged to Jonathan Gargill, whom she later marries. Edits National Academy of Sciences volume *BioDiversity*, contributing an opening chapter. Begins ongoing donation of his papers to the Library of Congress. Publishes *Success and Dominance in Ecosystems: The Case of the Social Insects*, and *The Ants*, with Bert Hölldobler; the latter wins the 1991 Pulitzer Prize for nonfiction.
1992	Receives the Levenson Prize, awarded by Harvard undergraduates, for his teaching skills. *The Diversity of Life* is published in October. Joins board of directors of the New York Botanical Garden.
1993	Receives the International Prize for Biology from the Japan Society for the Promotion of Science; after the ceremony, discusses natural history with Emperor Akihito, an expert in the fish family Gobiidae. *The Diversity of Life* is named a finalist for the National Book Critics Circle Award. Joins boards of directors of the American Museum of Natural History and The Nature Conservancy.

1994–95   Is appointed Pellegrino University Professor at Harvard in July 1994. Publishes *Journey to the Ants: A Story of Scientific Exploration* (with Bert Hölldobler), and a memoir, *Naturalist*, later named a finalist for the National Book Critics Circle Award.

1996–98   Publishes essay collection *In Search of Nature*. Retires from teaching duties at Harvard and from his curatorial position at the Museum of Comparative Zoology. Joins board of directors of Conservation International. *Consilience: The Unity of Knowledge* appears in March 1998.

1999–2002 Mother dies in Roswell, Georgia, on September 23, 1999, at age ninety-two. *American Scientist* names *The Insect Societies* (1971) one of the top one hundred science books of the century. In January 2002 publishes *The Future of Life*—a call to preserve the biosphere, with proposals for how to achieve this goal.

2003   Publishes the 794-page monograph *Pheidole in the New World: A Dominant, Hyperdiverse Ant Genus*, the culmination of almost two decades of taxonomic work in his home laboratory.

2004–5   Receives an honorary doctorate at Harvard's 2004 spring commencement. The E. O. Wilson Biodiversity Foundation is established in his honor the following year.

2006–9   Addresses people of faith in *The Creation: An Appeal to Save Life on Earth*, published in 2006: "I like to think that in this one life-and-death issue we have a common purpose." *The Superorganism: The Beauty, Elegance, and Strangeness of Insect Societies*, written with Bert Hölldobler, appears in 2009.

2010–12   In 2010, publishes *Anthill*, his first and only novel; *Kingdom of Ants: José Celestino Mutis and the Dawn of Natural History in the New World*, with José M. Gómez Durán, a study in the history of myrmecology; *The Leafcutter Ants: Civilization by Instinct*, with Bert Hölldobler; and controversial August *Nature* paper "The Evolution of Eusociality," with mathematicians Martin A. Nowak and Corina E. Tarnita. Makes first visit to Gorongosa National Park in Mozambique. *The*

Social Conquest of Earth, an expansive account of the evolution of social behavior, appears in April 2012 to laudatory reviews and becomes a best seller; President Bill Clinton calls it "a fabulous book." Collaborates with photographer Alex Harris on the photo-essay *Why We Are Here: Mobile and the Spirit of a Southern City.*

2013   *Letters to a Young Scientist*, published in April, offers advice and encouragement for students and young professionals: "The world needs you—badly."

2014–16   Attends March 2014 inauguration of the E.O. Wilson Biodiversity Laboratory in Gorongosa National Park, later publishing *A Window on Eternity: A Biologist's Walk through Gorongosa National Park*, about his time there. Brings an evolutionary approach to philosophical questions in *The Meaning of Human Existence*, published later that year; it is a best seller and subsequent finalist for the National Book Award. In *Half-Earth: Our Planet's Fight for Life*, published in 2016, proposes that half the planet be set aside in nature preserves to slow climate change and prevent ongoing mass extinction.

2017–19   Reflects on the beginnings of language, art, and narrative in *The Origins of Creativity*, published in 2017, and in 2019 offers a revisionist account of the evolution of altruism in *Genesis: The Deep Origin of Societies.*

2020   *Tales from the Ant World*, mixing scientific observations and personal recollections, is published in August. Lives in Lexington, Massachusetts, with his wife.

Note on the Texts

This volume contains three books by Edward O. Wilson: *Biophilia* (1984), *The Diversity of Life* (1992), and *Naturalist* (1994). Wilson is not known to have altered the texts of these works after their initial publication, though he contributed a new foreword and a new preface to *The Diversity of Life* in 1999 and 2010 respectively, and an afterword to *Naturalist* in 2006. The texts of all three works have been taken from first printings; Wilson's subsequent additions to *The Diversity of Life* and *Naturalist* have been included in the Notes.

Biophilia was first published by Harvard University Press on August 23, 1984. It has since been reprinted on at least twelve occasions, but no new editions are known to have appeared. The text of *Biophilia* in the present volume is that of the first printing.

The Diversity of Life was first published by Harvard University Press on October 6, 1992. The following year, new printings appeared from Allen Lane in London and W. W. Norton in New York, both of them reproducing the typesetting of the first edition. In 1999, Norton reissued the book with a new foreword, "Biodiversity at the Close of the Century"; this foreword was included in a Penguin edition published in London in 2001. Wilson contributed a "Preface to the 2010 Printing" for Harvard University Press, which reissued the book in that year. The text of *The Diversity of Life* in the present volume is that of the first Harvard University Press printing. In the book's Glossary, on pages 503–522 of the present volume, dates have been added to entries for individuals who have died since 1992. Wilson's 1999 foreword and 2010 preface are included in the Notes to the present volume.

Naturalist was first published by Island Press in Washington, D.C., on September 7, 1994. The book has since been published by Warner Books in New York (1995), Allen Lane in London (1995), Penguin in London (1996), and again by Island Press (2006), all in printings that reproduce the original typesetting, and the last with a new afterword by Wilson. A "25th Anniversary Edition" published by Island Press in 2009 is identical internally to its 2006 printing. The text of *Naturalist* in the present volume is that of the first printing; Wilson's 2006 afterword is included in the Notes.

This volume presents the texts of the original printings chosen for inclusion here, but it does not attempt to reproduce features of their design and layout. The texts are presented without change, except for the correction of typographical errors. Spelling, punctuation, and capitalization are often expressive features and they are not altered, even when inconsistent or irregular. The following is a list of errors corrected, cited by page and line number: 13.36, Frederick; 14.27, *Moby Dick*; 55.15, orginating; 76.29, venemous; 79.10, and half; 100.24, exhilirating; 126.27, Coca Cola; 143.18, wih; 191.11, makes; 332.15, have; 338.14, contract.; 375.27, Tay Sachs; 399.24, have; 421.21, rainforest; 431.18, *Menta*; 461.38, Sates,; 490.21. was a; 523.30, Kenenth; 583.38, tropical looking; 613.28, disciples) are; 642.6, De Voto,; 668.26, paperback,; 789.10, Wood's; 790.31, For five.

Notes

In the notes that follow, the reference numbers denote page and line of this volume (the line count includes chapter headings but not blank lines). References to Shakespeare are keyed to *The Riverside Shakespeare*, ed. G. Blakemore Evans (Boston: Houghton Mifflin, 1974), and references to the Bible to the King James Version.

Wilson's original endnotes to *Biophilia* and *The Diversity of Life* are included here verbatim, followed by an italic *W*; in a few instances, the original order of Wilson's notes has been rearranged to key them more closely to the text, and some longer notes have been split into two or more shorter ones.

BIOPHILIA

2.1–3 Soft, to your places . . . KINSELLA] "Soft, to Your Places," by Thomas Kinsella, in *Selected Poems, 1956–1968* (Dublin: Dolmen Press, 1973). *W*

5.11 biophilia] I first used the term *biophilia* in an article by that title in the *New York Times Book Review*, January 14, 1979, p. 43. *W*

9.33–34 "Go to the ant . . . her ways,"] Proverbs 6:6.

14.21–23 Edgar Allan Poe's . . . no man can discover."] See Poe's poem "Dream-Land," first published in *Graham's Magazine* in June 1844.

14.33–34 the historian Leo Marx . . . garden.] Leo Marx, *The Machine in the Garden: Technology and the Pastoral Ideal in America* (New York: Oxford University Press, 1964). *W*

15.10–11 the geographer Yi-Fu Tuan . . . world.] Yi-Fu Tuan, *Topophilia: A Study of Environmental Perception, Attitudes, and Values* (Englewood Cliffs: Prentice-Hall, 1974). *W*

16.18–20 Hobbes's definition . . . too late.] The quotation "Hell is truth seen too late" is conventionally attributed to English philosopher Thomas Hobbes (1588–1679).

17.20–35 In water films . . . life cycle.] The remarkable attack mechanism of the oomycete fungus *Haptoglossa mirabilis* was recently worked out in detail by E. Jane Robb and G. L. Barron, "Nature's Ballistic Missile," *Science*, 218:1221–1222 (1982). *W*

18.18–22 The abundance of organisms . . . bacteria.] The density of soil organisms is based on estimates given in John A. Wallwork, *The Distribution and Diversity of Soil Fauna* (New York: Academic Press, 1976), and Peter H. Raven, Ray F. Evert, and Helena Curtis, *Biology of Plants*, 3rd ed. (New York: Worth Publishers, 1981). *W*

18.28–39 The amount of information . . . *Britannica.*] The estimates of the number of nucleotide pairs in various kinds of organisms are given in the authoritative review by Ralph Hinegardner, "Evolution of Genome Size," pp. 179–199 in *Molecular Evolution*, ed. F. J. Ayala (Sunderland, Mass.: Sinauer Associates, 1976). Four permutations exist at the nucleotide pair sites, AT, TA, CG, and GC; and the amount of information present at each site can be roughly approximated as $\log_2 4 = 2$ bits. The number of bits per English word was derived from Henry Quastler in "A Primer on Information Theory," pp. 3–49 in *Symposium on Information Theory in Biology*, ed. H. P. Yockey (New York: Pergamon Press, 1958). *W*

22.27–30 Terry L. Erwin . . . tropical forests.] The upper estimate of 30 million living species of insects may seem improbable at first, but it has been carefully argued and documented by Terry L. Erwin, "Tropical Forest Canopies: The Last Biotic Frontier," *Bulletin of the Entomological Society of America*, 29:14–19 (1983). *W*

22.33–35 Charles Butler . . . 1609.] In 1609, Butler (1571–1647) published *The Feminine Monarchie*, the first English-language book about beekeeping.

22.38–23.1 All of man's troubles . . . to become.] See the epigraph to Vercors's novel *Les animaux dénaturés* (1952), as translated by Rita Barisse, under the title *You Shall Know Them*, in 1953. Vercors was a pseudonym of Jean Bruller (1902–1991).

25.6 the Minimum Critical Size Project] An authoritative account of the Critical Size Project is given by Sam Iker, "Islands of Life in a Forest Sea," *Mosaic* (National Science Foundation, Washington), 13:25–30 (September–October 1982). A nicely illustrated but very brief account is also presented by Peter T. White, "Tropical Rain Forests: Nature's Dwindling Treasures," *National Geographic*, 163:2–47 (January 1983). *W*

29.6–8 Somewhere a large tree . . . life.] I have never witnessed the spontaneous crash of a forest tree, but I have seen many giant rain forest trees brought down by the chain saw. That experience was used to reconstruct what must occur in the Amazonian forest. *W*

29.38–39 the leafcutter ant] The general biology of leafcutter ants is described by Edward O. Wilson, *The Insect Societies* (Cambridge: Harvard University Press, 1971), and Neal A. Weber, *Garden Ants: The Attines* (Philadelphia: American Philosophical Society, 1972). *W*

38.12–39.13 Louis Agassiz . . . stop this."] The account of the conversation between Louis Agassiz and Benjamin Peirce is in A. Hunter Dupree's *Asa Gray* (Cambridge: Harvard University Press, 1959). We know the subject but of course not the exact words. On the other hand, Agassiz's remark about Darwinism that evening ("We must stop this") is just as Gray later recalled it. *W*

43.17–44.10 AGASSIZ'S APPREHENSION . . . curiosity."] The exchanges between Agassiz, Darwin, and their friends are taken from David L. Hull's admirable *Darwin and His Critics: The Reception of Darwin's Theory of Evolution by the Scientific Community* (Cambridge: Harvard University Press, 1973). The protest by Agassiz is quoted from an article published posthumously in the *Atlantic Monthly*, 1874. W

44.19–20 Loren Graham . . . expansionists.] The characterization of restrictionists and expansionists is elaborated by Loren R. Graham in *Between Science and Values* (New York: Columbia University Press, 1981.) W

44.23–25 Bertrand Russell's . . . not know.] Bertrand Russell is quoted from an interview reprinted in *The Humanist*, November–December 1982, p. 39. W

45.3 great effects imply great causes.] Great effects do not imply great causes: I first heard the key conclusion of Darwin expressed in just this way by the philosopher John Passmore at a lecture at Cambridge University in 1982. W

45.16–30 Darwin wrote in his *N* Notebook . . . Locke."] The Darwin notebooks are cited by P. H. Barrett, *Metaphysics, Materialism, and the Evolution of Mind: Early Writings of Charles Darwin* (Chicago: University of Chicago Press, 1980). W

46.19 "Science grows . . . Tennyson.] See Tennyson's "Locksley Hall Sixty Years After," first published in 1886.

46.28–31 The romantic world . . . Thompson.] For examples of the modern restrictionist view of science, see John W. Bowker, "The Aeolian Harp: Sociobiology and Human Judgment," *Zygon*, 15:307–333 (1980); Theodore Roszak, "The Monster and the Titan: Science, Knowledge, and Gnosis," *Daedalus*, 103:17–32 (1974); and William Irwin Thompson, *The Time Falling Bodies Take to Light* (New York: St. Martin's Press, 1981). W

48.23–33 Its head is shaped . . . brown and black.] The description of the Emperor of Germany bird of paradise is based on my examination of specimens in the Museum of Comparative Zoology, Harvard University, and the excellent painting and biological summary by William T. Cooper and Joseph M. Forshaw in *The Birds of Paradise and Bower Birds* (Boston: David R. Godine, 1977). I never saw the species in the wild, even though in 1955 I walked over substantial parts of the Huon Peninsula inland from Finschhafen and Lae, and many *Paradisaea guilielmi* probably saw me. The reason is very simple: I was studying ants, encountering over 300 different kinds in this area alone, and almost always had my gaze focused on the ground. On one occasion I heard a sharp cry high in the treetops, and an Australian biologist nearby shouted "Bird of paradise!" But by the time I could adjust my eyeglasses and look up, the bird was gone. W

53.36–40 David Hilbert . . . independent development."] David Hilbert wrote on the vital importance of perpetual discovery in the preamble to his celebrated article presenting twenty-three fundamental problems in modern mathematics, "Sur les problemès futurs des mathématiques," in *Compte rendu du Deuxième Congrès International des Mathématiciens* (Paris, 1900), pp. 58–114. W

54.15–24 Einstein . . . we love him."] Einstein on Planck: "Principles of Research," *Ideas and Opinions by Albert Einstein*, based on *Mein Weltbild*, ed. Carl Seelig; rev. Sonja Bargmann (New York: Bonanza Books, 1954). W

55.20–26 P.A.M. Dirac . . . the beautiful."] Dirac wrote on the relationship between beauty and scientific truth in "The Evolution of the Physicist's Picture of Nature," *Scientific American*, 208:45–53 (May 1963). Weyl on aesthetics and truth: as quoted from a conversation with Freeman J. Dyson in an obituary essay, *Nature*, 177:457–458 (1956). W

56.6–16 David Hilbert . . . complete.] Hilbert's remarks are quoted by William N. Lipscomb in "Aesthetic Aspects of Science," pp. 1–24 in *The Aesthetic Dimension of Science*, ed. Deane W. Curtin (New York: Philosophical Library, 1982). W

56.21–26 The two vocations . . . is revealed.] I am indebted to the following sources for a more formal concept of art and the humanities used in my comparison with science: Richard W. Lyman et al., *The Humanities in American Life*, Report of the Commission on the Humanities (Berkeley: University of California Press, 1980); W. Jackson Bate, "The Crisis in English Studies," *Harvard Magazine*, September–October 1982, pp. 46–53; and Paul Oskar Kristeller, "The Humanities and Humanism," *Humanities Report*, January 1982, pp. 17–18. W

56.28–30 Roger Shattuck's . . . greatness."] Roger Shattuck on the autonomous tradition of art: "Humanizing the Humanities," *Change*, November 1974, pp. 4–5. W

56.34–38 In poetry . . . takes place.] T. S. Eliot wrote on the discipline of the poet in "Tradition and the Individual Talent" (1919), in *Selected Prose of T. S. Eliot* (New York: Harcourt Brace Jovanovich, 1975). W

57.4–5 Octavio Paz's . . . Waterjar"] Octavio Paz's "The Broken Waterjar" is translated by Lysander Kemp in *Early Poems, 1935–1955*. Copyright © 1963, 1973 by Octavio Paz and Muriel Rukeyser. Reprinted by permission of New Directions Publishing Corporation. W

58.11–26 Most scientists . . . mental operations.] Some of the best testimony concerning the creative process is to be found in lectures given by scientists and other scholars in the annual Nobel Conferences arranged by the faculty of Gustavus Adolphus College. The most pertinent are: *Creativity*, ed. John D. Roslansky (Amsterdam: North Holland, 1970); *The Aesthetic Dimension of Science*, ed. Deana W. Curtin (New York: Philosophical Library, 1982); and *Mind in Nature*, ed. Richard Q. Elvee (New York: Harper and Row, 1982). W

59.28–33   The great metallurgist . . . metal.] Cyril S. Smith recounts the origins of his love for metallurgy in *A Search for Structure: Selected Essays on Science, Art, and History* (Cambridge: MIT Press, 1981). *W*

59.33–37   Albert Camus . . . opened."] Camus characterizes the creative detour to rediscover the images of childhood in the preface to *The Wrong Side and the Right Side*, reprinted in *Lyrical and Critical Essays* (New York: Alfred A. Knopf, 1969). *W*

60.22–32   Hideki Yukawa . . . really creative.] Hideki Yukawa presented his view of the central role of analogy in *Creativity and Intuition: A Physicist Looks East and West*, trans. John Bester (Tokyo: Kodansha International, 1973). *W*

60.37   analogies] Einstein on analogies: "It is easy to find a superficial analogy which really expresses nothing. But to discover some essential feature hidden beneath the surface of external differences [and] to form on this basis a new successful theory is a typical example of the achievement of a successful theory by means of a deep and fortunate analogy." *The Evolution of Physics* (New York: Simon and Schuster, 1938). *W*

61.23–24   In 1962 . . . biogeography.] Robert H. MacArthur and I published our principal work in "An Equilibrium Theory of Insular Biogeography," *Evolution*, 17:373–387 (1963), and more fully in *The Theory of Island Biogeography* (Princeton: Princeton University Press, 1967). A more recent and comprehensive account of the theory and related topics is given by Mark Williamson in *Island Populations* (Oxford: Oxford University Press, 1981). *W*

67.10–16   Bishop Lowth . . . magnificent terms.] Bishop Lowth is quoted and the importance of his analysis examined by M. H. Abrams in *The Mirror and the Lamp* (New York: Oxford University Press, 1953), an authoritative review of the romantic tradition and the origins of literary criticism. *W*

67.19–20   Mankind . . . the poetic species.] Richard Rorty describes humanity as the poetic species in his superb review of philosophy of mind: "For beyond the vocabularies useful for prediction and control—the vocabulary of natural science—there are vocabularies of our moral and our political life and of the arts, of all those human activities which are not aimed at prediction and control but rather in giving us self-images which are worthy of our species. Such images are not true to the nature of species or false to it, for what is really distinctive about us is that we can rise above the question of truth or falsity. We are the poetic species, the one which can change itself by changing its behavior—and especially its linguistic behavior, the words it uses." "Mind as Ineffable," pp. 60–95 in *Mind in Nature*, ed. R. Q. Elvee (New York: Harper and Row, 1982). *W*

68.1–2   affirmation . . . Paleolithic cave art.] An excellent account of cave art and its possible use in the transmission of culture has been provided by John

E. Pfeiffer, *The Creative Explosion: An Inquiry into the Origins of Art and Religion* (New York: Harper and Row, 1982). W

68.21–27 Thomas Kinsella's . . . welcome us.] Kinsella's "Midsummer," *Selected Poems, 1956–1968* (Dublin: Dolmen Press, 1973). W

68.32–69.4 Richard Eberhart . . . memory belong.] Eberhart's stanza is from "Ultimate Song," *Collected Poems, 1930–1976* (New York: Oxford University Press, 1976). W

69.22–70.11 Cognitive psychology . . . long-term memory.] Some of the key reference works and textbooks on the mind and memory, including the node-link model, are *Cognitive Psychology and Its Implications*, by John R. Anderson (San Francisco: W. H. Freeman, 1980); *Mechanics of the Mind*, by Colin Blakemore (New York: Cambridge University Press, 1977); *Brainstorms: Philosophical Essays on Mind and Psychology*, by Daniel C. Dennett (Montgomery, Vt.: Bradford Books, 1978); *Psychology*, by Gardner Lindzey, C. S. Hall, and R. F. Thompson (New York: Worth Publishers, 1975); *Human Memory: The Processing of Information*, by G. R. and Elizabeth F. Loftus (Hillsdale: Lawrence Erlbaum Associates, 1976); *The Psychobiology of Mind*, by William R. Uttal (Hillsdale: Lawrence Erlbaum Associates, 1978); and *Cognitive Psychology*, by Wayne A. Wickelgren (Englewood Cliffs: Prentice-Hall, 1979). W

71.15–19 Eliot wrote . . . degenerate."] From "The Social Function of Poetry," first published in 1945 and collected in 1957 in *On Poets and Poetry*.

71.29–31 Gerda Smets . . . redundancy.] The measurement of varying brain arousal by different geometric designs was reported by Gerda Smets in *Aesthetic Judgment and Arousal: An Experimental Contribution to Psycho-physics* (Leuven, Belgium: Leuven University Press, 1973). W

73.12–21 Joseph Stella . . . my hopes.] Stella is quoted and his work analyzed in J. Gray Sweeney's *Themes in American Painting* (published under the auspices of the Grand Rapids Art Museum, Michigan, 1977). W

74.1 *The Serpent*] I have drawn most of the facts on the serpent in culture from Balaji Mundkur's *The Cult of the Serpent: An Interdisciplinary Survey of Its Manifestations and Origins* (Albany: State University of New York Press, 1983). This is a highly original and masterly work. Although I have long thought about our awe of the serpent, Mundkur has documented it in impressive detail from the history of art and literature. W

89.8–9 Among the early forms of Zeus . . . Meilikhios] A detailed and authoritative account of Zeus Meilikhios and the snake-Erinyes is given by Jane Ellen Harrison, *Prolegomena to the Study of Greek Religion*, 3rd ed. (Cambridge: Cambridge University Press, 1922). W

89.26–29 I will put enmity . . . their heel.] Genesis 3:15.

90.8   The combined biases . . . human nature.] The conception of biasing in mental development and its relation to human nature and culture is presented in greater detail in Charles J. Lumsden and Edward O. Wilson, *Promethean Fire* (Cambridge: Harvard University Press, 1983). *W*

91.19–24   The hunter-in-naturalist . . . inattentiveness."] José Ortega y Gasset, *Meditations on Hunting*, trans. Howard B. Wescott (New York: Charles Scribner's Sons, 1972). Other excellent discussions of the hunter's mystique are given by Paul Shepard, *The Tender Carnivore and the Sacred Game* (New York: Charles Scribner's Sons, 1973), and John G. Mitchell, *The Hunt* (New York: Alfred A. Knopf, 1980). *W*

91.31–33   pygmy . . . new species.] The pygmy desmognath salamander I collected was *Desmognathus chermocki*. It has since been formally combined with the more widespread *Desmognathus aeneus*, although I am informed by one of its discoverers, Barry D. Valentine, that its status remains problematical. In either case the field observations made in Alabama retain their significance with respect to the behavioral diversity of the desmognaths. *W*

92.14–18   William Mann . . . central Cuba.] Willliam Mann's account of ant collecting in Cuba is in "Stalking Ants Savage and Civilized," *National Geographic*, 66:171–192 (August 1934). *W*

93.15   a short technical article] "Behavior of the Cuban Lizard *Chamaelleolis chamaeleonitides* (Duméril and Bibron) in Captivity," *Copeia*, July 15, 1957.

95.17–19   Even colon bacteria . . . peculiar way.] The basic research on orientation and habitat selection in bacteria is ably summarized by Daniel E. Koshland, Jr., *Bacterial Chemotaxis as a Model Behavioral System* (New York: Raven Press, 1980). *W*

96.21–22   The archaeological evidence . . . original environment.] The evidence for the savanna habitat as the home of early man has been presented by several authors, including Karl W. Butzer, "Environment, Culture, and Human Evolution," *American Scientist*, 65:572–584 (1977), and Glynn Isaac, "Casting the Net Wide: A Review of Archaeological Evidence for Early Hominid Land-Use and Ecological Relations," pp. 114–134 in *Current Arguments on Early Man*, ed. L.-K. Königsson (New York: Pergamon Press, 1980). *W*

97.3   Gordon Orians] Gordon H. Orians developed the idea of the psychologically optimum human environment in "Habitat Selection: General Theory and Applications to Human Behavior," pp. 49–66 in *The Evolution of Human Social Behavior*, ed. Joan S. Lockard (New York: Elsevier North Holland, 1980). *W*

99.13–31   When Captain R. B. Marcy . . . dotting its surface."] The diary entries by Marcy and Parker are in Public Document 577 of the 31st Congress (1849); quoted by Orians. *W*

100.16–18 Melville wrote . . . miles to see it?"] The metaphor of the cataract of sand is given in the first chapter of *Moby-Dick*. Herman Melville understood as few other authors the innate aesthetic sense of the environment and especially the compelling attraction of open water: "Say, you are in the country, in some high land of lakes. Take almost any path you please, and ten to one it carries you down in a dale, and leaves you there by a pool in the stream. There is magic in it. Let the most absent-minded of men be plunged in his deepest reveries—stand that man on his legs, set his feet a-going, and he will infallibly lead you to water, if water there be in all that region." The yearning is of a very general kind, generating symbolism across many categories of thought. "It is the image of the ungraspable phantom of life; and this is the key to it all." *W*

102.3–25 Cyril Smith . . . *A Search for Structure*] Cyril S. Smith, *A Search for Structure: Selected Essays on Science, Art, and History* (Cambridge: MIT Press, 1981), p. 355. *W*

103.13 the actual colonization of space] On the colonization of space: the concept of self-contained stations was first brought into public discussion by Gerard K. O'Neill in an article for *Physics Today* (1974) and developed at length in his book *The High Frontiers: Human Colonies in Space* (New York: Bantam Books, 1976). An excellent popular exposition is also provided by T. A. Heppenheimer in *Colonies in Space* (Harrisburg: Stackpole Books, 1977). Extensions and criticisms, some of the latter quite severe, were written by physicists, ecologists, and others for *Space Colonies*, ed. Stewart Brand (New York: Penguin Books, 1977). *W*

106.26–30 Aldo Leopold . . . final outcome.] Aldo Leopold, "The Land Ethic," *A Sand County Almanac and Sketches Here and There* (New York: Oxford University Press, 1949). *W*

108.15 Extinction is accelerating] The acceleration of species extinction and its dangers for mankind have been ably documented by Norman Myers, *The Sinking Ark* (Elmsford: Pergamon Press, 1979), and Paul R. and Anne Ehrlich, *Extinction: The Causes and Consequences of the Disappearance of Species* (New York: Random House, 1981). They have been examined still further by Peter H. Raven and others in three National Research Council reports: *Conversion of Tropical Moist Forests* (1980); *Research Priorities in Tropical Biology* (1980); and *Ecological Aspects of Development in the Humid Tropics* (1982). *W*

109.16–17 the style of leadership . . . agitated moderation.] In his book *Physics and Politics* (1872), Walter Bagehot (1826–1877) discusses "animated moderation" as a quality in the temperament of great writers and of democratic political leaders.

111.39–112.1 In Israel . . . Tel Dan] For information on the rare flora of Tel Dan (Tel el Kadi), in the Hule Valley of Israel, I am grateful to Jehoshua Kugler and Eviatar Nevo. *W*

112.6–8   In the Western Ghats . . . flora and fauna.] The role of the sacred groves as unplanned nature reserves is explained by Madhav Gadgil and V. D. Vartak, "The Sacred Groves of Western Ghats in India," *Economic Botany*, 30:152–160 (1974). W

112.13   The modern practice of conservation] Perhaps the best historical review of the origin of the conservation ethic in the United States is Donald Fleming's "Roots of the New Conservation Movement," pp. 7–91 in *Perspectives in American History*, vol. 6, ed. Donald Fleming and Bernard Bailyn (Lunenburg: Stinehour Press, for the Charles Warren Center for Studies in American History, Harvard University, 1972). The concept of wilderness in particular is explored by Roderick Nash in his classic *Wilderness and the American Mind*, rev. ed. (New Haven: Yale University Press, 1973). W

112.23–24   Thus we favor . . . surrogate kin.] The idea of extended kinship contributing to the conservation ethic has been systematically examined by Gordon M. Burghardt and Harold A. Herzog, Jr., "Beyond Conspecifics: Is Brer Rabbit Our Brother?," *BioScience*, 30:763–768 (1980). W

113.20   pygmy chimpanzee.] The biology and status of the pygmy chimpanzee is described in "An Uncommon Chimp," by Paul Raeburn, *Science 83*, 4:40–48 (June 1983). W

115.25   Bishop Wilberforce's darkest thoughts] In a debate about evolution at Oxford University on June 30, 1860, soon after the publication of Darwin's *On the Origin of Species* (1859), Samuel Wilberforce (1805–1873), a bishop in the Church of England, is reported to have asked Thomas Huxley (1825–1895) whether it was through his grandfather or his grandmother that he claimed simian descent.

116.3–10   Peter Singer . . . *Have Standing?*] Peter Singer, *The Expanding Circle: Ethics and Sociobiology* (New York: Farrar, Straus and Giroux, 1981). Christopher D. Stone, *Should Trees Have Standing? Toward Legal Rights for Natural Objects* (Los Altos: William Kaufmann, 1974). W

116.40–117.1   the first law . . . Garrett Hardin] Garrett Hardin's tough-love approach to ethical philosophy is concisely expressed in *The Limits of Altruism: An Ecologist's View of Survival* (Bloomington: Indiana University Press, 1977). W

117.13–14   estimates . . . Norman Myers] My examples of edible tropical plants are taken from Norman Myers' important encyclopedic account, *A Wealth of Wild Species: Storehouse for Human Welfare* (Boulder: Westview Press, 1983). W

120.5–9   Thomas Eisner . . . book in a library.] Thomas Eisner compared species to a genetic looseleaf notebook in his testimony on the Endangered Species Act; his prepared statement was published in *The Congressional Record*, vol. 128 (April 1, 1982), and reprinted in the *Natural Areas Journal*, 2:31–32 (1982). W

126.15   Bernhardsdorp has changed strikingly] In December 1982 Richard
Prum, a young ornithologist working on the social behavior of birds in
Surinam, went to Bernhardsdorp at my request. He took detailed notes and
photographs and spoke with some of the residents. We subsequently met to
reconstruct the changes that occurred during the twenty years since my own
visit. The details of the recent political events in Surinam, especially the Bouterse takeover and executions of December 1982, are based on reports from
Amnesty International ("Urgent Action," December 13, 1982, January 11,
1983; *Amnesty International Report, 1983*, Amnesty International Publications,
London, 1983) and "A Country of Mutes," *Time*, May 30, 1983. The sources
of information used in the two accounts are at least partly independent, and
Amnesty International provided a full list of the names of the victims, as well
as details of exchanges it had with the Surinam government with reference
to the human-rights violations. The actions of the government have been in
one sense even-handed: the victims included Bram Behr, a journalist with
the communist weekly *Mokro*, as well as Bouterse's own local military commander. W

THE DIVERSITY OF LIFE

131.1   *The Diversity of Life*] This book was first published in 1992. Wilson
contributed a new foreword, "Biodiversity at the Close of the Century," for
a new printing in 1999, and a "Preface to the 2010 Printing" in 2010. These
forewords appear below:

BIODIVERSITY AT THE CLOSE OF THE CENTURY

Ten years have swept by since the multiple-authored book *BioDiversity*
introduced the title word into the English language, and six years since the
appearance of the work before you, *The Diversity of Life*. In that decade conservation biology, the new discipline explained by these works, has grown at
an explosive rate. By 1992 biodiversity was well enough known to be a central
concern of the Earth Summit held in Rio de Janeiro. The global Convention
on Biodiversity drafted at that conference has been ratified by almost 170
countries. Biodiversity courses are now taught at innumerable colleges and
universities in the United States and elsewhere. Natural history museums have
rewritten their agendas to focus on the study and conservation of ecosystems
and biotas. And conservation organizations routinely base their programs on
conservation biology; in response to the scientific content of the subject, they
count their successes not just by the salvaging of tigers and eagles but also the
protection of entire ecosystems that harbor such star species.

Humanity Versus the Natural World. What then is the status of biodiversity as this century ends? Without doubt, worldwide awareness of its
problems and promise has grown dramatically. Many countries, including
tropical "megadiversity" nations such as Brazil, Colombia, and Indonesia,

have added reserves and adjudicated—in principle at least—the practice of biological conservation as part of national policy. There is some justification for optimism, guarded by realism and shadowed by sensible apprehension, that the world is turning the corner in at least its attitude toward the rest of life.

Still, those who monitor the diversity of life are especially apprehensive because with all the good intentions of many scientists and policy makers, the growth of human population and the depletion of natural resources continues unabated. The prospects for biodiversity can be summarized by the following imagery of the bottleneck. The world's human population will increase by about a third before peaking within a century or so, then commence a slow decline. If the number at maximum is not much greater than 8 billion, everyone can, in theory at least, be housed and fed. However, the already intense pressures on the last remnants of wild biodiversity might easily grow fatal for a majority of the remaining ecosystems and their distressed species of plants and animals. The only way to carry biodiversity safely through the bottleneck of this critical period is by a combination of scientific and technological innovation, abatement of population growth, and environmental education, guided by a redirection of moral purpose.

Consider some of the basic facts and projections that define the bottleneck. The world population is now passing 6 billion. The average number of children produced per woman is about 2.6 and falling. To achieve zero population growth the average number of children per woman must drop to 2.1. (The extra tenth of a child, or one child per ten women, compensates for child mortality.) If the 2.1 level is reached very soon and held, there will be 7.7 billion people on Earth by 2050, leveling off at 8.5 billion in 2150. If a slight negative growth is reached, at 2.0 children per woman, the world population will peak at 7.8 billion and drop to 5.6 billion, a bit below the current level, also by 2150. To these estimates, unfortunately, must be added an even more worrisome estimate: if the replacement level is held even slightly above 2.1, the world population will continue to soar. Thus, only 2.2 children per woman will yield 12.5 billion by 2050, a potentially catastrophic doubling of present levels. In general, whatever the magnitude of increase, most of it will occur in the developing world, which is already economically impoverished.

Why would a doubling of the global population be so bad? After all, average per capita production is rising worldwide, a statistic often cited cheerfully by bottom-line economists. The reason for concern is that the increase is being achieved by an accelerating consumption of Earth's nonrenewable capital. Natural-resources experts generally agree that we are running out of arable land and water as a result of rising total demand, and that the finiteness of land and water represents a wall beyond which humanity cannot pass. There are no new continents to colonize, and displacement of other people by conquest and genocide is now generally frowned upon. As a species we are at long last approaching the Malthusian limit. The amount of cropland available per person, having dropped from 0.23 hectares in 1950 to 0.12 hectares in the mid-1990s, is continuing downward. The productivity per hectare has been elevated by fertilizer, Green-Revolution genetic improvement of crop strains, and other technological innovations, but the rate of improvement is diminishing, and its maintenance also depends on water, the other principal declining resource. Meanwhile, the population is shooting upward in most countries.

To the strain on the food and water resources can be added the aftershock of personal economic striving that follows the population explosion. Billions of people are not only producing children in excess of zero population growth but also taking every means available to improve their own and their family's personal security and welfare. The true measure of the human impact on Earth's environment is therefore not in pure population numbers but in the "ecological footprint," the amount of productive land needed to supply each person with food, water, energy, living space, commerce, transportation, and waste management. The footprint is far more than just land that each person uses directly. It is also composed of small pieces obtained from many places, some even from other continents—where food is grown, petroleum drawn, clothing manufactured, roads built, waste dumped, and public buildings raised. The ecological footprint of the average person in the developing countries is half a hectare; that of one citizen of the United States is 5 hectares. For the entire current world population to pull itself up to U.S. levels with existing technology would require two more planet Earths.

That amount of bootstrapping, of course, is not going to happen. The greatest challenge of the twenty-first century is to settle humanity down and accommodate 8 to 10 billion people with a decent standard of living before they wreck the planet. Meanwhile, the last remaining natural environments—the world's forests, grasslands, semideserts, and wetlands—are shrinking before the onslaught. Humanity's responsibility to the rest of life and to future generations is clear: bring with us as much of the environment and biodiversity through the bottleneck as possible.

Viewed this way, the study of biodiversity in all its aspects, from its origin, distribution, and extinction to its use, management, and preservation, is of paramount importance to human affairs. Here then is a brief review of some subjects in this field in which significant progress has been made since the 1992 publication of *The Diversity of Life*.

The Evolution of the Species Concept. Biologists still find it useful to divide living diversity into a hierarchy of three levels: ecosystem, species, and gene. An ecosystem is a local community of species (plants, animals, and microorganisms) plus their physical environment. Familiar examples include a pond, an estuary, and an alpine grassland atop a forested mountain. Because ecosystems are so often hard to delimit, and because genes are difficult to identify and count, the unit of choice in biodiversity studies remains the species, which is relatively easy to diagnose and has moreover been the central object of research for over two hundred years.

Throughout most of the century now ending, the prevailing definition of the species has been embodied in the "biological species concept," which holds that a species is most usefully defined as a population or series of populations of organisms capable of freely interbreeding with one another under natural conditions. In short, a species is a closed gene pool that perpetuates itself *in nature*. The production of hybrids in gardens and zoos does not justify combining two species in classification if they stay apart in nature. Unfortunately, as concrete as the biological species concept often appears in abstract, it often fails in practice. For example, it cannot be applied to the large number of organisms that reproduce asexually. During the 1990s, dissatisfaction with the concept has resulted in the promotion of an alternative category, the phylogenetic species. In this perspective, the most meaningful species is a genetically distinct population with a monophyletic lineage; that is, all members are derived from a single ancestral species. It is of little concern if the populations thus delimited indiscriminately breed with one another in nature. So long as the delimited population comprises individuals of the same coherent lineage that are distinguishable to a mutually agreed-upon amount from those of other populations, it can be ranked as a species by the purely phylogenetic definition.

The contest between the two opposing yet internally consistent species concepts creates a dilemma. Both criteria they respectively use are desirable. The advantage of the biological species concept is its recognition that closed gene pools are entities which for the most part have been irreversibly launched on an independent course of evolution. Given enough time, closed gene pools at first barely distinguishable from other closed pools are destined to diverge genetically and eventually to be very distinguishable. In contrast, the advantage of the phylogenetic species concept is that it reflects rigorously the past history of the lineages or groups of related species without reference to their reproductive status of hypothetical future history.

The cross-cutting criteria of interbreeding posed by the two concepts are actually not contradictory. They can be joined to create a synthetic species concept as follows: *a sexual species is a population that is both reproductively isolated in nature and monophyletic.* Suppose a given monophyletic sexual population is geographically—physically, not genetically—isolated from other, similar populations (as it would be, for example, if it were on an island) so that its ability to hybridize or to avoid hybridization is indecipherable. Its status is thus subjective. The population can be called a species if it is markedly different, or it can be called a subspecies (that is, a geographical race) if it is only slightly different. The same subjective criteria can be applied for populations that reproduce asexually. In short, the test of reproductive isolation is desirable but not crucial.

The new emphasis on monophyly provides greater precision and reliability in reconstructing the ancestry of species but does not constitute a fundamental shift away from the species concept already used by most practicing taxonomists. As a rule, specialists on the classification of organisms have embraced both reproductive isolation and monophyly for most of this century while conceding that the assignment of reproductive relationships among closely related but geographically isolated populations is little more than educated guesswork.

Why does all this semantic juggling matter? Because the current trend of the theory of classification is toward a higher degree of objectivity and consensus than existed in the past. Thanks to new techniques in molecular biology and statistical inference, a truly synthetic biological species concept, providing considerable information about each genetically distinguishable population, seems attainable. That goal is in turn pivotally important in ecology and conservation biology. How species are classified determines the number recognized, hence our sense of the magnitude of both local and global biodiversity. It affects the evaluation of the status of individual populations in conservation planning, in particular whether populations are ranked as species, subspecies, or unnamed entities. And finally, the refined species concept conforms more closely to the emerging picture of how biodiversity is created by evolution.

How many species? The estimates I made in 1986 and cited in *The Diversity of Life* (1992) put the number of recognized living species in the world—in other words, those formally described and bearing two-part scientific names—at approximately 1.4 million. About 13,000 additional "new" species are recognized each year. Thus in the decade since 1986 more than 100,000 species have been added, bringing the total as I counted it to 1.5 million. Meanwhile, the numbers in some important groups have been revised upward—in particular in the insects, the largest of all groups, from 751,000 in the 1980s to 865,000 in 1998. A similar elevation has been made for the fungi, from 47,000 to 69,000 species. A commonly cited total world figure in the late 1990s, suggested for example in the *Global Biodiversity Assessment* of the United Nations Environmental Programme (1995), is 1.75 million species. But this does not take into account the number of formal species names that have been erroneously applied to species named by earlier investigators, requiring an eventual reduction of the global number by 10 or even 20 percent.

The bottom line is that only a very rough estimate can be made of the number of recognized species in the world. It is reasonable on the basis of existing evidence to suppose that the figure lies somewhere between 1.5 and 1.7 million. Until a central register is established, however, with a census of the biodiversity of all of the groups of plants, animals, and microorganisms entered and constantly updated, the exact figure will remain elusive.

Even more distant, with a margin of uncertainty a hundred times greater, is the true number of species on Earth, both those described and the far larger number undescribed. As cited by the *Global Biodiversity Assessment* in 1995, estimates made during the past two decades range from 3,635,000, which is the sum of the lowest estimates for each group in turn by various authors, to 111,655,000, the sum of the highest estimates of each group by the same authors. A large part of the latter figure, 100 million, comes from the insects. Most entomologists now agree that this maximum number for insects is much too high, perhaps tenfold in excess of the actual number. "Working figures" for all groups, including insects, have tended to fall close to 10 million; the *Global Biodiversity Assessment* number, which constitutes no more than an educated guess and leans to conservative sentiment, is 13,620,000.

At first glance there may be a temptation to regard such counting and extrapolation as just numerology, playing with numbers for their own sake. But the ongoing attempt has an important practical purpose. It helps to identify the blank spaces on the biodiversity map, and it points the way to original and productive research. For example, as I write, about 98 percent of the living bird species of the world are believed to be known. In sharp contrast, one authority has reckoned that only 1.5 percent (15,000 out of 1 million) chromophyte algal species have been described. The past ten years have witnessed the discovery of large pools of species diversity in previously obscure or unknown groups, as illustrated by the following two examples:

- One third of all recognized species of organisms are parasites. Most higher organisms with larger bodies, including insects, have species that live in or on them. Further, a majority of the parasites are highly specialized for one or several host species. Because they are typically hidden as well, and often do not inflict obvious harm on the hosts, parasite species are seldom recognized until the host species themselves have been well studied. An extreme example, of many that can be cited, is provided by the mycoplasmas. In addition to being among the smallest of the bacteria, mycoplasmas lack cell walls, making them exceptionally hard to detect. Yet even a modest amount of research has yielded a wide array of species living in the tissues of insects and plants. Suppose that each free-living species had on average one parasite of such an obscure group limited to it, a modest assumption. Then the global count must be increased by many millions of undiscovered species.

- Bacteria continue to be the "black hole" of biodiversity, their depths unplumbed. That ignorance will not last indefinitely. The subject has been opened dramatically in the past decade by new molecular technology. Through most of the history of microbiology, perhaps fewer than 1 percent of bacterial species couple be cultured in large enough numbers to allow analysis by anatomical study and standard chemical tests. Now systematists are able to bypass the petri dish in favor of direct genetic analysis. A favored current method is to sequence the 16S rRNA segment, comprising about 1,500 base pairs whose variation allows not only the separation of the species but an estimation of their evolutionary relationships.

As a consequence of these advances, discoveries concerning the diversity of bacteria, as well as of archaea, the other major group of microbes, are now expanding swiftly. It turns out, for example, that while some bacterial species are widespread, especially those in the open sea, others are quite local in distribution in the manner typical of terrestrial animals and plants. Since bacteria and archaea are so small, with a million or more representing thousands of species to be found in a single gram of soil or marine silt, their potential variety is mind-boggling. Their prior obscurity was such that even common and widespread species are being recognized for the first time. For example, the extremely small members of the chlorophyll-bearing genus *Pro-chlorococcus*, discovered only in the late 1980s, are now recognized as one of the most abundant elements of oceanic plankton. Some newly found varieties of bacteria, called lithoautotrophs, survive on inorganic materials inside solid rock over 3 kilometers below the Earth's surface.

To summarize the prospect for biological exploration, Earth is biologically still a mostly unexplored planet. The opportunities for basic research, which promise us great scientific and practical rewards, seem limitless. Their pursuit will help make the next hundred years the Century of the Environment.

Extinction and Hot Spots. The extinction process as a biological phenomenon has assumed major importance in present-day studies of biodiversity. In 1995 the National Biological Service of the U.S. Department of the Interior (an agency subsequently dismantled to satisfy antienvironmental forces in the House of Representatives) released the first analysis of the status of ecosystems in the United States as a booklet, *Endangered Ecosystems of the United States: A Preliminary Assessment of Loss and Degradation*. Virtually all types of ecosystems, it reported, have been diminished by human activity. The greatest losses have occurred in wetlands and forests. Many of America's grasslands, savannas, and barrens communities have been cut to 2 percent or less of their original cover, with a corresponding threat to the native species living in them. Other ecosystem types reduced to an extreme degree are old-growth stands in the eastern deciduous forests, longleaf pine forests of the southeastern coastal plain, Atlantic white-cedar stands, and freshwater streams in the Mississippi alluvial plain. Surveys in other countries report comparable degrees of loss. By the late 1980s, for example, about three quarters of the world's original forest had been destroyed, including 50 percent of rain forests in both tropical and temperate regions.

Conservation biologists have continued to refine measures of endangerment in both ecosystems and species. The system for species introduced in 1994 by the International Union for Conservation of Nature and Natural Resources (IUCN), the source of the Red Data Books and Red Lists, is worth summarizing here for its authority and degree of documentation:

Extinct: There is no reasonable doubt that the last individual of the species (or subspecies) has died.

Extinct in the wild: The species survives only in cultivation, captivity, or in environments away from the species' native range.

Critically endangered: The species is at extremely high risk of extinction in the wild in the immediate future, with continued survival very unlikely.

Endangered: The species is at very high risk of extinction in the near future; in many cases, however, it is salvageable by conservation efforts.

Vulnerable: The species comprises shrinking populations, and extinction is a good possibility in the near or medium-range future.

Lower risk: The species is evidently safe for the time being, even if dependent on current conservation measures.

Around the world, biologists, like doctors in a critical-care ward, are monitoring the descent of selected individual species through the above categories on their road to extinction. Special interest is focused on the last survivors of vertebrate and plant species and how to save them. Among the many extremely rare, critically endangered species in the world can be counted the Catalina mountain mahogany, down to 6 adult trees in a gully on the southwest side of Santa Catalina Island, California; the Seychelles magpie robin, increasing slowly and at last report comprising 61 individuals on two islands; and four species of honeycreeper, among the most distinctive species of Hawaiian birds, limited to a scrap of forest on the Hawaiian island of Maui and represented by no more than several individuals, which by the time you read this have likely perished.

Experts generally agree that on a worldwide basis the causes of extinction, which are virtually all due to human activity, can be rank-ordered from the top down as follows: habitat destruction or degradation, the spread of exotic (nonnative) species, pollution, overharvesting, and disease. The data that measure the factors endangering U.S. species, as compiled by David S. Wilcove and his co-workers in 1998 (*BioScience*, 48(8): 607–615), are habitat loss, 88 percent; exotics, 46 percent; pollution, 20 percent; overharvesting, 14 percent; and disease, 2 percent. These numbers add up to more than 100 percent because the destructive factors seldom work alone. As habitat shrinks, for example, alien species and human hunters penetrate the interiors more easily, pollution reaches closer to the core populations, and disease strikes more easily to decimate the survivors.

What then is the rate of extinction? For local populations extinction can be total or nearly so. Virtually all the songbirds of Guam have been wiped out by the voracious brown tree snake, an exotic introduced from somewhere within its native range in the Solomon Islands, Papua New Guinea, and northern Australia shortly after the Second World War. Half of the 300 or more endemic cichlid fish species of Lake Victoria have been terminated by pollution and, especially, the introduction of the predatory Nile perch in midcentury and surface-choking water hyacinth decades earlier. On a global scale, the rates of extinction cannot yet be measured in absolute numbers except for exceptionally well-studied groups such as birds, mammals, and flowering plants. But the means exist to make rough indirect estimates of the *percentage* of ongoing extinctions. For example, studies conducted since 1988 by independent researchers using different techniques have variously estimated the rate of species extinction per decade to lie between 1 and 10 percent; the average of all the estimates combined is 6 percent.

Most local and global estimates, other than those taken from detailed direct observations of the fate of particular species, are based on the relation between the area of a habitat and the sustainable number of species inhabiting it. When the habitat is reduced in size, as when a forest is progressively clear-cut, the number of species declines in a predictable fashion. This reduction varies according to the kind of organism and habitat from about the sixth root to the third root of the reduction of area. A central and very commonly encountered value is a decline equal to the fourth root of area reduction. Its magnitude can be stated most clearly by the following rule of thumb: a loss of 90 percent in area of the habitat—that is, to 10 percent of the original cover—results in a reduction of 50 percent in the number of species that the habitat can indefinitely support.

Substantial reductions in habitat area, many beyond 90 percent, have already occurred in numerous places around the world. Recent studies have shown that they unfortunately tend to be concentrated in those regions where biodiversity is richest. The relation between shrinkage of habitat area and the loss of species, either through reduction of populations to critically endangered and endangered levels (hence "commitment" to early extinction) or outright extinction, has been well substantiated in many studies on the effects of deforestation and other habitat conversion beyond those described in *The Diversity of Life* in 1992. They include birds in the forests of the eastern United States, Indonesia, and the Philippines, and mammals in the national parks of western North America.

This area effect is not solely due to the loss of living space. Fragmentation alone—breaking available space into pieces—has its own impact, by increasing the length of the edge of the habitat patches relative to their area. Recent studies on the Amazonian rain forest, which is now being both clear-cut and fragmented at an alarming rate, have determined that the physical environment of the remaining forest is altered as much as 100 meters in from the edge, by drying and greater exposure to wind damage. Trees become more fragile, their biomass declines, and deep-forest plant and animal species disappear. Parallel monitoring in forests in the midwestern United States has revealed that populations of native birds decline as predators and parasites such as cowbirds penetrate more efficiently to their nesting places.

A major development in conservation biology in the 1990s has been the growing awareness and closer examination of the destructive effects of exotic species. A freshet of articles, together with appositely titled books such as *Strangers In Paradise* (Daniel Simberloff et al., eds., Island Press, 1997), *Alien Invasion* (Robert Devine, National Geographic Society, 1998), and *America's Least Wanted* (Bruce A. Stein and Stephanie R. Flack, eds., The Nature Conservancy, 1998), have brought the problem to the attention of a broad readership of environmental scientists and the general reading public. The problems caused by exotics for both native species and humanity are indeed severe, as these works suggest. The world's biodiversity is being homogenized at an ascending rate as more and more alien species invade even relatively natural environments. The United States, for example, is now home to red imported fire ants, ripsaw catfish, starlings, Asian tiger mosquitoes, green crabs, leafy spurge, Chinese tallow, Brazilian pepper, and hundreds of others, some beautiful and interesting but all pests. As they take hold over broad areas, they push native plants and animals back, some toward extinction, and as a consequence global biodiversity steadily declines. By the late 1990s, 11 percent of free-living plant species in the United States were exotics. The figure for Ontario was 28 percent, for the British Isles 43 percent, and for Hawaii 44 percent.

Those who consider such species potential assets should take another look. Here are a few of the environmental changes wrought by alien species in the American Great Lakes:

Sea lamprey: decline of native lake trout

Purple loosestrife: loss of waterfowl habitat and decline of native wetland plant species

Common carp: destruction of fish and waterfowl habitat

Eurasian watermilfoil: reduction of aquatic recreational resources

Zebra mussel: fouling of hard substrate, including water pipes, and the reduction of populations of native mollusks.

On the positive side, chinook and coho salmon, both introduced into the Great Lakes, have become valuable sports fishes. But even that advantage is bought at the price of reduction of the native species on which they prey. Overall, the costs of immigrants far outweigh the rather narrow and meager benefits they bring.

A sinister but poorly understood enhancement of the extinction of species is caused by inbreeding. The process is basically simple: when populations of breeding individuals drop to low levels—say, below 100—individuals are far more likely to mate with close relatives. The genetic similarity between kin makes it more probable that their offspring will be homozygous for genes (that is, possessing two of each kind instead of just one) that are deleterious or lethal. As a result of the increased mortality and reduced fertility, the entire population is more likely than otherwise to suffer a final descent to extinction.

Recent studies of song sparrows and white-footed mice in North America have demonstrated that inbred individuals do indeed fare less well in wild populations. However, mathematical models also suggest that in very small populations, random changes in the environment and in birth and death schedules are likely to cause extinction before inbreeding comes into play as an important factor. Empirical data are not yet adequate to settle the matter one way or the other. Still, in one field study of wild Finnish populations of a butterfly, the Glanville fritillary, more inbred populations vanished than outbred but otherwise comparable populations (*Nature*, 392: 442, 1998).

The principle that endangerment and extinction are rarely caused by a single factor is illustrated by the decline of frog populations, a trend that stirred worldwide publicity in the late 1990s. For a decade precipitous drops in populations of many species had been observed in Africa, Australia, the United States, and Central and South America. At least two species, the northern gastric brooding frog of Australia and the golden toad of Costa Rica (and probably many others yet to be verified), spiraled to extinction. The catastrophe raised alarm among environmentalists and the public. The question was asked: Are frogs our canary in the mine, whose death when carried there warns of silent but deadly changes that threaten our own existence?

A flurry of activity to find the reason for the frog die-off ensued, and multiple causes were quickly discovered. In the Midwest and Great Lakes region of the United States, a rise in multiple and missing legs and other abnormalities turned out to be due to water pollutants, as expected. In at least some other parts of the world, reduction of natural habitat has played a role. The increase in ultraviolet radiation by the thinning of the protective ozone layer, leading to aborted embryonic development, has also been suggested. Another, potentially decisive factor is the spread of a newly discovered species of chytrid fungus that causes a fatal skin disease. It appears to be especially important in frog populations of Australia and Central America.

All of these known and suspected sources of rising frog mortality are enhanced if not caused outright by human activity. Even the chytrid fungus is transported by humans from one continent to another, turning up in imported captive frogs, for example, in the United States.

Why should frogs be more vulnerable than other animal groups? The answer is believed to lie in their damp, porous skin, which allows easy access to pollutants and parasites. It may also be found in their amphibious existence. Most frog species live in the water as larvae, but on land as adults. Each habitat is subject to area loss, and each contains different pollutants and disease organisms, putting frogs in double jeopardy.

Hot Spots. In the 1990s systematists and biogeographers focused more intently on biodiversity mapping and preservation. Data were mined from the "focal" groups, those already well enough known, including birds, mammals, reptiles, amphibians, fishes, flowering plants, butterflies, tiger beetles, mollusks, crayfish, and a few other smaller groups, to allow sound analysis. The result has been an increase in the number recognized and a clearer definition of the hot spots. Some geographic areas contain ecosystems that are both threatened and the home of relatively larger numbers of species found nowhere else. In addition to those cited in 1992 in *The Diversity of Life*, others highlighted are the *cerrado* or savanna of central Brazil, the Great Lakes of East Africa, and the pantanal wetlands of the South American south-central heartland. Overdue attention is now also being paid to marine ecosystems, especially coral reefs, the widely threatened "rain forests of the sea."

Some important generalizations have emerged from these collective studies. One is the frequent discordance of hot spots for different groups of organisms. In the tropical forest wilderness it is a safe bet that local areas with the maximum diversity of trees also harbor a maximum diversity or at least very high numbers of species of shrubs, herbaceous plants, birds, ants, frogs, and most other principal groups of organisms, as well as the most biodiverse river systems and lakes. But in many other regions, such as the wildlands of Madagascar and northeastern Australia, which contain a mosaic of savanna, desert, and forest patches, hot spots differ according to groups of organisms. Bees and wasps, for example, are likely to be concentrated in the arid habitats and frogs and butterflies in the woodland. Nonetheless, when the biogeographical information is put together, it is still possible to delimit hot spots, and to do so with greater precision than in previous years.

Hot-spot mapping is a practical necessity. Those who wish to restrict conservation measures in deference to pure economic expansion are prone to claim that the mapping of endangered species imperils the "human" use of large tracts of land worldwide. Farmers are encouraged to believe that the discovery of a rare species on their woodlots will turn the land into a nature reserve.

This opposition is based on a gross misapprehension. In fact, hot spots cover only a very small fraction of Earth's surface. A 1997 survey by the staff of Conservation International, for example, revealed that the 17 hottest spots of the world occupy only 1.3 percent of the land surface yet contain a fourth of the planet's terrestrial vertebrate species and 40 percent of its plants. This study and others yielding comparable conclusions have resulted in a more precise and compelling policy guideline among conservation organizations: concentrate first on saving the hot spots and the remaining wilderness areas.

With these facts in hand, added to knowledge of the many benefits that biodiversity and natural environments bring, it should be possible to enlist a strong commitment to conservation from the countries most affected. After all, it is their own biological riches that they are being asked to protect.

Why care? It has been my experience that people tend to react to the evidences of species extinction with three stages of denial.

The *first stage* is cheerful resignation: "Why worry? Extinction is natural, isn't it?" And yes, species—as is often pointed out—have been dying through the more than 3.5 billion years of biological history without permanent harm to the biosphere. It has been estimated that 98 percent of the species that ever existed have vanished. The response then continues, "Evolution has always replaced extinct species with new ones, has it not?"

But these statements, while true, conceal a terrible twist. After the Mesozoic extinction spasm 65 million years ago, which ended the Age of Reptiles, and after each of the other four greatest previous spasms spaced over 400 million years at very roughly 100-million-year intervals, evolution required about 10 million years to restore the predisaster levels of diversity. That is an extremely long time for future generations to wait because of the damage we are inflicting on the environment within a few decades. Equally serious, evolution cannot perform as in previous ages if natural environments have been crowded out by artificial ones, the phenomenon known by biologists as "the death of birth." For example, Hawaii and most of the Polynesian islands have been either entirely deforested or largely covered by introduced plants.

In the *second stage* of denial, people ask, "Why do we need so many species anyway? Why care, especially since the vast majority are bugs, weeds, and fungi?" It is too easy to dismiss the creepy crawlies of the world, forgetting that less than a century ago and before the rise of the modern conservation movement native birds and mammals were dismissed with the same callous indifference. Now the value of the little things in the natural world has become compellingly clear. Recent experimental studies on whole ecosystems support what was long suspected: in most cases, the more species living in an ecosystem, the higher its productivity and the greater its ability to withstand drought and other kinds of environmental stress. Since we depend on an abundance of functioning ecosystems to cleanse our water, enrich our soil, and manufacture the very air we breathe, biodiversity is clearly not an inheritance to be discarded carelessly.

In addition to creating a habitable environment for human beings, wild species are the source of new pharmaceuticals, crops, fibers, and other products that help sustain our lives. The varied examples of such discoveries cited in *The Diversity of Life* continue to be multiplied. A new sense of opportunity is especially prominent in the search for novel antibiotics and other pharmaceuticals.

The clinching argument for the protection of species, however, may in the end prove to be moral. *Homo sapiens* is a brilliant and proud citizen of the biosphere, but Earth is where we originated and will stay. Who are we to destroy the planet's Creation? Each species around us is a masterpiece of evolution, exquisitely adapted to its environment. Species existing today are thousands to millions of years old. Their genes, having been tested by adversity over countless generations, engineer a staggeringly complex mix of biochemical devices that promote the survival and reproduction of the organisms carrying them.

Even if all these arguments are granted, the *third stage* of denial still emerges: "Why rush to save all the species now? We have more important things to do. Why not keep live specimens in zoos and botanical gardens and return them to the wild later?" The grim answer is that all the zoos in the world today can sustain tiny populations of at most 2,000 species of mammals, birds, reptiles, and amphibians out of a total of about 24,000 known to exist. The world's botanical gardens would be even more overwhelmed by the 250,000 plant species still alive on Earth. These refuges are invaluable in helping to save a few endangered species. So is freezing embryos in liquid nitrogen. But such stopgap measures cannot come close to solving the problem as a whole. To add to the difficulty, no one has even conceived a plan that might save the millions of kinds of insects, fungi, and other ecologically vital small organisms. Even if that prodigious feat could be accomplished, and scientists were poised to return species to independence, the forests, savannas, and other ecosystems in which many lived would no longer exist. To reassemble them would be like unscrambling an egg.

It is the universal consensus of conservation scientists and other environmental professionals that to protect the biosphere it will be necessary to maintain the natural environments in which wild species live. But given how rapidly these refuges are being destroyed, even that straightforward solution seems a daunting task. To keep the Creation, we need all the science, technology, and moral commitment that can be mustered in the service of ecology. Thus is the word ecology derived: *oikos*, our home; *logos*, its study and understanding.

E.O.W.
HARVARD UNIVERSITY
OCTOBER 20, 1998

PREFACE TO THE 2010 PRINTING

If any date can be fixed for the beginning of modern biodiversity studies, it would likely be September 21, 1986, when a National Forum on BioDiversity was held in Washington, D.C. Sponsored jointly by the National Research Council and the Smithsonian Institution, the three-day meeting brought together more than sixty leading biologists, economists, agricultural experts, philosophers, and representatives of assistance and leading agencies. When the contributions of this eclectic assemblage were published two years later under the title *BioDiversity*, the book became, by scientific publishing standards at least, an international bestseller. As its editor, I had already adopted in other writings the unified word "biodiversity," and that is where the term settled.

The material covered in *BioDiversity*, which was focused primarily on biology and written for a broad readership, became the subject of the present book, *The Diversity of Life*, first published in 1992.

Biodiversity, short for biological diversity, is defined as the totality of inherited variation in all the organisms of a selected area. The area can be a woodland or a system of forests or a pond or an ocean. It can be a political unit, say a state or a country. Or it can be the whole world. Once the location is selected, researchers study biodiversity at one or the other or all three levels of biological organization: first ecosystems, such as a patch of forest or a pond; then, next down, all the species, from microorganisms to trees and megafauna; and finally, at the third and lowest level, the genes that prescribe the traits of the species that make up the ecosystems.

It may well be asked at this point, what is so new about "modern" biodiversity studies? After all, record of humanity's attempt to identify all species of organisms dates back to Aristotle. The formalization of taxonomy to accomplish that goal was the consuming interest of Carl Linnaeus, one of the most influential scientists of the eighteenth century. Further, the discovery of the origin of species dates to Charles Darwin. The process of speciation and the multiplication of species, as opposed to the change of individual species through time, had been worked out by Alfred Russel Wallace by 1865, and elaborated to the level of chromosomes and genes during the first half of the twentieth century. The same can be said of biogeography, the mapping of species, and the deduction of phylogenies begun by Wallace and Lamarck—also in the nineteenth century.

The full maturation of biodiversity studies, however, can be dated to the era begun in the 1980s, and is based upon two new developments. The first is the re-igniting of the great Linnaean enterprise, based upon the recognition that despite more than two centuries of taxonomic labor, most of the biodiversity of the world remains unknown. The second development is the extension of the boundary of biodiversity studies in order to unite them with other branches of science and technology.

To say that the flora and fauna of the world remains largely unknown is no exaggeration. At the present time (2010), the numbers of known species of all kinds of organisms, defined as those discovered, characterized, and given a scientific name, is approximately 1.9 million. The actual number in existence on Earth has been variously estimated to be between 5 and more than 50 million; if microorganisms are added, the number would have to be increased dramatically, albeit to an entirely unknown degree.

In 2006–2007, approximately 18,000 new species of all kinds of non-microbial organisms were being described each year. Of these species, about 75 percent were invertebrates, 7 percent were vertebrates, and 11 percent were plants.* If we accept 10 million species as the total global number, a figure most biodiversity experts find conservative, it follows that at the present rate of species discovery it will be half another millennium, well into the twenty-seventh century, before the census of life on Earth is complete.

Nor are the still largely unknown groups of organisms trivial to the rest of life, including our own. The 100,000 species of fungi thus far discovered and named are but a tiny fraction of the roughly 1.5 million estimated to exist. Fewer than 25,000 species of nematodes, tiny wormlike creatures considered the most abundant animals on the planet, are known, whereas somewhat less than a half million have been estimated to await discovery. Some 14,000 known species of ants, the most abundant and ecologically dominant insects, probably represent no more than half of those in existence. The same rough figure—half or less—applies to the world's beetles and flies.

Science is at the dawn of a new age in the discovery of biodiversity. Yet it is clear that a much greater effort is needed from humanity if we wish to continue struggling along on a little-known planet. The technology exists to speed the effort to map the living world. DNA sequencing is now rapid enough for the mapping of complete genomes in days or even (in bacteria, for example) hours. Metagenomics, the sampling of DNA from assemblages of species in selected ecosystems, allows "quick and dirty" estimates of the amount of microbial diversity in soil and water. Once species of all kinds are separated and described by diagnostic traits, they can be quickly identified by DNA barcoding—the reading of selected parts of the genome.

As information accumulates, it is already being fed into databases available through single access on command. The most encompassing is the Encyclopedia of Life (http://www.eol.org), which was begun in 2005 and is well on its way to recording and making immediately accessible everything known about every species of organism, with updates on already known species and information on new species as they are discovered.

The first round of taxonomic discovery and archiving is only the beginning. Of the known species of organisms, only a very small fraction of the species, approximately 3 percent, have been studied well enough to evaluate their conservation status—whether they are abundant and widespread enough to

---

*These estimates are from Arthur D. Chapman, ed., *Numbers of Living Species in Australia and the World*, 2nd ed. (Australian Biodiversity Information Services, Department of the Environment, Water, Heritage and the Arts, Australia, September 2009).

be stable and safe, or liable in any degree to extinction. The 5,490 species of mammals and 9,998 species of birds thus far discovered have all been evaluated in this manner, but this knowledge is more than counterbalanced by our ignorance of plants (3.9 percent of 282,000 known species) and invertebrate animals (0.6 percent of 1.3 million species).*

Every species on Earth has been adapted by thousands to millions of years of evolution to the particularities of the environment in which it lives. Its genotype is different from that of all other species. The traits its genes prescribe are also unique, in biochemistry, anatomy, physiology, and behavior, and in the way it interacts with other species and serves the ecosystem it inhabits. Each species, in short, is a living encyclopedia of how to survive on planet Earth.

Humanity is in the very early stages of acquiring the knowledge biodiversity offers us. The impact of biodiversity studies on medicine, biotechnology, and agriculture is already substantial, and will in time become enormous. As the reach of general biology at all levels grows, the role of biodiversity studies will become disproportionately large. Its importance is foreordained by the fact that future biology as a whole is settling upon two laws. The first is that all living processes are ultimately obedient to the laws of physics and chemistry. This provides the foundation of molecular, cellular, and developmental biology. The second law is that all living processes have originated through evolution by natural selection. That perception is the foundation of evolutionary and environmental biology, much of which is devoted to biodiversity studies. In time, I believe, biology will be seen as a discipline that advances in coordinated manner along these two fronts.

Edward O. Wilson
20 May 2010

137.2 *Storm over the Amazon*] Parts of this chapter were modified from my earlier articles, "Storm over the Amazon," in Daniel Halpern, ed., *On Nature: Nature, Landscape, and Natural History* (San Francisco: North Point Press, 1987), pp. 157–159; and "Rain Forest Canopy: The High Frontier," *National Geographic*, 180:78–107 (December 1991). I covered the subject on which I reflected that night in a technical monograph, *Success and Dominance in Ecosystems: The Case of the Social Insects* (Oldendorf/Luhe, Germany: Ecological Institute, 1990). The precise call of the red howler monkey is based on the description by Louise H. Emmons in *Neotropical Rainforest Mammals: A Field Guide* (University of Chicago Press, 1990). W

142.11–12 to look for something . . . Ranges.] See Kipling's 1898 poem "The Explorer."

143.9–23 the great chemist Berzelius . . . attained.] The reflections of Jöns Jacob Berzelius are from his *Manual of Chemistry* (vol. 3, 1818), as quoted

*These estimates are from S. N. Stuart, E. O. Wilson, J. A. McNeely, R. A. Mittermeier, and J. P. Rodríguez, "The barometer of life," *Science* 328:177 (9 April 2010).

by Carl Gustaf Berhnard, "Berzelius, Creator of the Chemical Language," reprinted from the *Saab-Scania Griffin 1989/90* by the Royal Swedish Academy of Sciences. W

146.23–27   nototheniod fishes . . . cannot go.] The antifreeze technique of the nototheniod fishes is described in Joseph T. Eastman and Arthur L. DeVries, "Antarctic Fishes," *Scientific American*, 255:106–114 (November 1986). W

147.3–6   Archaebacteria . . . deep sea.] The archaebacteria or archaeotida, some of which exist in the most hostile environments on earth, have been most thoroughly documented by Carl R. Woese and his coresearchers. See Robert Pool, "Pushing the Envelope of Life," *Science*, 247:158–247 (1990). Some biologists, including Woese, consider these organisms to constitute a separate kingdom of life apart from true bacteria and other prokaryotic organisms comprising the kingdom Monera. W

149.2   *Krakatau*] The definitive account of the 1883 Krakatau eruptions, including personal accounts and research reports of the time, is provided by Tom Simkin and Richard S. Fiske, *Krakatau 1883: The Volcanic Eruption and Its Effects* (Washington, D.C.: Smithsonian Institution Press, 1983). Additional details on the tidal waves are given by Susanna Van Rose and Ian F. Mercer, *Volcanoes* (Cambridge: Harvard University Press, 1991). Analyses of the recolonization of Rakata are summarized in Robert H. MacArthur and Edward O. Wilson, *The Theory of Island Biogeography* (Princeton: Princeton University Press, 1967); Ian W. B. Thornton et al., "Colonization of the Krakatau Islands by Vertebrates: Equilibrium, Succession, and Possible Delayed Extinction," *Proceedings of the National Academy of Sciences*, 85:515–518 (1988); I.W.B. Thornton and T. R. New, "Krakatau Invertebrates: The 1980s Fauna in the Context of a Century of Recolonization," *Philosophical Transactions of the Royal Society of London*, ser. B, 322:493–522 (1988); and P. A. Rawlinson, A.H.T. Widjoya, M. N. Hutchinson, and G. W. Brown, "The Terrestrial Vertebrate Fauna of the Krakatau Islands, Sunda Strait, 1883–1986," *Philosophical Transactions of the Royal Society of London*, ser. B, 328:3–28 (1990). The estimate of the pumice temperatures following the explosion are by Noboru Oba of Kagoshima University, cited from personal communication by Thornton and New, "Krakatau Invertebrates." W

158.2   *The Great Extinctions*] I have drawn the details of the Cretaceous extinction episode and the meteorite-vulcanism debate from many sources, but especially Matthew H. Nitecki, ed., *Extinctions* (Chicago: University of Chicago Press, 1984); Steven M. Stanley, *Extinction* (New York: Scientific American Books, 1987), and "Periodic Mass Extinctions of the Earth's Species," *Bulletin of the American Academy of Arts and Sciences*, 40(8):29–48 (1987); David M. Raup, *Extinction: Bad Genes or Bad Luck?* (New York: Norton, 1991); Paul Whalley, "Insects and Cretaceous Mass Extinction," *Nature*, 327:562 (1987); Carl O. Moses, "A Geochemical Perspective on the Causes and Periodicity of Mass Extinctions," *Ecology* (70)4:812–823 (1989);

and William Glen, "What Killed the Dinosaurs?," *American Scientist*, 78 (4):354–370 (1990). Stanley, for example, has argued persuasively for the role of long-term cooling of the earth's climate as a principal factor in mass extinctions, including the Cretaceous. Whalley details the survival of the insects. The fate of the flowering plants is described by Andrew H. Knoll in Nitecki, *Extinctions*, pp. 21–68, and R. A. Spicer, "Plants at the Cretaceous-Tertiary Boundary," *Philosophical Transactions of the Royal Society of London*, ser. B, 325:291–305 (1989). The newest fossil evidence from the K-T boundary for many groups of organisms is summarized in *Evolution and Extinction*, special issue of the *Philosophical Transactions of the Royal Society of London*, ser. B, 325:239–488 (1989), ed. W. G. Chaloner and A. Hallam. Additional, unpublished details are reported by Richard A. Kerr in "Dinosaurs and Friends Snuffed Out?," *Science*, 251:160–162 (1991). W

160.27 Principle of Certainty] The certainty principle of opinion was stated by Robert H. Thouless, "The Tendency to Certainty in Religious Belief," *British Journal of Psychology*, 26(1):16–31 (1935). In both religious and secular opinion, Thouless wrote, "there is a real tendency amongst people for degree of belief to approach to certainty. Doubt and skepticism are for most people unusual and, I think, generally unstable states of mind." W

163.19–23 In 1990 . . . the first site.] Evidence for a giant meteor strike in the Caribbean area at the end of the Cretaceous is given by J.-M. Florentin, R. Maurrasse, and Gautam Sen, "Impacts, Tsunamis, and the Haitian Cretaceous-Tertiary Boundary Layer," *Science*, 252:1690–1693 (1991). They conclude that an immense meteorite impact near the Beloc formation of Haiti "produced microtektites that settled to form a nearly pure layer at the base. Vaporized materials with anomalously high extraterrestrial components settled last, along with carbonate sediments. The entire bed became sparsely consolidated. Subsequently, another major disruptive event, perhaps a giant tsunami, partly reworked the initial deposit . . . This process also may have caused further mixing of Cretaceous and Tertiary microfossils, as observed at Beloc and elsewhere." W

164.21–23 the large amounts of data . . . and others.] In 1984 David Raup and J. John Sepkoski Jr. proposed that large-scale extinctions have been periodic, occurring at intervals of about 26 million years. Their analysis was based on data from the families of marine animals. The Raup-Sepkoski hypothesis set off a round of speculation on possible extraterrestrial causes, such as meteorite or comet showers induced by cyclic approach or realignment of undiscovered celestial bodies. Most intriguing was the postulation of the role of a solar companion star, variously called "Nemesis" or "Death Star." But the whole idea has been effectively challenged by a combination of critiques of geological dating, statistical analysis, and taxonomic interpretation. The jury is still out on the matter, but an eventual negative verdict seems likely. For reviews of the subject see D. M. Raup, *The Nemesis Affair* (New York: Norton, 1986) and *Extinction: Bad Genes or Bad Luck?* (New York: Norton,

1991); S. M. Stanley, *Extinction* (New York: Scientific American Books, 1987); and a series of articles by paleobiologists in *Ecology*, 70(4):801–834 (1989), ed. Edward F. Connor. In 1991 Raup estimated that half of paleontologists best informed on the subject believed in periodicity; half did not. W

164.23–24   The loss of families . . . total destruction."] The extinction rates of families and species during the Permian crisis, based on rarefaction analysis, is given by David M. Raup, "Size of the Permo-Triassic Bottleneck and Its Evolutionary Implications," *Science*, 206:217–218 (1979). His statement on the near extinction of all higher life is given in a later review, "Diversity Crises in the Geological Past," in E. O. Wilson and Frances M. Peter, eds. *Biodiversity* (Washington, D.C.: National Academy Press, 1988), pp. 51–57. A separate examination of the evidence, which considers not only climatic cooling but also a regression (shrinking) of the shallow seas, is provided by Douglas H. Erwin in "The End-Permian Mass Extinction: What Really Happened and Did It Matter?," *Trends in Ecology and Evolution*, 4(8):225–229 (1989). W

165.4–5   massive volcanic eruptions . . . Permian extinctions] Evidence of massive volcanic eruptions at the time of the Permian extinction spasm was presented by Paul R. Renne and Asish R. Basu, "Rapid Eruption of the Siberian Traps Flood Basalts at the Permo-Triassic Boundary," *Science*, 253:176–179 (1991). W

172.4   the species concept] A thoroughgoing description and analysis of species concepts are given in Douglas J. Futuyma, *Evolutionary Biology*, 2nd ed. (Sunderland, Mass.: Sinauer, 1986); Alan R. Templeton, "The Meaning of Species and Speciation: A Genetic Perspective," in Daniel Otte and John A. Endler, eds., *Speciation and Its Consequences* (Sunderland: Sinauer, 1989), pp. 3–27; and Ernst Mayr and Peter D. Ashlock, *Principles of Systematic Zoology*, 2nd ed. (New York: McGraw-Hill, 1991). W

173.11–13   Tigers . . . Gir Forest.] The current status of wild tiger populations is documented in Lynn A. Maguire and Robert C. Lacy, "Allocating Scarce Resources for Conservation of Endangered Subspecies: Partitioning Zoo Space for Tigers," *Conservation Biology*, 4(2):157–166 (1990). W

176.26–178.2   *Anopheles* mosquitoes . . . from Europe.] The history of the link between malaria and the sibling species of *Anopheles maculipennis* is reviewed by Ernst Mayr, *Systematics and the Origin of Species* (New York: Columbia University Press, 1942). W

178.11   the biological-species concept] A history of the biological-species concept is given by Ernst Mayr, one of its later principal architects, in *Evolution and the Diversity of Life: Selected Essays* (Cambridge: Harvard University Press, 1976). W

178.14–17   each species . . . unique individual] The idea of a species as an individual unto itself has been argued forcefully by Michael T. Ghiselin, "Categories, Life, and Thinking," *Behavioral and Brain Sciences*, 4(2):269–313 (1981). W

178.31–38   At Cambridge University . . . ions mine.] The ions chorus of the Cavendish group, sung to the tune of "My Darling Clementine," is reported by Rutherford's student Samuel Devons in "Rutherford and the Science of His Day," *Notes and Records of the Royal Society of London*, 45(2):221–242 (1991). W

179.28–180.11   "semispecies," . . . ecological overlap.] Hybridization in oaks and the nature of semispecies are discussed by Alan T. Whittemore and Barbara A. Schaal, "Interspecific Gene Flow in Sympatric Oaks," *Proceedings of the National Academy of Sciences*, 88:2540–44 (1991). W

180.17–19   Tropical species . . . temperate zones.] The greater frequency of full reproductive isolation among tropical plant species has been remarked by Alwyn H. Gentry, "Speciation in Tropical Forests," in L. B. Holm-Nielsen, I. C. Nielsen, and H. Balslev, eds., *Tropical Forests: Botanical Dynamics, Speciation, and Diversity* (New York: Academic Press, 1989), pp. 113–134. W

180.28   species formation] General aspects of species formation are well reviewed by Douglas J. Futuyma, *Evolutionary Biology*, 2nd ed. (Sunderland, Mass.: Sinauer, 1986). Special topics are treated at a more advanced level in Daniel Otte and John A. Endler, eds., *Speciation and Its Consequences* (Sunderland: Sinauer, 1989). W

181.37   *Homo erectus*] The gradual evolution of the intermediate human species *Homo erectus* is described by Wu Rukang and Lin Shenglong, "Peking Man," *Scientific American*, 248:86–94 (June 1983). W

188.31–189.15   The giant silkworm moths . . . *heure d'amour.*] The data on mating times of giant silkworm moths are given by Phil and Nellie Rau, "The Sex Attraction and Rhythmic Periodicity in the Giant Saturniid Moths," *Transactions of the Academy of Science of St. Louis*, 26:83–221 (1929). W

189.20–190.3   Male jumping spiders . . . own species.] Details of the courtship displays of jumping spiders are given by Jocelyn Crane, "Comparative Biology of Salticid Spiders at Rancho Grande, Venezuela. Part 4: An Analysis of Display," *Zoologica* (New York), 34(4):159–214 (1949). W

199.35–200.18   Stephen O'Brien . . . native mammal.] The definition of formal subspecies for use in policy making was proposed by Stephen J. O'Brien and Ernst Mayr, "Bureaucratic Mischief: Recognizing Endangered Species and Subspecies," *Science*, 251:1187–88 (1991). They also reviewed the status of the Florida panther. W

201.19–34   leafroller moths . . . sex attractants.] The ultrasimple isolating mechanisms of leafroller moths are described by Wendell L. Roelofs and Richard L. Brown, "Pheromones and Evolutionary Relationships of Tortricidae," *Annual Review of Ecology and Systematics*, 13:395–422 (1982). W

205.2–3   *sympatric speciation*] Key review articles on various modes of sympatric speciation are Guy L. Bush, "Modes of Animal Speciation," *Annual*

Review of Ecology and Systematics, 6:339–364 (1975); Scott R. Diehl and G. L. Bush, "The Role of Habitat Preference in Adaptation and Speciation," in Daniel Otte and John A. Endler, eds., *Speciation and Its Consequences* (Sunderland: Sinauer, 1989), pp. 345–365; and Catherine A. and Maurice J. Taber, "Sympatric Speciation in Insects: Perception and Perspective," in Otte and Endler, *Speciation*, pp. 307–344. The theory of sympatric speciation by host races was developed principally by Guy Bush. A skeptical analysis is provided by Douglas J. Futuyma and Gregory C. Mayer, "Non-Allopatric Speciation in Animals," *Systematic Zoology*, 29(3):254–271 (1980). Futuyma and Mayer conclude that "the conditions under which host-associated sympatric speciation might occur are so exacting as to be met by very few species." W

209.7–12 At least five genes . . . skin texture.] The data on number of genes responsible for variation in simple traits are provided by Russell Lande, "The Minimum Number of Genes Contributing to Quantitative Variation Between and Within Populations," *Genetics*, 99(3,4):541–553 (1981). W

216.5–18 A dominant gene . . . single generation.] The rates of evolution due to change in frequencies of single genes are given by Daniel L. Hartl and Andrew G. Clark, *Principles of Population Genetics*, 2nd ed. (Sunderland: Sinauer, 1989). W

217.27–218.27 allometry . . . system emerges.] Allometric variation in the mandibles and horns of male beetles is reviewed by J. T. Clark, "Aspects of Variation in the Stag Beetle *Lucanus cervus* (L.) (Coleoptera: Lucanidae)," *Systematic Entomology*, 2(1):9–16 (1977). Allometry as the basis of caste differences in ants is presented in detail by Bert Hölldobler and Edward O. Wilson, *The Ants* (Cambridge: Harvard University Press, 1990). W

218.39–220.10 microevolution . . . macroevolution.] For examples of close linkage between microevolution and macroevolution, see: for the adaptive radiation of the Hawaii honeycreepers, Walter J. Bock, "Microevolutionary Sequences as a Fundamental Concept in Macroevolutionary Models," *Evolution*, 24(4):704–722 (1970), and reviewed here in Chapter 7; for the origin of a new vertebrate jaw type in bolyerine snakes on Round Island, Thomas H. Frazzetta, *Complex Adaptations in Evolving Populations* (Sunderland: Sinauer, 1975); and for the origin of new chromosome races and adaptive types in mole rats of the Middle East, Eviatar Nevo, "Speciation in Action and Adaptation in Subterranean Mole Rats: Patterns and Theory," *Bolletino Zoologia*, 52(1–2): 65–95 (1985). W

220.16 punctuated equilibrium] The punctuated equilibrium thesis was first presented by Niles Eldredge and Stephen J. Gould, "Punctuated Equilibria: An Alternative to Phyletic Gradualism," in T.J.M. Schopf, ed., *Models in Paleobiology* (San Francisco: Freeman, Cooper, 1972), pp. 82–115; and elaborated by Gould, "Is a New and General Theory of Evolution Emerging?," *Paleobiology*, 6(1):119–130 (1980), and by Eldredge, *Time Frames: The Rethinking of Darwinian Evolution and the Theory of Punctuated Equilibria*

(New York: Simon and Schuster, 1985). Among the more definitive critiques are Richard Dawkins, *The Blind Watchmaker* (New York: Norton, 1986); Max K. Hecht and Antoni Hoffman, "Why Not Neo-Darwinism? A Critique of Paleobiological Challenges," *Oxford Surveys in Evolutionary Biology*, 3:1–47 (1986); and Jeffrey Levinton, *Genetics, Paleontology, and Macroevolution* (New York: Cambridge University Press, 1988). The evidence originally presented by Eldredge and Gould has been found not to conform to the punctuated-equilibrium pattern; see William L. Brown Jr., "Punctuated Equilibrium Excused: The Original Examples Fail to Support It," *Biological Journal of the Linnean Society*, 31:383–404 (1987). W

221.33 species selection] The idea of species selection in the fossil record was first developed from fossil evidence by Steven M. Stanley, "A Theory of Evolution above the Species Level," *Proceedings of the National Academy of Sciences*, 72:646–650 (1975). The basic argument was expanded by Elisabeth S. Vrba and Stephen J. Gould, "The Hierarchical Expansion of Sorting and Selection," *Paleobiology*, 12(2):217–228 (1986). The basic genetic theory, however, had previously been worked out at the level of multiple competing populations of the same species, following essentially the same model as species selection. The key papers are Richard Levins, "Extinction," in M. Gerstenhaber, ed., *Some Mathematical Questions in Biology* (Providence: American Mathematical Society, 1970), pp. 77–107; and Scott A. Boorman and Paul R. Levitt, "Group Selection on the Boundary of a Stable Population," *Proceedings of the National Academy of Sciences*, 69(9):2711–13 (1972). W

223.15–23 Among insects . . . evolution.] The enhancement of species proliferation among insects by plant feeding was documented by Charles Mitter, Brian Farrell, and Brian Wiegmann, "The Phylogenetic Study of Adaptive Zones: Has Phytophagy Promoted Insect Diversification?," *American Naturalist*, 132(1):107–128. W

223.24–33 During the latter . . . mollusks.] The connection between species ranges and extinction rates among mollusks is from David Jablonski, "Heritability at the Species Level: Analysis of Geographic Ranges of Cretaceous Mollusks," *Science*, 288:360–363 (1987), and "Estimates of Species Duration: Response," *Science* 240:969 (1988). W

223.35 the *taxon cycle*] The taxon cycle was introduced by Edward O. Wilson, "The Nature of the Taxon Cycle in the Melanesian Ant Fauna," *American Naturalist*, 95:169–193 (1961); and most recently evaluated by James K. Liebherr and Ann E. Hajek, "A Cladistic Test of the Taxon Cycle and Taxon Pulse Hypotheses," *Cladistics*, 6:39–59 (1990). W

224.12–25 A process similar . . . circumstances.] Elisabeth Vrba has documented the turnover rates of African antelopes and other bovid mammals in "African Bovidae: Evolutionary Events since the Miocene," *South African Journal of Science*, 81:263–266 (1985), and "Mammals as a Key to Evolutionary Theory," *Journal of Mammalogy*, 73(1):1–28 (1992). W

872 NOTES

224.26–40 Desert plants . . . edge of extinction.] Finally, the example of countervailing organismic and species selection in desert plants was suggested by data from Delbert C. Wiens et al., "Developmental Failure and Loss of Reproductive Capacity in the Rare Paleoendemic Shrub *Dedeckera eurekensis*," *Nature*, 338:65–67 (1989); alternative hypotheses to explain the data are reviewed by Deborah Charlesworth, "Evolution of Low Female Fertility in Plants: Pollen Limitation, Resource Allocation and Genetic Load," *Trends in Ecology and Evolution*, 4(10):289–292 (1989). W

228.36 The 10,000 . . . Hawaii] The estimate of the number of endemic Hawaiian insect species is given by F. G. Howarth, S. H. Sohmer, and W. D. Duckworth, "Hawaiian Natural History and Conservation Efforts," *BioScience*, 38(4):232–238 (1988). W

229.35–231.3 More than half . . . unremembered dead.] The honeycreepers of Hawaii are reviewed by Walter J. Bock, "Microevolutionary Sequences as a Fundamental Concept in Macroevolutionary Models," *Evolution*, 24(4):704–722 (1970), and J. Michael Scott et al., "Conservation of Hawaii's Vanishing Avifauna," *BioScience*, 38(4):238–253 (1988). I have included additional information from Storrs L. Olson (personal communication), who with Helen F. James has pioneered in the study of subfossil species extinguished by the original Polynesian settlers of Hawaii. W

234.2–17 The details . . . its moorings.] The details of the force of the woodpecker's strike are from Philip R. A. May et al., "Woodpeckers and Head Injury," *Lancet*, February 28, 1976, pp. 454–455; and "Woodpecker Drilling Behavior: An Endorsement of the Rotational Theory of Impact Brain Injury," *Archives of Neurology*, 36:370–373 (1979). W

235.15–16 Darwin's finches . . . Geospizinae] The definitive work on the geospizine finches is Peter R. Grant, *Ecology and Evolution of Darwin's Finches* (Princeton: Princeton University Press, 1986). Another valuable, more popular account is Sherwin Carlquist, *Island Life: A Natural History of the Islands of the World* (Garden City: Natural History Press, 1965). W

237.35–36 the plant family Compositae] The best account of the evolution of composite herbs on islands is provided by Sherwin Carlquist in *Island Life* and in *Island Biology* (New York: Columbia University Press, 1974). More recent information on St. Helena flora is given in Mark Williamson, "St. Helena Ebony Tree Saved," *Nature*, 309:581 (1984). The data on St. Helena's mostly extinct beetles are from T. Vernon Wollaston's classic study, *Coleoptera Sanctae-Helenae* (London: John Van Voorst, 1877), updated by P. Basilewski and J. Decelle in their introduction to "La faune terrestre de l'Ile de Sainte-Hélène," *Annales, Musée Royale de l'Afrique Centrale, Tervuren, Belgium, Sciences Zoologiques*, 192:1–9 (1972). W

240.37–241.8 Most striking of all . . . personalized instruction.] The protean food habits of the Cocos Island finch were discovered by Tracey K. Werner and Thomas W. Sherry and reported in "Behavioral Feeding Specialization in *Pinaroloxias inornata*, the 'Darwin's Finch' of Cocos Island, Costa Rica," *Proceedings of the National Academy of Sciences*, 84:5506–10 (1987). W

241.24 the freshwater Cichlidae] Definitive reviews of cichlid fish evolution are presented in *Evolution of Fish Species Flocks*, ed. Anthony A. Echelle and Irv Kornfield (Orono: University of Maine Press, 1984), in articles by Wallace J. Dominey, P. Humphry Greenwood, Leslie S. Kaufman, Karel F. Liem, Kenneth R. McKaye, Richard E. Strauss, and Frans Witte.

242.31–244.15 All of the Lake Victoria . . . 200,000 years.] Molecular evidence of the origin of Lake Victoria cichlids is given by Axel Meyer et al., "Monophyletic Origin of Lake Victoria Cichlid Fishes Suggested by Mitochondrial DNA Sequences," *Nature*, 347:550–553 (1990). A recent analysis of the Lake Victoria species is given by F. Witte and M.J.P. van Oijen, "Taxonomy, Ecology and Fishery of Lake Victoria Haplochromine Trophic Groups," *Zoologische Verhandelingen*, 262:1–47 (1991). W

244.17 *species flocks*] The role of plasticity and behavior in macroevolution is emphasized by Mary Jane West-Eberhard, "Phenotypic Plasticity and the Origins of Diversity," *Annual Review of Ecology and Systematics*, 20:249–278 (1989). She provides many interesting examples to show how species formation can proceed rapidly during brief episodes of geographic isolation. A similar effect is posited by Wallace J. Dominey, "Effects of Sexual Selection and Life History on Speciation: Species Flocks in African Cichlids and Hawaiian *Drosophila*," in Echelle and Kornfield, *Evolution of Fish Species Flocks*. W

246.15–25 In Iceland's . . . thousand years ago.] The case of multiple forms in the arctic char is described by Skúli Skúlason, David L. G. Noakes, and Sigurdur S. Snorrason, "Ontogeny of Trophic Morphology in Four Sympatric Morphs of Arctic Charr *Salvelinus alpinus* in Thingvallavatn, Iceland," *Biological Journal of the Linnean Society*, 38:281–301 (1989). W

246.32–247.2 The cichlid fishes . . . have disappeared.] Estimates of extinction rates due to the Nile perch are made by C.D.N. Barel et al., "The Haplochromine Cichlids in Lake Victoria: An Assessment of Biological and Fisheries Interests," in M.H.A. Keenleyside, ed., *Cichlid Fishes: Behaviour, Ecology and Evolution* (London: Chapman and Hall, 1991), pp. 258–279. W

248.3 The archaic radiation of sharks] Parts of the description of adaptive radiation in sharks are adopted from my "In Praise of Sharks," *Discover*, 6(7):40–42, 48, 50–53 (1985). W

249.39–252.14 In 1976 . . . slack-jawed corpses.] The description of the great white shark by Hugh Edwards is in *Sharks*, ed. J. D. Stevens (New York: Facts on File, 1987), p. 212. Authoritative reviews of shark natural history are given by other authors in the same volume and by Victor G. Springer and Joy P. Gold, *Sharks in Question* (Washington, D.C.: Smithsonian Institution Press, 1989). W

252.29–31   Cookie-cutters . . . hydrophone arrays.] An account of the attacks by cookie-cutter sharks on nuclear submarines is provided by C. Scott Johnson, "Sea Creatures and the Problem of Equipment Damage," *U.S. Naval Institute Proceedings*, August 1978, pp. 106–107. W

253.22–254.17   the megamouth shark . . . for so long.] The rapidly developing story of the megamouth shark is summarized in Springer and Gold, *Sharks in Question*. An account of the vertical migration is given in the March 1991 *National Geographic*. I have benefited from background information provided in conversation with Robert J. Lavenberg of the Los Angeles County Natural History Museum, who studied the second California specimen live in its natural habitat. W

256.12–16   The third continental center . . . 2.5 million years ago.] A fourth adaptive radiation of mammals occurred on the great island of Madagascar, producing a wide array of lemurs, which are primitive primates resembling monkeys, and tenrecs, which are insectivores variously resembling shrews, moles, and hedgehogs. But the spread of major adaptive types in the fauna as a whole still fell far short of that on Australia, South America, and the World Continent. W

256.22–23   three major groups . . . Australia.] The classic survey of Australian mammals is Ellis Troughton, *Furred Animals of Australia* (London: Angus and Robertson, 1941). A more recent monograph on Australian mammals, with special attention to conservation, is Tim Flannery's beautifully illustrated *Australia's Vanishing Mammals* (Surry Hills, New South Wales: RD Press, 1990). W

258.26   the Great American Interchange] An excellent description of the Great American Interchange is given by George Gaylord Simpson, *Splendid Isolation: The Curious History of South American Mammals* (New Haven: Yale University Press, 1980). The most recent summary of the fossil and biogeographic evidence, on which my own account is largely based, is Larry G. Marshall et al., "Mammalian Evolution and the Great American Interchange," *Science*, 215:1351–57 (1982); and L. G. Marshall, "Land Mammals and the Great American Interchange," *American Scientist*, 76:380–388 (1988). A detailed description of the species that were extinguished is provided by Elaine Anderson, "Who's Who in the Pleistocene: A Mammalian Bestiary," in Paul S. Martin and Richard G. Klein, eds., *Quaternary Extinctions* (Tucson: University of Arizona Press, 1984), pp. 40–89. W

262.3   *A thousand ages . . . evening gone.*] See "Our God, Our Help in Ages Past," a 1709 hymn by Isaac Watts, drawn from Psalm 90.

265.22–40   success . . . risk of extinction.] I have treated the relation of diversity to longevity and dominance more formally in *Success and Dominance in Ecosystems: The Case of the Social Insects* (Oldendorf/Luhe: Ecology Institute, 1990). Longevity is defined as the duration through geological time of a species and all its descendants. But I will add a note now to say, more precisely, that the longevity of interest is the set of traits by which the species and its descendants are diagnosed, such as possession of a particular gland or bone structure or horn shape. The end of the group of species can then come either by absolute extinction, the death of all the populations, or by "chrono-taxon" extinction, in which populations of the species group evolve a new set of traits sufficiently different to rank the populations as a different genus or even higher ranking taxon. W

267.3–35   In 1983 . . . ocean's environment.] An incisive review of the phyla of organisms is Lynn Margulis and Karlene V. Schwartz, *Five Kingdoms: An Illustrated Guide to the Phyla of Life on Earth* (San Francisco: Freeman, 1982). That book just missed the description of loriciferans by R. M. Kristensen, "Loricifera, a New Phylum with Aschelminthes Characters from the Meiobenthos," *Zeitschrift für Zoologische Systematik und Evolutionsforschung*, 21(3):163–208 (1983). An updated account of loriciferans has been given by Richard C. and Gary J. Brusca, *Invertebrates* (Sunderland: Sinauer, 1990). W

268.34–269.2   With the help . . . live on earth.] The estimates of numbers of described species according to group come from my "The Current State of Biological Diversity," in E. O. Wilson and F. M. Peter, eds., *Biodiversity* (Washington, D.C.: National Academy Press, 1988), pp. 3–18. Detailed accounts of diversity and the degree of confidence in estimates of species numbers are provided for many groups individually in Sybil P. Parker, ed., *Synopsis and Classification of Living Organisms*, vols. 1 and 2 (New York: McGraw-Hill, 1982). That work supplied most of my estimates used to reach the world total of 1.4 million. In 1978 T.R.E. Southwood estimated 1.4 million for all described species except fungi, algae, and bacteria and other monerans, in Laurence A. Mound and Nadia Waloff, eds., *Diversity of Insect Faunas* (London: Blackwell, 1978), pp. 19–40. When the missing groups are added, Southwood's world total comes to 1.5 million. Nigel E. Stork, "Insect Diversity: Facts, Fiction, and Speculation," *Biological Journal of the Linnean Society*, 35:321–337 (1988), cites an unpublished estimate by N. M. Collins of 1.8 million species of plants and animals alone; if fungi and monerans are added, his figure would rise to about 1.9 million. I am inclined to believe this last number too high, while my own may be too low. W

269.25–273.2   So important are insects . . . animal life.] The speculative account of the fate of a world without insects was modified from my lecture, "The Little Things That Run the World," given at the National Zoological Park, Washington, D.C., May 7, 1987, and subsequently published in *Conservation Biology*, 1(4):344–346 (1987). W

273.10–277.11   In 1982 Terry Erwin . . . order of magnitude.] Terry Erwin's estimate of the diversity of rain-forest arthropods was first presented in "Tropical Forests: Their Richness in Coleoptera and Other Arthropod Species," *Coleopterists' Bulletin*, 36(1):74–75 (1982), and "Beetles and Other

Insects of Tropical Forest Canopies at Manaus, Brazil, Sampled by Insecticidal Fogging," in S. L. Sutton, T. C. Whitmore, and A. C. Chadwick, eds., *Tropical Rain Forest: Ecology and Management* (London: Blackwell, 1983), pp. 59–75. Evaluations of the estimate, with new analyses, are given by Robert M. May, "How Many Species Are There on Earth?," *Science*, 241:1441–49 (1988), and "How Many Species?," *Philosophical Transactions of the Royal Society of London*, ser. B, 330:293–304 (1990); Nigel Stork, "Insect Diversity," and personal communication; and Kevin J. Gaston, "The Magnitude of Global Insect Species Richness," *Conservation Biology*, 5(3):283–296 (1991). The account here is modified from my "Rain Forest Canopy: The High Frontier," *National Geographic*, 180:78–107 (December 1991). W

277.12–15   William Beebe . . . above it."] C. William Beebe wrote on the unexplored canopy of the tropical rain forest in *Tropical Wild Life in British Guiana*, by Beebe, G. Inness Hartley, and Paul G. Howes (New York: New York Zoological Society, 1917). W

278.5–9   J. Frederick Grassle . . . order of magnitude.] Our knowledge of biological diversity on the floor of the deep sea is summarized by J. Frederick Grassle, "Deep-Sea Benthic Biodiversity," *BioScience*, 41(7):464–469 (1991). W

279.31–281.14   Jostein Goksøyr . . . couple of centuries."] The estimation of diversity in soil bacteria by means of DNA strand matching is described in two articles by Jostein Goksøyr, Vigdis Torsvik, and their coworkers in *Applied and Environmental Microbiology*, 56(3):776–781, 782–787 (1990). I have benefited from additional unpublished manuscripts kindly provided by Jostein Goksøyr. W

280.37–281.2   The arbitrary standard . . . other species.] The 70 percent DNA matching criterion was proposed by the Ad Hoc Committee on Reconciliation of Approaches to Bacterial Systematics in *International Journal of Systematic Bacteriology*, 37:463–464 (1987). W

281.22–26   Recent drilling . . . early probes.] The new bacterial floras discovered by deep drilling are characterized by Carl B. Fliermans and David L. Balkwill, "Microbial Life in Deep Terrestrial Subsurfaces," *BioScience*, 39(6):370–377 (1989). W

281.27–28   other microorganisms] The diversity of fungi is another great unknown, its magnitude possibly approaching that of bacteria. In a recent assessment, David L. Hawksworth places the number of known species at 69,000 but the actual number on the earth as very likely 1.5 million. "The Fungal Dimension of Biodiversity: Magnitude, Significance, and Conservation," *Mycological Research*, 95(6):641–655 (1991). W

282.1–16   It has been described . . . in the plasma.] The symbiosis of scale insects, yeasts, and bacteria is described by Paul Buchner, *Endosymbiosis of Animals with Plant Microorganisms* (New York: Interscience Publishers, Wiley, 1965), pp. 271–272. W

283.23–28   In 1990 . . . this century."] Mittermeier is quoted in Philip Shabecoff, "New Species of Monkey Is Found in Populated Area, Brazilians Say," *New York Times*, June 21, 1990.

283.32–286.34   Not even the order Cetacea . . . separate species."] The information on the discovery of new species of whales and porpoises is drawn from W.F.J. Mörzer Bruyns, *Field Guide of Whales and Dolphins* (Amsterdam: C. A. Mees, 1971), and Katherine Ralls and Robert L. Brownell Jr., "A Whale of a New Species," *Nature*, 350:560 (1991). W

289.2–29   One is equitability . . . dependent on it."] A technical but clear account of equitability and other diversity measures of local faunas and floras is given by Anne E. Magurran, *Ecological Diversity and Its Measurement* (Princeton: Princeton University Press, 1988). W

293.1–4   At the very top . . . number:] The counts of kingdoms and phyla of living organisms are based on Lynn Margulis and Karlene V. Schwartz, *Five Kingdoms: An Illustrated Guide to the Phyla of Life on Earth* (San Francisco: Freeman, 1982). A case can be made for distinguishing the Archaebacteria as a sixth kingdom, but no consensus has been reached by systematists. W

293.27–295.11   *A goshawk* . . . number of species.] The natural history of the Black Forest hawks was drawn from Roger Tory Peterson, Guy Montfort, and P.A.D. Hollom, *A Field Guide to the Birds of Britain and Europe*, 2nd ed. (Boston: Houghton Mifflin, 1967). According to Hans Löhrl (personal communication through Ernst Mayr), the goshawk, a threatened species in parts of North America, not only continues to survive in the Black Forest but has increased its numbers, to the point of threatening the capercaillie, the huge grouse-like game bird of the region. W

295.39–298.3   By counting such variations . . . exists on earth] On the measurement of genetic diversity: the allozyme estimates are from Robert K. Selander, "Genetic Variation in Natural Populations," in F. J. Ayala, ed., *Molecular Evolution* (Sunderland: Sinauer, 1976), pp. 21–45; and "Genetic Variation in Natural Populations: Patterns and Theory," *Theoretical Population Biology*, 13(1):121–177 (1978). Other aspects of allozyme research and the newer measures of nucleotide diversity are given by Wen-Hsiung Li and Dan Graur, *Fundamentals of Molecular Evolution* (Sunderland: Sinauer, 1991), and R. K. Selander, Andrew G. Clark, and Thomas S. Whittam, *Evolution at the Molecular Level* (Sunderland: Sinauer, 1991). I am grateful to Russell Lande for providing valuable advice on the estimation of total genetic diversity based on this research. W

300.20–21   The most potent . . . *lutris*).] The role of sea otters as keystone species is detailed in David O. Duggins, "Kelp Beds and Sea Otters: An Experimental Approach," *Ecology*, 61(3):447–453 (1980). W

301.17–33   jaguars and pumas . . . produce these seeds.] The case for jaguars and pumas as keystone species is made by John Terborgh, "The Big Things

That Run the World—A Sequel to E.O. Wilson," *Conservation Biology*, 2(4):402–403 (1988). The estimate of a tenfold increase of coatis and rodents in the absence of jaguars and pumas on Barro Colorado Island is based on a comparison with the fauna at Cocha Cashu, Peru, where big cats still live. W

302.7–32   In a very different way . . . biological diversity.] The keystone role of large African mammals is documented by Norman Owen-Smith, "Megafaunal Extinctions: The Conservation Message from 11,000 Years B.P.," in *Conservation Biology*, 3(4):405–412 (1989). W

302.36–303.29   a colony of driver ants . . . ecosystem to bear.] The account of African driver ants is adapted from my *Success and Dominance in Ecosystems: The Case of the Social Insects* (Oldendorf/Luhe: Ecology Institute, 1990). W

306.8–32   One approach . . . higher up.] The assembly rules of community formation were inferred for the birds of New Guinea by Jared M. Diamond, "Assembly of Species Communities," in *Ecology and Evolution of Communities*, ed. M. L. Cody and J. M. Diamond (Cambridge: Harvard University Press, 1975), pp. 342–444. A critique of Diamond's approach based on statistical analysis is provided by Daniel Simberloff, "Using Island Biogeographic Distributions to Determine If Colonization Is Stochastic," *American Naturalist*, 112:713–726 (1978); "Competition Theory, Hypothesis Testing, and Other Community-Ecology Buzzwords," *American Naturalist*, 122:626–635 (1983). A general review across many groups of microorganisms and animals was recently provided by James A. Drake, "Communities as Assembled Structures: Do Rules Govern Pattern?," *Trends in Ecology and Evolution*, 5(5)159–164 (1990). Drake, who concludes that competition-based assembly rules do in fact exist, also employs the jigsaw-puzzle analogy to describe colonization sequences. Other, generally favorable approaches to the role of competition are provided by the authors of *Community Ecology*, ed. Jared Diamond and Ted J. Case (New York: Harper and Row, 1986). W

307.18–40   Fire ants . . . old haunts.] The fire-ant competition story is told in Hölldobler and Wilson, *The Ants*. W

308.15–309.3   Daphne Major . . . *character displacement*] Compression and release among competing Darwin's finches is described by Peter R. Grant, *Ecology and Evolution of Darwin's Finches* (Princeton: Princeton University Press, 1986). W

309.4–310.37   The classic example . . . avoid hybridization?] Character displacement in Darwin's finches was first suggested by David Lack in his classic *Darwin's Finches* (Cambridge: Cambridge University Press, 1947). It was then documented in convincing detail by Peter Grant in his own 1986 classic, *Ecology and Evolution of Darwin's Finches*. The plier analogy to bill structure was introduced by Robert I. Bowman as part of a detailed analysis in "Morphological Differentiation and Adaptation in the Galápagos Finches," *University of California Publications in Zoology*, 58:1–302 (1961). W

311.5–21 In a celebrated experiment . . . the competitors.] The relation
between predation and species numbers in intertidal mollusks is reported by
Robert T. Paine, "Food Web Complexity and Species Diversity," *American
Naturalist*, 100:65–75 (1966). W

312.19–28 forehead mites . . . skin crawl.] For showing me how to extract
and examine forehead mites, I am grateful to Michael Huben. W

315.17 a food web] For authoritative reviews of food webs see Joel E. Cohen,
Food Webs and Niche Space (Princeton: Princeton University Press, 1978); Joel
E. Cohen, Frédéric Briand, and Charles M. Newman, eds., *Community Food
Webs* (New York: Springer, 1990); and Stuart L. Pimm, John H. Lawton,
and Joel E. Cohen, "Food Web Patterns and Their Consequences," *Nature*,
350:669–674 (1991). W

317.21–32 On the west coast . . . prey of the other.] The strange reciprocal
predation of mosquito larvae and protozoans in the genus *Lambornella* was
reported by Jan O. Washburn et al., "Predator-Induced Trophic Shift of a
Free-Living Ciliate: Parasitism of Mosquito Larvae by Their Prey," *Science*,
240:1193–95 (1988). W

319.21–320.8 microbial mats . . . acidity.] Details of microbial mats and stro-
matolites are given by David J. Des Marais, "Microbial Mats and the Early
Evolution of Life," *Trends in Ecology and Evolution*, 5(5):140–144 (1990); and
lecture materials provided by J. William Schopf in Steve Olson, *Shaping the
Future: Biology and Human Values* (Washington, D.C.: National Academy
Press, 1989). W

320.9–326.15 Since the beginning . . . no bodies] The details of the history
of diversity have been taken from many sources, especially Andrew H. Knoll
and John Bauld, "The Evolution of Ecological Tolerance in Prokaryotes,"
Transactions of the Royal Society of Edinburgh, Earth Sciences, 80:209–223
(1989); lecture materials from J. William Schopf in Olson, *Shaping the Future*;
Philip W. Signor, "The Geologic History of Diversity," *Annual Review of
Ecology and Systematics*, 21:509–539 (1990); and Mark A. S. McMenamin,
"The Emergence of Animals," *Scientific American*, 256:94–102 (April 1987).
Reports of spores of earliest land plants and burrows of invertebrate animals
are by Gregory J. Retallack and Carolyn R. Feakes, "Trace Fossil Evidence
for Late Ordovician Animals on Land," *Science*, 235:61–63 (1987). I have also
benefited from an unpublished manuscript by A. H. Knoll and Heinrich D.
Holland, "Oxygen and Proterozoic Evolution: An Update." W

322.34 "evolutionary progress."] The concept of evolutionary progress given
here was first presented in my *Success and Dominance in Ecosystems: The Case
of the Social Insects* (Oldendorf/Luhe: Ecology Institute, 1990). W

323.15–17 overall average . . . numerous.] The idea of a moving average toward
larger and more complex animals through geological time is documented by
Geerat J. Vermeij, *Evolution and Escalation: An Ecological History of Life*

(Princeton: Princeton University Press, 1987); and by John Tyler Bonner, *The Evolution of Complexity, by Means of Natural Selection* (Princeton: Princeton University Press, 1988). *W*

323.27–28   the adjuration of C. S. Peirce . . . true.] See Peirce's essay "Some Consequences of Four Incapacities," first published in 1868 in *The Journal of Speculative Philosophy*: "Let us not pretend to doubt in philosophy what we do not doubt in our hearts."

324.5–6   As J. William Schopf . . . rusted.] See Schopf's *Cradle of Life: The Discovery of Earth's Earliest Fossils* (1987).

324.25–28   Approximately 540 million . . . history of life.] The age of the Cambrian period and hence of the entire Phanerozoic eon, 550 million years, is the consensus among geologists, according to Simon Conway Morris (personal communication). The connection between the rise of atmospheric oxygen and the origin of macroscopic animals in late Precambrian and early Cambrian times was first proposed as a theoretical model by Preston Cloud. *W*

328.3–7   end of the Paleozoic . . . to an end.] The estimates of extinction rates among marine organisms are based on many studies of the Permian and Triassic fossil record, which are reviewed in D. H. Erwin, "The End-Permian Mass Extinction," *Annual Review of Ecology and Systematics*, 21:69–91 (1990). *W*

328.12   Field of Bullets Scenario] The Field of Bullets scenario is given by David M. Raup, *Extinction: Bad Genes or Bad Luck?* (New York: Norton, 1991). It is derived from his techniques for estimating extinction rates according to taxonomic rank in "Taxonomic Diversity Estimation Using Rarefaction," *Paleobiology*, 1(4):333–342 (1975). I have given the scenario a military twist. *W*

329.10–330.33   Cambrian explosion . . . the Annelida.] An authoritative analysis of the Cambrian explosion in the evolution of marine animals is offered by S. Conway Morris, "Burgess Shale Faunas and the Cambrian Explosion," *Science*, 246:339–346 (1989). The affinities of several problematica, including *Hallucigenia*, to the living phylum Onychophora are suggested by L. Ramsköld and Hou Xianguang, "New Early Cambrian Animal and Onychophoran Affinities of Enigmatic Metazoans," *Nature*, 351:225–228 (1991). The diagnosis of the bizarre *Wiwaxia corrugata* was made by Nicholas A. Butterfield, "A Reassessment of the Enigmatic Burgess Shale Fossil *Wiwaxia corrugata* (Matthew) and Its Relationship to the Polychaete *Canadia spinosa* Walcott," *Paleobiology*, 16(3):287–303 (1990). *W*

331.10–11   continental land masses . . . species formation.] For Signor's summary of the correlation between continental geography and global biodiversity, see his "Geologic History of Diversity." *W*

332.7–9   These inhabited worlds . . . enrichment.] The trend toward enrichment of local faunas and floras is documented by J. John Sepkoski Jr. et al.,

"Phanerozoic Marine Diversity and the Fossil Record," *Nature*, 293:435–437 (1981); and Andrew H. Knoll, "Patterns of Change in Plant Communities through Geological Time," in Jared M. Diamond and Ted J. Case, eds., *Community Ecology* (New York: Harper and Row, 1986), pp. 126–141. W

332.25   latitudinal diversity gradient] Latitudinal species gradients. The numbers of breeding bird species are from Adrian Forsyth, *Portraits of the Rainforest* (Ontario: Camden House, Camden East, 1990). Raymond A. Paynter supplied me with the number of Colombian species. A list of publications documenting the latitudinal gradient in a wide range of plants and animals is given by George C. Stevens, "The Latitudinal Gradient in Geographical Range: How So Many Species Coexist in the Tropics," *American Naturalist*, 133(2):240–256 (1989). W

334.13–30   the vascular plants . . . Labrador.] Estimates of tropical versus temperate plant diversity are given by Peter H. Raven, "The Scope of the Plant Conservation Problem World-Wide," in David Bramwell, Ole Hamann, V. H. Heywood, and Hugh Synge, eds., *Botanic Gardens and the World Conservation Strategy* (New York: Academic Press, 1987), pp. 19–29. Alwyn H. Gentry's count of tree species in Peru, the world record, is given in "Tree Species Richness of Upper Amazonian Forests," *Proceedings of the National Academy of Sciences*, 85:156–159 (1988). Peter S. Ashton's unpublished estimates of Bornean tree diversity were provided in a personal communication. W

334.31–335.6   Butterflies . . . North Africa combined.] The butterfly diversity data from Peru and Brazil are cited by Gerardo Lamas, Robert K. Robbins, and Donald J. Harvey, "A Preliminary Survey of the Butterfly Fauna of Pakitza, Parque Nacional del Manu, Peru, with an Estimate of Its Species Richness," *Publicaciones del Museo de Historia Natural, Universidad Nacional Mayor de San Marcos, serie A Zoologia*, 40:1–19 (1991); and Thomas C. Emmel and George T. Austin, "The Tropical Rain Forest Butterfly Fauna of Rondonia, Brazil: Species Diversity and Conservation," *Tropical Lepidoptera*, 1(1):1–12 (1990). W

335.7–335.18   The ants . . . entire world.] The ants of a single tree in the Peruvian rain forest were analyzed by me in "The Arboreal Ant Fauna of Peruvian Amazon Forests: A First Assessment," *Biotropica*, 19(3):245–251 (1987). Terry L. Erwin estimated the number of beetle species in a Panamanian rain forest in "Tropical Forests: Their Richness in Coleoptera and Other Arthropod Species," *Coleopterist's Bulletin*, 36(1):74–75 (1982). Estimates of beetle diversity in North America and the world are given in Ross H. Arnett Jr., *American Insects: A Handbook of the Insects of America North of Mexico* (New York: Van Nostrand Reinhold, 1985). W

336.24–337.8   David Currie . . . more diversity.] David J. Currie's correlation of species richness of vertebrates and trees in North America with environmental variables is presented in "Energy and Large-Scale Patterns of Animal- and Plant-Species Richness," *American Naturalist*, 137(1):27–49 (1991). W

338.8–16   Rapoport's rule . . . temperate zones.] Rapoport's rule, as George Stevens has called it, was proposed by the Argentine ecologist Eduardo H. Rapoport in *Aerography: Geographical Strategies of Species*, English translation from the 1975 Spanish original (New York: Pergamon, 1982). Stevens himself, however, compiled the published data from many sources that sealed the point. He also made the connection between Rapoport's rule—that temperate species have wider latitudinal distributions—and the need for temperate species to occupy more variable local environments. The narrowing of altitudinal ranges on the sides of tropical mountains by the same effect, an idea essentially the same as Rapoport's rule, was introduced in 1967 by Daniel H. Janzen, "Why Mountain Passes Are Higher in the Tropics," *American Naturalist*, 101:233–249. W

339.2–6   Papua New Guinea . . . nematodes.] The New Guinea weevils with gardens of algae, lichens, and mosses on their backs were discovered by J. Linsley Gressitt, "Epizoic Symbiosis," *Entomological News*, 80(1):1–5 (1969). W

339.20   *Dynamine hoppi*] *Dynamine hoppi* and many other rare and beautiful butterfly species are described by Philip J. DeVries, *The Butterflies of Costa Rica and Their Natural History: Papilionidae, Pieridae, Nymphalidae* (Princeton: Princeton University Press, 1987). W

340.24–341.6   source-sink model . . . in the plot.] The source-sink model is evaluated by H. Ronald Pulliam, "Sources, Sinks, and Population Regulation," *American Naturalist*, 132(5):652–661 (1988). It has been especially well documented in the exhaustive study of tree diversity in Panama by Stephen Hubbell and Robin Foster: "Commonness and Rarity in a Neotropical Forest: Implications for Tropical Tree Conservation," in Michael E. Soulé, ed., *Conservation Biology: The Science of Scarcity and Diversity* (Sunderland: Sinauer, 1986), pp. 205–231. W

341.35–342.25   the epiphytes . . . the insects.] The account of epiphytes is modified from my "Rain Forest Canopy: The High Frontier," *National Geographic*, 180:78–107 (December 1991). W

344.1–15   The floor of the deep sea . . . degree.] The importance of deep-sea animals as support for environmental stability was first pointed out by Howard L. Sanders, "Marine Benthic Diversity: A Comparative Study," *American Naturalist*, 102:243–282 (1968). W

345.2–11   the size of an organism . . . deer.] Analyses of the effect of the size of organisms on biological diversity are given by D. R. Morse et al., "Fractal Dimension of Vegetation and the Distribution of Arthropod Body Lengths," *Nature*, 314:731–733 (1985); and Robert M. May, "How Many Species Are There on Earth?," *Science*, 241:1441–49 (1988). W

345.14–16   In 1959 . . . that length.] G. Evelyn Hutchinson and Robert H. MacArthur proposed the logarithmic rule of increasing biodiversity with decreasing organism size in "A Theoretical Ecological Model of Size

Distributions among Species of Animals," *American Naturalist*, 93:117–125 (1959). W

345.28–346.27 Nature is always . . . fractal dimensions.] The fractal analysis of niche size as a determinant of biodiversity was introduced by Morse et al., "Fractal Dimension." These ecologists measured actual vegetation surfaces to get the differences perceived by organisms of different sizes. W

348.1–19 the plumage of a bird . . . Carolina parakeet.] The feather-mite world in the plumage of parrots is described by Tila M. Pérez and Warren T. Atyeo, "Site Selection of the Feather and Quill Mites of Mexican Parrots," in D. A. Griffiths and C. E. Bowman, eds., *Acarology VI* (Chichester, Eng.: Ellis Horwood, 1984), pp. 563–570. Additional details were kindly provided to me by Tila Pérez in personal communication. W

348.19–23 Carolina parakeet . . . late 1930s.] The final days of the Carolina parakeet are described by Doreen Buscami, "The Last American Parakeet," *Natural History*, 87(4):10–12 (1978). W

348.24–27 Statistical studies . . . other resources.] The most thorough statistical studies of the factors affecting the numbers of animal species were recently conducted by Kenneth P. Dial and John M. Marzluff. See "Are the Smallest Organisms the Most Diverse?," *Ecology*, 69(5):1620–24 (1988); "Nonrandom Diversification within Taxonomic Assemblages," *Systematic Zoology*, 38(1):26–37 (1989); and "Life History Correlates of Taxonomic Diversity," *Ecology*, 72(2):428–439 (1990). W

348.27–349.9 The ultimate exemplars . . . perish without them.] The account of insect diversity and dominance given here is based on my "First Word," *Omni*, 12:6 (September 1990). W

349.10–23 Richard Southwood . . . newcomers.] The reasons for the great variety and ecological importance of insects are assessed by T.R.E. Southwood, "The Components of Diversity," in Laurence A. Mound and Nadia Waloff, eds., *Diversity of Insect Faunas* (London: Blackwell, 1978), pp. 19–40. W

349.37–350.10 Africa was an island . . . everything changed.] The description of adaptive radiation in African mammals is modified from Charles J. Lumsden and Edward O. Wilson, *Promethean Fire* (Cambridge: Harvard University Press, 1983). W

353.4–32 The New Zealand mistletoe . . . very end.] The account of the extinction of the New Zealand mistletoe is based on David A. Norton, "*Trilepidea adamsii*: An Obituary for a Species," *Conservation Biology*, 5(1):52–57 (1991). W

354.22–30 the duration of fish . . . nautiluses.)] The data on extinction rates of marine organisms are from David M. Raup, "Extinction: Bad Genes or Bad Luck?," *Acta geològica hispànica*, 16(1–2):25–33 (1981); and "Evolutionary

Radiations and Exctinction," in H. D. Holland and A. F. Trandall, eds., *Patterns of Change in Evolution* (Berlin: Dahlem Konferenzen, Abakon Verlagsgesellschaft, 1984), pp. 5–14. W

354.34–356.5   The probability of extinction . . . local events.] The approximate constancy of species extinction within a clade—and of clades within larger clades—was documented by Leigh van Valen, "A New Evolutionary Law," *Evolutionary Theory*, 1:1–30 (1973). An updated evaluation of longevity, confirming constancy but with a great many caveats, is provided by Jeffrey Levinton, *Genetics, Paleontology, and Macroevolution* (New York: Cambridge University Press, 1988). W

356.5–11   Clades of buffalos . . . African continent.] The recent history of African buffalos and antelopes, including a mass extinction episode 2.5 million years ago, is detailed by Elisabeth S. Vrba, "African Bovidae: Evolutionary Events since the Miocene," *South African Journal of Science*, 81:263–266 (1985). W

356.20–30   contemporary species formation . . . fifteen years.] Rapid species formation in Andean plants, especially orchids, is argued by Alwyn H. Gentry and Calaway H. Dodson, "Diversity and Biogeography of Neotropical Vascular Epiphytes," *Annals of the Missouri Botanical Garden*, 74:205–233 (1987). W

358.15   Surtsey] The birth of the island Surtsey on November 14, 1963, was followed by the colonization of plants and animals in a manner paralleling that of Krakatau (Chapter 2), though with many fewer species. The history of the island is detailed in Sturla Fridriksson, *Surtsey: Evolution of Life on a Volcanic Island* (New York: Halsted Press, Wiley, 1975). The Icelandic people have witnessed similar episodes many times. The tenth-century poem *Völuspá* transforms the eruptions into the rampages of the fire-giant Surtur the Black: "The hot stars down / from Heaven are whirled. / Fierce grows the steam / and the life-feeding flame. / Till fire leaps high / about Heaven itself." The name Surtsey means island of Surtur. W

358.37–39   This very simple model . . . 1963.] The theory of island biogeography was presented in 1963 by Robert H. MacArthur and Edward O. Wilson, "An Equilibrium Theory of Insular Zoogeography," *Evolution*, 17(4):373–387, and elaborated in our *The Theory of Island Biogeography* (Princeton: Princeton University Press, 1967). There have been many discussions and improvements of the idea, perhaps best presented by Mark Williamson in *Island Populations* (Oxford: Oxford University Press, 1981) and "Natural Extinction on Islands," *Philosophical Transactions of the Royal Society of London*, ser. B, 325:457–468 (1989). The rule that increasing the area of an island tenfold doubles the number of species was first suggested by Philip J. Darlington, *Zoogeography: The Geographical Distribution of Animals* (New York: Wiley, 1957). W

360.33–363.32   In the early 1960s . . . the islets.] The biogeographic experiment on the Florida Keys is reported in Daniel S. Simberloff and Edward O. Wilson, "Experimental Zoogeography of Islands: Defaunation and

Monitoring Techniques," *Ecology*, 50(2):267–278 (1969); and "Experimental Zoogeography of Islands: A Two-Year Record of Colonization," *Ecology*, 51(5):934–937 (1970). The theory of island biogeography, especially in its central proposition of a dynamic equilibrium in species numbers, has been tested by many other experiments using miniature systems, including diatoms suspended on slides in freshwater streams and microorganisms in bottles of water. Studies of turnover in patches of islands of varying areas have also contributed, as well as analyses of the postcatastrophe histories of Krakatau and Surtsey. W

363.37–365.39   In the late 1970s . . . biological reserves.] The early results of the Forest Fragments Project in Brazil are reported in Thomas E. Lovejoy et al., "Ecosystem Decay of Amazon Forest Remnants," in Matthew H. Nitecki, ed., *Extinction* (Chicago: University of Chicago Press, 1984), pp. 295–325; and Lovejoy et al., "Edge and Other Effects of Isolation on Amazon Forest Fragments," in Michael E. Soulé, ed., *Conservation Biology: The Science of Scarcity and Diversity* (Sunderland: Sinauer, 1986), pp. 257–285. The loss of beetle diversity was demonstrated by Bert C. Klein, "Effects of Forest Fragmentation on Dung and Carrion Beetle Communities in Central Amazonia," *Ecology*, 70(6):1715–25 (1989). W

366.20–26   The prediction . . . over time.] The theory of extinction probability, along with data from small British islands testing the theory, is presented in Stuart L. Pimm, H. Lee Jones, and Jared Diamond, "On the Risk of Extinction," *American Naturalist*, 132(6):757–785 (1988). In *The Theory of Island Biogeography* (1967), MacArthur and Wilson provide equations measuring the heavy dependence of the longevity of populations on population size and the birth and death rates of member organisms. W

367.19–369.35   endangered species of birds . . . West Indies.] Details on endangered North American bird species have been drawn from John W. Terborgh, "Preservation of Natural Diversity: The Problem of Extinction Prone Species," *BioScience*, 24(12):715–722 (1974); and *Where Have All the Birds Gone? Essays on the Biology and Conservation of Birds That Migrate to the American Tropics* (Princeton: Princeton University Press, 1989); David S. Wilcove and J. W. Terborgh, "Patterns of Population and Decline in Birds," *American Birds*, 38(1):10–13 (1984); and Russell Lande, "Genetics and Demography in Biological Conservation," *Science*, 241:1455–60 (1988). The diverse properties of rareness in organisms are classified by Deborah Rabinowitz, Sara Cairns, and Theresa Dillon, "Seven Forms of Rarity and Their Frequency in the Flora of the British Isles," in Michael E. Soulé, ed., *Conservation Biology: The Science of Scarcity and Diversity* (Sunderland: Sinauer, 1986), pp. 182–204. W

370.9–15   During the Paleozoic . . . platycerids.] The account of Paleozoic snails dwelling on the anuses of sea lilies is from Steven M. Stanley, "Periodic Mass Extinctions of the Earth's Species," *Bulletin of the American Academy of Arts and Sciences*, 40(8):29–48 (1987). W

372.7–30   Recently I studied ants . . . generalized species.] My study of extinction in ants of the West Indies was presented in "Invasion and Extinction in the West Indian Ant Fauna: Evidence from the Dominican Amber," *Science*, 229:265–267 (1985). W

372.31–374.4   The same trend . . . extinction.] Steven Stanley on the greater longevity of abundant mollusks in the fossil record: "Periodic Mass Extinctions," pp. 34–36. W

374.32–377.17   *inbreeding depression* . . . throughout the population.] The 50–500 rule of minimum population size was introduced by Ian Robert Franklin, "Evolutionary Change in Small Populations," in Michael E. Soulé and Bruce A. Wilcox, eds., *Conservation Biology: An Evolutionary-Ecological Perspective* (Sunderland: Sinauer, 1980), pp. 135–149. The lethal equivalents in the genetic makeup of zoo-animals are analyzed by John W. Senner, "Inbreeding Depression and the Survival of Zoo Populations," in Soulé and Wilcox, *Conservation Biology*, pp. 209–224; and by Katherine Ralls, Jonathan D. Ballou, and Alan Templeton, "Estimates of Lethal Equivalents and the Cost of Inbreeding in Mammals," *Conservation Biology*, 2(2):185–193 (1988). The 50–500 rule is reexamined by Otto Frankel and Michael E. Soulé, *Conservation and Evolution* (Cambridge: Cambridge University Press, 1981), and more critically by Russell Lande, "Genetics and Demography in Biological Conservation," *Science*, 241:1455–60 (1988). W

377.35–378.4   They include . . . New Mexico.] The tiny populations of the Frigate Island darkling beetle and Socorro sowbug are described in *The IUCN Invertebrate Red Data Book* (Old Woking: Unwin Brothers, 1983) and that of the hau kuahiwi tree of Kauai in *Plant Conservation* (Center for Plant Conservation), 3(4):1–8 (1988). W

378.10–33   the species as metapopulation . . . turns dark.] The metapopulation concept, originated by Richard Levins in 1970, has been most recently explored by Isabelle Olivieri et al., "The Genetics of Transient Populations: Research at the Metapopulation Level," *Trends in Ecology and Evolution*, 5(7):207–210 (1990); and in fine detail by authors in *Metapopulation Dynamics: Empirical and Theoretical Investigations*, ed. Michael Gilpin and Ilkka Hanski (New York: Academic Press, 1991), a book reprinted from the *Biological Journal of the Linnean Society*, 42(1–2) (1991). W

378.34–380.27   One metapopulation . . . or die.] Information on the Karner blue butterfly comes from "Minimum Area Requirements for Long-Term Conservation of the Albany Pine Bush and Karner Blue Butterfly: An Assessment," an unpublished report for the state of New York by Thomas J. Givnish, Eric S. Menges, and Dale F. Schweitzer, August 9, 1988; cited by permission of the authors. The Karner blue is one of a few scattered metapopulations classified as the eastern race of the Melissa blue, *Lycaeides melissa*. It was formally described by Vladimir Nabokov, the novelist and distinguished aurelian. W

380.31–382.13 a Brazilian parrot . . . expected.] The final days of Spix's macaw in the wild were reported by Jorgen B. Thomsen and Charles A. Munn, "*Cyanopsitta spixii*: A Non-Recovery Report," *Parrotletter*, 1(1):6–7 (1987) and in a news report, "Lone Macaw Makes a Vain Bid for Survival," *New Scientist*, August 18, 1990. I am indebted to Jorgen Thomsen for additional details on the status of the last surviving male. W

385.5–384.17 Centinela . . . species survived.] I am grateful to Alwyn H. Gentry for supplying me with the history of Centinela. Some of the characteristics of the flora are provided by Gentry in "Endemism in Tropical versus Temperate Plant Communities," in Michael E. Soulé, ed., *Conservation Biology: The Science of Scarcity and Diversity* (Sunderland: Sinauer, 1986), pp. 153–181. A history of deforestation in Ecuador is traced in Calway Dodson and Gentry, "Biological Extinction in Western Ecuador," *Annals of the Missouri Botanical Gardens*, 78(2):273–295 (1991). W

384.22–24 like the dead . . . unused.] See Thomas Gray's "Elegy Written in a Country Churchyard," first published in 1751.

384.26–386.27 During the past ten years . . . its zenith.] Mass extinction of Polynesian birds. The extinction of Hawaiian landbirds by the Polynesian colonists is described in Storrs L. Olson and Helen F. James, "Descriptions of Thirty-Two New Species of Birds from the Hawaiian Islands, Part 1: Non-Passeriformes," *Ornithological Monographs*, 45:1–88 (1991); and "Descriptions of Thirty-Two New Species of Birds from the Hawaiian Islands, Part 2: Passeriformes," *Ornithological Monographs*, 46:1–88 (1991). The destruction of the faunas in other parts of Polynesia are documented by David W. Steadman, "Extinction of Birds in Eastern Polynesia: A Review of the Record and Comparisons with Other Pacific Island Groups," *Journal of Archaeological Science*, 16:177–205 (1989); and Tom Dye and D. W. Steadman, "Polynesian Ancestors and Their Animal World," *American Scientist*, 78:207–215 (1990). The Henderson story is told by Steadman and Olson, "Bird Remains from an Archaeological Site on Henderson Island, South Pacific: Man-Caused Extinctions on an 'Uninhabited' Island," *Proceedings of the National Academy of Sciences*, 82:6191–95 (1985). W

386.31–390.8 In North America . . . toward people.] Ice Age extinctions. The definitive work on extinctions at the end of the last Ice Age, about 11,000 years ago, is the multiauthored *Quaternary Extinctions: A Prehistoric Revolution*, ed. Paul S. Martin and Richard G. Klein (Tucson: University of Arizona Press, 1984). The authors consulted here in order of appearance are David W. Steadman and Paul S. Martin (North American Pleistocene extinctions and late Pleistocene birds), Leslie F. Marcus and Rainer Berger (late Pleistocene megafauna as disclosed at Rancho La Brea), Larry D. Agenbroad (mammoths), Arthur M. Phillips III (ground sloths), C. Vance Haynes (Clovis culture and megafauna extinction), Jared M. Diamond (Iceland's bird fauna), James E. King and Jeffrey J. Saunders (mastodons), S. David Webb

(North American mammalian extinctions for the past 10 million years), and Donald K. Grayson (history of nineteenth-century explanations of Pleistocene extinctions). W

390.9–391.9   Before the coming . . . their own.] The extinction of the moas and other endemic birds on New Zealand is a story told by Michael M. Trotter, Beverley McCulloch, Atholl Anderson, and Richard Cassels, in Martin and Klein, *Quaternary Extinctions*; and more recently again by Anderson in *Prodigious Birds: Moas and Moa-hunting in Prehistoric New Zealand* (New York: Cambridge University Press, 1990). W

391.10–392.13   Madagascar . . . the continent's interior.] The fates of the Madagascan and Australian faunas are described by Robert E. Dewar, Peter Murray, Duncan Merrilees, and D. R. Horton in Martin and Klein, *Quaternary Extinctions*. W

392.14–394.17   In 1989 . . . feasts.] Jared Diamond's case identifying prehistoric man as destroyer of the world's megafauna is an improvement on that developed by Paul Martin and others, with important additions from Diamond's own research on birds of the Pacific region. It is presented in "Quaternary Megafaunal Extinctions: Variations on a Theme by Paganini," *Journal of Archaeological Science*, 16:167–175 (1989). W

394.17–19   As the Mexican . . . meat."] The demise of the imperial woodpecker in Mexico was reported by George Plimpton, "Un gran pedazo de carne," *Audubon Magazine*, 79(6):10–25 (1977). W

394.27   invasion of exotic animals] The origin and impact of exotic species are treated in Harold A. Mooney and James A. Drake, eds., *Ecology of Biological Invasions of North American and Hawaii* (New York: Springer, 1986). W

394.28–395.15   In the United States . . . forty years.] The status of extinct and vulnerable fish species in North America is reviewed by Jack E. Williams et al., "Fishes of North America. Endangered, Threatened, or of Special Concern: 1989," *Fisheries* (American Fisheries Society), 14(6):2–20 (1989); R. R. Miller et al., "Extinctions of North American Fishes During the Past Century," *Fisheries*, 14(6):22–38 (1989); and Jack E. Williams and Robert R. Miller, "Conservation Status of the North American Fish Fauna in Fresh Water," *Journal of Fish Biology*, 37(A):79–85 (1990). I am grateful to Karsten E. Hartel for sharing his unpublished analysis of data pertaining to species decline. W

396.25–397.13   a few anecdotes . . . deforestation continues.] The anecdotes of extinction of birds are based on Jared M. Diamond, "The Present, Past and Future of Human-Caused Extinction," *Philosophical Transactions of the Royal Society of London*, ser. B, 325:469–477 (1989); and John Terborgh, *Where Have All the Birds Gone? Essays on the Biology and Conservation of Birds That Migrate to the American Tropics* (Princeton: Princeton University Press, 1989). W

397.14–398.4   About 20 percent . . . riparian dwellers.] On the high rate of extinction of freshwater fishes, see Diamond, and Walter R. Courtenay Jr. and Peter B. Moyle, "Introduced Fishes, Aquaculture, and the Biodiversity Crisis," *Abstracts, 71st Annual Meeting, American Society of Ichthyologists and Herpetologists*, no pp.; and Irv Kornfield and Kent E. Carpenter, "Cyprinids of Lake Lanao, Philippines: Taxonomic Validity, Evolutionary Rates and Speciation Scenarios," in Anthony A. Echelle and Irv Kornfield, eds., *Evolution of Fish Species Flocks* (Orono: University of Maine Press, 1984). The total of 18 species accepted in the classical accounts of the Lake Lanao cyprinid species flock may be excessive, even though the Maranao people of the region recognize all of them. Some of the species may instead be morphs of very plastic species, as I described for the Mexican cichlid and arctic char in Chapter 7. However the matter is judged taxonomically, the adaptive radiation of the Lanao cyprinids is extreme for a single lake, and it has been almost completely erased during the past fifty years. The fate of the Lake Victoria fishes is described by Christopher G. Barlow and Allan Lisle, "Biology of the Nile Perch *Lates niloticus* (Pisces: Centropomidae) with Reference to Its Proposed Role as a Sport Fish in Australia," *Biological Conservation*, 39(4):269–289 (1987); Daniel J. Miller, "Introductions and Extinction of Fish in the African Great Lakes," *Trends in Ecology and Evolution*, 4(2):56–59 (1989); and C.D.N. Barel et al., "The Haplochromine Cichlids in Lake Victoria: An Assessment of Biological and Fisheries Interests," in M.H.A. Keenleyside, ed., *Cichlid Fishes: Behaviour, Ecology and Evolution* (London: Chapman and Hall, 1991), pp. 258–279. W

398.5–23   The United States . . . Coosa rivers.] The decline of freshwater mollusks is documented in *The IUCN Invertebrate Red Data Book* (Gland, Switzerland: International Union for Conservation of Nature and Natural Resources, 1983). W

398.27–399.4   The fate of the tree snails . . . endangered.] The Moorean tree snails have been the subject of classic studies of microevolution by Henry E. Crampton and Bryan C. Clarke. The snails' total destruction in the wild is described by James Murray, Elizabeth Murray, Michael S. Johnson, and Bryan Clarke, "The Extinction of *Partula* on Moorea," *Pacific Science*, 42(3,4):150–153 (1988); I am grateful to Bryan Clarke for supplying additional unpublished details on the episode. The loss of the Hawaiian tree snails is documented in *The IUCN Invertebrate Red Data Book* (1983). W

399.5–16   A recent survey . . . Miami.] The threatened plant species of the United States are counted by Linda R. McMahan, "CPC Survey Reveals 680 Native U.S. Plants May Become Extinct within 10 Years," *Plant Conservation* (Center for Plant Conservation), 3(4):1–2 (1988). The species already extinct were tabulated by Michael O'Neal and other CPC staff members in 1992 (personal communication). The account of the Puerto Rican endemic *Banara vanderbiltii* is based on John Popenoe, "One of the World's Rarest Species," *Plant Conservation*, 3(4):6 (1988). W

399.17–21   In western Germany . . . insect species.] The numbers of threatened and endangered invertebrate species of Europe were reported by Eladio Fernandez-Galiano in *IUCN Special Report Bulletin* (International Union for Conservation of Nature and Natural Resources), 18(7–9):7 (1987). In 1989, 501 insect species were listed as threatened under provisions of the U.S. Endangered Species Act. This represents only about 1 percent of the total known fauna, but it is also a gross underestimate owing to the poor state of taxonomic knowledge in all but a few groups. W

399.22–31   The fungi . . . find out.] The decline of European fungi is reviewed by John Jaenike, "Mass Extinction of European Fungi," *Trends in Ecology and Evolution*, 6(6):174–175 (1991). Similar studies have not yet been undertaken in North America. W

399.39–400.18   the case of the spotted owl . . . timber yield?"] The case of the northern spotted owl is discussed by Russell Lande, "Demographic Models of the Northern Spotted Owl (*Strix occidentalis caurina*)," *Oecologia*, 75(4):601–607 (1988), and "Genetics and Demography in Biological Conservation," *Science*, 241:1455–60 (1988). W

400.22–24   Among them . . . salamanders.] Rare frogs and salamanders of the Pacific Northwest forests are described by Hartwell H. Welsh Jr., "Relictual Amphibians and Old-Growth Forests," *Conservation Biology*, 4(3):308–319 (1990). W

401.20   a growing list of entire ecosystems] A catalogue of threatened and endangered habitats is provided in *The IUCN Invertebrate Red Data Book* (1983). W

402.27   "hot spots"] Norman Myers' eighteen hot spots were identified in two articles, "Threatened Biotas: 'Hot Spots' in Tropical Forests," *Environmentalist*, 8(3):187–208 (1988); and "The Biodiversity Challenge: Expanded Hot-Spots Analysis," *Environmentalist*, 10(4):243–256 (1990). W

406.11–12   Daniel Janzen . . . "living dead."] See "The Future of Tropical Ecology," *Annual Review of Ecology and Systematics* 17 (1986):305–24.

406.28–408.8   *Atlantic coast* . . . around it.] The present condition of the Brazilian Atlantic forest is detailed in Mark Collins, ed., *The Last Rain Forests: A World Conservation Atlas* (New York: Oxford University Press, 1990). This beautifully illustrated book, containing maps of the former and present extent of all the major tropical forests, is the best popular reference work of its kind. W

412.19–20   tropical deciduous forests] Among the concerns of ecologists, tropical deciduous forests have stood in the shadow of the rain forests, but they are in even greater peril. Because they occupy potentially prime agricultural and cattle land and are easily cleared, they are among the most heavily exploited of the world's land environments. In Central America they have been reduced to less than 10 percent of the original cover. Tropical deciduous forests are intermediate between rain forests and temperate deciduous forests in amount of diversity. A review is presented by Manuel Lerdau, Julie Whitbeck, and N. Michele Holbrook, "Tropical Deciduous Forest: Death of a Biome," *Trends in Ecology and Evolution*, 6(7):201–233 (1991). *W*

412.22–413.38 the coral reefs . . . warming trend.] The reduction of coral reefs by both natural and human-caused stress is reported in "Coral Reefs off 20 Countries Face Assaults from Man and Nature," *New York Times*, March 27, 1990; Peter W. Glynn, "Coral Reef Bleaching in the 1980s and Possible Connections with Global Warming," *Trends in Ecology and Evolution*, 6(6):175–179 (1991); and Leslie Roberts, "Greenhouse Role in Reef Stress Unproven," *Science*, 253:258–259 (1991). *W*

413.39–414.27 But the long-term . . . risk extinction.] The effects of climatic warming on biodiversity are predicted by Robert L. Peters and Joan D. S. Darling, "The Greenhouse Effect and Nature Reserves," *BioScience*, 35(11):707–717 (1985); Andy Dobson, Alison Jolly, and Dan Rubenstein, "The Greenhouse Effect and Biological Diversity," *Trends in Ecology and Evolution*, 4(3):64–68 (1989); and Robert L. Peters and Thomas E. Lovejoy, eds., *Global Warming and Biological Diversity* (New Haven: Yale University Press, 1992). The account given here is drawn from these sources and from my "Threats to Biodiversity," *Scientific American*, 260(9):108–116 (1989). *W*

414.28–37 In another arena . . . the sea.] The expected impact of the rise in sea level on biodiversity is examined by Walter V. Reid and Mark C. Trexler, *Drowning the National Heritage: Climate Change and U.S. Coastal Biodiversity* (Washington, D.C.: World Resources Institution, 1991). *W*

415.5–9 Our species appropriates . . . other species] The estimate of energy appropriated on the land by people was made by Peter M. Vitousek, Paul R. Ehrlich, Anne H. Ehrlich, and Pamela A. Matson, "Human Appropriation of the Products of Photosynthesis," *BioScience*, 36(6):368–373 (1986). The measure used by these authors was net primary production, the amount of energy left after subtracting the respiration of primary producers (mostly plants) from the total amount of energy (mostly solar) that is fixed biologically. The appropriation includes consumption of food, fiber, and timber; the productivity of all the land devoted exclusively to human needs, such as croplands (in addition to crops actually eaten); land burned over for clearing; and land devoted to dwellings or reduced to unproductive wastelands by overuse. The human appropriation of marine production remains relatively small. The relation of body size to population density and energy consumption among animal species is analyzed by James H. Brown and Brian A. Maurer, "Macroecology: The Division of Food and Space among Species on Continents," *Science*, 243:1145–50 (1989). *W*

415.14–18 In 1950 . . . 8 billion.] Global population trends were taken from *The Economist Book of Vital World Statistics* (New York: Times Books, 1990). *W*

415.32–417.9   the tropical rain forests . . . human intervention.] The account of the fragility of tropical rain forests is drawn from my "The Current State of Biological Diversity," in E. O. Wilson and F. M. Peter, eds., *Biodiversity* (Washington, D.C.: National Academy Press, 1988), pp. 3–18; from Christopher Uhl, "Restoration of Degraded Lands in the Amazon Basin," ibid., pp. 326–332; and from T. C. Whitmore, "Tropical Forest Nutrients: Where Do We Stand? A *Tour de Horizon*," in J. Proctor, ed., *Mineral Nutrients in Tropical Forest and Savanna Ecosystems* (Boston: Blackwell Scientific Publications, 1990), pp. 1–13. W

418.1–22   About 50,000 . . . the Atlantic.] Accounts of the record 1987 destruction of Amazonian forest are given by Mac Margolis, "Thousands of Amazon Acres Burning," *Washington Post*, September 8, 1988; Marlise Simons, "Vast Amazon Fires, Man-Made, Linked to Global Warming," *New York Times*, August 12, 1988; and "Amazon Holocaust: Forest Destruction in Brazil, 1987–88," *Briefing Paper*, Friends of the Earth (London, 1988). W

418.23–32   By 1989 . . . every year.] The estimates of annual tropical deforestation rates in 1989 were taken from the report by Norman Myers, *Deforestation Rates in Tropical Forests and Their Climatic Implications* (London: Friends of the Earth, 1989). They are based on data assembled country by country. Myers provides a summary of his study in "Tropical Deforestation: The Latest Situation," *BioScience*, 41(5):282 (1991). He defines tropical moist forests, roughly equated with tropical rain forests, as "evergreen or partly evergreen forests, in areas receiving not less than 100 mm of precipitation in any month for two out of three years, with mean annual temperature of 24-plus degrees Celsius, and essentially frost-free; in these forests some trees may be deciduous; the forests usually occur at altitudes below 1300 metres (though often in Amazonia up to 1,800 metres and generally in South-east Asia up to only 750 metres); and in mature examples of these forests, there are several more or less distinctive strata." In late 1991 the Food and Agriculture Organization of the United Nations released a preliminary report ("Second Interim Report on the State of Tropical Forests") that independently conforms to the assessment by Myers. The authors estimate that in 1981–1990 tropical forests were being removed at the rate of 170,000 square kilometers per year. The figure is 20 percent higher than Myers', but the FAO measurements included removal of thinner forests than those considered by Myers, as well as high bamboo stands. More precisely, forests were defined as collections of trees or bamboos with a minimum of 10 percent crown cover associated with wild floras and faunas and relatively undisturbed soil conditions. The extent of prehistoric forest cover is reviewed in Peter H. Raven, "The Scope of the Plant Conservation Problem World-Wide," in David Bramwell, Ole Hamann, V. H. Heywood, and Hugh Synge, eds., *Botanic Gardens and the World Conservation Strategy* (New York: Academic Press, 1987), pp. 20–29. The history of estimation of tropical deforestation rates from the 1970s to Myers' 1989 report is evaluated by J. A. Sayer and T. C. Whitmore, "Tropical Moist Forests: Destruction and Species Extinction," *Biological Conservation*,

55(2):199–213 (1991). They conclude that deforestation grew worse during the 1980s. They doubt that extinction was greatly increased as a result, but they make no reference to many of the data and models in the literature. W

419.8 z is what counts.] A comprehensive review of the large number of z values collected from faunas and floras around the world is provided in Mark Williamson, *Island Populations* (New York: Oxford University Press, 1981). W

419.28–420.22 In 1989 . . . mounts steeply.] Species extinction from loss of rain forest: projections similar to the ones I have made globally were obtained independently by Daniel S. Simberloff for plants and birds in the American tropics, "Are We on the Verge of a Mass Extinction in Tropical Rain Forests?," in David K. Elliott, ed., *Dynamics of Extinction* (New York: Wiley, 1986), pp. 165–180. Simberloff projects that with a halving of the original rain forest, expected by the end of this century (parallel to but not the same as cutting in half the amount left at this moment), 15 percent of the plant species—about 13,600 in all—will become extinct. If forests are saved only in existing parks and reserves, 66 percent will suffer extinction. For birds of the Amazon Basin, the figures are 12 and 70 percent respectively. W

420.36–38 When Cebu . . . joining them.] The extinction of the birds of Cebu is cited by Jared Diamond, "Playing Dice with Megadeath," *Discover*, April 1990, pp. 55–59. W

422.17–423.4 the exponential-decay model . . . originally present.] The use of land-bridge islands to estimate rates of species extinction was introduced by Jared Diamond, "Biogeographic Kinetics: Estimation of Relaxation Times for Avifaunas of Southwest Pacific Islands," *Proceedings of the National Academy of Sciences*, 69:3199–03 (1972), and "'Normal' Extinctions of Isolated Populations," in Matthew H. Nitecki, ed., *Extinction* (Chicago: University of Chicago Press, 1984), pp. 191–246; and by John Terborgh, "Preservation of Natural Diversity: The Problem of Extinction-Prone Species," *BioScience*, 24(12):715–722 (1974). The exponential-decay function in the decline of species is an assumption not yet proved, since extinction rates are difficult to track on single islands: Stanley H. Faeth and Edward F. Connor, "Supersaturated and Relaxing Island Faunas: A Critique of the Species-Age Relationship," *Journal of Biogeography*, 6(4):311–316 (1979). W

423.11–28 patches of forest . . . broken into fragments.] Bird extinction in isolated patches of Brazilian subtropical forest was reported by Edwin O. Willis, "The Composition of Avian Communities in Remanescent Woodlots in Southern Brazil," *Papéis avulsos de zoologia*, 33(1):1–25 (1979). The parallel study in the Bogor Botanical Garden was described by Jared M. Diamond, K. David Bishop, and S. Van Balen, "Bird Survival in an Isolated Javan Woodland: Island or Mirror?," *Conservation Biology*, 1(2):132–142 (1987). The decline of the bird fauna of southwestern Australia's wheatland was reported by D. A. Saunders, "Changes in the Avifauna of a Region, District and Remnant as a Result of Fragmentation of Native Vegetation: The Wheatbelt of Western Australia," *Biological Conservation*, 50(1–4):99–135 (1989). W

425.4–14   the maize species *Zea diploperennis*... machete and fire.] The discovery of a new species of perennial maize is reported by Hugh H. Iltis, John F. Doebley, Rafael Guzmán, and Batia Pazy, *"Zea diploperennis* (Gramineae): A New Teosinte from Mexico," *Science*, 203:186–188 (1979). The site of the wild population of perennial maize, together with surrounding land, totalling 139,000 hectares, has been set aside as the Sierra de Manantlán Biosphere Reserve by the Mexican government, specifically to protect the maize and other wild-crop relatives. It will also save many other native plant species as well as animals including ocelots and jaguars. W

427.7–13   An example ... cleared for agriculture.] The status of the Catharanthus periwinkles of Madagascar is described in Mark Plotkin et al., *Ethnobotany in Madagascar: Overview, Action Plan, Database* (Gland: International Union for Conservation of Nature and Natural Resources and World Wide Fund for Nature, 1985). Other details, including a discussion of the general promise of medicinal alkaloids, are provided in Thomas Eisner, "Prospecting for Nature's Chemical Riches," *Issues in Science and Technology*, 6(2):31–34 (1990). The alkaloid products of the rosy periwinkle have the following clinical record: vinblastine increases the ten-year survival rate for Hodgkin's disease from 2 percent to 58 percent, and vincristine increases the ten-year survival rate from 20 to 80 percent. The drugs are also effective against some other cancers, including Wilms' tumor, primary brain tumors, and testicular, cervical, and breast cancers. See Margery L. Oldfield, *The Value of Conserving Genetic Resources* (Sunderland: Sinauer, 1989). W

427.32–428.2   how much we already ... organism-derived.] Information on the natural origins of medicines used in the United States is provided in Chris Hails, *The Importance of Biological Diversity* (Gland: World Wild Fund for Nature, 1989). W

428.2–9   Yet these materials ... antidiuretic.] An authoritative account of pharmaceuticals harvested from plants, including a complete list of the 119 substances used in pure form, is provided by Norman R. Farnsworth, "Screening Plants for New Medicines," in E. O. Wilson and F. M. Peter, eds., *Biodiversity* (Washington, D.C.: National Academy Press, 1988), pp. 83–97. Additional perspectives are provided by D. D. Soejarto and N. R. Farnsworth, "Tropical Rain Forests: Potential Source of New Drugs?," *Perspectives in Biology and Medicine*, 32(2):244–256 (1989). W

428.12–24   The neem tree ... right."] The properties of the neem tree are described in Noel D. Vietmeyer, ed., *Neem: A Tree for Solving Global Problems* (Washington, D.C.: National Academy Press, 1992). W

428.37–430.9   The leech ... pit viper.] An account of leeches and the anticoagulant they produce is given by Paul S. Wachtel, "Return of the Bloodsucker," *International Wildlife*, September 1987, pp. 44–46. A news report of the new anticoagulants from vampire bats and pit vipers was published in *Science*, 253:621 (1991). W

430.23–432.12   a brief list . . . Anticancer] The list of pharmaceuticals derived from plants and fungi is drawn from Hails, *The Importance of Biological Diversity*; D. D. Soejarto and N. R. Farnsworth, "Tropical Rain Forests: Potential Source of New Drugs?," *Perspectives in Biology and Medicine*, 32(2):244–256 (1989); and Margery L. Oldfield, *The Value of Conserving Genetic Resources* (Sunderland: Sinauer, 1989). An impressive number of Amerindian natural products, few of which have been investigated to date, are described by Richard E. Schultes and Robert F. Raffauf, *The Healing Forest: Medicinal and Toxic Plants of the Northwest Amazonia* (Portland: Dioscorides Press, 1990). W

432.13–436.4   The same bright prospect . . . tropical countries.] The examples of food and forage plant species in early stages of economic development are taken in part from the much-esteemed "green book," *Underexploited Tropical Plants with Promising Economic Value*, published by the National Academy Press in 1975. This work is part of a series sponsored by the U.S. National Academy of Sciences under the direction of the Board on Science and Technology for International Development (BOSTID). Other studies in the series are *Tropical Legumes: Resources for the Future* (1979), *The Winged Bean: A High-protein Crop for the Tropics*, 2nd ed. (1981), *Amaranth: Modern Prospects for an Ancient Crop* (1983), and *Lost Crops of the Incas* (1989). Equally useful semitechnical reviews are found in Margery L. Oldfield, *The Value of Conserving Genetic Resources* (Sunderland: Sinauer, 1989), and Noel D. Vietmeyer, "Lesser-Known Plants of Potential Use in Agriculture and Forestry," *Science*, 232:1379–84 (1986). The best popular introductions, both influential in the development of this important subject, are Norman Myers, *A Wealth of Wild Species: Storehouse for Human Welfare* (Boulder: Westview Press, 1983), and the booklet compiled by Myers, *The Wild Supermarket* (Gland: World Wide Fund for Nature, 1990). W

436.5–29   From the mostly . . . consume them.] The potential of wild plant and animal species is detailed in the previously cited studies by Margery Oldfield, Norman Myers, and the authors in *Biodiversity*, as well as in Hails, *The Importance of Biological Diversity*. Inca agriculture is described in Hugh Popenoe, Noel D. Vietmeyer, and a panel of coauthors, *Lost Crops of the Incas* (Washington, D.C.: National Academy Press, 1989). W

436.30–438.11   Another premier native crop . . . amaranth.] The history of amaranth as an Amerindian crop is told by Jean L. Marx. "Amaranth: A Comeback for the Food of the Americas?," *Science*, 198:40 (1977). W

438.22–32   The Amazonian babassu palm . . . wild plant.] The stellar qualities of the babassu palm are detailed by Anthony B. Anderson, Peter H. May, and Michael J. Balick, *The Subsidy from Nature: Palm Forests, Peasantry, and Development on an Amazon Frontier* (New York: Columbia University Press, 1991). W

438.33–439.9 Another frontier . . . the world.] The promise of salt-tolerant plants is explored in two publications of the National Academy Press, prepared under the direction of the Board on Science and Technology for International Development: *Underexploited Tropical Plants with Promising Economic Value* (1975) and *Saline Agriculture: Salt-tolerant Plants for Developing Countries* (1990). An evaluation of the latter is provided by Susan Turner-Lewis, *National Research Council News Report*, May 1990, pp. 2–4. W

439.20–40 A good example . . . disastrous result.] The status and economic potential of the *Podocnemis* river turtles is described by Russell A. Mittermeier, "South American River Turtles: Saving Their Future," *Oryx*, 14(3):222–230 (1978). W

440.1–441.22 Similar advantages . . . practiced.] Chris Wille and Diane Jukofsky wrote on the green iguana in "Savory 'Chicken of the Trees' Could Play a Role in Saving Forests," *Canopy* (Rainforest Alliance), Summer 1991, p. 7. Dagmar Werner, who cheerfully calls himself the Iguana Mama, has provided a technical report on breeding and marketing the species, "The Rational Use of Green Iguanas," in J. G. Robinson and K. H. Redford, eds., *Neotropical Wildlife Use and Conservation* (Chicago: University of Chicago Press, 1991), pp. 181–201. W

441.22–442.32 Here in summary . . . profitably ranched.] The descriptions of wild animal species as potential food sources are based on *Little-known Asian Animals with a Promising Economic Future*, ed. Noel D. Vietmeyer (Washington, D.C.: National Academy Press, 1983), Oldfield, *The Value of Conserving Genetic Resources; Neotropical Wildlife Use and Conservation*, eds. John G. Robinson and Kent H. Redford (Chicago: University of Chicago Press, 1991); and *Microlivestock*, ed. Noel D. Vietmeyer (Washington, D.C.: National Academy Press, 1991). W

443.25–444.26 aquaculture . . . utilized diversity.] The account of aquaculture is based on Myers, *A Wealth of Wild Species.* W

444.29–445.6 The forests of the world . . . harvester.] New sources of pulp are recounted in Myers, *A Wealth of Wild Species.* W

445.7–15 Pulp . . . need for fertilizer.] Wood grass is described by Sinyan Shen in "Biological Engineering for Sustainable Biomass Production," in Wilson and Peter, *Biodiversity*, pp. 377–389. W

445.24–447.14 When they also learned . . . the industry.] The history of wild relatives and genetic diversity of crop plants is based principally on Erich Hoyt, *Conserving the Wild Relatives of Crops* (Rome and Gland: International Board for Plant Genetic Resources, International Union for Conservation of Nature and Natural Resources, and World Wide Fund for Nature, 1988); Hails, *The Importance of Biological Diversity*; Cary Fowler and Pat Mooney, *Shattering: Food, Politics, and the Loss of Genetic Diversity* (Tucson: University of Arizona Press, 1990); and "Bad Seed," a review of the Fowler-Mooney book by Ann Misch in *World-Watch*, 4(4):39–40 (1991). W

448.22–31 Thomas Eisner . . . other species.] The metaphor of a species as a loose-leaf book was used by Thomas Eisner, "Chemical Ecology and Genetic Engineering: The Prospects for Plant Protection and the Need for Plant Habitat Conservation," *Symposium on Tropical Biology and Agriculture* (St. Louis: Monsanto Company, July 15, 1985). W

449.23–451.1 In 1989 . . . of the soil.] The potential economic yield of Amazon rain forests is provided by Charles M. Peters, Alwyn H. Gentry, and Robert O. Mendelsohn, "Valuation of an Amazonian Rainforest," *Nature*, 339:655–656 (1989). The detailed ledger is from Charles M. Peters as quoted in the *New York Times*, July 4, 1989. W

451.8–14 Economists . . . Mishana tract.] Key contributions to the new interdisciplinary field of ecological economics include Herman E. Daly, *Steady-State Economics* (San Francisco: Freeman, 1977), and most recently, Robert Constanza, ed., *Ecological Economics: The Science and Management of Sustainability* (New York: Columbia University Press, 1991). An evaluation of the field from an environmentalist's perspective is provided by David W. Orr, "The Economics of Conservation," *Conservation Biology*, 5(4):439–441 (1991). A new journal devoted to the subject, *Ecological Economics*, was started by Elsevier (New York) in 1989. A related journal, *Ecological Engineering*, was inaugurated by the same publishers in 1992. W

451.15–26 "ecotourism." . . . save Rwanda.] Ecotourism is analyzed by Elizabeth Boo, *Ecotourism: The Potentials and Pitfalls* (Washington, D.C.: World Wildlife Fund, 1990). I am grateful to Gary Hartshorn and James Hirsch for information on ecotourism income in Costa Rica and to Elizabeth Boo for the most recent report from Rwanda. According to Hirsch, countryside ecotourism accounted for 7 percent, or $20 million, of the $275 million spent by visitors to Costa Rica in 1990. W

451.30–452.2 ecosystem services . . . agricultural heartland.] The possible consequences of deforestation of the Amazon forest on the region's climate was examined by J. Shukla, C. Nobre, and P. Sellers, "Amazon Deforestation and Climate Change," *Science*, 247:1322–25 (1990). W

452.3–21 When forests . . . slowed.] The role of tropical deforestation in the buildup of atmospheric carbon dioxide has been analyzed by many authors; the sources used here are Richard A. Houghton and George M. Woodwell, "Global Climatic Change," *Scientific American*, 260(4):36–44 (April 1989), and R. A. Houghton, "Emission of Greenhouse Gases," in Myers, *Deforestation Rates in Tropical Forests*, pp. 53–62. W

452.22–30 The very soils . . . croplands alike.] The genesis of soils by living organisms is described in Paul R. and Anne H. Ehrlich, *Healing the Planet:*

Strategies for Resolving the Environmental Crisis (Reading: Addison-Wesley, 1991). W

453.11–23 From a few key studies . . . fail.] The evidence for the role of biodiversity in the conservation and circulation of nutrients in forests is reviewed by Ariel E. Lugo, "Diversity of Tropical Species: Questions that Elude Answers," *Biology International* (International Union of Biological Sciences, Paris), special issue no. 19, 39 pp. (1988). W

454.11–455.6 "option value" . . . red ink.] Bryan G. Norton's assessment of the option value of species is given in "Commodity, Amenity, and Morality: The Limits of Quantification in Valuing Biodiversity," in Wilson and Peter, eds., *Biodiversity*, pp. 200–205. General aspects of economic analysis are explained by other authors in the same volume, including Nyle C. Brady, J. William Burley, Robert J. A. Goodland, and John Spears. They are also treated by Harold J. Morowitz, "Balancing Species Preservation and Economic Considerations," *Science*, 253:752–754 (1991). W

456.2 *Resolution*] In thinking about economic and moral foundations of conservation, I have been informed by the writings of ethical philosophers, including David Ehrenfeld, *The Arrogance of Humanism* (New York: Oxford University Press, 1978); Bryan Norton, "Commodity," and *Why Preserve Natural Variety?* (Princeton: Princeton University Press, 1987); Peter Singer, *The Expanding Circle: Ethics and Sociobiology* (New York: Farrar, Straus and Giroux, 1981); Holmes Rolston III, *Philosophy Gone Wild: Essays in Environmental Ethics* (Buffalo: Prometheus Books, 1986), and *Environmental Ethics: Duties to and Values in the Natural World* (Philadelphia: Temple University Press, 1988); Alan Randall, "The Value of Biodiversity," *Ambio*, 20(2):64–68 (1991); and the authors of *The Preservation of Species: The Value of Biological Diversity*, ed. Bryan G. Norton (Princeton: Princeton University Press, 1986). W

456.33–457.9 What is urgently . . . still unwritten.] The discussion of the conservation ethic is based in part on my *Biophilia* (Cambridge: Harvard University Press, 1984). The general definition of *ethic* comes from Aldo Leopold, *A Sand County Almanac and Sketches Here and There* (New York: Oxford University Press, 1949). W

457.17–20 a new discipline . . . humanity.] The definition of biodiversity studies given here and a discussion of its ramifications were presented in Paul R. Ehrlich and Edward O. Wilson, "Biodiversity Studies: Science and Policy," *Science*, 253:758–762 (1991). W

458.18–459.35 Three levels . . . in turn.] The three-level approach to surveying global biodiversity was developed in collaboration with Peter H. Raven. W

458.18 the RAP approach] The RAP search for hot spots is described by Sarah Pollock, "Biological SWAT Team Ranks for Diversity, Endemism," *Pacific Discovery*, 44(3):6–7 (1991). W

460.3–10   The prototype . . . economy.] An account of INBio, Costa Rica's National Institute of Biodiversity, is provided by Laura Tangley, "Cataloging Costa Rica's Diversity," *BioScience*, 40(9):633–636 (1990); and by Daniel H. Janzen, one of INBio's architects, in "How to Save Tropical Biodiversity," *American Entomologist*, 37(3):159–171 (1991). An equivalent institute for the United States is included in the National Biological Diversity Conservation and Environmental Research Act, which as of February 1992 remains to be passed by Congress. W

460.34–39   A working technology . . . *gap analysis.*] The use of Geographic Information Systems to map ecosystems is described by J. Michael Scott et al., "Species Richness: A Geographic Approach to Protecting Future Biological Diversity," *BioScience*, 37(11):782–788 (1987). On a much broader scale, essentially the same method has been applied by Eric Dinerstein and Eric D. Wikramanayake to assess reserves and parks in Asia and the western Pacific, in "Beyond 'Hotspots': How to Prioritize Investments in Biodiversity in the Indo-Pacific Region," *Conservation Biology*, 7(1):53–65 (1993). Techniques for mapping endangered species are given by many authors in Larry E. Morse and Mary Sue Henifin, eds., *Rare Plant Conservation: Geographical Data Organization* (New York: New York Botanical Garden, 1981). W

461.11–18   In the expanded . . . species and races.] The employment of landscape design to enhance biodiversity has been widely discussed. Summaries of key topics are provided in separate chapters by Bryn H. Green, Larry D. Harris (with John F. Eisenberg), and David Western, in Western and Mary C. Pearl, eds., *Conservation for the Twenty-First Century* (New York: Oxford University Press, 1989). W

461.19   "bioregions,"] The concept of bioregions, dating back to the 1800s and developed in modern form by Raymond F. Dasmann, Peter Berg, Charles H. W. Foster, and others, is reviewed by C.H.W. Foster, *Experiments in Bioregionalism: The New England River Basins Story* (Hanover: University Press of New England, 1984), and "Bioregionalism," *Renewable Resources Journal*, 4(3):12–14 (1986). W

463.7–31   Systematics . . . full-time basis.] The shortage of systematists is cited in my "The Biological Diversity Crisis: A Challenge to Science," *Issues in Science and Technology*, 2(1):20–29 (1985), and "Time to Revive Systematics," *Science*, 230:1227 (1985). W

464.37–465.3   GenBank . . . methods.] The progress of GenBank in recording DNA and RNA sequences is described by Christian Burks et al., in Russell F. Doolittle, ed., *Molecular Evolution: Computer Analysis of Protein and Nucleic Acid Sequences* (New York: Academic Press, 1990), pp. 3–22. W

465.34–37   As the Senegalese . . . taught."] Baba Dioum on knowledge and conservation is quoted by John Hopkins, "Preserving Native Biodiversity," Sierra Club special publication (San Francisco, 1991). W

465.38–466.15   A key enterprise . . . human health.] The concept of chemical prospecting was developed by Thomas Eisner during the late 1980s and presented in "Prospecting for Nature's Chemical Riches," *Issues in Science and Technology*, 6(2):31–34 (1990); and "Chemical Prospecting: A Proposal for Action," in F. Herbert Bormann and Stephen R. Kellert, eds., *Ecology, Economics, Ethics: The Broken Circle* (New Haven: Yale University Press, 1991), pp. 196–202. W

466.20–36   In 1991 . . . entire investment.] The 1991 agreement between Merck and Costa Rica's National Institute of Biodiversity was reported by William Booth, "U.S. Drug Firm Signs Up to Farm Tropical Forests," *Washington Post*, September 21, 1991. The cyclical nature of investment in natural products is described by Deborah Hay, "Pharmaceutical Industry's Renewed Interest in Plants Could Sow Seeds of Rainforest Protection," *The Canopy* (Rainforest Alliance), Spring 1991, pp. 1, 7. The use of wild species as sources of medicine is reviewed by Norman R. Farnsworth, "Screening Plants for New Medicines," in E. O. Wilson and F. M. Peter, eds., *Biodiversity* (Washington, D.C.: National Academy Press, 1988), pp. 83–97. W

467.21–468.14   the lore and traditional medicine . . . their cultures.] The data on pharmaceuticals discovered from folkloric medicine are reported in Farnsworth, "Screening Plants." Excellent brief accounts of traditional knowledge and the endangered status of indigenous people who possess it are given by Mark J. Plotkin, "The Outlook for New Agricultural and Industrial Products from the Tropics," in Wilson and Peter, *Biodiversity*, pp. 106–116, and by Eugene Linden, "Lost Tribes, Lost Knowledge," *Time*, September 23, 1991, pp. 46–56. The citation of Chinese traditional medicine was provided by Peter H. Raven (personal communication) and, for artemisinin, by Daniel L. Klayman, "*Qinghaosu* (Artemisinin): An Antimalarial Drug from China," *Science*, 228:1049–55 (1985), and by Xuan-De Luo and Chia-Chiang Shen, "The Chemistry, Pharmacology, and Clinical Applications of Qinghaosu (Artemisinin) and Its Derivatives," *Medicinal Research Reviews*, 7(1):29–52 (1987). W

468.25–33   A successful prototype . . . conservation.] The operations of the Tropical Agricultural Research and Training Center in Costa Rica are described by Laura Tangley in "Fighting Central America's Other War," *BioScience*, 37(11):772–777 (1987). W

469.12–14   Using an accountant's trick . . . expense.] The failure of governments, as well as of the United Nations Statistical Office and World Bank, to include deforestation and other natural resources in national depletion accounts is reported by Malcolm Gillis, "Economics, Ecology, and Ethics: Mending the Broken Circle for Tropical Forests," in Bormann and Kellert, *Ecology, Economics, Ethics*, pp. 155–179. W

469.35–470.12   the rubber tappers . . . timber.] The Brazilian rubber-tappers movement of the 1980s was bitterly opposed by some of the wealthy landowners of the western Amazon. On December 22, 1988, its leader Chico Mendes was killed by gunmen. The murder and the circumstances surrounding the struggle for control of the Amazonian environment are chronicled by Andrew Revkin, *The Burning Season* (Boston: Houghton Mifflin, 1990). W

470.13–32   Extractive reserves . . . forest managers."] The extractive reserves of the Amazon region are described by Walter V. Reid, James N. Barnes, and Brent Blackwelder, *Bankrolling Success: A Portfolio of Sustainable Development Projects* (Washington, D.C.: Environmental Policy Institute and National Wildlife Federation, 1989), and Philip M. Fearnside, "Extractive Reserves in Brazilian Amazonia," *BioScience*, 39(6):387–393 (1989). A critique of extractive reserves is presented by John O. Browder, "Extractive Reserves Will Not Save Tropics," *BioScience*, 40(9):626 (1990). W

470.35–471.15   The method of choice . . . the forest.] Strip logging as a sustainable industry is described by Carl F. Jordan, in "Amazon Rain Forests," *American Scientist*, 70:394–401 (1982), and by Gary S. Hartshorn, "Natural Forest Management by the Yanesha Forestry Cooperative in Peruvian Amazonia," in A. B. Anderson, ed., *Alternatives to Deforestation: Steps Toward Sustainable Use of the Amazon Rain Forest* (New York: Columbia University Press, 1990), pp. 128–137. W

471.24–473.2   Here are three . . . next century.] The examples of successful local sustainable development in Latin America are taken from Reid, Barnes, and Blackwelder, *Bankrolling Successes*. An account of local planning for sustainable tropical forest extraction is detailed in Leonard Berry et al., *Technologies to Sustain Tropical Forest Resources* (Washington, D.C.: Office of Technology Assessment, U.S. Congress, 1984). W

473.6–475.4   Today the poorest . . . international aid.] The impact of trade and subsidy policies of rich nations is described by Roger D. Stone and Eve Hamilton, *Global Economics and the Environment: Toward Sustainable Rural Development in the Third World* (New York: Council on Foreign Relations, 1991). W

476.26–37   Can extinct species . . . your hands.] The present status of research on DNA in fossils and archaeological remains is reviewed by Jeremy Cherfas, "Ancient DNA: Still Busy after Death," *Science*, 253:1354–56 (1991). W

477.16–26   The American Type Culture Collection . . . frogs.] The status of microbial preservation is described in "American Type Culture Collection Seeks to Expand Research Effort," *Scientist*, 4(16):1–7 (1990). W

478.11–479.6   One that works . . . in the wild.] Seed banks are reviewed by Erich Hoyt, *Conserving the Wild Relatives of Crops* (Rome and Gland: International Board for Plant Genetic Resources, etc., 1988); Jeffrey A. McNeely et al., *Conserving the World's Biological Diversity* (Gland and Washington, D.C.: International Union for Conservation of Nature and Natural Resources,

World Resources Institute, etc., 1990); Joel I. Cohen et al., "Ex Situ Conservation of Plant Genetic Resources: Global Development and Environmental Concerns," *Science*, 253:866–872 (1991). W

479.11–19   As of June 1991 . . . St. Helena.] The National Collection of Endangered Plants is the subject of a report in *Plant Conservation*, 6(1):6–7 (1991). W

479.21–480.4   Zoos . . . extinct in the wild.] The performance of zoos and other captive-animal facilities in maintaining diversity is described by William Conway, "Can Technology Aid Species Preservation?" in Wilson and Peter, *Biodiversity*, pp. 263–268; and by Colin Tudge, *Last Animals at the Zoo* (London: Hutchinson Radius, 1991). W

480.8–18   Conservation biologists . . . at risk.] The number of species of mammals facing extinction and requiring rescue is from Michael E. Soulé et al., "The Millennium Ark: How Long a Voyage, How Many Staterooms, How Many Passengers?," *Zoo Biology*, 5:101–114 (1986). William Conway is quoted on the limits of zoos by Edward C. Wolf, *On the Brink of Extinction: Conserving the Diversity of Life* (Washington, D.C.: Worldwatch Institute, 1987). W

481.8–12   The rescue . . . conservation.] A pioneering set of recommendations to save tropical ecosystems was advanced in 1980 by Peter H. Raven et al., *Research Priorities in Tropical Biology* (Washington, D.C.: National Academy Press, 1980). A review of ongoing efforts is provided by the authors in Wilson and Peter, *Biodiversity*; by McNeely et al., *Conserving*; by Janet N. Abramovitz, *Investing in Biological Diversity: U.S. Research and Conservation Efforts in Developing Countries* (Washington, D.C.: World Resources Institute; Gland: World Conservation Union; New York: United Nations Environment Program, 1992). W

481.18–34   The most inclusive . . . Endangered Species Act.] The provisions of the Endangered Species Act of the United States, as well as those of international regulatory protocols, are reviewed by Robert Boardman, *International Organization and the Conservation of Nature* (Bloomington: Indiana University Press, 1981); by Michael J. Bean, *The Evolution of National Wildlife Law* (New York: Praeger, 1983); and by Simon Lyster, *International Wildlife Law* (Cambridge, Eng.: Grotius, 1985). W

482.6–483.6   This means . . . hot spots.] The status of national parks and other reserves is documented by Walter V. Reid and Kenton R. Miller, *Keeping Options Alive: The Scientific Basis for Conserving Biodiversity* (Washington, D.C.: World Resources Institute, 1989); and by Michael E. Soulé, "Conservation: Tactics for a Constant Crisis," *Science*, 253:744–750 (1991). The percentage of the world's land surface under legal protection is from *1990 United Nations List of National Parks and Protected Areas*. W

483.7–32 debt-for-nature ... debt-for-nature swaps.] Debt-for-nature exchange schemes have been well explained by José Márcio Ayres, "Debt-for-Equity Swaps and the Conservation of Tropical Rain Forests," *Trends in Ecology and Evolution*, 4(11):331–332 (1989); and by Roger D. Stone and Eve Hamilton, *Global Economics and the Environment* (New York: Council on Foreign Relations, 1991). I have also used a master's thesis from University College, London, by Victoria C. Drake, "Debt-for-Nature Swaps: An Economic Appraisal." The Mexican trade was reported by Mark A. Uhlig, "Mexican Debt Deal May Save Jungle," *New York Times*, February 26, 1991. The idea of debt-for-nature was first proposed by Thomas Lovejoy of the Smithsonian Institution. W

484.30–485.20 the so-called SLOSS ... as large as possible.] The SLOSS controversy is examined, with different conclusions, by James F. Quinn and Alan Hastings, "Extinction in Subdivided Habitats," *Conservation Biology*, 1(3):198–208 (1987); and by Michael E. Gilpin, "A Comment on Quinn and Hastings: Extinction in Subdivided Habitats," *Conservation Biology*, 2(3):290–292 (1988). The advantages and disadvantages of corridors between small reserves are reviewed by William Stolzenburg, "The Fragment Connection," *Nature Conservancy*, July–August 1991, pp. 18–25. W

485.21–486.31 *Restore the wildlands* ... natural heritage.] The progress of ecosystem restoration in the United States can be followed in issues of *Restoration and Management Notes*, published by the University of Wisconsin Press since 1982. A recent account of prairie renewal and the general hopes and misgivings of restorationists is provided by William K. Stevens, "Green-Thumbed Ecologists Resurrect Vanished Habitats," *New York Times*, March 19, 1991. The creation of new dry tropical forest in Costa Rica's Guanacaste National Park is described by Reid et al., *Bankrolling Successes*. W

486.34–37 The return ... impoverished habitats.] The history of animal-species introduction into new environments is reviewed by Paul R. Ehrlich, "Which Animal Will Invade?," in Harold A. Mooney and James A. Drake, eds., *Ecology of Biological Invasions of North America and Hawaii* (New York: Springer, 1986), pp. 79–95. W

487.3–488.7 Most were exotics ... new communities together.] Ariel E. Lugo has spoken on behalf of exotic species in expanding local biodiversity. While conceding the high risk of introductions and the need to remove elements that endanger native fauna and flora, he notes that most such species are naturalized without creating ecological problems. "Exotics appear to do best in human-disturbed environments. Exotics can provide food and fiber without causing ecological havoc. For example, when managed properly, certain exotic trees grow well in highly degraded lands where they contribute to soil rehabilitation and reestablishment of native species." "Removal of Exotic Organisms," *Conservation Biology*, 4(4):345 (1990). W

489.27–33 Four splendid lines . . . the toil.] The Sibyl's advice to Aeneas is from the translation by Robert Fitzgerald, *The Aeneid: Virgil* (New York: Random House, 1983), book 6, p. 164. W

491.18–20 Tennyson . . . man is."] See Tennyson's poem "Flower in the Crannied Wall," first published in 1863 and collected in *The Holy Grail and Other Poems* (1870).

495.27–497.14 the particularities of human nature . . . the rest of life.] The innate affiliation of human beings with the natural world is elaborated in my *Biophilia* (Cambridge: Harvard University Press, 1984). The imagery of the serpent was drawn from Balaji Mundkur's masterful *The Cult of the Serpent: An Interdisciplinary Survey of Its Manifestations and Origins* (Albany: State University of New York Press, 1983). The concept of the idealized living place as a biological adaptation was developed by Gordon H. Orians, "Habitat Selection: General Theory and Applications to Human Behavior," in Joan S. Lockard, ed., *The Evolution of Human Social Behavior* (New York: Elsevier North Holland, 1980), pp. 46–66; and "An Ecological and Evolutionary Approach to Landscape Aesthetics," in Edmund C. Penning-Rowsell and David Lowenthal, eds., *Landscape Meanings and Values* (London: Allen and Unwin, 1986), pp. 3–22. W

497.5–7 In the United States . . . combined.] The greater attendance of people at zoos and aquariums than at professional sporting events (football, baseball, basketball, ice hockey) is cited in *Directory of the American Association of Zoological Parks and Aquaria*, ed. Linda Boyd (Wheeling, West Virginia: Ogle Bay Park, 1990–91). W

497.14–33 the idea of wilderness . . . generations to come.] Excellent histories of wilderness in the human imagination, especially in Europe and America, have been presented by Roderick Nash, *Wilderness and the American Mind* (New Haven: Yale University Press, 1967); and by Max Oelschlaeger, *The Idea of Wilderness: From Prehistory to the Age of Ecology* (New Haven: Yale University Press, 1991). W

NATURALIST

531.1 *Naturalist*] This book was first published in 1994. Wilson contributed an afterword for a new printing in 2006; it appears below.

AFTERWORD

For each of us, inborn propensities of temperament and talent screen childhood experience and weave the pattern of the adult mind. Through the twists and turns of my own development, I ended up a naturalist and made the practical decision to earn my living by research and teaching. Since then I have had many adventures, both those faithfully described in *Naturalist* and others, but none have altered me at the core. While more competent and cautious perhaps, and mercifully less roiled by passions, I remain the boy I was when my love of nature was awakened and the world beguiled me into thinking I could make something of my life.

For six decades, through my college student years and my later professional life, I have dwelled in basic science, and in particular my natural habitat of systematics, biogeography, evolutionary theory, ecology, and sociobiology, which all together I like to think of as scientific natural history. Halfway through this career I added nonfiction writing, because I had a talent for it, because well-wrought English prose had always inspired me, more even than music, and because I remained haunted by the cadences of the King James Bible and evangelical sermons of my childhood. A Southern writer had lived inside the biologist, growing impatient for release.

Would I have been as fulfilled had I taken up creative writing in college and only later turned to science for material? Perhaps, but I doubt it. I am now convinced that it is better to work from science into literature than to try the reverse, although many have done so with distinction. To understand the scientific culture deeply and, even more, to express the emotions that attend scientific exploration require that the writer inhabit science for a substantial part of his life, intent upon making important discoveries and placing them within the canon.

In the early 1970s I was summoned by a third altar call, so to speak. Earth's biodiversity, the Creation, which even as a secularist I regard with a spiritual reverence, is disappearing. Countless millions of years of evolution are being wiped out by the ignorant excesses of humankind. Biologists who most understand biodiversity and the causes of its peril should of all people, I felt, step forward in order to help save it.

In July 1994, the year *Naturalist* appeared, I was appointed Pellegrino University Professor, one of only fifteen university-wide professorships at Harvard University, intended to provide maximum latitude in teaching, although in fact I continued very much the same activities as before. I retired three years later, at the age of sixty-eight, in order to end teaching and administrative duties completely. I didn't have to quit: Harvard's mandatory age cap had been lifted by federal mandate. My wish was to turn more fully to research and writing, as well as to conservation activism. But, I confess, a deeper, more emotional reason for retiring was to shed a burden I had felt too long, that of juggling research and teaching in an effort to reach at least satisfactory achievement in both. A few years thereafter, at the 2004 spring commencement, I was awarded an honorary doctorate, a rare honor at Harvard, which grants only one such degree to a superannuated member of its own faculty on average every year or two. Because I had struggled much at this university, I felt deeply gratified by the double shoulder sword-touch.

Upon retirement, I was able to tip the balance toward research without dereliction of other duties and to intensify my efforts in global biodiversity conservation. Having served on the boards of The Nature Conservancy and the World Wildlife Fund, and as a chief advisor to the New York Botanical

Garden, I now focused my efforts on Conservation International, the young-est and in my view the most innovative member of the global conservation community. I was especially attracted by the two relatively young leaders of this organization: Russell Mittermeier, a distinguished conservation biolo-gist and primate expert of seemingly infinite energy, and Peter Seligmann, a genius at organizing such an enterprise, articulating a compelling vision of conservation thereby, and not least, fundraising. Conservation International has pioneered in finding new ways to establish and secure reserves in tropical, biodiversity-rich countries while promoting economic development in the surrounding regions.

All of the global conservation organizations seek for their boards of direc-tors leaders in business and the professions, as well as conservation experts and independently wealthy enthusiasts. CI has been notably successful in its recruitment efforts, managing at one time or another to attract, for example, the heads of Ford Motor Company, Gap, Intel, Starbucks, and Wal-Mart. Environmental purists undoubtedly would be dismayed by such a roster, but I have learned that there is no intrinsic antagonism between corporate and environmental leaders. Granted, corporations exist whose policies and leaders are wicked, or at least indifferent, but others are committed to protecting the environment and devote substantial time and money with minimal fanfare. If you want to save a tract of rain forest in Guyana, Liberia, or some other fractious developing country, it is often better to have such a person at your side than a diplomat or a professional environmentalist.

At Conservation International I served as a chairman of the Program Committee and as advisor in the founding of CI's Center for Applied Bio-diversity Science. The latter organization, aided by gifts totaling hundreds of millions of dollars from Gordon Moore, cofounder of Intel and chairman of the CI board, became the premier conservation-oriented research organiza-tion in the world.

Elsewhere, as an author and lecturer I tried to articulate the scientific argument for biodiversity conservation. *Consilience: The Unity of Knowledge* (1998), which makes the case for a return to the Enlightenment, closes with the chapter "To What End?"—meaning the ultimate goals of our species as a whole. The answer I suggested, and later spelled out at book length in *The Future of Life* (2002), is to bring humanity through the current bottleneck of overpopulation and rising per capita consumption in a manner that raises the quality of life for all, while ferrying through as much of the rest of life as science, technology, and an informed ethics make possible. A smaller, highly educated world population could, if supported cost-free by an independent, biologically rich natural world, turn Earth into a near-perfect habitat for humanity.

In 2005, I sent to press a new book tentatively entitled *The Creation*. Drawing on my own Southern Baptist upbringing for style and authenticity, I make the case that science and religion, the two most powerful social forces in the world, need to form an alliance in order to save Earth's biodiversity. All of life, I remind my hoped-for religious friends, is the living Creation of

Judeo-Christian sacred scripture. I press this appeal as a scientific humanist, with no argument from anything but our scientific knowledge of the state of the living environment and a rationalist argument to save the variety of life. I believe that this worldview can be annealed to that based on religious faith. Let us put aside our metaphysical differences, I say, and focus on a clearly defined issue that deeply concerns us all.

In addressing the issues of global conservation, I myself have focused in recent years on one particular major gap in scientific knowledge. The amazing fact is that we have discovered and analyzed only a small fraction of Earth's biodiversity. We lack a fully solid foundation for conservation science. As many as 90 percent of the species of organisms, mostly small invertebrates, protistans, and bacteria but also including even a few birds and mammals, remain unknown, hence with neither a diagnosis nor a scientific name. We have characterized about 180,000 of the flowering plants identified thus far out of a likely 230,000 species still living, as well as a large majority of land vertebrates (amphibians, reptiles, birds, and mammals), enough to advance conservation science and practices. But of the "little things that run the world," whose roles in maintaining healthy ecosystems are also crucial to our existence, we still know shockingly little. While the United States continues to spend billions to explore the rest of the solar system, the annual amount invested by public and private sources in the discovery and classification of biodiversity stays close to $200 million.

This stingy level of support has reduced taxonomy, the foundational discipline of biodiversity, to penury in comparison with the status of other biological disciplines. Many important groups of insects and other invertebrate animals, representing a large majority of known species, are addressed by no more than a dozen specialists, who tend to be aging and underpaid. In presentations and talks, I set out to put the following question before the public with some urgency: Why devote so much funding and attention to mapping the heavens, but so little to the home planet?

Since 2000, I have been pressing the issue to scientists and public analysts in other ways. I became principal advisor to the All Species Foundation, newly created to promote the completion of the world biodiversity survey. In the fall of 2001, I chaired a meeting of leaders in projects of this kind at the continental and global levels, sponsored by Harvard and Conservation International. We unanimously agreed that with funding at the level of the Human Genome Project, biologists could mostly complete the mapping of Earth's biodiversity within twenty-five years, in other words, a human generation. Both of these efforts have attracted a lot of favorable attention but little funding. At the beginning of the new century, plagued by terrorism and an economic recession, government and foundations have had other priorities.

Those of us arguing for the renaissance and public appreciation of biodiversity mapping have been hampered further by a common misconception, widespread within science itself, that taxonomy is just biological stamp collecting and not a serious part of modern biology. That is dangerously wrong. The foundational knowledge of ecology, the database of studies on biogeography and evolution, the substance of the tree of life over evolutionary time, the source of knowledge needed to save biodiversity in its entirety, all these great endeavors of biology depend upon a full taxonomy of the global fauna and flora.

In my inward struggle to frame concisely the reasons for the taxonomic mapping effort, I realized that the argument needed to be expressed as a concrete goal with a timetable for its completion, and in a manner that would make its significance obvious. The goal, I believe, should take the form of an electronic Encyclopedia of Life in which the species already known, now approaching 2 million in number, and, as they are found, the millions remaining to be discovered, would each have a page. The page would be indefinitely expandable and contain everything known about the species, including links to other databases, offering information ranging from its genome to its ecosystems function and, not least, its importance for humanity. The idea has been appealing enough to be adopted as a program of the U.S. National Museum of Natural History.

As taxonomists have soldiered on through the dark valley of their subject's neglect, a common complaint heard is that complete global mapping is too big a project to undertake. Not at all, I've contended, and to support my argument, I can point to a recent effort of my own: mastering the classification of *Pheidole*, the world's largest ant genus. There are so many species in this genus, spread out in warm climates around the world, that no one had ever attempted to sort and classify them. *Pheidole* species are also among the most abundant of all insects and hence are important ecological agents on the land. In trying to make sense of them, ecologists had been forced to designate the species with numbers—for example, *Pheidole* species 1, *Pheidole* species 2, and so on, up to as many as *Pheidole* species 50 from single locations. This procedure offers no hope of collating information from different localities without consulting the original voucher specimens placed in museum collections. *Pheidole*, the Mount Everest of ant taxonomy, had to be conquered.

In 1985 I set out to "climb" *Pheidole* for the New World, the center of its diversity. I had at my disposal in Harvard's ant collection (the world's largest) tens of thousands of specimens, collected by myself and others. I had all the necessary literature on the genus, much of it dating to the nineteenth century, and I managed to borrow and hold at Harvard almost all of the type specimens, on which the Latinized scientific names of previously described species had been based.

I worked on and off at odd times in my home laboratory, slowly threading my way through the museum collection while listening to classical and soft rock music. It was my version of knitting—relaxing and never boring, always edifying, and often exciting as I hit upon a species new to science. By 2001, after making about 6,000 measurements to an accuracy of 0.02 millimeter and more than 5,000 line drawings by my own hand, I came to the end. I had separated and diagnosed 624 species, including 337 new to science, which at that time composed 19 percent of all the known ant species of the Western Hemisphere and 6 percent of all the ant species of the world. My results, including analyses of the evolution of the genus and all that was known of the biology of each species, were published in 2003 in a 794-page book, *Pheidole in the New World: A Dominant, Hyperdiverse Ant Genus.*

Pheidole was thereby opened at last to study by scientists interested in its diversity and ecology. Of equal importance in my own mind, I had demonstrated that it is possible for just one person, working part-time, to master a substantial fraction of global biodiversity. If there are 20,000 species of ants in the world (about 12,000 are known to science as I write, in 2006), then only thirty or so specialists would be enough to achieve their discovery and analysis, and they would need perhaps no more than twenty years to do it. How many such experts would be required to diagnose and classify all organisms on Earth in the same amount of time? Leaving out for the moment the bacteria and archaea, of whose vast diversity we have almost no idea, and taking 10 million as a reasonable guess of the number of living species of other kinds on the planet, the number of specialists needed for this initial census could be under 20,000. That is a tiny fraction of the biologists currently employed in the world, and the multitudinous valuable discoveries they could make possible across the rest of biology would be beyond calculation.

In fact, new technology has turned even that time projection into an overestimate. As the *Pheidole* monograph approached publication, I learned of a recently developed method of illustration that has begun to revolutionize taxonomy. It is the combination of automontaged high-resolution digitized photography with Internet publishing. The automontage method entails making a series of photographs of the same specimen at different levels of the body, top to bottom and side to side (this can be done quickly by automation), and then combining them by computer to produce a three-dimensional image in perfect focus. The method allows very small objects, such as the type specimens of insects, to be viewed with greater clarity than when examined on the stage of a standard dissecting microscope. The images can be transmitted to others through the Internet, or collected together to create an electronic monograph or field guide.

I was introduced to the method by Piotr Naskrecki, one of the first biologists to use it in taxonomic practice. He kindly photographed the type specimens of *Pheidole* ants available in the Harvard collection and created a CD to be included in my book. This hybrid publication represents, to my view, the beginning of the end of the centuries-old technology of printed taxonomy, and its partial replacement by a CD is the start of a new, faster study and means of publication. I like to call my printed *Pheidole* book the "last of the great sailing ships." Henceforth, it should prove far easier to disseminate information about large and difficult genera of insects and other organisms.

To come to the final and most tumultuous track of my eclectic existence, the twelve years since the original publication of *Naturalist* have seen many changes in sociobiology, from which I have received, as its nominal founder, both anguish and satisfaction. Applied to ants and other animals, it has flourished. Applied to human social behavior it has also proliferated, but under the name "evolutionary psychology," now an academic subject with a life of its own. Evolutionary psychology has generated some excellent research and much else that is less than distinguished. Overall it has created an industry of popular books, with substantial combined impact, and become part of the popular culture. Criticism of the kind that followed the publication of my *Sociobiology: The New Synthesis* in 1975 has largely disappeared. However, attacks of the early era, which were heavily ideological in origin, have left a residue of misunderstanding not just about the content but about the very meaning of the term "sociobiology." It should be kept in mind that sociobiology is a discipline and, as such, is defined as the systematic study of the biological basis of all forms of social behavior. The thrust of criticism in the 1970s and 1980s, which arose from the now discredited conception of the human brain as a blank slate, was that sociobiology entails a belief in biological determinism. This was a canard, and one mischievously intended. Sociobiology is not a doctrine or a particular conclusion but a discipline, an open field of inquiry, allowing in theory for the human brain to be a blank slate (disproved), or completely hardwired (never claimed), or the product of interaction between genetic predisposition and environment (well established and now almost universally accepted).

Another outgrowth of the controversy was the widespread notion that sociobiology means the study of the genetic evolution of social behavior. However, as in all disciplines of biology, it comprises two approaches. The first is functional sociobiology, the study of how social systems are put together. It is this domain, including the theories of division of labor and of chemical communication, to which I made my principal contribution in the late 1950s to early 1970s. The second domain is the process of genetic evolution of social behavior, pioneered by J.B.S. Haldane and William D. Hamilton, among others, from the 1950s forward. In 1975, my *Sociobiology: The New Synthesis* brought functional sociobiology and evolutionary sociobiology together for the first time and established the boundaries of the subject. Thereafter, unfortunately, the public controversy was focused not on the discipline as such but on the application of its principles to the human species, and then only to the genetic interpretation of human social behavior. That is a sad and destructive misconstruction of an important scientific discipline.

Five years after the publication of *Naturalist*, my seventieth birthday came and went without a ripple in my mind. Now it recedes like a shoreline behind a departing ship, serenely, a shrinking abstract line of memory. Entering my late seventies as I write, in 2006, luck still holds: good health, good working conditions, creative capacities undiminished (of the last, granted I am not the best judge). I know better, but I press on as though I will live forever.

I am often asked, given the strong naturalism in my philosophical writings, to express my deepest convictions. They are simple, and I will give them here. Science is the global civilization of which I am a citizen. The spread of its democratic ethic and its unifying powers provides my faith in humanity. The astonishing depth of wonders in the universe, continuously revealed by science, is my temple. The capacity of the informed human mind, liberated at last by the understanding that we are alone and thus the sole stewards of

Earth, is my religion. The potential of humanity to turn this planet into a paradise for future generations is my afterlife.

You will understand, then, why I stay engaged with such purpose and optimism in all the subjects that have occupied me across six decades, from the natural history of ants through the labyrinth of behavioral and evolutionary biology to the great challenge that faces us all, citizen and scientist alike, in the decline of Earth's living environment. I am able still to continue field studies of ants island by island in the West Indies. In my brief visits there I am accompanied by younger myrmecologists, friends and colleagues in the study of ants. It is a time of joy, of entering habitats never before explored, discovering new species, learning and recording new facts of natural history, sharing with much hilarity war stories of earlier expeditions. The experience is primal. The true naturalist is a civilized hunter, and we are a happy hunter band. Thereby I revive the same emotions I experienced long ago as a teenage student at the University of Alabama, when the central ambition of my life was to be this kind of scientist. I am thus able to offer truth in testimony to the beautiful insight of Albert Camus:

> A man's work is nothing but this slow trek
> to rediscover through the detours of art
> those two or three great and simple images
> in whose presence his heart first opened.

Edward O. Wilson
January 15, 2006

539.1–7 When I get . . . Hughes] See Hughes's poem "Daybreak in Alabama," first published in 1940.

566.30–31 *Arrowsmith, The Sea Wolf,* and *Martin Eden*] The first of these three novels, by Sinclair Lewis (1885–1951), was published in 1925; the second and third, both by Jack London (1876–1916), first appeared in 1904 and 1909 respectively.

586.1–3 The vacant lot . . . observation.] See E. O. Wilson, and J. H. Eads, "A Report on the Imported Fire Ant *Solenopsis saevissima* var. *richteri* in Alabama," Special Report to the Alabama Department of Conservation, pp. 1–54 (1949).

586.26 Fibber McGee and Molly on the radio] A comedy series that was broadcast from 1935 to 1959.

588.29–33 Longfellow's invocation . . . the night.] See Longfellow's poem "The Ladder of St. Augustine," first published in 1855.

610.40–611.1 My observations . . . scientific papers.] See E. O. Wilson, T. Eisner, and B. D. Valentine, "The Beetle Genus *Paralimulodes* Bruch in North America, with Notes on Morphology and Behavior (Coleoptera: Limulodidae)", *Psyche* 61(4):154–61 (1954).

618.18–20   as Alfred North Whitehead . . . to discover.] See Whitehead's essay "Technical Education and Its Relation to Science and Literature," collected in *The Aims of Education, and Other Essays* (1917): "No man of science wants merely to know. He acquires knowledge to appease his passion for discovery. He does not discover in order to know, he knows in order to discover."

619.46   Barry Valentine . . . entomological journal.] See B. D. Valentine, and E. O. Wilson, "Records of the Order Zoraptera from Alabama," *Entomological News* 60:180–81 (1955).

620.39   a history . . . 1951] See "Variation and Adaptation in the Imported Fire Ant," *Evolution* 5:68–79 (March 1951).

648.20–23   *Belonopelta* and *Hylomyrma* . . . later publication] See "Ecology and Behavior of the Ant *Belonopelta deletrix* Mann (Hymenoptera: Formicidae)," *Psyche* 62:82–87 (1955).

664.2–4   three years later . . . army ants] See "The Beginnings of Nomadic and Group-Predatory Behavior in the Ponerine Ants," *Evolution* 12:24–36 (March 1958).

678.37–39   "Let the waters . . . heavens."] Genesis 1:20.

685.9   the White Tower] A hamburger restaurant, part of a chain, opened in Cambridge, Massachusetts, in 1932 and closed in 1973.

695.33–37   Shakespeare's muse . . . a name.] See *A Midsummer Night's Dream*, V.i.14–17.

700.7–9   As John Stuart Mill . . . field.] See Mill's *On Liberty* (1859).

721.32   the Matthew-Darlington cycle] Wilson explains this cycle in *Philip Jackson Darlington Jr., 1904–1983: A Biographical Memoir*, published by the National Academy of Sciences in 1991: "A theory of faunal dominance was first proposed by the American paleontologist William Diller Matthew in 1915. According to Matthew's scheme, new groups of vertebrates—such as the rhinoceros (Rhinocerotidae) and tapirs (Tapiridae)—had originated during the early Tertiary in the central Eurasian–North American land mass. Hardened by the cold and fickle climate of the north, they became better fitted for competition and survival everywhere and accordingly spread south through the southern land masses and archipelagos, pushing out the previous inhabitants. Rhinos and tapirs, for example, gave way before the great herds of antelopes, bovids, and other artiodactyls that still prevail in both the northern and southern hemispheres. Darlington realized, however, that Matthew had based his identification of the northern hemisphere as the key faunistic staging area on inadequate, and geographically biased, fossil data. Keeping Matthew's vision of dominance and cyclic replacement, he reconsidered the area of origin. With the aid of extensive new fossil evidence and close examination of the zoogeography of living cold-blooded vertebrates, he shifted the principal source area for dominant groups to the Old World tropics—more precisely, to central and northern Africa, southern Europe and the Middle East (tropical in climate during much of the Tertiary), and southern Asia. Dominant terrestrial groups originated more frequently within this domain than elsewhere. Their species tended to spread to parts of the world that Darlington now conceived to be peripheral: across the Bering land bridge to North America, and thence to Central and South America; across the Indonesian archipelagos to Australia and the Pacific archipelagos; and straight north into temperate Asia and south into southern Africa."

722.17–22   the conviction . . . be judged."] See Hardy's *A Mathematician's Apology* (1940).

723.35   two articles] See Robert H. MacArthur, "Fluctuations of Animal Populations and a Measure of Community Stability," *Ecology* 36(3):533–36 (1955), and "On the Relative Abundance of Bird Species," *Proceedings of the National Academy of Sciences* 43(3):293–95 (1957).

783.12–13   the enduring phrase . . . "selfish gene."] See Dawkins's 1976 book *The Selfish Gene.*

810.34–35   Marvin Harris] American anthropologist, author of *The Rise of Anthropological Theory: A History of Theories of Culture* (1968).

819.10–25   a pioneering study . . . if pursued] See note 97.3.

820.1–4   Balaji Mundkur] See note 74.1.

Index

Aboriginals, Australian, 76, 88, 176, 390–92

Abyssal benthos, 277–78, 344

Acacias, 98–99, 302, 496

Acadian flycatchers, 299

Acanthomyops ants, 577

Acarines, 312

Accipiter gentilis, 293–95

Acid rain, 160, 162

Acorn woodpeckers, 234

Acromyrmex ants, 34

Adaptation, 11, 77, 84, 113, 147, 190–91, 223–24, 240, 324, 337–38, 341, 376, 414, 694, 698, 713

Adaptive demography, 779–80

Adaptive radiation, 227–66, 268, 279, 309, 322, 329, 349–50, 387, 390–91, 397–98, 490

Adders, 83, 86

Aedes mosquitos, 662

Aedes sierrensis, 317

Aeolian plankton, 153–55

Aepyornis maximus, 391

Aesthetic criteria, 96–101, 103

Africa, 146, 173, 224, 228, 283, 343, 356, 391, 402, 483; acacias in, 98–99, 302, 496; agriculture in, 438, 444, 451; ants in, 302–4; chameleons in, 92–93; cichlid fishes in, 241–48, 255, 397–98; hominids in, 96, 113, 185–86, 350, 543, 631, 819; hot spots in, 408–9; as island continent, 349–50; and sickle-cell anemia, 211–12; snakes in, 83–85

African wild dogs, 264

Agassiz, Louis, 38–44, 53; "Essay on Classification," 43

Agency for International Development, 472

Aggression, 748, 781, 796

Agkistrodon piscivorus, 81–82

Agoutis, 301

Agriculture, 111, 126, 368, 423, 441, 457; Amazonian, 26–28, 30, 137, 335, 364, 384–87, 417; and conservation hot spots, 401, 403, 406, 408–12; early human, 16; food, 117–20, 432–38, 445–48, 450; monoculture, 432, 446; and sustainability, 426, 451, 461, 464–65, 468–76, 478, 482

Agriculture Department, U.S., 273, 762

Agrotechnology, 446, 468

Ailuropoda pandas, 292

Akepa, 231

Akiapolaau, 231, 233–37

Akihito, Emperor, 688–89

Alabama, 77, 80–82, 85, 91–92, 572–73, 579–626, 632, 814

Alabama Department of Conservation, 619–20

Albany Pine Bush, 378

Alder flycatchers, 94, 188

Aleutian islands, 300

Algae, 165, 170–71, 242, 246, 281, 314, 319, 321–22, 324, 326, 339, 346, 357, 413, 454, 596, 715

Alland, Alexander, Jr., 809–10

Allee, Warder Clyde, 777

Alleles, 212, 214–15, 218, 297

Allometry, 217–19, 779

Allopatric speciation, 205

All Species Foundation, 828–29

Alopias vulpinus, 251, 253

Alpacas, 263

Alpha diversity, 287–88

Alport, Gary, 754

Altmann, Stuart, 729, 775–78

Alvarez, Luis, 159–60

Amaranth, 436–38

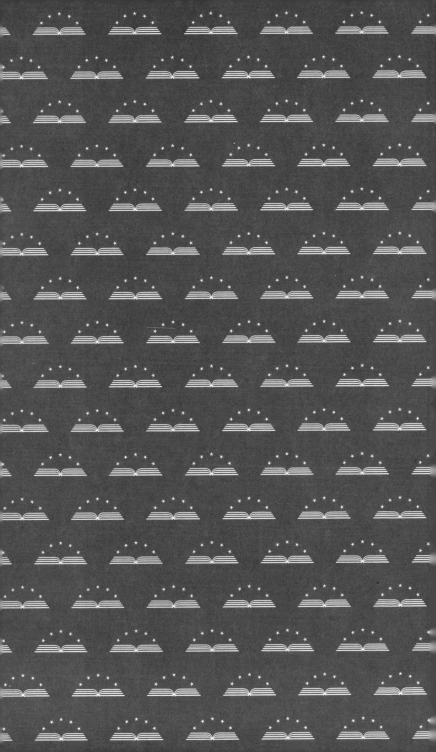